Advanced Undergraduate Quantum Mechanics

Lev I. Deych

Advanced Undergraduate Quantum Mechanics

Methods and Applications

 Springer

Lev I. Deych
Physics Department
Queens College and the Graduate Center
City University of New York
New York, NY, USA

ISBN 978-3-030-10073-5   ISBN 978-3-319-71550-6   (eBook)
https://doi.org/10.1007/978-3-319-71550-6

This Springer imprint is published by the registered company Springer International Publishing AG part
of Springer Nature.
The registered company address is: Gewerbestrasse 11, 6330 Cham, Switzerland

To Yelena, Daniil, Sergei, and my dog Xena

Preface

The market for undergraduate textbooks on quantum mechanics is oversaturated as one can literally choose from tens, if not hundreds, of different titles, with several of them being accepted as a standard. Writing yet another textbook on this subject might appear as a reckless and time-consuming adventure, and whoever got engaged in it can be rightfully suspected in arrogance and self-aggrandizement. Be that as it may, after 15 years of teaching an undergraduate quantum mechanics course at Queens College of the City University of New York, I finally came to realization that neither I nor my students are particularly happy with the existing standards. It also occurred to me that my approaches to teaching various topics in quantum mechanics, along with the lecture notes I have accumulated over these years, could form the foundation of a textbook that would be different from those I saw on the market. In many cases, students treat their physics textbooks as a reference source for formulas and postulates—used to solve problems rather than as actual reading material. I, on the other hand, dreamed about writing a book that would actually be read; and in order to achieve that, I have devoted a significant amount of time to the explanation of the most minute technical details of various technical derivations. The level of technical detail in the book would ideally allow students to use it without the need for a lecturer's explanations, allowing professors to use precious lecture time on something more productive and fun. But quantum mechanics is not just about derivations of formulas (though these might be fun, of course, for the mathematically inclined). The physical content of the derived results is immensely more important and interesting, if you ask me. Thus, I have included in the text extensive **qualitative** discussions of the physical significance of derived formulas and equations. Finally, to make reading even more enjoyable, I tried to preserve the informal, colloquial style of my lectures, addressing readers directly and avoiding the dry, impersonal manner found in too many formal scientific texts.

Most physics textbooks present ideas and concepts without more than a passing mention of the people who discovered them. Such an approach aims to emphasize the objectivity of the laws and principles that physics deals with. The essence of these laws does not depend on the personal traits and ideologies of their discoverers, which is not always the case in humanities. While I agree that such a cold, formal approach is justified by the objective nature of the laws of physics, I am still not sure that it is the best way to present the material to students. This approach dehumanizes physics, preventing people from relating to it on a personal level and seeing physics as part of the general human experience. In this book, I tried to break out of this tradition and introduced some personal details about the lives of those scientists who were responsible for developing quantum theory and changing our views of the universe along the way. I would like the readers to see that the complex technical and philosophical ideas in quantum mechanics were generated by mortal human beings with strengths and weaknesses—that they experienced struggles and made mistakes for which they had to bear full responsibility. Obviously, it would be impossible to talk about all these great (and sometimes not so great) men[1] in detail. But whenever possible, I have tried to provide bits and pieces about the personal lives of scientists whose names appear in the text.

Anyone writing a textbook on such an immense subject as quantum mechanics always struggles with the question of which topics to include and which to leave out. There are, of course, some standard concepts that cannot be avoided, but beyond those, the choice is always a function of the author's personal predilections. These predilections led me to include some topics that are not usually covered in under-graduate textbooks, such as the Heisenberg equations, the transfer-matrix method for one-dimensional problems, optical transitions in semiconductors, Landau levels, and the Hall conductivity. At the same time, I left out such popular topics as WKB and variational methods. The total contents of the book were chosen to satisfy the needs of a two-semester course for students who have already been exposed to some quantum ideas in a modern physics course or its equivalent. At the same time, the book can also be used for a one-semester course. While each instructor deciding to adopt this book can choose whatever material they prefer for their one-semester course, my suggestion would be to include Chaps. 1 through 9, as well as Chaps. 11 and 15, which can be taught almost independently of other chapters of the book. (While there are some cross-references between these and other chapters, they shouldn't impede the student's or instructor's ability to get through the arguments.) Finally, I would like to emphasize that the problems offered for solution by the students are an integral part of the text and must be treated as such.

Quantum mechanics is one of the ultimate triumphs of the human mind. I enjoyed writing this book, trying to convey the awesomeness of quantum ideas and of the people who contributed to their development, and I hope that you will enjoy reading

---

[1]Needless to say, at the time of birth of quantum physics, most of the scientists were, indeed, men. Even more remarkable are achievements of a few women who left their marks on the twentieth-century physics. Nobel Prize winners Marie and Irene Curie have since become household names, but other female scientists such as German mathematician Emmy Noether and Austrian-Swedish experimental physicist Lise Meitner also deserve to be remembered. Unfortunately, the contributions of Noether, who uncovered the intimate relationship between symmetries and conservation laws, and Meitner, who together with Otto Hahn discovered the fusion of uranium, are too advanced for an introductory quantum mechanics book, preventing me from talking about these scientists in more detail.

it. While physics students, unlike their brethren from humanity departments, are not accustomed to actually reading their assigned texts, I would like to encourage you to overcome the established habits and give this book a chance. Who knows, you might like it.

New York, NY, USA   Lev I. Deych

Acknowledgments

I would like to thank Drs. Mishaktul Bhattacharya and Alexey Tsvelik for reading and commenting on the portions of the manuscript and for general encouragement and Prof. Nathaniel Knight for making sure that my historical introduction did not deviate too much from actual historical facts. My special gratitude, however, goes to Prof. Lev Murokh of Queens College for test-driving the book in a quantum mechanics class he was teaching in fall 2017 and for discovering multiple misprints and quite a few errors, thereby saving me from a great deal of embarrassment. However, I take full responsibility for whatever mistakes might still remain in the text and will be immensely grateful to anyone pointing them out to me.

Contents

Part I
Language and Formalism
of Quantum Mechanics

Chapter 1
Introduction

1.1 The Rise of Quantum Physics and Its Many Oddities

In 1890, Scots-Irish physicist Lord Kelvin (born William Thomson, knighted in 1866 by Queen Victoria for his work on transatlantic telegraph, ennobled in 1892 for his achievements in thermodynamics, and became the first British scientist elevated to the House of Lords of the British Parliament) gave his famous speech identifying only two "clouds" in the clear sky of classical physics. One of them was the problem of luminous ether undetected in the series of experiments carried out between 1881 and 1887 by Americans Albert Michelson and Edward Morley, and the second one was the problem of the black-body radiation. Classical physics predicted that the amount of electromagnetic energy emitted by warm bodies increases with a decreased wavelength of radiation, making the total emitted energy infinite. This problem was dubbed an *ultraviolet catastrophe* by Paul Ehrenfest in 1911. As it turned out, both these little clouds spelled the end of the classical physics: the first of them resulted in the special relativity theory, and the solution of the second one achieved by German Physicist Max Planck laid the first stone in the foundation of quantum physics.

To explain the radiation of the black bodies, Planck had to introduce unusual entities—oscillators whose energy cannot be changed continuously, but must be a multiple of an elementary energy quantum $h\nu$, where $h = 6.55 \times 10^{-34}$ J s is a fundamental constant of nature introduced by Planck and ν is the classical frequency of the oscillator. (This expression is often written as $\hbar\omega$, where $\hbar$ (h-bar) is "the reduced" Planck's constant and ω is the angular frequency of the oscillator. Given that $\omega = 2\pi\nu$, you can easily find that $\hbar = h/2\pi$.) Using this assumption and the apparatus of classical statistical physics, Planck has derived his famous formula for the spectral energy density of the black-body radiation:

© Springer International Publishing AG, part of Springer Nature 2018
L.I. Deych, *Advanced Undergraduate Quantum Mechanics*,
https://doi.org/10.1007/978-3-319-71550-6_1

$$u(v, T) = \frac{8\pi h v^3}{c^3} \frac{1}{\exp\left(\frac{hv}{k_B T}\right) - 1},$$ (1.1)

where $u(v, T)dv$ is the energy density of electromagnetic waves with frequencies within the interval $[v, v + dv]$, k_B is the Boltzmann constant, and T is the absolute temperature.[1] The original value of the constant h which Planck derived by fitting his formula to the experimental data turned out to be quite close to the modern value, which is currently believed to be

$$h = 6.62607004 \times 10^{-34} \text{ J s}.$$

The idea that the energy of a classical particle cannot take an arbitrary continuously changing value seemed to Planck quite revolutionary, so much so that he refused to believe that his quantized oscillators are related to any real physical objects, such as atoms or molecules, thinking of them as of a purely mathematical trick, which works. It took Einstein's theory of photoelectric effect (1905), where Einstein explicitly postulated that electromagnetic energy propagates in the form of quantized and indivisible portions, his 1907 theory of specific heat in solids, and other developments, for Planck to finally declare in 1911 that the hypothesis of the quanta reflects physical reality and marks the beginning of a new era in physics.

From this point on, quantum ideas started rolling down the unsuspected physicists like an avalanche burying underneath the entire classical world of objective reality and certainty. Or so it seemed. Einstein opened the can of worms by proposing the notion of light quanta, which introduced into the conscience of physicists the idea of the wave–particle duality. In 1923 this idea was picked up by a young French Ph.D. student Louis de Broglie who in his dissertation suggested that not only light, normally thought to be a wave, can behave as a stream of particles, but also regular particles—electrons, neutrons, atoms, etc.— can manifest wavelike properties. He postulated that Einstein's relations connecting the wave characteristics of light (frequency, v, and wavelength, λ) with its particle's characteristics (energy E and momentum p)

$$E = hv$$ (1.2)

$$p = h/\lambda$$ (1.3)

---

[1]The actual story of this formula and Planck's contribution to quantum physics is not quite that simple. First, the ultraviolet catastrophe apparently did not motivate Planck at all as he did not think that it was an unavoidable logical consequence of classical physics. Second, Planck first obtained his formula empirically by trying to fit the experimental data and only after that found a theoretical "explanation" for it. Third, he did not believe that his quantized oscillators represent real atoms or molecules for quite some time and accepted the reality of energy quantization only very reluctantly. This story is described in more detail in the article by H. Kragh "Max Planck: the reluctant revolutionary" published in December 2000 in *Physics World*.

can be reversed and applied to electrons, protons, and other material particles. A significant difference between light and electrons, of course, is that the latter have a finite mass and obey a quadratic relation between energy and momentum, $E = p^2/2m_e$, where m_e is the mass of the particle, while the former is characterized by the linear relation $E = pc$, where c is the speed of light, following from the relativity theory for particles with zero rest mass. This difference as you will see later plays a crucial role in quantum theory.

The revolutionary *matter wave* idea of de Broglie was confirmed experimentally short 4 years later, in 1927, by American physicists Clinton Davisson and Lester Germer working at Bell Labs and independently by British scientist George Paget Thomson at the University of Aberdeen. Davisson and Germer observed diffraction of electrons propagating through crystalline nickel, while Thomson studied electrons propagating through a metal film.[2] These achievements resulted in the 1929 Nobel Prize for de Broglie and the shared 1937 Nobel prize for Davisson and Thomson. So, if you thought that the quantization of energy was a revolutionary idea, then the wave–particle dualism must really blow your mind: how can something be simultaneously a particle (localized, indivisible, countable entity) and a wave (extended, continuous, arbitrarily divisible excitation of a medium)? To wrap his mind around this oddity of the quantum world, Danish physicist Niels Bohr came up with his famed complementarity principle, which essentially states that all experiments that one can conduct with objects of atomic scale can be divided into two never-overlapping groups. In the experiments belonging to the first group, the material objects reveal their particle-like side, and in the experiments of the second group, they present to the world their wavelike quality, and it is impossible to design an experiment, in which both sides are manifested together. In the book *Evolution of Physics* by L. Infeld and A. Einstein, this principle was presented in the following way:

> But what is light really? Is it a wave or a shower of photons? There seems no likelihood for forming a consistent description of the phenomena of light by a choice of only one of the two languages. It seems as though we must use sometimes the one theory and sometimes the other, while at times we may use either. We are faced with a new kind of difficulty. We have two contradictory pictures of reality; separately neither of them fully explains the phenomena of light, but together they do.

Even though this quote refers to the properties of light, it can be repeated almost verbatim for the quantum description of any atomic object. Bohr's own formulation of the complementarity as presented in his famous 1927 lecture at a conference in the Italian town of Como is somewhat deeper even if more vague when taken out of the broader context:

> any given application of classical concepts precludes the simultaneous use of other classical concepts which in a different connection are equally necessary for the elucidation of the phenomena.

---

[2]Here is a historical irony for you: George Thomson, who got the Nobel Prize for proving that electrons are waves, is a son of J.J. Thomson (not to be confused with W. Thomson—Lord Kelvin), a prominent English physicist, who got the Nobel Prize for proving that an electron is a particle.

What Bohr is implying here is that by virtue of the macroscopic size of an observer (that would be us, humans), any measurement is necessarily conducted by a large macroscopic apparatus, which is supposed to be fully describable by the laws of classical physics. The results of the measurements, therefore, are interpreted in terms of classical concepts such as momentum, position, energy, time, etc. Then the complementarity principle poses that one cannot design a single experiment in which classical quantities belonging to complementary classes such as momentum and position, energy and time, etc. can be determined. Thus measuring a position of a quantum object, i.e., trying to localize it at a point in space and time, reveals this object's particle-like characteristics simultaneously destroying any traces of its wavelike behavior. The most frequently discussed illustration of this idea is a double-slit interference experiment beautifully analyzed, for instance, in the famous *The Feynman Lectures on Physics*, which are now freely available online at http://www.feynmanlectures.caltech.edu/, and I encourage you to go ahead and peruse them. The scale disparity between observers and atomic objects is what makes quantum theory so difficult to "understand"—our vocabulary developed to reflect the macroscopic world of our own scale fails when we try to apply it to the world of atoms and electrons. This is how Richard Feynman describes it:

> Because atomic behavior is so unlike ordinary experience, it is very difficult to get used to, and it appears peculiar and mysterious to everyone—both to the novice and to the experienced physicist. Even the experts do not understand it the way they would like to, and it is perfectly reasonable that they should not, because all of direct, human experience and of human intuition applies to large objects. We know how large objects will act, but things on a small scale just do not act that way. So we have to learn about them in a sort of abstract or imaginative fashion and not by connection with our direct experience.[3]

In 1927 Heisenberg found himself locked in the battle of ideas with Austrian physicist Erwin Schrödinger, the author of the famed Schrödinger equation, who took the de Broglie matter wave idea close to his heart and introduced a wave function satisfying a simple differential equation, which he tried to interpret quite literally as a quantity representing an actual electron smeared over some small but finite region of space. It differed strongly from Heisenberg's approach based upon the idea of "quantum jumps" introduced in Bohr's model of a hydrogen atom (I am sure you heard about Bohr's quantization postulates and his planetary model of an atom so I do not have to reproduce it here), which he described using abstract algebraic quantities later found by Heisenberg's collaborators and compatriots Ernst Pascual Jordan and Max Born to be matrices. Schrödinger's interpretation of the matter waves was very appealing, especially to older physicists, because it preserved most of the classical world view—continuous evolution of electron's charge density in space–time with no incomprehensible discontinuities introduced by quantum jumps. You might be amused by the following exchange between Schrödinger and Bohr that took place during Schrödinger's visit to Copenhagen in October of 1926[4]:

---

[3]R. Feynman, R. Leighton, M. Sands, *Feynman Lectures on Physics*, vol. 1, Ch. 37, online edition available at http://www.feynmanlectures.caltech.edu/.

[4]Quoted from the book by W. Moore, *Schrödinger. Life and Thought* (Cambridge University Press, Cambridge, 1989).

Schrödinger: "You surely must understand, Bohr, that the whole idea of quantum jumps necessarily leads to nonsense. ...the electron jumps from this orbit to another one and thereby radiates. Does this transition occur gradually or suddenly? If it occurs gradually, then the electron must gradually change its rotation frequency and energy. It is not comprehensible how this can give sharp frequencies for spectral lines. If the transition occurs suddenly, in a jump so to speak, ...one must ask how the electron moves in a jump. Why doesn't it emit a continuous spectrum? And what laws determine its motion in this jump?"

Bohr: "Yes, in what you say you are completely right. But that doesn't prove that there are no quantum jumps. It only proves that we can't visualize them, that means that the pictorial concepts we use to describe the events of everyday life and the experiments of the old physics do not suffice also to represent the process of a quantum jump. That is not surprising when one considers that the processes with which we are concerned here cannot be the subject of direct experience ...and our concepts do not apply to them".

Schrödinger: "I do not want to get into a philosophical discussion with you about the formation of concepts ...but I should simply like to know what happens in an atom. It's all the same to me in what language you talk about it. If there are electrons in atoms, which are particles, as we have so far supposed, they must also move about in some way. At the moment, it's not important to me to describe this motion exactly; but it must at least be possible to bring out how they behave in a stationary state or in a transition from one state to another. But one sees from the mathematical formalism of wave or quantum mechanics that it gives no rational answer to this questions. As soon, however, as we are ready to change the picture, so as to say that there are no electrons as particles but rather electron waves or matter waves, everything looks different. We no longer wonder about the sharp frequencies. The radiation of light becomes as easy to understand as the emission of radio waves by an antenna..."

Bohr: "No, unfortunately, that is not true. The contradictions do not disappear, they are simply shifted to another place...Think of the Planck radiation law. For the derivation of this law, it is essential that the energy of the atom have discrete values and change discontinuously ... You can't seriously wish to question the entire foundation of quantum theory."

Schrödinger: "Naturally, I do not maintain that all these relations are already completely understood ... but I think that the application of thermodynamics to the theory of matter waves may eventually lead to a good explanation of Planck's formula"

Bohr: "No, one cannot hope for that. For we have known for 25 years what the Planck formula means. And also we see the discontinuities, the jumps, quite directly in atomic phenomena, perhaps on the scintillation screen or in a cloud chamber ... You can't simply wave away these discontinuous phenomena as though they didn't exist."

Schrödinger: "If we are still going to have to put up with these damn quantum jumps, I am sorry that I ever had anything to do with quantum theory."

Bohr: "But the rest of us are very thankful for it—that you have—and your wave mechanics in its mathematical clarity and simplicity is a gigantic progress over the previous form of quantum mechanics."

By the end of this discussion, Schrödinger fell ill and for a few days had to stay as a guest in Bohr's house where Bohr's wife, Margrethe, was taking care of him. In the same year Schrödinger wrote:[5]

My theory was inspired by L de Broglie ...and by short but incomplete remarks by A Einstein. ...No genetic relation whatever with Heisenberg is known to me. I knew of his theory, of course, but felt discouraged not to say repelled, by the methods of transcendental algebra, which appeared very difficult to me and by the lack of visualizability.

---

[5]E. Schrödinger, On the relationship between the Heisenberg-Born-Jordan quantum mechanics and mine. Ann. Phys. **70**, 734 (1926).

Heisenberg fully understood the weakness of the matrix theory, writing in his 1925 paper, coauthored with Born and Jordan[6]:

> Admittedly, such a system of quantum-theoretical relations between observable quantities … would labor under the disadvantage that there can be no directly intuitive geometrical interpretation because the motion of electrons cannot be described in terms of the familiar concepts of space and time.

The popularity of Schrödinger's matter waves not only spelled for him purely scientific troubles as a competing theory but also jeopardized his career: he was ripe to search for a permanent university position, and fondness of professors in charge of the academic appointments for Schrödinger's views created problems for Heisenberg. Thus, in 1927, Heisenberg undertook concentrated efforts to remedy the perceived weaknesses of his approach to quantum theory, which resulted in the celebrated Uncertainty Paper,[7] where his now famous uncertainty principle

$$\Delta x \Delta p \geq \frac{\hbar}{2} \tag{1.4}$$

was shown to be an unavoidable consequence of the inner structure of his quantum formalism (Δx and Δp are loosely defined "uncertainties" of the position and momentum). An equally important portion of the paper was devoted to expounding the connections between the formalism and the real world. Heisenberg made a point to emphasize that in the end the goal of the theory is to predict values of those quantities, which can be observed by a clearly defined physical process. Considering operational and physical details of an observation of the coordinate, Heisenberg showed that it is impossible to carry out such an observation without disturbing the velocity of the electron, and therefore, such concepts as trajectory or continuous time dependence of a particle's coordinate shall have no place in the theory of the electron's properties. He illustrated these ideas with a thought experiment involving observation of an electron using a gamma-ray microscope. The idea of the experiment is that in order to actually observe ("see") a position of the electron, one has to shine light on it and detect the reflected (or scattered if you prefer) rays using a microscope. The accuracy in measuring the position is limited by the microscope's resolution (electron can be anywhere within a region resolved by the microscope), which is proportional to the wavelength of light: the shorter the wavelength, the smaller region of space it is able to resolve. Thus to determine the precise position of the electron, one must use light of a very short wavelength (hence, the gamma-ray microscope in Heisenberg's paper). The problem is, however, that light of very short wavelength has very large momentum (see Eq. 1.3), which it transfers to the electron changing its momentum by the amount which is inversely proportional to the wavelength. The precise value of the electron's new momentum and its direction

---

[6]M. Born, W. Heisenberg, P. Jordan, Zeitschr. Phys. **35**, 557 (1926).

[7]W. Heisenberg, The physical content of quantum kinematics and mechanics. Zeitschr. Phys. **43**, 172 (1927).

are completely unknown, and, therefore, trying to improve our knowledge about electron's position, we destroy our ability to know its momentum.

Heisenberg's uncertainty principle and the idea that it is the process of measurement that defines what can be known about the system contributed to Bohr's thinking about his complementarity principle. Taken together with the uncertainty principle and Born's interpretation of the results of quantum measurements in terms of probabilities,[8] it eventually resulted in the complete mathematical and conceptual framework of quantum theory known as the Copenhagen interpretation. According to this interpretation, the observable quantities describing a quantum system do not have any definite value before they are actually measured and can randomly take one of the allowed values only after a measurement is performed. The act of measurement abruptly changes the system bringing it in the state corresponding to the realized value of the measured quantity.

Not everyone was happy with the Copenhagen interpretation, including one of the originators of the quantum revolution, Albert Einstein. He couldn't reconcile himself with the loss of the classical determinism—the idea that the true laws of nature must provide us with complete and precise information about all essential characteristics of material objects and that given the right tools, we can always experimentally measure them. Einstein wrote to Max Born in 1926[9]:

> Quantum mechanics is certainly imposing. But an inner voice tells me that it is not yet the real thing. The theory says a lot, but does not really bring us any closer to the secret of the "old one." I, at any rate, am convinced that **He does not throw dice**.

"He" in this context is a direct reference to the Creator, who appears explicitly in a catchier and better known expression of the same idea recorded in the book *Einstein and the Poet* by William Hermann: "God does not play dice with the world." Leaving aside the issue of religious beliefs of Einstein, you need to understand why this particular feature of the quantum theory bothered him that much. After all, Einstein had no objections to using probability in classical statistical physics and successfully used probabilistic concepts himself in his work on Brownian motion. The difference of course comes from the fact that the probabilities in classical statistical physics reflect the objective lack of information about positions and velocities of individual molecules, and in the absence of this information, the language of the distribution functions and probabilities is the best way to deal with this situation. In the Copenhagen interpretation of quantum mechanics, the statistical uncertainty was made inherent to the nature of the universe, and Einstein had difficulty with this idea. For many years, Einstein and Bohr participated in public discussions and exchanged letters, in which Einstein tried to find a counter-example to the Copenhagen interpretation and Bohr debunked all of them. These exchanges between Einstein and Bohr are the finest example of the scientific oral and epistolary debates.

---

[8]M. Born, On the quantum mechanics of collisions. Zeitschr. Phys. **37**, 863 (1926).

[9]Letter to Max Born (4 December 1926); The Born-Einstein Letters (translated by Irene Born) (Walker and Company, New York, 1971).

In this book, I will rely on the traditional Copenhagen interpretation of quantum theory and will present it taking the "bull by the horns": in spirit of Heisenberg's views that only those quantities which can be really observed in a particular set of experiments shall be used to describe quantum systems, I will start with the abstract concept of the quantum state defined by a set of such observable quantities. Then I will place this concept within a mathematical structure of a linear vector space and will proceed from there. The Schrödinger wave function will appear much later in the text (as compared to most other undergraduate-level textbooks) as one of many alternative and equivalent ways of describing the states of quantum systems.

However, before plunging into the depth of the quantum formalism and its application, I would like to provide you with the historical background on the times when quantum theory was born. While it will not help you understand the physics better, it will help you see that advances in physics were not an isolated incident but a part of a general trend in the movement of humanity toward modernity. You will also be able to appreciate that people responsible for the birth of quantum theory are not just abstract great names—they are real people with different views on life, different political preferences, and moral principles, people forced to make difficult choices outside of their professional lives and bearing responsibility for their choices.

1.2   Brief Overview of the Historical Background

The history of quantum theory from earlier childhood to maturity covers the time period between 1900 and 1930 and involves such countries as Germany, Austria, France, Italy, Denmark, the UK, and the USA. During this time, not only the facade of classical physics crumbled, but the entire world changed drastically. It was the time of great upheavals and great discoveries, unendurable misery, and unbeatable achievements in science, architecture, arts, music, and literature. The period between 1900 and 1914 was the time of inter-European and even transcontinental integration (or what we would call now globalization) with free flow of goods, people, and capital across the borders and the time of liberalization and democratization, when even monarchic regimes of Germany, Austria-Hungary, and Russia provided more political rights and more freedom to their citizens. The liberalization and the growth of wealth that accompanied it were beneficial for developments in science, arts, music, and literature. Physics was not an exception— the free travel between countries and free exchange of ideas created fertile ground for new developments, which included relativity theory and quantum mechanics.

At the same time, the wealth and prosperity weren't shared by all, and the socialist ideas started attracting a significant number of followers. The Austro-Hungarian Empire proved to be the less stable and vulnerable to the rising nationalism of smaller nations comprising it. On the surface, the everyday life of people in European countries appeared to follow the normal "business like usual" path; underneath of this placidity the tensions were rising, especially in the Balkans. This was the calm before the storm as everything came crushing down on June

28, 1914, with the assassination of the heir to Austro-Hungarian throne, Archduke Franz Ferdinand, at the hands of a Serbian teenage nationalist. On July 28 Austria declared war on Serbia, and on August 1 and 3, Germany declared war on Russia and France correspondingly and invaded Belgium. On August 4 Britain declared war on Germany and World War I was in full swing.

In response to the war, the nationalist and militarist fever was rising even among intellectuals: scientists, artists, and poets. Such prominent German physicists as Nobel Prize winners Max Planck, Philipp Lenard (photoelectric effect), Walther Nernst (third law of thermodynamics), Wilhelm Roentgen (X-rays), and Wilhelm Wien (black-body radiation) signed the infamous *Manifesto of 93* declaring unequivocal support of the German occupation of Belgium, the actions known in history as the Rape of Belgium. The war devastated Europe: 18 million died (11 million military personnel and 7 million civilians), and 23 million were wounded. This war witnessed the first widespread use of chemical weapons on the battle field, most effectively by the German Army but also by the Austrians, French, and British. The chemical terms chlorine and mustard gas became household names.

During the war, not surprisingly, there was a lull in the development of physics in general and quantum theory in particular as many scientists on all sides participated in the war efforts. In addition to direct deaths, suffering, and destruction, the war started a chain of events that led to the even greater horrors of World War II. The Austrio-Hungarian Empire disintegrated, leaving in its wake a multitude of smaller countries in Central and Eastern Europe (Austria, Hungary, Yugoslavia, Czechoslovakia, Poland); the Russian Empire went through the bloody "Bolshevik" revolution, which gave birth to a cruel totalitarian regime under the guise of the "Dictatorship of Proletariat." Germany lost the war and was forced to sign the harsh Versailles peace treaty. By the terms of the treaty, Germany and her allies took all responsibility for all the losses and damages during the war, had to pay heavy reparations, and lost significant territory, especially Alsace-Lorraine, and output of coalmines of Saar to France, Upper Silesia, Eastern Pomerania, and part of Eastern Prussia to Poland. The treaty left Germany humiliated and fostered feelings of resentment and a desire for revenge among large segments of the German population.

The German Empire disintegrated and was replaced by the weak and ineffective Weimar Republic, which survived until 1933, when Hitler, newly appointed as Chancellor, suspended democratic procedures and declared himself Fuhrer of the Third German Reich. In Germany and to a lesser extent in Austria, the period of Weimar Republic was the time of high unemployment, hyperinflation, and intense and often violent political struggles between socialists, liberal democrats, communists, conservatives, and national socialists. But it was also the time of unprecedented explosion of creativity in science, architecture, literature, film, music, and arts. Berlin became a thriving cosmopolitan city, the center of attraction for artistic and scientific elites. In addition to quantum theory, this time produced significant advancements in nuclear physics and radioactivity, and engineering, but also gave rise to eugenics and radical racial theories.

Fig. 1.1 From back to front and from left to right: Auguste Piccard, Émile Henriot, **Paul Ehrenfest**, Édouard Herzen, Théophile de Donder, **Erwin Schrödinger**, Jules-Émile Verschaffelt, **Wolfgang Pauli**, **Werner Heisenberg**, Ralph Howard Fowler, Léon Brillouin, Peter Debye, Martin Knudsen, William Lawrence Bragg, Hendrik Anthony Kramers, **Paul Dirac**, Arthur Compton, **Louis de Broglie**, **Max Born**, **Niels Bohr**, Irving Langmuir, **Max Planck**, Marie Skłodowska Curie, Hendrik Lorentz, **Albert Einstein**, Paul Langevin, Charles-Eugène Guye, Charles Thomson Rees Wilson, Owen Willans Richardson

Quantum theory reached its maturity between 1923 and 1930, thanks to the resumption of international contacts and free flow of information. A big role in fostering the progress was played by Solvay conferences that took place in Brussels and financed by Belgian industrialist Ernest Solvay. The first two conferences took place in 1911 and 1913 and resumed in 1921 after an 8-year interruption due to the war. These conferences were attended by all the main players in physics and chemistry of the times. The photograph above (Fig. 1.1) shows the attendees of the fifth conference that took place in 1927 and was a culmination of a struggle between Bohr and Einstein's views on the interpretation of quantum mechanics. Bohr won, and from that time on, the Copenhagen interpretation dominated physicists thinking about quantum theory.

By the end of the 1920s, the economic situation in Germany started improving, but the market crash of 1929 and the start of Great Depression in the USA interrupted the recovery. The economics of Weimar Republic fell into the abyss facilitating the rise of Nazis to power in 1933. The intellectual atmosphere in Germany and Austria had already begun to deteriorate in the beginning of the 1930s with the rise of the movement for Aryan physics spearheaded by such luminaries as Philipp Lenard and Johannes Stark, but after 1933 the situation for German physicists of Jewish origin became intolerable. Albert Einstein visited the USA in 1933 and never returned; the same year, Max Born was suspended from his position at the University of Göttingen and immigrated first to England and then to the USA. Nazis drove away Jewish scientists not only from Germany and Austria but also from all of Central and Eastern Europe. Here is the incomplete list of the refugee physicists: Hans Bethe, John von Neumann, Leo Szilard, James Franck, Edward Teller, Rudolf Peierls, and Klaus Fuchs. Enrico Fermi, whose wife was Jewish, fled fascist Italy. Ironically, while trying to preserve the racial purity of their science, Nazis destroyed German predominance in physics (as well as in other areas of intellectual and artistic pursuit) and made America into a science powerhouse.

Not only physicists of Jewish origin bore the wrath of the Nazis in Germany and Austria. Erwin Schrödinger, who was known to oppose Nazism, was ordered not to leave the country after Hitler declared *Anschluss* (union) between Germany and Austria in 1938. Luckily, he managed to escape to Italy, from where he moved to the UK and finally settled in Dublin as the Head of the Institute for Advanced Studies. During this time, he tried to develop a unified field theory, but his most important work of this period was the book *What Is Life*, where he introduced the idea that complex molecules can contain genetic information. He returned to Vienna in 1956.

However, not all physicists fled, and some even joined the National Socialist Party. Among the most prominent Nazi physicists was the aforementioned Philipp Lenard, who contributed to the discovery of photoelectric effect and won the Nobel Prize in 1905. He was an ardent anti-Semite and dismissed Einstein's works as "Jewish science." Lenard lived through the war and was demoted from his emeritus position at Heidelberg University in 1945 by Allied forces.

Especially sad is the case of Born's student and collaborator Pascual Jordan. He joined the Nazi party in 1933 and even became the member of its SA unit[10] and enlisted in Luftwaffe (German Air Force) in 1939. During World War II, he attempted to interest party officials in various weapon schemes but was deemed politically unreliable. Indeed, to his honor, he refused to condemn Einstein and other Jewish physicists. After the war, Wolfgang Pauli interceded on Jordan's behalf declaring him rehabilitated. This allowed Jordan to continue his academic career and even secure a tenured position in 1953. Still, flirting with Nazism costed him the Nobel Prize, which he would have probably shared with Born.

I also feel obliged to mention that some German physicists who remained in Germany during the Nazi era, while not actively opposing the regime (which would be akin to signing a death sentence), still behaved in a very noble way. Arnold Sommerfeld, who was nominated 84 times for the Nobel Prize, but never won, and was among the earlier contributors to quantum theory, was an admirer of Einstein and made special efforts to help his Jewish students and assistants, such as Rudolf Peierls and Hans Bethe, find employment outside of Germany. Another patriarch

---

[10]Sturmabteilung, or SA—the original paramilitary wing of the Nazi party—played a significant role in Hitler's rise to power.

of German science, Otto Hahn, who made significant contributions to physics and chemistry of radioactivity and won the 1944 Nobel Prize, was an opponent of national socialism. Einstein wrote that Hahn was "one of the very few who stood upright and did the best he could in these years of evil." For instance, he fostered a longtime collaboration with an Austrian physicist with Jewish roots, Lise Meitner. In 1938, Hahn helped her escape from Berlin to the Netherlands, giving her his mother's diamond ring to bribe the frontier guards if needed. After the war, Hahn became the founding president of the Max Planck Society in the new Federal Republic of Germany and one of the most respected citizens of the new country.

Werner Heisenberg, on the other hand, found himself in a very difficult situation. While not a Nazi, he thought of himself as a German patriot and believed that Hitler was a necessary evil to save Germany. So, he stayed put during the Nazi period justifying it by the desire to preserve German physics. He also agreed to lead the Nazi nuclear program. Obviously, his relationship with former friends became very strained. It is known that he visited Bohr in 1941 while Denmark was under Nazi occupation. The content of the meeting remains a mystery, and Bohr refused to provide his account of what transpired, but the meeting did not go too well and Bohr was visibly upset. Bohr wrote down his account of the meeting, but it was sealed in his personal papers by the decision of his family. The mystery of this meeting inspired an award-winning play *Copenhagen* by Michael Frayn, and of the film by the same name, where the role of Heisenberg was played by Daniel Craig (future James Bond in the last three installations of the series: *Casino Royale*, *Quantum of Solace*, and *Skyfall*). I personally found Craig very convincing as Heisenberg, even when he talked about scientific matters. After the war, Heisenberg was cleared of accusations in Nazi collaboration in the course of the denazification process and was allowed to continue his scientific career. However, his behavior during the Nazi time isolated him from other European and American physicists.

Chapter 2
Quantum States

2.1 Classical and Quantum States

The task of any physical theory is to develop the means to predict the results of
the measurements of a physical quantity sometime in the future based upon the
information about the current state of the system under study and knowledge about
interaction of this system with its environment. The term "state" has many different
uses in physics—it is used to describe different **states** of matter (solids, liquids,
etc.)—in thermodynamics we derive various **equations of state**, relations between
various thermodynamic parameters. We can also talk about a particular **state** of a
system, meaning specific values of a set of quantities important for a problem at
hand. In this book I will consider, for the most part, systems consisting of a small
number of particles, placed in a variety of different environments. The states, which
I will be dealing with here, are "mechanical states," but the precise meaning of this
term depends upon the choice of a conceptual framework, classical or quantum, with
which the problem is approached.

In classical physics a mechanical state is most completely described by specifying coordinates and velocities of the particles at any given time. As Laplace said:

> Given for one instant an intelligence which could comprehend all the forces by which nature
> is animated and the respective positions of the beings which compose it, if moreover this
> intelligence were vast enough to submit these data to analysis, it would embrace in the same
> formula both the movements of the largest bodies in the universe and those of the lightest
> atom; to it nothing would be uncertain, and the future as the past would be present to its
> eyes.

In modern language it means that acting forces and initial coordinates and velocities
determine future positions of all bodies in the universe with any accuracy limited
only by the accuracy of the available experimental and computational instruments.
Mathematically, the evolution of classical states of a single particle is described
by ordinary differential equations of the second order, where given values of
coordinates and velocities form a set of initial conditions necessary to find their

© Springer International Publishing AG, part of Springer Nature 2018
L.I. Deych, *Advanced Undergraduate Quantum Mechanics*,
https://doi.org/10.1007/978-3-319-71550-6_2

unique solution. I am ignoring here complications arising in the cases of nonlinear chaotic dynamics, when a uniquely existing solution becomes unstable with respect to small variations in initial conditions, so that practical predictability of the equations of motion is lost. These situations are excluded from consideration in this book.

Finding states of a classical particle can be significantly simplified if the system under consideration possesses conserving quantities, called integrals of motion, such as energy or angular momentum. These quantities do not change in the course of the time evolution of the system, and, therefore, their values are determined by the initial conditions. Knowledge of these quantities makes solving the equations of motion easier: for instance, a differential equation of the second order can be reduced to the equation of the first order, or a three-dimensional problem can be converted into its one-dimensional equivalent. At the same time, while simplifying the technical task of solving equations of motion, the existence of integrals of motion does not change the fundamental nature of the classical description of the system.

In a quantum world, we are forced to give up the ability to have complete knowledge of all desirable parameters (coordinates, velocities, energy, etc.) and, therefore, have to redefine the meaning of term "state." To develop the concept of quantum state, I first introduce the idea of an observable defined as any quantity whose numerical value can be experimentally measured. The list of possible observables is essentially the same as the list of classical parameters: it can include coordinates, momentums, energies, angular momentums, etc. The main difference between quantum and classical descriptions appears, however, when you recognize that in the quantum world, according to the complementarity principle discussed in the Introduction, not all observables can be measured within the same set of experiments. For instance, reinterpreting the Heisenberg's uncertainty principle, you may say that if a system is in a state with precisely known coordinate, its momentum cannot be prescribed any definite value and vice versa—if the system is in the state with precisely known momentum, its coordinate remains completely undefined. However, you can imagine that there might exist more than one observable, whose values can be found with certainty for the same state of the system (an obvious example of such observables is three components of the position vector or three components of the vector of momentum). Such observables, called mutually consistent, play an important role in quantum theory. The largest set of such observables is called a **complete set of mutually consistent observables**. Mutually consistent observables often correspond to classical integrals of motion such as energy, angular momentum, etc., which in quantum theory are of much greater importance than in classical mechanics. The complete set of mutually consistent observables is all the information which we can have about a quantum state of a system, and therefore, it seems reasonable to define a quantum state simply as a collection of values of these observables.

To proceed I need to translate the words into some kind of mathematical formulas or at least something which looks like mathematical formulas. So, embrace yourself as I am going to hit you with some heavy formal stuff. Let's say that the complete set of compatible observables contains N_{max} elements $\{q^{(i)}\}$, $1 \leq i \leq N_{max}$. Let $q_k^{(i)}$ represent a k-th value of i-th observable, and let us assume that any given observable q_i can either change continuously or take a discrete set of values. In the former case, we say that the observable has a continuous spectrum, while in the latter case, observables are said to have a discrete spectrum. A British scientist Paul Dirac (I will tell you more about him later in the book), one of the founding fathers of quantum theory who shared the 1933 Nobel Prize in Physics with Schrödinger, suggested a rather picturesque and intuitively appealing way of representing a quantum state as

$$\left| q_k^{(1)}, q_m^{(2)}, \cdots q_p^{(N_{max})} \right\rangle. \tag{2.1}$$

For now this is just a fancy notation, but as I will continue to develop quantum formalism, its utility will become more and more apparent. Information contained in Eq. 2.1 is quite transparent: this expression tells us that if the system is in the state given by Eq. 2.1 and we measure observable $q^{(i)}$, the result of the measurement will be value $q_k^{(i)}$. States, in which at least one of the observables has different values, are mutually exclusive, in a sense that if you repeat the measurement over and over, you will never observe two distinct values of the same observable as long as the measurement is performed on the system in the state described by Eq. 2.1.

The states of the type presented in Eq. 2.1, in which all mutually consistent observables have definite values, are the simplest, but not the only possible states of quantum systems. Actually, as you already know, the whole brouhaha about quantum mechanics and Einstein's rejection of it was because of the fact that frequently observables would not have definite values, so that one could not predict with certainty an outcome of a measurement. If a measurement is performed on an ensemble of identical systems, all placed in such a state (same for all systems), different members of the ensemble would generate different outcomes in an unpredictable manner. A similar situation arises in the case of repeating measurements on the same system provided it is returned back to its initial state after each measurement. From a theoretical point of view, states with uncertain outcomes of a measurement can be described as a linear superposition of the simple states as discussed in the next section.

2.2 Quantum States and Hilbert Vector Space

2.2.1 Superposition Principle

Experimental evidence of existence of quantum superposition states comes from observations of interference of electrons or neutrons described in any textbook dealing with the basic ideas of quantum mechanics. To explain the connection between interference and superposition, we usually draw on an analogy with electromagnetic waves, where interference is a result of the addition of two spatially overlapping coherent waves. Depending upon the phase difference between the waves at any particular point in space, one can observe either bright interference fringes (constructive interference resulting from the phase difference $2\pi n$, where n is an integer) or dark fringes (destructive interference resulting at points where the phase difference is $(2n+1)\pi$). Thus, the argument goes, the interference of quantum particles must also result from the superposition of something having wavelike properties. The particle–wave duality was briefly described in the Introduction, and it is also discussed in lots of other textbooks including the already quoted *The Feynman Lectures on Physics*, so I will not dwell any longer on that. I will, however, use the existence of quantum interference to justify introducing superposition of the states defined in Eq. 2.1. While interference experiments provide convincing but indirect evidence for quantum superposition, recently, superposition states of few photons and atoms have been observed directly.

Let me assume that a quantum system can be in one of two states $|q_1\rangle$ and $|q_2\rangle$ characterized by different values q_1 and q_2 of some observable, q, with a discrete spectrum. According to the superposition principle, this system can also be in a state formed by a linear superposition of these two states:

$$|superposition\rangle = a_1 |q_1\rangle + a_2 |q_2\rangle \tag{2.2}$$

where a_1 and a_2 are, in general, complex numbers. Equation 2.2 can be interpreted verbally by saying that in order to form a superposition state $|superposition\rangle$, one must "multiply" each of the states $|q_1\rangle$ and $|q_2\rangle$ by complex numbers a_1 and a_2, respectively, and then "add" the results. The problem here is, of course, that since we have no idea about the mathematical nature of the object I call "quantum state" (is it a function, or a number, or some other mathematical object?), I cannot actually tell you how to perform operations I called multiplication and addition and what they actually mean (this is why I surrounded "add" and "multiply" by the quotation marks). Is it possible to make sense of Eq. 2.2 without first assigning some more concrete mathematical meaning to these "quantum states"? It might surprise you, but the answer is yes, it is possible, and mathematicians do this all the time. It is not really necessary to know either the mathematical nature of the state objects or the meaning of algebraic operations we want to perform with them. All what we need is to postulate that these objects and operations exist and possess certain properties. If this sounds too abstract for you, consider this. Modern object-oriented computer languages such as C++ are based exactly on this idea—they introduce an "object," which can be anything (a number, a matrix, a word, a figure), and define various operations with them such as addition, multiplication, etc. All these operations have well-defined properties, but their concrete meaning depends upon an object to which they are applied. Thus, symbol "+" between two numbers means one thing, while the same symbol between words or matrices mean something completely different, but it still is the same operation because it has the same basic properties.

2.2.2 Linear Vector Spaces

I am afraid that now I have to get a bit more abstract with you than you would have probably liked, but do not get too frustrated about it. After all you probably have passed a bunch of calculus classes taught by math professors forcing you to learn proofs of all those theorems of existence and uniqueness. The stuff I am feeding you here is just small potatoes compared to that. Anyway, I will begin by postulating that all quantum states can be represented by special objects belonging to a certain class or "space" defined in such a way that if objects given by Eq. 2.1 belong to this "space," then the objects defined by Eq. 2.2 also belong to it. (Note that the word "space" here has a completely different meaning than in our everyday language or even in the language of the introductory physics courses and replaces such words as class or set of objects with special properties.) I will also assume that there exists a *null* object $|0\rangle$ such that $|q\rangle + |0\rangle = |q\rangle$, $0 \cdot |q\rangle = |0\rangle$ (where 0 in the last expression is just a number zero unlike $|0\rangle$, which is used to designate the null state object, and a dot $\cdot$ means multiplication by a number). I will also postulate a few distributive and associative properties (ditching the dot $\cdot$ for the sake of compactness and out of habit) such as

$$a\left(|q_1\rangle + |q_2\rangle\right) = a\,|q_1\rangle + a\,|q_2\rangle \tag{2.3}$$

$$a_1\,|q_i\rangle + a_2\,|q_i\rangle = (a_1 + a_2)\,|q_i\rangle \tag{2.4}$$

$$a_1\left(a_2\,|q_i\rangle\right) = (a_1 a_2)\,|q_i\rangle \tag{2.5}$$

which allows carrying out standard algebraic operations with the quantum state objects.

A familiar example of objects, which possess all the properties specified in Eqs. 2.2–2.5 provided that the coefficients a_i are confined to the set of real numbers, is given by the usual three-dimensional vectors, such as displacement, or velocity vectors used in elementary mechanics. All operations used in Eqs. 2.2–2.5 have in this case specific definitions: multiplication by a number is defined as multiplication of a vector's length by a number while keeping its direction intact for positive coefficients and reversed for negative coefficients, and the addition of vectors is defined by a triangle or parallelogram rule. It is a matter of simple geometry and algebra to prove the distributive and associative properties of these operations as presented in Eqs. 2.3–2.5.

Abstract objects satisfying the abovementioned quantities are also called vectors and are said to belong to a linear vector space. I will use notation based on Greek letters such as $|\alpha\rangle$, $|\beta\rangle$, etc. to represent generic elements of the vector space (not necessarily vectors based on values of certain physical observables). Even though these abstract vectors are quite different from what you are used to calling vectors in classical mechanics (e.g., they do not point to any direction in our regular three-dimensional space and can have infinitely many components, represented by complex numbers), I am going to use some of what you know about regular vectors to help you infer additional properties of our abstract vector objects.

For instance, you know that regular three-dimensional vectors can be presented as a sum of three mutually perpendicular unit vectors, whose directions are predetermined by an arbitrary choice of three coordinate axes (X, Y, and Z). These unit vectors form what is called a *basis* in the three-dimensional space—a set of vectors which cannot be expressed as linear combinations of one another but can be used to present any other vector in this space as their linear combination: $A = A_x e_x + A_y e_y + A_z e_z$, where e_x, e_y, e_z are the unit vectors in the direction of X, Y, and Z axes, respectively. In order to expand the idea of basis to the arbitrary space of abstract linear vectors, I first have to acquaint you with the concept of linear independence of vectors. A set of vectors $|q_1\rangle$, $|q_2\rangle$, $\cdots |q_N\rangle$ is called linearly independent if none of the vectors in the set can be presented as a linear combination of others. A set of linearly independent vectors is complete if adding any other distinct vector to the set makes it linearly dependent. The number of linearly independent vectors in the complete set determines the dimension of the space. This number does not need to be finite—there are spaces with infinite number of linearly independent vectors. From the definition of the complete set of linearly independent vectors, it follows that any other vector in a given space can be presented as their linear combination:

$$|\alpha\rangle = \sum_i a_i |q_i\rangle \tag{2.6}$$

where summation is over all vectors in the complete set. Such a set of vectors is called a basis in a given space. Apparently, representation of an arbitrary vector in the form of Eq. 2.6 is a formal generalization of the physical superposition principle. I can also identify the set of states characterized by different values of the complete set of mutually consistent observables with a set of linearly independent vectors. Indeed, these states are mutually exclusive and correspond to certain values of the corresponding observables so that they cannot be presented as a superposition since the superposition generates states with uncertain values of the corresponding observables. In addition, since I am assuming that I am dealing with a complete set of consistent observables, I cannot add any more states to the set, which means that the set is linearly independent and complete, i.e., can be considered as a basis. Starting with a basis based on the complete set of mutually consistent observables, I can, in principle, form such linear combinations, which would remain linearly independent among themselves even though they would not any longer correspond to definite values of any physical observables. From a purely mathematical standpoint, this set still forms a basis, so that the choice of the basis is not unique.

In addition to generalizing the concept of the basis, I can use the example of three-dimensional geometrical vectors to introduce one more operation involving vectors. You, of course, know that two three-dimensional vectors can be combined to form a "dot" or "scalar" product, which plays an important role in physics. Consider, for instance, the vector of force F and position vector r. Assuming that the force is constant, you can define its work, W, as the scalar product of the force and the position vector: $W = \boldsymbol{F}{\cdot}\boldsymbol{r}$. You know how to compute this dot product, actually you know even two different equivalent ways of doing so. The dot product can be computed either as

$$W = |\boldsymbol{F}|\,|\boldsymbol{r}|\cos\theta \tag{2.7}$$

where $|\boldsymbol{F}|$ or $|\boldsymbol{r}|$ are magnitudes of the force and of the position vector, while θ is the angle between these two vectors, or as

$$W = F_x x + F_y y + F_z z \tag{2.8}$$

where $F_{x,y,z}$ are components of the force along X, Y, and Z axes of a specified coordinate system correspondingly, while x, y, and z are corresponding components of the position vector (or coordinates, if you wish). The magnitudes of the force and position vectors can be defined via the scalar products of the vectors with themselves, e.g., $|\boldsymbol{F}| = \sqrt{\boldsymbol{F}{\cdot}\boldsymbol{F}}$. The magnitudes of the vectors and their dot products possess certain important properties expressed by inequalities listed below:

$$|\boldsymbol{A}| \geq 0 \tag{2.9}$$

$$|\boldsymbol{A}{\cdot}\boldsymbol{B}| \leq |\boldsymbol{A}|\,|\boldsymbol{B}| \tag{2.10}$$

$$|\boldsymbol{A}{+}\boldsymbol{B}| \leq |\boldsymbol{A}| + |\boldsymbol{B}|\,. \tag{2.11}$$

The first of these inequalities as well as the statement that the equality in it is only reached for a null vector is quite trivial for regular vectors. Equation 2.10 follows from the limitation on the values of the cosine function, $\cos\theta \leq 1$, and the last of these inequalities expresses a well-known geometrical fact that the sum of any two sides of a triangle is always larger than the third side. The equality in this case can only be reached for degenerate triangles, in which all its sides are aligned along a single line.

Since the magnitude and the dot product play such an important role in application of regular vectors, you would be correct to think that it is a clever idea to introduce the similar operation for the abstract vectors as well. The problem is that since I do not know what my abstract vectors actually are, I cannot give the magnitude and the dot product an operational definition (meaning a prescription how to compute them similar to Eq. 2.7 or 2.8). What I can and will do is to postulate that these operations exist and define them by requiring that whatever they are operationally, they must have the same properties as those given in Eqs. 2.9–2.11. When talking about abstract vectors, however, it is customary to use the term "norm" instead of the magnitude and "inner product" instead of the dot product. Notation usually used for the norm of an abstract vector $|\alpha\rangle$ is $\|\alpha\|$, so that Eq. 2.9 takes the form of

$$\|\alpha\| \geq 0 \tag{2.12}$$

which obviously implies that the norm is necessarily real-valued (as opposed to complex-valued). However, an attempt at defining the norm as an inner product of the vector with itself, as well as defining the inner product by simply extending Eq. 2.8 to an arbitrary number of components, results in a problem.

It turns out that unlike the case of regular three-dimensional vectors, in general it is not possible to define the inner product using only vectors belonging to the same vector space. To see why this is so, consider an example of single-column matrices, k, with N rows ($1 \times N$ matrix or a column vector) with complex-valued elements. Multiplication by a complex number $a \cdot k$ in this case is defined obviously as

$$a \cdot k \equiv a \begin{bmatrix} k_1 \\ k_2 \\ \vdots \\ k_{N-1} \\ k_N \end{bmatrix} = \begin{bmatrix} ak_1 \\ ak_2 \\ \vdots \\ ak_{N-1} \\ ak_N \end{bmatrix} \tag{2.13}$$

and produces another column vector. The addition of two column vectors $k + p$ is defined as

$$k + p \equiv \begin{bmatrix} k_1 \\ k_2 \\ \vdots \\ k_{N-1} \\ k_N \end{bmatrix} + \begin{bmatrix} p_1 \\ p_2 \\ \vdots \\ p_{N-1} \\ p_N \end{bmatrix} = \begin{bmatrix} k_1 + p_1 \\ k_2 + p_2 \\ \vdots \\ k_{N-1} + p_{N-1} \\ k_N + p_N \end{bmatrix} \tag{2.14}$$

and also produces a column vector. Obviously, column vectors form a linear space. Now, in the regular matrix algebra, you can define two types of products, inner product and outer product, but neither of them can be introduced using only column vectors. To introduce either of these two operations, you need to combine a column vector with a row vector (matrix $N \times 1$). Now, if you place the row vector to the left of the column vector and use regular matrix multiplication rules (*row by the column*), you can convert the two vector objects into a number:

$$\begin{bmatrix} k_1 & k_2 & \cdots & k_{N-1} & k_N \end{bmatrix} \begin{bmatrix} p_1 \\ p_2 \\ \vdots \\ p_{N-1} \\ p_N \end{bmatrix} = \sum_{i=1}^{N} k_i p_i. \tag{2.15}$$

This is the inner product of the row and the column vectors. If you swap the two objects placing the column to the left of the row, you can generate a $N \times N$ matrix using the following rule:

$$\begin{bmatrix} p_1 \\ p_2 \\ \vdots \\ p_{N-1} \\ p_N \end{bmatrix} \begin{bmatrix} k_1 & k_2 & \cdots & k_{N-1} & k_N \end{bmatrix} = \begin{bmatrix} p_1k_1 & p_1k_2 & \cdots & p_1k_N \\ p_2k_1 & p_2k_2 & \cdots & p_2k_N \\ \vdots & \ddots & \ddots & \vdots \\ p_Nk_1 & p_Nk_2 & \cdots & p_Nk_N \end{bmatrix}. \tag{2.16}$$

What you get here is the so-called outer or tensor product of the row and the column vectors.

The row vectors do not belong to the same vector space as column vectors (you cannot add a column and a row): they form their own space called *adjoint* space. In order to establish a proper relationship between the row space and the column space, we need to determine how to convert a column vector into a row vector. It appears that the answer is almost trivial—one can do it using matrix operation known as transposition, which transforms a column vector k into a respective row vector k^T. However, it is easy to see that by using simple transposition, you will not be able to generate the inner product and the norm satisfying the conditions presented in Eq. 2.12. Indeed, generalizing Eq. 2.9, you can try introducing the norm using an inner product of a row obtained by transposition of the initial column and the column itself. This procedure will yield

$$k^T \cdot k \equiv \begin{bmatrix} k_1 & k_2 & \cdots & k_{N-1} & k_N \end{bmatrix} \begin{bmatrix} k_1 \\ k_2 \\ \vdots \\ k_{N-1} \\ k_N \end{bmatrix} = \sum_{i=1}^{N} (k_i)^2. \tag{2.17}$$

If k_i are complex-valued quantities, the result of this multiplication is not necessarily real-valued in clear contradiction with the required property of the norm. The problem, however, can be fixed if in addition to transposing a column vector, you would also complex conjugate its elements. The resulting operation is called Hermitian transposition or Hermitian conjugation, and it turns a column vector k into its adjoint or Hermitian conjugate row vector $k^\dagger$. Using Hermitian conjugation rather than simple transposition turns Eq. 2.17 into

$$k^\dagger \cdot k \equiv \begin{bmatrix} k_1^* & k_2^* & \cdots & k_{N-1}^* & k_N^* \end{bmatrix} \begin{bmatrix} k_1 \\ k_2 \\ \vdots \\ k_{N-1} \\ k_N \end{bmatrix} = \sum_{i=1}^{N} |k_i|^2 \tag{2.18}$$

where $|k_i|$ means the absolute value of the respective complex number. Now you can define the inner product of two column vectors, k and p (k, p), as an operation which involves Hermitian conjugation of a vector on the left and forming an inner product of the resulting row with the parent column: $(k, p) = k^\dagger \cdot p$. It is obvious that the inner product defined this way is not commutative, meaning that $(k, p) \neq (p, k)$. Actually, it is easy to see that $(k, p) = (p, k)^*$. Using Hermitian conjugation one can also define a new kind of the outer product as well, but I will leave the discussion of the latter till later.

A linear vector space with a defined inner product and a norm becomes something which mathematicians call Hilbert space. (There are some mathematical niceties and details concerning the exact definition of the Hilbert space, but they are of no concern to us here.) I hope that the example with column and row vectors helps you to realize that in order to define the norm and the inner product for our abstract vectors $|\alpha\rangle$, you need complimentary vectors inhabiting an adjoint space. Again following Dirac I will designate a vector adjoint to $|\beta\rangle$ as $\langle\beta|$ and use notation $\langle\beta|\alpha\rangle$ for the inner product. In the case of abstract vectors, we do not really know how to actually compute the inner product, but whatever its operational definition might be, we require that it obeys the following condition:

$$\langle\beta|\alpha\rangle = (\langle\alpha|\beta\rangle)^* . \tag{2.19}$$

This property ensures that the norm defined as

$$\|\alpha\| \equiv \sqrt{\langle\alpha|\alpha\rangle} \tag{2.20}$$

is real and nonnegative. Indeed, applying Eq. 2.20 to the case when $|\beta\rangle = |\alpha\rangle$, you have $\langle\alpha|\alpha\rangle = (\langle\alpha|\alpha\rangle)^*$, proving that $\langle\alpha|\alpha\rangle$ is real-valued.

To distinguish between adjoint vectors in speech, we call $|\alpha\rangle$ a *ket vector* and $\langle\alpha|$—a *bra vector*. These terms have been introduced by Paul Dirac together with the respective notation. Their origin can be traced to word "bracket" split in two halves "bra - ket" just like angular brackets in $\langle\alpha|\alpha\rangle$ are split in two vectors (what happened to letter "c" in the process is anybody's guess).

So far, we have no operational prescription on converting a single generic ket into a respective bra: all what you need to do is just to change the orientation and position of the angular bracket from $|\rangle$ to $\langle|$. However, an important question which we need to figure out now is how to do this conversion in the case of such expressions as $a|\alpha\rangle$ or even more complex expressions of the kind of Eq. 2.2. What is clear is that the adjoint of this expression must look like

$$(a|\alpha\rangle)^\dagger = \widetilde{a}\langle\alpha| ,$$

where I used symbol $\dagger$ to designate conversion to the adjoint space (or performing Hermitian conjugation just like in Eq. 2.18). Now I need to find how coefficients $\widetilde{a}$ are related to a. To this end let me compute the norm $\|a\alpha\|$:

$$\|a\alpha\|^2 = \widetilde{a}a\langle\alpha|\alpha\rangle = \widetilde{a}a\|\alpha\|^2 .$$

For this expression to be real-valued for all possible coefficients a, I have no choice but to require that $\widetilde{a} = a^*$. This yields a simple rule for Hermitian conjugation of complex expressions involving ket vectors: convert all kets into bras by simply replacing one angular bracket with the other and complex conjugate all numerical coefficients appearing in the original expression. Here are a few examples of application of this rule:

Example 1 (Hermitian Conjugation) Perform Hermitian conjugation of the following expression:

$$|\alpha\rangle = (2i + 3)\,|q_1\rangle + 5i\,|q_2\rangle.$$

Solution

$$\langle\alpha| = (-2i + 3)\,\langle q_1| - 5i\,\langle q_2|.$$

Example 2 (Norm Calculation) If $\|q_1\| = 2$, $\|q_2\| = 4$, and $\langle q_1| q_2\rangle = i - 1$, find the norm of $|\alpha\rangle$ defined in the previous example.

Solution

$$\|\alpha\|^2 = ((-2i + 3)\,\langle q_1| - 5i\,\langle q_2|)\,((2i + 3)\,|q_1\rangle + 5i\,|q_2\rangle) =$$
$$(-2i + 3)\,(2i + 3)\,\|q_1\|^2 + 5i\,(-5i)\,\|q_2\|^2 +$$
$$5i\,(-2i + 3)\,\langle q_1| q_2\rangle - 5i\,(2i + 3)\,\langle q_2| q_1\rangle =$$
$$13 \times 4 + 25 \times 16 + (10 + 15i)\,(i - 1) + (10 - 15i)\,(-i - 1) = 402$$
$$\|\alpha\| = \sqrt{402}.$$

It will be shown in the future development of the formalism that vectors $|\alpha\rangle$ and $a\,|\alpha\rangle$, where a is an arbitrary complex coefficient, describe the same quantum state, i.e., contain the same information about the system. Therefore, it is often convenient to deal only with the states whose norm is equal to unity (in a way such states are analogous to unit vectors in a regular three-dimensional space). Such vectors can be produced via normalization procedure, which consist in replacing an original vector $|\alpha\rangle$ with the vector $|\alpha\rangle / \|\alpha\|$. In the future, we shall assume that all abstract vectors used in calculations are normalized, and those which are not will have to be normalized before any calculations with them are performed.

Example 3 (Normalization) Normalize the following state:

$$|\alpha\rangle = 2\,|q_1\rangle + 3i\,|q_2\rangle - \frac{2}{3}\,|q_3\rangle,$$

assuming that the norm of all vectors $|q_i\rangle$ is unity and that inner products involving pairs of different vectors are all equal to zero.

Solution

1st step: Hermitian conjugation

$$\langle\alpha| = 2\langle q_1| - 3i\langle q_2| - \frac{2}{3}\langle q_3|.$$

2nd step: Forming the inner product

$$\|\alpha\|^2 = \langle\alpha|\alpha\rangle = \left(2\langle q_1| - 3i\langle q_2| - \frac{2}{3}\langle q_3|\right)\left(2|q_1\rangle + 3i|q_2\rangle - \frac{2}{3}|q_3\rangle\right) =$$

$$4 + 9 + \frac{4}{9} = \frac{121}{9}.$$

3rd step: The normalization

$$\|\alpha\| = \sqrt{\frac{121}{9}} = \frac{11}{3}$$

$$|\alpha\rangle_N = \frac{3}{11}\left(2|q_1\rangle + 3i|q_2\rangle - \frac{2}{3}|q_3\rangle\right).$$

I will complete this discussion of vector states in their ket and bra reincarnations by considering another important example of vector spaces and respective inner products and norms. Consider a class of complex functions, $\psi(x)$, of a single variable x defined over domain $x \in [-\infty, \infty]$, which also satisfy the condition that $\int_{-\infty}^{\infty} |\psi(x)|^2\, dx < \infty$. It is very easy to show that linear combinations of such functions also belong to the same class, so they do form a linear vector space. The inner product of two functions $\psi(x)$ and $\varphi(x)$ defined as

$$(\psi, \varphi) = \int_{-\infty}^{\infty} \psi^*(x)\varphi(x)dx \tag{2.21}$$

satisfies condition presented by Eq. 2.19, and the respective norm

$$\|\psi\| = \sqrt{\int_{-\infty}^{\infty} |\psi(x)|^2\, dx}$$

is obviously real-valued and nonnegative. Thus, these functions, called square-integrable, do form a Hilbert vector space, in which for each ket vector $|\alpha\rangle \equiv \psi(x)$, there is an adjoint bra vector $\langle\alpha| \equiv \psi^*(x)$ with an inner product and a norm defined as respective integrals.

So, I gave you two very different concrete realizations of abstract Hilbert space: column vectors and square-integrable functions with two different operational definitions of the inner product. Despite the difference in the operational meaning of inner product in these two cases, they all had the same defining properties.

2.2.3 Superposition Principle and Probabilities

States characterized by definite values of the complete set of mutually consistent observables have an important mathematical property, which I cannot yet prove (it will be done later), but it is important for the argument I am trying to present, so you will have to trust me on this for now. So, here is the property:

$$\left\langle q_l^{(1)}, q_n^{(2)}, \cdots q_r^{(N_{max})} \,\Big|\, q_k^{(1)}, q_m^{(2)}, \cdots q_p^{(N_{max})} \right\rangle = \delta_{l,k}\delta_{n,m}\cdots\delta_{r,p}, \tag{2.22}$$

where I assumed that the state vectors are normalized ($\delta_{l,k}$ is Kronecker delta, which is equal to unity for coinciding indexes, and zero otherwise). This equation surely looks rather mysterious, but its actual content is rather simple. To see this imagine that you are dealing with a system described by a single observable, say, energy. To make the matter even simpler, assume also that the energy can only take two values 0 and 1. Then you are dealing with only two states with definite values of energy $|1\rangle$ and $|0\rangle$. Equation 2.22 in this case simply states that

$$\langle 0\,|1\rangle = 0, \ \langle 0\,|0\rangle = \langle 1\,|1\rangle = 1.$$

Abstract vectors, whose inner product is equal to zero, are called orthogonal in the obvious generalization of the concept of orthogonality for regular perpendicular three-dimensional vectors, whose dot product is equal to zero. What I am driving at here is the connection between the mathematical concept of orthogonality and the physical notion of mutual exclusivity discussed in Sect. 2.1: Eq. 2.22 does say that mutually exclusive states are also orthogonal. If the basis is constructed of mutually orthogonal and normalized vectors, it is called orthonormalized. Such bases are particularly convenient to work with, and, therefore, they are used almost exclusively in practical calculations. Here is an example of a system of vectors forming an orthonormal basis, which you are quite familiar with.

Example 4 (Fourier Series: Examples of a Discrete Orthonormal Basis) Consider a set of functions of a single variable x defined on a finite interval $[0, L]$. These functions obviously form a linear space with an inner product defined as

$$\langle f|\,g\rangle = \int_0^L f^*(x)g(x)dx.$$

It is well known that these functions can be expended into a Fourier series[1]

$$f(x) = \sqrt{\frac{1}{L}} \sum_{-\infty}^{\infty} a_n e^{i2\pi nx/L}$$

with expansion coefficients given by

$$a_n = \sqrt{\frac{1}{L}} \int_0^L f(x) e^{-2i\pi nx/L} dx.$$

One can identify $|\chi_n\rangle \equiv \sqrt{\frac{1}{L}} e^{i2\pi nx/L}$ with vectors of an orthonormal basis so that the Fourier series expansion of the function can be presented in the form of Eq. 2.6. Indeed, it is easy to see that

$$\langle \chi_m | \chi_n \rangle = \frac{1}{L} \int_0^L dx e^{-2i\pi mx/L} e^{2i\pi nx/L} = \begin{cases} 1 & m = n \\ 0 & m \neq n \end{cases}.$$

Mutual exclusivity can be more formally expressed in terms of probabilities: if a quantum system is in a state with prescribed values $q_l^{(1)}, q_n^{(2)}, \cdots q_r^{(N_{max})}$ of a complete set of mutually consistent observables, then the measurement of the observables $q^{(s)}$ will produce value $q_p^{(s)}$ appearing in the definition of the state with a probability equal to 1, while the probability of any other value is equal to zero. Now, let's turn our attention to the superposition state of the type presented in Eq. 2.2 and ask a question: what should we expect if we measure observable q and the system is in the state given by this equation? You already know that the exact result of the measurement cannot be predicted, but it is intuitively clear that it must be either q_1 or q_2. The only question you need to ponder on now is that if the measurement is repeated multiple times with the system always brought back to the same initial state (or if, instead of one system, you got your hands on an ensemble of identical systems all in the same state), what are the fractions of measurement outcomes yielding q_1 and q_2? It seems reasonable to assume that it is coefficients $a_{1,2}$ in the superposition which will determine an answer to this question. Indeed, if I set either a_1 or a_2 to zero, I will take you back to the state in which all measurements yield either q_1 or q_2, respectively, while its counterpart is never observed. It is natural to describe this situation in terms of probabilities and assume that these probabilities are determined

---

[1]There are some conditions on the required smoothness of the functions for their Fourier series actually to represent them accurately, but leave it to mathematicians to worry about these details. Here I just want to mention that the representation as a Fourier series actually works even for functions, which are not necessarily continuous.

by the coefficients a_i. However, it is clear that I cannot identify the coefficients themselves with the probabilities because a_i are not necessarily positive or even real (if this were not the case, we could not describe both constructive interference and destructive interference just like in the case of electromagnetic waves). At the same time, you can recall that in the case of wave interference, it is not the amplitudes of the waves but their intensities proportional to the absolute values of the squared amplitudes that determine the brightness of the interference fringes. Thus, we can surmise that in the case of quantum superposition, respective probabilities are given by $|a_i|^2$:

$$p(q_i) = |a_i|^2 . \tag{2.23}$$

Multiplying Eq. 2.2 by $\langle q_1|$ or $\langle q_2|$ (the bra counterparts of respective vectors $|q_i\rangle$) from the left and using the orthogonality condition, Eq. 2.22, I can derive for the coefficients a_i

$$\langle q_1| \alpha \rangle = \langle q_1| (a_1 |q_1\rangle + a_2 |q_2\rangle) = a_1 \|q_1\|^2 \Rightarrow a_1 = \langle q_1| \alpha \rangle$$

$$\langle q_2| \alpha \rangle = \langle q_2| (a_1 |q_1\rangle + a_2 |q_2\rangle) = a_1 \|q_2\|^2 \rightarrow a_2 = \langle q_2| \alpha \rangle, \tag{2.24}$$

where I took into account the convention that all vectors describing quantum states are presumed to be normalized. Expressions derived in Eq. 2.24 allow presenting Eq. 2.23 for probability in a more generic form:

$$p(q_i) = |\langle q_i| \alpha \rangle|^2 . \tag{2.25}$$

Applying Eq. 2.25 to the case of $|\alpha\rangle = |q_j\rangle$, you find $p(q_i) = \delta_{ij}$ establishing formal correspondence between notions of mutual exclusivity and orthogonality. Computation of the norm of the state $|\alpha\rangle$ yields

$$\|\alpha\|^2 = |a_1|^2 + |a_2|^2 \equiv p_1 + p_2.$$

If $\|\alpha\| = 1$, i.e., the state $|\alpha\rangle$ is normalized as presumed, then you obtain relation $p_1 + p_2 = 1$ in complete agreement with what is expected of the probabilities. This result reinforces my (well, actually Max Born's) suggestion to interpret $|\langle q_i| \alpha \rangle|^2$ as a probability that the measurement of observable q on a system in state $|\alpha\rangle$ will produce q_i.

It is important to emphasize that any uncertainty in the result of the measurement of the observable q exists only *before* the measurement has taken place and that the probability referred to in this discussion describes the number of given outcomes in the series of such measurements. After the measurement is carried out, and one of the values of the observable is actually observed, all uncertainty has disappeared. We now know that the measurement yielded a particular value q_i, which, according to our earlier proposition, is only possible if the system is in the respective state

$|q_i\rangle$. Thus we have to conclude that the act of measurement has destroyed the initial state $|\alpha\rangle$ and "collapsed" it into state $|q_i\rangle$. The most intriguing question, of course, is what determines the state in which the system collapses into. This question has been debated during the entire 100+ year-long history of quantum mechanics and is being debated still. The orthodox Copenhagen interpretation of quantum mechanics essentially leaves this question without an answer claiming that the choice of the final (after measurement) state is completely random.[2] I propose that you accept this interpretation as quite sufficient for most practical purposes, while it does leave people with a philosophical state of mind somewhat unsatisfied.

Equation 2.2 describes superposition of only two states. It is not too difficult to imagine that it can be extended to the case of the arbitrary number of states $\left|q_k^{(1)}, q_m^{(2)}, \cdots q_p^{(N_{max})}\right\rangle$ generated by mutually consistent observables with discrete spectrum:

$$|\alpha\rangle = \sum_{k,m\cdots,p} a_{k,m\cdots,p} \left|q_k^{(1)}, q_m^{(2)}, \cdots q_p^{(N_{max})}\right\rangle. \tag{2.26}$$

This sum, in principle, can contain any number of terms, including an infinite amount. In the latter case, of course, one have to start worrying about its convergence, but I will leave these worries to mathematicians. Coefficients $a_{k,m\cdots,p}$ appearing in this equation have the same meaning as the coefficients in the two-state superposition expressed by Eq. 2.25, in which $|q_i\rangle$ is replaced with a more general state appearing in Eq. 2.26.

---

[2] At a talk given at the physics department of Queens College in New York in 2014, British mathematician J.H. Conway (currently Professor Emeritus of Mathematics at Princeton University) dismissed the randomness postulate of the Copenhagen interpretation as a "cop-out" and also because the use of probabilities only makes sense when one deals with a well-defined ensemble of events or particles, which is not true in the case of a single electron or photon. At the same time, he and S. Kochen (Canada) proved a mathematical theorem asserting that the entire structure of quantum mechanics is inconsistent with the idea of existence of some unknown characteristics of quantum systems, which would, shall we find them, provide deterministic description of the system. In this sense, they proved completeness of the existing structure of quantum theory and buried the idea of "hidden variables"—unknown elements of reality, which could restore determinism to the quantum world—provided that we are unwilling to throw away the entire conceptual structure of quantum mechanics, which, so far, gave excellent quantitative explanation of a vast number of the experimental data. The Conway and Kochen theorem is called "free will theorem" because it can be interpreted as an assertion that electrons, just like humans, have "free will," which in strict mathematical sense means that electron's future behavior might not be a deterministic function of its past. The description of the theorem can be found here: https://en.wikipedia.org/wiki/Free_will_theorem.

2.3   States Characterized by Observables with Continuous Spectrum

In the previous section, I considered only states generated by observables with discrete spectrum. As a result, even though the number of states in Eq. 2.26 can be infinite, they are still countable (one can enumerate them using natural numbers $1, 2, 3, \cdots$). Some observables, however, have continuous spectrum, meaning that they can take values from a continuous (finite or infinite) interval of values. One of such important observables is a particle's position, measured by its position vector r or a set of Cartesian coordinates (x, y, z), defined in a particular coordinate system. It is interesting to note in this regard that while in classical mechanics, descriptions using Cartesian coordinates are largely equivalent to those relying on spherical or polar coordinates, it is not so in quantum description, where angular coordinates in spherical or cylindrical systems do not easily submit to quantum treatment. This comment obviously appears somewhat cryptic here, but its meaning will be clarified in the subsequent chapters. Another peculiarity of the position observable is the need to carefully distinguish between coordinates being characteristics of a particle's position and coordinates being markers of various points in space, needed to describe position dependence of various mathematical and physical quantities.

Other observables, such as energy or momentum, might have either continuous or discrete spectrum depending upon the environment, in which a particle finds itself, or might have mixed spectrum, where an interval of discretely defined values crosses over into an interval of continuously distributed values.

Two main peculiarities of states characterized by observables with continuous spectrum are that (1) they cannot be normalized in a regular sense of the word and (2) the concept of probability as defined by Eq. 2.25 loses its meaning because in the case of continuous random variables, the probability can only be defined for an interval (which might be infinitesimally small) of values, but not for any particular value of the variable. These two features are not independent and are related to each other as it will be seen from the future analysis.

I will illustrate the properties of states corresponding to observables with continuous spectrum using the position of a particle as an example. Assuming that there are no other observables mutually consistent with the position, I will present a state in which this observable has a definite value r as $|r\rangle$. In order to construct a superposition state using these vectors, I have to replace the sum in Eq. 2.26 with an integral over all possible values of position vector r introducing instead of coefficients a_k with discrete indexes a function $\psi(r)$ of a continuous variable:

$$|\alpha\rangle = \int d^3 r \psi(r) |r\rangle . \tag{2.27}$$

Now I want to compute the norm $\|\alpha\|$ of the superposition state given by Eq. 2.27. The Hermitian conjugation of Eq. 2.27 produces the respective bra vector

$$\langle \alpha | = \int d^3 r \psi^*(r) \langle r| \tag{2.28}$$

so that the norm becomes

$$\|\alpha\|^2 \equiv \langle \alpha | \alpha \rangle = \iint d^3 r_1 d^3 r_2 \psi^* (r_1) \psi (r_2) \langle r_1 | r_2 \rangle, \tag{2.29}$$

where I had to rename the integration variables in order to be able to replace the product of integrals with a double integral (note that r_1 appears in those parts of Eq. 2.29, which originate from the bra vector of Eq. 2.28, and r_2 appears in the ket-related parts of the integral). States $|r_1\rangle$ and $|r_2\rangle$ remain mutually exclusive even in the case of continuous spectrum as long as $r_1 \neq r_2$. Thus I can write based on the discussion in the previous sections that $\langle r_1 | r_2 \rangle = 0$ for $r_1 \neq r_2$. If I now require that $\langle r_1 | r_2 \rangle = 1$ for $r_1 = r_2$, which would correspond to the "regular" normalization condition, I will end up with an integral, in which the integrand is zero everywhere with exception of one point, where it is finite. Clearly such an integral would be zero in contradiction with the properties of the norm (it can only be zero for null-vector). To save the situation, something has to give, and I have to reject one of the assumptions made when evaluating Eq. 2.29. The mutual exclusivity of $|r_1\rangle$ and $|r_2\rangle$ and the related requirement that $\langle r_1 | r_2 \rangle = 0$ for $r_1 \neq r_2$ are connected with the basic ideas discussed in Sect. 2.2.3 and, therefore, appears untouchable. So, the only choice left to me is to reject the assumption that $\langle r_1 | r_1 \rangle$ is equal to unity or takes any other finite value. As a result we are left with the following requirements on $\langle r_1 | r_2 \rangle$: this expression must be zero for unequal values of its arguments while producing a non-zero result when being integrated with any "normal" function.

These requirements are satisfied by an object called Dirac's delta-function. Dirac introduced notation $\delta (x)$ for its simplest single-variable version and presented most of its properties in the form useful for physicists in his influential 1930 book *The Principles of Quantum Mechanics*, which since then has been reissued many times (it is the same Paul Dirac who introduced the bra-ket notation for quantum states). It seems a bit unfair to name this object after him because it was already known to such mathematicians as Poisson and Fourier in the nineteenth century, but physicists learned about it from Dirac, so we stick to our guns and call it a Dirac's function. The first thing one needs to understand about the delta-function is that it is not a function in any reasonable sense of the word. Therefore, the meaning of such operations as integration or differentiation involving this object cannot be defined following the standard rules of regular calculus. Nevertheless, physicists keep working with this object as though nothing is wrong (giving nightmares to rigor-sensitive mathematicians) with the only requirements that the results of all performed operations must make sense (that is from a physicist's perspective). Mathematicians call such objects "distributions" or treat them as examples of "functionals." Below I supply you with all properties of the delta-functions you will need to know.

The main defining property of the delta-function of a single variable is

$$\int_{x_1}^{x_2} f(x)\delta(x)dx = \begin{cases} f(0), & 0 \in [x_1, x_2] \\ 0 & 0 \notin [x_1, x_2] \end{cases} \tag{2.30}$$

with its immediate generalization to

$$\int_{x_1}^{x_2} f(x)\delta(x-x_0)dx = \begin{cases} f(x_0), & x_0 \in [x_1,x_2] \\ 0 & x_0 \notin [x_1,x_2]. \end{cases} \qquad (2.31)$$

These equations express the main property of the delta-function—it acts as a selector singling out the value of function $f(x)$ at $x = x_0$, where the argument of the delta-function vanishes. In a particular case of $f(x) = 1$, Eq. 2.31 yields another important characteristics of the delta-function:

$$\int_{x_1}^{x_2} \delta(x-x_0)dx = \begin{cases} 1, & x_0 \in [x_1,x_2] \\ 0 & x_0 \notin [x_1,x_2], \end{cases}$$

which expresses the idea that while the "width" of the delta-function is zero and its "height" is infinite, the area covered by it is equal to unity. An example of actual limiting procedure producing a delta-function out of a regular function based on this idea can be found in the exercises in this chapter.

One can also define the delta-function of a more complex argument such as $\delta[g(x)]$, where $g(x)$ is an arbitrary function. If $g(x)$ has only one zero at $x = x_0$, I can define $\delta[g(x)]$ by replacing $g(x)$ with the first term of its Taylor expansion around x_0: $g(x) \approx b(x-x_0)$, where $b \equiv (dg/dx)_{x-x_0}$, and making a substitution of variable $\tilde{x} = b(x-x_0)$, which yields

$$\int_{x_1}^{x_2} f(x)\delta[g(x)]\,dx = \frac{1}{|b|} \int_{g(x_1)}^{g(x_2)} f\left(\frac{\tilde{x}}{b}+x_0\right)\delta(\tilde{x})\,d\tilde{x} = \frac{1}{|b|}f(x_0), \qquad (2.32)$$

The expansion of $g(x)$ in the Taylor series is justified here because the value of the integral is determined by the behavior of this function in the immediate vicinity of x_0.

If the function $g(x)$ has multiple zeroes within the interval of integration, then we must isolate each zero and perform the procedure described above for each of them. The result will look something like this:

$$\int_{x_1}^{x_2} f(x)\delta[g(x)]\,dx = \sum_i \frac{1}{|b_i|}f(x_0^{(i)}) \qquad (2.33)$$

where b_i is the value of the derivative of $g(x)$ at the respective i-th zero $x_0^{(i)}$. To illustrate this procedure, consider an example.

Example 5 (Delta-function With Two Zeros) Consider

$$g(x) = x^2 - x_0^2.$$

In this case the method outlined above yields

$$\int_{x_1}^{x_2} f(x)\delta\left[x^2 - x_0^2\right] dx \equiv \frac{1}{2x_0}\left[\int_{x_1}^{x_2} f(x)\delta(x - x_0)\,dx + \int_{x_1}^{x_2} f(x)\delta(x + x_0)\,dx\right]$$

$$(2.34)$$

where I assumed that both x_0 and $-x_0$ belong to the interval between x_1 and x_2. I can also define a derivative of the delta-function using integration by parts and assuming that integral of df/dx is still equal to $f(x)$ even if $f(x) \equiv \delta(x)$. This is how it goes:

$$\int_{x_1}^{x_2} f(x)\delta'(x - x_0)dx = f(x)\delta(x - x_0)|_{x_1}^{x_2} - \int_{x_1}^{x_2} \delta(x - x_0)f'(x)dx$$

$$= \begin{cases} -\left.\frac{df}{dx}\right|_{x=x_0}, & x_0 \in [x_1, x_2] \\ 0 & x_0 \notin [x_1, x_2]. \end{cases} \qquad (2.35)$$

Similarly one can define higher derivatives of the delta-function.

We will also need an important representation of the delta-function as a Fourier transform:

$$\delta(x) = \frac{1}{2\pi} \int_{-\infty}^{\infty} e^{ikx}dk. \qquad (2.36)$$

To demonstrate that this representation of the delta-function actually makes sense, consider direct and inverse Fourier transforms:

$$f(x) = \frac{1}{\sqrt{2\pi}} \int_{-\infty}^{\infty} \tilde{f}(k)e^{ikx}dk \qquad (2.37)$$

$$\tilde{f}(k) = \frac{1}{\sqrt{2\pi}} \int_{-\infty}^{\infty} f(x)e^{-ikx}dx. \qquad (2.38)$$

Substituting Eq. 2.38 into Eq. 2.37, I get

$$f(x) = \frac{1}{2\pi} \int_{-\infty}^{\infty} f(x_1)e^{-ikx_1}e^{ikx}dkdx_1 = \frac{1}{2\pi} \int_{-\infty}^{\infty} dx_1 f(x_1) \int_{-\infty}^{\infty} e^{ik(x-x_1)}dk$$

and the only way to make this into an identity for any function is to accept Eq. 2.36 for the integral over k.

Finally, you will need a generalization of the delta-function to the case of several variables. For instance, delta-function involving position vectors in Cartesian coordinates can be defined as

$$\delta \left(r_1 - r_2\right) \equiv \delta \left(x_1 - x_2\right) \delta \left(y_1 - y_2\right) \delta \left(z_1 - z_2\right), \tag{2.39}$$

in which case its representation in the form of a Fourier transform becomes

$$\delta \left(r_1 - r_2\right) = \frac{1}{(2\pi)^3} \int_{-\infty}^{\infty} e^{ik_x(x_1-x_2)} e^{ik_y(y_1-y_2)} e^{ik_z(z_1-z_2)} dk_x dk_y dk_z =$$

$$\frac{1}{(2\pi)^3} \int_{-\infty}^{\infty} e^{ik\cdot(r_1-r_2)} d^3k. \tag{2.40}$$

Now back to the calculation of the norm, i.e., to Eq. 2.29. To complete this calculation, I will introduce a generalized, so-called delta-function normalization condition for states $|r\rangle$ by requiring that

$$\langle r_1 | r_2 \rangle = \delta \left(r_1 - r_2\right). \tag{2.41}$$

Substituting Eq. 2.41 into Eq. 2.29 and using the properties of the delta-function, I finally arrive at

$$\|\alpha\|^2 = \int d^3r \psi^* \left(r\right) \psi \left(r\right). \tag{2.42}$$

Now, in order to ensure correct normalization of the state $|\alpha\rangle$, you only need to require that function $\psi \left(r\right)$ is chosen to be normalized such that

$$\int d^3r \left| \psi \left(r\right) \right|^2 = 1. \tag{2.43}$$

Example 6 (Normalization of a Wave Function) Normalize the state $|\alpha\rangle$ presented by the following function.

$$\psi(x) = e^{ikx} e^{-ax^2/2}.$$

Solution

Using the definition of the norm, Eq. 2.42, I have

$$\|\alpha\|^2 = \int\limits_{-\infty}^{\infty} \psi^*(x)\psi(x)dx = \int\limits_{-\infty}^{\infty} e^{-ax^2} dx = \sqrt{\frac{\pi}{a}}$$

where I used the substitution of variables $y = \sqrt{a}x$ and the well-known integral

$$\int\limits_{-\infty}^{\infty} \exp\left(-y^2\right) dy = \sqrt{\pi}.$$

Thus, the normalized form of the state can be written as

$$|\alpha\rangle = \left(\frac{a}{\pi}\right)^{1/4} \int\limits_{-\infty}^{\infty} e^{ikx} e^{-ax^2/2} |x\rangle .$$

Function $\psi(r)$ in these expressions is called the wave function and is often cited in quantum mechanics textbooks as the descriptor of a quantum state. You can see now that this is not quite the case—the definition of the wave function involves two different states: the actual state of the system $|\alpha\rangle$ and a state, $|r\rangle$, in which a particle would have a definite position. You can think of the wave function as a projection of $|\alpha\rangle$ on $|r\rangle$. If the position r could only take discrete values, we would have interpreted $|\psi(r)|^2$ as a probability. In the continuous case, however, we can only ask about a probability that the measurement of position would produce a result within a certain (possibly infinitesimally small) volume around some central point r. The answer to this question is well expressed in terms of differential probability

$$dP(r) \equiv d^3r \, |\psi(r)|^2 ,$$

where $|\psi(r)|^2$ can be interpreted as the position probability density. The probability that the measured position vector belongs to some finite volume V is given by expression

$$P(V) = \iiint\limits_{V} d^3r \, |\psi(r)|^2 , \tag{2.44}$$

while the normalization condition 2.43 simply states the fact that the measurement of the particle's position will produce some value within the entire volume available to the particle with probability equal to one.

2.4 Problems

Problem 1 Consider two states:

$$|\psi_1\rangle = |\phi_1\rangle + i\,|\phi_2\rangle - 2\,|\phi_3\rangle$$
$$|\psi_2\rangle = -\,|\phi_1\rangle + 2\,|\phi_2\rangle - i\,|\phi_3\rangle\,,$$

where $|\phi_{1,2,3}\rangle$ are all normalized and orthogonal to each other.

1. Normalize states $|\psi_1\rangle$ and $|\psi_2\rangle$.
2. Find adjoint counterparts of these states $\langle\psi_1|$ and $\langle\psi_2|$.
3. Compute inner products $\langle\psi_1|\,\psi_2\rangle$ and $\langle\psi_2|\,\psi_1\rangle$ and verify that $\langle\psi_1|\,\psi_2\rangle = \langle\psi_2|\,\psi_1\rangle^*$.
4. Find a linear combination of states $|\psi_1\rangle$ and $|\psi_2\rangle$ that would be orthogonal to $|\psi_1\rangle$.
5. Compute $(\langle\psi_1| + \langle\psi_2|)(|\psi_1\rangle + |\psi_2\rangle)$. Do it in two ways: (a) by computing the sums first and then taking the inner product and (b) using the distributive property of the inner product, remove the parenthesis and compute the inner products of the resulting individual terms.

Problem 2 Determine if the following sets of vectors, defined by their components in some basis, are linearly dependent or independent:

1. $(2, 2, 0)$, $(1, 0, 1)$, $(0, i, -1)$
2. $(0, 0, 1)$, $(i, 0, 0)$, $(0, 0, -1)$
3. $(1, i, 2)$, $(1, i, -1)$, $(i, -i, 2i)$

Problem 3 Consider the set of functions

$$f_n(x) = A \sin \frac{\pi n x}{L},$$

where L is a positive quantity.

1. Prove that these functions form an orthogonal system with the inner product defined as

$$\langle f_n | f_m \rangle = \int\limits_0^L f_n(x) f_m(x)\,dx.$$

2. Normalize these functions.
3. Find an expression for coefficients c_n in the expansion

$$\psi(x) = \sum_{n=0}^{\infty} c_n f_n(x),$$

where $f_n(x)$ is given by the expression from the first part of the problem with amplitude A replaced by the normalization coefficient found in Part 2.

Problem 4 Repeat problem 3 with the set of functions

$$\varphi_n(x) = \exp\left(i\frac{2\pi nx}{L}\right),$$

and the inner product defined as

$$\langle\varphi_n|\,\varphi_m\rangle \equiv \int\limits_0^L \varphi_n^*(x)\varphi_m(x)dx.$$

Problem 5 Consider functions $g_1(x) = x$, $g_2(x) = x^2$, and $g_3(x) = x^3$ defined on an interval $x \in [-1, 1]$ with inner product defined as

$$\langle g_n|\,g_m\rangle = \int\limits_{-1}^1 dxg_n(x)g_m(x).$$

1. Which of these three functions are mutually orthogonal, and which are not?
2. Consider linear combination of functions $g_1(x)$ and $g_3(x)$: $ag_1(x) + bg_3(x)$ and find coefficients a and b which would make this function orthogonal to $g_1(x)$.
3. Find a different linear combination of the same functions, which would be orthogonal to $g_3(x)$.
4. Are these two new functions orthogonal to each other?

Problem 6 Consider a wave function of the form

$$\psi(x) = Ae^{ikx}xe^{-x^2/2}.$$

1. Normalize this function using the standard definition of the inner product for the square-integrable functions.
2. Find the probability that a measurement of the x-coordinate of the particle will produce a value between $0 \le x \le \sqrt{2}$.
3. Find the probability that a measurement of the x-coordinate of the particle will produce a value such that $x > 2$. Use mathematical tables or any available computational tools to obtain the numerical values.

Problem 7 Consider a function of the form

$$f(x) = \begin{cases} \frac{1}{\Delta} & |x| < \Delta/2 \\ 0 & |x| > \Delta/2. \end{cases}$$

Show that this function turns into a Dirac's δ-function in the limit $\Delta \to 0$.

Problem 8 Compute the following integrals:

1.

$$\int_0^1 \left(x^3 + 5x\right) \delta\left(x - 2\right) dx$$

2.

$$\int_0^3 \left(\sin 2x + 2 \tan 3x\right) \delta\left(x^2 - 5x + 4\right) dx$$

3.

$$\int_{-\infty}^{\infty} xe^{-x^2} \delta'\left(x + 5\right) dx$$

Problem 9 Evaluate the following expression:

$$\int_{-\infty}^{\infty} dx f\left(x\right) \int_{-\infty}^{\infty} dk k e^{ik(x-x')}.$$

Hint: Use the representation of the delta-function as a Fourier integral to figure out the integral with respect to k.

Chapter 3
Observables and Operators

3.1 Hamiltonian Formulation of Classical Mechanics

The version of classical mechanics based on forces and Newton's laws resists any meaningful reformation into a quantum theory because it depends critically on such concepts (trajectory, acceleration, etc.) that do not correspond to any observable reality in the quantum world. More productive for finding links between classical and quantum realms is an alternative formulation, where energy rather than force takes the central role. There are two essential elements in this formulation of classical mechanics. One is the idea of canonical coordinates in the so-called phase space (as opposed to regular three-dimensional configuration space), and the other is the concept of Hamiltonian.

Points in the phase space represent classical states of the system, characterized, for instance, by its coordinates x_i and components of the momentum vector p_i. For a single particle moving along a straight line (one-dimensional motion), the phase space is two-dimensional; for the fully three-dimensional motion, the phase space is six-dimensional; and for a three-dimensional motion of N particles, the dimension of the phase space is $6N$. Each point in the phase space represents the most complete information about a classical system—its coordinates and velocities. When particles move, their coordinates and momentums change, drawing a phase trajectory of the system in the phase space. For a single particle allowed to move only along a straight line, this trajectory is a curve in two-dimensional space. If the motion of the particle is conservative, i.e., its energy is a conserving quantity, each phase trajectory is an equienergetic line—each point on the trajectory corresponds to the state of the system with exactly the same energy (energy does not change, while particles change their position and momentum). Using the phase space, we effectively put space coordinates and momentum of the particles on equal footing without imposing any a priori relationships between them (as opposed to elementary mechanics, when the momentum is defined via the time derivative of coordinates). You shall see that

© Springer International Publishing AG, part of Springer Nature 2018
L.I. Deych, *Advanced Undergraduate Quantum Mechanics*,
https://doi.org/10.1007/978-3-319-71550-6_3

the relationship between coordinate and momentum, called canonically conjugate variables, arising within this framework is much closer to its quantum version than it would have been in the Newtonian approach.

The Hamiltonian is essentially the energy of a conservative system expressed in terms of coordinates and momentum $H(\boldsymbol{p},\boldsymbol{r})$, which in the case of a single particle takes the form of

$$H(\boldsymbol{p},\boldsymbol{r}) = \frac{p^2}{2m} + V(r), \tag{3.1}$$

where $\boldsymbol{p}$ is the momentum vector[1] and $V(r)$ is the potential energy of the particle in the external field. The Hamiltonian occupies a special place in classical mechanics (as compared, for instance, to angular momentum, which can also be a conserving quantity under certain circumstances) because it determines system's dynamics via Hamiltonian equations, which can be formulated as

$$\frac{dp_i}{dt} = -\frac{\partial H}{\partial r_i} \tag{3.2}$$

$$\frac{dr_i}{dt} = \frac{\partial H}{\partial p_i}, \tag{3.3}$$

where r_i and p_i ($i = 1, 2, 3$) are Cartesian components of the position and momentum vectors x, y, z and p_x, p_y, p_z, respectively. Hamiltonian equations can be rewritten in another interesting form using so-called Poisson brackets $\{f, g\}$ defined for two arbitrary functions of canonical variables:

$$\{f,g\} = \sum_{i=1}^{N} \left(\frac{\partial f}{\partial r_i} \frac{\partial g}{\partial p_i} - \frac{\partial f}{\partial p_i} \frac{\partial g}{\partial r_i} \right). \tag{3.4}$$

Summation in Eq. 3.4 is over all relevant canonical conjugated pairs of coordinates. It is easy to see that the Poisson brackets for momentum and corresponding coordinates are

$$\{r_i, p_j\} = \delta_{i,j}. \tag{3.5}$$

This form of Poisson brackets is called canonical: any pair of variables possessing Poisson brackets of this form form a canonically conjugated pair and satisfy Hamiltonian equations 3.2 and 3.3.

Applying the definition of the Poisson brackets, Eq. 3.4, to the pair of functions p_i, H and r_i, H, you can find (check it out!)

---

[1] p^2 is defined as usual as the square of the magnitude of the vector in Cartesian coordinates $p_x^2 + p_y^2 + p_z^2$.

$$\{p_i, H\} = -\frac{\partial H}{\partial r_i},$$

$$\{r_i, H\} = \frac{\partial H}{\partial p_i},$$

so that Hamiltonian equations 3.2 and 3.3 can be rewritten even in a more symmetric form:

$$\frac{dp_i}{dt} = \{p_i, H\}, \tag{3.6}$$

$$\frac{dr_i}{dt} = \{r_i, H\}. \tag{3.7}$$

Finally, the time derivative of an arbitrary function of canonical coordinates can be expressed in terms of Poisson brackets involving Hamiltonian. I illustrate this statement for a function of only one pair of coordinates $f(x, p, t)$:

$$\frac{df}{dt} = \frac{\partial f}{\partial t} + \frac{\partial f}{\partial p}\frac{dp}{dt} + \frac{\partial f}{\partial x}\frac{dx}{dt} = \frac{\partial f}{\partial t} - \frac{\partial f}{\partial p}\frac{\partial H}{\partial x} + \frac{\partial f}{\partial x}\frac{\partial H}{\partial p} = \frac{\partial f}{\partial t} + \{f, H\}. \tag{3.8}$$

3.2 Operators in Quantum Mechanics

3.2.1 General Definitions

The main task of quantum theory is to be able to predict (or explain) results of experiments conducted with quantum systems. All such experiments involve taking a system in some initial state, subjecting it to external influences, which change its environment, and observing a reaction of the system to these changes. A theoretician in me would say that by doing all these manipulations and measurements, experimentalists change the quantum state of the system, but so far the formalism I have at my disposal does not have any theoretical representation of all these turning knobs and dials, lasers, which go on and off, magnets, thermostats, and all other real material objects in the arsenal of an experimentalist. I need additional mathematical tools, which would allow me to describe theoretically all these changes inflicted upon an unsuspecting system by the men in lab coats. Since the quantum states are presented in the theory by vectors of a linear vector space, what I need are objects that can change these vectors. Such objects are known to mathematicians—they call them operators. The role of operators in quantum theory is twofold. On one hand, they are used to describe transformations of state vectors, and on the other hand, they provide the theoretical means to predict the outcomes of measurements of observables.

From the mathematical standpoint, an operator is a rule prescribing how to change one abstract vector of a linear vector space, say, $|\alpha\rangle$, into another abstract vector, say, $|\beta\rangle$ of the same or a different vector space. Symbolically this can be represented as,

$$|\beta\rangle = \hat{T}|\alpha\rangle,\tag{3.9}$$

where the "hat"ˆabove a capital letter (in this case T) signifies that $\hat{T}$ represents such a rule, or an operator, "acting" on $|\alpha\rangle$ and converting it into $|\beta\rangle$). Note that the symbol of the operator appears in Eq. 3.9 next to the vertical line marking the "tail" of the ket $|\alpha\rangle$.

The special role in quantum mechanics and other applications is played by *linear operators*—the class of rules satisfying the following condition:

$$\hat{T}(a_1|\alpha_1\rangle + a_2|\alpha_2\rangle) = a_1\hat{T}|\alpha_1\rangle + a_2\hat{T}|\alpha_2\rangle.\tag{3.10}$$

Here are a few examples of linear operators:

1. Differentiation operator d/dx converting a function $f(x)$ into its derivative $g(x) = (d/dx)f \equiv df/dx$ (note how the operator symbol appears on the left of the function)
2. Gradient operator $\vec{\nabla} = e_x\partial/\partial x + e_y\partial/\partial y + e_z\partial/\partial z$, where $e_{x,y,z}$ are unit vectors in the directions of the respective coordinate axes, converting a scalar function of three spatial variables into a vector:

$$\vec{\nabla}f(x,y,z) = e_x\partial f/\partial x + e_y\partial f/\partial y + e_z\partial f/\partial z$$

3. Integration operator, $\hat{K}$, which is defined by its kernel $K(x_1, x_2)$ and converts one function to another as

$$|g\rangle = \hat{K}|f\rangle \Longleftrightarrow g(x_1) = \int_{-\infty}^{\infty} K(x_1, x_2)f(x_2)\,dx_2$$

4. Rotation operator $\hat{R}$, which changes the orientation of a vector without changing its length

Linearity of the first three operators is evident from linearity of differentiation and integration, and the proof of linearity of rotations is a simple exercise in geometry and is left to the readers to perform.

Equation 3.9 defines an operator by its action on a ket vector. It is also possible to define an operator acting on bra vectors. One can, for instance, perform formal Hermitian conjugation of Eq. 3.9 and introduce Hermitian conjugate operator $\hat{T}^\dagger$:

$$\langle\beta| = \langle\alpha|\hat{T}^\dagger.\tag{3.11}$$

Notice that now operator $\hat{T}^\dagger$ stands to the right of the respective bra vector but still next to its tail, "acting" to the left. Thus, Hermitian conjugation in this case involves also the change in the order, in which the participating objects are written, as well as the "direction" of "action" of the operators from right to left.

In order to help you develop intuition regarding transition between Eqs. 3.9 and 3.11, consider a linear space of column vectors—$1 \times N$ matrices. For operators you can take $N \times N$ matrices and define its action on a vector as regular matrix multiplication. For this definition to make sense from the point of view of matrix multiplication rules, the matrix must be placed to the left of the column vector. The result of this operation is another column vector:

$$
\begin{bmatrix}
t_{11} & t_{12} & \cdots & t_{1N} \\
t_{21} & t_{22} & \cdots & t_{2N} \\
\vdots & \vdots & \ddots & \vdots \\
t_{N1} & t_{N2} & \cdots & t_{NN}
\end{bmatrix}
\begin{bmatrix}
a_1 \\
a_2 \\
\vdots \\
a_N
\end{bmatrix}
=
\begin{bmatrix}
b_1 \\
b_2 \\
\vdots \\
b_N
\end{bmatrix}.
\tag{3.12}
$$

The Hermitian conjugate of a column vector is a row vector ($N \times 1$ matrix) with complex-conjugated elements. Equation 3.12 contains two column vectors, and its Hermitian conjugated version must describe the relation between two rows $[a_1^* \; a_2^* \; \cdots \; a_N^*]$ and $[b_1^* \; b_2^* \; \cdots \; b_N^*]$. However, in order to be able to multiply a square matrix and a row vector, I must place the former to the left of the latter:

$$
[a_1^* \; a_2^* \; \cdots \; a_N^*]
\begin{bmatrix}
t_{11}^\dagger & t_{12}^\dagger & \cdots & t_{1N}^\dagger \\
t_{21}^\dagger & t_{22}^\dagger & \cdots & t_{2N}^\dagger \\
\vdots & \vdots & \ddots & \vdots \\
t_{N1}^\dagger & t_{N2}^\dagger & \cdots & t_{NN}^\dagger
\end{bmatrix}
= [b_1^* \; b_2^* \; \cdots \; b_N^*],
\tag{3.13}
$$

where $t_{ij}^\dagger$ represents elements of a Hermitian conjugate operator matrix $\hat{T}^\dagger$. If I want (and I certainly do) that the relation between elements a_i^* and b_i^* expressed by Eq. 3.13 reproduce complex-conjugated relations given by Eq. 3.12, I must require that the rows of the matrix in Eq. 3.12 coincide with the complex-conjugated columns of the matrix in Eq. 3.13: $t_{ij}^\dagger = t_{ji}^*$. This gives me an operational (not just formal) rule for performing the Hermitian conjugation of the matrix operator: it consists in regular matrix transposition and complex conjugation of all matrix elements. This example serves two important purposes: first, it demonstrates why reversal of the order in which vectors and operators appear after Hermitian conjugation makes sense, and, second, it yields a rule for Hermitian conjugation of a matrix.

In a general case, Eq. 3.11 does not give us any clue on how to actually generate Hermitian conjugate operators. In order to derive such a rule, I need to relate both (initial and Hermitian conjugate) operators to a quantity, which I know how to transform and which does not depend on any concrete realization of the vector space or an operator. The only quantity of this kind, which I know of, is an inner product, and I, in order to get to it, will multiply Eq. 3.9 by a bra vector $\langle\beta|$ from the left. This will leave me with expression $\langle\beta|\hat{T}|\alpha\rangle$, which can be understood as a product of a bra vector $\langle\beta|$ and a ket vector $\hat{T}|\alpha\rangle$. Complex conjugating this expression and applying Eq. 2.19, I get

$$\langle\beta|\hat{T}|\alpha\rangle^* = \langle\alpha|\hat{T}^\dagger|\beta\rangle, \tag{3.14}$$

where I also used Eq. 3.11 to convert ket $\hat{T}|\alpha\rangle$ into a corresponding bra $\langle\alpha|\hat{T}^\dagger$. Equation 3.14 can be used to find a Hermitian conjugate of any particular operator as illustrated by the following examples.

Example 7 (Hermitian Conjugation) Consider differentiation operator $\hat{D}$ acting on differentiable square-integrable functions as

$$\hat{D}|f\rangle \equiv \frac{df}{dx},$$

Using the definition of the inner product defined by Eq. 2.21, you can present the expression in Eq. 3.14 as

$$\langle g|\hat{D}|f\rangle \equiv \int_{-\infty}^{\infty} dx g^*(x)\frac{df}{dx}.$$

Integration by parts converts this expression into the following form:

$$\left(\langle g|\hat{D}|f\rangle\right)^* = \left(\int_{-\infty}^{\infty} dx g^*(x)\frac{df}{dx}\right)^* = g(x)f^*(x)\big|_{-\infty}^{\infty} - \int_{-\infty}^{\infty} dx f^*(x)\frac{dg}{dx} =$$

$$- \int_{-\infty}^{\infty} dx f^*(x)\frac{dg}{dx},$$

where I took into account that any square-integrable functions must vanish at both positive and negative infinities. Presenting this result in the form of the right side of Eq. 3.14 $\langle f|\hat{D}^\dagger|g\rangle$, you can identify $\hat{D}^\dagger$ as $\hat{D}^\dagger = -d/dx$.

If an operator and its Hermitian conjugate coincide

$$\langle\beta|\hat{T}|\alpha\rangle^* = \langle\alpha|\hat{T}|\beta\rangle \tag{3.15}$$

or $\hat{T} = \hat{T}^\dagger$, the respective operator is called Hermitian or self-adjoint operator. Hermitian operators have a number of important properties, which will be discussed in more detail in Sect. 3.3. Here I shall note just one important property of Hermitian operators, which trivially follows from Eq. 3.15: a quantity defined as $\langle\alpha|\hat{T}|\alpha\rangle$ is a real-valued number for any choice of state $|\alpha\rangle$. Expressions of this type are called *expectation values of the operator in a given state*. The origin of this name will become clear in Sect. 3.3. A few examples of Hermitian operators follow below.

Example 8 (Hermitian Operators) Let me prove that operator $i\hat{D}$, where $\hat{D}$ is the differentiation operator, introduced in the previous example, is Hermitian. To this end I just need to repeat computations from Example 7:

$$\left(\langle g|\, i\hat{D}\,|f\rangle\right)^* = \left(i \int_{-\infty}^{\infty} dx\, g^*(x) \frac{df}{dx}\right)^* = g(x) f^*(x)\big|_{-\infty}^{\infty} + i \int_{-\infty}^{\infty} dx\, f^*(x) \frac{dg}{dx} \equiv$$

$$\langle f|\, i\hat{D}\,|g\rangle .$$

Example 9 (Hermitian Operators) As a second example of the Hermitian operator, I consider a 3×3 matrix M acting on vectors in a three-dimensional vector space:

$$M = \begin{bmatrix} 1 & i & 2 \\ -i & 1 & 4i \\ 2 & -4i & 0 \end{bmatrix} .$$

I will demonstrate that this matrix is Hermitian by directly remembering that Hermitian conjugation of matrices consists of transposition and complex conjugation. Consequently carrying out these operations, you can convince yourselves that they yield the same matrix M:

$$\begin{bmatrix} 1 & i & 2 \\ i & 1 & 4i \\ 2 & -4i & 0 \end{bmatrix} \rightarrow \begin{bmatrix} 1 & -i & 2 \\ i & 1 & -4i \\ 2 & 4i & 0 \end{bmatrix} \rightarrow \begin{bmatrix} 1 & i & 2 \\ -i & 1 & 4i \\ 2 & -4i & 0 \end{bmatrix} .$$

You can also compute expression $a^\dagger \cdot M \cdot a$, where a is an arbitrary column vector and $a^\dagger$ its Hermitian conjugate:

$$\begin{bmatrix} a_1^* & a_2^* & a_3^* \end{bmatrix} \begin{bmatrix} 1 & i & 2 \\ -i & 1 & 4i \\ 2 & -4i & 0 \end{bmatrix} \begin{bmatrix} a_1 \\ a_2 \\ a_3 \end{bmatrix} = \begin{bmatrix} a_1^* & a_2^* & a_3^* \end{bmatrix} \begin{bmatrix} a_1 + ia_2 + 2a_3 \\ -ia_1 + a_2 + 4ia_3 \\ 2a_1 - 4ia_2 \end{bmatrix} =$$

$$a_1^* a_1 + ia_1^* a_2 + 2a_1^* a_3 - ia_2^* a_1 + a_2^* a_2 + 4ia_2^* a_3 + 2a_1 a_3^* - 4ia_2 a_3^* =$$

$$|a_1|^2 + |a_2|^2 + |a_3|^2 + 2\left(a_1^* a_3 + a_1 a_3^*\right) + i\left(a_1^* a_2 - a_2^* a_1 + 4a_2^* a_3 - 4a_2 a_3^*\right) .$$

It is obvious that the final expression is real-valued as promised.

3.2.2 Commutators, Functions of Operators, and Operator Identities

In addition to Hermitian conjugation, you will need to perform on operators other, less exotic, operations, such as multiplication. The product of two operators $\hat{T}_1$ and $\hat{T}_2$ is defined as consecutive action of the operators. If you consider action on a ket vector, the first operator to do the work is the one on the right:

$$\left(\hat{T}_2\hat{T}_1\right)|\alpha\rangle \equiv \hat{T}_2\left(\hat{T}_1|\alpha\rangle\right).$$

In the case of operators acting on the bra vector, the order is opposite: the first to act is the leftmost operator:

$$\langle\alpha|\left(\hat{T}_2\hat{T}_1\right) \equiv \left(\langle\alpha|\hat{T}_2\right)\hat{T}_1.$$

The most important property of the operator multiplication is actually the absence of a property: multiplication of operator is not, in general, commutative[2]:

$$\hat{T}_2\hat{T}_1 \neq \hat{T}_1\hat{T}_2.$$

The non-commutative nature of operator multiplication is of extreme importance in quantum mechanics, and as you will see, it is the main mathematical feature responsible, for instance, for the uncertainty relation. For the same reason, sets of operators that do commute with each other also play an important role in the quantum formalism.

The non-commutativity of operator multiplication is expressed quantitatively via the notion of a commutator. The commutator of two operators $\left[\hat{T}_1, \hat{T}_2\right]$ is defined as

$$\left[\hat{T}_1, \hat{T}_2\right] = \hat{T}_1\hat{T}_2 - \hat{T}_2\hat{T}_1. \tag{3.16}$$

The knowledge of the commutator or, as it is sometimes called, a commutation relation between two operators is essential and, often, the most important information about operators that you can have. You will see throughout the course how the commutation relations of different operators are used in a variety of applications and calculations.

Commutators have a few important properties, the most frequently used of which are the following:

---

[2]We all are used to deal with commutative multiplication of numbers: the result does not depend on the order, in which multiplication is performed. The lack of commutativity of multiplication was one of the features of the Heisenberg theory, which especially freaked out Schrödinger.

$$\left[\hat{T}_1, \hat{T}_2\right] = -\left[\hat{T}_2, \hat{T}_1\right] \tag{3.17}$$

$$\left[\hat{T}_1 + \hat{T}_2, \hat{T}_3\right] = \left[\hat{T}_1, \hat{T}_3\right] + \left[\hat{T}_2, \hat{T}_3\right] \tag{3.18}$$

$$\left[c_1\hat{T}_1, c_2\hat{T}_2\right] = c_1 c_2 \left[\hat{T}_1, \hat{T}_2\right]. \tag{3.19}$$

The proof of all these identities is quite obvious, and I shall leave it for you as an exercise.

Having defined a product of two operators, I can introduce a power function for the operators: $\hat{T}^n$ simply means applying the same operator n times. The power function is important because it allows defining other, more complex, functions of the operators. In general, expression $f\left(\hat{T}\right)$, where $f(x)$ is an arbitrary function, which has infinitely many derivatives at $x = 0$, can be expended in the infinite Taylor series. Using this series one can define the operator function $f\left(\hat{T}\right)$ by simply substituting the operator instead of x in the series:

$$f\left(\hat{T}\right) = \sum_{n=0}^{\infty} \frac{1}{n!} \frac{d^n f}{dx^n}\bigg|_{x=0} \hat{T}^n.$$

However, a number of important functions, which you are used to dealing with routinely, cannot be defined this way and, therefore, do not make sense for operators. Among them are $\sqrt{\hat{T}}$, $\ln\left(\hat{T}\right)$ and other similar functions with singularities at zero. An important exception is function $\hat{T}^{-1}$, called inverse operator, which is defined by equation

$$\hat{T}\hat{T}^{-1} = \hat{T}^{-1}\hat{T} = \hat{I}, \tag{3.20}$$

where $\hat{I}$ is a unity operator, i.e., an operator which does not change a vector it acts upon. The meaning of the inverse operator can be illustrated by the following expressions:

$$\hat{T}|\alpha\rangle = |\beta\rangle$$
$$\hat{T}^{-1}|\beta\rangle = |\alpha\rangle,$$

where the second line is obtained from the first one by multiplying both sides of the latter by $\hat{T}^{-1}$. Finding inverse operators is usually a difficult task and often amounts to solving an entire problem. If an operator has a form of a matrix, its inverse can be found according to standard rules for inverting matrices.

Finding inverse operators is significantly simplified for a special class of operators called *unitary operators*. These operators, defined by the condition

$$\hat{U}^\dagger = \hat{U}^{-1},$$

play an extremely important role in quantum theory (we value them, of course, not just because their inverse is easy to find). The main property of unitary operators is that they do not change the norm of the vectors or their inner products. Indeed, consider vectors $|\alpha\rangle$ and $|\beta\rangle$, and define new vectors $|\tilde{\alpha}\rangle = \hat{U}|\alpha\rangle$ and $\left|\tilde{\beta}\right\rangle = \hat{U}|\beta\rangle$, where $\hat{U}$ is a unitary operator. Direct computation of $\left\langle\tilde{\alpha}\right|\left.\tilde{\beta}\right\rangle$ proves this statement:

$$\langle\tilde{\alpha}| = \langle\alpha|\,\hat{U}^\dagger \Rightarrow \left\langle\tilde{\alpha}\right|\left.\tilde{\beta}\right\rangle = \langle\alpha|\,\hat{U}^\dagger\hat{U}\,|\beta\rangle = \langle\alpha|\,\hat{U}^{-1}\hat{U}\,|\beta\rangle = \langle\alpha|\,\beta\rangle\,.$$

Unitary operators are a generalization of the rotation operator acting on regular three-dimensional vectors: rotation of two vectors by the same angle does not change their lengths as well as an angle between them. As a result, the dot product of these vectors also does not change. Here is an example of a unitary operator based on the two-dimensional rotation matrix.

Example 10 (Unitary Operators) Consider the well-known matrix used to relate the coordinates of a two-dimensional vector rotated by an angle θ from its initial position:

$$R = \begin{bmatrix} \cos\theta & -\sin\theta \\ \sin\theta & \cos\theta \end{bmatrix}$$

Its Hermitian conjugate is

$$R^\dagger = \begin{bmatrix} \cos\theta & \sin\theta \\ -\sin\theta & \cos\theta \end{bmatrix}.$$

Simple computation shows that product $R^\dagger R$ is a unity matrix:

$$R^\dagger R = \begin{bmatrix} \cos\theta & \sin\theta \\ -\sin\theta & \cos\theta \end{bmatrix}\begin{bmatrix} \cos\theta & -\sin\theta \\ \sin\theta & \cos\theta \end{bmatrix} =$$

$$\begin{bmatrix} \cos^2\theta + \sin\theta & -\cos\theta\sin\theta + \cos\theta\sin\theta \\ -\cos\theta\sin\theta + \cos\theta\sin\theta & \cos^2\theta + \sin\theta \end{bmatrix} = \begin{bmatrix} 1 & 0 \\ 0 & 1 \end{bmatrix}.$$

This proves, of course, that $R^\dagger = R^{-1}$.

An important example of an operator function is an exponential function defined as

$$\exp\left(\hat{T}\right) \equiv \sum_{n=0}^{\infty}\frac{1}{n!}\hat{T}^n. \tag{3.21}$$

Some of the familiar properties of this function remain valid even when its argument is an operator. For instance, the derivative of the expression $f(\lambda) = \exp\left(\lambda\hat{T}\right)$ with respect to the parameter λ is calculated as though $\hat{T}$ were a regular number:

$$df/d\lambda = \hat{T}\exp\left(\lambda\hat{T}\right).$$

You should be warned, however, that a very convenient property of exponential functions

$$\exp(x+y) = \exp(x)\exp(y) \tag{3.22}$$

does not hold for operator arguments. One way to understand the reason for this unfortunate circumstance is to notice that if two operators $\hat{T}_1$ and $\hat{T}_2$ in the argument of the exponential function $\exp\left(\hat{T}_1 + \hat{T}_2\right)$ do not commute, expressions $\exp\left(\hat{T}_1\right)\exp(\hat{T}_2)$ and $\exp\left(\hat{T}_2\right)\exp(\hat{T}_1)$ are not equivalent, so they both cannot be equal to the exponential of the sum of these operators. Generalization of Eq. 3.22 to the case of operator arguments is, in general, very complicated and will not be considered here. There is, however, one case, when such a generalization has a relatively simple form and can be derived without too much efforts, while some work is still required, of course. This simplification takes place when the commutator of the operators $\hat{T}_1$ and $\hat{T}_2$ commutes with both of them. In most cases, this means that the commutator is a regular number, but it does not have to be.

So, suppose that the commutator of two operators $\hat{T}_1$ and $\hat{T}_2$ is $\left[\hat{T}_1, \hat{T}_2\right] = \hat{C}$, where $\hat{C}$ is such that $\left[\hat{T}_1, \hat{C}\right] = \left[\hat{T}_2, \hat{C}\right] = 0$. This assumption appears to be quite restrictive, but in reality, it is fulfilled in a great many pairs of operators that are important for quantum mechanics. In order to derive the promised generalization of Eq. 3.22, I have to, first, prove two intermediate identities, which, however, are useful in their own right. Let me begin by computing the following expression:

$$\left[\hat{T}_1, e^{\lambda\hat{T}_2}\right] = \sum_{n=0}^{\infty}\frac{1}{n!}\left[\hat{T}_1, \lambda^n\hat{T}_2{}^n\right] = \sum_{n=0}^{\infty}\frac{\lambda^n}{n!}\left[\hat{T}_1, \hat{T}_2{}^n\right]. \tag{3.23}$$

To proceed I need to prove the following identity for the commutators:

$$\left[\hat{T}_1, \hat{T}_2{}^n\right] = n\hat{C}\hat{T}_2{}^{n-1}. \tag{3.24}$$

The easiest way to do it is to use the method of mathematical induction. For those who have forgotten how this method works, the first step is to prove the statement for the first nontrivial value of the index ($n = 2$ in this case). After that you assume that the statement is correct for $n = k$ and, using this assumption, prove it for $n = k+1$. Thus, the first step—consider $n = 2$:

$$\left[\hat{T}_1, \hat{T}_2^{\ 2}\right] = \hat{T}_1\hat{T}_2^{\ 2} - \hat{T}_2^{\ 2}\hat{T}_1 = \hat{T}_1\hat{T}_2\hat{T}_2 - \hat{T}_2\hat{T}_1\hat{T}_2 + \hat{T}_2\hat{T}_1\hat{T}_2 - \hat{T}_2\hat{T}_2\hat{T}_1$$

$$= \left(\hat{T}_1\hat{T}_2 - \hat{T}_2\hat{T}_1\right)\hat{T}_2 + \hat{T}_2\left(\hat{T}_1\hat{T}_2 - \hat{T}_2\hat{T}_1\right) = 2\hat{C}\hat{T}_2.$$

(Note that it works because $\hat{C}$ commutes with $\hat{T}_2$.) Next, $n = k$ assumption:

$$\left[\hat{T}_1, \hat{T}_2^{\ k}\right] = k\hat{C}\hat{T}_2^{\ k-1}$$

The final step—proof for $n = k + 1$:

$$\left[\hat{T}_1, \hat{T}_2^{\ k+1}\right] = \hat{T}_1\hat{T}_2^{\ k+1} - \hat{T}_2^{\ k+1}\hat{T}_1 = \hat{T}_1\hat{T}_2^{\ k+1} - \hat{T}_2\hat{T}_1\hat{T}_2^{\ k} + \hat{T}_2\hat{T}_1\hat{T}_2^{\ k} - \hat{T}_2^{\ k+1}\hat{T}_1$$

$$= \left[\hat{T}_1, \hat{T}_2\right]\hat{T}_2^{\ k} + \hat{T}_2\left[\hat{T}_1, \hat{T}_2^{\ k}\right] = \hat{C}\hat{T}_2^{\ k} + k\hat{C}\hat{T}_2^{\ k} = (k+1)\hat{C}\hat{T}_2^{\ k}.$$

Using this identity I can transform Eq. 3.23 into

$$\left[\hat{T}_1, e^{\lambda\hat{T}_2}\right] = \hat{C}\sum_{n=0}^{\infty} \frac{\lambda^n n}{n!}\hat{T}_2^{\ n-1} = \hat{C}\lambda\sum_{n=1}^{\infty} \frac{\lambda^{n-1}}{(n-1)!}\hat{T}_2^{\ n-1} = \lambda\hat{C}e^{\lambda\hat{T}_2} \qquad (3.25)$$

This result can be used to derive another important identity. Multiply Eq. 3.25 by $e^{-\lambda\hat{T}_2}$ from the left:

$$e^{-\lambda\hat{T}_2}\left[\hat{T}_1, e^{\lambda\hat{T}_2}\right] = e^{-\lambda\hat{T}_2}\lambda\hat{C}e^{\lambda\hat{T}_2}.$$

The right-hand side of this expression simplifies to $\hat{C}\lambda$: $e^{-\lambda\hat{T}_2}e^{\lambda\hat{T}_2} = e^{-\lambda\hat{T}_2+\lambda\hat{T}_2} = \hat{I}$, since Eq. 3.22 is applicable for any commuting operators and any operator commutes with itself. Now you can expand the commutator on the left of the expression above to get

$$e^{-\lambda\hat{T}_2}\hat{T}_1 e^{\lambda\hat{T}_2} - \hat{T}_1 = \hat{C}\lambda$$

or

$$e^{-\lambda\hat{T}_2}\hat{T}_1 e^{\lambda\hat{T}_2} = \hat{T}_1 + \hat{C}\lambda. \qquad (3.26)$$

Now I am ready to approach my main target and to prove that

$$e^{\hat{T}_1+\hat{T}_2} = e^{\hat{T}_1}e^{\hat{T}_2}e^{-\frac{1}{2}[\hat{T}_1,\hat{T}_2]}. \qquad (3.27)$$

The proof of this identity is more involved than the two previous derivations. Direct proof (for instance, by using series expansions of the exponential functions on both sides of Eq. 3.27) results in expressions too cumbersome to allow for fruitful analysis. Therefore, I am going to use an indirect approach, which was invented by Harvard Professor Roy Glauber, winner of the 2005 Nobel Prize for his contribution in quantum optics. Glauber considered function $f(x) = e^{x\hat{T}_1} e^{x\hat{T}_2}$, for which he derived a differential equation by computing its derivative:

$$\frac{df}{dx} = \hat{T}_1 e^{x\hat{T}_1} e^{x\hat{T}_2} + e^{x\hat{T}_1} \hat{T}_2 e^{x\hat{T}_2}.$$

Note how operators $\hat{T}_1$ and $\hat{T}_2$ are placed in this expression: $\hat{T}_1$ appears in front of the exponent containing $\hat{T}_2$ because it originates from the exponential function of $\hat{T}_1$ positioned to the left of $e^{x\hat{T}_2}$. At the same time, $\hat{T}_2$ appears behind $e^{x\hat{T}_1}$ following the respective position of $e^{x\hat{T}_2}$. Relative positions of $e^{x\hat{T}_i}$ and respective $\hat{T}_i$ are not important because these operators commute (any operator commutes with any function of the same operator). Now the derivative can be rewritten in the following way:

$$\frac{df}{dx} = e^{x\hat{T}_1} e^{x\hat{T}_2} e^{-x\hat{T}_2} e^{-x\hat{T}_1} \left(\hat{T}_1 e^{x\hat{T}_1} e^{x\hat{T}_2} + e^{x\hat{T}_1} \hat{T}_2 e^{x\hat{T}_2} \right).$$

It is not too difficult to see that the expression in front of the brackets is equal to unity so writing it there does not change anything. Continue

$$\frac{df}{dx} = f(x) \left(e^{-x\hat{T}_2} \hat{T}_1 e^{x\hat{T}_2} + \hat{T}_2 \right) = f(x) \left(\hat{T}_1 + x\left[\hat{T}_1, \hat{T}_2\right] + \hat{T}_2 \right),$$

where the identity given by Eq. 3.26 is used and $\hat{C}$ is replaced with $\left[\hat{T}_1, \hat{T}_2\right]$. This differential equation can now be solved for function $f(x)$:

$$\int \frac{df}{f} = \int dx \left(\hat{T}_1 + \hat{T}_2 + x\left[\hat{T}_1, \hat{T}_2\right] \right) \Rightarrow$$

$$\ln \frac{f}{f_0} = x \left(\hat{T}_1 + \hat{T}_2 \right) + \frac{1}{2}x^2\left[\hat{T}_1, \hat{T}_2\right],$$

where integration constant f_0 is chosen to satisfy the obvious initial condition: $f(0) = 1$. With this in mind, function f can be written as

$$f = e^{x(\hat{T}_1 + \hat{T}_2) + \frac{1}{2}x^2[\hat{T}_1, \hat{T}_2]}.$$

Setting $x = 1$ in this expression and multiplying it by $e^{-\frac{1}{2}[\hat{T}_1, \hat{T}_2]}$, Eq. 3.27 is finally obtained, completing the proof.

I want to finish this section with two important technical statements about Hermitian operators. The first one is concerned with Hermitian conjugation of a product of two Hermitian operators. It can be shown that

$$\left(\hat{T}_1\hat{T}_2\right)^\dagger = \hat{T}_2\hat{T}_1. \tag{3.28}$$

This statement can be proven as follows. By definition

$$\langle\alpha|\left(\hat{T}_1\hat{T}_2\right)^\dagger|\beta\rangle = \left(\langle\beta|\hat{T}_1\hat{T}_2|\alpha\rangle\right)^*.$$

Introducing $\langle\beta|\hat{T}_1 = \langle\tilde{\beta}|, \hat{T}_2|\alpha\rangle = |\tilde{\alpha}\rangle$ and using Eq. 2.19, you can write the right-hand side of this expression as

$$\left(\langle\tilde{\beta}|\tilde{\alpha}\rangle\right)^* = \langle\tilde{\alpha}|\tilde{\beta}\rangle.$$

Rules for Hermitian conjugation yield $|\tilde{\beta}\rangle = \hat{T}_1^\dagger|\beta\rangle$ and $\langle\tilde{\alpha}| = \langle\alpha|\hat{T}_2^\dagger$, which allows to proceed as follows:

$$\left(\langle\beta|\hat{T}_1\hat{T}_2|\alpha\rangle\right)^* = \left(\langle\tilde{\beta}|\tilde{\alpha}\rangle\right)^* = \langle\tilde{\alpha}|\tilde{\beta}\rangle = \langle\alpha|\hat{T}_2^\dagger\hat{T}_1^\dagger|\beta\rangle.$$

By the way, you may have noticed that I had actually proved a more general statement. Indeed, the last equation means that

$$\left(\hat{T}_1\hat{T}_2\right)^\dagger = \hat{T}_2^\dagger\hat{T}_1^\dagger,$$

which is valid for any linear, not necessarily Hermitian, operator. Equation 3.28 follows from this result if $\hat{T}_1$ and $\hat{T}_2$ are Hermitian. An immediate corollary of this result is the following:

$$\left[\hat{T}_1, \hat{T}_2\right]^\dagger = -\left[\hat{T}_2, \hat{T}_1\right]. \tag{3.29}$$

Operators which change sign upon Hermitian conjugation are called anti-Hermitian, so the commutator of two Hermitian operators is also anti-Hermitian. It is now easy to demonstrate that a commutator of two Hermitian operators can be presented as

$$\left[\hat{T}_1, \hat{T}_2\right] = i\hat{A}, \tag{3.30}$$

where $\hat{A}$ is Hermitian. If the commutator is a number, Eq. 3.30 is reduced to

$$\left[\hat{T}_1, \hat{T}_2\right] = ic, \tag{3.31}$$

where c is real.

3.2.3 Eigenvalues and Eigenvectors

When an operator acts on a generic vector, the result is a different vector. For instance, differentiation operator acting on function e^{-x^2}: $\hat{D}e^{-x^2} = -2xe^{-x^2}$ — produces a different function. If, however, you apply the same operator to function $e^{\kappa x}$, the result will be the same function, multiplied by a number: $\hat{D}e^{\kappa x} = \kappa e^{\kappa x}$. This example illustrates a general phenomenon: among many vectors that are changed by operators in completely different vectors, there are some that are only being multiplied by a number. This special class of vectors, called eigenvectors, plays an important role in the application of operators in quantum physics. The number, which appears as a factor in front of an eigenvector, is specific for each vector (or a limited subset thereof) and is called an eigenvalue. The formal definition of an eigenvector and an eigenvalue is as follows: vector $|\alpha\rangle$ is an eigenvector of operator $\hat{T}$ with a respective eigenvalue λ_α if

$$\hat{T}|\alpha\rangle = \lambda_\alpha |\alpha\rangle. \tag{3.32}$$

For each eigenvector there might be one and only one corresponding eigenvalue, but the opposite of this statement is not always true. If for each eigenvalue there exists only a single eigenvector, we describe this eigenvalue as non-degenerate. If an opposite happens, and several eigenvectors "belong" to the same eigenvalue, the respective eigenvalue is naturally called "degenerate." In the non-degenerate case, an eigenvalue describes a respective eigenvector with an accuracy to a constant factor (a vector appearing in Eq. 3.32 can be multiplied by any number without destroying the equation). If we, however, require that all eigenvectors be normalized, then the eigenvalue will define the respective eigenvector uniquely (with accuracy to an arbitrary phase factor, which cannot be fixed by normalization but which does not affect any physical results) so that I can designate it simply as $|\lambda\rangle$.

To distinguish between different eigenvectors belonging to the same eigenvalue, I need an additional index so that Eq. 3.32 becomes

$$\hat{T}|\lambda, \mu\rangle = \lambda |\lambda, \mu\rangle. \tag{3.33}$$

The physical meaning of the additional index will become clear later, but for now, it is just a way to distinguish between different eigenvectors belonging to the same eigenvalue. An important property of degenerate eigenvectors is that any linear combination of these vectors is again an eigenvector belonging to the same eigenvalue. Indeed, consider a vector

$$|\alpha\rangle = a_{\mu_1} |\lambda, \mu_1\rangle + a_{\mu_2} |\lambda, \mu_2\rangle$$

and apply operator $\hat{T}$ to it

$$\hat{T} |\alpha\rangle = \hat{T} \left(a_{\mu_1} |\lambda, \mu_1\rangle + a_{\mu_2} |\lambda, \mu_2\rangle \right) = a_{\mu_1} \lambda |\lambda, \mu_1\rangle + a_{\mu_2} \lambda |\lambda, \mu_2\rangle = \lambda |\alpha\rangle$$

where I used Eq. 3.33. Using mathematical lingo, you can say that eigenvectors belonging to a degenerate eigenvalue form a subspace of the total linear space because by forming any linear combination thereof you remain within the same set of vectors in complete agreement with the definition of a vector space.

Now I shall prove an important theorem concerning eigenvectors of commuting operators and discuss its consequences.

Theorem 1 (Eigenvectors of Commuting Operators) *Consider two operators $\hat{T}_1$ and $\hat{T}_2$ such that $\hat{T}_1 \hat{T}_2 = \hat{T}_2 \hat{T}_1$. Also assume that λ_{T_1} is a non-degenerate eigenvalue of $\hat{T}_1$ with eigenvector $|\lambda_{T_1}\rangle$. Then, this vector is also an eigenvector of the operator $\hat{T}_2$.*

Proof Consider

$$\hat{T}_2 \hat{T}_1 |\lambda_{T_1}\rangle = \lambda_{T_1} \hat{T}_2 |\lambda_{T_1}\rangle = \hat{T}_1 \hat{T}_2 |\lambda_{T_1}\rangle$$

where at the last step I used the commutative property of the operators. The obtained result means that $\hat{T}_2 |\lambda_{T_1}\rangle$ is also an eigenvector of $\hat{T}_1$ with the same eigenvalue λ_{T_1}. However, since it was assumed that λ_{T_1} is non-degenerate, this new eigenvector might differ from $|\lambda_{T_1}\rangle$ only by a constant factor:

$$\hat{T}_2 |\lambda_{T_1}\rangle = \lambda_{T_2} |\lambda_{T_1}\rangle \,,$$

which means that $|\lambda_{T_1}\rangle$ is an eigenvector of $\hat{T}_2$.

The non-degenerate nature of the eigenvalue of $\hat{T}_1$ is essential for this proof to work. Thus, if eigenvalues of $\hat{T}_1$ are degenerate, not all eigenvectors of $\hat{T}_1$ will also be eigenvectors of $\hat{T}_2$. However, it can be proven (though the proof is much more involved and will not be reproduced here) that one can always form such a linear combination of these degenerate eigenvectors which will become an eigenvector of $\hat{T}_2$, with its own eigenvalue λ_{T_2}. In this case, assigning eigenvalues of both $\hat{T}_1$ and $\hat{T}_2$ might provide a unique characterization of a vector, which is a simultaneous eigenvector of both operators and can be notated as $|\lambda_{T_1}, \lambda_{T_2}\rangle$. Comparing this notation to Eq. 3.33, one can see that the index μ in that equation can be understood as an eigenvalue of a commuting partner operator. If there exists a third operator, $\hat{T}_3$, commuting with both $\hat{T}_1$ and $\hat{T}_2$, one can find common eigenvectors for all three operators, in which case a full unique characterization of such a state would require specifying three eigenvalues: $|\lambda_{T_1}, \lambda_{T_2}, \lambda_{T_3}\rangle$, where

$$\hat{T}_1 |\lambda_{T_1}, \lambda_{T_2}, \lambda_{T_3}\rangle = \lambda_{T_1} |\lambda_{T_1}, \lambda_{T_2}, \lambda_{T_3}\rangle$$
$$\hat{T}_2 |\lambda_{T_1}, \lambda_{T_2}, \lambda_{T_3}\rangle = \lambda_{T_2} |\lambda_{T_1}, \lambda_{T_2}, \lambda_{T_3}\rangle$$
$$\hat{T}_3 |\lambda_{T_1}, \lambda_{T_2}, \lambda_{T_3}\rangle = \lambda_{T_3} |\lambda_{T_1}, \lambda_{T_2}, \lambda_{T_3}\rangle .$$

In general, in order to fully uniquely characterize an eigenvector of an operator with degenerate eigenvalues, one needs to find the complete set of commuting operators (CSCO), i.e., all operators which commute with each other.

To help you visualize these rather abstract concepts, I will illustrate them with a simple example involving commuting matrices, but you have to be prepared for some lengthy computations. So, embrace yourself! This example will also illustrate the process of finding eigenvalues and eigenvectors of operators in a matrix form.

Example 11 (Eigenvectors of Commuting Matrices) Consider two 3×3 matrices

$$M1 = \begin{bmatrix} \frac{5}{4} & \frac{1}{2\sqrt{2}} & \frac{1}{4} \\ \frac{1}{2\sqrt{2}} & \frac{3}{2} & \frac{1}{2\sqrt{2}} \\ \frac{1}{4} & \frac{1}{2\sqrt{2}} & \frac{5}{4} \end{bmatrix} ; \quad M2 = \begin{bmatrix} 1 & -\frac{1}{\sqrt{2}} & -1 \\ -\frac{1}{\sqrt{2}} & 0 & -\frac{1}{\sqrt{2}} \\ -1 & -\frac{1}{\sqrt{2}} & 1 \end{bmatrix} . \tag{3.34}$$

It does not take much effort to compute their products (you can use symbolic computational platform such as Mathematica or Maple if you are too lazy to do it yourself) and to see that the matrices, indeed, commute:

$$M1 \cdot M2 = M2 \cdot M1 = \begin{bmatrix} \frac{3}{4} & -\frac{3}{2\sqrt{2}} & -\frac{5}{4} \\ -\frac{3}{2\sqrt{2}} & -\frac{1}{2} & -\frac{3}{2\sqrt{2}} \\ -\frac{5}{4} & -\frac{3}{2\sqrt{2}} & \frac{3}{4} \end{bmatrix} .$$

Vectors in this case are single columns with three elements:

$$|\alpha\rangle = \begin{bmatrix} u_1 \\ u_2 \\ u_3 \end{bmatrix} .$$

and the eigenvector equation 3.32 takes the form of a matrix equation. For $M1$ this equation is

$$\begin{bmatrix} \frac{5}{4} & \frac{1}{2\sqrt{2}} & \frac{1}{4} \\ \frac{1}{2\sqrt{2}} & \frac{3}{2} & \frac{1}{2\sqrt{2}} \\ \frac{1}{4} & \frac{1}{2\sqrt{2}} & \frac{5}{4} \end{bmatrix} \begin{bmatrix} u_1 \\ u_2 \\ u_3 \end{bmatrix} = \lambda \begin{bmatrix} u_1 \\ u_2 \\ u_3 \end{bmatrix} .$$

It is convenient to collect all terms on one side and present this equation in the form

$$\begin{bmatrix} \frac{5}{4} - \lambda & \frac{1}{2\sqrt{2}} & \frac{1}{4} \\ \frac{1}{2\sqrt{2}} & \frac{3}{2} - \lambda & \frac{1}{2\sqrt{2}} \\ \frac{1}{4} & \frac{1}{2\sqrt{2}} & \frac{5}{4} - \lambda \end{bmatrix} \begin{bmatrix} u_1 \\ u_2 \\ u_3 \end{bmatrix} = 0. \tag{3.35}$$

What we have here is a matrix form of a system of three linear homogeneous equations, which always has at least one solution: $u_1 = u_2 = u_3 = 0$. This solution, however, is not what I had in mind when introducing the concept of eigenvectors. We need non-zero solutions, but they might exist only if the determinant of the matrix representing coefficients of this equation is equal to zero. (Cramer's rule of linear algebra, anyone?) Computing the determinant and setting it to zero, I arrive at the following equation:

$$\lambda^3 - 4\lambda^2 + 5\lambda - 2 = 0,$$

which has three solutions, $\lambda_{1,2} = 1$; $\lambda_3 = 2$, two of which coincide signifying that the matrix does have degenerate eigenvalues. (These solutions can be found by factoring the determinant as $(\lambda - 1)^2 (\lambda - 2)$.)

Now, for each eigenvalue, I will find a respective eigenvector, beginning with a non-degenerate eigenvalue $\lambda_3 = 2$. Substituting this eigenvalue in Eq. 3.35, I reduce it to

$$\begin{bmatrix} -\frac{3}{4} & \frac{1}{2\sqrt{2}} & \frac{1}{4} \\ \frac{1}{2\sqrt{2}} & -\frac{1}{2} & \frac{1}{2\sqrt{2}} \\ \frac{1}{4} & \frac{1}{2\sqrt{2}} & -\frac{3}{4} \end{bmatrix} \begin{bmatrix} u_1^{(3)} \\ u_2^{(3)} \\ u_3^{(3)} \end{bmatrix} = 0.$$

where added upper index in $u_i^{(3)}$ indicates that this eigenvector belongs to the third eigenvalue. Expanding the matrix equation in an explicit system of linear equations yields

$$-\frac{3}{4}u_1^{(3)} + \frac{1}{2\sqrt{2}}u_2^{(3)} + \frac{1}{4}u_3^{(3)} = 0 \Rightarrow -3u_1^{(3)} + \sqrt{2}u_2^{(3)} + u_3^{(3)} = 0$$

$$\frac{1}{2\sqrt{2}}u_1^{(3)} - \frac{1}{2}u_2^{(3)} + \frac{1}{2\sqrt{2}}u_3^{(3)} = 0 \Rightarrow u_1^{(3)} - \sqrt{2}u_2^{(3)} + u_3^{(3)} = 0$$

$$\frac{1}{4}u_1^{(3)} + \frac{1}{2\sqrt{2}}u_2^{(3)} - \frac{3}{4}u_3^{(3)} = 0 \Rightarrow u_1^{(3)} + \sqrt{2}u_2^{(3)} - 3u_3^{(3)} = 0.$$

Combining the last two equations, I get $2u_1^{(3)} - 2u_3^{(3)} = 0 \Rightarrow u_1^{(3)} = u_3^{(3)}$. Then, the first two equations are reduced to two identical equations:

$$-2u_1^{(3)} + \sqrt{2}u_2^{(3)} = 0$$

$$2u_1^{(3)} - \sqrt{2}u_2^{(3)} = 0,$$

which means that the value for one of the coefficients $u_{1,2}^{(3)}$ can be chosen arbitrarily. For instance, you can express these coefficients in terms of yet undefined $u_1^{(3)}$: $u_2^{(3)} = \sqrt{2}u_1^{(3)}$; $u_3^{(3)} = u_1^{(3)}$. Using notation $|2\rangle$ to designate this eigenvector (2 in this notation refers to the value of the respective eigenvalue), I can write

$$|2\rangle = u_1^{(3)} \begin{bmatrix} 1 \\ \sqrt{2} \\ 1 \end{bmatrix}.$$

The value of the remaining coefficient can be fixed (if the undefined coefficients make you nervous) by requiring that the vector is normalized:

$$\left| u_1^{(3)} \right|^2 [1 \ \sqrt{2} \ 1] \begin{bmatrix} 1 \\ \sqrt{2} \\ 1 \end{bmatrix} = \left| u_1^{(3)} \right|^2 4 = 1 \Rightarrow u_1^{(3)} = \frac{1}{2}.$$

Thus, the normalized eigenvector belonging to the eigenvalue $\lambda = 2$ is found to be

$$|2\rangle = \frac{1}{2} \begin{bmatrix} 1 \\ \sqrt{2} \\ 1 \end{bmatrix}. \tag{3.36}$$

Now let me deal with degenerate eigenvalue $\lambda_{1,2} = 1$. In this case, the eigenvector equation becomes

$$\begin{bmatrix} \frac{1}{4} & \frac{1}{2\sqrt{2}} & \frac{1}{4} \\ \frac{1}{2\sqrt{2}} & \frac{1}{2} & \frac{1}{2\sqrt{2}} \\ \frac{1}{4} & \frac{1}{2\sqrt{2}} & \frac{1}{4} \end{bmatrix} \begin{bmatrix} u_1^{(1,2)} \\ u_2^{(1,2)} \\ u_3^{(1,2)} \end{bmatrix} = 0$$

or in the expanded form

$$\frac{1}{4}u_1^{(1,2)} + \frac{1}{2\sqrt{2}}u_2^{(1,2)} + \frac{1}{4}u_3^{(1,2)} = 0 \Rightarrow u_1^{(1,2)} + \sqrt{2}u_2^{(1,2)} + u_3^{(1,2)} = 0$$

$$\frac{1}{2\sqrt{2}}u_1^{(1,2)} + \frac{1}{2}u_2^{(1,2)} + \frac{1}{2\sqrt{2}}u_3^{(1,2)} = 0 \Rightarrow u_1^{(1,2)} + \sqrt{2}u_2^{(1,2)} + u_3^{(1,2)} = 0$$

$$\frac{1}{4}u_1^{(1,2)} + \frac{1}{2\sqrt{2}}u_2^{(1,2)} + \frac{1}{4}u_3^{(1,2)} = 0 \Rightarrow u_1^{(1,2)} + \sqrt{2}u_2^{(1,2)} + u_3^{(1,2)} = 0.$$

In this case, all three equations coincide, meaning that I can choose arbitrarily two coefficients, e.g., $u_1^{(1,2)}$ and $u_3^{(1,2)}$, while expressing the remaining coefficients as

$$u_2^{(1,2)} = -\left(u_1^{(1,2)} + u_3^{(1,2)}\right)/\sqrt{2}.$$

Choosing different values of the remaining coefficients, I can generate different eigenvectors all belonging to the same eigenvalue. For instance, choosing $u_3^{(2)} = 0$ and $u_1^{(1)} = 0$, I generate distinct vectors:

$$|1\rangle_1 = u_3^{(2)} \begin{bmatrix} 0 \\ -\frac{1}{\sqrt{2}} \\ 1 \end{bmatrix} ; \ |1\rangle_2 = u_3^{(1)} \begin{bmatrix} 1 \\ -\frac{1}{\sqrt{2}} \\ 0 \end{bmatrix}, \tag{3.37}$$

which can also be normalized. Any linear combination of these vectors will also be an eigenvector.

Now I turn my attention to matrix $M2$. Again computing the determinant

$$\begin{Vmatrix} 1-\lambda & -\frac{1}{\sqrt{2}} & -1 \\ -\frac{1}{\sqrt{2}} & -\lambda & -\frac{1}{\sqrt{2}} \\ -1 & -\frac{1}{\sqrt{2}} & 1-\lambda \end{Vmatrix}$$

and setting it to zero, I end up with the equation

$$\lambda^3 - 2\lambda^2 - \lambda + 2 = 0$$

which again can be solved by factorization and yields $\lambda_1 = 2$, $\lambda_2 = -1$, $\lambda_3 = 1$. Each of these eigenvalues (which, by the way, are non-degenerate) has its own eigenvector, which can be found in the same way as above. I will leave the actual calculations as an exercise and present here only the final answers for the normalized eigenvectors:

$$|2\rangle = \frac{1}{\sqrt{2}} \begin{bmatrix} -1 \\ 0 \\ 1 \end{bmatrix}, \ |-1\rangle = \frac{1}{2} \begin{bmatrix} 1 \\ \sqrt{2} \\ 1 \end{bmatrix}, \ |1\rangle = \frac{1}{2} \begin{bmatrix} 1 \\ -\sqrt{2} \\ 1 \end{bmatrix}, \tag{3.38}$$

where eigenvectors are again labeled by their respective eigenvalues. Now, it is obvious that eigenvector $|-1\rangle$ of matrix $M2$ is also an eigenvector of $M1$, so I only need to check the remaining vectors:

$$
\begin{bmatrix} \frac{5}{4} & \frac{1}{2\sqrt{2}} & \frac{1}{4} \\ \frac{1}{2\sqrt{2}} & \frac{3}{2} & \frac{1}{2\sqrt{2}} \\ \frac{1}{4} & \frac{1}{2\sqrt{2}} & \frac{5}{4} \end{bmatrix} \begin{bmatrix} -1 \\ 0 \\ 1 \end{bmatrix} = \begin{bmatrix} -1 \\ 0 \\ 1 \end{bmatrix},
$$

so this vector is an eigenvector of $M1$ with eigenvalue $\lambda = 1$. Note that the elements of this vector obey condition $u_2^{(1,2)} = -\left(u_1^{(1,2)} + u_3^{(1,2)}\right)/\sqrt{2}$ derived for the degenerate eigenvectors of $M1$ with $u_2 = 0$, $u_1 = -u_3 = 1$. Now, for the remaining eigenvector of $M2$, I have

$$
\begin{bmatrix} \frac{5}{4} & \frac{1}{2\sqrt{2}} & \frac{1}{4} \\ \frac{1}{2\sqrt{2}} & \frac{3}{2} & \frac{1}{2\sqrt{2}} \\ \frac{1}{4} & \frac{1}{2\sqrt{2}} & \frac{5}{4} \end{bmatrix} \begin{bmatrix} 1 \\ \sqrt{2} \\ 1 \end{bmatrix} = \begin{bmatrix} 1 \\ -\sqrt{2} \\ 1 \end{bmatrix},
$$

i.e., this is also an eigenvector of $M1$ with the same eigenvalue. For this vector I also have $u_2 = -(u_1 + u_3)/\sqrt{2}$ with $u_1 = u_3 = 1$. Thus, I can present the system of common eigenvectors of these two matrices, in which degenerate eigenvectors become uniquely defined by the virtue of their belonging to the eigenvalues of a second commuting matrix. Now, all these eigenvectors can be designated as $|1, 2\rangle$, $|1, 1\rangle$, and $|2, -1\rangle$, where the first and second numbers refer to the eigenvalues of $M1$ and $M2$, respectively.

3.3 Operators and Observables

3.3.1 Hermitian Operators

One might notice a striking similarity between CSCO and the concept of the complete set of mutually consistent observables discussed in Sect. 2.1. Also, the state vectors characterized by definite values of compatible observables look like eigenvectors of operators characterized by eigenvalues of commuting operators. It appears reasonable, therefore, to expect that one can establish a connection between physical observables and quantum states characterized by the values of the observables on one hand and the mathematical concepts of operators and their eigenvalues and eigenvectors on the other hand. This connection is indeed established by the following postulates laying down the foundation of formalism of quantum mechanics.

> **Postulate 1 (Observables and Hermitian Operators)** Every observable is represented in quantum theory by a Hermitian operator.

Postulate 2 Eigenvalues of operators constructed to represent an observable determine values, which a measurement of the observable might yield, and eigenvectors define states, in which a measurement of the observable represented by the operator will with certainty produce the corresponding value.

The first question which might pop up in someone's mind after reading the first of these postulates is, why does it single out Hermitian operators? The fact of the matter is that Hermitian operators possess a number of special properties, which make them practically suitable for their intended use as representative of physical observables. These properties can be formulated in the form of several theorems.

Theorem 2 (Theorem of the Eigenvalues) *Eigenvalues of Hermitian operators with discrete spectrum are necessarily real-valued.*

Proof Let $|\lambda_n\rangle$ be an eigenvector of a Hermitian operator $\hat{T}$ corresponding to eigenvalue λ_n:

$$\hat{T}|\lambda_n\rangle = \lambda_n|\lambda_n\rangle.$$

Premultiplying this expression by $\langle\lambda_n|$, I get

$$\langle\lambda_n|\hat{T}|\lambda_n\rangle = \lambda_n\langle\lambda_n|\lambda_n\rangle.$$

Performing complex conjugation of this expression and using the definition of the Hermitian conjugate operator, Eq. 3.14, I derive

$$\left(\langle\lambda_n|\hat{T}|\lambda_n\rangle\right)^* = \langle\lambda_n|\hat{T}^\dagger|\lambda_n\rangle = \lambda_n^*\langle\lambda_n|\lambda_n\rangle$$

where it is assumed that the norm of the vector exists and is a real-valued quantity. For Hermitian operators $\hat{T}^\dagger = \hat{T}$, in which case left-hand sides of the last two equations coincide yielding $\lambda_n^* = \lambda_n$, which means, of course, that λ_n is a real number.

The importance of this theorem for association between physical observables and operators is obvious—results of any measurements are always expressed by real numbers, and the theorem guarantees that the mathematical constructs (eigenvalues) used to connect the formalism with the real world of experiments and observations are consistent with this natural requirement. The assumption that the norm of the respective eigenvectors exists, which is a critical element of the proof of the theorem, can be rigorously validated only for Hermitian operators with discrete spectrum.[3]

Eigenvectors of operators with continuous spectrum are not normalizable in the usual sense (see Sect. 2.3), so this theorem does not apply to them. At the same time, we need such continuous spectrum operators as momentum or coordinate

---

[3]I borrowed this fact without proof from the branch of mathematics called functional analysis that studies the properties of linear operators.

to describe physical reality, so we have to find a way to avoid having to deal with unrealistic complex eigenvalues. Leaving the mathematical intricacies of this problem to mathematicians, I solve it here by a sleight of hand. I simply postulate that only real eigenvalues and their corresponding eigenvectors of such operators can be used to represent quantum states and the results of measurements. It can be shown that the eigenvectors corresponding to real eigenvalues of Hermitian operators with continuous spectrum can be normalized in the sense of Eq. 2.41. To illustrate the last point, consider operator id/dx that I have previously proved to be Hermitian. The eigenvectors of this operator have the form of e^{-ikx}, with k being an eigenvalue:

$$id(e^{-ikx})/dx = ke^{-ikx}.$$

If I force k to be a real number, I can use the properties of the delta-function to write

$$\int dx e^{ix(k-k_1)} = 2\pi\delta(k-k_1),$$

which is the orthonormalization requirement for the eigenvectors belonging to continuous spectrum. You may want to notice that the integral in this expression is reduced to the delta-function only for real-valued k.

Theorem 3 (Theorem of Eigenvectors) *Eigenvectors of Hermitian operators with discrete spectrum belonging to different eigenvalues are necessarily orthogonal.*

Proof Consider two different eigenvalues λ_1 and λ_2 of a Hermitian operator $\hat{T}$ together with their eigenvectors $|\lambda_1\rangle$ and $|\lambda_2\rangle$:

$$\hat{T}|\lambda_1\rangle = \lambda_1|\lambda_1\rangle$$
$$\hat{T}|\lambda_2\rangle = \lambda_2|\lambda_2\rangle.$$

Premultiply first of these equations by $\langle\lambda_2|$ and the second one by $\langle\lambda_1|$:

$$\langle\lambda_2|\hat{T}|\lambda_1\rangle = \lambda_1\langle\lambda_2|\lambda_1\rangle$$
$$\langle\lambda_1|\hat{T}|\lambda_2\rangle = \lambda_2\langle\lambda_1|\lambda_2\rangle.$$

Complex conjugate the second of these equations, use Eq. 3.14 (which defines the Hermitian conjugate operator), and take into account that $\hat{T}$ is Hermitian. This yields

$$\langle\lambda_2|\hat{T}|\lambda_1\rangle = (\lambda_2)^*\langle\lambda_1|\lambda_2\rangle^*.$$

so that the pair of equations from above can be written as

$$\langle\lambda_2|\hat{T}|\lambda_1\rangle = \lambda_1\langle\lambda_2|\lambda_1\rangle$$
$$\langle\lambda_2|\hat{T}|\lambda_1\rangle = (\lambda_2)^*\langle\lambda_1|\lambda_2\rangle^*$$

Taking into account that the eigenvalues of the Hermitian operators are real and that according to the property of the inner product $\langle \lambda_2 | \lambda_1 \rangle = \langle \lambda_1 | \lambda_2 \rangle^*$, you finally obtain

$$\lambda_1 \langle \lambda_2 | \lambda_1 \rangle = \lambda_2 \langle \lambda_2 | \lambda_1 \rangle .$$

If $\lambda_1 \neq \lambda_2$, you have no choice but to conclude that $\langle \lambda_2 | \lambda_1 \rangle = 0$.

In the case of Hermitian operators with degenerate spectrum, the situation is more complex because, as we saw in the matrix example in Sect. 3.2.3, one can generate multiple sets of linearly independent vectors belonging to the same eigenvalue, and they do not have to be orthogonal. At the same time, we also saw that one can always find such a set, in which eigenvectors are orthogonal. These special sets of orthogonal vectors belonging to the degenerate eigenvalues are usually also eigenvectors of another operator from the respective CSCO. Thus, you can be rest assured that for any Hermitian operator, there exists a set of mutually orthogonal eigenvectors. I already mentioned that the physical meaning of the mathematical concept of orthogonality is mutual exclusivity of values of the observables used to characterize the states, and this comment essentially completes our identification of mutually exclusive states characterized by a set of mutually consistent set of observables with eigenvectors of operators belonging to a complete set of mutually commuting operators.

Theorem 4 (Completeness of Eigenvectors) *The set of eigenvectors of Hermitian operators is complete in a sense that any state in the respective Hilbert vector space can be presented as a linear combination of these eigenvectors.*

The completeness property gives a rigorous mathematical justification to the generalization of the superposition principle expressed by Eq. 2.26. This property essentially states that eigenvectors of Hermitian operators with discrete spectrum form a countable basis in the Hilbert vector space. It can also be expressed in the form of a so-called completeness or "closure" relation, which can be presented as a useful operator identity. To derive it, I, first, rewrite Eq. 2.26 in a more compact form as

$$|\alpha\rangle = \sum_n a_n |\chi_n\rangle , \tag{3.39}$$

where index n enumerates the eigenvectors and each eigenvector $|\chi_n\rangle$, which is assumed to be normalized, is characterized by all available eigenvalues of the respective CSCO. Expansion coefficients a_n in this expression can be found as $a_n = \langle \chi_n | \alpha \rangle$ as established in Eq. 2.24. After substitution of this expression back into Eq. 3.39, the latter becomes

$$|\alpha\rangle = \sum_n |\chi_n\rangle \langle \chi_n | \alpha \rangle \equiv \left(\sum_n |\chi_n\rangle \langle \chi_n| \right) |\alpha\rangle . \tag{3.40}$$

In the last expression here, I split off ket vector $|\alpha\rangle$ from the bra $\langle\chi_n|$ and combined the latter with another ket $|\chi_n\rangle$. The ket and bra vectors enclosed in the brackets are in unusual positions: the bra is on the left of the ket, which is opposite to their regular positions in the standard inner product. As you can guess, expression $\hat{P}^{(n)} = |\chi_n\rangle\langle\chi_n|$ is not an inner product, but does it have any sensible meaning at all? In the matrix example of the vectors, this expression corresponds to the situation in which the column vector is written down to the left of the row vector— the arrangement used to form the outer or tensor product mentioned in the previous section. Respectively, in the case of abstract generic ket and bra vectors, $|\chi_n\rangle\langle\chi_n|$ can be understood as an outer product of two vectors. Naturally, just as the outer product of rows and columns yields a matrix, the outer product of bras and kets generates an operator: indeed, if you bring the split-off ket vector back, you can construct the following expression:

$$\hat{P}^{(n)}|\alpha\rangle = |\chi_n\rangle\langle\chi_n|\alpha\rangle. \tag{3.41}$$

i.e., the result of the action of $\hat{P}^{(n)}$ on $|\alpha\rangle$ is vector $|\chi_n\rangle$ multiplied by a number. If $|\alpha\rangle$ and $|\chi_n\rangle$ were a regular three-dimensional vector and one of the unit vectors specifying a particular direction correspondingly, you could say that $\hat{P}^{(n)}$ projects $|\alpha\rangle$ on $|\chi_n\rangle$ and generates a component of $|\alpha\rangle$ in the direction specified by $|\chi_n\rangle$. It is customary to maintain the same terminology and call operator $\hat{P}^{(n)}$ a projection operator.

Example 12 (Projection Operators) To get accustomed to working with operators of the form $\hat{P}^{(n)} = |\chi_n\rangle\langle\chi_n|$, let me prove the main property of the projection operators, $\left[\hat{P}^{(n)}\right]^2 = \hat{P}^{(n)}$:

$$\left[\hat{P}^{(n)}\right]^2 = |\chi_n\rangle\langle\chi_n|\chi_n\rangle\langle\chi_n|.$$

The expression in the middle looks like an inner product of a basis vector with itself, and as such it is equal to unity. Thus, we have

$$\left[\hat{P}^{(n)}\right]^2 = |\chi_n\rangle\langle\chi_n| = \hat{P}^{(n)}.$$

The expression inside the parentheses in Eq. 3.40 is a sum of projection operators, but most importantly, it is easy to see that this sum is identical to a unity operator: it acts on vector $|\alpha\rangle$ and generates the same vector. This statement can be written as the following identity:

$$\sum_n |\chi_n\rangle\langle\chi_n| = \hat{I}, \tag{3.42}$$

which is the completeness or closure relation. This is a useful operator identity, which will be frequently used in what follows.

Not all vector spaces used in quantum mechanics can be described by a discrete basis, and sometimes we have to use as a basis eigenvectors of operators with continuous spectrum. I have already discussed this possibility in Sect. 2.3 using states characterized by a definite value of particle's position $|r\rangle$. Now you can associate these states with eigenvectors of a position operator $\hat{r}$. In general, if $|q\rangle$ is an eigenvector of some Hermitian operator with continuous spectrum and q is the respective eigenvalue, you can present an arbitrary state $|\alpha\rangle$ as an integral instead of a sum:

$$|\alpha\rangle = \int dq\psi(q)\,|q\rangle . \tag{3.43}$$

Premultiplying Eq. 3.43 by bra $\langle q_1|$ and using the orthogonality condition for continuous spectrum, Eq. 2.41, you will obtain

$$\langle q_1\,|\alpha\rangle = \int dq\psi(q)\,\langle q_1\,|q\rangle = \int dq\psi(q)\delta\,(q_1 - q) = \psi(q_1). \tag{3.44}$$

Replacing $\psi(q)$ in Eq. 3.43 with its expression derived in Eq. 3.44, you end up with

$$|\alpha\rangle = \int dq\,|q\rangle\,\langle q\,|\alpha\rangle .$$

Considering expression $\int dq\,|q\rangle\,\langle q|$ as an operator, you can, similarly to the case of discrete basis, write

$$\int dq\,|q\rangle\,\langle q| = \hat{I}. \tag{3.45}$$

Equation 3.45 constitutes a completeness condition for eigenvectors of operators with continuous spectrum.

Example 13 (Expansion in Terms of Continuous Basis) To illustrate Eq. 3.43, consider again a linear vector space of integrable functions of a single variable: $|\alpha\rangle \equiv f(x)$. The Fourier transform of this function can be defined as

$$f(x) = \frac{1}{\sqrt{2\pi}}\int\limits_{-\infty}^{\infty} dk\tilde{f}(k)e^{ikx},$$

where the "coefficient" function $\tilde{f}(k)$ is defined via the inverse transform

$$\tilde{f}(k) = \frac{1}{\sqrt{2\pi}}\int\limits_{-\infty}^{\infty} dxf(x)e^{-ikx}.$$

The role of the continuous basis is played here by functions

$$|k\rangle \equiv \frac{1}{\sqrt{2\pi}} e^{ikx},$$

which are eigenvectors of Hermitian operator $-id/dx$ with continuous spectrum consisting of real numbers k. These eigenvectors are orthogonal and delta-function normalized:

$$\frac{1}{2\pi} \int\limits_{-\infty}^{\infty} dx e^{i(k_1-k)x} = \delta\left(k - k_1\right).$$

The completeness condition, Eq. 3.45, for these functions takes the form of

$$\frac{1}{2\pi} \int\limits_{-\infty}^{\infty} dk e^{i(x_1-x)k} = \delta\left(x - x_1\right),$$

with delta-function $\delta\left(x - x_1\right)$ playing the role of the identity operator $\hat{I}$ in this space:

$$\int f(x)\delta\left(x - x_1\right) dx = f(x_1).$$

Some operators have a mixed spectrum: it is discrete for one range of eigenvalues and continuous for another range. Completeness relation in this case will be a combination of Eq. 3.42 and Eq. 3.45 with sum over all discrete eigenvectors and the integral over the continuous one.

3.3.2 Quantization Postulate

Most physical observables can be constructed from just two elements: position vector r and momentum p. I have already introduced states with definite values of the position vector, $|r\rangle$, which are supposed to be eigenvectors of a respective Hermitian operator $\hat{r}$. Similarly, I can introduce states with definite values of momentum $|p\rangle$, which are supposed to be eigenvectors of the Hermitian momentum operator $\hat{p}$. The first question, of course, which you shall want to know is what these operators do to quantum states. You could have guessed the answer for the states represented by eigenvectors of respective operators: $\hat{r}|\tilde{r}\rangle = \tilde{r}|\tilde{r}\rangle$, $\hat{p}|\tilde{p}\rangle = \tilde{p}|\tilde{p}\rangle$, where I placed $\sim$ above r and p to better distinguish between symbols of respective operators and their eigenvalues and eigenvectors. Using these results I can compute expressions like $\hat{r}|\alpha\rangle$ or $p|\alpha\rangle$ by expanding the state $|\alpha\rangle$ in terms of eigenvectors of the respective operators. For instance, by presenting

$$|\alpha\rangle = \int d\tilde{r}\psi\left(\tilde{r}\right)|\tilde{r}\rangle,$$

I can find

$$\hat{r}\,|\alpha\rangle = \int d\tilde{r}\psi\,(\tilde{r})\,\hat{r}\,|\tilde{r}\rangle = \int d\tilde{r}\psi\,(\tilde{r})\,\tilde{r}\,|\tilde{r}\rangle\,.$$

Similar treatment for the momentum operator yields

$$\hat{p}\,|\alpha\rangle = \int d\tilde{p}\varphi\,(\tilde{p})\,\hat{p}\,|\tilde{p}\rangle = \int d\tilde{p}\varphi\,(\tilde{p})\,\tilde{p}\,|\tilde{p}\rangle\,.$$

The problem arises when both position and momentum operators appear in the same expression and we have to figure out how to operate, say, $\hat{p}$, on a state expanded in terms of eigenvectors of $\hat{r}$ or vice versa. I will discuss this issue later in the book, in the section devoted to "representations" of the state vectors and operators. For now I would just like to say that the solution to this problem depends on the fundamental assumptions about commutation relations involving position and momentum operators. Essentially, the quantization procedure, i.e., the rules determining how to replace classical observables with their representation as quantum operators, consists in the postulation of these commutation relations. You will see many times in this text that the knowledge of the commutators of various operators is all what you need to know to perform quantum mechanical calculations. So, please meet the fundamental commutation relations of quantum mechanics.

Postulate 3 (Quantization Postulate) Operators, corresponding to various Cartesian components of position vector and momentum, obey the following commutation relations:

$$[\hat{r}_i, \hat{r}_j] = 0; \ [\hat{p}_i, \hat{p}_j] = 0 \qquad (3.46)$$

$$[\hat{r}_i, \hat{p}_j] = i\hbar\delta_{i,j}, \qquad (3.47)$$

where subindexes take values $1, 2, 3$ indicating x, y, z Cartesian components of the position and momentum vectors, respectively.

The first of the commutators in Eq. 3.46 indicates that the Cartesian components of the position vectors are mutually consistent observables. In other words, it means that if a system is in the state with a certain position, all three components of the position vector are well-defined. The same is true for the vector of momentum as expressed by the second of the commutators in Eq. 3.46. These commutators reflect our desire born out of empirical experience for the position and momentum of the quantum systems to be genuinely well-defined quantities, at least when measured independently of each other.

The commutators presented in Eq. 3.47 are often called canonical commutation relations, and they also express our empiric experience, namely, the fact that the same Cartesian components of position and momentum vectors of a quantum system are not mutually consistent observables and cannot, therefore, be described by commuting operators. The actual form of the commutator is chosen to reproduce Heisenberg's uncertainty principle, which is discussed in the next section. You will also see later that the empirical foundation for this form of the commutator can be traced to the de Broglie relation, Eq. 1.3. It is interesting to note a striking similarity between commutators given in Eqs. 3.46 and 3.47 and canonical Poisson brackets of classical mechanics, Eq. 3.5. This similarity lies in the foundation of the so-called canonical quantization rule: any classical conjugated quantities satisfying Eq. 3.5 in quantum theory are promoted to quantum operators obeying the canonical commutation relation 3.47. Therefore, canonically conjugated variables never belong to the same class of mutually consistent observables and are found on the opposite sides of the Bohr complementarity principle.

3.3.3 Constructing the Observables: A Few Important Examples

Using coordinate and momentum operators, I can construct operators for other observables, which is done according to the standard quantization rule.

Quantization Rule To turn a classical observable into an operator, replace all coordinate and momentums appearing in its classical definition with corresponding operators respecting the requirements of hermiticity and the order of multiplication, when necessary.

In many situations, the issues related to hermiticity or to the multiplication order of observables are resolved automatically, but in some cases one needs to pay special attention to them. To have you started, consider several simplest examples.

Kinetic Energy
Kinetic energy of a single particle with mass m_e is described by operator

$$\hat{K} = \frac{\hat{p}^2}{2m_e},$$

which is obtained from the corresponding classical expression by replacing classical momentum with the momentum operator. The eigenvectors of this operator coincide with the eigenvectors of the momentum operator, and its eigenvalues, which form a continuous spectrum, provide values of kinetic energy that can be observed for a system under study.

Potential Energy
Potential energy is obtained from the respective classical potential energy function by replacing classical coordinate argument of the function with its operator equivalent: $U(r) \rightarrow U(\hat{r})$. It is assumed here, of course, that the potential energy function can be presented as a series of positive and negative powers of r, in

which case the corresponding operator expression would have an easily identifiable meaning. Examples of such transformations are one-dimensional harmonic potential $(kx^2 \rightarrow k\hat{x}^2)$ and Coulomb potential $(k/r \rightarrow k\hat{r}^{-1})$, where r is the absolute value of the position vector.[4] The eigenvectors of this operator are the same as of the position operator, and the respective eigenvalues determine the possible values of the potential energy of the system.

Hamiltonian

Hamiltonian, which in classical mechanics is defined as the energy of the system expressed in terms of canonically conjugated coordinate and momentum, in quantum mechanics becomes, in a single particle case, an operator of the form

$$\hat{H} = \frac{\hat{p}^2}{2m} + U(\hat{r}).\tag{3.48}$$

Since position and momentum operators do not commute, the eigenvectors of the Hamiltonian are usually different from the eigenvectors of both position and momentum operators. Eigenvalues of Hamiltonian can belong to discrete, continuous, or mixed spectrum and determine the values of energy, which the system can have in the given environment. This is the most important operator in all of the quantum physics: just like classical Hamiltonian, its quantum counterpart controls the dynamics of the quantum objects.

Angular Momentum

Angular momentum is a very special kind of an observable. Classical angular momentum is a vector defined as a cross product of the position and momentum operators $L = r \times p$. The quantization rule requires that the quantum mechanical angular momentum operator is constructed by promoting position and momentum vectors to the corresponding operators:

$$\hat{L} = \hat{r} \times \hat{p}.\tag{3.49}$$

However, since this expression involves the product of the potentially non-commuting operators, one has to be careful with the order of the multiplication. One also needs to make sure that the resulting operator is Hermitian. To address both these concerns, I will expand the angular momentum vector in its Cartesian components:

$$\hat{L}_x = \hat{y}\hat{p}_z - \hat{z}\hat{p}_y\tag{3.50}$$

---

[4]This transformation is not as trivial as it might seem since taking absolute value of a vector involves operation of square root, which is not well defined for operators. Practically it is not a problem, however, because usually one works in the basis of the eigenvectors of the position operator, in which case $\hat{r}^{-1}$ becomes simply $1/r$. If you are not concerned with any of this, this note is not for you. I mention it here simply in order to avoid accusations in sweeping something under the rug.

$$\hat{L}_y = \hat{z}\hat{p}_x - \hat{x}\hat{p}_z \tag{3.51}$$

$$\hat{L}_z = \hat{x}\hat{p}_y - \hat{y}\hat{p}_x. \tag{3.52}$$

(One can use as a useful mnemonic device representation of the vector product as a determinant:

$$r \times p \equiv \begin{Vmatrix} e_x & e_y & e_z \\ x & y & z \\ p_x & p_y & p_z \end{Vmatrix},$$

where the first line is formed by unit vectors defining corresponding axes of a Cartesian coordinate system.)

The first thing to notice in Eqs. 3.50–3.52 is that operators that are actually being multiplied correspond to commuting components of the position and momentum vectors; thus, the order, in which you place these operators, is not important. Next, you need to verify that each of the components of the angular momentum operator is a Hermitian operator. Hermitian conjugation, e.g., on x-component yields

$$\hat{L}_x^\dagger = (\hat{y}\hat{p}_z)^\dagger - (\hat{z}\hat{p}_y)^\dagger = \hat{p}_z\hat{y} - \hat{p}_y\hat{z} = \hat{L}_x$$

proving hermiticity of this operator. Similarly, you can demonstrate the Hermitian nature of two other components. The most unusual property of the angular momentum, however, is that different components of the angular momentum do not commute. To illustrate this point, compute commutator $\left[\hat{L}_x, \hat{L}_y\right]$:

$$\left[\hat{L}_x, \hat{L}_y\right] = (\hat{y}\hat{p}_z - \hat{z}\hat{p}_y)(\hat{z}\hat{p}_x - \hat{x}\hat{p}_z) - (\hat{z}\hat{p}_x - \hat{x}\hat{p}_z)(\hat{y}\hat{p}_z - \hat{z}\hat{p}_y) =$$

$$\hat{y}\hat{p}_z\hat{z}\hat{p}_x + \hat{z}\hat{p}_y\hat{x}\hat{p}_z - \hat{p}_z\hat{y}\hat{z}\hat{p}_x - \hat{y}\hat{p}_z\hat{x}\hat{p}_z - \hat{z}\hat{p}_x\hat{y}\hat{p}_z - \hat{x}\hat{p}_z\hat{z}\hat{p}_y + \hat{z}\hat{p}_x\hat{z}\hat{p}_y + \hat{x}\hat{p}_z\hat{y}\hat{p}_z =$$

$$\hat{y}\hat{p}_x\hat{p}_z\hat{z} + \overline{\hat{p}_y\hat{x}\hat{z}\hat{p}_z} - \hat{z}^2\hat{p}_y\hat{p}_x - \hat{p}_z^2\hat{x}\hat{y} - \hat{y}\hat{p}_x\hat{z}\hat{p}_z - \overline{\hat{p}_y\hat{x}\hat{p}_z\hat{z}} + \hat{z}^2\hat{p}_y\hat{p}_x + \hat{p}_z^2\hat{x}\hat{y} =$$

$$\hat{y}\hat{p}_x(\hat{p}_z\hat{z} - \hat{z}\hat{p}_z) + \hat{p}_y\hat{x}(\hat{z}\hat{p}_z - \hat{p}_z\hat{z}) = i\hbar(\hat{p}_y\hat{x} - \hat{y}\hat{p}_x) = i\hbar\hat{L}_z, \tag{3.53}$$

where, when transitioning from the second line to the third, I took into account that different components of the coordinate and momentum operators do commute, so that their order can be changed at will. Similarly, you will find (do it!)

$$\left[\hat{L}_z, \hat{L}_x\right] = i\hbar\hat{L}_y \tag{3.54}$$

$$\left[\hat{L}_y, \hat{L}_z\right] = i\hbar\hat{L}_x. \tag{3.55}$$

These results indicate that the vector of the angular momentum in quantum theory is quite different from regular classical vectors as well as from vector operators of position and momentum: different components of this vector do not belong to the same group of mutually commuting operators and do not represent mutually consistent observables, meaning that this vector is not really well-defined. More specifically, if a quantum system is in a state in which one of the Cartesian components of the angular momentum is known with certainty, measurements of two other components will produce statistically uncertain results. This conclusion, in addition to making the direction of the angular momentum vector uncertain, also raises a question about its magnitude. Indeed, the magnitude of a generic classical 3-D vector is defined as $|A| = \sqrt{A_x^2 + A_y^2 + A_z^2}$. Formal quantization of this expression is not possible because the square root of an operator $\sqrt{\hat{A}_x^2 + \hat{A}_y^2 + \hat{A}_z^2}$ is not a well-defined object. In the case of position and momentum operators, this problem did not arise because different components of these operators are commuting so that one can always choose a coordinate system in which all but one component of the position or momentum operators are equal to zero. The possible values of the remaining non-zero component will define the magnitude of the entire vector. This approach is not possible in the case of angular momentum because of the incompatibility of its components. This problem is circumvented by choosing the operator of the square of angular momentum defined as

$$\hat{L}^2 = \hat{L}_x^2 + \hat{L}_y^2 + \hat{L}_z^2 \tag{3.56}$$

to represent its magnitude. Computing commutators $\left[\hat{L}^2, L_{x,y,z}\right]$ you will find that all three commutators vanish. (The proof of this statement is left to you as an exercise.) This means that operators of the square of the angular momentum and one (any) component of the angular momentum are compatible observables, so that a quantum system can be created in a state in which one of the components and the magnitude of the angular momentum are known with certainty. Obviously such a state would be a common eigenvector of $\hat{L}^2$ and $\hat{L}_z$.

Quantization of $p \cdot r$

As a last example, consider a classical expression of the form $\boldsymbol{p} \cdot \boldsymbol{r}$, which appears in some applications. An attempt to directly transform this expression in the quantum form by promoting the momentum and position vectors to operators faces two obstacles. First, the operators in this expression do not commute, and so it is unclear what is the correct order of multiplication. Second, even if I arbitrarily impose a particular order, say, $\hat{\boldsymbol{p}} \cdot \hat{\boldsymbol{r}}$, the resulting operator is not Hermitian because $(\hat{\boldsymbol{p}} \cdot \hat{\boldsymbol{r}})^\dagger = \hat{\boldsymbol{r}} \cdot \hat{\boldsymbol{p}} \neq \hat{\boldsymbol{p}} \cdot \hat{\boldsymbol{r}}$. To carry out the quantization procedure in this case, you need to come up with an expression, which would coincide with its original classical version but would not depend on the order of the operators, and be Hermitian. One way to achieve this is to introduce operator

$$\frac{1}{2} \left(\hat{\boldsymbol{p}} \cdot \hat{\boldsymbol{r}} + \hat{\boldsymbol{r}} \cdot \hat{\boldsymbol{p}} \right)$$

which satisfies all these conditions. However, this quantization procedure is not unique, and it might (and does) create problems down the road, but luckily for us this is not the road I choose for us to travel.

3.3.4 Eigenvalues of the Angular Momentum

The operators of the angular momentum play an extraordinary role in quantum theory, both on the fundamental level and for applications. The fundamental role of the angular momentum is derived from its relation to the rotation operator and rotational symmetry of quantum systems, but discussion of this topic is well above your pay grade. Those interested in the topic are free to consult any graduate level quantum mechanics text. From the point of view of applications, the importance of the angular momentum stems from the fact that many fundamental interactions in nature are described by so-called central potentials. The potential energy of such interactions depends only on the absolute value of the distance between two interacting particles, but not on the orientation of the vector of their relative position. This text is mostly concerned with quantum mechanics of a single particle in an external potential (a two-particle problem can often be presented in this form as well). If the external potential belongs to the class of central potentials, it can be shown that the Hamiltonian of such a system commutes with all components of the angular momentum as well as with operator $\hat{L}^2$. The proof of this statement requires proving it separately for kinetic energy operator (essentially for operator $\hat{p}^2$) and for the potential energy operator $V(\hat{r})$. I believe that the readers of this text are already equipped to prove that $\left[\hat{L}_{x,y,z}, \hat{p}^2\right] = \left[\hat{L}^2, \hat{p}^2\right] = 0$, so I leave it to you as an exercise. As far as the commutators with the potential energy operator go, this proof will have to be left till later.

Vanishing of the commutators of angular momentum operators and the Hamiltonian means that the Hamiltonian, $\hat{L}^2$, and one of the components of the angular momentum form a system of commuting operators and that the eigenvectors of $\hat{L}^2$ and, say, $\hat{L}_z$ are also eigenvectors of the Hamiltonian. This fact can significantly simplify finding eigenvalues and eigenvectors of the Hamiltonian.

It is also remarkable that the eigenvalues of $\hat{L}^2$ and, for instance, $\hat{L}_z$ can be found using only commutation relations given by Eqs. 3.53–3.55. The choice of the z-component here is a random historical occasion and does not have any physical significance. By choosing this particular component, which, you shall understand, is attached to a particular choice of the coordinate system, we essentially say to the experimentalists that if the quantum system is in a state described by the eigenvectors of $\hat{L}_z$ as defined by this coordinate system, then a measurement of a component of the angular momentum in the same direction will produce results corresponding to the respective eigenvalue with certainty, while measurements of any other component of the angular momentum will have quantum uncertainty.

I begin the search for the eigenvalues by introducing abstract vectors $|\lambda_L, \lambda_z\rangle$ defined as common eigenvectors of operators $\hat{L}^2$ and $\hat{L}_z$ characterized by some yet unknown eigenvalues λ_L and λ_z:

$$\hat{L}^2 |\lambda_L, \lambda_z\rangle = \lambda_L |\lambda_L, \lambda_z\rangle, \tag{3.57}$$

$$\hat{L}_z |\lambda_L, \lambda_z\rangle = \lambda_z |\lambda_L, \lambda_z\rangle. \tag{3.58}$$

It is convenient to present these eigenvalues as $\lambda_L = \hbar^2 p$ and $\lambda_z = \hbar m$. Pulling out factors $\hbar^2$ and $\hbar$ from eigenvalues of $\hat{L}^2$ and $\hat{L}_z$, respectively, makes the remaining quantities p and m dimensionless since the dimension of the angular momentum is the same as that of Planck's constant. Apparently, I will need to invoke, somehow, two remaining components of the angular momentum. It is not right away obvious how to do it, but let's say that I have had a divine intervention or premonition that the following two new operators might be useful:

$$\hat{L}_+ = \hat{L}_x + i\hat{L}_y, \tag{3.59}$$

$$\hat{L}_- = \hat{L}_x - i\hat{L}_y. \tag{3.60}$$

The first thing I need to do with these operators is to compute their commutators with operators $\hat{L}^2$ and $\hat{L}_z$:

$$\left[\hat{L}_+, \hat{L}_z\right] = \left[\hat{L}_x, \hat{L}_z\right] + i\left[\hat{L}_y, \hat{L}_z\right] = -i\hbar\hat{L}_y - \hbar\hat{L}_x = -\hbar\hat{L}_+, \tag{3.61}$$

$$\left[\hat{L}_-, \hat{L}_z\right] = \left[\hat{L}_x, \hat{L}_z\right] - i\left[\hat{L}_y, \hat{L}_z\right] = -i\hbar\hat{L}_y + \hbar\hat{L}_x = \hbar\hat{L}_-. \tag{3.62}$$

It is also easy to see that commutators $\left[\hat{L}^2, \hat{L}_\pm\right]$ vanish. Indeed, $\hat{L}^2$ commutes with all component operators and, therefore, with $\hat{L}_\pm$, which are combinations of $\hat{L}_x$ and $\hat{L}_y$. Now, the new operators for a theoretician are like new toys for a child, and I am eager to play with them and see what they can do. So, to satisfy the urge, and in hopes to learn something new, I want to apply operators $\hat{L}_\pm$ to Eq. 3.57:

$$\hat{L}_\pm \hat{L}^2 |\lambda_L, \lambda_z\rangle = \hat{L}^2 \hat{L}_\pm |\lambda_L, \lambda_z\rangle = \hbar^2 p \hat{L}_\pm |\lambda_L, \lambda_z\rangle, \tag{3.63}$$

where I used $\hat{L}^2 \hat{L}_\pm = \hat{L}_\pm \hat{L}^2$. OK, and what did we learn from this exercise? Well, I know now that if $|\lambda_L, \lambda_z\rangle$ is the eigenvector of $\hat{L}^2$ with eigenvalue $\hbar^2 p$, then vector $\hat{L}_\pm |\lambda_L, \lambda_z\rangle$ is still the eigenvector of $\hat{L}^2$ with the same eigenvalue, which is not really surprising because $L_\pm$ do commute with $\hat{L}^2$. So far, it is not much, and you would be right to say that so far operators $\hat{L}_\pm$ have not given us any particular advantages because we would have gotten the same result with operators $\hat{L}_{x,y}$. But let's not jump the gun—always a bad idea—while patience and persistence are virtues. Instead, let me play another game and apply $\hat{L}_+$ to Eq. 3.58:

$$\hat{L}_+ \hat{L}_z |\lambda_L, \lambda_z\rangle = \hbar m \hat{L}_+ |\lambda_L, \lambda_z\rangle,$$

$$\left(\hat{L}_z\hat{L}_+ - \hbar\hat{L}_+\right)|\lambda_L, \lambda_z\rangle = \hbar m\hat{L}_+|\lambda_L, \lambda_z\rangle,$$

$$\hat{L}_z\hat{L}_+|\lambda_L, \lambda_z\rangle = \hbar(m+1)\hat{L}_+|\lambda_L, \lambda_z\rangle, \tag{3.64}$$

where I used commutation relation 3.61 to make the transition from the first to the second line. Now, the last line in Eq. 3.64 tells us that $\hat{L}_+|\lambda_L, \lambda_z\rangle$ is an eigenvector of $\hat{L}_z$ with eigenvalue $\hbar m + \hbar$. This is a quite exciting result: it means that if I start with some eigenvector with a known eigenvalue, I can generate new eigenvectors with progressively increasing eigenvalues: $\hbar m + \hbar$, $\hbar m + 2\hbar$, $\hbar m + 3\hbar \ldots$. This is already something new, which we could not have gotten without the operator $\hat{L}_+$. The secret of this operator lies in its commutator with $\hat{L}_z$, which is proportional to $\hat{L}_+$ itself. The same is true for operator $\hat{L}_-$, so it is worth looking into what this operator can do:

$$\hat{L}_-\hat{L}_z|\lambda_L, \lambda_z\rangle = \hbar m\hat{L}_-|\lambda_L, \lambda_z\rangle,$$

$$\left(\hat{L}_z\hat{L}_- + \hbar\hat{L}_-\right)|\lambda_L, \lambda_z\rangle = \hbar m\hat{L}_-|\lambda_L, \lambda_z\rangle,$$

$$\hat{L}_z\hat{L}_-|\lambda_L, \lambda_z\rangle = \hbar(m-1)\hat{L}_-|\lambda_L, \lambda_z\rangle. \tag{3.65}$$

When deriving Eq. 3.65, I again applied commutator from Eq. 3.62 to its first line. The final result of this calculation indicates that operator $\hat{L}_-$ also generates new eigenvectors of $\hat{L}_z$ but with progressively *decreasing* eigenvalues. Not surprisingly operators $\hat{L}_+$ and $\hat{L}_-$ are called raising and lowering ladder operators.

Now, the question arises: will this process of generating new eigenvectors and eigenvalues ever stop? In other words, can operator $\hat{L}_z$ have arbitrary large and arbitrary small eigenvalues? Intuitively, it is clear that the answer to this question must be negative and that the possible eigenvalues of $\hat{L}_z$ must be limited both from above and from below. Indeed, these eigenvalues represent possible results of the measurement of one component of a vector, while eigenvalues of $\hat{L}^2$ represent possible experimentally observable values of the squared magnitude of the same vector. It is difficult to imagine that the component of a vector can be larger than the magnitude of the same vector, and therefore one should expect that there must be some kind of a relation between these two eigenvalues, e.g., something like this $m^2 < p$. In order to see if such a relation, indeed, exists, consider the following expression:

$$\langle\lambda_L, \lambda_z|\hat{L}^2|\lambda_L, \lambda_z\rangle = \langle\lambda_L, \lambda_z|\hat{L}_x^2|\lambda_L, \lambda_z\rangle + \langle\lambda_L, \lambda_z|\hat{L}_y^2|\lambda_L, \lambda_z\rangle$$

$$+ \langle\lambda_L, \lambda_z|\hat{L}_z^2|\lambda_L, \lambda_z\rangle.$$

Taking into account Eqs. 3.57 and 3.58, this can be written as

$$\hbar^2 p = \langle\lambda_L, \lambda_z|\hat{L}_x^2|p, \lambda_z\rangle + \langle\lambda_L, \lambda_z|\hat{L}_y^2|p, \lambda_z\rangle + \hbar^2 m^2. \tag{3.66}$$

Since the expectation values of operators $\hat{L}_x^2$ and $\hat{L}_y^2$ in any state are positive quantities, Eq. 3.66 yields that $p > m^2$. This means that there exists the smallest m, which I will designate as $\bar{l}$, and there exists the largest m, for which I will use symbol l. Now, assume that you are dealing with the eigenvector $|\lambda_L, \hbar\bar{l}\rangle$ and applying operator $\hat{L}_-$ to it. Generally speaking, this operator must lower the eigenvalue, but we assumed that this eigenvalue is already the lowest. The only way to reconcile Eq. 3.65 with this assumption is to require that

$$\hat{L}_- |\lambda_L, \hbar\bar{l}\rangle = 0. \tag{3.67}$$

In order to figure out how to use this important piece of information, I again need a bit of divine inspiration, or I can just notice that the product of operators $\hat{L}_+\hat{L}_-$ can be expressed in terms of operators $\hat{L}^2$ and $\hat{L}_z$:

$$\hat{L}_+\hat{L}_- = \hat{L}_x^2 + \hat{L}_y^2 + i\hat{L}_y\hat{L}_x - i\hat{L}_x\hat{L}_y = \hat{L}^2 - \hat{L}_z^2 + \hbar\hat{L}_z.$$

Rewriting this expression as

$$\hat{L}^2 = \hat{L}_z^2 - \hbar\hat{L}_z + \hat{L}_+\hat{L}_-, \tag{3.68}$$

and applying it to vector $|\lambda_L, \hbar\bar{l}\rangle$ while taking into account Eq. 3.67, I obtain

$$\hat{L}^2 |\lambda_L, \hbar\bar{l}\rangle = \hat{L}_z^2 |\lambda_L, \hbar\bar{l}\rangle - \hbar\hat{L}_z |\lambda_L, \hbar\bar{l}\rangle + \hat{L}_+\hat{L}_- |\lambda_L, \hbar\bar{l}\rangle \Rightarrow$$

$$\hbar^2 p |\lambda_L, \hbar\bar{l}\rangle = \hbar^2\bar{l}^2 |\lambda_L, \hbar\bar{l}\rangle - \hbar^2\bar{l} |\lambda_L, \hbar\bar{l}\rangle \Rightarrow$$

$$p = \bar{l}^2 - \bar{l}. \tag{3.69}$$

Now, consider the state characterized by the largest values of m, $|\lambda_L, \hbar l\rangle$. Attempting to act on this vector with operator $\hat{L}_+$ leaves you with the same conundrum encountered when discussing vector $|\lambda_L, \hbar\bar{l}\rangle$, but by now you know the way out: you must require that

$$\hat{L}_+ |\lambda_L, \hbar l\rangle = 0. \tag{3.70}$$

The derivation of Eq. 3.69 based on Eq. 3.67 was successful because the lowering operator $\hat{L}_-$ appears in this equation after operator $\hat{L}_+$. Consequently, when the product $\hat{L}_+\hat{L}_-$ is made to act on $|\lambda_L, \hbar\bar{l}\rangle$, the resulting expression vanishes. In order to achieve the same effect with state $|\lambda_L, \hbar l\rangle$ and Eq. 3.70, I need to modify Eq. 3.68 in such a way that it would contain combination $\hat{L}_-\hat{L}_+$ instead of $\hat{L}_+\hat{L}_-$. To achieve this, consider

$$\hat{L}_-\hat{L}_+ = \hat{L}_x^2 + \hat{L}_y^2 - i\hat{L}_y\hat{L}_x + i\hat{L}_x\hat{L}_y = \hat{L}^2 - \hat{L}_z^2 - \hbar\hat{L}_z \tag{3.71}$$

which can be rewritten in the desired form

$$\hat{L}^2 = \hat{L}_z^2 + \hbar\hat{L}_z + \hat{L}_-\hat{L}_+.$$ (3.72)

Now applying $\hat{L}^2$ to $|\lambda_L, \hbar l\rangle$ and using Eqs. 3.72 and 3.70, I get

$$p = l^2 + l.$$ (3.73)

Comparing Eq. 3.69 with Eq. 3.73, I infer that smallest and largest eigenvalues of $\hat{L}_z$ are related to each other as

$$l^2 + l = \bar{l}^2 - \bar{l}.$$

It is easy to see (one can always just solve the quadratic equation for $\bar{l}$) that this relation implies that $\bar{l} = -l$ or $\bar{l} = l + 1$. The latter solution contradicts to the assumption that $\bar{l}$ is the smallest eigenvalue and l is the largest; thus the only possibility which makes sense is $\bar{l} = -l$.

Now imagine that you have found the smallest eigenvalue $-l$ and you start applying operator $\hat{L}_+$ to state $|\lambda_L, -\hbar l\rangle$. After each application of the operator, the eigenvalue of $\hat{L}_z$ increases by one, so that after applying it N times, you end up with eigenvalue $-l + N$. Eventually you must reach the largest eigenvalue l, at which point you will have $-l + N = l \Rightarrow 2l = N$. N is apparently an integer number, so l can be either integer, if N is even, or half-integer, if N is odd.

Now, let us gather our thoughts and try to summarize what it is that we have got:

1. The eigenvalue of operator $\hat{L}^2$ is equal to $\hbar^2 l(l + 1)$, where l determines the maximum eigenvalue of the operator $\hat{L}_z$, $\hbar l$.
2. l can take either integer or half-integer values, forming two non-overlapping series of allowed values: $0, 1, 2, 3 \cdots$ or $1/2, 3/2, 5/2 \cdots$.
3. Allowed values of m start at $-l$ and advance increasing by one until it reaches l. For instance, for $l = 0$, the only possible value of m is zero; for $l = 1/2$, m can be $-1/2, 1/2$; and for $l = 1$, we can have states with $m = -1, 0, 1$. In general for a state characterized by the same eigenvalue of operator $\hat{L}^2$, $\hbar^2 l(l + 1)$, there are $2l + 1$ possible states with different eigenvalues of $\hat{L}_z$.

It is interesting to note that if I were talking about a classical vector, the maximum magnitude of its component along an axis would simply equal to the length of the vector. If we interpret expression $\hbar l$ as such a component's length, then the squared length of the entire vector would have been $\hbar^2 l^2$, which is different from the quantum result $\hbar^2 l^2 + \hbar^2 l$. One can see that the "extra" contribution to the "length" comes from fluctuations of two other components of the angular momentum. Indeed, using what you have learned from Eq. 3.66, you can write

$$\hbar^2 l^2 + \hbar^2 l = \hbar^2 l^2 + \langle l, l| \hat{L}_x^2 |l, l\rangle + \langle l, l| \hat{L}_y^2 |l, l\rangle \Rightarrow$$

$$\langle l, l| \hat{L}_x^2 |l, l\rangle + \langle l, l| \hat{L}_y^2 |l, l\rangle = \hbar^2 l \Rightarrow$$

$$\langle l, l| \hat{L}_x^2 |l, l\rangle = \langle l, l| \hat{L}_y^2 |l, l\rangle = \hbar^2 l/2.$$

In the last expression, I introduced a shortcut notation for the common eigenvectors of operators $\hat{L}^2$ and $\hat{L}_z$, which in general looks like $|l, m\rangle$ with the first number indicating that this vector belongs to the eigenvalue $\hbar^2 l(l+1)$ of $\hat{L}^2$ and the second number pointing at the eigenvalue $\hbar m$ of $\hat{L}_z$. For brevity, l is often referred to as the "angular momentum," and m is often called a "magnetic" quantum number. The origin of this name will become clear later, when we get to consider the behavior of atoms in the magnetic field.

Finally, let me note that even though we know now that ladder operators $\hat{L}_\pm$ generate eigenvectors of $\hat{L}_z$, there is no guarantee that the resulting eigenvectors will be normalized even if the initial vector is. So, in order to finalize the rule for obtaining *normalized* eigenvectors using ladder operators, we have to analyze their action more carefully. First, it is easy to see that they are Hermitian conjugates of each other:

$$\hat{L}_- = \hat{L}_+^\dagger. \tag{3.74}$$

Assuming that vectors $|l, m\rangle$ and $|l, m+1\rangle$ are normalized and introducing yet unknown normalization coefficient, I can write

$$\hat{L}_+ |l, m\rangle = A_{l,m} |l, m+1\rangle.$$

The Hermitian conjugation of this expression yields

$$\langle l, m| \hat{L}_- = \langle l, m+1| A_{l,m}^*.$$

Multiplying the left-hand side of this equation by the left-hand side of the previous one and doing the same to their right-hand sides yields

$$\langle l, m| \hat{L}_- \hat{L}_+ |l, m\rangle = A_{l,m}^* A_{l,m} \langle l, m+1| |l, m+1\rangle.$$

Since it was assumed that all ket vectors are normalized, I now immediately have for $|A_{lm}|^2$:

$$|A_{lm}|^2 = \langle l, m| \hat{L}_- \hat{L}_+ |l, m\rangle.$$

Taking into account Eq. 3.71, and the fact that kets in this expression are eigenvectors of $\hat{L}^2$ and $\hat{L}_z$, I find

$$|A_{lm}|^2 = \hbar^2 \left[l(l+1) - m(m+1) \right]$$

which allows to establish the final rule for the generation of new eigenvectors from the known ones:

$$\hat{L}_+ |l, m\rangle = \hbar \sqrt{l(l+1) - m(m+1)} |l, m+1\rangle. \tag{3.75}$$

I will leave it to you to show that

$$\hat{L}_- |l, m\rangle = \hbar \sqrt{l(l+1) - m(m-1)} |l, m-1\rangle. \tag{3.76}$$

To conclude this section, let me just emphasize once again that we were able to find eigenvalues for the system of operators, as well as a rule for generating their eigenvectors, using nothing but their commutation relations. The key to successful completion of this task was the existence of the ladder operators with their very special commutation relations given by Eqs. 3.61 and 3.62.

3.3.5 Statistical Interpretation

In Chap. 2 I have already introduced the relation between coefficients in the superposition states and probabilities of various outcomes of the measurements on quantum systems. This time I will elaborate those ideas in a more precise way by formulating two postulates introducing statistical interpretation to the formalism of quantum mechanics.

Postulate 4 (Born's Rule) A measurement of an observable can only yield a value from the set of the eigenvalues of the operator representing the measured observable. If a system before the measurement is not in a state described by one of the eigenvectors of this operator, the result of the measurement cannot be predicted a priori. Only a probability (or probability density for observables with continuous spectrum) of a particular outcome can be known. If the measured eigenvalue is not degenerate, this probability is given by

$$p_n = |\langle \alpha | \lambda_n \rangle|^2, \tag{3.77}$$

where $|\alpha\rangle$ represents a state of the system before the measurement, λ_n is one of the eigenvalues, and $|\lambda_n\rangle$ is the corresponding eigenvector. If the eigenvalue is degenerate, the probabilities given by Eq. 3.77 must be summed up with other degenerate states belonging to this eigenvalue. In the case of observables with continuous spectrum, the probability is replaced with probability density $p(q)$:

$$p(q) = |\langle \alpha | q \rangle|^2,$$

which determines a differential probability dP that the measured value of the observable lies within interval of values $[q, q + dq]$ as $dp = p(q)dq$.

Postulate 5 Regardless of the state in which the system was before an observable is measured, immediately after the measurement, the system will be in a state represented by the eigenvector of the corresponding operator belonging to the observed non-degenerate eigenvalue. If the measured eigenvalue is degenerate, all we can state is that after the measurement the system will be in a state in the subspace of eigenvectors belonging to this eigenvalue.

Both these postulates are essentially more accurate restatements of the propositions already discussed in Sect. 2.2.3, where somewhat vague notion of "the state with definite values of an observable" is replaced with its mathematical representation as an eigenvector of a respective operator. This more formal approach allows carrying out a more comprehensive exploration of the statistical interpretation of quantum mechanical formalism.

I begin by considering an expression of the form $\langle \alpha | \hat{T} | \alpha \rangle$, where $|\alpha\rangle$ is an arbitrary state and $\hat{T}$ is a Hermitian operator representing a certain observable. I have already mentioned that this expression is often referred to as "expectation value," but now I can demonstrate what it actually means. Expanding this state into eigenvectors of $\hat{T}$ (Eq. 3.39), I can present $\langle \alpha | \hat{T} | \alpha \rangle$ as

$$\langle \alpha | \hat{T} | \alpha \rangle = \sum_n \sum_m a_n^* a_m \langle \chi_n | \hat{T} | \chi_m \rangle =$$

$$\sum_n \sum_m \lambda_m a_n^* a_m \langle \chi_n | \chi_m \rangle = \sum_n \lambda_n |a_n|^2, \tag{3.78}$$

where I first took advantage of the fact that $|\chi_m\rangle$ is an eigenvector of $\hat{T}$ with eigenvalue λ_m: $\hat{T} |\chi_m\rangle = \lambda_m |\chi_m\rangle$ and then used orthonormalization condition for the eigenvectors, $\langle \chi_n | \chi_m \rangle = \delta_{nm}$. According to Born's rule, $|a_n|^2$ is the probability that the measurement of the observable will produce λ_n. Then, it becomes clear that the final result in Eq. 3.78 has the meaning of the average value of the observable, which one would "expect" to find if the same measurement is repeated multiple times or if an experimentalist carries out the measurement on multiple identical copies of the same system. The simplest measure of the statistical uncertainty of such measurements would be the standard deviation, which in regular probability theory would be defined as

$$\sigma_T = \sqrt{\overline{\lambda^2} - \overline{\lambda}^2}$$

where the bar above the letters means statistical averaging with probabilities given by $p_n = |a_n|^2$: $\overline{\lambda^2} = \sum_n p_n \lambda_n^2$; $\overline{\lambda}^2 = \left(\sum_n p_n \lambda_n \right)^2$. In the context of quantum theory, the measure of uncertainty of a measurement can be described as

$$\sigma_T = \sqrt{\langle \alpha | \hat{T}^2 | \alpha \rangle - \left(\langle \alpha | \hat{T} | \alpha \rangle \right)^2}. \tag{3.79}$$

Indeed,

$$\langle \alpha | \hat{T}^2 | \alpha \rangle = \sum_n \sum_m a_n^* a_m \langle \chi_n | \hat{T}\hat{T} | \chi_m \rangle = \sum_n \sum_m \lambda_m a_n^* a_m \langle \chi_n | \hat{T} | \chi_m \rangle$$

$$= \sum_n \sum_m \lambda_m^2 a_n^* a_m \langle \chi_n | \chi_m \rangle = \sum_n \lambda_n^2 |a_n|^2 .$$

This shows that the measure of uncertainty expressed by Eq. 3.79 does agree with the probabilistic definition of the standard deviation. If state $|\alpha\rangle$ is one of the eigenvectors $|\chi_{n_0}\rangle$, all coefficients a_n are zeroes, with the exception of $a_{n_0} = 1$. In this case, we have $\langle \alpha | \hat{T}^2 | \alpha \rangle = \lambda_{n_0}^2 = \left(\langle \alpha | \hat{T} | \alpha \rangle \right)^2$, and uncertainty σ_T vanishes. This justifies calling states represented by eigenvectors determinant states or states in which the observable has a definite value. If there are several mutually consistent observables represented by commuting operators, we can have a state, which is a common eigenvector of all operators, in which all observables will have definite values.

If two observables are not mutually consistent and are described by operators $\hat{T}_1$ and $\hat{T}_2$ that do not commute, one can derive the following inequality for uncertainties of these operators σ_{T_1} and σ_{T_2}:

$$\sigma_{T_1} \sigma_{T_2} \geq \frac{1}{2} \langle \alpha | \left| \left[\hat{T}_1, \hat{T}_2 \right] \right| | \alpha \rangle \tag{3.80}$$

which is valid for an arbitrary state $|\alpha\rangle$. This is the so-called generalized uncertainty principle. Using canonical commutation relations 3.47, I can immediately reproduce the Heisenberg inequality

$$\sigma_x \sigma_p \geq \frac{1}{2}\hbar \tag{3.81}$$

which now becomes a particular case of a more general result presented by Eq. 3.80. It is interesting that using Heisenberg uncertainty principle, Eq. 1.4, as an empiric formula and combining it with Eq. 3.80, I can "derive" or justify, if you want, the canonical commutator between the coordinate and momentum operators. Indeed, since Eq. 3.81 is valid for an arbitrary state, in order to reconcile Eq. 3.81 with Eq. 3.80, I have to admit that the commutator of coordinate and momentum operators must be a regular number (only in this case the right-hand side of Eq. 3.81 becomes proportional to $\langle \alpha | \alpha \rangle = 1$, so that the dependence on the state vanishes). The absolute value of this number must obviously be equal to $\hbar$, but recalling that if the commutator of two Hermitian operators is a number, it must be an imaginary number (see Eq. 3.31), I can conclude that $[\hat{x}, \hat{p}_x] = i\hbar$, which is the canonical commutation relation given in Eq. 3.47. Of course, these arguments are not sufficient to show if this commutator is $+i\hbar$ or $-i\hbar$, but the choice of the sign is, actually, the matter of convention, and the standard agreement is to write this commutator as given in Eq. 3.47.

To illustrate all these rather abstract postulates, I will finish this section with an example, in which, to save time, I will again use matrices $M1$ and $M2$ defined by Eq. 3.37.

Example 14 (Probabilities of Measurements) Assume that these matrices represent two observables of some quantum system and that you intend to measure these observables. It is given that the system is prepared in the state $|\eta\rangle$ represented by the column

$$|\eta\rangle = \frac{1}{\sqrt{7}} \begin{bmatrix} 2i \\ 1 \\ 1-i \end{bmatrix}$$

and you are asked to predict the results of the different sequence of measurements of observables $M1$ and $M2$. The first step you have to do is to verify that your initial state is normalized, which is just a good housekeeping habit. The norm of this vector is (do not forget to do complex conjugation when converting ket into a bra—for some reason even good students keep forgetting about it)

$$\|\eta\| = \frac{1}{7} \begin{bmatrix} -2i & 1 & 1+i \end{bmatrix} \begin{bmatrix} 2i \\ 1 \\ 1-i \end{bmatrix} =$$

$$\frac{1}{7} \left((-2i)(2i) + 1 + (1+i)(1-i) \right) =$$

$$\frac{1}{7} (4 + 1 + 2) = 1.$$

Once normalization is verified, you are ready for the next step. Let's say you first want to measure the observable represented by $M2$. We found earlier that the eigenvalues of this matrix are $\lambda_1 = 2$, $\lambda_2 = 1$, and $\lambda_3 = -1$. Thus, these are the values that you can expect to see on the dial of your measuring device (more or less, experimental errors are unavoidable, of course). The actual issue is to find the corresponding probabilities. Using Born's rule, Eq. 3.77, and the corresponding eigenvectors given in Eq. 3.38, you can find for each of the eigenvalues

$$p_{\lambda_1} = |\langle \eta | 2 \rangle|^2 = \left| \frac{1}{\sqrt{2}} \frac{1}{\sqrt{7}} \begin{bmatrix} -2i & 1 & 1+i \end{bmatrix} \begin{bmatrix} -1 \\ 0 \\ 1 \end{bmatrix} \right|^2 =$$

$$\frac{1}{14} |2i + 1 + i|^2 = \frac{5}{7}$$

$$p_{\lambda_2} = |\langle \eta| -1\rangle|^2 = \left| \frac{1}{2}\frac{1}{\sqrt{7}} \begin{bmatrix} -2i & 1 & 1+i \end{bmatrix} \begin{bmatrix} 1 \\ \sqrt{2} \\ 1 \end{bmatrix} \right|^2 =$$

$$\frac{1}{28}\left| -2i + \sqrt{2} + 1 + i \right|^2 = \frac{\left(1+\sqrt{2}\right)^2 + 1}{28} = \frac{2+\sqrt{2}}{14}$$

and

$$p_{\lambda_3} = |\langle \eta| 1\rangle|^2 = \left| \frac{1}{2}\frac{1}{\sqrt{7}} \begin{bmatrix} -2i & 1 & 1+i \end{bmatrix} \begin{bmatrix} 1 \\ -\sqrt{2} \\ 1 \end{bmatrix} \right|^2 =$$

$$\frac{1}{28}\left| -2i - \sqrt{2} + 1 + i \right|^2 = \frac{\left(1-\sqrt{2}\right)^2 + 1}{28} = \frac{2-\sqrt{2}}{14}.$$

It is always a good idea to run a quick check:

$$p_{\lambda_1} + p_{\lambda_2} + p_{\lambda_3} = \frac{5}{7} + \frac{2+\sqrt{2}}{14} + \frac{2-\sqrt{2}}{14} = \frac{5}{7} + \frac{2}{7} = 1,$$

as it should be. So far so good. The expectation value of $M2$ can be computed in two different ways. First, I will use the standard probabilistic definition of the average

$$\langle M2\rangle = \left(p_{\lambda_1}\lambda_1 + p_{\lambda_2}\lambda_2 + p_{\lambda_3}\lambda_3 \right) =$$

$$2 \times \frac{5}{7} + (-1)\frac{2+\sqrt{2}}{14} + 1\frac{2-\sqrt{2}}{14} = \frac{10-\sqrt{2}}{7}.$$

And I will also compute this quantity using quantum-mechanical definition:

$$\langle M2\rangle \equiv \langle \eta| \widehat{M2} |\eta\rangle =$$

$$\frac{1}{7}\begin{bmatrix} -2i & 1 & 1+i \end{bmatrix} \begin{bmatrix} 1 & -\frac{1}{\sqrt{2}} & -1 \\ -\frac{1}{\sqrt{2}} & 0 & -\frac{1}{\sqrt{2}} \\ -1 & -\frac{1}{\sqrt{2}} & 1 \end{bmatrix} \begin{bmatrix} 2i \\ 1 \\ 1-i \end{bmatrix} =$$

$$\frac{1}{7}\begin{bmatrix} -2i & 1 & 1+i \end{bmatrix} \begin{bmatrix} 2i - \frac{1}{\sqrt{2}} - 1 + i \\ -\frac{2i}{\sqrt{2}} - \frac{1-i}{\sqrt{2}} \\ -2i - \frac{1}{\sqrt{2}} + 1 - i \end{bmatrix} =$$

$$\frac{1}{7}\begin{bmatrix} -2i & 1 & 1+i \end{bmatrix} \begin{bmatrix} 3i - \frac{1}{\sqrt{2}} - 1 \\ -\frac{1+i}{\sqrt{2}} \\ -3i - \frac{1}{\sqrt{2}} + 1 \end{bmatrix} =$$

$$\frac{1}{7}\left(6 + \frac{2i}{\sqrt{2}} + 2i - \frac{1+i}{\sqrt{2}} - 2i - \frac{1}{\sqrt{2}} + 4 - \frac{i}{\sqrt{2}}\right) = \frac{1}{7}\left(10 - \sqrt{2}\right),$$

again, exactly as promised. If immediately after measuring $M2$ you will attempt to measure $M1$ and are interested in probabilities of various outcomes (now you are talking about outcomes consisting of pairs of measurements, which are given by all nine possible pairs of eigenvalues $(\lambda_i^{(M2)}, \lambda_j^{(M2)})$), you have to take into account that after the first measurement, the system is no longer in the initial state $|\eta\rangle$. Depending on the outcome of the first measurement, it will be in a state presented by one of the eigenvectors of $M2$. However, since these two matrices commute, and the eigenvectors of $M1$ are also eigenvectors of $M2$, the outcomes of the second measurement are completely determined by the outcome of the first, and there are only three possible results. For instance, if the first measurement produced for $M2$ value -1 (probability $(2 - \sqrt{2})/14$), the measurement of $M1$ will be guaranteed to yield 2 (the state corresponding to eigenvalue -1 of matrix $M2$ is described by the same vector as the eigenvector of $M1$ belonging to its eigenvalue 2). Thus, the probability of getting the pair $(-1, 2)$ is still $(2 - \sqrt{2})/14$.

If you measure $M1$ first, the situation is a bit more complex since $M1$ has degenerate eigenvalues. So, if you want, for instance, to find the probability of getting 1 after measuring $M1$, you have to compute two probabilities—one for each degenerate state—and sum them up. To do that you can use the corresponding orthogonal and normalized vectors given in Eq. 3.38, which are common eigenvectors of both $M1$ and $M2$. This will yield

$$p_1 = \left| \frac{1}{\sqrt{2}} \frac{1}{\sqrt{7}} \begin{bmatrix} -2i & 1 & 1+i \end{bmatrix} \begin{bmatrix} -1 \\ 0 \\ 1 \end{bmatrix} \right|^2 +$$

$$\left| \frac{1}{2} \frac{1}{\sqrt{7}} \begin{bmatrix} -2i & 1 & 1+i \end{bmatrix} \begin{bmatrix} 1 \\ -\sqrt{2} \\ 1 \end{bmatrix} \right|^2 = \frac{10}{14} + \frac{4 - 2\sqrt{2}}{28} = \frac{12 - \sqrt{2}}{14}.$$

At this point a question might pop up in your head, if this result is unique. Indeed, you already know that degenerate eigenvalues can be characterized by an infinite number of different normalized and orthogonal eigenvectors. It would be nice if the probability would not depend on this arbitrary choice, but is it really so? I will give you a chance to answer this question as an exercise.

Finally let me compute the uncertainty of the observable $M2$ in this experiment. For this computation I need to first find $M2^2$, which is

$$M2^2 = \begin{bmatrix} \frac{5}{2} & 0 & -\frac{3}{2} \\ 0 & 1 & 0 \\ -\frac{3}{2} & 0 & \frac{5}{2} \end{bmatrix}.$$

Now you can compute

$$\langle \eta | \widehat{M2^2} | \eta \rangle = \frac{1}{7} \begin{bmatrix} -2i & 1 & 1+i \end{bmatrix} \begin{bmatrix} \frac{5}{2} & 0 & -\frac{3}{2} \\ 0 & 1 & 0 \\ -\frac{3}{2} & 0 & \frac{5}{2} \end{bmatrix} \begin{bmatrix} 2i \\ 1 \\ 1-i \end{bmatrix} =$$

$$\frac{1}{7} \begin{bmatrix} -2i & 1 & 1+i \end{bmatrix} \begin{bmatrix} -\frac{3}{2} + \frac{13}{2}i \\ 1 \\ -\frac{5}{2} - \frac{11}{2}i \end{bmatrix} = \frac{17}{7}$$

so that the uncertainty σ_{M2}^2 is found to be

$$\sigma_{M2}^2 = \langle \eta | \widehat{M2^2} | \eta \rangle - \langle \eta | \widehat{M2} | \eta \rangle^2 = \frac{17}{7} - \frac{1}{49} \left(10 - \sqrt{2} \right)^2 = \frac{17 + 20\sqrt{2}}{49}.$$

3.4 Problems

Section 3.1

Problem 10 A constant force F is acting on a particle of mass m. Derive an expression for the potential energy associated with this force, write down the Hamiltonian of the system, and derive Hamiltonian equations.

Problem 11 Consider a particle moving in a central potential field with Hamiltonian

$$H = \frac{p^2}{2m} + V(|r|).$$

Compute the following Poisson bracket:

$$\{L_x, H\}, \ \{L_y, H\}, \ \{L_z, H\},$$

where $L_{x,y,z}$ are Cartesian coordinates of angular momentum of the particle in some arbitrarily chosen coordinate system. Interpret the results.

Section 3.2.1

Problem 12 Which of the following is a linear operator?

1. Inversion operator $\hat{P}$, which acts on functions of coordinates according to the rule $\hat{P}f(r) = f(-r)$.
2. Square operator $\hat{S}$ defined as $\hat{S}f = f^2$.
3. Determinant operator $\widehat{Det}$, which when applied to a square matrix turns it into the matrix's determinant.
4. Exchange operator $\hat{E}$ acting on functions of two variables as $\hat{E}f(x_1, x_2) = f(x_2, x_1)$.
5. Trace operator $\widehat{Tr}$, which acts on a matrix and turns it into the sum of its diagonal elements.

Problem 13 Prove the linearity of the rotation operator.

Problem 14 Find a Hermitian conjugate for the integral operator $\hat{K}$ acting on integrable functions of a single variable and defined by kernel $K(x_1, x_2)$:

$$\hat{K}f = \int_{-\infty}^{\infty} K(x_1, x_2)f(x_2).$$

The inner product is defined in a regular way: $\langle g | f \rangle = \int_{-\infty}^{\infty} g^*(x)f(x)dx$. Determine under which condition on the kernel this operator is Hermitian.

Problem 15 Expression $\hat{P} = |\alpha\rangle \langle\beta|$ can be understood as an operator acting in the following way:

$$\hat{P}|\sigma\rangle \equiv |\alpha\rangle \langle\beta| \sigma\rangle.$$

Find its Hermitian conjugate.

Section 3.2.2

Problem 16 Specify the condition that must be obeyed by an operator so that it is both unitary and Hermitian.
Consider the following matrices:

$$\begin{bmatrix} 1 & 0 \\ 0 & -1 \end{bmatrix}, \begin{bmatrix} 0 & 1 \\ 1 & 0 \end{bmatrix}, \begin{bmatrix} 0 & i \\ -i & 0 \end{bmatrix}.$$

Do they satisfy this condition?

Problem 17 For three operators $\hat{A}$, $\hat{B}$, and $\hat{C}$, prove the following identity (known as Jacobi identity):

$$\left[\left[\hat{A},\hat{B}\right],\hat{C}\right] + \left[\left[\hat{C},\hat{A}\right],\hat{B}\right] + \left[\left[\hat{B},\hat{C}\right],\hat{A}\right] = 0.$$

Problem 18 Which of the following matrices are Hermitian?

1.

$$\begin{bmatrix} 3i & 5i & 7 \\ -5i & 2 & 3 \\ 7 & 3 & 0 \end{bmatrix}$$

2.

$$\begin{bmatrix} 1 & i & 2i \\ -i & 0 & 3 \\ -2i & 3 & 2 \end{bmatrix}$$

3.

$$\begin{bmatrix} \sqrt{2} & 1 & -2 \\ -1 & 2 & 4\sqrt{5} \\ 7 & -4\sqrt{5} & \sqrt{3} \end{bmatrix}$$

4.

$$\begin{bmatrix} 7 & 4 & 2 \\ 4 & 2 & 1 \\ 2 & 1 & 4 \end{bmatrix}$$

Problem 19 Prove the identity

$$\left(\hat{A}\hat{B}\right)^{-1} = \hat{B}^{-1}\hat{A}^{-1}.$$

Problem 20 Prove the following properties of the commutators:

$$\left[\hat{T}_1,\hat{T}_2\right] = -\left[\hat{T}_2,\hat{T}_1\right]$$

$$\left[\hat{T}_1 + \hat{T}_2,,\hat{T}_3\right] = \left[\hat{T}_1,\hat{T}_3\right] + \left[\hat{T}_2,\hat{T}_3\right]$$

$$\left[c_1\hat{T}_1,c_2\hat{T}_2\right] = c_1c_2\left[\hat{T}_1,\hat{T}_2\right].$$

Problem 21 If operator $\hat{D}$ is defined as

$$\hat{D}f(x) = \frac{df}{dx},$$

what would be an inverse of this operator?

Problem 22 Find an inverse of the following matrices:

1.

$$\begin{bmatrix} 1 & i & 2i \\ -i & 0 & 3 \\ -2i & 3 & 2 \end{bmatrix}$$

2.

$$\begin{bmatrix} 0 & i & 2 \\ -i & 0 & 1 \\ -i & i & 0 \end{bmatrix}$$

Problem 23 Consider an operator $\hat{\sigma}$ characterized by the following property: $\hat{\sigma}^2 = \hat{I}$, where $\hat{I}$ is a unity operator. Using power series expansion, find the closed-form expression (not in the form of a series) for the operator $\exp{(i\hat{\sigma}t)}$.

Problem 24 Prove that if the commutator of two Hermitian operators is a number, this number is necessarily imaginary.

Problem 25 Given that $[\hat{x}, \hat{p}] = i\hbar$, compute

$$\left[\hat{x}^2, \hat{p}^2\right].$$

Section 3.2.3

Problem 26 Consider matrices $\begin{bmatrix} 0 & i \\ -i & 0 \end{bmatrix}$ and $\begin{bmatrix} 0 & 1 \\ 1 & 0 \end{bmatrix}$.

1. Find the eigenvalues and normalized eigenvectors of these matrices.
2. Check orthogonality of the found vectors.

Problem 27 Consider two matrices:

$$A_1 = \begin{bmatrix} 1 & 0 & 0 \\ 0 & -1 & 0 \\ 0 & 0 & -1 \end{bmatrix} ; A_2 = \begin{bmatrix} 1 & 0 & 0 \\ 0 & 0 & 1 \\ 0 & 1 & 0 \end{bmatrix}.$$

1. Show that these operators commute.
2. Find a set of eigenvectors common for both of them.

Problem 28 Find eigenvalues and normalized eigenvectors of the following matrix:

$$
\begin{bmatrix}
1 & -\frac{1}{\sqrt{2}} & -1 \\
-\frac{1}{\sqrt{2}} & 0 & -\frac{1}{\sqrt{2}} \\
-1 & -\frac{1}{\sqrt{2}} & 1
\end{bmatrix}.
$$

Problem 29 Consider the following matrix:

$$
A = \begin{bmatrix}
0 & 0 & -1 \\
0 & 1 & 0 \\
-1 & 0 & 0
\end{bmatrix}.
$$

1. Find its eigenvalues. Are there degenerate ones?
2. Construct a system of normalized and orthogonal eigenvectors.
3. Show that

$$
e^{xA} = \cosh x + A \sinh x.
$$

Section 3.3.1

Problem 30 Consider an operator defined as

$$
\hat{A} = |\phi_1\rangle \langle \psi_1| + |\phi_2\rangle \langle \phi_2| + |\phi_3\rangle \langle \phi_3| -
$$

$$
i |\phi_1\rangle \langle \phi_2| - |\phi_1\rangle \langle \phi_3| + i |\phi_2\rangle \langle \phi_1| - |\phi_3\rangle \langle \phi_1|
$$

where $|\phi_1\rangle$, $|\phi_2\rangle$, and $|\phi_3\rangle$ form an orthonormalized basis.

1. Check if this operator is Hermitian by computing $\hat{A}^\dagger$.
2. Compute $\hat{A}^2$.
3. What are the possible values an experimentalist can observe when measuring an observable represented by this operator?
4. Find states in which the system will be immediately after the measurement for each of the possible outcomes. Verify that the states are presented by orthogonal vectors.

Problem 31 Show that if $\hat{P}$ is a projection operator, $\hat{I} - \hat{P}$ is also a projection operator.

Section 3.3.2

Problem 32 Derive the commutation relations

$$\left[\hat{L}_z, \hat{L}_x\right] = i\hbar \hat{L}_y$$

$$\left[\hat{L}_y, \hat{L}_z\right] = i\hbar \hat{L}_x.$$

Problem 33 Prove that the commutator of the operator of the square of angular momentum $\hat{L}^2$ commutes with all components of the angular momentum operator, $\hat{L}_{x,y,z}$.

Problem 34 Compute commutators

$$\left[\hat{L}_z, \hat{x}\right], \ \left[\hat{L}_z, \hat{y}\right], \ \left[\hat{L}_z, \hat{z}\right]$$

$$\left[\hat{L}_z, \hat{p}_x\right], \ \left[\hat{L}_z, \hat{p}_y\right], \ \left[\hat{L}_z, \hat{p}_z\right]$$

$$\left[\hat{L}^2, \hat{x}\right], \ \left[\hat{L}^2, \hat{y}\right], \ \left[\hat{L}^2, \hat{z}\right]$$

$$\left[\hat{L}^2, \hat{p}_x\right], \ \left[\hat{L}^2, \hat{p}_y\right], \ \left[\hat{L}^2, \hat{p}_z\right].$$

Problem 35 Prove that

$$\left[\hat{L}_{x,y,z}, \hat{p}^2\right] = \left[\hat{L}^2, \hat{p}^2\right] = 0.$$

Section 3.3.4

Problem 36 Prove that

$$\hat{L}_- |l, m\rangle = \hbar \sqrt{l(l+1) - m(m-1)} \, |l, m-1\rangle.$$

Problem 37 Compute the following expressions:

$$\langle l, m' | \hat{L}_- |l, m\rangle$$

$$\langle l, m' | \hat{L}_+ |l, m\rangle.$$

For $l = 1$ present the results as a matrix.

Problem 38 Compute

$$\langle l, m' | \hat{L}_x^2 | l, m \rangle .$$

Hint: Use the representation of $\hat{L}_x$ in terms of raising and lowering ladder operators.

Section 3.3.5

Problem 39 An observable A represented by an operator $\hat{A}$ can be in two mutually exclusive states represented by eigenvectors of $\hat{A}$ $|a_1\rangle$ and $|a_2\rangle$, where $a_{1,2}$ are corresponding eigenvalues. The second observable B represented by an operator $\hat{B}$ also can be in two mutually exclusive states represented by eigenvectors of $\hat{B}$ $|b_1\rangle$ and $|b_2\rangle$, where $b_{1,2}$ are corresponding eigenvalues. These eigenvectors can be related to each other as

$$|a_1\rangle = \frac{1}{5}(3|b_1\rangle + 4|b_2\rangle)$$

$$|a_2\rangle = \frac{1}{5}(4|b_1\rangle - 3i|b_2\rangle) .$$

1. If observable A is measured and value a_1 is obtained, what is the state of the system immediately after the measurement?
2. If now B is measured, what are the possible outcomes, and what are their probabilities?
3. Right after B was measured, A is measured again. What is the probability of getting a_1 for different possible outcomes of the first measurement?

Problem 40 A quantum system is in a state described by a vector

$$|\alpha_1\rangle = \frac{i}{\sqrt{3}}|\chi_1\rangle + \frac{\sqrt{2}}{\sqrt{3}}|\chi_2\rangle .$$

Find the probability that a measurement of some observable will bring the system to state described by a vector

$$|\alpha_2\rangle = \frac{1+i}{\sqrt{3}}|\chi_1\rangle + \frac{1}{\sqrt{6}}|\chi_2\rangle + \frac{1}{\sqrt{6}}|\chi_3\rangle$$

where $|\chi_{1,2,3}\rangle$ form an orthonormalized basis.

Problem 41 Consider a quantum system in a state described by a column vector

$$|\psi\rangle = \frac{1}{\sqrt{5}}\begin{bmatrix} -i \\ 2 \\ 0 \end{bmatrix} .$$

The system is characterized by two observables T_1 and T_2 presented by matrices

$$T_1 = \begin{bmatrix} 1 & i & 1 \\ -i & 0 & 0 \\ 1 & 0 & 0 \end{bmatrix} ; \ T_2 = \begin{bmatrix} 3 & 0 & 0 \\ 0 & 1 & i \\ 0 & -i & 0 \end{bmatrix}.$$

1. If T_1 is measured first and T_2 immediately afterward, what is the probability of obtaining -1 for T_1 and 3 for T_2?
2. What are the probabilities of getting the same values if the order of measurements is reversed? Discuss the result in terms of commutation properties of the two matrices.

Problem 42 Consider a system described by the Hamiltonian

$$H = \frac{1}{\sqrt{2}} \begin{bmatrix} 0 & -i & 0 \\ i & 3 & 3 \\ 0 & 3 & 0 \end{bmatrix}$$

placed in a quantum state described by a column vector

$$|\psi\rangle = \begin{bmatrix} 4-i \\ -2+5i \\ 3+2i \end{bmatrix}.$$

1. Find the expectation value of energy in this state.
2. Find the uncertainty of energy in this state.
3. Find the possible values of energy measurements and their probabilities.
4. Use the results of the previous task to calculate the expectation value and uncertainty of energy again. Compare the results with results of tasks 1 and 2.

Problem 43 Go back to Example 14 at the end of the chapter, and using a different set of orthogonal and normalized eigenvectors of $M1$ (you will have to find it first, of course), compute the probability of getting the degenerate eigenvalue of $M1$. Is the result the same?

Problem 44 Consider a system described by a Hamiltonian

$$\hat{H} = -\frac{1}{2} \frac{d^2}{dx^2} + \frac{1}{2} x^2$$

presented by an operator acting on square-integrable functions of a single variable x forming a Hilbert space with an inner product defined in Sect. 2.1. This system is prepared in state

$$|\psi\rangle = \frac{1}{\sqrt{3}} |\phi_1\rangle + \frac{\sqrt{2}}{\sqrt{3}} |\phi_2\rangle$$

where vectors $|\phi_{1,2}\rangle$ are defined as the following functions:

$$|\phi_1\rangle = \exp\left(-\frac{x^2}{2}\right); \quad |\phi_2\rangle = (1 - 2x^2) \exp\left(-\frac{x^2}{2}\right).$$

1. Verify that these functions are eigenvectors of the Hamiltonian, determine the respective eigenvalues, and normalize the eigenvectors.
2. Rewrite the expression for the state $|\psi\rangle$ in terms of normalized versions of the vectors $|\phi_{1,2}\rangle$.
3. If the energy of the system is measured, what are the possible outcomes, and what are their probabilities?
4. Find expectation values and uncertainties of the operators

$$\hat{\pi}f(x) = -i\frac{df}{dx}; \quad \hat{x}f(x) = xf(x)$$

in state $|\psi\rangle$.

Chapter 4
Unitary Operators and Quantum Dynamics

In the previous section, I explained how one can dig out experimentally relevant information using states of a quantum system and operators representing the observables. The remaining burning question, however, is how can we find these states so that we could use these methods. In a typical experiment, an experimentalist begins by "preparing" a quantum system in some state, which they believe they know.[1] After that they smash the system with a hammer, or hit it by a laser light, or subject it to an electric or magnetic field, wait for some time, and measure new values of the selected observables. In order to predict the results of new measurements, you must be able to describe how the quantum system changes between the time of preparation and the time of subsequent measurement, or, speaking more scientifically, you must know its dynamics. As it has been made clear in the previous section, you need two objects to predict the results of a measurement: a state of the system and the operator assigned to the measured observable. Now you can ask an interesting question: "When the quantum system evolves in time, what is actually changing—the state or the operator?" To make this question more specific, consider an expectation value of an observable described by operator $\hat{T}$: $\langle \alpha | \hat{T} | \alpha \rangle$. When your system evolves, this expectation value becomes a function of time. The question is, which element of the expression for the expectation value, $\hat{T}$ or $|\alpha\rangle$, must be considered as a time-dependent quantity to describe the dynamics of the expectation value? It turns out that time dependence can be ascribed to either of these two elements, and depending on the choice, it will generate two different but equivalent *pictures* of quantum mechanics. In the so-called Schrödinger picture, the state vectors are treated as time-dependent quantities, while operators remain fixed rules transforming the states. In the Heisenberg picture, the state vector is considered as a constant, and all the dynamics of the system is ascribed to the time-dependent operators. The origins of these two pictures can be found in the earlier

---

[1]Preparation of a quantum system in a predefined state usually consists in carrying out a measurement, but it is not an easy task to prepare a system in a state we want.

© Springer International Publishing AG, part of Springer Nature 2018
L.I. Deych, *Advanced Undergraduate Quantum Mechanics*,
https://doi.org/10.1007/978-3-319-71550-6_4

days of quantum theory with the Heisenberg matrix mechanics competing against Schrödinger's matter wave theory. The first attempt to prove equivalence of the two pictures was undertaken by Schrödinger as early as in 1926, but the rigorous mathematical proof of the equivalence did not exist until John von Neumann published in 1932 his definitive book *Mathematical Foundations of Quantum Mechanics*.

Von Neumann was one of the major figures in mathematics and mathematical physics of the twentieth century. Born to a rich Jewish family in Hungary, which was elevated to nobility by Austro-Hungary Emperor Franz Joseph (hence the prefix von in his name), he was a child prodigy, got his Ph.D. in mathematics at the age of 23, and became the youngest privatdocent at the University of Berlin. In 1929 he got an offer from Princeton University and moved to the USA. He brought his entire family to America in 1938 saving them from certain death. In addition to laying rigorous mathematical foundation to quantum theory, von Neumann is famous for his role in the Manhattan Project and developing the concept of digital computers (among other things).

After this brief historical detour, I begin presentation of quantum dynamics starting with the Schrödinger picture.

4.1 Schrödinger Picture

4.1.1 Time-Evolution Operator and Schrödinger Equation

The statistical interpretation of quantum mechanical formalism makes sense only if all vectors describing states of quantum system remain normalized at all times. I will begin digging deeper into this issue by computing the norm of a generic vector $\|\alpha\|$ using Eq. 3.39. First, I need the corresponding bra vector:

$$\langle \alpha | = \sum_n a_n^* \langle \chi_n |$$

so that I can write for the norm

$$\|\alpha\|^2 = \langle \alpha | \alpha \rangle = \sum_m \sum_n a_m a_n^* \langle \chi_n | \chi_m \rangle = \sum_m \sum_n a_m a_n^* \delta_{nm} = \sum_n |a_n|^2 .$$

(4.1)

According to the postulate 4 in Sect. 3.3.5, $|a_n|^2$ is equal to probability p_n that the respective eigenvalue will be observed. Equation 4.1 in this case can be interpreted as a statement that the norm of a generic vector is equal to the sum of probabilities of all possible measurement outcomes. The latter must obviously be equal to unity $\sum_n p_n = 1$ regardless of the time dependence of state $|\alpha\rangle$. This result has quite a profound consequence. Indeed, time dependence of a state vector can be considered as a transformation of a vector $|\alpha(t_0)\rangle$ defined at some initial instant of time t_0 into another vector $|\alpha(t)\rangle$ at time t under the action of an operator:

$$|\alpha(t)\rangle = \hat{U}(t, t_0)|\alpha(t_0)\rangle. \tag{4.2}$$

In order to keep the norm of the vector unchanged, the operator $\hat{U}(t, t_0)$ must be unitary, which significantly limits the class of operators that can be used to describe the dynamics of quantum states. It also must obey an obvious condition:

$$\hat{U}(t_0, t_0) = \hat{I}. \tag{4.3}$$

Now, consider an evolution of the system from state $|\alpha(t_0)\rangle$ to state $|\alpha(t_1)\rangle$ and then to state $|\alpha(t_f)\rangle$, which can be described as

$$|\alpha(t_1)\rangle = \hat{U}(t_1, t_0)|\alpha(t_0)\rangle$$
$$|\alpha(t_f)\rangle = \hat{U}(t_f, t_1)|\alpha(t_1)\rangle.$$

I can also describe a system's dynamics from the initial state to the final, bypassing the intermediate state:

$$|\alpha(t_f)\rangle = \hat{U}(t_f, t_0)|\alpha(t_0)\rangle.$$

Comparing this with the first two lines of the previous equation, you can infer an important property of the time-evolution operator $\hat{U}$:

$$\hat{U}(t_f, t_0) = \hat{U}(t_f, t_1)\hat{U}(t_1, t_0). \tag{4.4}$$

An important corollary of Eq. 4.4 is obtained by setting $t_f = t_0$, which yields

$$\hat{U}(t_0, t_1)\hat{U}(t_1, t_0) = \hat{I} \Rightarrow \hat{U}(t_0, t_1) = \hat{U}^{-1}(t_1, t_0) \tag{4.5}$$

where I also used Eq. 4.3. In other words, the reversal of time in quantum dynamics is equivalent to replacing the time-evolution operator with its inverse. This idea can also be expressed by saying that by inverting the time-evolution operator, you describe the evolution of the system from present to the past. This property can also be described as reversibility of quantum dynamics: taking a system from t_0 to t_f and back brings the system in its original state completely reversing its initial evolution.

Now, let me consider the action of $\hat{U}(t_1, t_0)$ over an infinitesimally small time interval $t_0, t_1 - t_0 + dt$. Expanding this operator over the small interval dt and using Eq. 4.3, I can write:

$$\hat{U}(t_0 + dt, t_0) = \hat{I} + \hat{G}dt \tag{4.6}$$

where $\hat{G} \equiv d\hat{U}/dt\big|_{t=t_0}$ is an operator obtained by differentiating the time-evolution operator with respect to time. Inverse to the operator defined by Eq. 4.6 can be found by expanding function $(1 + x)^{-1}$ with respect to x and keeping only linear in x terms: $(1 + x)^{-1} \simeq 1 - x$. Applying this to operator $\left(\hat{I} + \hat{G}dt\right)^{-1}$, I get

$$\hat{U}^{-1}(t_0 + dt, t_0) = \hat{I} - \hat{G}dt.$$

At the same time, Hermitian conjugation of Eq. 4.6 returns

$$\hat{U}^{\dagger}(t_0 + dt, t_0) = \hat{I} + \hat{G}^{\dagger}dt.$$

Since the time-evolution operator is unitary ($\hat{U}^{-1} = \hat{U}^{\dagger}$), operator $\hat{G}$ has to be anti-Hermitian: $\hat{G}^{\dagger} = -\hat{G}$, so that it can be presented as $\hat{G} = -i\hat{H}/\hbar$ (see Eq. 3.30), where $\hat{H}$ is a Hermitian operator and $\hbar$ is introduced to ensure that $\hat{H}$ has the dimension of energy. Indeed, since the time-evolution operator is dimensionless, it is clear that operator $\hat{G}$ has the dimension of inverse time. The dimension of the Planck's constant is that of energy multiplied by time, so it is clear that $\hat{H}$ has indeed the dimension of energy. This simple analysis leads the way to the next postulate of quantum theory.

Postulate 6 Hermitian operator $\hat{H}$ in the expansion of the time-evolution operator is the operator version of Hamiltonian function of classical mechanics.

Thus, Eq. 4.6 can now be rewritten as

$$\hat{U}(t_0 + dt, t_0) = \hat{I} - i\frac{\hat{H}}{\hbar}dt. \tag{4.7}$$

Taking advantage of the composition rule, Eq. 4.4, I can write:

$$\hat{U}(t + dt, t_0) = \hat{U}(t + dt, t)\,\hat{U}(t, t_0) =$$

$$\left(\hat{I} - i\frac{\hat{H}}{\hbar}dt\right)\hat{U}(t, t_0) = \hat{U}(t, t_0) - i\frac{\hat{H}}{\hbar}\hat{U}(t, t_0)\,dt$$

where I also used Eq. 4.7. The main difference between this last expression and Eq. 4.7 is that t in the latter can be separated from t_0 by a finite interval. The last equation can be rewritten in the form of differential equation:

$$\frac{d\hat{U}(t, t_0)}{dt} = -i\frac{\hat{H}}{\hbar}\hat{U}(t, t_0). \tag{4.8}$$

Applying Eq. 4.7 to Eq. 4.2, I can also derive:

$$|\alpha(t + dt)\rangle = \left(\hat{I} - \frac{i}{\hbar}\hat{H}dt\right)|\alpha(t)\rangle \Rightarrow \frac{|\alpha(t + dt)\rangle - |\alpha(t)\rangle}{dt} = -\frac{i}{\hbar}\hat{H}|\alpha(t)\rangle$$

which can be rewritten in a standard form

$$i\hbar \frac{d\,|\alpha\rangle}{dt} = \hat{H}\,|\alpha\rangle \tag{4.9}$$

called Schrödinger equation. As any differential equation, Eq. 4.9 has to be complemented by an initial condition specifying the state of the system at an arbitrary chosen initial time. Given the role of Hamiltonian in classical mechanics discussed in Sect. 3.1, it is not very surprising that the same quantity (in its operator reincarnation) determines the dynamics of quantum systems as well.

4.1.2 Stationary States

If Hamiltonian does not contain explicit time dependence, which might appear, for instance, if an atom interacts with a time-dependent electric field $E(t)$, Eq. 4.9 has a very simple formal solution:

$$|\alpha(t)\rangle = \exp\left(-i\frac{\hat{H}}{\hbar}t\right)|\alpha_0\rangle, \tag{4.10}$$

where $|\alpha_0\rangle$ is the state of the system at time $t = 0$. For practical calculations, however, this solution is not very helpful because the action of the exponent of an operator on an arbitrary vector in general is not easy to compute. Situation becomes much simpler if the initial state is presented by one of the eigenvectors of the Hamiltonian. If $|\alpha_0\rangle = |\chi_n\rangle$, where $|\chi_n\rangle$ is an eigenvector of $\hat{H}$ with respective eigenvalue E_n

$$\hat{H}\,|\chi_n\rangle = E_n\,|\chi_n\rangle, \tag{4.11}$$

the right-hand side of Eq. 4.10 can be computed as follows:

$$|\alpha(t)\rangle = \exp\left(-i\frac{\hat{H}}{\hbar}t\right)|\chi_n\rangle = \sum_{m=0}^{\infty}\frac{1}{m!}\left(\frac{-it}{\hbar}\right)^m \hat{H}^m\,|\chi_n\rangle =$$

$$\sum_{m=0}^{\infty}\frac{1}{m!}\left(\frac{-it}{\hbar}\right)^m E_n^m\,|\chi_n\rangle = \exp\left(-i\frac{E_n}{\hbar}t\right)|\chi_n\rangle, \tag{4.12}$$

where I used the definition of the exponential function of an operator, Eq. 3.21, and the fact that

$$\hat{H}^m\,|\chi_n\rangle = E_n^m\,|\chi_n\rangle,$$

which is easily proved. (You will have a chance to prove it when doing your homework.) Thus, if a system is initially in a state represented by an eigenvector of the Hamiltonian, it remains in this state forever and ever. The time-dependent factor in this case is a complex number with absolute value equal to unity (pure phase as physicists like to say) and does not affect, therefore, any measurable quantities. Indeed, consider, for instance, an expectation value of some generic operator $\hat{T}$ when a system is in the state described by Eq. 4.12:

$$\langle \alpha(t) | \, \hat{T} \, | \alpha(t) \rangle = \exp\left(i \frac{E_n}{\hbar} t \right) \langle \chi_n | \, \hat{T} \, | \chi_n \rangle \exp\left(-i \frac{E_n}{\hbar} t \right) = \langle \chi_n | \, \hat{T} \, | \chi_n \rangle \,.$$

The eigenvector equation 4.11 is often called time-independent Schrödinger equation, and it's solutions represent the very same stationary states, which were postulated by Bohr, whose existence was proven in Davisson–Germer experiments mentioned in the Introduction, and which became the main object of the Schrödinger wave mechanics. The corresponding eigenvalues are called energy levels or simply energies. Here I will use the term stationary states to designate solutions of the time-independent Schrödinger equation with exponential time dependence attached to the eigenvectors and given by Eq. 4.12. As you just saw, this time dependence does not affect experimentally observable quantities, which remain independent of time, justifying the name "stationary" for these states.

I need to point out at a general ambiguity of relation between quantum states and vectors representing them: the latter are always defined with accuracy to a phase, meaning that all vectors can be multiplied by a complex number of unit magnitude without affecting any physical results. This is obvious from Eq. 4.10, which does not change upon multiplying the state vector by any constant factor. But since it is required that the states are normalized, this constant factor is limited to have a magnitude equal to unity, i.e., to be a pure phase. However, in the case presented in Eq. 4.12, the multiplying factor is not constant and, therefore, cannot be simply dismissed making it physically significant. This significance manifests itself, however, only when we have to deal with several stationary states. Indeed, since energy is always defined with an accuracy up to a constant factor, one can always make the energy eigenvalue corresponding to any one of stationary states to vanish killing thereby the time dependence of the corresponding stationary state. This vanishing trick, however, can be achieved only for one state, while all others will retain their exponential factor albeit with different energy values equal to the difference between their initial values and the one you chose to be equal to zero.[2] In order to demonstrate that this general property of energies retain its meaning in quantum theory as well, I will consider a state evolving from a superposition of two eigenvectors of a Hamiltonian with different eigenvalues. Thus, assume that the initial state of the system is

$$|\alpha_0\rangle = a_1 \, |\chi_1\rangle + a_2 \, |\chi_2\rangle \,.$$

---

[2]Technically this can be achieved by subtracting one of the energy eigenvalues from the potential appearing in the Hamiltonian, which is equivalent to simple change of the zero level of the energies.

Using linearity of the time-evolution operator and results from Eq. 4.12, I can easily compute:

$$|\alpha(t)\rangle = a_1 \exp\left(-i\frac{\hat{H}}{\hbar}t\right)|\chi_1\rangle + a_2 \exp\left(-i\frac{\hat{H}}{\hbar}t\right)|\chi_2\rangle =$$

$$a_1 \exp\left(-i\frac{E_1}{\hbar}t\right)|\chi_1\rangle + a_2 \exp\left(-i\frac{E_2}{\hbar}t\right)|\chi_2\rangle =$$

$$\exp\left(-i\frac{E_1}{\hbar}t\right)\left(a_1 |\chi_1\rangle + a_2 \exp\left(-i\frac{E_2 - E_1}{\hbar}t\right)|\chi_2\rangle\right). \qquad (4.13)$$

Equation 4.13 shows that an initial vector in the form of a superposition of two eigenvectors of a Hamiltonian evolves by "dressing up" each of the initial state with the exponential time factor containing the energy eigenvalue corresponding to the respective eigenvector. However, the absolute values of these energy eigenvalues are again not important as the dynamics of the state is determined by the difference between them. I emphasized this point in the last line of Eq. 4.13 by factoring out one of the time-dependent exponential factors. It is clear that the overall phase factor will again disappear from all experimentally relevant expressions, and the entire time dependence will be determined by $\exp\left(-i\frac{E_2 - E_1}{\hbar}t\right)$. Apparently, it would not matter for this dynamic if I factored out the other exponential factor. To illustrate this point, I will now compute an expectation value of some generic operator with the state described by Eq. 4.13:

$$\langle\alpha(t)|\hat{T}|\alpha(t)\rangle = \exp\left(i\frac{E_1}{\hbar}t\right)\left(a_1^* \langle\chi_1| + a_2^*\left\langle\chi_2 \exp\left(i\frac{E_2 - E_1}{\hbar}t\right)\right|\right)$$

$$\hat{T}\left(a_1 |\chi_1\rangle + a_2 \exp\left(-i\frac{E_2 - E_1}{\hbar}t\right)|\chi_2\rangle\right)\exp\left(-i\frac{E_1}{\hbar}t\right)$$

$$|a_1|^2 T_{11} + |a_2|^2 T_{22}+$$

$$T_{12}a_1^* a_2 \exp\left(-i\frac{E_2 - E_1}{\hbar}t\right) + T_{21}a_2^* a_1 \exp\left(i\frac{E_2 - E_1}{\hbar}t\right)$$

where $T_{ij} = \langle\chi_i|\hat{T}|\chi_j\rangle$. Taking into account that for Hermitian operators diagonal elements are real-valued and nondiagonal are complex conjugates of their transposed elements ($T_{ij} = T_{ji}^*$), this can be written down as

$$\langle\alpha(t)|\hat{T}|\alpha(t)\rangle = |a_1|^2 T_{11} + |a_2|^2 T_{22}+$$

$$2|T_{12}||a_1||a_2|\cos\left(\frac{E_2 - E_1}{\hbar}t + \delta_{T_{21}} + \delta_{a_1} - \delta_{a_2}\right), \qquad (4.14)$$

where δs are phases of elements appearing in the corresponding subindexes. This expression is explicitly real and is periodic with frequency dependent on the difference of energies $(E_2 - E_1)/\hbar$. I will leave it to you as an exercise to demonstrate that this result wouldn't change if you factor out $\exp\left(i\frac{E_2}{\hbar}t\right)$ instead of $\exp\left(i\frac{E_1}{\hbar}t\right)$.

In a general case, expanding an arbitrary initial state vector in the basis of the eigenvectors of the Hamiltonian, you can see that a time dependence of the vector representing the state of the system is obtained by adding the corresponding exponential factors $\exp\left(-i\frac{E_n}{\hbar}t\right)$ in front of each $|\chi_n\rangle$ term in this expansion:

$$|\alpha(t)\rangle = \sum_n a_n \exp\left(-i\frac{E_n}{\hbar}t\right)|\chi_n\rangle. \tag{4.15}$$

Expansion coefficients a_n are determined by the initial state $|\alpha_0\rangle$ with the help of Eq. 2.24. Equation 4.15 essentially solves the problem of quantum dynamics provided one knows eigenvalues and eigenvectors of the system's Hamiltonian. For this reason solving the time-independent Schrödinger equation is one of the main technical problems in quantum theory. Respectively, much of this text as well as of all other books on quantum mechanics will be devoted to devising various ways of doing so.

If the Hamiltonian has a continuous spectrum of energy, the same idea for generating the time-dependent state from an initial state still works. One only needs to replace the sum over the discrete index in Eq. 4.15 by an integral over a relevant continuous quantity k labeling states of the system to get

$$|\alpha(t)\rangle = \int dk a(k) \exp\left(-i\frac{E_k}{\hbar}t\right)|\chi_k\rangle. \tag{4.16}$$

Coefficients $a(k)$ are again determined by an initial state in exactly the same way as in the discrete case (you will be well advised to remember though that the operational definitions of the inner product can be very different in the discrete and continuous cases).

4.1.3   Ehrenfest Theorem and Correspondence Principle

I want to finish the discussion of the Schrödinger picture by deriving the so-called Ehrenfest theorem, which is concerned with the dynamics of the expectation value of a generic Hermitian operator $\hat{A}(t)$, which might have its own explicit time dependence. Assuming that the system is in state $|\alpha(t)\rangle$, I will derive a differential equation for quantity $\left\langle \hat{A}(t)\right\rangle = \langle\alpha(t)|\hat{A}|\alpha(t)\rangle$, where $\left\langle \hat{A}(t)\right\rangle$ is a frequently used shortened notation for the expectation values. This expression can be differentiated using standard rules for differentiation of a product of several functions:

$$\frac{d\left\langle \hat{A}(t)\right\rangle}{dt} = \frac{d\left\langle \alpha(t)\right|}{dt}\hat{A}\left|\alpha(t)\right\rangle + \left\langle \alpha(t)\right|\frac{\partial \hat{A}}{\partial t}\left|\alpha(t)\right\rangle + \left\langle \alpha(t)\right|\hat{A}\frac{d\left|\alpha(t)\right\rangle}{dt} =$$

$$\frac{i}{\hbar}\left\langle \alpha(t)\right|\hat{H}\hat{A}\left|\alpha(t)\right\rangle - \frac{i}{\hbar}\left\langle \alpha(t)\right|\hat{A}\hat{H}\left|\alpha(t)\right\rangle + \left\langle \alpha(t)\right|\frac{\partial \hat{A}}{\partial t}\left|\alpha(t)\right\rangle =$$

$$-\frac{i}{\hbar}\left\langle \left[\hat{A},\hat{H}\right]\right\rangle + \left\langle \frac{\partial \hat{A}}{\partial t}\right\rangle. \tag{4.17}$$

In the Schrödinger picture, the operators are devoid of their own dynamics. The time derivative in the last term of the Ehrenfest theorem takes into account a possibility of an external time dependence of an operator, which is not related to their internal dynamics. This explicit time dependence is a reflection of the changing environment of the system, such as a time-dependent electromagnetic field interacting with an atom. If operator $\hat{A}$ does not have such an externally imposed time dependence, then the last term in Eq. 4.17 vanishes, and the dynamics of the expectation value of the observable represented by $\hat{A}$ is completely determined by its commutator with the Hamiltonian.

There is a special class of observables, whose operators commute with Hamiltonian. You already know that such observables are compatible with Hamiltonian, i.e., they would have a definite value if the system is in one of its stationary states. Ehrenfest theorem shows that such observables have an additional property— regardless of the state of the system, their expectation values do not depend on time. In other words, the expectation values of observables whose operators commute with the Hamiltonian are conserving quantities.

Finally, I would like you to note a remarkable similarity between the Ehrenfest theorem and Eq. 3.8 expressing time derivative of a classical function of coordinate and momentum in terms of its Poisson brackets with the classical Hamiltonian: the two equations become identical if one makes a substitution $\{\cdots\} \rightarrow -(i/\hbar)[\cdots]$. This similarity is physically significant as illustrated by the following example. Let me apply Ehrenfest theorem to a very special but extremely important case of coordinate and momentum operators of a single particle described by a time-independent Hamiltonian, like the one given in Eq. 3.48. For simplicity I will limit myself to a one-dimensional case, so that I will only need to consider one component of the position and momentum operators and can treat the potential energy as a function of a single coordinate only. The Ehrenfest theorem involves commutators of the respective operators with the Hamiltonian. In the case under consideration, I have to compute $[\hat{x}, \hat{H}]$ and $[\hat{p}_x, \hat{H}]$ for $\hat{H}$ given by the one-dimensional version of Eq. 3.48, which I reproduce below for your convenience:

$$\hat{H} = \frac{\hat{p}_x^2}{2m} + V(\hat{x}).$$

The easiest commutator to compute is $[\hat{x}, \hat{H}]$:

$$\left[\hat{x}, \hat{H}\right] = \left[\hat{x}, \frac{\hat{p}_x^2}{2m}\right] = \frac{i\hbar}{m}\hat{p}_x \tag{4.18}$$

where I used the fact that $\hat{x}$ commutes with $V(\hat{x})$ as well as identity 3.24 and canonical commutation relation, Eq. 3.47. It takes a bit more labor to compute $[\hat{p}, \hat{H}] = [\hat{p}, V(\hat{x})]$. In order to evaluate this commutator, I first present the potential energy as a power series:

$$V(\hat{x}) = \sum_{n=0}^{\infty} \frac{1}{n!} \frac{d^n V}{dx^n} \hat{x}^n$$

so that I can write

$$[\hat{p}_x, V(\hat{x})] = \sum_{n=0}^{\infty} \frac{1}{n!} \frac{d^n V}{dx^n} [\hat{p}_x, \hat{x}^n].$$

Again using identity 3.24 to evaluate commutator $[\hat{p}, \hat{x}^n]$, I get

$$[\hat{p}, \hat{x}^n] = -i\hbar n \hat{x}^{n-1},$$

substitution of which into the previous equation yields

$$[\hat{p}_x, V(\hat{x})] = -i\hbar \sum_{n=1}^{\infty} \frac{1}{(n-1)!} \frac{d^n V}{dx^n} \hat{x}^{n-1} \equiv -i\hbar \sum_{n=0}^{\infty} \frac{1}{n!} \frac{d^{n+1} V}{dx^{n+1}} \hat{x}^n.$$

Here I took into account that $n = 0$ term of the initial series is a constant and vanishes upon the differentiation. Correspondingly the summation in the middle expression above begins with $n = 1$. In the next step, I changed the dummy index of summation $n - 1 \rightarrow n$, turning $n = 1$ term into $n = 0$, $n = 2$ into $n = 1$, and so on. This process naturally forces to replace $n - th$ derivative with $n + 1$th and $\hat{x}^n$ with $\hat{x}^{n+1}$. All what is now left is to recognize that the final resulting power series is the expansion of the derivative of function $V(x)$:

$$\frac{dV}{dx} = \sum_{n=0}^{\infty} \frac{1}{n!} \frac{d^{n+1} V}{dx^{n+1}} \hat{x}^n.$$

Thus, I can proudly present

$$[\hat{p}_x, V(\hat{x})] = -i\hbar \frac{dV}{dx}. \tag{4.19}$$

Now, the Ehrenfest theorem for these two operators becomes

$$\frac{d \langle \hat{x} \rangle}{dt} = \frac{\langle \hat{p}_x \rangle}{m}$$

$$\frac{d \langle \hat{p}_x \rangle}{dt} = -\left\langle \frac{dV}{dx} \right\rangle.$$

Repeating the same calculations for all three components of the position and the momentum vectors, you can easily obtain the three-dimensional version of these equations:

$$\frac{d \langle \hat{r} \rangle}{dt} = \frac{\langle \hat{p} \rangle}{m} \tag{4.20}$$

$$\frac{d \langle \hat{p} \rangle}{dt} = -\langle \nabla V \rangle \tag{4.21}$$

where ∇V (in case you forgot) is a gradient of V defined in the Cartesian coordinates with unit vectors e_x, e_y, and e_z in the direction of the corresponding axes X, Y, and Z as

$$\nabla V = e_x \frac{\partial V}{\partial x} + e_y \frac{\partial V}{\partial y} + e_z \frac{\partial V}{\partial z}.$$

The obtained equations *resemble* classical Hamiltonian equations, but it actually would be wrong to say (as many textbooks do) that Ehrenfest equations make expectation values of position and momentum operators to behave like corresponding classical quantities. In reality these equations do not even constitute a closed system of equations, which becomes almost obvious once you realize that generally speaking $\langle V(\hat{x}) \rangle \neq V(\langle \hat{x} \rangle)$. Equality here is realized only if the potential energy is either a linear or a quadratic function of the coordinates. In the former case $-d\hat{V}/dx = F = const$, so that the Ehrenfest equations have a simple solution:

$$\langle \hat{p} \rangle = p_0 + Ft;$$

$$\langle \hat{x} \rangle = x_0 + (p_0/m)\, t + (1/2)\,(F/m)\, t^2,$$

reproducing classical equations for a particle moving with constant acceleration. In the case of a quadratic potential (harmonic oscillator)

$$-d\hat{V}/dx = k\hat{x}$$

so that

$$-\left\langle d\hat{V}/dx \right\rangle = k\, \langle \hat{x} \rangle$$

reducing the Ehrenfest equations to classical equations describing a harmonic oscillator. To illustrate the difficulty arising in a more general situation, consider $V(x) = ax^3/3$. In this case Ehrenfest equations become

$$\frac{d\langle\hat{x}\rangle}{dt} = \frac{\langle\hat{p}\rangle}{m}$$

$$\frac{d\langle\hat{p}\rangle}{dt} = -a\langle x^2\rangle.$$

Since $\langle x^2\rangle \neq \langle x\rangle^2$, the resulting system of equations is not complete, because now you need to derive a separate equation for $\langle x^2\rangle$. Trying to do so (see the exercises) will appear as a recurring nightmare—you will end up with new variables at each step, and this process will never end. You might wonder, of course, that maybe it is possible to find such a state for which $\langle\hat{x}^n\rangle = \langle\hat{x}\rangle^n$, in which case Ehrenfest equations will literally coincide with the Hamiltonian equations. It is not very difficult to prove that the only state in which this might be true is the state represented by the eigenvector of the coordinate operator. Unfortunately, even if at some time $t = 0$ you can create a system in such a state, it will lose this property as it evolves in time. To see that this is indeed the case, imagine a state $|x(t)\rangle$ such that $\hat{x}|x(t)\rangle = x(t)|x(t)\rangle$ and try to plug it in Eq. 4.9 describing the dynamics of quantum states. You will immediately see that since the coordinate and momentum operators do not commute, this state cannot be a solution of the time-dependent Schrödinger equation.

There is, however, another way to make Ehrenfest equations identical to their classical Hamiltonian counterparts. All what you need to do is to neglect quantum uncertainty of coordinate and momentum. Since technically these uncertainties arise from the canonical commutation relation, you can do away with them by passing to the limit $\hbar \to 0$. The emergence of Hamiltonian equations, in this so-called classical limit, is a very attractive and soothing feature of quantum formalism indicating that the developed theory adheres to the correspondence principle formulated (again!) by Niels Bohr. This principle played an important heuristic and philosophical role in the development of quantum theory. It states that the quantum theory must reproduce the results of classical physics in situations where classical physics is known to be valid. Even though the concrete mathematical expressions defining situations when the quantum description must reduce to the classical one vary from phenomenon to phenomenon, they all involve taking the limit $\hbar \to 0$. In this limit, for instance, the quantum of energy $\hbar\omega$ introduced by Planck vanishes, or de Broglie wavelength $\lambda = h/p$ goes to zero, and quantum uncertainties of various observables, which prevented you from replacing $\langle\hat{x}^n\rangle \to \langle\hat{x}\rangle^n$, disappear.

4.2 Heisenberg Picture

As I mentioned in the beginning of this chapter, quantum dynamics can be described by imposing a time dependence on operators rather than on the states. This approach, a version of which was designed by Heisenberg, Born, and Jordan, was historically first, is directly connected with classical Hamiltonian equations, and is quite popular in current research literature on quantum mechanics, especially in quantum optics. However, for some reasons it rarely appears in undergraduate texts on quantum theory. Probably, it is believed that the idea of time-dependent operators is too complicated for infirm minds of undergraduate physics majors to comprehend, but I personally do not see why this must be the case. So, let's try to remove the veil of mystery from this alternative version of quantum theory, called the Heisenberg picture.

At first glance, the idea of time-dependent operators seems indeed quite strange: if an operator is, e.g., a prescription to differentiate a function, how can this rule change with time? The best way to answer this type of question is to first develop a formal way to describe the time dependence of operators and, then, to illustrate it using a few simple examples.

I begin by considering an expectation value of some arbitrary operator $\hat{A}$ in state $|\alpha(t)\rangle$: $\langle \alpha(t)| \hat{A} |\alpha(t)\rangle$. Using the time-evolution operator defined in Eq. 4.2, I can present this expectation value as

$$\langle A(t)\rangle = \langle \alpha_0| \hat{U}^\dagger(t,0)\hat{A}(t)\hat{U}(t,0) |\alpha_0\rangle . \tag{4.22}$$

Lumping together all three operators appearing between the bra and ket vectors in Eq. 4.22 into a new operator

$$\hat{A}_H(t) = \hat{U}^\dagger(t,0)\hat{A}(t)\hat{U}(t,0) \tag{4.23}$$

yields a time-dependent Heisenberg representation of the initial operator. This time dependence has, in general, two sources: an external time dependence of the initial Schrödinger operator discussed in the previous section and an internal time dependence responsible for the quantum dynamics of the system represented by the time-evolution operators. Differentiating this equation with respect to time, I obtain

$$\frac{d\hat{A}_H(t)}{dt} = \frac{d\hat{U}^\dagger(t,0)}{dt}\hat{A}(t)\hat{U}(t,0)+$$

$$\hat{U}^\dagger(t,0)\hat{A}(t)\frac{d\hat{U}(t,0)}{dt} + \hat{U}^\dagger(t,0)\frac{\partial\hat{A}(t)}{\partial t}\hat{U}(t,0).$$

Using Eq. 4.8 this can be rewritten as

$$\frac{d\hat{A}_H(t)}{dt} = \frac{i}{\hbar} \hat{U}^\dagger(t,0)\hat{H}\hat{A}(t)\hat{U}(t,0) -$$

$$\frac{i}{\hbar}\hat{U}^\dagger(t,0)\hat{A}(t)\hat{H}\hat{U}(t,0) + \hat{U}^\dagger(t,0)\frac{\partial \hat{A}(t)}{\partial t}\hat{U}(t,0).$$

Taking advantage of the unitarity of the time-evolution operator, I insert combination $\hat{U}\hat{U}^\dagger \equiv \hat{I}$ between the Hamiltonian and operator $\hat{A}$ in the first two terms of the equation above. This procedure yields

$$\frac{d\hat{A}_H(t)}{dt} = \frac{i}{\hbar} \underbrace{\hat{U}^\dagger(t,0)\hat{H}\hat{U}} \, \underbrace{\hat{U}^\dagger\hat{A}(t)\hat{U}(t,0)} -$$

$$\frac{i}{\hbar} \underbrace{\hat{U}^\dagger(t,0)\hat{A}(t)\hat{U}} \, \underbrace{\hat{U}^\dagger\hat{H}\hat{U}(t,0)} + \underbrace{\hat{U}^\dagger(t,0)\frac{\partial \hat{A}(t)}{\partial t}\hat{U}(t,0)}$$

where each of the bracketed terms defines, according to Eq. 4.23, a Heisenberg representation of the corresponding operator. Therefore, the last equation can be rewritten as

$$\frac{d\hat{A}_H(t)}{dt} = -\frac{i}{\hbar}\left[\hat{A}_H, \hat{H}_H\right] + \frac{\partial \hat{A}_H(t)}{\partial t}. \tag{4.24}$$

The resulting equation is called the Heisenberg equation for time-dependent operators. It looks very much like the Ehrenfest theorem, and just like the latter, it resembles the classical Eq. 3.8. However, unlike Ehrenfest theorem, the Heisenberg equation describes the time evolution of operators rather than of the expectation values and, therefore, does not suffer from perpetual emergence of new variables. The Ehrenfest theorem can be obtained from Eq. 4.24 by computing the expectation values of both sides of this equation with an initial state $|\alpha_0\rangle$. The initial condition for the Heisenberg equation can be easily ascertained from Eq. 4.23: setting $t = 0$ in this equation immediately yields that the initial conditions for Heisenberg operators are given by the corresponding Schrödinger operators, establishing an intimate connection between the two pictures. Now you can answer the question posed in the beginning of this section: how can a rule representing an operator change with time? The time dependence comes from combinations of various immutable rules with time-dependent coefficients. You will see the example of this a few paragraphs below.

Hamiltonian appearing in Eq. 4.24 is the Heisenberg representation of the regular Schrödinger Hamiltonian and must be evaluated before Heisenberg equations can be used. However, in the special important case of a time-independent Schrödinger Hamiltonian, one can easily show that $\hat{H}_H \equiv \hat{H}$. Indeed, one can easily infer from

Eq. 4.10 that the time-evolution operator for time-independent Hamiltonian is

$$\hat{U}(t,0) = \exp\left(-i\hat{H}t/\hbar\right).$$ (4.25)

This operator obviously commutes with Hamiltonian, and as a result we have

$$\hat{H}_H = \hat{U}^\dagger \hat{H} \hat{U} = \hat{U}^\dagger \hat{U} \hat{H} = \hat{H}.$$

The same arguments apply to any Schrödinger operator commuting with Hamiltonian, so all such operators remain independent of time. This result also obviously follows from Eq. 4.24. Thus, one can say that operators commuting with Hamiltonian represent quantum conserving observables not only at the level of expectation values as in the Ehrenfest theorem but at a deeper level of operators themselves.

In the case of Hamiltonians with explicit time dependence, you can no longer claim that the Heisenberg representation of the Hamiltonian coincides with the Schrödinger one. The Heisenberg picture in this case loses its immediate appeal, and people often prefer a picture, intermediate between Schrödinger's and Heisenberg's, called *interaction representation*. In this representation the Hamiltonian is divided into time-independent and time-dependent parts:

$$\hat{H}(t) = \hat{H}_0 + \hat{V}(t).$$

The transition to new operators is carried out now using the time-evolution operator $\hat{U}(t,0) = \exp\left(-i\hat{H}_0 t/\hbar\right)$. As a result one ends up with both operators and the state of the system displaying dependence of time: the dynamics of the operators is defined by operator $\hat{H}_0$ and the dynamics of the states by $\hat{V}(t)$. However, something tells me that continuing with this line of thought would bring me way over the line allowed in the undergraduate course. So, consider this as a teaser and preview of things to come if you decide to deepen your knowledge of quantum theory.

Now back to the time-independent Hamiltonians. The same calculations as in the case of Ehrenfest equations yield the following Heisenberg equations for the one-dimensional motion of a quantum particle:

$$\frac{d\hat{x}}{dt} = \frac{\hat{p}_x}{m_e}$$ (4.26)

$$\frac{d\hat{p}_x}{dt} = -\frac{d\hat{V}}{dx}$$ (4.27)

which coincide with the respective Hamiltonian equations of classical mechanics. Again just like in the case of Ehrenfest theorem, these equations can be easily generalized for the three-dimensional case:

$$\frac{d\hat{r}}{dt} = \frac{\hat{p}}{m_e} \tag{4.28}$$

$$\frac{d\hat{p}}{dt} = -\nabla \hat{V}. \tag{4.29}$$

To illustrate how Heisenberg equations work, let me consider the case of a one-dimensional harmonic oscillator—a particle moving in a quadratic potential of the form $V = \frac{1}{2}m\omega^2 x^2$, in which case Eq. 4.27 becomes

$$\frac{d\hat{p}_x}{dt} = -m\omega^2 \hat{x}. \tag{4.30}$$

Differentiation of this equation with respect to time yields a differential equation of the second order:

$$\frac{d^2\hat{p}_x}{dt^2} = -\omega^2 \hat{p}_x$$

where the term $d\hat{x}/dt$ is replaced with $\hat{p}/m$ with the help of Eq. 4.26. It is straightforward to verify that the equation for the momentum operator is solved by

$$\hat{p}(t) = \hat{\pi}_1 \cos \omega t + \hat{\pi}_2 \sin \omega t \tag{4.31}$$

while an expression for the coordinate operator is obtained from Eq. 4.30 by simple differentiation:

$$\hat{x}(t) = \frac{\hat{\pi}_1}{m\omega} \sin \omega t - \frac{\hat{\pi}_2}{m\omega} \cos \omega t. \tag{4.32}$$

Unknown operators $\hat{\pi}_{1,2}$ in Eqs. 4.31 and 4.32 are to be determined from the initial conditions. (General solution of any linear differential equation is a combination of particular solutions with undefined constant coefficients. Since we are dealing with operator equations, these unknown coefficients must also be operators.) Substituting $t = 0$ in the found solutions, you can see that

$$\hat{\pi}_1 = \hat{p}_{x0}$$

$$\hat{\pi}_2 = -m\omega \hat{x}_0$$

so that, just as I advertised, the time-dependent momentum and coordinate operators are expressed as linear combinations of Schrödinger operators $\hat{p}_{x0}$ and $\hat{x}_0$ with time-dependent coefficients

$$\hat{p}(t) = \hat{p}_{x0} \cos \omega t - m\omega \hat{x}_0 \sin \omega t$$

$$\hat{x}(t) = \frac{\hat{p}_{x0}}{m\omega} \sin \omega t + \hat{x}_0 \cos \omega t. \tag{4.33}$$

The obtained solution for the Heisenberg operators looks identical to the solution of the classical harmonic oscillator problem, but do not get deceived by this similarity. For instance, in classical case one can envision initial conditions such that either x_o or p_{x0} (not both, of course) are zeroes. In quantum case these are operators and cannot be set to zero. At this point you do not know enough about properties of the quantum harmonic oscillator to analyze this result any further, so I will postpone doing this till later. Still, we can have a bit of fun and, as an exercise, calculate the commutator between operators $\hat{p}(t)$ and $\hat{x}(t)$, taken not necessarily at the same time.

Example 15 (Commutator of Heisenberg Operators for Harmonic Oscillator)

$$[\hat{x}(t_1), \hat{p}(t_2)] = \sin \omega t_1 \cos \omega t_2 \left[\frac{\hat{p}_{x0}}{m\omega}, \hat{p}_{x0} \right] + \cos \omega t_1 \cos \omega t_2 \, [\hat{x}_0, \hat{p}_{x0}] -$$

$$\sin \omega t_1 \sin \omega t_2 \, [\hat{p}_{x0}, \hat{x}_0] - \sin \omega t_2 \cos \omega t_1 \, [\hat{x}_0, m\omega \hat{x}_0] =$$

$$i\hbar \cos \omega t_1 \cos \omega t_2 + i\hbar \sin \omega t_1 \sin \omega t_2 = i\hbar \cos \left[\omega \left(t_1 - t_2 \right) \right]$$

It is interesting to note that this commutator depends only on the time interval $t_1 - t_2$ and not on t_1 and t_2 separately. The equal time commutator ($t_1 = t_2$) coincides with the canonical commutator for Schrödinger coordinate and momentum operators.

It is also fun to think about eigenvectors and eigenvalues of the Heisenberg operators in Eq. 4.33, but I will let you play this game as an exercise.

4.3 Problems

Section 4.1.1

Problem 45 Consider a Hamiltonian presented by a 2×2 matrix

$$\hat{H} = \hbar \omega \begin{bmatrix} \cos \theta & \sin \theta \exp (i\varphi) \\ \sin \theta \exp (-i\varphi) & -\cos \theta \end{bmatrix}.$$

1. Find Hermitian conjugate and inverse matrix and convince yourself that this operator is simultaneously Hermitian and unitary.
2. Using representation of an exponential function as a power series, evaluate the time-evolution operator for this Hamiltonian.

3. Assume that the initial state of the system is given by a vector

$$|\alpha_0\rangle = \frac{1}{\sqrt{2}} \begin{bmatrix} 1 \\ 1 \end{bmatrix}$$

and find $|\alpha(t)\rangle$ using the time-evolution operator.

Section 4.1.2

Problem 46 Prove that if $\hat{H}|\chi_n\rangle = E_n|\chi_n\rangle$, then $\hat{H}^m|\chi_n\rangle = E_n^m|\chi_n\rangle$.

Problem 47 Consider a system with Hamiltonian

$$\hat{H} = \frac{\hat{p}^2}{2m_e} + V(r).$$

Assume that you know its eigenvalues and eigenvectors $|\chi_n\rangle$ and E_n. Show that if you change the potential in this Hamiltonian to $V(r) - E_0$, all the eigenvectors will stay the same, the eigenvalue E_0 will become equal to zero, and all other eigenvalues will become $E_n - E_0$.

Problem 48 Re-derive Eq. 4.14 factoring out $\exp\left(i\frac{E_2}{\hbar}t\right)$ instead of $\exp\left(i\frac{E_1}{\hbar}t\right)$, and demonstrate that this result does not change.

Problem 49 Consider a system described by a Hamiltonian

$$\hat{H} = E_0 \begin{bmatrix} 1 & ia \\ -ia & 1 \end{bmatrix}.$$

1. Find stationary states of this Hamiltonian.
2. Assuming that at $t = 0$ the system is in the state

$$|\alpha_0\rangle = \frac{1}{\sqrt{2}} \begin{bmatrix} 1 \\ i \end{bmatrix},$$

find $|\alpha(t)\rangle$ using stationary states of the Hamiltonian.

Section 4.1.3

Problem 50 Prove that $\langle\alpha|\hat{x}^2|\alpha\rangle = (\langle\alpha|\hat{x}|\alpha\rangle)^2$ if and only if $\hat{x}|\alpha\rangle = x|\alpha\rangle$. It is not very difficult to prove that $\hat{x}|\alpha\rangle = x|\alpha\rangle$ implies $\langle\alpha|\hat{x}^2|\alpha\rangle = (\langle\alpha|\hat{x}|\alpha\rangle)^2$, but the proof of the opposite statement requires a bit more ingenuity. You can try to prove it by demonstrating that if $\hat{x}\,|\alpha\rangle \neq x\,|\alpha\rangle$, then $\langle\alpha|\,\hat{x}^2\,|\alpha\rangle$ cannot be equal to $(\langle\alpha|\,\hat{x}\,|\alpha\rangle)^2$.

Problem 51 Go back to the problem involving a one-dimensional motion of a particle in the cubic potential $\hat{V} = a\hat{x}^3/3$ discussed in Sect. 3.3.1. It has been shown in the text that the Ehrenfest equation for $\langle\hat{p}\rangle$ involves the expectation value of $\langle x^2\rangle$. Derive the Ehrenfest equation for this quantity. Do you see expectation values of any new operators or a combination of operators in the equation for $\langle x^2\rangle$? Derive Ehrenfest equations for those new quantities. Comment on the results.

Section 4.2

Problem 52 You will learn in the following section that quantum states can be described by functions of coordinates—wave functions, in which case Schrödinger momentum operator becomes

$$\hat{p}_{x0}\psi(x) = -i\hbar\frac{d\psi}{dx}$$

while the coordinate operator becomes simple multiplication by the coordinate $\hat{x}_0\psi(x) = x\psi(x)$. Using this form of time-independent operators, find the functions representing eigenvectors of their time-dependent Heisenberg counterparts:

$$\hat{p}(t) = \hat{p}_{x0}\cos\omega t - m\omega\hat{x}_0\sin\omega t$$

$$\hat{x}(t) = \frac{\hat{p}_{x0}}{m\omega}\sin\omega t + \hat{x}_0\cos\omega t.$$

Analyze the behavior of these eigenvectors as functions of time; especially, consider limits $t = \pi n$ and $t = \pi/2 + \pi n$, where $n = 0, 1, 2\cdots$. Hint: Time-dependent terms here are just parameters, and their time dependence does not affect how you shall solve the respective differential equations.

Problem 53 Derive Heisenberg equations for operators $\hat{a}$, $\hat{a}^\dagger$ and $\hat{b}$, $\hat{b}^\dagger$ appearing in the following Hamiltonian:

$$\hat{H} = \hbar\omega\hat{a}^\dagger\hat{a} + \hbar\Omega\hat{b}^\dagger\hat{b} + \kappa\left(\hat{a}^\dagger\hat{b} + \hat{b}^\dagger\hat{a}\right).$$

Commutation relations for these operators are as follows:

$$\left[\hat{a},\hat{a}^\dagger\right] = 1;\ \left[\hat{b},\hat{b}^\dagger\right] = 1;\ \left[\hat{a},\hat{b}^\dagger\right] = 0;\ \left[\hat{a},\hat{b}\right] = 0.$$

Chapter 5
Representations of Vectors and Operators

5.1 Representation in Continuous Basis

We have managed to get through four chapters of this text without specifying any concrete form of the state vectors, and treating them as some abstractions defined only by the rules of the games that we could play with them. This approach is very convenient and rewarding from a theoretical point of view as it emphasizes the generality of quantum approach to the world and allows to derive a number of important general results with relative ease. However, when it comes to responding to experimentalists' requests to explain/predict their quantitative experimental results, we do need to have something a bit more concrete and tangible than the idea of an abstract vector. The similar situation actually arises also in the case of our regular three-dimensional geometric vectors. It is often convenient to think of them as purely geometrical objects (arrows, for instance) and derive results independent of any choice of coordinate system. However, at some point, eventually, you will need to get to some "down-to-earth" computations, and to carry them out, you will have to choose a coordinate system and replace the "arrows" with a set of numbers—the vector components.

In the case of abstract vectors that live in an abstract linear vector space, you can use the same idea to get a more concrete and handy representation of the quantum states. All these representations require that we use a basis in our abstract space. It seems more logical to begin with representations based on discrete bases, but in reality, we are somewhat forced to start with continuous bases. The reason for this is that two main observables in quantum mechanics, from which almost all of them can be constructed, are the position and the momentum (see Sect. 3.3.2). Operators corresponding to these observables have continuous spectrum, and, therefore, you will have to learn how to represent these operators using continuous bases.

© Springer International Publishing AG, part of Springer Nature 2018
L.I. Deych, *Advanced Undergraduate Quantum Mechanics*,
https://doi.org/10.1007/978-3-319-71550-6_5

5.1.1   Position and Momentum Operators in a Continuous Basis

Let me begin with some abstract operator $\hat{C}$ and a continuous basis formed by orthonormalized vectors $|q\rangle$:

$$\langle q| \, q'\rangle = \delta\left(q - q'\right),$$

where q is a continuously changing parameter. You already know from Sect. 2.3 that an abstract ket vector $|\alpha\rangle$ can be presented as an integral:

$$|\alpha\rangle = \int dq \varphi_\alpha(q) \, |q\rangle . \tag{5.1}$$

Hermitian conjugation of this expression produces its bra counterpart:

$$\langle \alpha| = \int dq \varphi_\alpha^*(q) \, \langle q| . \tag{5.2}$$

I can now write down the inner product between vectors $|\alpha\rangle$ and $|\beta\rangle$ as

$$\langle \alpha| \, \beta\rangle = \int dq \int dq' \, \varphi_\alpha^*(q) \varphi_\beta(q') \, \langle q| \, q'\rangle =$$

$$\int dq \int dq' \, \varphi_\alpha^*(q) \varphi_\beta(q') \delta\left(q - q'\right) = \int dq \varphi_\alpha^*(q) \varphi_\beta(q). \tag{5.3}$$

Subindexes α and β in these expressions indicate the correspondence between an abstract vector and the respective function appearing in the superposition given by Eq. 5.1. Applying Eq. 5.3 to the case when $|\alpha\rangle = |\beta\rangle$, I reproduce Eq. 2.42, and by recalling that all state vectors must be normalized, I end up with condition

$$\int dq \varphi_\alpha^*(q) \varphi_\alpha(q) = 1 \tag{5.4}$$

generalizing Eq. 2.43, which was originally derived only for the functions of coordinates. It should be noted here that while I am using a single variable q as an argument of the functions $\varphi\,(q)$, you must understand that it is just a convenient notation, and in reality, q can represent several variables. For instance, eigenvectors of the position operator depend on three components of the position vector, but we have been using the single symbol r to designate them all.

As long as we all agree on the choice of the basis, and do not change it in the middle of a conversation (or calculations), we have a one-to-one correspondence between abstract vectors and the respective superposition coefficients. This function $\varphi_\alpha(q)$ provides a complete description of the corresponding vector and can be, therefore, considered as its faithful representation. It can be expressed in terms of vector $|\alpha\rangle$ and the basis vectors by premultiplying Eq. 5.1 by $\langle q'|$ and using the orthonormality condition:

$$\varphi_\alpha(q) = \langle q|\,\alpha\rangle\,. \tag{5.5}$$

This essentially completes the discussion of the representation of vectors, but this was an easy part. You also need to learn how to find representation of operators appropriate for the developed representation of vectors, which is the hard part. For starters, I need to explain to you what it means to represent an operator. Consider an expression

$$|\beta\rangle = \hat{Q}\,|\alpha\rangle$$

where two abstract vectors are related to each other by abstract operator $\hat{Q}$. It seems reasonable to define a representation of the operator as such an object that would yield the same relation between functions $\varphi_\alpha(q)$ and $\varphi_\beta(q)$ representing the corresponding abstract vectors $|\alpha\rangle$ and $|\beta\rangle$. In order to figure it out, let me try to insert the completeness condition for the continuous spectrum, Eq. 3.45, formed with basis vectors $|q\rangle$ into three places in $|\beta\rangle = \hat{Q}\,|\alpha\rangle$, in front of vector $|\beta\rangle$, in front of operator $\hat{Q}$, and between the operator and $|\alpha\rangle$:

$$\int dq\,|q\rangle\,\langle q\,|\beta\rangle = \int dq \int dq'\,|q\rangle\,\langle q|\hat{Q}\,|q'\rangle\langle q'\,|\alpha\rangle\,. \tag{5.6}$$

(This amounts to inserting unity operators in all places, so, obviously, I have not changed anything.) Using Eq. 5.5 I get

$$\int dq\,|q\rangle\,\varphi_\beta(q) = \int dq \int dq'\,|q\rangle\,\langle q|\hat{Q}\,|q'\rangle\varphi_\alpha(q')\,|q\rangle\,,$$

which can be rewritten as

$$\int dq\,|q\rangle\,\left[\varphi_\beta(q) - \int dq'\,|q\rangle\,\langle q|\hat{Q}\,|q'\rangle\varphi_\alpha(q')\right] = 0.$$

Since the vectors of the basis are linearly independent, the integral in this expression can be zero only if the integrand is zero, which yields

$$\varphi_\beta(q) = \int dq'\,\langle q|\hat{Q}\,|q'\rangle\varphi_\alpha(q') \equiv \int dq'Q(q,q')\varphi_\alpha(q') \tag{5.7}$$

where I introduced $Q(q,q') = \langle q|\hat{Q}\,|q'\rangle$. Thus, the abstract operator $\hat{Q}$ in the continuous basis takes the form of an integral operator with kernel $\langle q|\hat{Q}\,|q'\rangle$. If we know how this operator acts on the basis vectors, we can determine the kernel and replace the abstract relation $|\beta\rangle = \hat{Q}\,|\alpha\rangle$ by an integral relation given by Eq. 5.7.

For instance, you can easily find the kernel if the basis is formed by eigenvectors of operator $\hat{Q}$. Indeed, if you know that $\hat{Q}|q\rangle = q|q\rangle$, then $Q(q, q') = \langle q|\hat{Q}|q'\rangle = q\delta(q - q')$ and Eq. 5.7 simplifies to

$$\varphi_\beta(q) = \int dq' Q(q, q')\varphi_\alpha(q') = q\varphi_\alpha(q), \tag{5.8}$$

i.e., the integral operator is reduced to simple multiplication by a corresponding eigenvalue, and what can be simpler? Unfortunately, life is always more complicated, and most cases, you will have to deal simultaneously with at least two non-commuting operators, so that eigenvectors of one operator will not be the eigenvectors of the other. To deal with this situation, you have to learn how to represent an operator in a basis formed by vectors that are not its eigenvectors.

Most of the bases used for practical calculations are formed by eigenvectors of some Hermitian operator, and recognition of this fact can be quite useful. So, in addition to the original operator $\hat{Q}$ with eigenvectors $|q\rangle$ and eigenvalues q, let me introduce another operator $\hat{S}$ with eigenvectors $|s\rangle$ and eigenvalues s. The goal now is to find an integral kernel representing operator $\hat{Q}$ in the basis of vectors $|s\rangle$. In other words, I need to rewrite expressions $\langle s|\hat{Q}|s'\rangle$ in the basis formed by vectors $|q\rangle$. It can be done by exploiting (twice) again the same old trick with the completeness relation expressed in terms of these vectors:

$$\langle s|\hat{Q}|s'\rangle = \int dqdq' \langle s|q\rangle \langle q|\hat{Q}|q'\rangle\langle q'|s'\rangle.$$

Now, taking into account that kernel $Q(q, q')$ in the basis of its own eigenvectors is $Q(q, q') = q\delta(q - q')$, I can simplify the above expression into

$$Q(s, s') \equiv \langle s|\hat{Q}|s'\rangle = \int dqq \langle s|q\rangle \langle q|s'\rangle \tag{5.9}$$

which gives me exactly what I have been looking for. For this expression to be useful, however, you would need to know functions $\chi_q(s) = \langle s|q\rangle$ and $\chi_s(q) = \langle q|s\rangle$. The first of them can be interpreted as the representation of eigenvector $|q\rangle$ in the basis of vectors $|s\rangle$, and the second one is clearly a representation of $|s\rangle$ in the basis of $|q\rangle$. These functions are related to each other by the property of the inner product described by Eq. 2.19: $\chi_q(s) = \chi_s^*(q)$. So, the whole business of finding the kernel $Q(s, s')$ is now reduced to finding representation of eigenvectors of $\hat{Q}$ in terms of those of $\hat{S}$ (or vice versa).

Finding function $\chi_s(q)$ is impossible without bringing in some additional information. It can be, for instance, a commutator between operators $\hat{Q}$ and $\hat{S}$, or just outright expression for $\chi_s(q)$ obtained empirically or heuristically, on the ground of some physical arguments, or just by divine insight. Whatever method you chose, you need to specify now which operators we are dealing with. Of biggest interest are, of course, operators of position and momentum, so let's agree to identify operator $\hat{Q}$ with x-component $\hat{P}_x$ of the momentum operator and $\hat{S}$ with operator

$\hat{X}$—x-component of the position operator. In this case function $\chi_q(s)$ becomes the coordinate representation $\chi_{p_x}(x)$ of the eigenvectors of the momentum operator (its x-component, of course).

This function represents a state of the particle with definite momentum p_x, which, according to de Broglie hypothesis, corresponds to motion of a free particle and is described by a harmonic wave with the wave vector with x-component $k_x = p_x/\hbar$. Disregarding the time-dependent portion of such a wave, I can write its coordinate-dependent part as $\chi_{p_x}(x) = a\exp(ik_xx) = a\exp(ip_xx/\hbar)$. Choice $a = 1/\sqrt{2\pi\hbar}$ generates, according to Eq. 2.36, a delta-normalized function:

$$\chi_{p_x}(x) = \frac{1}{\sqrt{2\pi\hbar}}\exp(ip_xx/\hbar). \tag{5.10}$$

Indeed,

$$\frac{1}{2\pi\hbar}\int_{-\infty}^{\infty}e^{i(p_x-p_x')x/\hbar}dx = \frac{1}{2\pi}\int_{-\infty}^{\infty}e^{i(p_x-p_x')\tilde{x}}d\tilde{x} = \delta(p_x - p_x') \tag{5.11}$$

where $\tilde{x} = x/\hbar$. Similarly, you can also find that

$$\frac{1}{2\pi\hbar}\int_{-\infty}^{\infty}e^{i(x-x')p_x/\hbar}dp_x = \delta(x - x'). \tag{5.12}$$

Now I can write Eq. 5.9 as

$$P_x(x,x') = \frac{1}{2\pi\hbar}\int_{-\infty}^{\infty}dp_xp_xe^{ip_xx/\hbar}e^{-ip_xx'/\hbar} = \frac{1}{2\pi\hbar}\int_{-\infty}^{\infty}dp_xp_xe^{i(x-x')p_x/\hbar}.$$

This integral might look puzzling, because it is naturally diverging. Applying a magic trick, however, I can turn it into something that actually makes sense. The trick is quite popular, so it is useful to have it up your sleeve. Differentiation of Eq. 5.12 with respect to x produces

$$\frac{d\delta(x-x')}{dx} = \frac{i}{2\pi\hbar^2}\int_{-\infty}^{\infty}dp_xp_xe^{i(x-x')p_x/\hbar}.$$

I hope that you have recognized the integral of interest on the right-hand side of this equation, so that you can find for $P_x(x,x')$:

$$P_x(x,x') = \frac{\hbar}{i}\frac{d\delta(x-x')}{dx}. \tag{5.13}$$

Substituting Eq. 5.13 into Eq. 5.7 with variable q replaced with s (in that equation q was just a generic variable not yet identified with eigenvalues of the operator), I derive a relation between functions $\varphi_\alpha(x)$ and $\varphi_\beta(x)$ indicating states $|\alpha\rangle$ and $|\beta\rangle$ in the representation of the eigenvectors of the position operator (position representation for brevity):

$$\varphi_\beta(x) = \frac{\hbar}{i} \int dx' \frac{d\delta(x-x')}{dx} \varphi_\alpha(x') = \frac{\hbar}{i}\frac{d}{dx} \int dx' \varphi_\alpha(x')\delta(x-x') = -i\hbar \frac{d\varphi_\alpha(x)}{dx}.$$

Thus, we see that the $\hat{P}_x$ operator in the position (coordinate) representation is equivalent to a differential operator:

$$\hat{p}_x = -i\hbar \frac{d}{dx} \qquad (5.14)$$

where I used the lowercase letter for a particular representation of the operator as opposed to the uppercase used for abstract operators. The coordinate operator in the coordinate representation is obviously just an operator of multiplication by the coordinate's eigenvalue.

You can turn this analysis around and identify $\hat{S}$ with a component of the momentum operator and $\hat{Q}$ with x-coordinate. Then $\chi_q(s)$ becomes the momentum representation of the eigenvector of coordinate:

$$\chi_x(p_x) = \langle p | x\rangle = (\langle x | p\rangle)^* = \frac{1}{\sqrt{2\pi\hbar}} \exp\left(-ip_x x/\hbar\right). \qquad (5.15)$$

Repeating all the same manipulation as before, you will end up with the coordinate representation of the coordinate operator in the form of a differential operator:

$$\hat{x} = i\hbar \frac{d}{dp_x}. \qquad (5.16)$$

Equations 5.14 and 5.16 are obtained from each other by interchanging $x \leftrightarrows p_x$ and complex conjugating the result. The complex conjugation bit is, of course, in sync with the fact that coordinate representation of the momentum's eigenvectors and momentum representation of the coordinate's eigenvectors are complex conjugates of each other. Obviously, all the same arguments can be carried out for any other Cartesian component of the position and momentum operators, which brings us to the following conclusion. The position representation of the momentum operator is given by the coordinate gradient operator $\vec{\nabla}$ as

$$\hat{p} = -i\hbar \vec{\nabla}, \qquad (5.17)$$

while the momentum representation of the position operator is

$$\hat{r} = i\hbar \vec{\nabla}_p \qquad (5.18)$$

where $\vec{\nabla}_p$ is defined as

$$\vec{\nabla}_p = e_x \frac{\partial}{\partial p_x} + e_y \frac{\partial}{\partial p_y} + e_z \frac{\partial}{\partial p_z} \tag{5.19}$$

and $e_{x,y,z}$ are unit vectors of Cartesian coordinate system with axes X, Y, and Z. The functions representing the eigenvectors of the 3-D momentum operator in the position representation and of the 3-D position operator in the momentum representation are, obviously, obtained by multiplying their respective one-dimensional counterparts:

$$\chi_r(p) = \frac{1}{(2\pi h)^{3/2}} e^{-i p \cdot r / \hbar} \tag{5.20}$$

$$\chi_p(r) = \frac{1}{(2\pi \hbar)^{3/2}} e^{i p \cdot r / \hbar}. \tag{5.21}$$

You might be wondering how we ended up with $\hat{P}$ and $\hat{R}$ being represented by differential rather than by integral operators as was my original intention. It happened thanks to a singular nature of the kernels, $\langle r' | \hat{P} | r \rangle$ and $\langle p' | \hat{R} | p \rangle$, which turned out to be proportional to the derivative of the delta-function. And the delta-function derivatives are quite capable of turning integrals into derivatives, as it happened in this particular case.

Having found representations of the coordinate and momentum operators, you can easily compute their commutator. For instance, for the x-components in the coordinate representation, you will easily find

$$\left[\hat{X}, \hat{P}_x\right] f(x) = -i\hbar x \frac{d}{dx} f(x) + i\hbar \frac{d}{dx} (xf(x)) = i\hbar f(x) \Rightarrow \left[\hat{X}, \hat{P}_x\right] = i\hbar. \tag{5.22}$$

Since this commutator is just a number, it must not depend on the particular representation (this is why I returned capital letters for the operators). Indeed, the calculations carried out in the momentum representation yield

$$\left[\hat{X}, \hat{P}_x\right] f(p) = i\hbar \frac{d}{dp} (pf(p)) - i\hbar p \frac{d}{dp} f(p) = i\hbar f(p) \Rightarrow \left[\hat{X}, \hat{P}_x\right] = i\hbar$$

as expected.

I will finish this section by deriving a relation between functions $\varphi_\alpha(r)$ and $\tilde{\varphi}_\alpha(p)$ representing the same state $|\alpha\rangle$ correspondingly in the coordinate and momentum representations. To achieve this, I am again resorting to the magic of the completeness relation based upon the eigenvectors of momentum. Substitution of this relation into Eq. 5.5 adapted for the eigenvectors of the position operator gives

$$\varphi_\alpha(r) = \langle r | \alpha \rangle = \int dp \, \langle r | p \rangle \, \langle p | \alpha \rangle = \int dp \, \chi_r(p) \tilde{\varphi}_\alpha(p) \qquad (5.23)$$

$$= \frac{1}{(2\pi\hbar)^{3/2}} \int d^3p \, e^{-i p \cdot r / \hbar} \tilde{\varphi}_\alpha(p)$$

where I used Eq. 5.20 for the momentum representation of the position operator's eigenvector. One can easily invert Eq. 5.23 using Fourier representation of the delta-function, Eq. 2.36, to obtain

$$\tilde{\varphi}_\alpha(p) = \frac{1}{(2\pi\hbar)^{3/2}} \int d^3r \, e^{i p \cdot r / \hbar} \varphi_\alpha(r). \qquad (5.24)$$

5.1.2 Parity Operator

In this section I want to make a slight detour and define an important operator closely related to the eigenvectors of the position operator used to introduce the position representation of the state vectors. This operator, called parity operator, is often used to classify wave functions arising in this representation, so it seems quite appropriate to talk about it here.

The parity operator is defined by its action on the eigenvectors of the position operator $|r\rangle$ as

$$\hat{\Pi} |r\rangle = |-r\rangle , \qquad (5.25)$$

the operation often called *inversion*. It is easy to see that this operator is Hermitian:

$$\langle r' | \hat{\Pi} | r \rangle = \langle r' | -r \rangle = \delta \left(r' + r \right)$$

$$\left(\langle r | \hat{\Pi} | r' \rangle \right)^* = \left(\langle r | -r' \rangle \right)^* = \delta \left(r' + r \right)$$

and that it is equal to its inverse, $\hat{\Pi}^2 |r\rangle = \hat{\Pi} |-r\rangle = |r\rangle \Rightarrow \hat{\Pi}^2 = \hat{I}$, where $\hat{I}$ is the identity operator. It follows immediately from the last expression that $\hat{\Pi} = \hat{\Pi}^{-1}$. It also means that this operator is unitary. The action of this operator on an arbitrary state can be defined using its position representation:

$$\langle r | \hat{\Pi} | \psi \rangle = \langle -r | \psi \rangle = \psi \left(-r \right)$$

where $\psi(r) = \langle r | \psi \rangle$. It is also important to know how this operator acts on the eigenvectors of the momentum operator, which can be found out using again the coordinate representation:

$$\hat{\Pi}\,|p\rangle = \int d\boldsymbol{r}\,\langle\boldsymbol{r}|\,p\rangle\,\hat{\Pi}\,|\boldsymbol{r}\rangle = \int d\boldsymbol{r}\,\langle\boldsymbol{r}|\,p\rangle\,|-\boldsymbol{r}\rangle = \int d\boldsymbol{r}\,\langle-\boldsymbol{r}|\,p\rangle\,|\boldsymbol{r}\rangle\,.$$

To derive this result I used coordinate representation of $|p\rangle$ and changed the integration variable $\boldsymbol{r}$ to $-\boldsymbol{r}$ in the last integral. Finally, using Eq. 5.21 I can write $\langle-\boldsymbol{r}|\,p\rangle = \langle\boldsymbol{r}|\,-p\rangle$, which results in the following transformation rule for $|p\rangle$:

$$\hat{\Pi}\,|p\rangle = |-p\rangle\,. \tag{5.26}$$

Parity operator has only two eigenvalues: 1 or -1. Indeed, assume that $|\pi\rangle$ is an eigenvector with eigenvalue σ : $\hat{\Pi}\,|\pi\rangle = \sigma\,|\pi\rangle$. Apply the parity operator to this relation again: $\hat{\Pi}^2\,|\pi\rangle = \sigma\,\hat{\Pi}\,|\pi\rangle \iff |\pi\rangle = \sigma^2\,|\pi\rangle \Rightarrow \sigma = \pm1$. Accordingly, eigenvectors of $\hat{\Pi}$ represent those states that either do not change upon inversion (we can call them even states) or those that change their sign (odd states). Obviously all even and all odd functions are coordinate representations of the eigenvectors of the parity operator.

Parity operator is one of the simplest *symmetry* operators, which means that it can be used to determine that a Hamiltonian or another operator corresponding to a quantum observable does not change, when a system is transformed in a certain way. In fancy language, this property is called invariance with respect to a certain transformation. To see why such invariance can be important, consider the time-independent Schrödinger equation:

$$\hat{H}\,|\alpha\rangle = E\,|\alpha\rangle$$

and assume that there is an operator (usually a unitary one), which can be used to describe a transformation of the system. Parity operator is one such example: it generates spatial inversion of the system with respect to an origin of the coordinate system. Rotations with respect to an axis or a point provide other examples of transformations described by unitary operators. In what follows I will use notation for the parity operator for the sake of concreteness, but most conclusions in the next paragraph will be applicable to any symmetry operator.

So, let me apply operator $\hat{\Pi}$ to the time-independent Schrödinger equation. In addition, I will also insert expression $\hat{\Pi}^{-1}\hat{\Pi}$, which is obviously equal to the identity operator, between $\hat{H}$ and $|\alpha\rangle$:

$$\hat{\Pi}\hat{H}\hat{\Pi}^{-1}\hat{\Pi}\,|\alpha\rangle = E\hat{\Pi}\,|\alpha\rangle\,.$$

The Schrödinger equation preserves its form when rewritten in terms of new vector $|\tilde{\alpha}\rangle = \hat{\Pi}\,|\alpha\rangle$ and new Hamiltonian $\hat{H}' = \hat{\Pi}\hat{H}\hat{\Pi}^{-1}$. This exercise demonstrates that a relation between vectors and operators is preserved if a transformation of a vector is accompanied by the corresponding transformation of the operator. I can now give a formal definition of the invariance of a system, which I earlier loosely described by saying that "the system does not change" upon certain operation, i.e., the system is invariant under a transformation if its Hamiltonian obeys the following condition: $\hat{H}' = \hat{\Pi}\hat{H}\hat{\Pi}^{-1} = \hat{H}$. One of the immediate consequences of this condition is that if $|\alpha\rangle$ is the eigenvector of the Hamiltonian, then $|\tilde{\alpha}\rangle = \hat{\Pi}\,|\alpha\rangle$ is also an eigenvector. This is an important conclusion, but I cannot dwell on it for too long as it will bring us way outside of our comfort zone. What is more important for us is that condition $\hat{\Pi}\hat{H}\hat{\Pi}^{-1} = \hat{H}$ implies that the Hamiltonian and the transformation operator commute $\hat{\Pi}\hat{H} = \hat{\Pi}\hat{H}$. This information can be immediately put to use because we already know what this means—the transformation operator and the Hamiltonian have a common set of eigenvectors. Usually eigenvectors of the former are known, and this knowledge makes finding of the eigenvectors of the latter easier. For instance, if I prove that my Hamiltonian is invariant with respect to parity transformation, I can immediately conclude that all eigenvectors of the Hamiltonian are presented by either even or odd functions, which, as you will see in Sect. 6.2, significantly simplify their computation.

Hamiltonian is not the only operator whose behavior under parity transformation is of interest. Other operators worthy of our consideration are position and momentum operators. Let me begin with a position operator defined, as you well know, by $\hat{r}\,|r\rangle = r\,|r\rangle$. Performing the same manipulation with this expression as the one to which I just subjected Hamiltonian, I will have

$$\hat{\Pi}\hat{r}\hat{\Pi}^{-1}\hat{\Pi}\,|r\rangle = r\hat{\Pi}\,|r\rangle\,.$$

Using Eq. 5.25 I transform this into

$$\hat{\Pi}\hat{r}\hat{\Pi}^{-1}\,|-r\rangle = r\,|-r\rangle$$

which only makes sense if

$$\hat{\Pi}\hat{r}\hat{\Pi}^{-1} = -\hat{r}.$$

This result demonstrates that the position operator changes its sign upon inversion, which, after some reflection, appears as almost obvious. Operators which have this property are called "odd" as opposed to "even" operators, which do not change upon parity transformation. Obviously, the inversion-invariant operators are by definition "even." I will leave it for you as an exercise to prove that the momentum operator is also "odd."

5.1.3 Schrödinger Equation in the Position Representation

The position representation is the most popular in practical applications of quantum theory. This is the representation in which the original de Broglie matter waves were described and in which Schrödinger wrote his equation. Much of the classical physics deals with processes occurring in space and time, so it is not surprising that the wave functions written in the position representation hold a special place in our hearts.[1] It is also important, of course, that the potential energy operator, which might have quite elaborate position dependence, looks the simplest in the position representation. The momentum operator, on the other hand, does not have a significant multiplicity of the forms appearing mostly in kinetic energy as $\hat{p}^2$ term, whose coordinate representation looks quite tolerable.

To derive the coordinate representation of the Hamiltonian, I need first to resolve a few technical questions. In particular, I need to know how to generate a representation of the product of two operators from representations of individual factors. Consider, for instance, operator expression $\hat{Q}\hat{S}$, whose integral kernel in some basis $|\lambda\rangle$ is $\langle \lambda'| \hat{Q}\hat{S} |\lambda\rangle$. Inserting a completeness relation (again!) between the operators, I obtain

$$\langle \lambda'| \hat{Q}\hat{S} |\lambda\rangle = \int d\lambda'' \langle \lambda'| \hat{Q} |\lambda''\rangle \langle \lambda''| \hat{S} |\lambda\rangle = \int d\lambda'' Q(\lambda', \lambda'') S(\lambda'', \lambda). \quad (5.27)$$

An important example is operator $\hat{P}_x^2$, whose position representation would be useful to know. The integral kernel for $\hat{P}_x$ was found in the previous section as $Q(x', x'') = i\hbar \delta'(x' - x'')$, where $'$ on the delta-function signifies differentiation with respect to the first argument. Substitution of these expressions into Eq. 5.27 yields

$$\langle x'| \hat{P}_x^2 |x\rangle = -\hbar^2 \int dx'' \frac{d\delta(x' - x'')}{dx'} \frac{d\delta(x'' - x)}{dx''} =$$

$$\hbar^2 \left. \frac{d^2\delta(x' - x'')}{dx' dx''} \right|_{x''=x} = -\hbar^2 \frac{d^2\delta(x' - x)}{dx'^2} = -\hbar^2 \frac{d^2\delta(x - x')}{dx^2}$$

where in the last line, I used evenness of the delta-function to switch from $\delta(x' - x)$ to $\delta(x - x')$ and the chain differentiation rule to change the differentiation variable. If you plug this result into expression

$$\varphi_\beta(x') = \int \langle x'| \hat{P}_x^2 |x\rangle \, \varphi_\alpha(x) dx,$$

you will get

$$\varphi_\beta(x') = -\hbar^2 \int \frac{d^2\delta(x - x')}{dx^2} \varphi_\alpha(x) dx = -\hbar^2 \frac{d^2\varphi_\alpha(x)}{dx^2} \quad (5.28)$$

which means that the coordinate representation of $\hat{P}_x^2$ operator is just the square of $-i\hbar d/dx$ operator (could have guessed this, of course, but this derivation was

---

[1]You might remember that the lack of spatial-temporal picture was the main complaints Schrödinger leveled against Heisenberg's "transcendental" algebraic approach.

a nice exercise, wasn't it?). Obviously the same result can be obtained for two other components of the momentum, which means that operator $\hat{P}^2$ in the position representation is given by

$$\hat{p}^2 = -\hbar^2 \nabla^2, \tag{5.29}$$

where $\nabla^2 = \vec{\nabla} \cdot \vec{\nabla}$ is the Laplacian operator. Using Eq. 5.19, one can easily derive

$$\nabla^2 = \frac{\partial^2}{\partial x^2} + \frac{\partial^2}{\partial y} + \frac{\partial^2}{\partial z^2}. \tag{5.30}$$

Since the action of the position operator in the position representation amounts to the simple multiplication by position vector r, the position representation of the potential energy operator $V(\hat{r})$ amounts to multiplication by $V(r)$. Thus the action of the entire Hamiltonian in the position representation can now be described as

$$\hat{H}_r \Psi_\alpha(r) \equiv \int dr' \langle r| \hat{H} |r' \rangle \Psi_\alpha(r') =$$

$$\frac{1}{2m_e} \int dr' \langle r|\hat{p}^2 |r' \rangle \Psi_\alpha(r') + \int dr' \langle r| \hat{V} |r' \rangle \Psi_\alpha(r') =$$

$$-\frac{\hbar^2}{2m_e} \nabla^2 \Psi_\alpha(r) + \int dr' V(r') \langle r| r' \rangle \Psi_\alpha(r') =$$

$$-\frac{\hbar^2}{2m_e} \nabla^2 \Psi_\alpha(r) + \int dr' V(r') \delta(r - r') \Psi_\alpha(r') =$$

$$-\frac{\hbar^2}{2m_e} \nabla^2 \Psi_\alpha(r) + V(r) \Psi_\alpha(r), \tag{5.31}$$

where $\Psi_\alpha(r)$ stands for $\langle r |\alpha \rangle$. Correspondingly, I can write down the Hamiltonian in the position representation simply as

$$\hat{H}_r = -\frac{\hbar^2}{2m} \nabla^2 + V(r, t) \tag{5.32}$$

which acts on functions $\Psi(r, t)$ realizing the position representation of the corresponding quantum states.

The time-dependent Schrödinger equation in the coordinate representation is obtained from Eq. 4.9 by premultiplying it with the basis bra vector $\langle r|$ and using the completeness relation:

$$i\hbar \frac{d \langle r |\alpha \rangle}{dt} = \int dr' \langle r| \hat{H} |r' \rangle \langle r' |\alpha \rangle.$$

The left-hand side of this equation is simply $\Psi(r, t)$ (I will drop the subindex α from now on), while the right-hand side was evaluated just a few lines above in Eq. 5.31. Thus, I can write the position representation of Eq. 4.9 as

$$i\hbar \frac{\partial \Psi(r, t)}{\partial t} = \left[-\frac{\hbar^2}{2m} \nabla^2 + V(r, t) \right] \Psi(r, t). \tag{5.33}$$

This is what most of quantum mechanics textbooks call the celebrated time-dependent Schrödinger equation governing quantum dynamics of a single-particle quantum state represented by wave function $\Psi(r, t)$. If the potential function in Eq. 5.33 does not depend on time, one can separate time and coordinate dependence of the wave function as

$$\Psi(r, t) = \exp\left(-i\frac{E}{\hbar} t \right) \psi(r) \tag{5.34}$$

where $\psi(r)$ obeys equation

$$\left[-\frac{\hbar^2}{2m} \nabla^2 + V(r) \right] \psi(r) = E\psi(r) \tag{5.35}$$

often called time-independent Schrödinger equation. Rewritten in the form $\hat{H}_r \psi(r) = E\psi(r)$, where subindex r points to the position representation, it becomes reminiscent of Eq. 4.11 defining eigenvalues and eigenvectors of the Hamiltonian. Obviously, Eq. 5.35 produces eigenvectors of the Hamiltonian in the position representation.

This equation, which is a linear differential equation of the second order, has to be complemented by boundary conditions specifying behavior of the wave functions at infinity. They depend on the type of spectrum (discrete or continuous) the respective wave functions belong to. If the eigenvalue E belongs to a discrete spectrum, we know from the discussion in Sect. 2.2 that the corresponding states are square-integrable, which means that integral $\int |\psi(r)|^2 \, dr$ taken over the entire volume (it defines the norm of the state vector in the coordinate representation; see Eq. 2.43 or 5.4) is finite. Only functions which tend to zero fast enough when $|r| \to \infty$ will satisfy this requirement. Thus, the boundary condition for the wave functions of discrete spectrum can be formulated as

$$\lim_{|r| \to \infty} |\psi(r)| = 0. \tag{5.36}$$

The existence of a discrete spectrum depends on the behavior of the potential function $V(r)$ and is closely related to the type of classical motion at a given energy. Imagine, for instance, that there exists a closed surface in space separating regions where $E > V(r)$ from the regions where $E < V(r)$. A classical particle can only exist in the latter region, because the former would correspond to negative values of kinetic energies. Regions where classical kinetic energy would be positive are called *classically allowed*, while regions where kinetic energy turns negative are called *classically forbidden*. The boundary between these two regions, where $E = V(r)$, forms a surface, which a classical particle cannot cross. Such motion of a classical particle is called bound motion. In the quantum mechanical case, Schrödinger Eq. 5.35 has solutions in both regions, which, however, have a completely different behavior. An analysis in the most generic three-dimensional case is mathematically too involved to attempt it here, so I shall illustrate this difference considering a one-dimensional model, with the wave function and the potential depending on a single coordinate, e.g., x. For a classically bound motion to take place in this case, there must exist an interval of coordinates $x_1 < x < x_2$, where $E > V(x)$, while everywhere else $E < V(x)$. The terminal points of this interval are so-called turning points, where a classical particle would momentarily stop before reversing its velocity.

It is convenient to analyze this situation quantum mechanically by rewriting the Schrödinger equation as

$$\frac{d^2\psi(x)}{dx^2} = \frac{2m}{\hbar^2} \left[V(x) - E\right] \psi(x). \tag{5.37}$$

In the classically forbidden regions, which extend to infinity in both positive and negative directions of the coordinate axes, the second derivative of the wave function always has the same sign as the wave function itself. It is easier to discuss the meaning of this result assuming that the wave function and, respectively, its second derivative, in the classically forbidden region, are positive. In this case, if the first derivative is positive (wave function grows), it becomes even more positive so that the wave function bends upward growing even faster with increasing x. If, however, the first derivative is negative (wave function decreases), it is becoming less and less negative approaching zero. The wave function in this case must also asymptotically approach zero without ever changing its sign. This wave function would obviously satisfy the boundary condition given in Eq. 5.36 and, therefore, correspond to the eigenvalue from the discrete spectrum. If the wave function is negative, all the same arguments work, and the wave function is either monotonically decreasing, becoming even more negative, or increasing approaching zero from the negative side. In the classically allowed region, the second derivative is negative, and the solution to the equation does not have to be monotonic. The main conclusion following from these arguments is that the energy eigenvalues belonging to an interval corresponding to a classically bound motion form, in quantum description, a discrete spectrum.

Wave functions corresponding to the continuous spectrum of energy usually appear in the situations when the potential approaches a constant finite value at infinity. If energy E exceeds this limiting value of the potential, than asymptotically for large values of x, Eq. 5.37 takes the following form:

$$\frac{d^2\psi(x)}{dx^2} = -\frac{2m}{\hbar^2} \left[E - V(\infty)\right] \psi(x)$$

which has two possible solutions $\psi(x) \propto \exp(ikx)$ or $\psi(x) \propto \exp(-ikx)$, where $k = \sqrt{2m[E - V(\infty)]}/\hbar$. Any one of these asymptotic forms can be chosen as a boundary condition at infinity: the actual choice is determined by the physical problem at hand. This situation often appears in so-called scattering problems, when one is interested in the behavior of a stream of particles incident on the potential from infinity and being registered by a detector on the opposite side of the potential. For this reason, wave functions with asymptotic behavior of this kind are called scattering wave functions. I will talk much more about this situation in subsequent chapters of the book.

Finally, you need to learn about the continuity properties of the wave functions. This issue arises only if the potential $V(r)$ is not everywhere continuous (if the potential is continuous, the wave functions are automatically continuous). We require that the wave function remains continuous regardless of the discontinuity of the potential. The physical foundation for such a requirement can be given as follows. A discontinuity of wave function means that its first derivative becomes infinite at the point of discontinuity, which creates a whole bunch of problems, e.g., the expectation value of the momentum of the particle at this point becomes infinite.

However, the continuity of the first derivative of the wave function is not necessarily guaranteed. In one-dimensional case, one can show that if the discontinuities of the potential only occur in the form of finite "jumps," the first derivative of the wave function remains continuous (provided that the mass of the particle remains the same on both sides of the "step in the potential"). To see this one simply needs to integrate Eq. 5.37 over an infinitesimal interval surrounding the point of discontinuity of the potential, x_d:

$$\lim_{\varepsilon \to 0} \int_{x_d - \varepsilon}^{x_d + \varepsilon} \frac{d^2\psi(x)}{dx^2} dx = \lim_{\varepsilon \to 0} \left(\frac{d\psi(x)}{dx}\bigg|_{x_d + \varepsilon} - \frac{d\psi(x)}{dx}\bigg|_{x_d - \varepsilon} \right) =$$

$$\lim_{\varepsilon \to 0} \frac{2m}{\hbar^2} \int_{x_d - \varepsilon}^{x_d + \varepsilon} [V(x) - E] \psi(x) dx \Rightarrow$$

$$\lim_{\varepsilon \to 0} \left(\frac{d\psi(x)}{dx}\bigg|_{x_d + \varepsilon} - \frac{d\psi(x)}{dx}\bigg|_{x_d - \varepsilon} \right) = \frac{2m}{\hbar^2} [(V_2 - V_1)\psi(x_d)\varepsilon - E\psi(x_d)\varepsilon] = 0 \Rightarrow$$

$$\frac{d\psi(x)}{dx}\bigg|_{x_d + 0} = \frac{d\psi(x)}{dx}\bigg|_{x_d - 0} \tag{5.38}$$

where $V_1 = V(x_d - 0)$ and $V_2 = V(x_d + 0)$. In some semiconductor heterostructures (alternating planar layers of different semiconductors), Eq. 5.37 is sometimes used to describe the behavior of charged particles in the so-called effective mass approximation. In this approximation the periodic potential of ions felt by electrons is approximately taken into account by modifying the mass of the electrons from their normal "free electron" value. The new "effective" masses are usually different in different materials, and if the discontinuity of the potential occurs due to an electron passing from one semiconductor to another, its effective mass also changes. Repeating previous derivation taking into account the possibility of discontinuity of the mass, you can derive a generalized derivative continuity condition:

$$\frac{1}{m_1} \frac{d\psi(x)}{dx}\bigg|_{x_d+0} = \frac{1}{m_2} \frac{d\psi(x)}{dx}\bigg|_{x_d-0} \tag{5.39}$$

where $m_{1,2}$ are values of the effective mass on both sides of the potential step.

The position representation allows for a useful and conceptually important generalization of the idea of probability conservation expressed by the normalization condition 2.43. Consider the following quantity:

$$P(t) = \int_v |\Psi(r,t)|^2 \, d^3r,$$

which yields a probability that a measurement of the particle's position will find it within the integration volume v. Computing the time derivative of this quantity and utilizing Schrödinger's equation, Eq. 5.33, you get

$$\frac{\partial P}{\partial t} = \int_v \left[\Psi(r,t) \frac{\partial \Psi^*(r,t)}{\partial t} + \Psi^*(r,t) \frac{\partial \Psi(r,t)}{\partial t} \right] d^3r =$$

$$\frac{1}{i\hbar} \int_v \left\{ \Psi^*(r,t) \left[-\frac{\hbar^2}{2m} \nabla^2 + V(r,t) \right] \Psi(r,t) \right.$$

$$\left. - \Psi(r,t) \left[-\frac{\hbar^2}{2m} \nabla^2 + V(r,t) \right] \Psi^*(r,t) \right\} d^3r =$$

$$\frac{i\hbar}{2m} \int_v \left\{ \Psi^*(r,t) \nabla^2 \Psi(r,t) - \Psi(r,t) \nabla^2 \Psi^*(r,t) \right\} d^3r. \tag{5.40}$$

To proceed you will need the following vector identity:

$$\Psi^*(r,t) \nabla^2 \Psi(r,t) - \Psi(r,t) \nabla^2 \Psi^*(r,t) \equiv$$

$$\nabla \cdot \left[\Psi^*(r,t) \nabla \Psi(r,t) - \Psi(r,t) \nabla \Psi^*(r,t) \right]$$

which is easily proved by working it out from the right to the left. What is important is that the expression on the right has a form of a divergence of a vector so that Eq. 5.40 can be written as

$$\int_v \frac{\partial}{\partial t} |\Psi(r,t)| \, d^3r + \int_v \nabla \cdot j \, d^3r = 0, \tag{5.41}$$

where I introduced a vector called probability current density

$$j = \frac{i\hbar}{2m_e} \left[\Psi(r, t) \nabla \Psi^*(r, t) - \Psi^*(r, t) \nabla \Psi(r, t) \right]. \tag{5.42}$$

One important property of this quantity is that it vanishes for the wave functions representing a stationary state if its time-independent part is real. Indeed, substituting

$$\Psi(r, t) = \exp(-iEt/\hbar)\, \psi(r),$$

you can see that the product of time-dependent factors yields unity, and, if $\psi(r)$ is real, the remaining two terms simply cancel each other. Equation 5.42 can be rewritten in a more illuminating form: introducing a velocity operator

$$\hat{v} \equiv \frac{\hat{p}}{m_e} = -\frac{i\hbar}{m_e} \nabla$$

it can be presented as

$$j = \frac{1}{2} \left(\Psi^* \hat{v} \Psi + \Psi (\hat{v}\Psi)^* \right). \tag{5.43}$$

If you do not see an immediate usefulness of bringing out the velocity operator in the definition of j (besides a purely aesthetic fact that Eq. 5.43 is more pleasant to the eye), let me point out that it highlights the connection between quantum and classical concepts of the current density. As you may remember from introductory physics course, the current density for any flowing quantity in classical physics can be written down as ρv, where ρ is the density of whatever does the flowing (charge, mass, etc.) and v is the velocity of the flow. This connection becomes even more direct for a free propagating particle with wave function:

$$\Psi(r, t) = A \exp(-iEt/\hbar + ipr/\hbar).$$

Substituting this wave function into Eq. 5.42 or 5.43, you will find for the quantum j:

$$j = |A|^2 p/m,$$

which is an exact reproduction of the classical expression if you identify $|A|^2$ with ρ.

Using Gauss' theorem (google it, if you do not remember!), I can rewrite Eq. 5.41 as

$$\int_v \frac{\partial}{\partial t} |\Psi(r, t)|^2 \, d^3r = - \int_\Sigma j \cdot n dS \tag{5.44}$$

where n is a unit vector normal to surface Σ enclosing volume v (directed outward). The right-hand side of Eq. 5.44 has a meaning of a flux (just like electric field flux in electromagnetism) characterizing the "flow" of probability across a boundary encompassing the volume. This equation simply states that the probability "to locate" a particle within a given volume decreases if the probability "flows" outside of the volume and increases if the flow of probability is reversed. In this sense, this equation is the statement of conservation of probability, just like a similar statement in electromagnetism would mean conservation of charge, and in hydrodynamics, conservation of mass. An alternative expression of this statement can be obtained if you drop the volume integration in Eq. 5.41 and introduce probability density $\rho(r, t) \equiv |\Psi(r, t)|^2$:

$$\frac{\partial}{\partial t} \rho(r, t) + \nabla \cdot j = 0. \tag{5.45}$$

This equation is called probability continuity equation, and it looks very much like any other continuity equation: in the electrodynamic context, ρ is the charge density, and j is the current density; in hydrodynamics, ρ is a density of a fluid, and j is the mass flux; in thermodynamics, ρ is local energy density, and j is energy flux; etc. While in quantum mechanics this equation does not describe the flow of anything material, such as charge or mass, it has very similar empirical significance. Probability current density, for instance, determines such experimentally observable characteristics as scattering cross-sections or reflection and transmission coefficients.

5.1.4 Orbital Angular Momentum in Position Representation

5.1.4.1 Operators

When I first introduced angular momentum operators in Sect. 3.3.4, I emphasized that the importance of the angular momentum is derived from the fact that it commutes with the Hamiltonian of a particle in a central field. At that time I did not have the tools to prove this fact as well as to study eigenvectors of the angular momentum operators in any particular detail. Using position representation for these operators, I can eliminate some of those gaps. This representation is generated by substituting Eq. 5.17 for position representation of the momentum operator to Eqs. 3.50–3.52 with additional understanding that the action of the position operator is reduced to mere multiplication by r. This procedure generates the following expressions for the Cartesian components of the angular momentum defined with respect to some coordinate axes:

$$\hat{L}_x = -i\hbar y \frac{\partial}{\partial z} + i\hbar z \frac{\partial}{\partial y} \tag{5.46}$$

$$\hat{L}_y = -i\hbar z \frac{\partial}{\partial x} + i\hbar x \frac{\partial}{\partial z} \tag{5.47}$$

$$\hat{L}_z = -i\hbar x \frac{\partial}{\partial y} + i\hbar y \frac{\partial}{\partial x}. \tag{5.48}$$

Fig. 5.1 Spherical
coordinate system

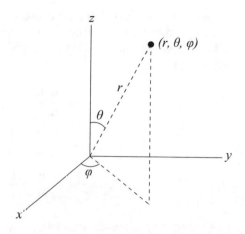

These expressions imply that operators $\hat{L}_{x,y,z}$ act on wave functions defined in terms
of Cartesian coordinates x, y, and z of a position vector $\mathbf{r}$. However, Cartesian
coordinates are not the only way to characterize a position of a point in space.
Spherical coordinates, for instance, can do the same job, and in some instances,
we might want to have operators acting on functions $\psi(r, \theta, \varphi)$, where r, θ, and φ
are radial, polar, and azimuthal spherical coordinates (see Fig. 5.1). To make sure
that there is no confusion left, let me reiterate: I am using spherical coordinates to
describe position dependence of the wave functions in the coordinate representation,
but I keep using Cartesian coordinate system to introduce components of the vector
of the angular momentum and respective operators. It is important that the two
coordinate systems are mutually dependent: the spherical angles θ and φ are defined
with respect to the same axes, which are used to define Cartesian components of the
angular momentum.

To proceed with my plan, I need to remind you the well-known relations between
Cartesian and spherical coordinates:

$$z = r\cos\theta \tag{5.49}$$

$$x = r\sin\theta\cos\varphi \tag{5.50}$$

$$y = r\sin\theta\sin\varphi \tag{5.51}$$

and

$$r = \sqrt{x^2 + y^2 + z^2} \tag{5.52}$$

$$\theta = \arccos\left(\frac{z}{\sqrt{x^2 + y^2 + z^2}}\right) \tag{5.53}$$

$$\varphi = \arctan\left(\frac{y}{x}\right). \tag{5.54}$$

To make the transition from the operators defined in space of functions $f(x, y, z)$ to the operators acting on functions $f(r, \theta, \varphi)$, I shall use the regular chain rule for differentiation of the functions of several variables, which in this case takes the following form:

$$\frac{\partial}{\partial x} = \frac{\partial r}{\partial x}\frac{\partial}{\partial r} + \frac{\partial \theta}{\partial x}\frac{\partial}{\partial \theta} + \frac{\partial \varphi}{\partial x}\frac{\partial}{\partial \varphi}$$

$$\frac{\partial}{\partial y} = \frac{\partial r}{\partial y}\frac{\partial}{\partial r} + \frac{\partial \theta}{\partial y}\frac{\partial}{\partial \theta} + \frac{\partial \varphi}{\partial y}\frac{\partial}{\partial \varphi}$$

$$\frac{\partial}{\partial z} = \frac{\partial r}{\partial z}\frac{\partial}{\partial r} + \frac{\partial \theta}{\partial z}\frac{\partial}{\partial \theta} + \frac{\partial \varphi}{\partial z}\frac{\partial}{\partial \varphi}.$$

I will illustrate this transition deriving expression for $\hat{L}_z$ in the spherical coordinates. According to Eq. 5.48, I need derivative operators $\partial/\partial x$ and $\partial/\partial y$. Using Eqs. 5.52–5.54, as well as Eqs. 5.49–5.51, I get

$$\frac{\partial r}{\partial x} = \frac{x}{\sqrt{x^2 + y^2 + z^2}} = \sin\theta \cos\varphi.$$

To compute derivative $\partial\theta/\partial x$, it is more convenient to transform Eq. 5.53 into

$$\cos\theta = \frac{z}{\sqrt{x^2 + y^2 + z^2}}$$

and differentiate it with respect to x:

$$-\sin\theta\frac{\partial\theta}{\partial x} = -\frac{zx}{(x^2 + y^2 + z^2)^{3/2}}.$$

This expression can now be transformed into

$$\frac{\partial\theta}{\partial x} = \frac{r^2 \cos\theta \sin\theta \cos\varphi}{r^3 \sin\theta} = \frac{\cos\theta \cos\varphi}{r}.$$

Similarly, starting with $\tan\varphi = y/x$, I find

$$\frac{1}{\cos^2\varphi}\frac{\partial\varphi}{\partial x} = -\frac{y}{x^2} = -\frac{\sin\varphi}{r \sin\theta \cos^2\varphi} \Rightarrow \frac{\partial\varphi}{\partial x} = -\frac{\sin\varphi}{r \sin\theta}.$$

Gathering all these results together, I finally have

$$y\frac{\partial}{\partial x} = r\sin^2\theta \cos\varphi \sin\varphi\frac{\partial}{\partial r} + \sin\theta \sin\varphi \cos\theta \cos\varphi\frac{\partial}{\partial \theta} - \sin^2\varphi\frac{\partial}{\partial \varphi}. \qquad (5.55)$$

Now I need to repeat these calculations for $x\partial/\partial y$ contribution to Eq. 5.48:

$$\frac{\partial r}{\partial y} = \frac{y}{\sqrt{x^2 + y^2 + z^2}} = \sin\theta\sin\varphi$$

$$-\sin\theta\frac{\partial\theta}{\partial y} = -\frac{zy}{(x^2 + y^2 + z^2)^{3/2}} \rightarrow \frac{\partial\theta}{\partial y} = \frac{\cos\theta\sin\varphi}{r}$$

$$\frac{1}{\cos^2\varphi}\frac{\partial\varphi}{\partial y} = \frac{1}{x} = \frac{1}{r\sin\theta\cos\varphi} \rightarrow \frac{\partial\varphi}{\partial y} = \frac{\cos\varphi}{r\sin\theta}$$

$$x\frac{\partial}{\partial y} = r\sin^2\theta\cos\varphi\sin\varphi\frac{\partial}{\partial r} + \sin\theta\sin\varphi\cos\theta\cos\varphi\frac{\partial}{\partial\theta} + \cos^2\varphi\frac{\partial}{\partial\varphi}. \tag{5.56}$$

Finally, combining Eqs. 5.55 and 5.56, I am getting my reward for all this hard work because the derived expression for $\hat{L}_z$ is so remarkably simple:

$$\hat{L}_z = -i\hbar x\frac{\partial}{\partial y} + i\hbar y\frac{\partial}{\partial x} = -i\hbar\frac{\partial}{\partial\varphi}. \tag{5.57}$$

This result justifies going into all these troubles involved in transitioning to spherical coordinates. One can also derive similar expressions for x- and y-components of the angular momentum, but they are not that pretty:

$$\hat{L}_x = i\hbar\left(\sin\varphi\frac{\partial}{\partial\theta} + \cot\theta\cos\varphi\frac{\partial}{\partial\varphi}\right) \tag{5.58}$$

$$\hat{L}_y = i\hbar\left(-\cos\varphi\frac{\partial}{\partial\theta} + \cot\theta\sin\varphi\frac{\partial}{\partial\varphi}\right). \tag{5.59}$$

The remarkable simplicity of $\hat{L}_z$ expressed in terms of the derivative with respect to spherical coordinates is the main reason why it became customary to consider the pair $\hat{L}_z$, $\hat{L}^2$ as a set of commuting operators, when dealing with the angular momentum. Derivation of the expression for operator $\hat{L}^2$ in terms of spherical coordinates is quite straightforward, and while it is excruciatingly tedious, it does lead to a really awesome answer:

$$\hat{L}^2 = -\hbar^2\left[\frac{1}{\sin\theta}\frac{\partial}{\partial\theta}\left(\sin\theta\frac{\partial}{\partial\theta}\right) + \frac{1}{\sin^2\theta}\frac{\partial^2}{\partial\varphi^2}\right]. \tag{5.60}$$

However, in order to appreciate its awesomeness, you might have to google "Laplacian operator" unless, of course, you are also awesome and remember how

Fig. 5.2 Breaking down a
classical momentum in
components

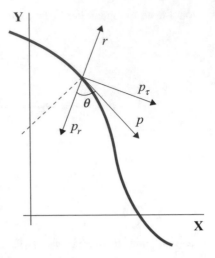

it looks like in spherical coordinates (in Cartesian coordinates it was defined in
Eq. 5.30). For your convenience I will present it here:

$$\nabla^2 = \frac{1}{r^2} \frac{\partial}{\partial r}\left(r^2 \frac{\partial}{\partial r}\right) + \frac{1}{r^2}\left[\frac{1}{\sin\theta}\frac{\partial}{\partial\theta}\left(\sin\theta\frac{\partial}{\partial\theta}\right) + \frac{1}{\sin^2\theta}\frac{\partial^2}{\partial\varphi^2}\right] \qquad (5.61)$$

hoping that you notice that the angular part of the Laplacian (expression in square
brackets) is identical to $-\hat{L}^2/\hbar^2$. And this fact is not left without important consequences. Recall that the Laplacian operator defines the coordinate representation of
the kinetic energy operator $\hat{K} = -\hbar^2\nabla^2/2m_e$ which now can be written down in
spherical coordinates as

$$\hat{K} = -\frac{\hbar^2}{2m_e r^2}\frac{\partial}{\partial r}\left(r^2\frac{\partial}{\partial r}\right) + \frac{\hat{L}^2}{2m_e r^2}. \qquad (5.62)$$

This presentation of the kinetic energy makes it plainly obvious that $\left[\hat{K},\hat{L}^2\right] = 0$.

Indeed, the radial part of kinetic energy commutes with $\hat{L}^2$ because they contain
derivatives with respect to different coordinates, and the angular part is simply
proportional to $\hat{L}^2$, which obviously commutes with itself. To get an even better
appreciation of Eq. 5.62, it is interesting to consider a classical kinetic energy
rewritten in terms of two mutually perpendicular components of the momentum:
p_τ, which is normal to the particle's position vector, and p_r, which is aligned with
it. Taking into account that the momentum is tangential to the particle's trajectory,
you can see (Fig. 5.2) that $p_\tau = p\sin\vartheta$, where ϑ is the angle between the vector
of momentum and the position vector at a given point. In terms of these two
components, the kinetic energy can be presented as

$$K = \frac{p_\tau^2}{2m_e} + \frac{p_r^2}{2m_e} = \frac{p^2\sin^2\vartheta}{2m_e} + \frac{p_r^2}{2m_e}.$$

Now, let me play a bit with the first of these terms multiplying its numerator and denominator by r^2:

$$\frac{p^2 \sin^2 \vartheta}{2m_e} = \frac{p^2 r^2 \sin^2 \vartheta}{2m_e r^2}.$$

I am sure you recognize now that the numerator of this expression is $|\mathbf{r} \mathbf{x} \mathbf{p}|^2$, which is nothing, but the classical angular momentum $\mathbf{L} = \mathbf{r} \mathbf{x} \mathbf{p}$. Thus, the classical kinetic energy can be presented as

$$K = \frac{p_r^2}{2m_e} + \frac{L^2}{2m_e r^2}$$

where the last term "miraculously" reproduces a similar term in quantum mechanical Eq. 5.62. Isn't it true that physics (and math) work in mysterious ways?

Now I can fulfill my promise made in Sect. 3.3.4 and prove that the operators of the angular momentum commute with the Hamiltonian if the particle's potential energy belongs to the class of central potentials. Actually, Eq. 5.60 makes the proof quite trivial: the angular momentum operators in the position representation contain only derivatives with respect to angular variables, so that if the potential energy $V(r)$ depends only on the radial coordinate $V(r)$ (definition of the central potential!), then neither $\hat{L}_z$ nor $\hat{L}^2$ affects $V(r)$ so that $\hat{L}^2 V(r)\psi (r, \theta, \varphi) = V(r)\hat{L}^2 \psi (r, \theta, \varphi)$, and the same is obviously true for $\hat{L}_z$ operator. Since I already showed that the angular momentum commutes with the kinetic energy, the last remark completes the required proof.

The direct consequence of vanishing commutators $\left[\hat{H}, \hat{L}_z\right]$ and $\left[\hat{H}, \hat{L}^2\right]$ is that the common eigenvectors of $\hat{L}^2$ and $\hat{L}_z$ are also eigenvectors of Hamiltonians with a central potential, which makes the task of finding these eigenvectors especially important. And this is what, without further ado, I am going to do now.

5.1.4.2 Eigenvectors

First of all, let me remind you that we are looking for the functions, which represent common eigenvectors of operators $\hat{L}^2$ and $\hat{L}_z$. This means that these functions must simultaneously obey both equations:

$$\hat{L}_z \psi_{lm} (\theta, \varphi) = \hbar m \psi_{lm} (\theta, \varphi) \qquad (5.63)$$

and

$$\hat{L}^2 \psi_{lm} (\theta, \varphi) = \hbar^2 l(l + 1)\psi_{lm} (\theta, \varphi). \qquad (5.64)$$

I begin with operator $\hat{L}_z$ whose eigenvectors in the coordinate representation are particularly easy to find. First, let me notice that this operator only contains derivatives with respect to φ, so that the angular variable θ plays here the role of "silent" parameter, a constant, as far as operator $\hat{L}_z$ is concerned. In formal language it means that dependence of θ may appear in function $\psi_{lm}(\theta, \varphi)$ only as a factor in front of the "main" function dependent only of φ:

$$\psi_{lm}(\theta, \varphi) = P_l^m(\theta) \Phi_m(\varphi). \tag{5.65}$$

Substituting this form into Eq. 5.63, you can see that $P_l^m(\theta)$ indeed behaves as a constant and can be discarded. The resulting equation for the remaining function

$$-i\hbar \frac{\partial \Phi_m(\varphi)}{\partial \varphi} = \hbar m \Phi_m(\varphi)$$

has an obvious solution

$$\Phi_m(\varphi) = \frac{1}{\sqrt{2\pi}} \exp(im\varphi). \tag{5.66}$$

Now consider how function $\Phi_m(\varphi)$ evolves when the position vector rotates around the axis Z. After one complete rotation, which corresponds to the change of φ by 2π, the position vector returns to the initial position. It would have been weird if the wave function would not return to its initial value as well. In a somewhat more sophisticated language, it means function $\Phi_m(\varphi)$ is expected to be periodic in φ. This can be only achieved if you allow only for integer values of m: $m = 0, \pm 1, \pm 2 \cdots$. This is only half of the eigenvalues of the operator $\hat{L}_z$ found by algebraic methods in Sect. 3.3.4. The eigenvalues corresponding to half-integer values of m result in the solutions that change its sign upon rotation by 2π and shall be discarded. It does not mean, of course, that half-integer values m have no place in quantum theory; it only means that they cannot correspond to eigenvectors permitted the position representation. The factor $1/\sqrt{2\pi}$ in Eq. 5.66 ensures that the wave function $\Phi_m(\varphi)$ is normalized with respect to the inner products defined as

$$\langle \Phi_{m_1} | \Phi_{m_2} \rangle \equiv \int_0^{2\pi} \Phi_{m_1}^*(\varphi) \Phi_{m_2}(\varphi) \, d\varphi = \frac{1}{2\pi} \int_0^{2\pi} \exp\left[i(m_2 - m_1)\varphi\right] d\varphi. \tag{5.67}$$

It is obvious that with this definition of the inner product, the functions representing the eigenvectors are not only normalized but also orthogonal. The integral in Eq. 5.67 is a part of a surface integral carried out over the surface of a sphere, which in spherical coordinates has the following form:

$$\langle \psi_1 | \psi_2 \rangle = \int_0^\pi \int_0^{2\pi} d\theta \, d\varphi \sin\theta \, \psi_1^*(\theta, \varphi) \psi_2(\theta, \varphi) \tag{5.68}$$

where $d\theta \, d\varphi \sin\theta$ is a spherical area element. The remaining integration over polar angle θ defines the inner product for yet unknown functions $P_l^m(\theta)$:

$$\langle P_{l_1}^{m_1} \mid P_{l_2}^{m_2} \rangle = \int\limits_0^\pi d\theta \, \sin\theta \left[P_{l_1}^{m_1}(\theta) \right]^* P_{l_2}^{m_2}(\theta). \tag{5.69}$$

These functions are found by substituting $\psi_{lm}(\theta, \varphi) = P_l^m(\theta) \exp(im\varphi)$ into Eq. 5.64, which results in the following equation:

$$-\left[\frac{1}{\sin\theta} \frac{\partial}{\partial\theta} \left(\sin\theta \frac{\partial}{\partial\theta} \right) + \frac{1}{\sin^2\theta} \frac{\partial^2}{\partial\varphi^2} \right] P_l^m(\theta) \exp(im\varphi) = l(l+1)P(\theta) \exp(im\varphi).$$

Carrying out the differentiation with respect to φ and canceling the exponential factor results in the following equation for $P_l^m(\theta)$:

$$\frac{1}{\sin\theta} \frac{\partial}{\partial\theta} \left(\sin\theta \frac{\partial P_l^m}{\partial\theta} \right) - \frac{m^2}{\sin^2\theta} P_l^m + l(l+1)P_l^m = 0.$$

Do you see now why I kept both indexes l and m in the notation for P_l^m? By introducing the new variable $x = \cos\theta$, this equation can be rewritten as

$$\frac{d}{dx}\left[(1-x^2) \frac{dP_l^m}{dx} \right] + \left[l(l+1) - \frac{m^2}{1-x^2} \right] P_l^m = 0 \tag{5.70}$$

where I used relation $d/d\theta = (dx/d\theta) \, d/dx = -\sin\theta \, d/dx$ and replaced $\sin^2\theta$ with $1 - \cos^2\theta = 1 - x^2$.

This equation is very well known in mathematical physics as *general Legendre equation*, whose solutions can be presented in the form of *associated Legendre functions*, $P_l^m(x) \equiv P_l^m(\cos\theta)$. As is clear from the relation between variables x and $\cos\theta$, functions $P_l^m(x)$ are defined on the interval $x \in [-1, 1]$, where they are orthogonal with the inner product defined as $\int_{-1}^1 P_{l_1}^m(x)P_l^m(x)dx$:

$$\int\limits_{-1}^1 P_{l_1}^m(x)P_l^m(x)dx = \frac{2(l+m)!}{(2l+1)(l-m)!} \delta_{l,l_1}. \tag{5.71}$$

You may want to notice that the substitution of the integration variable $x = \cos\theta$ converts this integral into the form identical to the integral in Eq. 5.69.

The proof of orthogonality of the Legendre functions is fairly standard for differential equations of this kind, and you will benefit from learning how to carry it out. First, copy Eq. 5.70 for $P_{l_1}^m$:

$$\frac{d}{dx}\left[(1-x^2) \frac{dP_{l_1}^m}{dx} \right] + \left[l_1(l_1+1) - \frac{m^2}{1-x^2} \right] P_{l_1}^m = 0. \tag{5.72}$$

Now, multiply Eq. 5.70 by $P_{l_1}^m$ and Eq. 5.72 by P_l^m, and integrate the resulting expressions from -1 to 1:

$$\int_{-1}^{1} P_{l_1}^m \frac{d}{dx}\left[(1-x^2)\frac{dP_l^m}{dx}\right] dx + l(l+1)\int_{-1}^{1} P_{l_1}^m(x)P_l^m(x)dx-$$

$$m^2 \int_{-1}^{1} \frac{P_{l_1}^m(x)P_l^m(x)}{1-x^2}dx = 0$$

$$\int_{-1}^{1} P_l^m \frac{d}{dx}\left[(1-x^2)\frac{dP_{l_1}^m}{dx}\right] dx + l_1(l_1+1)\int_{-1}^{1} P_{l_1}^m(x)P_l^m(x)dx-$$

$$m^2 \int_{-1}^{1} \frac{P_{l_1}^m(x)P_l^m(x)}{1-x^2}dx = 0.$$

Integration of the first terms in both equations by parts yields

$$-\int_{-1}^{1} (1-x^2)\frac{dP_l^m}{dx}\frac{dP_{l_1}^m}{dx}dx + l(l+1)\int_{-1}^{1} P_{l_1}^m(x)P_l^m(x)dx$$

$$- m^2 \int_{-1}^{1} \frac{P_{l_1}^m(x)P_l^m(x)}{1-x^2}dx = 0$$

$$-\int_{-1}^{1} (1-x^2)\frac{dP_l^m}{dx}\frac{dP_{l_1}^m}{dx}dx + l_1(l_1+1)\int_{-1}^{1} P_{l_1}^m(x)P_l^m(x)dx$$

$$- m^2 \int_{-1}^{1} \frac{P_{l_1}^m(x)P_l^m(x)}{1-x^2}dx = 0,$$

and by subtracting these two expressions, you get

$$[l(l+1) - l_1(l_1+1)]\int_{-1}^{1} P_{l_1}^m(x)P_l^m(x)dx = 0.$$

It is quite obvious now that for $l \neq l_1$ this equality can only hold if

$$\int_{-1}^{1} P_{l_1}^m(x)P_l^m(x)dx = 0.$$

The derivation of the normalization coefficient in Eq. 5.71 requires a bit more effort, and I shall leave it for the most curious readers to discover it for themselves (google it!). You can also notice that in the case of functions with equal l and different m, the same line of reasoning results in a different orthogonality condition:

$$\int_{-1}^{1} \frac{P_l^{m_1}(x) P_l^m(x)}{1 - x^2} dx = \begin{cases} 0 & m \neq m_1 \\ \frac{(l+m)!}{m(l-m)!} & m = m_1 \neq 0 \\ \infty & m = m_1 = 0 \end{cases} \tag{5.73}$$

where, again, derivation of the normalization integral lies outside the scope of this text.

The associated Legendre polynomials can be computed using the following expression:

$$P_l^m(x) = (-1)^m \left(1 - x^2\right)^{m/2} \frac{d^{l+m}}{dx^{l+m}} \left(x^2 - 1\right)^l \tag{5.74}$$

where factor $(-1)^m$ is known as Condon-Shortley phase and is sometimes excluded from the definition of $P_l^m(x)$. Equation 5.74 makes sense and gives non-zero results if and only if l and m are integers and $0 \leq l + m \leq 2l \Leftrightarrow -l \leq m \leq l$. The integer part of this statement is obvious—derivatives of fractional order are not something that we can live with at this point. The second part of this statement, which reiterates what we have already learned about the relation between these two quantum numbers in Sect. 3.3.4, can be understood by noticing that function $\left(x^2 - 1\right)^l$ is a polynomial of the order $2l$ and, therefore, can be differentiated no more than $2l$ times before it starts producing zeroes.

Legendre equation 5.70 is invariant (does not change) if you replace m to $-m$. This means that solutions of this equation characterized by m and $-m$ must be proportional to each other. Indeed, one can show that functions defined by Eq. 5.74 satisfy the following important relation:

$$P_l^{-m}(x) = (-1)^m \frac{(l-m)!}{(l+m)!} P_l^m(x). \tag{5.75}$$

Finally, combining Eqs. 5.66 and 5.74 and adding corresponding normalization coefficients, we end up with a set of functions $\psi_{lm} \equiv Y_l^m(\theta, \varphi)$ known as spherical harmonics and defined as

$$Y_l^m(\theta, \varphi) = (-1)^m \sqrt{\frac{2l + 1}{4\pi} \frac{(l-m)!}{(l+m)!}} P_l^m(\cos \theta) e^{im\varphi}. \tag{5.76}$$

In light of the results presented above, the spherical harmonics are obviously orthogonal and normalized:

$$\int\limits_{0}^{\pi} \int\limits_{0}^{2\pi} \left[Y_l^m(\theta, \varphi) \right]^* Y_{l_1}^{m_1}(\theta, \varphi) \sin \theta d\theta d\varphi = \delta_{ll_1} \delta_{mm_1} \qquad (5.77)$$

providing us with the position representation of normalized common eigenvectors of operators $\hat{L}^2$ and $\hat{L}_z$.

I will conclude this section with a brief description of main qualitative properties of the spherical harmonics. Numerous identities and recursion relations involving associated Legendre functions are well documented and are easily available in the literature and on the Internet. However, it is important to have a qualitative understanding of how spherical harmonics behave off the top of one's head.

The first thing to notice is the symmetry of the spherical harmonics upon inversion of the position vector on the sphere with respect to the origin of the coordinate system: $r \to -r$. This corresponds to the transformation of the angular spherical coordinates $\theta \to \pi - \theta$, $\varphi \to \varphi + \pi$. Upon this transformation $\exp(im\varphi) \to (-1)^m \exp(im\varphi)$, while the argument $x = \cos\theta$ of the associated Legendre function, $P_l^m(\cos\theta)$, transforms as $\cos\theta \to \cos(\pi - \theta) = -\cos\theta$, i.e., we are dealing here with inversion $x \to -x$. Associated Legendre functions have a definite parity: they are either even (do not change) or odd (change the sign) when their argument changes the sign. This is quite obvious from Eq. 5.74: replacing $x \to -x$ does not change the function being differentiated or the factor preceding differentiation, while the derivatives with respect to x change their sign with each differentiation. It is obvious, therefore, that

$$P_l^m(-x) = (-1)^{l+m} P_l^m(x). \qquad (5.78)$$

Combining this result with the transformation property of $\exp(im\varphi)$, we have for the spherical harmonics

$$Y_l^m(\pi - \theta, \varphi + \pi) = (-1)^l Y_l^m(\theta, \varphi) \qquad (5.79)$$

which means that the spherical harmonics have definite parity: they are either even with respect to inversion (for even values of l) or odd, if l is an odd number. This behavior is consistent with the fact that the operator of the orbital angular momentum $\hat{L} = \hat{r} \times \hat{p}$ is invariant with respect to the parity transformation, and, therefore, its eigenvectors must also be eigenvectors of the parity operator, i.e., have a definite parity.

It is also important to have a picture of dependence of the spherical harmonics upon its arguments. Dependence on azimuth angle φ is trivial: the real and imaginary parts of the spherical harmonics oscillate with frequency m, but these oscillations are not really significant, unless we are dealing with a superposition state comprised of several spherical harmonics with different azimuthal numbers. For a single spherical harmonics, relevant properties are often described by its absolute values $\left| Y_l^m(\theta, \varphi) \right|^2$, which lose all dependence on φ. Dependence on polar angle contained in $P_l^m(\cos\theta)$ is a more interesting matter and is determined by values of both quantum numbers l and m separately, as well as by their difference $l - m$. For instance, for $m = l$, it is easy to see that

$$P_l^l(\cos\theta) \propto \sin^l\theta,$$

which takes zero values at $\theta = 0, \pi$ (two poles of the sphere) and has a single maximum at the equator $\theta = \pi/2$. The width of the maximum (loosely defined) becomes smaller with increasing l (the function decreases more rapidly away from equator for larger l). If one likes pseudoclassical mind helpers (I would not even call them "analogies"), one can think about a particle rotating around the equator with its angular momentum pointing in the polar direction. The larger the angular momentum is, the more torque would be required to turn it away from the poles, which can be kind of loosely interpreted as a smaller probability for a particle to deviate from the equatorial trajectory. But, please, do not take these pseudoclassical mumbo jumbo too seriously.

The case of $m = 0$ corresponds to the classical angular momentum lying in the equatorial plane, while respective spherical harmonics are reduced to regular Legendre polynomials:

$$P_l(x) = \frac{d^l}{dx^l}(x^2 - 1)^l.$$

These are the only spherical harmonics which do not have zeroes at the poles of the sphere ($x = \pm 1$, or $\theta = 0, \pi$), but it has zeroes between the poles, whose number is equal to the orbital number l. Obviously, the number of minimums and maximums of these functions is always equal to $l - 1$ (the only exception is $l = 0$, when we are dealing with a constant). In the case of generic values of $m \neq 0$, the spherical harmonics vanish at the poles, and the number of their nods in the polar direction is equal to $l - m$. In my opinion, mastering the provided information will help you not only to have a qualitative feeling for various expressions and phenomena involving spherical harmonics but also to make quite an impression at a cocktail party. To help you with visualizing these properties, I plotted graphs of the associated Legendre polynomials with $l = 3$ in Fig. 5.3. To make the picture prettier, I normalized all functions in the plot to bring their maximum values closer to each other; obviously this procedure did not change their qualitative behavior.

5.2 Representations in Discrete Basis

Now let's talk about the representation of abstract vectors in discrete bases. Equation 3.39, which represents vector $|\alpha\rangle$ in a basis $|\chi_n\rangle$, establishes a one-to-one correspondence between the vector and a set of coefficients a_n. These coefficients are a discrete analog of functions representing vectors in continuous

Fig. 5.3 Graphs of
associated Legendre
polynomials with $l = 3$ and
$0 \leq m \leq 3$

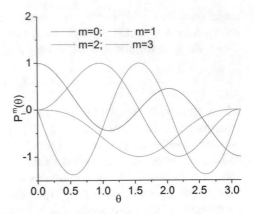

basis introduced in the previous section and can be arranged in the form of a
column vector. Thus, in this case we are representing the abstract vector space by
a space of column vectors with all the rules of matrix addition and multiplication
defined for these objects. The Hermitian conjugation in this space was discussed in
Sect. 2.2.2 and includes transitioning to the adjoint space inhabited by row vectors
with complex-conjugated elements:

$$\langle \alpha | = \sum_n a_n^* \langle \chi_n |.$$

The inner product now becomes a standard matrix multiplication between a row
vector on the left and a column vector on the right:

$$\langle \alpha | \beta \rangle = \begin{bmatrix} a_1^* & a_2^* & \cdots & a_N^* & \cdots \end{bmatrix} \begin{bmatrix} b_1 \\ b_2 \\ \vdots \\ b_N \\ \vdots \end{bmatrix} = \sum_{i=1}^{\infty} a_i^* b_i, \tag{5.80}$$

where b_n are coefficients in the expansion of ket $|\beta\rangle$ in the same basis, while the
outer or tensor product $|\alpha\rangle \langle\beta|$ is represented by a matrix formed according to the
rules of the matrix tensor product, Eq. 2.16:

$$S_{nm}^{(\alpha,\beta)} = \begin{bmatrix} a_1 \\ a_2 \\ \vdots \\ a_N \\ \vdots \end{bmatrix} \begin{bmatrix} b_1^* & b_2^* & \cdots & b_N^* & \cdots \end{bmatrix} = \begin{bmatrix} a_1 b_1^* & a_1 b_2^* & \cdots & a_1 b_N^* & \cdots \\ a_2 b_1^* & \ddots & \ddots & a_2 b_N^* & \ddots \\ \vdots & \ddots & \ddots & \vdots & \ddots \\ a_N b_1^* & a_N b_2^* & \cdots & a_N b_N^* & \ddots \\ \vdots & \ddots & \ddots & \ddots & \cdots \end{bmatrix}. \tag{5.81}$$

Due to the normalization requirement accepted for the state vectors, the expansion coefficients obey the following obvious "sum" rule:

$$\sum_n |a_n|^2 = 1 \qquad (5.82)$$

which is, again, a discrete analog of Eq. 5.4.

If a space of states of a given quantum system can be fully described by a discrete basis, these states can always be presented in the form of column and row vectors reducing the problem to that of a matrix algebra (remember Heisenberg's matrix mechanics—this is where it finds its roots). The main difference, of course, is that in standard linear algebra problems, the dimension of the space is always finite, while normally spaces of quantum mechanical states have infinite dimensionality. This creates a number of technical problems of mathematical nature, but we shall let mathematicians to worry about them. At any rate, in most practical applications of quantum theory, you wouldn't have to deal with the entire infinitely dimensional space of states. Usually, it is possible to find a way to restrict attention to a much smaller (sometimes just two-dimensional) subspace using certain physically meaningful assumptions about hierarchy of interactions relevant for the problem under study.

To have you started, consider this simplest of simplest example, which, however, often gives students a headache.

Example 16 (A Basis Vector in Its Own Basis) This example deals with the following question: what is a representation of a vector in a basis to which this vector itself belongs? In other words, if $|\chi_{\tilde{n}}\rangle$ is one of the set of orthogonal normalized vectors $|\chi_n\rangle$, $n = 1, 2 \cdots$, which column vector will represent it in the basis formed by these vectors? Even though the answer to this question is almost trivial, it never fails to confuse students. Assume, for instance, that $\tilde{n} = 1$. In this case I have $|\chi_1\rangle = 1\,|\chi_1\rangle + 0\,|\chi_2\rangle + 0\,|\chi_3\rangle + \cdots$. Obviously, the corresponding column vector is

$$\begin{bmatrix} 1 \\ 0 \\ 0 \\ \vdots \end{bmatrix}.$$

Considering $\tilde{n} = 2$, I will similarly find that the column representing this vector contains unity in the second position and zeroes everywhere else. This pattern, of course, repeats itself for all other elements of the basis: any basis vector $|\chi_{\tilde{n}}\rangle$ is represented in the basis it is the element of, by a column, where all components but one are zeroes, and the only component in the $\tilde{n}$-th place is unity.

Now, if column vectors can represent vector states, it is almost obvious that operators must be represented by matrices, in which case the word "act" would mean matrix multiplication. Matrices multiply column vectors from the left and row vectors from the right. The main question is how to construct a matrix representing a given operator in a chosen basis. To answer this question, I will again rely on the completeness relation (its discrete basis reincarnation, Eq. 3.42) for the basis vectors $|\chi_n\rangle$. Insertion of this relation into $|\beta\rangle = \hat{T}|\alpha\rangle$ yields

$$\sum_{n=0}^{\infty} |\chi_n\rangle \langle \chi_n |\beta\rangle = \sum_{n=0}^{\infty} |\chi_n\rangle \langle \chi_n| \hat{T} \sum_{m=0}^{\infty} |\chi_m\rangle \langle \chi_m |\alpha\rangle$$

$$= \sum_{n=0}^{\infty} \sum_{m=0}^{\infty} \langle \chi_n| \hat{T} |\chi_m\rangle \langle \chi_m |\alpha\rangle |\chi_n\rangle .$$

Taking into account that coefficients b_n are given by $b_n = \langle \chi_n |\beta\rangle$ and coefficients a_n are $a_n = \langle \chi_m |\alpha\rangle$, I transform the previous equation into

$$\sum_{n=0}^{\infty} b_n |\chi_n\rangle = \sum_{n=0}^{\infty} \sum_{m=0}^{\infty} \langle \chi_n| \hat{T} |\chi_m\rangle a_m |\chi_n\rangle .$$

Now, thanks to the linear independence of the basis vectors, I can simply equate the coefficients in front of each of $|\chi_n\rangle$ separately:

$$b_n = \sum_{m=0}^{\infty} \langle \chi_n| \hat{T} |\chi_m\rangle a_m. \tag{5.83}$$

This expression can be rewritten in the matrix form as

$$b = T \cdot a$$

where I am using bold Latin letters to denote columns and matrices representing vectors and operators in a given discrete basis. This means that the required matrix representation of the given operator is

$$T_{nm} = \langle \chi_n| \hat{T} |\chi_m\rangle . \tag{5.84}$$

To illustrate an application of this result, consider the matrix of the operator $\hat{S} = |\alpha\rangle \langle \beta|$ obtained by the outer product of two vectors. Using Eq. 5.84, you immediately find

$$S_{mn} = \langle \chi_n |\alpha\rangle \langle \beta| \chi_m\rangle = a_m b_n^*$$

with full agreement with the result obtained using standard matrix definition of the outer product, Eq. 5.81.

The equation for eigenvectors $\hat{T}|\alpha\rangle = \lambda |\alpha\rangle$ in the matrix representation is reduced to the matrix equation:

$$\sum_{m=0}^{\infty} T_{nm} a_m = \lambda a_n$$

which can be rewritten as

$$\sum_{m=0}^{\infty} (T_{nm} - \lambda \delta_{nm}) a_m = 0. \tag{5.85}$$

This is essentially a shortcut notation for the system of uniform linear equations, which has nontrivial (meaning non-zero) solutions, only if the determinant $\|T_{nm} - \lambda \delta_{nm}\|$ vanishes (Cramer's rule that had already been mentioned earlier). If you go back to Sect. 3.2.3, you will find examples of eigenvector and eigenvalue calculations with matrices.

The matrix representation of operators is practically useful only if you know how the operator acts on the vectors of the chosen basis. If an operator in question is built out of basis vectors, the problem is resolved almost trivially. For instance, the projection operators introduced in Eq. 3.41 have a simple matrix representation in the same basis in which it is defined:

$$P_{km}^{(n)} = \langle \chi_k | \chi_n \rangle \langle \chi_n | \chi_m \rangle = \delta_{kn} \delta_{nm}$$

which is a matrix with a single non-zero element $k = m = n$ on a main diagonal. You can find other examples of matrix representations for operators of this kind, in the exercises in this chapter.

In most cases the issue of finding how operators act on the basis vectors is not that trivial. Often it is resolved by using position representation of operators and the basis vectors. This approach works especially well for the class of operators, which can be presented as a combination of position and momentum operators. This class includes many important operators, but not all of them.

5.2.1 Discrete Representation from a Continuous One

To illustrate this point, let me consider an example of a single particle of mass m_e allowed to move freely along a linear segment of finite length L. The probability that the particle's position can be anywhere outside of this segment is assumed to be zero. This condition is most naturally expressed in the position representation, where Hamiltonian, which contains only kinetic energy term, takes the form of

$$\hat{H} = \frac{\hat{p}_x^2}{2m} = -\frac{\hbar^2}{2m} \frac{d^2}{dx^2}. \tag{5.86}$$

The confinement of the particle inside the specified linear segment is formally expressed by the requirement that the wave function $\psi(x)$ representing states of the system is equal to zero outside of the allowed interval. The continuity of the wave function then requires that it also vanishes at the terminal points of this interval. Choosing the origin of a coordinate system at the left end of the allowed interval, and assigning coordinate $x = L$ to its right end, I can express the confinement conditions by requiring that the wave function vanished at both ends of the interval:

$$\psi(0) = \psi(L) = 0. \tag{5.87}$$

It is easy to check that Schrödinger equation 5.37 for a free particle ($V(x) = 0$) has two linearly independent solutions $\psi_+(x) = \exp(ikx)$ and $\psi_-(x) = \exp(-ikx)$, where $k = \sqrt{2mE}/\hbar$. Now I need to construct a linear combination of these functions obeying the confinement conditions, Eq. 5.87. Beginning with a general solution

$$\psi(x) = Ae^{ikx} + Be^{-ikx},$$

I find that requirement $\psi(0) = 0$ yields $A + B = 0$, which allows me to write the wave function as

$$\psi(x) = A \sin kx.$$

(I used Euler's formula $\sin kx = (\exp(ikx) - \exp(-ikx))/2i$ and incorporated constant $2i$ into coefficient A.) The condition at $x = L$ yields

$$A \sin kL = 0$$

with two possible ways to fulfill it. One is to make $A = 0$, in which case the entire wave function vanishes, and we definitely do not want this to happen. Thus, you are stuck with the only other option, namely, to require that

$$kL = \pi n, \ n = 1, 2 \cdots .$$

This result means that the states of the system considered in this example can only be presented by a discrete set of wave functions:

$$\psi_n(x) = A \sin k_n x$$

characterized by parameter

$$k_n = \frac{\pi n}{L} \tag{5.88}$$

with corresponding discrete energy levels

$$E_n = \frac{\hbar^2 \pi^2 n^2}{2mL^2}.$$
(5.89)

The appearance of the discrete spectrum is not surprising here, of course, since the classical motion in this example is clearly bound. The remaining unknown coefficient A remains unknown at this point—it cannot be fixed by the boundary condition, which is a fairly typical situation in problems of this kind. I, however, have one additional weapon at my disposal—the normalization condition, which in this case reads as (using standard definition of the inner product for the square-integrable functions)

$$\int_{-\infty}^{\infty} |\psi(x)|^2 \, dx = |A|^2 \int_0^L \sin^2 k_n x \, dx = 1 \Rightarrow A = \sqrt{\frac{2}{L}}$$

where at the last step, I chose A to be a real positive quantity. This choice while pleasing to the eye does not make any difference since normalization condition only defines A up to an arbitrary phase factor of the form $\exp(i\varphi)$ in alignment with the already mentioned general principle that vectors representing quantum states are always defined only up to a phase.

The system of wave functions

$$\psi_n(x) = \sqrt{\frac{2}{L}} \sin \frac{\pi n x}{L}$$
(5.90)

forms a normalized orthogonal basis, which can be used to present any other wave function defined on the interval $x \in [0, L]$ (one can recognize here just a Fourier series expansion for a function defined on a finite interval). This basis can also be used to represent various operators acting on such functions. For instance, Hamiltonian, Eq. 5.86, in this basis is represented by an infinite diagonal matrix:

$$H_{mn} = -\frac{\hbar^2}{2m} \frac{2}{L} \int_0^L \sin \frac{\pi m x}{L} \frac{d^2}{dx^2} \sin \frac{\pi n x}{L} dx =$$

$$\frac{\hbar^2 \pi^2 n^2}{2mL^2} \frac{2}{L} \int_0^L \sin \frac{\pi m x}{L} \sin \frac{\pi n x}{L} dx = \frac{\hbar^2 \pi^2 n^2}{2mL^2} \delta_{mn}.$$
(5.91)

Now, assume that the particle that you follow is also subjected to an external uniform electric field (with all other conditions and limitations intact). This will add a potential energy term to the Hamiltonian of the form $V(x) = eFx$, where e is the absolute value of the particle's charge, presumed to be negative, and F is the magnitude of the field. Now, I want you to try to present the new Hamiltonian of the particle:

$$\hat{H} = \frac{\hat{p}_x^2}{2m} + eFx$$

in the same basis of functions $\psi_n(x)$ defined in Eq. 5.90. The resulting matrix would have the diagonal part given in Eq. 5.91, and the part, which can be written as eFx_{mn}, where

$$x_{mn} = \frac{2}{L} \int_0^L x \sin \frac{\pi n x}{L} \sin \frac{\pi m x}{L} dx =$$

$$\frac{1}{L} \int_0^L x \left[\cos \frac{\pi (n-m) x}{L} - \cos \frac{\pi (n+m) x}{L} \right] dx = \tag{5.92}$$

$$\frac{1}{L} \frac{L}{\pi} \left[\frac{x}{n-m} \sin \frac{\pi (n-m) x}{L} - \frac{x}{n+m} \sin \frac{\pi (n+m) x}{L} \right]_0^L -$$

$$\frac{1}{L} \frac{L}{\pi} \left[\frac{1}{n-m} \int_0^L \sin \frac{\pi (n-m) x}{L} dx - \frac{1}{n+m} \int_0^L \sin \frac{\pi (n+m) x}{L} dx \right] =$$

$$\frac{L}{\pi^2} \frac{1}{(n-m)^2} \cos \frac{\pi (n-m) x}{L} \Big|_0^L - \frac{L}{\pi^2} \frac{1}{(n+m)^2} \cos \frac{\pi (n+m) x}{L} \Big|_0^L =$$

$$\frac{L}{\pi^2} [(-1)^{n-m} - 1] \frac{4nm}{(n^2 - m^2)^2}, \quad n \neq m. \tag{5.93}$$

The diagonal element of this matrix, which is just an expectation value of the coordinate, is easily found to be (from the first line of Eq. 5.93) $x_{nn} = L/2$, which has an obvious physical meaning. The total Hamiltonian in the representation based on functions defined in Eq. 5.90 is now an infinite nondiagonal matrix:

$$H_{mn} = \left(\frac{\hbar^2 \pi^2 n^2}{2mL^2} + \frac{eFL}{2} \right) \delta_{mn} + \frac{eFL}{\pi^2} [(-1)^{n-m} - 1] \frac{4nm}{(n^2 - m^2)^2} \tag{5.94}$$

where the second term contributes only to the nondiagonal elements. The electric field-related correction to the diagonal elements of the Hamiltonian is just a constant and can be eliminated by choosing a different zero level for the energies, for instance, by writing the electric field potential as $eF(x - L/2)$.

This example illustrates a rather general situation: often in order to find a representation of an operator in one basis, we have to use its known representation in a different basis. This approach works especially well with observables that can be expressed as combinations of position and momentum operators, whose representations in continuous bases were discussed above. This approach often leads to nondiagonal matrices, and finding the eigenvalues and eigenvectors of the operator of interest is reduced to finding eigenvalues and eigenvectors of the resulting matrix. In many cases this cannot be done exactly because the dimensionality of the resulting matrices can be infinite, but it is often possible to truncate them and solve the problem approximately. How this is done practically will be discussed in a separate chapter.

5.2.2 Transition from One Discrete Basis to Another

Quite often you will find yourself in a situation when having found (or being given) matrix representation of an operator in one discrete basis, you will need to find an equivalent matrix representing this operator in a different basis. Here I will show how this can be done.

So, let's assume that you have an operator $\hat{T}$ and a system of basis vectors $\left|\chi_m^{(old)}\right\rangle$. The representation of this operator in this basis, as we have already established, is given by a matrix:

$$T_{mn}^{(old)} = \left\langle \chi_m^{(old)} \middle| \hat{T} \middle| \chi_n^{(old)} \right\rangle.$$

However, I would like to re-derive this expression in a slightly different way. Let me multiply the operator $\hat{T}$ by two unity operators expressed by the completeness relation, Eq. 3.42, formed with the vectors of this basis:

$$\hat{T} = \sum_{n=0}^{\infty} \sum_{m=0}^{\infty} \left|\chi_m^{(old)}\right\rangle \left\langle \chi_m^{(old)} \middle| \hat{T} \middle| \chi_n^{(old)} \right\rangle \left\langle \chi_n^{(old)} \right| =$$

$$\sum_{n=0}^{\infty} \sum_{m=0}^{\infty} T_{mn}^{(old)} \left|\chi_m^{(old)}\right\rangle \left\langle \chi_n^{(old)} \right|. \qquad (5.95)$$

This representation of an operator in terms of a matrix and operators $\left|\chi_m^{(old)}\right\rangle \left\langle \chi_n^{(old)} \right|$ is akin to the expansion of a vector into a linear combination of basis vectors. Now, let me assume that I have another basis $\left|\chi_m^{(new)}\right\rangle$, and I want to relate the matrix of the operator in this basis $T_{mn}^{(new)}$ to matrix $T_{mn}^{(old)}$. To achieve this goal, let me express the matrix $T_{mn}^{(new)}$ using Eq. 5.95:

$$T_{kl}^{(new)} = \sum_{n=0}^{\infty} \sum_{m=0}^{\infty} T_{mn}^{(old)} \left\langle \chi_k^{(new)} \middle| \chi_m^{(old)} \right\rangle \left\langle \chi_n^{(old)} \middle| \chi_l^{(new)} \right\rangle.$$

This can be rewritten with the help of two new matrices:

$$U_{nl} = \left\langle \chi_n^{(old)} \middle| \chi_l^{(new)} \right\rangle \tag{5.96}$$

and

$$\tilde{U}_{km} = \left\langle \chi_k^{(new)} \middle| \chi_m^{(old)} \right\rangle$$

as

$$T_{kl}^{(new)} = \sum_{n=0}^{\infty} \sum_{m=0}^{\infty} \tilde{U}_{km} T_{mn}^{(old)} U_{nl}.$$

(Note the position of the indexes in this expression, which adhere to the regular rule for the matrix multiplication.) Now I need to figure out how to construct matrices U and $\tilde{U}$ and their relation to each other. Let me first perform complex conjugation of each element of matrix $\tilde{U}_{km}$ and take advantage of the main property of the inner product, Eq. 2.19:

$$\tilde{U}_{km}^* = \left\langle \chi_k^{(new)} \middle| \chi_m^{(old)} \right\rangle^* = \left\langle \chi_m^{(old)} \middle| \chi_k^{(new)} \right\rangle = U_{mk}.$$

Thus I can see that matrix $\tilde{U}$ can be obtained from U by complex conjugation and transposition, or, expressing this in fewer words, $\tilde{U}$ is a Hermitian conjugate of U: $\tilde{U} = U^\dagger$, and the matrix transformation rule can be presented as

$$T_{kl}^{(new)} = \sum_{n=0}^{\infty} \sum_{m=0}^{\infty} U_{km}^\dagger T_{mn}^{(old)} U_{nl}. \tag{5.97}$$

Now, let me focus on one particular column of matrix U, say, column l_0. Then you can easily recognize that quantities $\left\langle \chi_n^{(old)} \middle| \chi_{l_0}^{(new)} \right\rangle$ are nothing but coefficients of expansion of the new basis vector $\left| \chi_{l_0}^{(new)} \right\rangle$ in the old basis:

$$\left| \chi_{l_0}^{(new)} \right\rangle = \sum_n \left| \chi_n^{(old)} \right\rangle \left\langle \chi_n^{(old)} \middle| \chi_{l_0}^{(new)} \right\rangle,$$

which gives a simple recipe for preparing matrix U: find representation of the nth vector of a new basis in the old one and use the corresponding coefficients as a n-th column of matrix U. Let me illustrate this rule with a simple example.

Example 17 (Transformation to a New Basis) Consider a Hermitian matrix:

$$\begin{bmatrix} 1 & i \\ -i & -1 \end{bmatrix}$$

and rewrite it in the basis of its own eigenvectors.

Solution

First I need to find these eigenvectors, which are given by equation

$$\begin{bmatrix} 1 & i \\ -i & 1 \end{bmatrix} \begin{bmatrix} a_1 \\ a_2 \end{bmatrix} = \lambda \begin{bmatrix} a_1 \\ a_2 \end{bmatrix}.$$

The corresponding eigenvalues are found from

$$(1 - \lambda)^2 - 1 = 0 \Rightarrow$$
$$-2\lambda + \lambda^2 = 0 \Rightarrow$$
$$\lambda_1 = 0, \ \lambda_2 = 2.$$

Now I can find two eigenvectors:
For $\lambda_1 = 0$, I have

$$a_1 + ia_2 = 0 \Rightarrow |0\rangle = \frac{1}{\sqrt{2}} \begin{bmatrix} 1 \\ i \end{bmatrix}$$

where I used $|0\rangle$ as a notation for a normalized eigenvector belonging to $\lambda_1 = 0$. You can verify that this vector is indeed normalized.
For $\lambda_2 = 2$, the eigenvector equations become

$$a_1 + ia_2 = 2a_1 \Rightarrow |2\rangle = \frac{1}{\sqrt{2}} \begin{bmatrix} 1 \\ -i \end{bmatrix}.$$

What you need to realize now (quite obvious, but always gives students a shudder) is that the numbers in these columns are the coefficients in the representation of the new basis (vectors $|0\rangle$ and $|2\rangle$) in terms of the vectors of the old basis. Thus, the transformation matrix U can be generated as

$$U = \frac{1}{\sqrt{2}} \begin{bmatrix} 1 & 1 \\ i & -i \end{bmatrix}$$

and its Hermitian conjugate matrix as

$$U^\dagger = \frac{1}{\sqrt{2}} \begin{bmatrix} 1 & -i \\ 1 & i \end{bmatrix}.$$

Plugging these matrices in the transformation rule, Eq. 5.97, I get

$$\frac{1}{2} \begin{bmatrix} 1 & -i \\ 1 & i \end{bmatrix} \begin{bmatrix} 1 & i \\ -i & 1 \end{bmatrix} \begin{bmatrix} 1 & 1 \\ i & -i \end{bmatrix} =$$

$$\frac{1}{2}\begin{bmatrix} 1 & -i \\ 1 & i \end{bmatrix}\begin{bmatrix} 0 & 2 \\ 0 & -2i \end{bmatrix} = \begin{bmatrix} 0 & 0 \\ 0 & 2 \end{bmatrix},$$

which is exactly what you should have expected: a matrix in the basis of its own eigenvectors is diagonal with eigenvalues along the main diagonal.

Sometimes the transformation rule connecting representation of operators in different bases is presented in an alternative form

$$T_{kl}^{(new)} = \sum_{n=0}^{\infty}\sum_{m=0}^{\infty} \tilde{U}_{km} T_{mn}^{(old)} \tilde{U}_{nl}^{\dagger} \tag{5.98}$$

with matrix $\tilde{U}_{nl}$ defined as

$$\tilde{U}_{nl} = \langle \chi_n^{(new)} | \chi_l^{(old)} \rangle. \tag{5.99}$$

Complex conjugation of Eq. 5.99 yields

$$\tilde{U}_{nl}^* = \langle \chi_n^{(new)} | \chi_l^{(old)} \rangle^* = \langle \chi_l^{(old)} | \chi_n^{(new)} \rangle = U_{ln}$$

Performing matrix transposition and recalling that complex conjugation plus transposition yields Hermitian conjugation, you can see that

$$\tilde{U} = U^{\dagger}$$

and that Eq. 5.98 and Eq. 5.97 are equivalent to each other.

The transformation matrix in the form of Eq. 5.99 appears naturally when one is looking for the transformation between components of the same vector written in two different bases. Indeed, consider a vector $|\alpha\rangle$ represented in two different bases as

$$|\alpha\rangle = \sum_l a_l^{(old)} \left| \chi_l^{(old)} \right\rangle = \sum_l a_l^{(new)} \left| \chi_l^{(new)} \right\rangle.$$

The simplest way to express coefficients $a_l^{(new)}$ in terms of coefficients $a_l^{(old)}$, which is the goal of this exercise, is to premultiply the expression above by the bra-vector $\langle \chi_m^{(new)} |$ and take advantage of the orthogonality of the basis vectors. This yields

$$a_m^{(new)} = \sum_l \langle \chi_m^{(new)} | \chi_l^{(old)} \rangle a_l^{(old)} = \sum_l \tilde{U}_{ml} a_l^{(old)} = \sum_l U_{ml}^{\dagger} a_l^{(old)}.$$

What is left for me to do now is to show that matrix U defined by Eq. 5.96 is unitary. To this end I need to compute the product of two matrices U_{nm} and $U_{ml}^{\dagger}$, using standard matrix multiplication rule:

$$\left(UU^{\dagger}\right)_{nl} = \sum_{m} U_{nm} U_{ml}^{\dagger}.$$

Substituting here Eq. 5.96 I can write

$$\left(UU^{\dagger}\right)_{nl} = \sum_{m} \left\langle \chi_{n}^{(old)} \middle| \chi_{m}^{(new)} \right\rangle \left\langle \chi_{m}^{(new)} \middle| \chi_{l}^{(old)} \right\rangle = \left\langle \chi_{n}^{(old)} \middle| \chi_{l}^{(old)} \right\rangle = \delta_{nl},$$

where I replaced the sum over m with a unity operator because it is again just a completeness condition and used the orthonormalization of the vectors of the basis to replace their inner product with Kronecker's delta-symbol. This calculation reveals that $U^{\dagger} = U^{-1}$, which is the definition of the unitary matrix. One should not be surprised that the transformation of the vector components from one basis to another is provided by a unitary matrix. Indeed, such a transformation clearly should not change the norm of the vector, and it is, indeed, one of the important properties of the unitary operators.

5.2.3 Spin Operators

The approach to generating representation of operators outlined in the previous section would not work for operators which cannot be built out of momentum and position. If, however, you somehow know the eigenvalues of the operator in question (most likely this knowledge comes by distilling empirical facts), you can construct the matrix of this operator in the basis of its own eigenvectors. Indeed, if $|\chi_m\rangle$ is the eigenvector of $\hat{T}$, corresponding to eigenvalue t_m, i.e., $\hat{T}|\chi_m\rangle = t_m|\chi_m\rangle$, then Eq. 5.84 immediately gives

$$T_{nm} = \langle \chi_n| \hat{T} |\chi_m \rangle = t_m \langle \chi_n| \chi_m \rangle = t_m \delta_{nm}.$$

Thus, any operator in the basis of its own eigenvectors is presented by a diagonal matrix with eigenvalues along the main diagonal. Unfortunately, we often have to deal with a set of non-commuting operators, only one of which can be presented by a diagonal matrix. The question then remains how to generate a matrix representation of other non-commuting operators in the same basis. Fortunately, in all practical situations, this problem can be solved if one knows commutation relations between relevant operators. I will illustrate this approach by considering representation of angular momentum operators in the situation when eigenvalues of the z-component of the angular momentum can only take two values $+\hbar/2$ and $-\hbar/2$. Quantum numbers m and l introduced in Sect. 3.3.4 take in this case values $\pm1/2$ and $1/2$, respectively. In Sect. 5.1.4, you saw that the orbital angular momentum, which is constructed of position and momentum operators, admits only integer values for these numbers. The suggested half-integer values, which are allowed by the algebraic properties of these operators, can, therefore, correspond only to a very special angular momentum of electrons not related to their orbital motion. This intrinsic angular momentum is known as *spin*. Leaving more detailed discussion of this quantity till later, here let's just accept its existence and use it to illustrate a method of generating matrix representation of operators, which do not have a position or momentum representation.

To distinguish between spin and orbital angular momentum, I will introduce special notations for the former designating the respective operators as $\hat{S}_x$, $\hat{S}_y$, and $\hat{S}_z$, which have the same meaning as operators $\hat{L}_x$, L_y, and L_z of Sect. 3.3.4. Accordingly, I will replace the quantum number l with s and m with m_s. It is important to realize from the outset that, while orbital quantum number l is allowed to take any integer values, the value of the respective spin number s is fixed at $1/2$ and cannot be changed—it is an intrinsic property of electrons just like its mass or charge. Thus, the only quantum number which can be used to distinguish between different spin states is m_s.

Since m_s takes on only two distinct values, there exist only two respective states described by eigenvectors of operator $\hat{S}_z$. Thus, the space occupied by different spin states is two-dimensional, and the respective vectors are represented by 2×1 column vectors, and operators are represented by 2×2 matrices. In the basis of its own eigenvectors, $\hat{S}_z$ is simply a diagonal matrix:

$$S_z = \begin{bmatrix} \frac{\hbar}{2} & 0 \\ 0 & -\frac{\hbar}{2} \end{bmatrix}, \tag{5.100}$$

while the states take the form of columns

$$|1/2\rangle = \begin{bmatrix} 1 \\ 0 \end{bmatrix}, \ |-1/2\rangle = \begin{bmatrix} 0 \\ 1 \end{bmatrix}, \tag{5.101}$$

where I chose to numerate the state corresponding to the positive eigenvalue as first. (This choice determines the positions of negative and positive elements in the matrix S_z and the ones and zeroes in the corresponding columns.) An arbitrary state in the space of spin states can be written down as a linear combination of the basis vectors:

$$|\chi\rangle = a \begin{bmatrix} 1 \\ 0 \end{bmatrix} + b \begin{bmatrix} 0 \\ 1 \end{bmatrix}. \tag{5.102}$$

The result expressed by Eq. 5.100 is somewhat obvious, and our main task is to find matrices realizing a representation of two remaining components of the spin angular momentum, $\hat{S}_x$ and $\hat{S}_y$ in this basis. (Operator $\hat{S}^2$ in this instance is trivial— it is diagonal with identical diagonal elements equal to $\hbar^2 s(s + 1) = 3\hbar^2/4$, so it is proportional to an identity matrix.)

I begin solving this problem by focusing on operators $\hat{S}_\pm = \hat{S}_x \pm i\hat{S}_y$, which are spin analogs of the ladder operators $\hat{L}_\pm$ introduced in Eqs. 3.64 and 3.65. Since I postulated that spin operators obey the same commutation relations as operators

$\hat{L}_{x,y,z}$, I can use the results obtained for these operators, in particular Eq. 3.75 describing how operator $\hat{L}_+$ acts on eigenvectors of $\hat{L}_z$. Adapting this equation to the case of spin states, I can write

$$\hat{S}_+ |s, m_s\rangle = \hbar \sqrt{\frac{3}{4} - m_s (m_s + 1)} |s, m_s + 1\rangle \qquad (5.103)$$

where I took into account that $s = 1/2$. Applying this equation to the only two existing states $|1/2\rangle$ and $|-1/2\rangle$ (I dropped quantum number s, because it never changes), I have

$$\hat{S}_+ |1/2\rangle = 0$$
$$\hat{S}_+ |-1/2\rangle = \hbar |1/2\rangle \qquad (5.104)$$

from which you can immediately infer that the matrix representation of $\hat{S}_+$ is

$$S_+ = \hbar \begin{bmatrix} 0 & 1 \\ 0 & 0 \end{bmatrix}. \qquad (5.105)$$

The matrix representation for the lowering operator $\hat{S}_-$ can be derived in a similar way using Eq. 3.76, which yields

$$\hat{S}_- |1/2\rangle = \hbar |-1/2\rangle$$
$$\hat{S}_- |-1/2\rangle = 0, \qquad (5.106)$$

but it is much faster simply to recall that $\hat{S}_- = \hat{S}_+^\dagger$, so that the respective matrix is obtained by matrix transposition and complex conjugation of Eq. 5.105:

$$S_- = \hbar \begin{bmatrix} 0 & 0 \\ 1 & 0 \end{bmatrix}. \qquad (5.107)$$

Now, using the definition of the ladder operators, you can write for $\hat{S}_x$ and $\hat{S}_y$:

$$\hat{S}_x = \frac{1}{2} \left(\hat{S}_+ + \hat{S}_- \right) \qquad (5.108)$$

$$\hat{S}_y = \frac{1}{2i} \left(\hat{S}_+ - \hat{S}_- \right) \qquad (5.109)$$

which together with Eqs. 5.105 and 5.107 generate the required matrices:

$$S_x = \frac{\hbar}{2} \begin{bmatrix} 0 & 1 \\ 1 & 0 \end{bmatrix} \qquad (5.110)$$

$$S_y = \frac{\hbar}{2} \begin{bmatrix} 0 & -i \\ i & 0 \end{bmatrix}. \qquad (5.111)$$

Equations 5.110 and 5.111 provide the solution to the problem of finding the matrix representation for operators, which cannot be reduced to combinations of the position and momentum. As you can see, the commutation relations played the crucial role in solving this problem.

5.3   Problems

Section 5.1.1

Problem 54 Reproduce calculations leading to Eq. 5.16 for the momentum representation of the coordinate.

Problem 55 Derive Eq. 5.20 generalizing the approach that led to Eq. 5.15 in Sect. 5.1.1.

Problem 56 Derive Eq. 5.24 using the same method which I used deriving Eq. 5.23 (do not attempt to simply invert the previous equation).

Problem 57 Assuming that function $\chi_s(q)$ presenting eigenvectors of operator $\hat{S}$ in the basis of operator $\hat{Q}$ is given by

$$\chi_s(q) = Ae^{isq-s^2q^2},$$

find the integral representation of the operator $\hat{S}$ in this basis.

Section 5.1.2

Problem 58

1. Prove that the momentum operator is "odd" (changes its sign upon the parity transformation).
2. Prove that the operator of the angular momentum is invariant with respect to the parity transformation ("even").

Section 5.1.3

Problem 59 Which of the following can be used as wave functions describing states of the discrete spectrum:

1. $e^{x^2/2}$
2. $\left(2x^2 - x^4/3\right) e^{-x^2/2}$
3. $A \sin kx$
4. $B \exp(ikx) + C \exp(-ikx)$
5. $xe^{-|x|}$
6. $Ae^{-x^2} \cos kx$

Problem 60 It is known that a potential energy of a quantum particle exhibits a finite discontinuity at point $x = 0$. It is also known that for $x < 0$ and $x > 0$, the wave functions of the particle are presented by

$$\psi(x) = \begin{cases} A\left(x^2 + 2\right) \exp\left(-\alpha_1 x^2\right) & x < 0 \\ B \exp\left(-\alpha_2 x^2\right) & x > 0. \end{cases}$$

Using continuity of the wave function and its derivative, establish relations between parameters A, B, α_1, and α_2.

Problem 61 Prove the following identity:

$$\Psi^*(r, t) \nabla^2 \Psi(r, t) - \Psi(r, t) \nabla^2 \Psi^*(r, t) \equiv \nabla \cdot \left[\Psi^*(r, t) \nabla \Psi(r, t) - \Psi(r, t) \nabla \Psi^*(r, t)\right].$$

Problem 62 Compute probability current densities for a particle in the states described by the following wave functions:

1. $\psi(x) = A \exp(ikz) + B \exp(-ikz)$
2. $\psi(x) = A \cos kx$
3. $\psi(r) = \frac{A}{r} \exp(ikr) + \frac{B}{r} \exp(-ikr)$, where $r = \sqrt{x^2 + y^2 + z^2}$
4. $\psi(r) = A \exp(i\,k \cdot r) + C \exp(-i\,k \cdot r)$

Section 5.1.4

Problem 63 Derive expressions for operators $\hat{L}_x$ and $\hat{L}_y$ in spherical coordinates presented in Eqs. 5.58 and 5.59.

Problem 64 Derive the orthogonalization condition for the associated Legendre functions with $l_1 = l_2$ and $m_1 \neq m_2$ (Eq. 5.73). Do not attempt to obtain the normalization coefficient.

Problem 65 Consider a function of polar and azimuthal angles θ and φ defined as

$$\psi(\theta, \varphi) = \sin \theta \left(1 - \cos \theta\right) \cos \varphi.$$

1. Normalize this function.
2. Present this function as a linear combination of spherical harmonics $Y_l^m(\theta, \varphi)$.

3. If the observables presented by operators $\hat{L}^2$ and $\hat{L}_z$ are measured when a particle is in the state presented by this wave function, what would be the possible outcomes and their probabilities?
4. Find the expectation values and uncertainties of these observables in this state.

Problem 66 Repeat the previous problem for a following function:

$$\psi(\theta, \varphi) = \frac{3}{2} \sin 2\theta \exp(-i\varphi) + 2 \sin^2 \theta \sin 2\varphi,$$

but do not attempt to normalize it before rewriting it as a combination of spherical harmonics.

Problem 67 Find the coordinate representation for lowering and raising ladder operators introduced in Sect. 3.3.4, and using found expressions, find $Y_l^l(\theta, \varphi)$ and $Y_l^{-l}(\theta, \varphi)$.

Problem 68 Find all zeroes of the angular probability distribution for a particle in angular states described by spherical harmonics $Y_3^0(\theta, \varphi)$, $Y_3^1(\theta, \varphi)$, $Y_3^2(\theta, \varphi)$, and $Y_3^3(\theta, \varphi)$.

Problem 69 Find energy values for the system described by Hamiltonian:

$$H = \frac{\hat{L}_x^2 + \hat{L}_y^2}{2I_1} + \frac{\hat{L}_z^2}{2I_2}.$$

Section 5.2

Problem 70 Using eigenvectors $|1\rangle$ and $|2\rangle$ of matrix

$$\begin{bmatrix} 0 & i \\ -i & 0 \end{bmatrix}$$

as a basis, construct the matrix representation of operators $|1\rangle \langle 1|$ and $|2\rangle \langle 2|$ and verify the closure (or completeness) condition:

$$|1\rangle \langle 1| + |2\rangle \langle 2| = \hat{I}$$

where $\hat{I}$ is the unity operator.

Problem 71 Find the matrix of the operator:

$$\hat{A} = |\phi_1\rangle \langle \phi_1| + |\phi_2\rangle \langle \phi_2| + |\phi_3\rangle \langle \phi_3| -$$
$$i|\phi_1\rangle \langle \phi_2| - |\phi_1\rangle \langle \phi_3| + i|\phi_2\rangle \langle \phi_1| - |\phi_3\rangle \langle \phi_1|$$

in the basis formed by orthonormalized vectors $|\phi_1\rangle$, $|\phi_2\rangle$, and $|\phi_3\rangle$.

Problem 72 Write down expressions for spherical harmonics with orbital quantum number $l = 1$. You can consider them as a basis in the subspace of eigenvectors of operator $\hat{L}^2$ belonging to this eigenvalue in the sense that any linear combination of them will also be an eigenvector of this operator.

1. Prove that this is indeed the case, i.e., that any linear combination of spherical harmonics with $l = 1$ represents an eigenvector of $\hat{L}^2$ with the same eigenvalue.
2. Find the matrix representation of operator $\hat{L}_x$ in this basis, and using the obtained matrix, find the representation of this operator's eigenvectors in this basis.
3. The found eigenvectors can also be considered as yet another basis. Find the representation of operators $\hat{L}_x$ and $\hat{L}_z$ in this basis.

Problem 73 Present operator id/dx in the basis of spherical harmonics with $l = 1$.

Problem 74 Consider the matrix:

$$A = \begin{bmatrix} 0 & 0 & -1 \\ 0 & 1 & 0 \\ -1 & 0 & 0 \end{bmatrix}.$$

Transform this matrix to the basis of its eigenvectors. Verify that the elements along the diagonal of the resulting matrix are eigenvalues of A.

Problem 75 Consider two matrices:

$$A = \begin{bmatrix} 1 & i & 1 \\ -i & 0 & 0 \\ 1 & 0 & 0 \end{bmatrix}$$

and

$$B = \begin{bmatrix} 3 & 0 & 0 \\ 0 & 1 & i \\ 0 & -i & 0 \end{bmatrix}.$$

Rewrite matrix B in the basis formed by eigenvectors of matrix A.

Section 5.2.3

Problem 76 Using the same approach, which was used in Sect. 5.2.3 for spin $1/2$, find the matrix representation for the operators of the spin $s = 3/2$. Hint: What is the dimension of the space that contains the vectors representing states of this spin?

Part II
Quantum Models

In this part of the book, we will play with some of the toys, which physicists created in order to get better insight into a variety of new and unusual properties exhibited by real systems obeying laws of quantum mechanics. These toys rarely represent real systems and this is why we call them *models*. Still, in many instances they provide necessary first experience and conceptual understanding required to deal with reality in all its complexity. Using models allows us to focus on those properties of real world, which appear to be of the most significance at least for the class of problems we are interested in. One can think of models in quantum mechanics as of impressionist or postimpressionist paintings, when instead of painstaking attention to details, the main focus is on capturing "the essence" of the object, whatever this might mean. Quantum mechanical models develop physical intuition about the phenomena under study and can often be used as a first iteration of an approximation scheme yielding more accurate and quantitative description of nature.

Chapter 6
One-Dimensional Models

One-dimensional models might appear in quantum mechanics in two, in a way, diametrically opposite situations. In one case, you can pretend that the potential energy of a particle changes only in one direction, such as a potential energy of a uniform electric field. Classically, this would mean a motion characterized by acceleration in one direction and constant velocity in perpendicular directions. By choosing an appropriate inertial coordinate system, you can always eliminate the constant velocity component and consider this motion as straightlinear. Quantum mechanically, this situation has to be described in the coordinate representation, and the respective coordinate wave function can be presented in the form

$$\psi\left(r\right) = e^{i(k_x x + k_y y)}\varphi(z) \tag{6.1}$$

where Z-axis of the coordinate system is arbitrarily chosen to lie along the direction, in which the potential energy changes. The behavior of the wave function in two perpendicular directions (X and Y) is that of a free particle with conserving components of momentums being $p_x = \hbar k_x$ and $p_y = \hbar k_y$. Substituting this expression into Eq. 5.35, where the potential energy is taken to have the form of $V\left(r\right) \equiv V(z)$, and canceling the exponential factors on both sides of the equation, you will end up with the following one-dimensional equation:

$$-\frac{\hbar^2}{2m_e}\frac{d^2\varphi}{dz^2} + V(z)\varphi\left(z\right) = E_z\varphi\left(z\right) \tag{6.2}$$

where

$$E_z = E - \frac{p_x^2}{2m_e} - \frac{p_y^2}{2m_e}$$

Fig. 6.1 A schematic of a
semiconductor
heterostructure, in which the
motion of electrons in the
direction perpendicular to the
planes of the layers can be
described by the
one-dimensional model

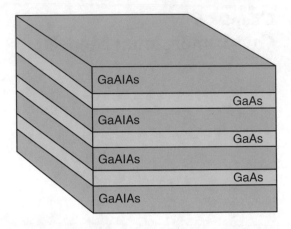

is the contribution of the motion in z direction to the total energy of the system E.
Values of p_x and p_y are determined by the initial state of the system, which may
or may not be one of the eigenvectors of the Hamiltonian with given values of p_x
and p_y. In the latter case, the particular solution of the time-dependent Schrödinger
equation satisfying the initial conditions will be given by the linear combination of
the functions presented in Eqs. 6.1 and 6.2, but for now I will focus only on the
stationary states, which correspond to initial conditions with definite p_x and p_y.

While for a long time this type of one-dimensional model was used mostly
in classrooms to illustrate basic quantum effects to unsuspecting students, the
technological advances of the last 50 years made this model quite relevant as
a stepping stone to understanding properties of practically important artificial
structures made of planar layers of several different semiconductors arranged in an
alternating order (see Fig. 6.1). It can be shown (way above your pay grade though)
that the motion of electrons in such structures can be approximately described by a
potential energy, which only changes in the direction perpendicular to the plane of
the layers (growth direction).

The second situation, in which the one-dimensional model can have at least some
relation to reality, is the case of potentials confining the motion of the particle in all
directions but one. One can imagine a particle moving inside of a cylindrical tube
with impenetrable walls. The motion perpendicular to the axis of the cylinder is
characterized by discrete allowed values of energy (I will show it later, for now
you will have to trust me on that), and if the radius of the tube is small enough, the
distance between adjacent energy levels can be sufficiently large that for all practical
purposes only one of this energy levels can be taken into account. In this case, the
transverse (as in perpendicular to the axes of the cylinder) motion is completely
"frozen," and one is again left with pure one-dimensional motion. In all cases, we
are dealing with the Schrödinger equation in the form of Eq. 6.2, which is the main
object of study in this chapter.

6.1 Free Particle and the Wave Packets

Before taking on quantum states of electrons in one-dimensional piecewise potentials such as wells or barriers, it is useful to consider the simplest quantum
mechanical model—a freely propagating, i.e., not interacting with anything, particle. In classical physics, as we all know, such a particle would move with a constant
velocity, v, and can be characterized by conserving kinetic energy $K = mv^2/2$ and
momentum $p = mv$. In quantum mechanics, states of a free particle are the solution
of the Schrödinger equation with zero potential

$$i\hbar \frac{\partial |\Psi\rangle}{\partial t} = \frac{\hat{P}^2}{2m_e} |\Psi\rangle .$$ (6.3)

It is quite easy to see that the stationary states of a free particle are eigenvectors of
the momentum operator:

$$|\Psi\rangle = \exp\left(-i\frac{E_p}{\hbar}t\right) |p\rangle$$ (6.4)

where $|p\rangle$ is defined by $\hat{P} |p\rangle = p |p\rangle$. Substitution of Eq. 6.4 into Eq. 6.3 yields

$$E_p = \frac{p^2}{2m_e}$$ (6.5)

which is an expected classical relation between energy and momentum of a free
particle, often called dispersion relation. Historically, Schrödinger equation was
devised to make sure that the quantum theory respects this relation between energy
and momentum. Indeed, in the position representation, the Schrödinger equation
becomes

$$i\hbar \frac{\partial \Psi(r, t)}{\partial t} = -\frac{\hbar^2 \nabla^2}{2m_e} \Psi(r, t)$$ (6.6)

with stationary state solutions of the form

$$\Psi(r, t) = \frac{1}{(2\pi\hbar)^{3/2}} \exp\left(-i\frac{E_p}{\hbar}t + i\frac{p \cdot r}{\hbar}\right)$$ (6.7)

where I used δ-function normalized eigenvectors of momentum operator given in
Eq. 5.21. Now, one can argue that the Schrödinger equation contains the first-order
time derivative because the dispersion relation, Eq. 6.5, is linear in E, while the
derivative over coordinates must be of the second order to reproduce the term p^2 in
Eq. 6.5. Further, one can argue that since the time derivative in the Schrödinger

equation is only of the first order, the corresponding wave function must be represented by a complex exponential function rather than by a real trigonometric function, which, in turn, makes the factor i in front of the time derivative necessary to compensate for the similar factor in the argument of the wave function.

The wave function of the form given in Eq. 6.7 has been conceived at the early days of quantum mechanics as a mean to reconcile particle and wavelike properties of quantum objects. However, it was clear from the very beginning that regardless of the chosen interpretation (statistical due to Born or Schrödinger's pilot wave), there are several problems with assigning this function to represent quantum states of real particles. First, its absolute value is uniform in space, which can hardly represent an actual localized particle regardless of the chosen interpretation. Also, the motion of the wave represented by Eq. 6.7 is characterized by phase velocity $v_{ph} = \omega/k = E/p = p/(2m_e)$, which is half of the corresponding classical velocity $v_{cl} = p/m_e$ making it difficult to associate it with the motion of a particle.

To get around this conundrum, it was suggested that actual states of the particles (in either interpretation) are presented not by stationary states but by their superposition, which still will solve the Schrödinger equation 6.6. It is quite easy to show that by choosing an appropriate superposition, it is possible, for instance, to localize a particle within an arbitrarily small region solving at least one of the listed problems. To see how this comes about, consider a wave function at time $t = 0$ and form a superposition of the form

$$\psi(\boldsymbol{r}) = \frac{1}{(2\pi\hbar)^{3/2}} \int d^3p A(\boldsymbol{p}) \exp\left(i\frac{\boldsymbol{p} \cdot \boldsymbol{r}}{\hbar}\right). \tag{6.8}$$

In Sect. 5.1.1, it was shown that Eq. 6.8 can be inverted to yield

$$A(\boldsymbol{p}) = \frac{1}{(2\pi\hbar)^{3/2}} \int d^3r \psi(\boldsymbol{r}) \exp\left(-i\frac{\boldsymbol{p} \cdot \boldsymbol{r}}{\hbar}\right) \tag{6.9}$$

so that by choosing an appropriate $A(\boldsymbol{p})$, I can "generate" an initial ($t = 0$) wave function with an arbitrary degree of localization. Now all what I need is to consider the time dependence of this initial superposition to see if other problems outlined above can also be circumvented by the superposition states, which, in the case of free propagating particles, are often called *wave packets*.

Here I will focus on just one particular example of the wave packets, which despite its relative simplicity will help me to illustrate most of the relevant ideas. First of all, I will simplify the consideration by limiting it to the case of one-dimensional motion described by a wave function, which depends on a single coordinate, say, z. Integrals in Eqs. 6.8 and 6.9 are in this case reduced to one-dimensional form

$$\psi(z) = \frac{1}{\sqrt{2\pi\hbar}} \int\limits_{-\infty}^{\infty} dp_z A(p_z) \exp\left(i\frac{p_z z}{\hbar}\right) \tag{6.10}$$

$$A(p_z) = \frac{1}{\sqrt{2\pi\hbar}} \int\limits_{-\infty}^{\infty} dz \psi\,(z) \exp\left(-i\frac{p_z z}{\hbar}\right). \tag{6.11}$$

Next, I will assume that the initial state of the particle is described by function

$$\psi\,(z) = C \exp\left[-\frac{(z-\bar{z})^2}{(2\Delta z_0)^2}\right] \exp\left(i\frac{\bar{p}_z}{\hbar}z\right) \tag{6.12}$$

where constant C is found from the normalization condition

$$\int\limits_{-\infty}^{\infty} |\psi\,(z)|^2\,dz = |C|^2 \int\limits_{-\infty}^{\infty} \exp\left[-\frac{(z-\bar{z})^2}{2\,(\Delta z_0)^2}\right] dz =$$

$$|C|^2\,\sqrt{2}\Delta z_0 \int\limits_{-\infty}^{\infty} \exp\left(-x^2\right) dx = |C|^2\,\Delta z_0 \sqrt{2\pi} = 1 \Longrightarrow$$

$$C = \frac{1}{\sqrt{\Delta z_0 \sqrt{2\pi}}}.$$

In the course of computing the normalization integral, I introduced a new integration variable $x = (z - \bar{z})/\left(\sqrt{2}\Delta z_0\right)$ and used a well-known integral $\int_{-\infty}^{\infty} \exp\left(-x^2\right) dx = \sqrt{\pi}$. Thus, my initial state is represented by the normalized wave function, where the amplitude of the plane wave $\exp\left(i\bar{p}_z z/\hbar\right)$ is modulated by the so-called Gaussian function

$$\psi\,(z) = \frac{1}{\sqrt{\Delta z \sqrt{2\pi}}} \exp\left[-\frac{(z-\bar{z})^2}{(2\Delta z_0)^2}\right] \exp\left(i\frac{\bar{p}_z}{\hbar}z\right). \tag{6.13}$$

The probability distribution corresponding to this wave function is peaked at $z = \bar{z}$ and falls off from its maximum value as z moves away from $\bar{z}$. Parameter Δz_0 determines how fast the decrease of the probability takes place: the larger Δz_0, the larger deviation from $\bar{z}$ is required to decrease the probability density by e. Varying Δz_0, one can control the degree of particle localization—a smaller Δz_0 corresponds to better localized particles (see Fig. 6.2). Formally speaking, one can define $\bar{z}$ and Δz_0 as expectation value and uncertainty of the coordinate in the state described by this wave function. Indeed, I can easily compute

Fig. 6.2 Normalized Gaussian wave functions with different values of the width parameter Δz_0: with decreasing Δz_0 the function narrows, while its maximum grows such that the total area under the curve remains equal to unity

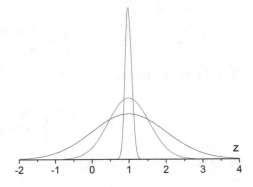

$$\langle z \rangle = \frac{1}{\Delta z \sqrt{2\pi}} \int_{-\infty}^{\infty} dz\, z \exp\left[-\frac{(z-\bar{z})^2}{2(\Delta z_0)^2}\right] =$$

$$\frac{1}{\sqrt{\pi}} \int_{-\infty}^{\infty} \left(x\sqrt{2}\Delta z_0 + \bar{z}\right) \exp\left(-x^2\right) dx = \bar{z}$$

where I took into account that normalization integral computed earlier is equal to unity and the fact that the integral of an odd function over a symmetric interval is zero. The uncertainty takes a bit more work:

$$\langle z^2 \rangle = \frac{1}{\Delta z \sqrt{2\pi}} \int_{-\infty}^{\infty} dz\, z^2 \exp\left[-\frac{(z-\bar{z})^2}{(\Delta z_0)^2}\right] =$$

$$\frac{1}{\sqrt{\pi}} \int_{-\infty}^{\infty} \left(x\sqrt{2}\Delta z_0 + \bar{z}\right)^2 \exp\left(-x^2\right) dx =$$

$$\bar{z}^2 + \frac{2(\Delta z_0)^2}{\sqrt{\pi}} \int_{-\infty}^{\infty} x^2 \exp\left(-x^2\right) dx = \bar{z}^2 + (\Delta z_0)^2$$

where I used another well-known integral $\int_{-\infty}^{\infty} x^2 \exp\left(-x^2\right) dx = \sqrt{\pi}/2$. Subtracting $\bar{z}^2$ from $\langle z^2 \rangle$, you can convince yourself that Δz_0 is, indeed, the uncertainty of the coordinate.

It shall be noticed that these arguments do not contradict to Schrödinger's pilot wave interpretation, according to which the wave presented by the wave packet is a real material object accompanying a particle and whose width defines the degree of particle localization.

Now I can find the appropriate amplitudes $A(p_z)$ in Eq. 6.10, which would reproduce the wave function given by Eq. 6.13. Substitution of this equation into Eq. 6.11 yields

$$A(p_z) = \frac{1}{\sqrt{2\pi\hbar}} \frac{1}{\sqrt{\Delta z_0 \sqrt{2\pi}}} \int_{-\infty}^{\infty} dz \exp\left[-\frac{(z-\bar{z})^2}{(2\Delta z_0)^2}\right] \exp\left(-i\frac{(p_z-\bar{p}_z)z}{\hbar}\right) =$$

$$\frac{2\Delta z_0}{\sqrt{2\pi\hbar}\sqrt{\Delta z_0\sqrt{2\pi}}} \int_{-\infty}^{\infty} dx \exp\left[-x^2 - i\frac{(p_z-\bar{p}_z)}{\hbar}(2x\Delta z_0 + \bar{z})\right] =$$

$$\frac{\sqrt{2\Delta z_0}}{\sqrt{\pi\hbar}(2\pi)^{1/4}} \exp\left(-i\frac{p_z-\bar{p}_z}{\hbar}\bar{z}\right) \int_{-\infty}^{\infty} dx \exp\left[-x^2 - i2\Delta z_0\frac{p_z-\bar{p}_z}{\hbar}x\right] =$$

$$\frac{\sqrt{2\Delta z_0}}{\sqrt{\pi\hbar}(2\pi)^{1/4}} \exp\left(-i\frac{p_z-\bar{p}_z}{\hbar}\bar{z}\right)$$

$$\times \int_{-\infty}^{\infty} dx \exp\left[-x^2 - i2\Delta z_0\frac{p_z-\bar{p}_z}{\hbar}x - \left(i\frac{(p_z-\bar{p}_z)\Delta z_0}{\hbar}\right)^2 + \left(i\frac{(p_z-\bar{p}_z)\Delta z_0}{\hbar}\right)^2\right]$$

$$= \frac{\sqrt{2\Delta z_0}}{\sqrt{\pi\hbar}(2\pi)^{1/4}} \exp\left(-i\frac{p_z-\bar{p}_z}{\hbar}\bar{z}\right) \times$$

$$\exp\left[-\frac{(p_z-\bar{p}_z)^2(\Delta z_0)^2}{\hbar^2}\right] \int_{-\infty}^{\infty} dx \exp\left[-\left(x + i\frac{(p_z-\bar{p}_z)\Delta z_0}{\hbar}\right)^2\right] =$$

$$\frac{\sqrt{2\Delta z_0}}{\sqrt{\hbar}(2\pi)^{1/4}} \exp\left(-i\frac{p_z-\bar{p}_z}{\hbar}\bar{z}\right) \exp\left[-\frac{(p_z-\bar{p}_z)^2(\Delta z_0)^2}{\hbar^2}\right].$$

This was a long calculation, but it is worth the efforts to carefully peruse it. Some of the tricks that I used in its course were the substitution of variable $x = (z - \bar{z})/(2\Delta z_0)$, presenting an expression of the form $a^2 + 2ba$ as a complete square, $a^2 + 2ba + b^2 - b^2 = (a + b)^2 - b^2$, and finally the fact that integral $\int_{-\infty}^{\infty} dx \exp\left[-(x - x_0)^2\right]$ still equals to $\sqrt{\pi}$ regardless of the value of x_0. Before continuing, I will set $\bar{z} = 0$, which amounts to the choice of the zero of the coordinate z, and introduce new parameter $\Delta p = \hbar/(2\Delta z_0)$. Then the expression for $A(p_z)$ becomes

$$A(p_z) = \frac{1}{\sqrt{\Delta p}(2\pi)^{1/4}} \exp\left[-\frac{(p_z-\bar{p}_z)^2}{(2\Delta p)^2}\right], \tag{6.14}$$

and it is easy to verify (do it!) that as expected $\int_{-\infty}^{\infty} |A(p)|^2 dp = 1$, while $\langle p_z \rangle = \bar{p}_z$, and parameter Δp determines the uncertainty of the particle's momentum. Recalling the definition of this parameter in terms of the uncertainty of coordinates, you can see that these two parameters obey the minimum version of the Schrödinger uncertainty principle:

$$\Delta p \Delta z_0 = \frac{\hbar}{2}. \tag{6.15}$$

This is the special property of the Gaussian distribution: for all other initial states, the product of the uncertainties would be larger than $\hbar/2$.

Having found $A(p_z)$, I can now find the time dependence of the initial wave function by considering the superposition of the stationary states at an arbitrary time t:

$$\Psi(z,t) = \frac{1}{\sqrt{2\pi\hbar\Delta p}\,(2\pi)^{1/4}} \int\limits_{-\infty}^{\infty} dp_z \exp\left[-\frac{(p_z - \bar{p}_z)^2}{(2\Delta p)^2}\right] \exp\left[-i\frac{p_z^2}{2\hbar m_e}t + i\frac{p_z}{\hbar}z\right],$$

which at $t = 0$ is obviously reduced to the function given in Eq. 6.12. I begin evaluating this integral, again, by introducing a dimensionless variable

$$x = \frac{p_z - \bar{p}_z}{2\Delta p}$$

and transforming this integral into

$$\Psi(z,t) =$$

$$\frac{2\Delta p}{\sqrt{2\pi\hbar\Delta p}\,(2\pi)^{1/4}} \int\limits_{-\infty}^{\infty} dx \exp\left(-x^2\right) \exp\left[-it\frac{1}{2\hbar m_e}(\bar{p}_z + 2\Delta px)^2 + i\frac{z}{\hbar}(\bar{p}_z + 2\Delta px)\right]$$

$$= \frac{2\sqrt{\Delta p}}{\sqrt{2\pi\hbar}\,(2\pi)^{1/4}} \exp\left(-it\frac{\bar{p}_z^2}{2\hbar m_e} + i\frac{z}{\hbar}\bar{p}_z\right) \times$$

$$\int\limits_{-\infty}^{\infty} dx \exp\left[-x^2 - it\frac{2\bar{p}_z\Delta p}{\hbar m_e}x - it\frac{2(\Delta p)^2}{\hbar m_e}x^2 + 2i\frac{z\Delta p}{\hbar}x\right] =$$

$$\frac{2\sqrt{\Delta p}}{\sqrt{2\pi\hbar}\,(2\pi)^{1/4}} \exp\left(-it\frac{\bar{p}_z^2}{2\hbar m_e} + i\frac{z}{\hbar}\bar{p}_z\right) \times$$

$$\int\limits_{-\infty}^{\infty} dx \exp\left[-x^2\left(1 + it\frac{2(\Delta p)^2}{\hbar m_e}\right) - 2ix\frac{\Delta p}{\hbar}\left(\frac{\bar{p}_z}{m_e}t - z\right)\right].$$

Before continuing, let me brush up this expression a bit, first, by replacing $\bar{p}_z/m$, which corresponds to the classical velocity of the particle with momentum $\bar{p}_z$ with v_{gr}, and second by introducing the notation

$$\alpha = \sqrt{1 + it\frac{2(\Delta p)^2}{\hbar m_e}} = \sqrt{1 + it\frac{\hbar}{2m_e(\Delta z_0)^2}} \qquad (6.16)$$

where in the second expression I replaced Δp with Δz_0 using Eq. 6.15. As a result, the expression for the wave function now takes a somewhat less cumbersome form:

$$\Psi(z, t) = \frac{2\sqrt{\Delta p}}{\sqrt{2\pi\hbar}(2\pi)^{1/4}} \exp\left(-it\frac{\bar{p}_z^2}{2\hbar m_e} + i\frac{z}{\hbar}\bar{p}_z\right) \times$$

$$\int_{-\infty}^{\infty} dx \exp\left[-x^2\alpha^2 - 2ix\frac{\Delta p}{\hbar}(v_{gr}t - z)\right].$$

Performing integral over x (using all the same tricks as before, substitution $\tilde{x} = \alpha x$ and completion of the square), I will obtain the final expression for the wave function $\Psi(z, t)$

$$\Psi(z, t) =$$

$$\frac{\sqrt{2\Delta p}}{\sqrt{\hbar}(2\pi)^{1/4}\alpha} \exp\left(-it\frac{\bar{p}_z^2}{2\hbar m_e} + i\frac{z}{\hbar}\bar{p}_z\right) \exp\left[-\left(\frac{\Delta p}{\alpha\hbar}\right)^2 (v_{gr}t - z)^2\right] =$$

$$\frac{1}{\alpha\sqrt{\Delta z_0}(2\pi)^{1/4}} \exp\left[-\frac{(v_{gr}t - z)^2}{(2\alpha\Delta z_0)^2}\right] \exp\left(-it\frac{\bar{p}_z^2}{2\hbar m_e} + i\frac{z}{\hbar}\bar{p}_z\right) \qquad (6.17)$$

where at the last step I used Eq. 6.15 to replace Δp with Δz_0. Not surprisingly, at $t = 0$ Eq. 6.17 is reduced to $\psi(z)$ as given by Eq. 6.13.

It is quite educational to inspect various factors in this expression separately. The last factor is a regular plane wave with a wave number determined by the expectation value of the momentum $\bar{p}_z$ and corresponding frequency $\bar{\omega} = \bar{p}_z^2/(2\hbar m_e)$. This wave propagates with standard phase velocity $v_{ph} = \hbar\bar{\omega}/\bar{p}_z$, but its amplitude, defined by the second exponential factor, is also time and coordinate dependent. For any given instant t, there is coordinate $z_{max} = v_{gr}t$, when the amplitude is the largest, and decreases when z deviates away from it in any direction. One can say that the amplitude factor modulates the initial plane wave turning it into a wave packet more or less localized within a finite coordinate region. This localization region obviously changes its position with time as is evident from the definition of z_{max}, and this motion of the localization region occurs with velocity $v_{gr} = z_{max}/t$. This velocity is called group velocity of the wave packet because it characterizes the motion of the entire group of waves participating in the superposition forming the packet, while the phase velocity describes the motion of each separate wave component of this superposition. These two velocities are different because the phase velocity $v_{ph} = p/2m_e$ depends on the momentum p and is, therefore, different for each member of the group. To illustrate all these points, I plotted the real part of the wave function

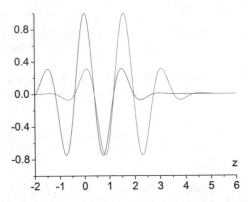

Fig. 6.3 The real part of the wave function representing a wave packet as a function of the coordinate for two different instances. You can see suppression of oscillations away from the main maximum as well as the displacement of the main maximum. The time interval is chosen to be equal to the period of the oscillating factor so that the magnitude of the main maximums remains the same for both instances. For other time intervals, it does not have to be the case because the decrease of the cos function can damp the maximum's magnitude. Also, this plot does not account for the fact that the parameter α in Eq. 6.17 is complex-valued and depends on time. See discussion of the role of this parameter further in the text

presented in Eq. 6.17 for two distinct time instances as shown in Fig. 6.3. It is easy to see that the expression for the group velocity $v_{gr} = \bar{p}/m_e$ can be obtained from the dispersion relation $E(p) = p_z^2/2m_e$ of the free particle as

$$v_{gr} = \left. \frac{dE}{dp_z} \right|_{p_z=\bar{p}}. \tag{6.18}$$

Equation 6.18 can also be generalized to the case of three-dimensional propagation, in which case the derivative is replaced by a gradient $v_{gr} = \nabla E(p)|_{p=\bar{p}}$ and also to more exotic cases of particles whose dispersion relation is different from Eq. 6.5. If you are wondering where on earth you can find free particles with dispersion different from standard quadratic form, here are two examples for you: (1) relation between energy and momentum of relativistic particles is $E = \sqrt{m_e^2 c^4 + p^2 c^2}$ and (2) electrons in semiconductors can be in many practically important cases approximated as "free" particles with modified dispersion relation $E(p)$. In general, the picture of a wave packet propagating with the group velocity can be generalized to non-Gaussian wave packets as long as they can be described by a momentum wave function $A(p_z)$ with a single and relatively narrow maximum.

So far, the behavior of the wave packet appears to be consistent not only with traditional Copenhagen interpretation of quantum mechanics but also with Schrödinger's pilot wave picture. However, when discussing the role of the amplitude modulating factor in Eq. 6.17, I so far ignored an obvious "elephant in the room," which makes this discussion somewhat more nuanced. The parameter α, which sits "quietly" in the denominator of the modulating factor (as well as in a normalization pre-factor), is complex-valued and time dependent. The first of these circumstances makes the expression

$$B(z,t) = \frac{1}{\alpha\sqrt{\Delta z_0}\,(2\pi)^{1/4}}\exp\left[-\frac{(v_{gr}t-z)^2}{(2\alpha\,\Delta z_0)^2}\right] \tag{6.19}$$

not quite a pure amplitude because it also has a phase attached to it. This phase, however, is of little interest, and in order to focus on the actual amplitude part of this expression, I will consider its squared absolute value, $|B(z,t)|^2$, which, of course, coincides with $|\Psi(z,t)|^2$ and in the Copenhagen interpretation yields the probability distribution $P(z,t)$ for the coordinates of the particle at any given time in the state described by the wave packet $\Psi(z,t)$:

$$P(z,t) = \frac{1}{\sqrt{2\pi}\,\Delta z_0\,|\alpha|^2}\exp\left[-\frac{(v_{gr}t-z)^2}{(2\Delta z_0)^2}\left(\frac{1}{\alpha^2}+\frac{1}{(\alpha^2)^*}\right)\right]. \tag{6.20}$$

Now I define

$$\left(\frac{1}{\Delta z}\right)^2 = \left(\frac{1}{\Delta z_0}\right)^2\left(\frac{1}{\alpha^2}+\frac{1}{(\alpha^2)^*}\right)$$

and using the definition of α from Eq. 6.16 calculate

$$\left(\frac{1}{\Delta z}\right)^2 = \frac{1}{2\,(\Delta z_0)^2}\left(\frac{1}{1+it\frac{\hbar}{2m_e(\Delta z_0)^2}}+\frac{1}{1-it\frac{\hbar}{2m_e(\Delta z_0)^2}}\right) =$$

$$\frac{1}{2\,(\Delta z_0)^2}\frac{2}{1+t^2\frac{\hbar^2}{4m_e^2(\Delta z_0)^4}} = \frac{(\Delta z_0)^2}{(\Delta z_0)^4+\frac{\hbar^2 t^2}{4m_e^2}}. \tag{6.21}$$

I can also find

$$\frac{1}{|\alpha|^2} = \frac{1}{\sqrt{\left(1+it\frac{\hbar}{2m_e(\Delta z_0)^2}\right)\left(1-it\frac{\hbar}{2m_e(\Delta z_0)^2}\right)}} =$$

$$\frac{1}{\sqrt{1+t^2\frac{\hbar^2}{4m_e^2(\Delta z_0)^4}}} = \frac{\Delta z_0}{\Delta z}. \tag{6.22}$$

Substitution of Eqs. 6.21 and 6.22 to Eq. 6.20 converts the expression for the probability density in the following nice looking form:

$$P(z,t) = \frac{1}{\sqrt{2\pi}\,\Delta z}\exp\left[-\frac{(v_{gr}t-z)^2}{2\,(\Delta z)^2}\right]. \tag{6.23}$$

Comparing this result with Eq. 6.13, it becomes clear that $z_{max} = v_g t$ is the expectation value of the coordinate in the state described by the wave packet, while Δz represents its uncertainty. Rewriting Eq. 6.21, I can present this uncertainty in a more illuminating form:

$$\Delta z = \Delta z_0 \sqrt{1 + \frac{\hbar^2 t^2}{4 m_e^2 (\Delta z_0)^4}} \tag{6.24}$$

which shows that the localization range of the wave packet increases with time with the rate (roughly defined as derivative $d(\Delta z)^2 / dt^2$) inversely proportional to the initial uncertainty Δz_0. In other words, the tighter you try to squeeze your particle into a smaller volume, the faster the localization volume of the particle increases with time. This phenomenon of the wave packet spreading is what kills Schrödinger's pilot wave interpretation: the broadening of the wave packets would make such a pilot wave unstable. To get an intuitive feeling for how fast this spreading takes place, assume that an electron is initially localized in a region of atomic dimensions with $\Delta z_0 \simeq 10^{-10}$ m. Substituting the values of the Planck constant and the electron's mass into Eq. 6.24, you will get for Δz:

$$\Delta z = 10^{-10} \sqrt{1 + 1.32 \times 10^{33} t^2} \text{ m,}$$

which reaches the value of 10^3 m in just about 3 ms!

This essentially completes the discussion of the free particle wave packets, but I cannot pass an opportunity to play with the Heisenberg picture whenever it is possible, and this is one of the simplest situations to showcase it. You can consider it as a reward to you for being such a good sport and wading with me through the tedious analysis of the Gaussian wave packet.

Recalling that the Hamiltonian of a free particle is just

$$\hat{H} = \frac{\hat{P}^2}{2m_e},$$

you can easily derive the Heisenberg equations for the components of the position and momentum operators

$$\frac{d\hat{r}}{dt} = \frac{\hat{P}}{m}$$

$$\frac{d\hat{P}}{dt} = 0$$

with obvious solution

$$\hat{r} = \hat{r}_0 + \frac{\hat{P}_0}{m} t \tag{6.25}$$

where $\hat{r}_0$ and $\hat{P}_0$ are as usual Schrödinger picture's operators setting initial conditions in the Heisenberg picture. Assuming that the particle is in some arbitrary state $|\chi\rangle$, which does not change with time in the Heisenberg picture, I can immediately derive for the expectation value of the position operator:

$$\langle \hat{r} \rangle = \langle \hat{r}_0 \rangle + \frac{\langle \hat{P}_0 \rangle}{m} t$$

which is, of course, a three-dimensional version of the expression for the expectation value of the coordinate found from Eq. 6.23 with z_0 set to zero. Now, squaring Eq. 6.25 I get

$$\hat{r}^2 = \hat{r}_0^2 + \frac{\hat{P}_0^2}{m^2} t^2 + \frac{t}{m}\left(\hat{r}_0 \hat{P}_0 + \hat{P}_0 \hat{r}_0\right).$$

Using position representation for the operators $\hat{r}_0$, $\hat{P}_0$, and Eq. 6.13 to represent state $|\chi\rangle$, you can demonstrate by direct computations that

$$\langle \chi | \hat{r}_0 \hat{P}_0 + \hat{P}_0 \hat{r}_0 | \chi \rangle = 0$$

so that one has for the uncertainty of the position Δr^2:

$$\Delta r^2 = \Delta r_0^2 + \frac{\Delta p^2}{m^2} t^2, \tag{6.26}$$

where Δp^2 is again the uncertainty of the momentum computed with an initial state of the particle. If this initial state is Gaussian, and limiting Eq. 6.26 to just a single coordinate, you can use Eq. 6.15 to replace the momentum uncertainty with Δz_0, which will yield Eq. 6.24 for the spreading of the wave packet. In addition to this, however, Eq. 6.26 demonstrates that the phenomenon of wave packet spreading is not limited only to one-dimensional Gaussian packets and is a general feature of free propagation of quantum particles.

6.2   Rectangular Potential Wells and Barriers

6.2.1   Potential Wells: Systems with Mixed Spectrum

The first important model, which I am going to introduce in this section, is characterized by a potential profile shown in Fig. 6.4, and which can be described as

$$V(z) = \begin{cases} V_w & |z| < d/2 \\ V_b & |z| > d/2 \end{cases} \tag{6.27}$$

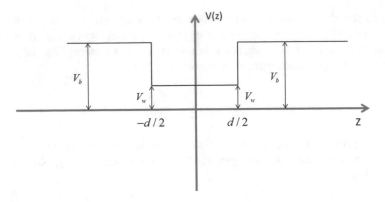

Fig. 6.4 Rectangular potential well

where I assumed for concreteness that $V_b > V_w$. Such a potential profile is called a *potential well*. If one chooses to count the energy from the bottom of the well, then $V_w \rightarrow 0$, and $V_b \rightarrow V_b - V_w$. The energy levels in this potential must be separated in two different regions $V_w < E_z < V_b$ and $E > V_b$ with distinctly different types of behavior (states with $E < V_w$ do not exist). In the former case, a classical particle would have been confined between the two "walls" of this potential well at $|z| = d/2$ bouncing back and forth, while in the latter, classical motion is unbounded (the particle can be anywhere along the Z-axis). As was already discussed in Sect. 5.1.3, two different types of classical behavior translate in to different quantum behaviors as well.

Bound States: Discrete Spectrum
I will begin with spectral region $V_w < E_z < V_b$, which corresponds to bound classical motion, and where you should expect to see discrete spectrum of energy eigenvalues. The potential we are dealing with is a piecewise continuous function with finite jumps at $z = \pm d/2$. For the range of energies under consideration, spatial regions defined by $|z| > d/2$ are classically forbidden. Therefore, as it was discussed in Sect. 5.1.3, Eq. 5.37 must be complemented by the boundary conditions requiring that the wave function vanishes at $z \rightarrow \pm\infty$ and by continuity conditions at $|z| = d/2$.

However, before I start dirtying my hands and digging into the boring business of actually writing down the wave functions and matching the boundary conditions and all that, I want to play with the problem a little bit more and see if I can make this task a bit less boring. The nice thing about this particular potential is that it is symmetric with respect to inversion of coordinate z: $V(-z) = V(z)$, and I hope that you recognize here your old acquaintance from Sect. 5.1.2—the parity transformation, $\hat{\Pi}V(z) = V(-z)$. And not only the potential is symmetric, but the boundary conditions are also symmetric: $\varphi(z) \rightarrow 0$, when $|z| \rightarrow \infty$. Since the kinetic energy was shown earlier to be always parity invariant (does not change upon the parity transformation), you can confidently conclude that the entire

Hamiltonian of this system is symmetric with respect to this transformation. And as I have already explained in Sect. 5.1.2, it means that the Hamiltonian commutes with the parity operator, $\hat{\Pi}$, so that the wave functions representing eigenvectors of this Hamiltonian also represent eigenvectors of $\hat{\Pi}$. Wherefore, solutions of the Schrödinger equation 5.37 with potential given by Eq. 6.27 can be classified into even $(\varphi(-z) = \varphi(z))$ and odd $(\varphi(-z) = -\varphi(z))$ functions with the immediate consequence that you only need to deal with boundary and continuity conditions at $z > 0$. Indeed, the definite parity of the solutions, even or odd, ensures that the conditions for $z < 0$ are satisfied simultaneously with those at $z > 0$. Here is the power of the symmetry to you: I just cut the number of equations to be solved to satisfy the continuity conditions by half without even breaking a sweat! Using the symmetry arguments for such a simple problem might seem a bit as an overkill—it is not too difficult to solve it by simply using the brute force. I still wanted to show it to you so that you would be better prepared to understand the implications of symmetry in more "sanity-threatening situations."

Because of the discontinuity of the potential at $|z| = d/2$, solutions for intervals $|z| < d/2$ and $|z| > d/2$ must be found independently and stitched afterward using continuity conditions.

1. $|z| < d/2$. The Schrödinger equation for this interval takes the form

$$\frac{d^2\varphi}{dz^2} = -\frac{2m_e}{\hbar^2} (E_z - V_w) \, \varphi \, (z)$$

where $E_z - V_w > 0$. This equation is similar to that of the free particle with positive energy $E_z - V_w$, and its most general solution has, therefore, the form

$$\varphi(z) = Ae^{ikz} + Be^{-ikz} = \tilde{A} \sin kz + \tilde{B} \cos kz$$

where

$$k = \sqrt{2m_e \, (E_z - V_w)}/\hbar \tag{6.28}$$

is a real quantity. The choice of the exponential or trigonometric functions to represent this solution is a matter of one's taste and/or convenience: the expressions are equivalent with $\tilde{A} = i(A - B)$ and $\tilde{B} = A + B$. However, since we know that $\varphi(z)$ must have a definite parity, the trigonometric form is more convenient to take advantage of this insight. Indeed, in order to generate an even solution I can simply make $\tilde{A} = 0$, while an odd solution is obtained by choosing $\tilde{B} = 0$:

$$\varphi_e(z) = B \cos kz, \quad |z| < d/2 \tag{6.29}$$

$$\varphi_o(z) = A \sin kz, \quad |z| < d/2. \tag{6.30}$$

Notation for the remaining coefficients is irrelevant, so I dropped the tildes above the letters.

2. $|z| > d/2$. In this case, the right-hand side of the corresponding Schrödinger equation

$$\frac{d^2\varphi}{dz^2} = \frac{2m_e}{\hbar^2} (V_b - E_z) \, \varphi \, (z)$$

is positive for the range of energies under consideration ($V_b - E_z > 0$) so that the general solution of this equation is given by

$$\varphi(z) = Ce^{\kappa z} + De^{-\kappa z} \tag{6.31}$$

where now

$$\kappa = \sqrt{2m_e \, (V_b - E_z)}/\hbar \tag{6.32}$$

is a real quantity. As has already been mentioned, the solution of the Schrödinger equation in the region $z > d/2$ must vanish at $z \to +\infty$ (Eq. 5.36). The function presented by Eq. 6.31 satisfies this requirement only if the exponentially growing term is gotten rid of, which I achieve by simply requiring that $C = 0$. Thus, the wave function for $z > d/2$ becomes

$$\varphi(z) = De^{-\kappa z}, \, z > d/2 \tag{6.33}$$

for both even and odd solutions. For negative values of coordinates $z < -d/2$, Eq. 6.33 would produce for even and odd solutions correspondingly:

$$\varphi_e(z) = De^{\kappa z} \tag{6.34}$$

$$\varphi_o(z) = -De^{\kappa z}. \tag{6.35}$$

Before continuing with stitching the wave functions at $z = d/2$, let me point out that the solutions in the classically allowed region $|z| < d/2$ are presented by oscillating functions. Upon crossing to the classically forbidden region $|z| > d/2$, the oscillating character of the solutions turns into a monotonic decrease. This example illustrates generic properties discussed in Sect. 5.1.3 and can be used to formulate a general rule of thumb applied to any piecewise constant potential: in the classically allowed regions, the wave function is represented by combination of trigonometric function, while in classically forbidden regions, the solution is given by combination of exponential functions with a real argument. However, I need to warn you to pay attention to the fact that I was able to eliminate the exponentially growing terms in Eqs. 6.33–6.35 only because the classically forbidden region extended all the way to positive or negative infinities. If, as it might happen in certain problems, the potential would have another jump and a classically forbidden region would have crossed over to a classically allowed region, you would have to keep both growing and decreasing exponential functions because the conditions at infinity can only be used within a region of coordinates extending, well, to infinity.

Equation 6.33 must be stitched with either Eq. 6.29 or 6.30 to generate continuous solution describing the wave function in the entire domain of the coordinate z. This must be done separately for even and odd solutions. In the latter case, the continuity of the wave function and of its derivative at $z = d/2$ requires that

$$B \cos \frac{kd}{2} = De^{-\kappa d/2} \tag{6.36}$$

$$-Bk \sin \frac{kd}{2} = -\kappa De^{-\kappa d/2}. \tag{6.37}$$

For arbitrary values of E_z, which appears in these equations via parameters k and κ, Eqs. 6.36 and 6.37 can have only trivial solution $B = D = 0$. Obviously, this is not what we want. However, if I insist on having non-zero solutions, I must impose a special condition on the allowed values of k and κ. One way to derive this condition is to divide Eq. 6.36 by Eq. 6.37 (this is allowed because we require that $B, D \neq 0$) yielding

$$k \tan \frac{kd}{2} = \kappa. \tag{6.38}$$

Taking into account Eqs. 6.28 and 6.32, you can recognize Eq. 6.38 is a transcendental equation for energy E_z. Solutions of this equation determine the values of energy permitting the existence of non-zero coefficients B and D and, hence, of the wave functions satisfying all the boundary conditions. The solutions of Eq. 6.38 are obviously eigenvalues of the Hamiltonian also called allowed energy values or energy levels.

Equation 6.38 does not submit to an analytical solution, but it still can be qualitatively analyzed to help you to determine, at least, the number of solutions it might have. To this end, it is convenient to rewrite this equation introducing dimensionless variables, such as $kd/2$ and $\kappa d/2$. To facilitate the transition to these variables, I first compute $k^2 + \kappa^2$ using Eqs. 6.28 and 6.32:

$$k^2 + \kappa^2 = \frac{2m_e (V_b - V_w)}{\hbar^2}.$$

Multiplying this expression by $d^2/4$ and introducing dimensionless ε for $kd/2$, I have

$$\varepsilon^2 + \frac{\kappa^2 d^2}{4} = \frac{m_e (V_b - V_w) d^2}{2\hbar^2} \Rightarrow$$

$$\frac{\kappa d}{2} = \sqrt{\varepsilon_0^2 - \varepsilon^2},$$

where I introduced another dimensionless parameter ε_0 defined as

$$\varepsilon_0 = \frac{d}{\hbar} \sqrt{\frac{m_e \left(V_b - V_w\right)}{2}}.$$

Multiplying both sides of Eq. 6.38 by $d/2$, I can rewrite it now as

$$\tan \varepsilon = \frac{\sqrt{\varepsilon_0^2 - \varepsilon^2}}{\varepsilon}. \tag{6.39}$$

You can see that ε_0 incorporates all relevant parameters of the system, solely determining the allowed energy values. This is an excellent illustration of the power of dimensionless variables: four different parameters have collapsed into a single one, which rules them all. Without even solving the equation, I know now that all rectangular potential wells with different values of m_e, $V_b - V_w$ and d will have the same dimensionless energy levels as long as all these parameters correspond to the same ε_0.

Obviously, Eq. 6.39 only makes sense for $\varepsilon \leq \varepsilon_0$, so it is important to understand the physical meaning of this condition. Substituting all necessary definitions, you can see that $\varepsilon = \varepsilon_0$ turns into

$$\frac{m_e \left(E_z - V_w\right) d^2}{2\hbar^2} = \frac{m_e \left(V_b - V_w\right) d^2}{2\hbar^2} \Rightarrow E_z = V_b,$$

i.e., at the point $\varepsilon = \varepsilon_0$ an energy crosses over the potential barrier, where assumptions used to derive Eq. 6.39 lose their validity.

In order to understand solutions to Eq. 6.39, it is useful to visualize graphs of its left-hand and right-hand sides. As ε increases from zero, the function on the right decreases from positive infinity to zero at $\varepsilon = \varepsilon_0$, where it terminates. The left-hand side is a tangent, which grows from zero at $\varepsilon = 0$ and reaches its asymptotic behavior at $\varepsilon = \pi/2$, where it jumps all the way to negative infinity, starts its climb toward the next zero at π, goes to infinity at $3\pi/2$, and so on. If $\varepsilon_0 < \pi$, the two functions will cross only once because the right-hand side will end before the left-hand side manages to get to the positive territory again. Once ε_0, however, crosses the π threshold, the second crossing becomes possible, and one more when ε_0 exceeds 2π, and so on. An important point here is that at least one even solution always exists no matter how small ε_0 becomes. Another important qualitative point one may take home is that the magnitude of ε_0 depends on two main parameters: the depth of the well $V_b - V_w$ and its geometric width d; the wider and deeper wells would be able to accommodate a large number of allowed energy levels, and in order to decrease the number of energy eigenvalues belonging to the discrete spectrum, one can either make the well narrower or shallower. It is important to remember though that the geometric width of the well affects not only values of allowed energies but also the difference between adjacent energy levels, which are closer to each other in wider wells.

The stitching conditions for the odd wave functions take the form

$$A \sin \frac{kd}{2} = De^{-\kappa d/2} \tag{6.40}$$

$$Ak \cos \frac{kd}{2} = -\kappa De^{-\kappa d/2} \tag{6.41}$$

resulting to a different equation for allowed energy values

$$\cot \varepsilon = -\frac{\sqrt{\varepsilon_0^2 - \varepsilon^2}}{\varepsilon}. \tag{6.42}$$

The right-hand side of this equation changes from negative infinity to zero at $\varepsilon = \varepsilon_0$, while the left-hand side begins at positive infinity at $\varepsilon = 0$ and crosses to the negative territory only for $\varepsilon > \pi/2$. Thus, if $\varepsilon_0 < \pi/2$, this equation has no solutions. In this case, the only allowed eigenvalue of energy corresponds to a single even solution for the wave function. Increasing ε_0 beyond $\pi/2$ will produce the first odd solution, and it will happen before the second even solution appears. Following this line of reasoning, you can see that with increasing ε_0, eigenvalues corresponding to odd and even wave functions appear in an alternating manner.

The state corresponding to the lowest energy is called the ground state, and as you just saw, it is always represented by an even function. The next energy corresponds to an odd solution, then you have again an energy level corresponding to the even solution, then to an odd one again, and this pattern repeats until the last allowed eigenvalue is reached, which can be either odd or even.

All solutions of Eqs. 6.39 and 6.42 can be enumerated as ε_n, where $n = 1$ corresponds to the ground state (even) solution, $n = 2$ to the lowest in energy odd solution, and so on and so forth. Similar enumeration can be applied to the wave functions

$$\varphi_n(z) = \begin{cases} B_n \cos k_n z & |z| < d/2 \\ B_n \cos (k_n d/2) \, e^{\kappa_n d/2} e^{-\kappa_n |z|} & |z| > d/2 \end{cases}, \quad n = 1, 3, 5 \ldots \tag{6.43}$$

and

$$\varphi_n(z) = \begin{cases} A_n \sin k_n z & |z| < d/2 \\ A_n \sin (k_n d/2) \, e^{\kappa_n d/2} e^{-\kappa_n z} & z > d/2 \\ -A_n \sin (k_n d/2) \, e^{\kappa_n d/2} e^{\kappa_n z} & z < -d/2 \end{cases}, \quad n = 2, 4, 6 \ldots \tag{6.44}$$

Here, $k_n = 2\varepsilon_n/d$, $\kappa_n = 2\sqrt{\varepsilon_0^2 - \varepsilon_n^2}/d$, while the corresponding values of energy E_{zn} are

$$E_{zn} = V_w + \frac{2\hbar^2}{md^2} \varepsilon_n^2. \tag{6.45}$$

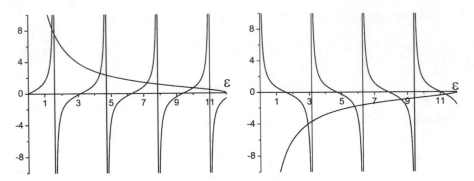

Fig. 6.5 Graphic solution of the eigenvalue equation for even and odd wave functions

You may notice that wave functions in Eqs. 6.43 and 6.44 still contain undefined coefficients B_n and A_n correspondingly, while coefficient D was eliminated using Eqs. 6.36 and 6.40. This is, however, normal because all eigenvectors and representing their functions are always defined only up to a constant factor (I have said that already and not once, right?). As usual, values of these coefficients can be fixed by the normalization condition

$$\int_{-\infty}^{\infty} \varphi_n^2(z)dz = 1.$$

Figure 6.5 illustrates the process of emergence of the even and odd solutions described above. The graph on the left refers to Eq. 6.39 for energies of even states, and the graph on the right corresponds to Eq. 6.44 for energies of the odd states. One can see that the crossing points on the graphs signifying values of the dimensionless energy parameter ε alternate in their values between even and odd states: the lowest energy value comes from the graph on the left, the second lowest appears in the graph on the right, and this alternation continues throughout all the ten energy values depicted in these plots. It is also instructive to plot the wave functions corresponding to a few lowest energy eigenvalues. Graphs in Fig. 6.6 present (from left to right) the ground state and the, first and second excited states. In addition to clearly demonstrating the even and odd nature of the respective states, these graphs reveal an important phenomenon—a transition to a higher energy level always adds an extra zero to the corresponding wave function. This behavior is actually a manifestation of a mathematical theorem valid for any one-dimensional problems with discrete spectrum: the number of zeroes of a wave function corresponding to n-th energy level ($n = 1$ corresponds to the ground state) is always equal to $n - 1$. Since a rigorous proof of this statement is beyond our reach, I will illustrate this point considering a limiting case of a very deep well, such that $\varepsilon_0 \gg 1$. In the limit $\varepsilon_0 \to \infty$, Eq. 6.39 has solutions $\varepsilon_n = \pi n/2$, $n = 1, 3, 5 \cdots$, while solutions of Eq. 6.42 are $\varepsilon_n = \pi n/2$, $n = 2, 4, 6 \cdots$. The corresponding energy values from Eq. 6.45 coincide with those given in Eq. 5.89 (if one replaces L with d) for a

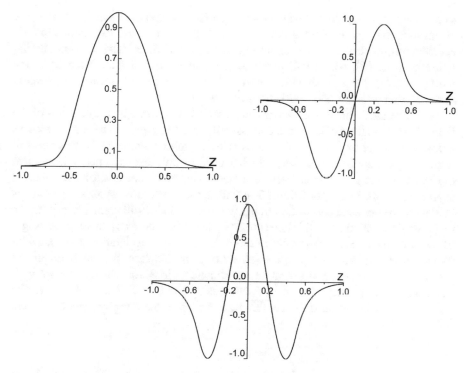

Fig. 6.6 Wave functions corresponding to the first three lowest energy eigenvalues of a rectangular potential well

particle, whose motion is confined in a finite region of the total length d. Obviously, this confinement corresponds to the limit of the potential well with infinitely high barriers. The wave function in this case becomes

$$\varphi_n(z) = \begin{cases} B_n \cos \frac{\pi n z}{d} & n = 1, 3, 5 \cdots \\ A_n \sin \frac{\pi n z}{d} & n = 2, 4, 6 \cdots \end{cases}$$

within the well $|z| < d/2$, and it is exact zero outside of the well (κ_n goes to infinity and vanquishes the exponential terms $\exp \kappa_n (-z + d/2)$ for all $z > d/2$ and $\exp \kappa_n (z + d/2)$ for all $z < -d/2$.) Now, one can clearly see how the increase of n by 1 transforms cos into sin adding an extra zero to the function when z changes from $-d/2$ to $d/2$.

Unbound (Scattering) States: Continuous Spectrum

The range of energies satisfying the condition $E_z > E_b$ corresponds to an unbound classical motion, where the entire domain $-\infty < z < \infty$ becomes classically allowed. The motion of the classical particle depends on the combination of its initial position and initial velocity: if a particle is initially at $z < -d/2$ with velocity directed to the left or at $z > d/2$ with positive velocity, it will keep moving with the same velocity—the potential well would not affect its motion at all. If, however, the initial motion of the particle is directed toward the well, it will experience infinite acceleration (or deceleration) for infinitesimally short time interval when passing points $z = \pm d/2$, which will result in finite increase and then decrease of the particle's speed. After passing the region of the well, the particle resumes its straight linear motion with the same velocity as before.

Quantum mechanical behavior of the particle is described by the solution of the Schrödinger equation, which in the classically allowed region can be presented as a combination of exponential functions with complex arguments. In this spectral region, the symmetry arguments, which I used to find discrete energy levels and the corresponding wave functions, are no longer valid because of the inherent asymmetry in the initial conditions. You will see soon that this asymmetry, which is evident in the classical description of the unbound motion, will manifest itself in the quantum description as well. Therefore, it is no longer necessary to keep the origin of the coordinate axis at the center of the well, and the consideration becomes a bit more convenient if I move it to the left by $d/2$. In this case, the left boundary of the well corresponds to $z = 0$, and the coordinate regions, for which different wave functions must be written, are now defined as $z < 0$, $0 < z < d$, and $z > d$. The most general solution for the wave function in each of these regions can be written down as

$$\varphi(z) = \begin{cases} A_1 e^{ik_1 z} + B_1 e^{-ik_1 z} & z < 0 \\ A_2 e^{ik_2 z} + B_2 e^{-ik_2 z} & 0 < z < d \\ A_3 e^{ik_1 z} + B_3 e^{-ik_1 z} & z > d \end{cases} \tag{6.46}$$

where $k_1 = \sqrt{2m(E_z - V_b)}$ and $k_2 = \sqrt{2m(E_z - V_w)}$. This expression for the wave function has to be complemented by four stitching conditions—two at each of the points of discontinuity. Requiring continuity of the function and its derivative at $z = 0$ and $z = d$, I get

$$A_1 + B_1 = A_2 + B_2 \tag{6.47}$$

$$k_1 (A_1 - B_1) = k_2 (A_2 - B_2) \tag{6.48}$$

$$A_2 e^{ik_2 d} + B_2 e^{-ik_2 d} = A_3 e^{ik_1 d} + B_3 e^{-ik_1 d} \tag{6.49}$$

$$k_2 \left(A_2 e^{ik_2 d} - B_2 e^{-ik_2 d} \right) = k_1 \left(A_3 e^{ik_1 d} - B_3 e^{-ik_1 d} \right) \tag{6.50}$$

where Eqs. 6.47 and 6.49 ensure continuity of the wave function at $z = 0$ and $z = d$ correspondingly, while Eqs. 6.48 and 6.50 do the same for its derivative. Simple counting of the number of unknown coefficients and comparing it with the number of equations tells me that I have got a problem here: there are only four equations for six unknowns, which is one unknown too many. However, I have not yet specified a desirable behavior of the wave function at infinity (a boundary condition), which can be useful in eliminating extra unknowns. Unlike the case of the discrete spectrum, where the behavior of the wave functions at infinity is uniquely prescribed, here I have an array of choices reflecting different physical situations for which the problem at hand is being used.

Before digging into the issue of the boundary conditions at infinity for this problem, it might be useful to get a better physical understanding of the terms appearing in the expressions for $\varphi(z)$. To this end, let me dust off some results from Sect. 5.1.3, namely, the concept of the probability current, Eq. 5.42, which in its one-dimensional reincarnation takes the form of

$$j = \frac{i\hbar}{2m_e} \left(\varphi \frac{d\varphi^*}{dz} - \varphi^* \frac{d\varphi}{dz} \right). \tag{6.51}$$

Substituting a generic form of the wave function $A_i e^{ik_i z} + B_i e^{-ik_i z}$ into Eq. 6.51, you find

$$j = \frac{i\hbar}{2m_e} \left[-ik_i \left(A_i e^{ik_i z} + B_i e^{-ik_i z} \right) \left(A_i^* e^{-ik_i z} - B_i^* e^{ik_i z} \right) \right.$$

$$\left. -ik_i \left(A_i^* e^{-ik_i z} + B_i^* e^{ik_i z} \right) \left(A_i e^{ik_i z} - B_i e^{ik_i z} \right) \right] =$$

$$\frac{\hbar k_i}{2m_e} \left(|A_i|^2 - |B_i|^2 + B_i A_i^* e^{-2ik_i z} - B_i^* A_i e^{2ik_i z} + \right.$$

$$\left. + |A_i|^2 - |B_i|^2 - B_i A_i^* e^{-2ik_i z} + B_i^* A_i e^{2ik_i z} \right) =$$

$$\frac{\hbar k_i}{m_e} |A_i|^2 - \frac{\hbar k_i}{m_e} |B_i|^2 .$$

The first term in this expression describes a positive (directed in positive z direction) probability current associated with term $A_i e^{ik_i z}$ in the wave function, while the second, negative, term describes a probability current in the opposite, negative z direction and is associated with the term $B_i e^{-ik_i z}$ in the wave function. Now, imagine a classical beam of particles of mass m_e all moving with the same speed v in the positive direction of the z-axis. The current of particles in this beam (a number of particles crossing a plane perpendicular to the flow per unit time per unit area of the cross section of the beam) is easily found to be Nv, where N is the number of particles in the beam per beam's unit volume. This expression coincides with the quantum mechanical probability current if you replace v with p/m_e, p with $\hbar k$ and identify $|A|^2$ with N. This comparison allows interpreting the terms of the wave function containing e^{ikz} as corresponding to the beam of particles propagating from left to right and terms with e^{-ikz} as describing particles propagating in the opposite direction.

A typical experiment involving particles with energies in the continuous segment of the spectrum consists in sending particles created by some source, positioned far away from the potential well, toward the well and counting the number of particles in the beam behind the well (transmitted particles) or the number of particles in front of the well but propagating in the negative z direction (reflected particles). In this case, the asymptotic behavior of the wave function at negative infinity must contain both left- and right-propagating currents, while the wave function at the positive infinity only contains the right-propagating particles. This gives us one of the possible boundary conditions at infinity corresponding to this particular experimental situation: $\varphi(z \to \infty) = A_3 e^{ik_1 z}$. For the wave function to have this form, coefficient B_3 in Eq. 6.46 must be set to zero. As a result, I end up with five unknown coefficients and the same four equations, and what is left to realize is that the term $A_1 e^{ik_1 z}$ describes the current of particles created by the source, which is external to the Schrödinger equation and is determined by an experimentalist controlling the concentration of particles in the outgoing beam. Thus, A_1 shall be treated as a free parameter, while all remaining coefficients must be expressed in its terms.

Quantities actually measured in the experiment, i.e., the fraction of particles reflected by the potential or the fraction of particles transmitted past the potential, can be interpreted quantum mechanically as probabilities of reflection $R = j_r/j_{inc}$ and transmission $T = j_{tr}/j_{inc}$, where I introduced notations for the reflected current $j_r = \hbar k_1 |B_1|^2 / m_e$, the incident current $j_{inc} = \hbar k_1 |A_1|^2 / m_e$, and transmitted current $j_r = \hbar k_3 |A_3|^2 / m_e$. Wave number $k_3 = \sqrt{2m_e(E_z - V_\infty)}$ is determined by the value of the potential at $z \to \infty$, V_∞. In the particular case I am dealing with now, the potentials at $z < 0$ and $z > d$ are the same, so that $V_\infty = V_b$ and $k_3 = k_1$. One should realize, however, that it is not always the case, so that one has to be careful when defining the transmission probability. The most general expressions for the reflection and transmission probabilities are

$$R = \frac{|B_1|^2}{|A_1|^2} \tag{6.52}$$

$$T = \frac{k_3 |A_3|^2}{k_1 |A_1|^2}. \tag{6.53}$$

Now it becomes clear that in order to obtain experimentally relevant form for the scattering wave function, you need to solve the following system of equations expressing all unknown coefficients in terms of amplitude of the incident particles A_1:

$$1 + r = A_2 + B_2, \tag{6.54}$$

$$k_1(1 - r) = k_2(A_2 - B_2), \tag{6.55}$$

$$A_2 e^{ik_2 d} + B_2 e^{-ik_2 d} = t e^{ik_1 d}, \tag{6.56}$$

$$k_2\left(A_2 e^{ik_2 d} - B_2 e^{-ik_2 d}\right) = k_1 t e^{ik_1 d}. \tag{6.57}$$

Here, I introduced the amplitude reflection and transmission coefficients $r = B_2/A_1$ and $t = A_3/A_1$ correspondingly and redefined amplitudes A_2 and B_2 as $A_2/A_1 \rightarrow A_2$, and $B_2/A_1 \rightarrow B_2$. Combining the first two equations, I obtain

$$\left(1 + \frac{k_1}{k_2}\right) + \left(1 - \frac{k_1}{k_2}\right) r = 2A_2,$$

$$\left(1 - \frac{k_1}{k_2}\right) + \left(1 + \frac{k_1}{k_2}\right) r = 2B_2,$$

while the other two yield

$$\left(1 + \frac{k_1}{k_2}\right) t e^{ik_1 d} = 2A_2 e^{ik_2 d},$$

$$\left(1 - \frac{k_1}{k_2}\right) t e^{ik_1 d} = 2B_2 e^{-ik_2 d}.$$

Expressing A_2 and B_2 from the last pair of equations and substituting it in the first ones, I get

$$\left(1 + \frac{k_1}{k_2}\right) + \left(1 - \frac{k_1}{k_2}\right) r = t\left(1 + \frac{k_1}{k_2}\right) e^{ik_1 d} e^{-ik_2 d},$$

$$\left(1 - \frac{k_1}{k_2}\right) + \left(1 + \frac{k_1}{k_2}\right) r = t\left(1 - \frac{k_1}{k_2}\right) e^{ik_1 d} e^{ik_2 d},$$

which after some brushing up yields

$$\left[1 + r\frac{k_2 - k_1}{k_2 + k_1}\right] e^{-ik_1 d} e^{ik_2 d} = t,$$

$$\left[1 + r\frac{k_2 + k_1}{k_2 - k_1}\right] e^{-ik_1 d} e^{-ik_2 d} = t.$$

Equating the left-hand sides of these two equations gives

$$e^{-ik_1 d} e^{ik_2 d}\left(1 + \frac{k_2 - k_1}{k_2 + k_1} r\right) = e^{-ik_1 d} e^{-ik_2 d}\left(1 + \frac{k_2 + k_1}{k_2 - k_1} r\right) \rightarrow$$

$$r\left(\frac{k_2 + k_1}{k_2 - k_1} e^{-2ik_2 d} - \frac{k_2 - k_1}{k_2 + k_1}\right) = 1 - e^{-2ik_2 d}.$$

Finally, some simple algebraic manipulations, which I hope you can reproduce yourselves, yield

$$r = \frac{\left(k_1^2 - k_2^2\right) \sin k_2 d}{\left(k_2^2 + k_1^2\right) \sin k_2 d + 2ik_2 k_1 \cos k_2 d},$$ (6.58)

$$t = \frac{2ik_2 k_1}{\left(k_2^2 + k_1^2\right) \sin k_2 d + 2ik_2 k_1 \cos k_2 d}.$$ (6.59)

Now, you can easily find two remaining coefficients A_2 and B_2:

$$A_2 = \frac{1}{2}\left(1 + \frac{k_1}{k_2}\right) e^{i(k_1 - k_2)d} \frac{2ik_2 k_1}{\left(k_2^2 + k_1^2\right) \sin k_2 d + 2ik_2 k_1 \cos k_2 d},$$ (6.60)

$$B_2 = \frac{1}{2}\left(1 - \frac{k_1}{k_2}\right) e^{i(k_1 + k_2)d} \frac{2ik_2 k_1}{\left(k_2^2 + k_1^2\right) \sin k_2 d + 2ik_2 k_1 \cos k_2 d}.$$ (6.61)

Now, once the expressions for the coefficients of the wave function are found in terms of A_1, you might wonder if it is possible and/or necessary to fix the value of the latter. Generally speaking, this is again a question of normalization of the wave function, and according to our general understanding, we must be able to normalize this function using the delta-function. However, when the wave function is not just a plane wave, the procedure becomes rather cumbersome and requires careful evaluation of diverging integrals. From a practical point of view, it does not make much sense to jump from all these hoops to achieve the normalization, which would matter only if you plan to use the resulting functions as a basis, and this almost never happens. Thus, if you are only concerned with obtaining experimentally relevant quantities, you will be happy leaving A_1 undetermined and use Eqs. 6.58 and 6.59 to find the transmission and reflection probabilities from Eqs. 6.52 and 6.53:

$$R = \frac{\left(k_1^2 - k_2^2\right)^2 \sin^2 k_2 d}{\left(k_2^2 + k_1^2\right)^2 \sin^2 k_2 d + 4k_2^2 k_1^2 \cos^2 k_2 d}$$ (6.62)

$$T = \frac{4k_2^2 k_1^2}{\left(k_2^2 + k_1^2\right)^2 \sin^2 k_2 d + 4k_2^2 k_1^2 \cos^2 k_2 d}.$$ (6.63)

The denominator of these expressions can be rewritten in the following form:

$$\left(k_2^2 + k_1^2\right)^2 \sin^2 k_2 d + 4k_2^2 k_1^2 \cos^2 k_2 d = 4k_2^2 k_1^2 + \left(k_1^2 - k_2^2\right)^2 \sin^2 k_2 d.$$

Thanks to this rearrangement, you can realize two important facts. First, you can immediately see that

$$R + T = 1$$ (6.64)

and, second, that the transmission, considered as a function of energy, oscillates between its maximum value equal to unity, achieved at $k_2 d = \pi n$, $n = 1, 2, 3 \cdots$, and its minimum value

$$T_{min} = \frac{4 k_2^2 k_1^2}{\left(k_2^2 + k_1^2 \right)^2}$$

which occurs at $k_2 d = \pi/2 + \pi n$. For large values of energy $E_z \gg V_b$, when k_1 and k_2 become close to each other, the minimum value of transmission differs little from unity, so that the transmission remains close to one for almost all energies. The reflection probability, in this case, becomes correspondingly small for all energies as well. This is the behavior close to what you would expect from a classical particle, so that the higher energy limit means transition to the classical regime. This behavior is illustrated in Fig. 6.7.

Equation 6.64 is an important expression of the conservation of probability— it simply states that since transmission and reflection are the only two mutually exclusive events that can occur when a particle is incident on the potential, the sum of their probabilities must be equal to unity. Even though this relation was derived here for the particular case of the rectangular well, it is valid for a generic potential asymptotically approaching a constant value at $z \to \pm\infty$. The validity of Eq. 6.64 serves in reality as a test on correctness of Eqs. 6.62 and 6.63. Using the definitions of transmission and reflection coefficients in terms of the probability currents, I can rewrite Eq. 6.64 as

$$\frac{j_r}{j_{inc}} + \frac{j_{tr}}{j_{inc}} = 1 \iff j_{inc} - j_r = j_{tr}. \tag{6.65}$$

This equation establishes that the total probability current on the left of the potential well is equal to the probability current on its right, which is just a general statement of the conservation of probability, which can also be interpreted as a continuity of the probability current across any finite discontinuity of the potential.

Fig. 6.7 Transmission probability for the rectangular potential well

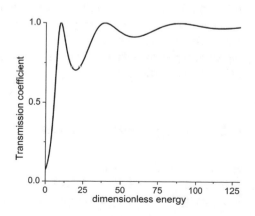

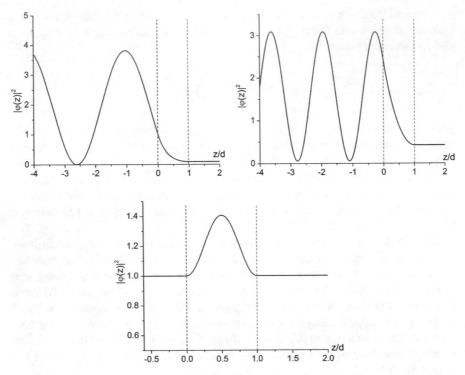

Fig. 6.8 Spatial dependence of $|\varphi(z)|^2$ for three different values of energy: low energy, high energy, and resonance energy where transmission goes to one and reflection to zero. The absence of reflection in the last plot is evidenced by the absence of oscillations of the probability density due to interference of the incident and reflected waves. The vertical lines delineate the edges of the well

The behavior of the wave function also changes with energy. Figure 6.8 illustrates this point plotting the spatial dependence of the respective probability density $|\varphi(z)|^2$ at three different energies, including the one which corresponds to zero reflection. In the latter case, the probability distribution becomes flat at both $z < -d/2$ and $z > d/2$ signaling the absence of interference between incident and reflected waves. One can also notice the decrease in the period of the oscillations for higher energies as it should be expected because higher energy means large wave number and shorter wavelength.

6.2.2 Square Potential Barrier

Square potential barrier is a potential well turned upside down, when the higher value of the potential energy V_b is limited to the finite interval $|z| < d/2$, while the lower energy V_w corresponds to the semi-infinite regions $|z| > d/2$ outside of this interval. The first principal difference between this situation and the one considered in the previous section is that there are no energies corresponding to a classically bound motion in this potential, and, therefore, there are no states corresponding to discrete energy levels. In both cases, $E_z < V_b$ and $E_z > V_b$, classical motion is unbound, and quantum mechanical states belong to the continuous spectrum (there are no states with $E_z < V_w$). The difference between these energy regions is that in the former case, the interval $|z| < d/2$ is classically forbidden, while in the latter, the entire domain of z-coordinate is classically allowed. Respectively, there are two different types of wave functions: when $V_w < E_z < V_b$

$$\varphi(z) = \begin{cases} A_1 e^{ik_2 z} + B_1 e^{-ik_2 z} & z < -d/2 \\ A_2 e^{\kappa_1 z} + B_2 e^{-\kappa_1 z} & -d/2 < z < d/2 \\ A_3 e^{ik_2 z} & z > d/2 \end{cases} \tag{6.66}$$

where k_2 is defined as in the previous section, while $\kappa_1 = \sqrt{2m (V_w - E_z)}$ is related to k_1 as $k_1 = -i\kappa_1$. I already mentioned it once, but I would like to emphasize again—you cannot eliminate either of the real exponential functions in the second line of Eq. 6.66 because the requirement for the wave function to decay at infinity can be used only when the classically forbidden region expands to infinity. In the case at hand, it is limited to the region $|z| < d/2$, so the exponential growth of the wave function does not have enough "room" to become a problem. For energies $E_z > V_b$, the wave function has the form of

$$\varphi(z) = \begin{cases} A_1 e^{ik_2 z} + B_1 e^{-ik_2 z} & z < -d/2 \\ A_2 e^{ik_1 z} + B_2 e^{-ik_1 z} & -d/2 < z < d/2 \\ A_3 e^{ik_2 z} & z > d/2. \end{cases} \tag{6.67}$$

This wave function is essentially equivalent to the one considered in the case of the potential well, so you can simply copy Eqs. 6.58 and 6.59 while exchanging k_1 and k_2:

$$B_1 = \frac{e^{-ik_2 d} \left(k_2^2 - k_1^2\right) \sin k_1 d}{\left(k_2^2 + k_1^2\right) \sin k_1 d + 2ik_2 k_1 \cos k_1 d} \tag{6.68}$$

$$A_3 = \frac{2ik_2 k_1 e^{-ik_2 d}}{\left(k_2^2 + k_1^2\right) \sin k_1 d + 2ik_2 k_1 \cos k_1 d}. \tag{6.69}$$

Transmission and reflection coefficients in this case have all the same properties as in the case of the potential well, which I am not going to repeat again.

Going back to the case $V_w < E_z < V_b$, it might appear that here I would have to carry out all the calculations from scratch because now I have to deal with real exponential functions. But fear not, you still can use the previous result by replacing k_1 with $k_1 = -i\kappa_1$. The negative sign in this expression is important—it ensures that the coefficient A_2 in Eq. 6.67 goes over to the same coefficient A_2 in Eq. 6.66 (the same obviously applies to coefficients B_2). In order to finish the transformation of Eqs. 6.68 and 6.69 for the under-the-barrier case, you just need to recall that $\sin(ix) = i\sinh x$ and $\cos(ix) = \cosh x$. With these relations in mind, you easily obtain

$$B_1 = -\frac{ie^{-ik_2d}\left(k_2^2 + \kappa_1^2\right)\sinh \kappa_1 d}{-i\left(k_2^2 - \kappa_1^2\right)\sinh \kappa_1 d + 2k_2\kappa_1 \cosh \kappa_1 d} \tag{6.70}$$

$$A_3 = \frac{2k_2\kappa_1 e^{-ik_2d}}{-i\left(k_2^2 - \kappa_1^2\right)\sinh \kappa_1 d + 2k_2\kappa_1 \cosh \kappa_1 d}. \tag{6.71}$$

The respective transmission and reflection coefficients become

$$R = \frac{\left(k_2^2 + \kappa_1^2\right)^2 \sinh^2 \kappa_1 d}{\left(k_2^2 - \kappa_1^2\right)^2 \sinh^2 \kappa_1 d + 4k_2^2\kappa_1^2 \cosh^2 \kappa_1 d} \tag{6.72}$$

$$T = \frac{4k_2^2\kappa_1^2}{\left(k_2^2 - \kappa_1^2\right)^2 \sinh^2 \kappa_1 d + 4k_2^2\kappa_1^2 \cosh^2 \kappa_1 d}. \tag{6.73}$$

Even though I derived Eqs. 6.70 and 6.71 by merely extending Eqs. 6.68 and 6.69 to the region of imaginary k_1 (for mathematically sophisticated—this procedure is a simple example of what is known in mathematics as analytical continuation), the properties of the reflection and transmission coefficients given by Eq. 6.72 are very different from those derived for the over-the-barrier transmission case $E_z > V_b$. Gone are their periodic dependence on the energy and d, as well as special values of energy, when the transmission turns to unity and reflection goes to zero. What do we have instead? Actually quite a boring picture: transmission is exponentially decreasing with increasing width of the barrier d and slowly approaches unity as the energy swings between V_w and V_b because $\kappa_1 = \sqrt{2m_e (V_b - E_z)}$ vanishes at $E_z = V_b$. To illustrate the exponential dependence of the transmission on d, I will consider a case of a "thick" barrier, which in mathematical language means $\kappa_1 d \gg 1$. To find the required approximate expression for T and R, I need to remind you a simple property of hyperbolic functions $\cosh x$ and $\sinh x$: for large values of their argument x, these functions can be approximated by a simple exponential, $\sinh x \simeq \cosh x \simeq \frac{1}{2}\exp x$. Taking this into account, I can derive

$$R \simeq 1 \tag{6.74}$$

$$T \simeq \frac{4k_2^2\kappa_1^2}{\left(k_2^2 + \kappa_1^2\right)^2}e^{-4\kappa_1 d}. \tag{6.75}$$

When deriving the expression for the reflection coefficient, I "lost" the exponentially small term, which is supposed to be subtracted from unity to ensure conservation of probability. At the same time, this term makes the main contribution to the transmission coefficient and, therefore, survives. A better approximation for the reflection coefficient can be found simply by writing it down as $R = 1 - T$. Obviously, the same results can be derived directly from Eq. 6.72 by being a bit more careful and keeping leading exponentially small terms.

What is surprising here is, of course, not the fact that the transmission is small, but that it is not exactly equal to zero. Because what it means is that there exists a non-zero probability for the particle to travel across a classically forbidden region, emerge on the other side, and keep moving as a free particle. This phenomenon, which is a quantum mechanical version of "walking through the wall," is called tunneling, and you can hear physicists saying that the particle tunnels through the barrier. The exponential nature of the dependence upon d is very important, because exponential function is one of the fastest changing functions appearing in mathematical description of natural processes. It means that a small change in d results in a substantial change in transmission. This effect has vast practical importance and is used in many applications such as tunneling diodes, tunneling microscopy, flush memory, etc.

6.3   Delta-Functional Potential

In this section, I will present a rather peculiar model potential, which does not really have direct analogies in the real world. I can justify spending some time on it by making three simple points: (a) it is easily solvable, so considering it would not take too much of our time, (b) it is useful as an illustration of a situation when the derivative of the wave function loses its continuity property, and (c) in the case of shallow potential wells, which are able to hold only a single bound state, it can provide a decent qualitative understanding of real physical situations. This utterly unrealistic potential has the form of a delta-function

$$V = -\varsigma \delta(z), \tag{6.76}$$

where the negative sign signifies that the potential is attractive and that the states with negative energies are possible and must belong to the discrete spectrum. Indeed, the entire region of z except of a single point $z = 0$ is classically forbidden, so the motion of a classical particle, if one can imagine being localized to a single point as a motion, is finite. Parameter ς in this expression represents a "strength" of the potential, but one needs to understand that the dimension of this parameter is *energy × length*, so it should not be interpreted as a "magnitude" of the potential. It becomes obvious if one integrates Eq. 6.76: $\varsigma = \int V(z)dz$, so it is clear that ς is the area under the potential. If one thinks of the delta-function as a limiting case of a rectangular potential of depth V_w and width d, with $V_w \to \infty$ and $d \to 0$, in such a way that $\varsigma = V_w d$ remains constant, the meaning of this parameter becomes even more transparent.

The main peculiarity of this model is that the discontinuity of the potential in this case involves more than just a finite jump, so that my previous arguments concerning the continuity of the derivative of the wave function are no longer applicable. Actually, this derivative is not continuous at all, and the first matter of business is to figure out how to "stitch" derivatives of the wave function defined at $z < 0$ with those defined at $z > 0$. To solve this puzzle, let me start with the basics—the Schrödinger equation

$$-\frac{\hbar^2}{2m_e}\frac{d^2\varphi}{dz^2} - \varsigma\delta(z)\varphi(z) = E\varphi(z). \tag{6.77}$$

Integrating this equation over infinitesimally small interval $-\epsilon, \epsilon$ and taking into account that an integral of a continuous function over such an interval is zero (in the limit $\epsilon \to 0$), I get

$$-\frac{\hbar^2}{2m_e}\left(\frac{d\varphi}{dz}\bigg|_{z=\epsilon} - \frac{d\varphi}{dz}\bigg|_{z=-\epsilon}\right) - \varsigma\varphi(0) = 0.$$

This yields the derivative stitching rule:

$$\frac{d\varphi}{dz}\bigg|_{z=\epsilon} - \frac{d\varphi}{dz}\bigg|_{z=-\epsilon} = -\frac{2m_e}{\hbar^2}\varsigma\varphi(0). \tag{6.78}$$

Now, all what I need is to solve the Schrödinger equation with zero potential and negative energy separately for $z < 0$ and $z > 0$ and stitch the solutions. Since both these regions are classically forbidden for a particle with $E < 0$, the solutions have the form of real-valued exponential functions:

$$\varphi(z) = \begin{cases} A_1 e^{\kappa z} & z < 0 \\ A_2 e^{-\kappa z} & z > 0 \end{cases}$$

where $\kappa = \sqrt{-2m_e E}/\hbar$, and I discarded the contributions which would grow exponentially at positive and negative infinities to satisfy the boundary conditions. A continuity of the wave function at $z = 0$ requires that $A_2 = A_1$, and Eq. 6.78 yields

$$-2\kappa A = -\frac{2m_e}{\hbar^2}\varsigma A.$$

Assuming that A is non-zero (naturally) and taking into account the definition of κ, I find that this expression is reduced to the equation for allowed energy levels:

$$E = -\frac{m_e\varsigma^2}{2\hbar^2}. \tag{6.79}$$

Obviously, Eq. 6.79 shows that there is only one such energy, which is why this model can only be useful for description of shallow potential wells with a single discrete energy level.

Solutions with positive energies can be constructed in the same way as it was done for the rectangular potential well or barrier

$$\varphi(z) = \begin{cases} A_1 e^{ikz} + B_1 e^{-ikz} & z < 0 \\ A_2 e^{ikz} & z > 0 \end{cases} \tag{6.80}$$

where $k = \sqrt{2m_e E}/\hbar$, and the continuity of the wave function at $z = 0$ yields

$$A_1 + B_1 - A_2.$$

The derivative stitching condition, Eq. 6.78, generates the following equation:

$$ikA_2 - ik(A_1 - B_1) = -\frac{2m_e}{\hbar^2} \varsigma A_2.$$

Solving these two equations for B_1 and A_2, one can obtain

$$\frac{A_2}{A_1} = \frac{1}{1 - i\chi},$$

$$\frac{B_1}{A_1} = \frac{i\chi}{1 - i\chi},$$

where I introduced a convenient dimensionless parameter χ defined as

$$\chi = \frac{m_e \varsigma}{k\hbar^2}.$$

The amplitude transmission and reflection coefficients $t = A_2/A_1$ and $r = B_1/A_1$ are complex numbers, which can be presented in the exponential form using Euler formula as

$$r = \sqrt{R} e^{i\theta_r}; \ t = \sqrt{T} e^{i\theta_t}$$

where reflection and transmission probabilities $R = |r|^2$ and $T = |t|^2$ and corresponding phases θ_r and θ_t are given by

$$R = \frac{\chi^2}{1 + \chi^2}; \ T = \frac{1}{1 + \chi^2} \tag{6.81}$$

$$\theta_r = -\arctan\frac{1}{\chi}; \ \theta_t = \arctan\chi. \tag{6.82}$$

While the phase of the amplitude reflection and transmission coefficients do not affect the probabilities, they still play an important role and can be observed. The reflected wave function interferes with the function describing incident particles and determines the spatial distribution of relative probabilities of position measurements. These phases also define the temporal behavior of the particles in the situations involving nonstationary states, but discussion of this situation is outside of the scope of this book.

6.4 Problems

Problems for Sect. 6.2.1

Problem 77 Derive Eq. 6.42.

Problem 78 Find the reflection and transmission coefficients for the potential barrier shown in Fig. 6.9. Show that $R + T = 1$.

Problem 79 In quantum tunneling, the penetration probability is sensitive to slight changes in the height and/or width of the barrier. Consider an electron with energy $E = 15\,\text{eV}$ incident on a rectangular barrier of height $V = 7\,\text{eV}$ and width $d = 1.8\,\text{nm}$. By what factor does the penetration probability change if the width is decreased to $d = 1.7\,\text{nm}$?

Problem 80 Consider a step potential

$$V(z) = \begin{cases} 0 & x < 0 \\ V_0 & x > 0. \end{cases}$$

Calculate the reflection and transmission probabilities for two cases $0 < E < V_0$ and $E > V_0$.

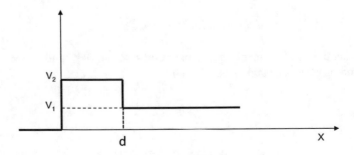

Fig. 6.9 Potential barrier with an asymmetric potential

Problem 81 Find an equation for the energy levels of a particle of mass m_e moving in a potential of the form

$$V(z) = \begin{cases} \infty & x < -a \\ 0 & -a < x < -b \\ V_0 & -b < x < b \\ 0 & b < x < a \\ \infty & x > a. \end{cases}$$

Consider even and odd wave functions separately. Using any graphic software, find the approximate values of the two lowest values of the energy if $m - 1.78 \times 10^{-27}$ kg, $a = 0.12$ nm, $b = 0.42$ nm, $V_0 = 1.5$ eV. Sketch the respective wave functions for each of the found eigenvalues.

Problem 82 Consider a particle moving in a potential comprised of two attractive delta-functional potentials separated by a distance d:

$$V(x) = -\varsigma\delta(x + d/2) - \varsigma\delta(x - d/2).$$

1. Derive an equation for discrete energy levels in this potential, and solve it if possible. How many discrete energy levels does this potential have? Analyze the behavior of these energy levels when the distance d between the wells increases.
2. Find the wave functions corresponding to the continuous segment of the spectrum, and determine the respective transmission and reflection probabilities.

Chapter 7
Harmonic Oscillator Models

It is as difficult to overestimate the role of harmonic oscillator models in physics in general and in quantum mechanics in particular as the influence of Beatles and Led Zeppelin on modern popular music. Harmonic oscillators are ubiquitous and appear every time when one is dealing with a system that has a state of equilibrium in the vicinity of which it can oscillate, i.e., in a vast majority of physical systems—atoms, molecules, solids, electromagnetic field, etc. It also does not hurt their popularity that the harmonic oscillator is one of the very few models which can be solved exactly.

Consider a particle moving in a potential $V(x, y, z)$, which has a minimum at some point $x = y = z = 0$. Mathematically speaking, this means that at this point $\partial V/\partial x = \partial V/\partial y = \partial V/\partial z = 0$, while the matrix of the second derivatives $L_{ij} \equiv \partial^2 V/\partial r_i \partial r_j|_{x=y=z=0}$, where $r_1 \equiv x$, $r_2 \equiv y$, and $r_3 \equiv z$, is positive definite. If you still remember the connection between the potential energy and the force in classical mechanics, you should recognize that in this situation, point $x = y = z = 0$ corresponds to the particle being in the state of stable equilibrium. Stable in this context means that a particle removed from the equilibrium by a small distance will be forced to move back toward it rather than away from it. Expanding potential energy in a power series in the vicinity of the equilibrium and keeping only the first nonvanishing terms, you will get

$$V(x, y, z) \approx \frac{1}{2} \sum_{i,j} L_{i,j} r_i r_j.$$

Respective classical Hamiltonian equations 3.2 and 3.3 yield for this potential:

$$\frac{dp_i}{dt} = -\sum_j L_{ij} r_j \tag{7.1}$$

© Springer International Publishing AG, part of Springer Nature 2018
L.I. Deych, *Advanced Undergraduate Quantum Mechanics*,
https://doi.org/10.1007/978-3-319-71550-6_7

$$\frac{dr_i}{dt} = \frac{p_i}{m_e} \tag{7.2}$$

($i = x, y, z$). They can be converted into Newton's equations by differentiating Eq. 7.2 (with respect to time) and eliminating the resulting time derivative of the momentum using Eq. 7.1:

$$\frac{dr_i^2}{dt^2} = -\frac{1}{m_e} \sum_j L_{i,j} r_j. \tag{7.3}$$

The presence in matrix L_{ij} of nondiagonal elements indicates that the particle's motion in the direction of any of the chosen axes X, Y, or Z is not independent of its motion in other directions. In layman's terms, it means that it is impossible to arrange for this particle to move purely in the direction of any of the axes. Nevertheless, solutions of these equations still can be presented in the standard time-harmonic form $r_i = a_i \exp(i\omega t)$ with amplitudes a_i and frequency ω obeying equations:

$$\frac{1}{m_e} \sum_j L_{i,j} a_j = \omega^2 a_i, \tag{7.4}$$

which is an eigenvalue equation for the matrix $L_{i,j}/m_e$. It is obvious that this is symmetric ($L_{i,j} = L_{j,i}$), real-valued, and, therefore, Hermitian matrix. Thus, based on the eigenvalue theorems discussed in Sect. 3.3.1, this matrix is guaranteed to have real eigenvalues and corresponding orthogonal eigenvectors. The equation for the eigenvalues is found by requiring that Eq. 7.4 has nontrivial solutions:

$$det\left(m_e\omega^2\delta_{i,j} - L_{i,j}\right) = 0$$

and in general has three solutions ω_n^2, where $n = 1, 2, 3$. Substituting each of these frequencies back in Eq. 7.4, you can find amplitudes $a_x^{(n)}$, $a_y^{(n)}$ $a_z^{(n)}$, which form corresponding eigenvectors. These eigenvectors are regular three-dimensional vectors defining three mutually orthogonal directions in space. Oscillations in each of these directions, called normal modes, are characterized by their unique frequencies ω_n^2 and can occur independently of each other. Indeed, these three vectors can be used as a new basis, which in this particular case amounts to introducing new coordinate axes along the directions of the normal modes. The matrix $L_{i,j}/m_e$ transformed to this basis becomes diagonal, and introducing notation ξ_1, ξ_2, and ξ_3, to represent coordinates along these new directions, Eq. 7.3 will take a form of three independent differential equations:

$$\frac{d^2\xi_n}{dt^2} = -\omega_n^2\xi_n. \tag{7.5}$$

One can also show (those interested in details are welcome to read any of many text-books on classical mechanics, or molecular oscillations, or a combination thereof) that Hamiltonian written in terms of these new coordinates and corresponding conjugated momentums π_n takes the form

$$H = \sum_n \left(\frac{\pi_n^2}{2m_n} + m_n \omega_n^2 \xi_n^2 \right)$$

which is the sum of three independent one-dimensional Hamiltonians. The transition to this form is not so trivial, and the mass parameter m_n does not have to coincide with the actual mass of the particle. Nevertheless, as long as π_n and ξ_n are a canonically conjugated pair characterized by the standard for the coordinate and momentum Poisson brackets, Eq. 3.5, we can treat them as such for all practical purposes, including quantization.

Thus, using the concept of normal modes, one can always reduce a problem involving harmonic oscillations to a simple combination of one-dimensional problems. This is actually true even in the case involving oscillations of several particles such as multi-atom molecules. Therefore, using the one-dimensional model to describe harmonic oscillations is even more justified than the one-dimensional models described in the previous chapter. And so, the one-dimensional model of the quantum harmonic oscillator is what I am going to consider next.

7.1 One-Dimensional Harmonic Oscillator

7.1.1 Stationary States (Eigenvalues and Eigenvectors)

Classical mechanics of the one-dimensional harmonic oscillator is described by Hamiltonian:

$$H = \frac{p^2}{2m_e} + \frac{1}{2}m_e \omega^2 x^2, \tag{7.6}$$

and respective Hamiltonian equations for momentum p and coordinate x are obtained by specializing Eqs. 7.1 and 7.2 to the one-dimensional situation:

$$\frac{dp}{dt} = -m_e \omega^2 x \tag{7.7}$$

$$\frac{dx}{dt} = \frac{p}{m_e} \tag{7.8}$$

where I replaced the corresponding diagonal element of matrix $L_{i,j}$ as $L_{xx} \equiv m_e \omega^2$. These equations are, of course, easy to solve, and the solution is well known:

$$x = x_0 \cos \omega t + \frac{p_0}{m_e \omega} \sin \omega t$$

$$p = -m_e \omega x_0 \sin \omega t + p_0 \cos \omega t, \tag{7.9}$$

where x_0 and p_0 are initial values of the coordinate and momentum of the particle. Equation 7.9 describes a familiar harmonic time dependence, which can be presented in terms of amplitude A and initial phase δ:

$$x(t) = A \sin (\omega t + \delta).$$

Both A and δ are determined by the initial conditions: the amplitude—by the total energy E of the oscillator, which, as you know, is a conserving quantity—and phase, by the ratio of the initial coordinate and momentum. Recalling that at the maximum displacement E takes entirely the form of the potential energy, you can write

$$\frac{1}{2} m_e \omega^2 A^2 = E = \frac{p_0^2}{2m_e} + \frac{1}{2} m_e \omega^2 x_0^2 \Rightarrow$$

$$A = \sqrt{\frac{2E}{m_e \omega^2}} = \sqrt{x_0^2 + \frac{p_0^2}{m_e^2 \omega^2}}. \tag{7.10}$$

The phase of the oscillator can be found by expanding $\sin (\omega t + \delta) = \sin \omega t \cos \delta + \cos \omega t \sin \delta$ and equating the resulting terms with their counterparts in Eq. 7.9. This yields

$$A \cos \delta = \frac{p_0}{m_e \omega}$$

$$A \sin \delta = x_0$$

and subsequently

$$\tan \delta = \frac{x_0 m_e \omega}{p_0}.$$

Obviously, the motion of a harmonic oscillator is bounded with the maximum deviation from the equilibrium position given by its amplitude A, Eq. 7.10. The coordinate x becomes equal to A at two turning points, where the velocity of the oscillator and, respectively, its kinetic energy turn to zero. The relation between total, potential, and kinetic energies of the harmonic oscillator can be illustrated by a diagram shown in Fig. 7.1, where vertical lines show the turning points of the classical motion.

Even though you all have known the solution to the harmonic oscillator problem almost since the elementary school, you might find it useful to play with its Hamiltonian a bit more. Let me, for instance, factorize the Hamiltonian, taking advantage of its $u^2 + v^2$ form, which can be presented as $(u + iv)(u - iv)$:

Fig. 7.1 Energy diagram for
a classical oscillator. The
horizontal line corresponds to
its total energy E, and vertical
dashed lines indicate the
turning points and
coordinates corresponding to
two maximum displacements

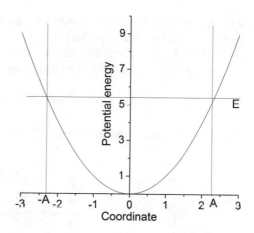

$$H = \left(\frac{p}{\sqrt{2m_e}} + i\sqrt{\frac{m_e}{2}}\omega x \right) \left(\frac{p}{\sqrt{2m_e}} - i\sqrt{\frac{m_e}{2}}\omega x \right). \tag{7.11}$$

Now, on a whim, I am going to compute the Poisson bracket involving these factors. Designating the first of them as u,

$$u = \frac{p}{\sqrt{2m_e}} + i\sqrt{\frac{m}{2}}\omega x,$$

and the second one as u^*,

$$u^* = \frac{p}{\sqrt{2m_e}} - i\sqrt{\frac{m_e}{2}}\omega x,$$

I find, using Eq. 3.4 for Poisson bracket,

$$\{u, u^*\} = \frac{\partial u}{\partial x}\frac{\partial u^*}{\partial p} - \frac{\partial u}{\partial p}\frac{\partial u^*}{\partial x} = i\omega.$$

Using this result as a hint, I now introduce new variables:

$$b = -\frac{i}{\sqrt{\omega}}u = \sqrt{\frac{m_e\omega}{2}}x - i\frac{p}{\sqrt{2m_e\omega}} \tag{7.12}$$

$$b^\dagger = \frac{1}{\sqrt{\omega}}u^* = -i\sqrt{\frac{m_e\omega}{2}}x + \frac{p}{\sqrt{2m_e\omega}} \tag{7.13}$$

whose Poisson bracket, by design, of course, is

$$\{b, b^\dagger\} = 1.$$

This means that b and $b^\dagger$ constitute a canonically conjugated pair (if you have already forgotten what I am talking about, check Sect. 3.1), with b playing the role of the coordinate and $b^\dagger$ pretending to be the momentum. Computing $bb^\dagger$ (do it!), you will see that the Hamiltonian can be presented as

$$H = i\omega bb^\dagger.$$

The corresponding Hamiltonian equations are

$$\frac{db}{dt} = \frac{\partial H}{\partial b^\dagger} = i\omega b \qquad (7.14)$$

$$\frac{db^\dagger}{dt} = -\frac{\partial H}{\partial b} = -i\omega b^\dagger. \qquad (7.15)$$

The advantage of these equations as compared to initial Eqs. 7.7 and 7.8 is that they are independent first-order differential equations, which can be easily solved:

$$b = b_0 e^{i\omega t}; \; b^\dagger = b_0^\dagger e^{-i\omega t}. \qquad (7.16)$$

Initial coordinate and momentum can be expressed in terms of b and $b^\dagger$ by inverting Eqs. 7.12 and 7.13, but I will leave it for you as an exercise.

Transition between pairs x, p and $b, b^\dagger$ is an example of a so-called canonical transformation of variables, and the only reason I decided to bother you with it is that it paves a way to better understanding its quantum analog, which is of crucial importance. According to the quantization rules discussed in Sect. 3.3.2, transition from classical to quantum description consists in promoting classical variables to quantum operators, and the coordinate-momentum dyad plays a crucial role in the process, namely, because it is a canonical pair. The operators replacing classical variables are, to a large extent, defined by their commutation relations, and in the case of canonical pairs, the commutator is directly linked to the respective Poisson brackets, as I have already mentioned previously. In the case of the coordinate-momentum pair, the corresponding commutator is obtained from the Poisson bracket by multiplying it by $i\hbar$. As well as any pair of variables characterized by the canonical Poisson bracket, which play the role similar to coordinate and momentum in classical mechanics, any quantum mechanical pair of operators with canonical commutator $i\hbar$ will have properties similar to those of the coordinate and momentum. For instance, if I know that two Hermitian operators $\hat{\upsilon}$ and $\hat{\sigma}$ have a commutator $[\hat{\upsilon}, \hat{\sigma}] = i\hbar$, I can without any doubts claim that the operator $\hat{\sigma}$ in the representation based on eigenvectors of $\hat{\upsilon}$ is $\hat{\sigma}_\upsilon = -i\hbar\partial/\partial\upsilon$ similar to the momentum operator in the coordinate representation. I have to emphasize, however, the requirement that the operators must be Hermitian. Therefore, if I were to promote b and $b^\dagger$ to operators, it would not work, because they would not be Hermitian. Nevertheless, operators similar to b and $b^\dagger$ (while not exactly like them) do play an important role in quantum theory (and not just for harmonic oscillators).

Now I am ready to get down to our main business and start developing quantum theory of harmonic oscillators. The goal is to develop the theory as far as possible without resorting to any particular representation for momentum and coordinate operators. Such an approach will produce the most general results, independent of a representation, offer important insights into the quantum properties of oscillators, and create a formal framework for extending this theory beyond pure mechanical harmonic oscillators.

I start by "factorizing" the quantum Hamiltonian in a way similar to factorization of the classic Hamiltonian in Eq. 7.11. However, to make sure that this factorization works for operators, I would like to review the origin of the identity:

$$u^2 + v^2 = (u + iv)(u - iv).$$

Removing the parentheses on its right-hand side, I have

$$(u + iv)(u - iv) = u^2 + v^2 + ivu - iuv.$$

If u and v are regular variables, the last two terms in this expression cancel, but if they are non-commuting operators, it is quite obvious that the original factorization rule is no longer true and must be corrected:

$$\hat{u}^2 + \hat{v}^2 = (\hat{u} + i\hat{v})(\hat{u} - i\hat{v}) + i[\hat{u}, \hat{v}]. \tag{7.17}$$

The order of the terms in the parentheses on the right-hand side of this expression can be changed, which will result in an alternative form of the identity:

$$\hat{u}^2 + \hat{v}^2 = (\hat{u} - i\hat{v})(\hat{u} + i\hat{v}) - i[\hat{u}, \hat{v}]. \tag{7.18}$$

Identifying $\hat{u}$ and $\hat{v}$ as

$$\hat{u} = \sqrt{\frac{m_e}{2}}\omega\hat{x} \tag{7.19}$$

$$\hat{v} = \frac{\hat{p}}{\sqrt{2m_e}}, \tag{7.20}$$

I find

$$[\hat{u}, \hat{v}] = \frac{1}{2}\omega[\hat{x}, \hat{p}] = \frac{1}{2}i\hbar\omega. \tag{7.21}$$

Operators $\hat{u}$ and $\hat{v}$ have a dimension of $\sqrt{energy}$, while their commutator, proportional to $\hbar\omega$, obviously has the dimension of energy. If I am not mistaken, I have already remarked that it is often quite beneficial to work with dimensionless quantities. Thus, taking a clue from Eqs. 7.17 and 7.18 and the experience gained working with classical Hamiltonian, I will try to generate dimensionless operators such as

$$\hat{a} = \frac{1}{\sqrt{\hbar\omega}}(\hat{u} + i\hat{v}) = \sqrt{\frac{m_e\omega}{2\hbar}}\hat{x} + i\frac{\hat{p}}{\sqrt{2m_e\hbar\omega}} \qquad (7.22)$$

$$\hat{a}^\dagger = \frac{1}{\sqrt{\hbar\omega}}(\hat{u} - i\hat{v}) = \sqrt{\frac{m_e\omega}{2\hbar}}\hat{x} - i\frac{\hat{p}}{\sqrt{2m_e\hbar\omega}}. \qquad (7.23)$$

The commutator of these operators is

$$[\hat{a}, \hat{a}^\dagger] = -i\sqrt{\frac{m_e\omega}{2\hbar}}\frac{1}{\sqrt{2m_e\hbar\omega}}[\hat{x}, \hat{p}] + i\sqrt{\frac{m_e\omega}{2\hbar}}\frac{1}{\sqrt{2m_e\hbar\omega}}[\hat{p}, \hat{x}] =$$

$$-\frac{i}{2\hbar}[\hat{x}, \hat{p}] + \frac{i}{2\hbar}[\hat{p}, \hat{x}] = 1.$$

Due to a special importance of this result, I will reproduce it as a separate numbered formula:

$$[\hat{a}, \hat{a}^\dagger] = 1. \qquad (7.24)$$

The operators $\hat{a}$ and $\hat{a}^\dagger$ are clearly not Hermitian: performing Hermitian conjugation of Eqs. 7.22 and 7.23, you can immediately see that they are actually Hermitian conjugates of each other, hence the notation $\hat{a}^\dagger$. It will also be useful to express coordinate and momentum operators in terms of $\hat{a}$ and $\hat{a}^\dagger$. Adding and subtracting Eqs. 7.22 and 7.23, I can invert these equations to get

$$\hat{x} = \sqrt{\frac{\hbar}{2\omega m_e}}\left(\hat{a} + \hat{a}^\dagger\right), \qquad (7.25)$$

$$\hat{p} = i\sqrt{\frac{\hbar\omega m_e}{2}}\left(\hat{a}^\dagger - \hat{a}\right). \qquad (7.26)$$

Using the operator factorization identities 7.17 or 7.18 with $\hat{u}$ and $\hat{v}$ defined in Eqs. 7.19 and 7.20, I can derive two alternative forms of the Hamiltonian:

$$\hat{H} = \hbar\omega\left(\frac{1}{2} + \hat{a}^\dagger\hat{a}\right) = \hbar\omega\left(-\frac{1}{2} + \hat{a}\hat{a}^\dagger\right).$$

These two expressions differ by the order of the operators in it and by the sign in front of $1/2$. Formally they are absolutely equivalent, and one can be reduced to another using commutation relation 7.24. However, from a practical point of view (and you will have to trust me on this for now), the first of these expressions is much more convenient to use than the other. Thus, in what follows, I will rely on the representation of the Hamiltonian in the form

$$\hat{H} = \hbar\omega \left(\frac{1}{2} + \hat{a}^\dagger \hat{a} \right). \tag{7.27}$$

Our first task is to find the eigenvalues and eigenvectors of this Hamiltonian, i.e., the stationary states of the harmonic oscillator. Since the classical motion in the harmonic potential is bound for all values of energy, it should be expected that the entire spectrum of the Hamiltonian is discrete so that yet unknown energy eigenvalues can be labeled by a discrete index as E_n and the respective eigenvectors as $|E_n\rangle$:

$$\hat{H} |E_n\rangle = E_n |E_n\rangle. \tag{7.28}$$

Since I am not allowed to use any particular representation for the coordinate and momentum operators, all what I have to go on with are the commutation relations. This invites me to use the same purely algebraic technique, which I successfully used previously when searching for eigenvalues of the operators of the angular momentum in Sect. 3.3.4. However, in the role of the angular momentum ladder operators $\hat{L}_\pm$, I am going to cast operators $\hat{a}$ and $\hat{a}^\dagger$, which appear to have some similarities with $\hat{L}_\pm$: they are also non-Hermitian and are Hermitian conjugates of each other. You might remember that operators $\hat{L}_\pm$ applied to an eigenvector of operator $\hat{L}_z$ generate other eigenvectors with decreased or increased eigenvalue. Will you be surprised if it turns out that operators $\hat{a}$ and $\hat{a}^\dagger$ are doing the same to the eigenvectors of the harmonic oscillator? Probably not.

The first step is to note that eigenvectors of the Hamiltonian coincide with those of the operator $\hat{N} = \hat{a}^\dagger \hat{a}$, which is called the number operator and is obviously Hermitian. Indeed, once you rewrite the Hamiltonian as

$$\hat{H} = \hbar\omega \left(\frac{1}{2} + \hat{N} \right),$$

this statement becomes pretty obvious. Moreover, you can immediately see that if λ_n is the eigenvalue of $\hat{N}$: $\hat{N} |E_n\rangle = \lambda_n |E_n\rangle$, then

$$E_n = \hbar\omega \left(\frac{1}{2} + \lambda_n \right). \tag{7.29}$$

Therefore, I can focus my attention on finding the eigenvalues and eigenvectors of the number operator $\hat{N}$. To this end, I first compute the commutator $\left[\hat{N}, \hat{a} \right]$ (if you want to know what prompted me to do so, the only excuse I can offer is that there isn't much more for me to do, so why not do that?):

$$\left[\hat{N}, \hat{a} \right] = \hat{a}^\dagger \hat{a}^2 - \hat{a}\hat{a}^\dagger \hat{a} = \left(\hat{a}^\dagger \hat{a} - \hat{a}\hat{a}^\dagger \right) \hat{a} = -\hat{a}, \tag{7.30}$$

where I took advantage of Eq. 7.24. Carrying out Hermitian conjugation of this result, and remembering to change the order of the operators in their product after Hermitian conjugation, I immediately obtain

$$\left[\hat{N}, \hat{a}^\dagger\right] = \hat{a}^\dagger. \tag{7.31}$$

In the next step, I consider $\hat{N}\hat{a}\left|E_n\right\rangle$ and use the commutation relation 7.30 to get

$$\hat{N}\hat{a}\left|E_n\right\rangle = -\hat{a}\left|E_n\right\rangle + \hat{a}\hat{N}\left|E_n\right\rangle =$$
$$\lambda_n\hat{a}\left|E_n\right\rangle - \hat{a}\left|E_n\right\rangle = (\lambda_n - 1)\,\hat{a}\left|E_n\right\rangle.$$

This result shows that $\hat{a}\left|E_n\right\rangle$ is an eigenvector of $\hat{N}$ with eigenvalue $\lambda_n - 1$, i.e., the operator $\hat{a}$ generates eigenvectors of $\hat{N}$ with eigenvalues decreasing by one with each application of the operator. Not surprisingly, this operator is called *lowering operator*. The questions, which naturally pop up at this point, are how far down in energy one can go and how one knows when the bottom is reached. The answer to the first question is obvious—that energy eigenvalues of the harmonic oscillator can never be negative, and thus $\lambda_n > -1/2$. The second question is answered by recycling arguments that I have already used when discussing the angular momentum—the only way to reconcile the ability of $\hat{a}$ to keep decreasing λ_n every time it is applied and the requirement that there must exist a smallest λ_n is to impose on the eigenvector corresponding to this minimum value condition:

$$\hat{a}\left|E_{min}\right\rangle = 0. \tag{7.32}$$

Another useful relation is obtained by performing Hermitian conjugation of this equation:

$$\left\langle E_{min}\right|\hat{a}^\dagger = 0. \tag{7.33}$$

Now you are going to appreciate the wisdom of writing the Hamiltonian in the form of Eq. 7.27 and of introducing operator $\hat{N}$. Indeed Eq. 7.32 used in $\hat{N}\left|E_{min}\right\rangle$ gives $\hat{N}\left|E_{min}\right\rangle = 0$, which means that the minimum value $\lambda_{min} = 0$, and $E_{min} = \hbar\omega/2$. So, behold the power of the lowering operator—we found the bottom, the lowest possible energy of a harmonic oscillator, its ground state!

Just like in other examples, the lowest energy is not zero, which is, of course, the consequence of the uncertainty principle: zero energy would require that both kinetic and potential energies are equal to zero, which would mean that both coordinate and momentum operators would have certain values of zero, which is impossible. The ground state energy $\hbar\omega/2$ is one of the clearest examples of the energy associated with so-called quantum fluctuations.

The contribution of these fluctuations to the energy can be quantified by computing the expectation values of $\hat{p}^2$ and $\hat{x}^2$, which determine the quantum uncertainties of the respective observables. Using Eqs. 7.25 and 7.26 that express operators $\hat{p}$ and $\hat{x}$ in terms of operators $\hat{a}$ and $\hat{a}^\dagger$ in conjunction with Eqs. 7.32 and 7.33, you can immediately see that

$$\langle E_{min}|\hat{x}|E_{min}\rangle = \langle E_{min}|\hat{p}|E_{min}\rangle = 0,$$

so that the uncertainties Δp and Δx are $\Delta p = \sqrt{\langle \hat{p}^2 \rangle}$ and $\Delta x = \sqrt{\langle \hat{x}^2 \rangle}$. Squaring Eqs. 7.25 and 7.26, and computing these expectation values with state $|E_{min}\rangle$, you will get

$$\langle E_{min}|\hat{x}^2|E_{min}\rangle = \frac{\hbar}{2m_e\omega}\left(\langle E_{min}|\hat{a}^2|E_{min}\rangle + \langle E_{min}|\hat{a}^{\dagger 2}|E_{min}\rangle + \right.$$

$$\left. \langle E_{min}|\hat{a}^\dagger\hat{a}|E_{min}\rangle + \langle E_{min}|\hat{a}\hat{a}^\dagger|E_{min}\rangle\right).$$

The first three terms in this expression vanish, thanks to Eqs. 7.32 and 7.33. However, the last term requires some more efforts because the order of operators $\hat{a}$ and $\hat{a}^\dagger$ in it is "wrong" in the sense that it is not conducive to the immediate application of Eqs. 7.32 and 7.33. The situation, however, can be quite easily rectified by using the commutation relations 7.24 to change this order and rewrite this term as

$$\langle E_{min}|\hat{a}\hat{a}^\dagger|E_{min}\rangle = \langle E_{min}|1 + \hat{a}^\dagger\hat{a}|E_{min}\rangle = 1,$$

where I, as usual, assumed that whatever the state vector $|E_{min}\rangle$ is, it is normalized. Thus, finally, I find

$$\langle E_{min}|\hat{x}^2|E_{min}\rangle = \frac{\hbar}{2m_e\omega}. \tag{7.34}$$

Similarly,

$$\langle E_{min}|\hat{p}^2|E_{min}\rangle = -\frac{m_e\omega\hbar}{2}\left(\langle E_{min}|\hat{a}^2|E_{min}\rangle + \langle E_{min}|\hat{a}^{\dagger 2}|E_{min}\rangle - \right.$$

$$\left. \langle E_{min}|\hat{a}^\dagger\hat{a}|E_{min}\rangle - \langle E_{min}|\hat{a}\hat{a}^\dagger|E_{min}\rangle\right) =$$

$$\frac{m_e\omega\hbar}{2}\langle E_{min}|\hat{a}\hat{a}^\dagger|E_{min}\rangle = \frac{m_e\omega\hbar}{2}. \tag{7.35}$$

Using Eqs. 7.34 and 7.35 in expressions for kinetic and potential energies, $\hat{p}^2/2m_e$ and $m_e\omega^2 x^2/2$, I immediately find that the ground state expectation values $\langle \hat{p}^2 \rangle/2m_e$ and $m_e\omega^2 \langle \hat{x}^2 \rangle/2$ are both equal to $\hbar\omega/4$. Isn't it remarkable that while the ground state of harmonic oscillator is characterized by a certain value of energy $\hbar\omega/2$, it is formed by two fluctuating quantities, kinetic and potential energies, contributing equal amounts? One can actually see here a certain analogy with classical harmonic oscillator, whose energy, while being time independent, includes contributions from kinetic and potential energies, whose time dependencies totally compensate each other yielding a constant sum.

OK, by finding the energy of the ground state, I took you down to the very bottom of the energy valley. Now it is time to climb back up, and we are going to do it with the assistance of … wait for it…, of course, the operator $\hat{a}^\dagger$! Actually, there is not much surprise or suspense here because this is exactly what happened with angular momentum operators: we used $\hat{L}_-$ to find the lowest eigenvalue and operator $\hat{L}_+$ to move up from there. My next step is pretty obvious now—consider $\hat{N}\hat{a}^\dagger\,|E_n\rangle$:

$$\hat{N}\hat{a}^\dagger\,|E_n\rangle = \hat{a}^\dagger\,|E_n\rangle + \hat{a}^\dagger\hat{N}\,|E_n\rangle =$$

$$\lambda_n\hat{a}^\dagger\,|E_n\rangle + \hat{a}^\dagger\,|E_n\rangle = (\lambda_n + 1)\,\hat{a}^\dagger\,|E_n\rangle$$

where this time I used commutation relation from Eq. 7.31. So, as expected, $\hat{a}^\dagger\,|E_n\rangle$ is an eigenvector of the number operator with eigenvalue $\lambda_n + 1$, i.e., operator $\hat{a}^\dagger$ does generate eigenvectors with eigenvalues increasing by one for each application of the operator. Starting with the ground state, for which $\lambda_{min} = 0$, operator $\hat{a}^\dagger$ will generate eigenvectors with eigenvalues of $\hat{N}$ equal to $1, 2, 3\cdots$. In other words, the eigenvalues of the number operator are all natural numbers n starting with 0, which make energy levels of quantum harmonic oscillator, according to Eq. 7.29, equal to

$$E_n = \hbar\omega\left(\frac{1}{2} + n\right),\ n = 0, 1, 2\cdots. \qquad (7.36)$$

What is left for us now is to find the corresponding eigenvectors, for which, from now on, I will use the simplified notation $|n\rangle$. All what I know at this point is that if $|n\rangle$ is an eigenvector corresponding to the eigenvalue of the number operator n, then $\hat{a}^\dagger\,|n\rangle$ is an eigenvector corresponding to the eigenvalue $n + 1$. But I cannot guarantee that this new eigenvector will be normalized even if $|n\rangle$ is. Therefore, reserving the bra and ket notation only for normalized vectors, the best I can write for now is

$$\hat{a}^\dagger\,|n\rangle = c_n\,|n + 1\rangle, \qquad (7.37)$$

where $|n + 1\rangle$ is assumed normalized and c_n is yet an unknown normalization factor. Again, I cannot help but remind you that we encountered exactly the same situation when discussing eigenvectors of $\hat{L}^2$. To find c_n I, first, write down a Hermitian conjugated version of Eq. 7.37:

$$\langle n|\,\hat{a} = c_n^*\,\langle n + 1|. \qquad (7.38)$$

Then, multiplying left-hand and right-hand sides of Eqs. 7.37 and 7.38, I get

$$\langle n|\,\hat{a}\hat{a}^\dagger\,|n\rangle = |c_n|^2\,\langle n + 1|\,n + 1\rangle.$$

Using commutation relation 7.24 and taking into account that all vectors are now assumed normalized, I have

$$\langle n| \hat{N} + 1 |n \rangle = |c_n|^2 \Rightarrow |c_n|^2 = n + 1.$$

Taking advantage of the freedom in the choice of the phase of the normalization factor, I choose c_n to be real positive. Now I have the rule for generating new normalized eigenvectors:

$$|n + 1\rangle = \frac{1}{\sqrt{n+1}} \hat{a}^\dagger |n\rangle .$$

Applying this rule sequentially starting with the ground state, I end up with the following expression for an arbitrary eigenvector $|n\rangle$:

$$|n\rangle = \frac{1}{\sqrt{n!}} \left(\hat{a}^\dagger \right)^n |0\rangle , \tag{7.39}$$

where $|0\rangle$ stands for the eigenvector corresponding to the ground state. One can also show that

$$\hat{a} |n\rangle = \sqrt{n} |n - 1\rangle , \tag{7.40}$$

but I will leave a proof of this relation as an exercise.

Equation 7.39 relates eigenvectors describing excited stationary states of the oscillator to its ground state. The latter, however, might appear to you to be undetermined, which is true if by "determining" it you mean expressing it in terms of some known vectors or functions. However, for most purposes, all information that you need about the ground state is contained in Eq. 7.32, and in this sense, this equation is the definition of the ground state. You can use it to find answers to any specific question pertaining to this state. For instance, if you are interested in a function representing this state in coordinate representation, you can use the coordinate representation of the momentum and coordinate operators to turn Eq. 7.32 into an easy-to-solve differential equation for $\varphi_0(x) \equiv \langle x| E_{min}\rangle$:

$$\left(\sqrt{\frac{m_e \omega}{2\hbar}} x + \sqrt{\frac{\hbar}{2m_e \omega}} \frac{d}{dx} \right) \varphi_0(x) = 0 \Rightarrow$$

$$\frac{d\varphi_0(x)}{dx} = -\frac{m_e \omega}{\hbar} x \varphi_0(x) \Rightarrow$$

$$\varphi_0 = C \exp\left(-\frac{x^2}{2\xi^2} \right) . \tag{7.41}$$

Parameter ξ appearing in this equation is defined as

$$\xi = \sqrt{\frac{\hbar}{m_e \omega}} \tag{7.42}$$

and has the dimension of length. It specifies the characteristic scale of the spatial dependence of the wave function: for $x \ll \xi$ the wave function is almost constant, while for $x \gg \xi$ its behavior crosses over to a steep descent. It is easy to see that this parameter characterizes a transition between classically allowed and classically forbidden regions of coordinates for the harmonic oscillator. Indeed, the substitution of quantum ground state energy $E = \hbar\omega/2$ to Eq. 7.10 for the amplitude A of classical oscillator yields $A = \xi$, which means that for the ground state of the oscillator $x < \xi$ corresponds to the classically allowed region, and the region $x > \xi$ is classically forbidden.

Integration constant C in Eq. 7.41 is found from the normalization condition:

$$C^2 \int_{-\infty}^{\infty} \exp\left(-\frac{x^2}{\xi^2}\right) dx = C^2 \xi \int_{-\infty}^{\infty} \exp\left(-\tilde{x}^2\right) d\tilde{x} = C^2 \sqrt{\pi}\xi = 1$$

where I computed the integral by introducing a dimensionless variable $\tilde{x} = x/\xi$ and using a known value of the Gaussian integral $\int_{-\infty}^{\infty} \exp\left(-y^2\right) dy = \sqrt{\pi}$. Thus, the normalized version of the oscillator ground state wave function becomes

$$\varphi_0 = \frac{1}{\sqrt{\sqrt{\pi}\xi}} \exp\left(-\frac{x^2}{2\xi^2}\right). \tag{7.43}$$

Having found the normalized ground state wave function in the coordinate representation, I can now use the raising operator (also rewritten in the coordinate representation) to generate wave functions representing an arbitrary stationary state of the Hamiltonian:

$$\varphi_n(x) = \frac{1}{\sqrt{2^n n! \xi \sqrt{\pi}}} \left(\tilde{x} - \frac{d}{d\tilde{x}}\right)^n \left[\exp\left(-\frac{\tilde{x}^2}{2}\right)\right].$$

Here I used the coordinate representation for the raising operator expressed in terms of dimensionless variable $\tilde{x}$:

$$\hat{a}^\dagger = \sqrt{\frac{m_e \omega}{2\hbar}} x - \frac{\hbar}{\sqrt{2m_e\hbar\omega}} \frac{d}{dx} =$$

$$\frac{x}{\sqrt{2}\xi} - \frac{\xi}{\sqrt{2}} \frac{d}{dx} = \frac{1}{\sqrt{2}} \left(\tilde{x} - \frac{d}{d\tilde{x}}\right)$$

substituted in Eq. 7.39. You can easily convince yourselves that expression

$$\left(\tilde{x} - \frac{d}{d\tilde{x}}\right)^n \left[\exp\left(-\frac{\tilde{x}^2}{2}\right)\right]$$

generates polynomials multiplied by an exponential function $\exp\left(-\tilde{x}^2/2\right)$. Pulling out this exponential factor, you end up with so-called Hermite polynomials $\mathcal{H}_n(\tilde{x})$ defined as

$$\mathcal{H}_n(\tilde{x}) = \exp\left(\frac{\tilde{x}^2}{2}\right)\left(\tilde{x} - \frac{d}{d\tilde{x}}\right)^n \left[\exp\left(-\frac{\tilde{x}^2}{2}\right)\right]$$

so that the oscillator's wave function takes the form

$$\varphi_n(x) = \frac{1}{\sqrt{2^n n! \xi \sqrt{\pi}}} \exp\left(-\frac{\tilde{x}^2}{2}\right) \mathcal{H}_n(\tilde{x}). \tag{7.44}$$

Hermitian polynomials are well known in mathematical physics and can be computed from the following somewhat simpler expression:

$$\mathcal{H}_n(\tilde{x}) = (-1)^n e^{\tilde{x}^2} \frac{d^n}{d\tilde{x}^n}\left(e^{-\tilde{x}^2}\right). \tag{7.45}$$

The properties of these polynomials are well documented (google it!), so I will only emphasize one point: these polynomials and, therefore, the entire wave function have a definite parity—it is even for $n = 0, 2, 4 \cdots$, and it is odd for $n = 1, 3, 5 \cdots$. Obviously this fact is the result of the symmetry of the harmonic oscillator potential with respect to inversion and is an agreement with our previous discussions of the connection between this symmetry and the parity of the quantum states. Figure 7.2 presents graphs of wave functions representing states with $n = 0, 1, 2, 3$, from which you can see that another general rule is also fulfilled here: the number of zeroes of the wave function coincides with the number of the respective energy level n. Note that n is counted here starting from zero; therefore, the number of zeroes of the wave function is n instead of $n - 1$.

Coordinate representation is, obviously, not the only possible way to present eigenvectors of the harmonic oscillator. As a second example, I want to discuss a representation based on eigenvectors of the Hamiltonian $|n\rangle$. The eigenvectors themselves in this representation are presented, as all basis vectors, by columns with a single entry, equal to unity, in the row corresponding to the number of the respective basis vector. The Hamiltonian in this basis is presented by a diagonal matrix $H_{nm} = E_n \delta_{nm}$, where E_n are energy eigenvalues given by Eq. 7.36. Less trivial is the representation of coordinate and momentum operators, and to find it I must first compute the matrix elements of the lowering (or raising—does not really matter) operator, $a_{mn} = \langle m| \hat{a} |n\rangle$:

$$a_{mn} = \langle m| \hat{a} |n\rangle = \sqrt{n} \langle m| n-1\rangle = \sqrt{n}\delta_{m,n-1},$$

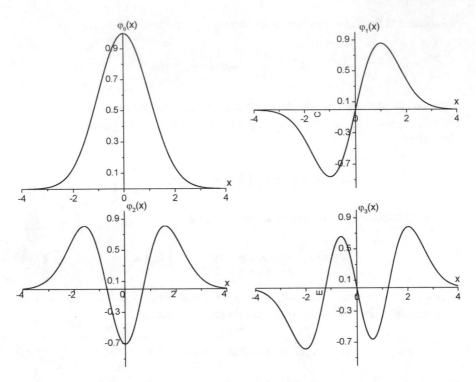

Fig. 7.2 Wave functions representing states of harmonic oscillators with $n = 0$ (upper left graph), $n = 1$ (upper right graph), $n = 2$ (lower left graph), and $n = 3$ (lower right graph)

where I used Eq. 7.40 and the fact that all eigenvectors $|n\rangle$ are orthonormal. To visualize this matrix correctly, it is important to remember that index n in Eq. 7.40 starts counting from zero, and it is convenient to keep it this way when numerating matrix elements. In this case the first row is given by a_{0n}, second by $a_{1,n}$, and so on. Respectively, the first column is given by a_{m0}. Non-zero elements in matrix a_{mn} are characterized by column index exceeding the respective row index by one, i.e., $a_{0,1}, a_{1,2}$, etc.—they go parallel to the main diagonal but one element above it:

$$a_{mn} = \begin{bmatrix} 0 & \sqrt{1} & 0 & 0 & \cdots \\ 0 & 0 & \sqrt{2} & 0 & \cdots \\ 0 & 0 & 0 & \sqrt{3} & \cdots \\ \vdots & \vdots & \vdots & \ddots & \vdots \\ 0 & 0 & 0 & 0 & \ddots \end{bmatrix}. \tag{7.46}$$

The matrix for the raising operator

$$a^\dagger_{mn} = \langle m| \hat{a}^\dagger |n\rangle = \sqrt{n+1}\, \langle m|\, n+1\rangle = \sqrt{n+1}\delta_{m,n+1}$$

is obtained from Eq. 7.46 by simple transposition:

$$a_{mn}^\dagger = \begin{bmatrix} 0 & 0 & 0 & 0 & \cdots \\ \sqrt{1} & 0 & 0 & 0 & \cdots \\ 0 & \sqrt{2} & 0 & 0 & \cdots \\ \vdots & \vdots & \ddots & \ddots & \vdots \\ 0 & 0 & 0 & 0 & \ddots \end{bmatrix}. \tag{7.47}$$

Obtaining matrices for coordinate and momentum operators is now as easy as adding two matrices. Using Eqs. 7.25 and 7.26, I find

$$x_{mn} = \sqrt{\frac{\hbar}{2m_e\omega}} \begin{bmatrix} 0 & \sqrt{1} & 0 & 0 & \cdots \\ \sqrt{1} & 0 & \sqrt{2} & 0 & \cdots \\ 0 & \sqrt{2} & 0 & \sqrt{3} & \cdots \\ \vdots & \vdots & \vdots & \ddots & \vdots \\ 0 & 0 & 0 & 0 & \ddots \end{bmatrix} \tag{7.48}$$

$$p_{mn} = i\sqrt{\frac{m_e\hbar\omega}{2}} \begin{bmatrix} 0 & -\sqrt{1} & 0 & 0 & \cdots \\ \sqrt{1} & 0 & -\sqrt{2} & 0 & \cdots \\ 0 & \sqrt{2} & 0 & -\sqrt{3} & \cdots \\ \vdots & \vdots & \vdots & \ddots & \vdots \\ 0 & 0 & 0 & 0 & \ddots \end{bmatrix}. \tag{7.49}$$

Both matrices are obviously Hermitian, but for the matrix representing the momentum operator, one must remember to do complex conjugation in addition to matrix transposition.

7.1.2 Dynamics of Quantum Harmonic Oscillator

When talking (or thinking) about a harmonic oscillator, we are intuitively looking for a quantity that changes periodically with time—oscillates. However, the stationary states, which I presented to you in the preceding section, are not very helpful in satisfying our intuitive subconscious desire to see a pendulum or at least something oscillating. Stationary states even though they have non-zero energies associated with them do not describe any dynamics and any physically relevant time dependence. Any expectation values computed with stationary states are time-independent, and those for coordinate and momentum are zeroes, not only for the ground state but for any stationary state. This is, of course, obvious from the coordinate and momentum matrices presented in Eqs. 7.48 and 7.49, but one can also make a symmetry-based argument explaining this result. Even though this is a detour from the main goal of this section, I will take it because symmetry-based arguments are important in many areas of quantum mechanics and also because they are cool.

The Hamiltonian of the harmonic oscillator is invariant with respect to the inversion operator $\hat{\Pi}$ (see Sect. 6.2.1), and, therefore, its eigenvectors have a definite parity, as it was already mentioned. Any expectation value involves a bra and ket pair of them, and, therefore, either they are odd or even, their overall contribution is invariant with respect to $\hat{\Pi}$ (a product of two odd functions is even). At the same time, coordinate and momentum operators are odd with respect to parity transformation: $\hat{\Pi}^{-1}\hat{x}\hat{\Pi} = -\hat{x}$, $\hat{\Pi}^{-1}\hat{p}\hat{\Pi} = -\hat{p}$, as it was shown in Sect. 5.1.2. Thus, on one hand, expectation values are supposed to change sign upon inversion, but on the other hand, they must not because they represent a property of the system invariant with respect to inversion. Thus, ponder this: you have a situation when a quantity must simultaneously change its sign while remaining the same. Clearly, there is only one quantity capable of this Houdini trick, and it is the great invention of Hindu mathematicians—the zero.

Now, back to the main topics. It is clear that the only way a quantum harmonic oscillator can actually oscillate is by being in a nonstationary state. We have discussed two approaches to dealing with nonstationary phenomena—the Schrödinger picture (operators are time-independent, state vectors are time-dependent) and the Heisenberg picture (operators depend on time, and state vectors do not). I will treat the dynamics of the harmonic oscillator using both pictures beginning with the Heisenberg approach.

Heisenberg equations 4.24 can be derived for any operator, and in Sect. 4.2 I did that for the momentum and coordinate operators. Equation 4.33 provides you with the complete solution of the respective Heisenberg equations and essentially with all what you might need to describe the dynamics of any experimentally relevant quantity. However, I would like to revisit the problem of finding time-dependent position and momentum operators, but this time I will do it by solving the Heisenberg equations for lowering and raising operators. The corresponding equations are

$$\frac{d\hat{a}_H}{dt} = -\frac{i}{\hbar}\left[\hat{a}_H, \hat{H}\right]$$

$$\frac{d\hat{a}_H^\dagger}{dt} = -\frac{i}{\hbar}\left[\hat{a}_H^\dagger, \hat{H}\right].$$

The expression for the Hamiltonian in terms of Heisenberg operators $\hat{a}_H$, $\hat{a}_H^\dagger$ is the same as in terms of Schrödinger operators:

$$e^{\frac{i}{\hbar}\hat{H}t}\hat{H}e^{-\frac{i}{\hbar}\hat{H}t} = \hat{H} = \hbar\omega\left(e^{\frac{i}{\hbar}\hat{H}t}\hat{a}^\dagger e^{-\frac{i}{\hbar}\hat{H}t}e^{\frac{i}{\hbar}\hat{H}t}\hat{a}e^{\frac{-i}{\hbar}}\hat{H}\right)$$

$$\hbar\omega\left(\frac{1}{2} + \hat{a}_H^\dagger\hat{a}_H\right),$$

and, therefore, all the commutation relations, which we calculated for Schrödinger operators, remain the same. In particular, using Eqs. 7.30 and 7.31, I can find $\left[\hat{a}_H, \hat{H}\right] = \hbar\omega\left[\hat{a}_H, \hat{N}\right] = \hbar\omega\hat{a}_H$, and $\left[\hat{a}_H^\dagger, \hat{H}\right] = \hbar\omega\left[\hat{a}_H^\dagger, \hat{N}\right] = -\hbar\omega\hat{a}_H^\dagger$. Substituting it in the Heisenberg equations, I obtain the following nice-looking equations:

$$\frac{d\hat{a}_H}{dt} = - i\omega\hat{a}_H \tag{7.50}$$

$$\frac{d\hat{a}_H^\dagger}{dt} = i\omega\hat{a}_H^\dagger. \tag{7.51}$$

Unlike equations for the momentum and coordinate operators, Eqs. 7.50 and 7.51 are not coupled, so they can be solved independently—in that they have a striking resemblance to classical Eqs. 7.14 and 7.13. Solutions to these equations are easy to write:

$$\hat{a}_H = \hat{a}e^{-i\omega t}; \; \hat{a}_H^\dagger = \hat{a}^\dagger e^{i\omega t}, \tag{7.52}$$

where $\hat{a}$ and $\hat{a}^\dagger$ are Schrödinger operators that play the role of initial conditions for the Heisenberg equations. Equations 7.25 and 7.26 are obviously valid for Heisenberg operators as well so that one can obtain for time-dependent coordinate and momentum operators:

$$\hat{x}_H = \sqrt{\frac{\hbar}{2m_e\omega}} \left(\hat{a}e^{-i\omega t} + \hat{a}^\dagger e^{i\omega t}\right) \tag{7.53}$$

$$\hat{p}_H = i\sqrt{\frac{m_e\hbar\omega}{2}} \left(\hat{a}^\dagger e^{i\omega t} - \hat{a}e^{-i\omega t}\right). \tag{7.54}$$

Using the Euler relation for the exponential functions, I can rewrite this result in the form previously derived in Eq. 4.33:

$$\hat{x}_H = \sqrt{\frac{\hbar}{2m_e\omega}} \left[\left(\hat{a} + \hat{a}^\dagger\right)\cos\omega t + i\left(\hat{a}^\dagger - \hat{a}\right)\sin\omega t\right] \tag{7.55}$$

$$\hat{p}_H = i\sqrt{\frac{m_e\hbar\omega}{2}} \left[\left(\hat{a}^\dagger - \hat{a}\right)\cos\omega t + i\left(\hat{a} + \hat{a}^\dagger\right)\sin\omega t\right], \tag{7.56}$$

which agrees with Eq. 4.33 after one recognizes that at $t = 0$ these equations reproduce coordinate and momentum operators in the Schrödinger representation. Either of Eqs. 7.53–7.56 can be used, for instance, to compute the expectation values of coordinate and momentum for an arbitrary initial state $|\chi_0\rangle$. This task can be facilitated by using the basis of the eigenvectors $|n\rangle$ to represent this state:

$$|\chi_0\rangle = \sum_{n=0}^{\infty} c_n |n\rangle. \tag{7.57}$$

It is a bit more convenient to carry out these calculations using the exponential form of time dependence as in Eqs. 7.53 and 7.54:

$$\langle x \rangle = \sqrt{\frac{\hbar}{2m_e\omega}} \left[e^{-i\omega t} \sum_{n,m=0}^{\infty} c_m^* c_n \langle m| \hat{a} |n\rangle + e^{i\omega t} \sum_{n,m=0}^{\infty} c_m^* c_n \langle m| \hat{a}^\dagger |n\rangle \right] =$$

$$\sqrt{\frac{\hbar}{2m_e\omega}} \left[e^{-i\omega t} \sum_{n,m=0}^{\infty} c_m^* c_n \sqrt{n}\delta_{m,n-1} + e^{i\omega t} \sum_{n,m=0}^{\infty} c_m^* c_n \sqrt{n+1}\delta_{m,n+1} \right] =$$

$$\sqrt{\frac{\hbar}{2m_e\omega}} \left[e^{-i\omega t} \sum_{m=0}^{\infty} c_m^* c_{m+1} \sqrt{m+1} + e^{i\omega t} \sum_{m=0}^{\infty} c_{m+1}^* c_m \sqrt{m+1} \right], \tag{7.58}$$

where I used previously derived matrix elements for the lowering and raising operators. Now, we are getting something familiar: the expectation value of coordinate does indeed oscillate with the frequency of the harmonic oscillator ω, and what is interesting, this behavior does not depend on the actual initial state, as long as it has contributions from at least two adjacent stationary states so that both c_m and c_{m+1} are different from zeroes. This requirement, of course, excludes initial stationary states, which would have only one nonvanishing coefficient c_m, as well as nonstationary states with a definite parity, which would contain only coefficients c_m with either odd or even m. With a bit of imagination, you can recognize in Eq. 7.58 the typical for a classical harmonic oscillator behavior which can be described as

$$\langle x \rangle = A \cos (\omega t + \phi), \tag{7.59}$$

where the amplitude and phase of the oscillations are determined by the initial conditions (as they also are in the classical case):

$$A = \sqrt{\frac{\hbar}{2m_e\omega}} \left| \sum_{m=0}^{\infty} c_m^* c_{m+1} \sqrt{m+1} \right|;$$

$$\phi = \arctan \frac{Im\left[\sum_{m=0}^{\infty} c_{m+1}^* c_m \sqrt{m+1} \right]}{Re\left[\sum_{m=0}^{\infty} c_{m+1}^* c_m \sqrt{m+1} \right]}. \tag{7.60}$$

Similar calculations for the momentum operator produce

$$\langle p \rangle = i\sqrt{\frac{m_e\hbar\omega}{2}} \left[e^{i\omega t} \sum_{m=0}^{\infty} c_{m+1}^* c_m \sqrt{m+1} - e^{-i\omega t} \sum_{m=0}^{\infty} c_m^* c_{m+1} \sqrt{m+1} \right], \tag{7.61}$$

which can be rewritten using the same amplitude and phase as

$$\langle p \rangle = -m_e \omega A \sin(\omega t + \phi) \tag{7.62}$$

in full agreement with the Ehrenfest theorem, Eq. 4.17.

Before shifting attention to the Schrödinger picture, let me consider a few more examples of the application of the Heisenberg equations.

Example 18 (Uncertainties of Coordinate and Momentum of a Harmonic Oscillator) Assume that the harmonic oscillator is initially in a state described by an equal superposition of its ground and the first excited states:

$$|\alpha_0\rangle = \frac{1}{\sqrt{2}} (|0\rangle + |1\rangle).$$

Compute uncertainties of the coordinate and momentum operators at an arbitrary time t and demonstrate, using the Heisenberg picture, that the uncertainty relation is fulfilled at all times.

Using Eqs. 7.59, 7.62, and 7.60 with $c_0 = c_1 = 1/\sqrt{2}$ and $c_m = 0$ for $m > 1$, I find for the expectation values

$$\langle x \rangle = \frac{1}{2} \sqrt{\frac{\hbar}{2m_e\omega}} \cos \omega t$$

$$\langle p \rangle = -\frac{1}{2} \sqrt{\frac{\hbar \omega m_e}{2}} \sin \omega t.$$

To find the uncertainties, I first have to compute $\langle p^2 \rangle$ and $\langle x^2 \rangle$. I begin by computing

$$\hat{x}^2 = \frac{\hbar}{2m_e\omega} \left(\hat{a}e^{-i\omega t} + \hat{a}^\dagger e^{i\omega t} \right)^2 =$$

$$\frac{\hbar}{2m_e\omega} \left(\hat{a}^2 e^{-2i\omega t} + \hat{a}\hat{a}^\dagger + \hat{a}^\dagger\hat{a} + \left(\hat{a}^\dagger \right)^2 e^{2i\omega t} \right),$$

$$\hat{p}^2 = -\frac{m_e\hbar\omega}{2} \left(\hat{a}e^{-i\omega t} - \hat{a}^\dagger e^{i\omega t} \right)^2 =$$

$$-\frac{m_e\hbar\omega}{2} \left(\hat{a}^2 e^{-2i\omega t} - \hat{a}\hat{a}^\dagger - \hat{a}^\dagger\hat{a} + \left(\hat{a}^\dagger \right)^2 e^{2i\omega t} \right).$$

Now, remembering that $\hat{a}|0\rangle = 0$, $\hat{a}|1\rangle = |0\rangle$, $\hat{a}|2\rangle = \sqrt{2}|1\rangle$, $\hat{a}^\dagger|0\rangle = |1\rangle$, $\hat{a}^\dagger|1\rangle = \sqrt{2}|2\rangle$, $\hat{a}^\dagger|2\rangle = \sqrt{3}|3\rangle$, I get

$$\hat{x}^2 |\alpha_0\rangle = \frac{\hbar}{2\sqrt{2}m_e\omega} \left(|0\rangle + \sqrt{2}e^{2i\omega t} |2\rangle + 2|1\rangle + |1\rangle + \sqrt{6}e^{2i\omega t} |3\rangle \right),$$

$$\hat{p}^2 |\alpha_0\rangle = -\frac{m_e\hbar\omega}{2\sqrt{2}} \left(-|0\rangle + \sqrt{2}e^{2i\omega t} |2\rangle - 2|1\rangle - |1\rangle + \sqrt{6}e^{2i\omega t} |3\rangle \right),$$

and, finally,

$$\langle \alpha_0 | \hat{x}^2 | \alpha_0 \rangle = \frac{\hbar}{4m_e\omega} (1 + 3) = \frac{\hbar}{m_e\omega}$$

$$\langle \alpha_0 | \hat{p}^2 | \alpha_0 \rangle = \frac{m_e\hbar\omega}{4} (1 + 3) = m_e\hbar\omega.$$

Now I can find the uncertainties:

$$\Delta x = \sqrt{\langle \hat{x}^2 \rangle - \langle \hat{x} \rangle^2} = \sqrt{\frac{\hbar}{m_e\omega} \left(1 - \frac{1}{8} \cos^2 \omega t \right)}$$

$$\Delta p = \sqrt{\langle \hat{p}^2 \rangle - \langle \hat{p} \rangle^2} = \sqrt{m_e\hbar\omega \left(1 - \frac{1}{8} \sin^2 \omega t \right)}$$

$$\Delta x \Delta p = \frac{\hbar}{2\sqrt{2}} \sqrt{7 + \frac{1}{32} \sin^2 2\omega t} > 0.93\hbar$$

in agreement with the uncertainty principle.

There also exists an alternative approach to computing time-dependent averages of various observables using the Heisenberg picture, which allows to establish their dependence of time in a more generic way. To develop such an approach, let me first rewrite Eqs. 7.55 and 7.56 using Eqs. 7.25 and 7.26 for Schrödinger versions of the coordinate and momentum operators, which I will designate here as $\hat{x}_0$ and $\hat{p}_0$ to emphasize the fact that they serve as initial values for the Heisenberg equations:

$$\hat{x}_H = \hat{x}_0 \cos \omega t + \frac{1}{m_e\omega} \hat{p}_0 \sin \omega t \qquad (7.63)$$

$$\hat{p}_H = \hat{p}_0 \cos \omega t - m_e\omega \hat{x}_0 \sin \omega t. \qquad (7.64)$$

Now, let's say I want to compute the uncertainty of the coordinate for an arbitrary state $|\alpha\rangle$. The expectation values of the coordinate and momentum in this state, using Eqs. 7.63 and 7.64, can be written as

$$\langle x \rangle = \langle \hat{x}_0 \rangle \cos \omega t + \frac{1}{m_e\omega} \langle \hat{p}_0 \rangle \sin \omega t$$

$$\langle p \rangle = \langle \hat{p}_0 \rangle \cos \omega t - m_e\omega \langle \hat{x}_0 \rangle \sin \omega t,$$

where $\langle \hat{x}_0 \rangle$ and $\langle \hat{p}_0 \rangle$ are time-independent "Schrödinger" expectation values that can be computed for a given state using any of the representations for the Schrödinger coordinate and momentum operators. Similarly, I can find for the expectation values of the squared operators

$$\langle \hat{x}^2 \rangle = \langle \hat{x}_0^2 \rangle \cos^2 \omega t + \frac{1}{2m_e\omega} \left(\langle \hat{x}_0 \hat{p}_0 \rangle + \langle \hat{p}_0 \hat{x}_0 \rangle \right) \sin 2\omega t + \frac{1}{m_e^2\omega^2} \langle \hat{p}_0^2 \rangle \sin^2 \omega t$$

$$\langle \hat{p}^2 \rangle = \langle \hat{p}_0^2 \rangle \cos^2 \omega t - \frac{1}{2}m_e\omega \left(\langle \hat{x}_0 \hat{p}_0 \rangle + \langle \hat{p}_0 \hat{x}_0 \rangle \right) \sin 2\omega t + m_e^2\omega^2 \langle \hat{x}_0^2 \rangle \sin^2 \omega t$$

which will yield the following for the uncertainties:

$$(\Delta x)^2 = (\Delta \hat{x}_0)^2 \cos^2 \omega t + \frac{1}{m_e^2\omega^2} (\Delta \hat{p}_0)^2 \sin^2 \omega t +$$

$$\frac{1}{2m_e\omega} \left[(\langle \hat{x}_0 \hat{p}_0 \rangle + \langle \hat{p}_0 \hat{x}_0 \rangle - 2 \langle \hat{x}_0 \rangle \langle \hat{p}_0 \rangle) \right] \sin 2\omega t$$

$$(\Delta p)^2 = (\Delta \hat{p}_0)^2 \cos^2 \omega t + m_e^2\omega^2 (\Delta \hat{x}_0)^2 \sin^2 \omega t -$$

$$\frac{1}{2}m_e\omega \left[(\langle \hat{x}_0 \hat{p}_0 \rangle + \langle \hat{p}_0 \hat{x}_0 \rangle - 2 \langle \hat{x}_0 \rangle \langle \hat{p}_0 \rangle) \right] \sin 2\omega t.$$

I already mentioned it once, but it is worth emphasizing again: all expectation values in this expression refer to Schrödinger operators and can be computed using any of the representations for the latter. Let me illustrate this point by considering the following example.

Example 19 (Harmonic Oscillator with Shifted Minimum of the Potential) Consider a harmonic oscillator with mass m_e and frequency ω in the ground state. Suddenly, without disruption of the oscillator's state, the minimum of the potential shifts by d along the axes of oscillations and the stiffness of the potential changes such that it is now characterized by a new classical frequency Ω. Find the expectation value and uncertainty of coordinate and momentum of the electron in the potential with the new position of its minimum.

It is convenient to solve this problem using coordinate representation for the initial state and for the Schrödinger operators $\hat{x}$ and $\hat{p}$. First of all, let's agree to place the origin of the X-axis at the new position of the minimum. Then, the initial wave function, which is the ground state wave function of the oscillator with potential in the original position, is

$$\psi_0(x) = \left(\frac{m_e\omega}{\pi\hbar} \right)^{1/4} \exp\left(-\frac{m_e\omega}{2\hbar} (x+d)^2 \right),$$

where x is counted from the new position of the potential. The expectation values of the Schrödinger operators $\langle \hat{x}_0 \rangle$ and $\langle \hat{p}_0 \rangle$ are

$$\langle \hat{x}_0 \rangle = \sqrt{\frac{m_e\omega}{\pi\hbar}} \int\limits_{-\infty}^{\infty} x \exp\left(-\frac{m_e\omega}{\hbar} (x+d)^2 \right) =$$

$$\sqrt{\frac{m_e\omega}{\pi\hbar}} \int_{-\infty}^{\infty} (x-d)\exp\left(-\frac{m_e\omega}{\hbar}x^2\right) = -d$$

where I made a substitution of variables and took into account that the wave function of the initial state is even. Similarly, I can find

$$\langle \hat{p}_0 \rangle = -i\hbar\sqrt{\frac{m_e\omega}{\pi\hbar}} \int_{-\infty}^{\infty} \exp\left(-\frac{m_e\omega}{2\hbar}(x+d)^2\right) \times$$

$$\frac{d}{dx}\left[\exp\left(-\frac{m_e\omega}{2\hbar}(x+d)^2\right)\right]dx =$$

$$i\hbar\sqrt{\frac{m_e\omega}{\pi\hbar}}\frac{m_e\omega}{\hbar} \int_{-\infty}^{\infty} \exp\left(-\frac{m_e\omega}{\hbar}(x+d)^2\right)(x+d)\,dx = 0.$$

Thus, I have for the time-dependent expectation values

$$\langle x \rangle = -d\cos\Omega t$$

$$\langle p \rangle = m_e\Omega d \sin\Omega t.$$

(Obviously, the dynamics of raising and lowering operators is defined by new frequency Ω.) In order to find the respective uncertainties, I need to compute $\Delta\hat{x}_0$, $\Delta\hat{p}_0$, and $\langle\hat{x}_0\hat{p}_0\rangle$. The uncertainties of the regular Schrödinger coordinate and momentum operators do not depend on the position of the potential minimum with respect to the origin of the coordinate axes, so I can simply recycle the results from Eqs. 7.34 and 7.35:

$$\Delta\hat{x}_0 = \sqrt{\frac{\hbar}{2m_e\omega}};\ \Delta\hat{p}_0 = \sqrt{\frac{\hbar m_e\omega}{2}}.$$

For the last expectation value, $\langle\hat{x}_0\hat{p}_0\rangle$, I will actually have to do some work:

$$\langle\hat{x}_0\hat{p}_0\rangle = -i\hbar\left(\frac{m_e\omega}{\pi\hbar}\right)^{1/2} \times$$

$$\int_{-\infty}^{\infty} x\exp\left(-\frac{m_e\omega}{2\hbar}(x+d)^2\right)\frac{d}{dx}\left[\exp\left(-\frac{m_e\omega}{2\hbar}(x+d)^2\right)\right]dx =$$

$$i\hbar\left(\frac{m_e\omega}{\pi\hbar}\right)^{1/2}\left(\frac{m_e\omega}{\hbar}\right) \int_{-\infty}^{\infty} x(x+d)\exp\left(-\frac{m_e\omega}{\hbar}(x+d)^2\right)dx =$$

$$i\hbar \left(\frac{m_e\omega}{\pi\hbar}\right)^{1/2} \left(\frac{m_e\omega}{\hbar}\right) \int_{-\infty}^{\infty} x\,(x-d)\exp\left(-\frac{m_e\omega}{\hbar}x^2\right) dx$$

$$= i\hbar \left(\frac{m_e\omega}{\hbar}\right) \left(\frac{\hbar}{2m_e\omega}\right) = \frac{i\hbar}{2}.$$

In the last line of this expression, I took into account that the integral with the linear in x factor vanishes because of the oddness of the integrand, while the integral containing x^2 together with the normalization factor of the wave function reproduces the uncertainty of the coordinate $(\triangle \hat{x}_0)^2$. If you are spooked by the imaginary result here, you shouldn't. Operator $\hat{x}_0\hat{p}_0$ is not Hermitian, and its expectation value does not have to be real. To complete this calculation, I would have to compute $\langle \hat{p}_0\hat{x}_0\rangle$, but I will save us some time and use the canonical commutation relation $[\hat{x}, \hat{p}] = i\hbar$ to find

$$\langle \hat{x}_0\hat{p}_0\rangle + \langle \hat{p}_0\hat{x}_0\rangle = 2\,\langle \hat{x}_0\hat{p}_0\rangle - i\hbar = 0.$$

Oops, so much efforts to get zero in the end? Feeling disappointed and a bit cheated? Well, you should be, because we could have guessed that the answer here is zero without any calculations. Indeed, the momentum operator contains imaginary unity in it, and with the wave function being completely real, this imaginary factor is not going anywhere. But, on the other hand, the result must be real because $\hat{x}_0\hat{p}_0 + \hat{p}_0\hat{x}_0$ is a Hermitian operator. So, the only conclusion a reasonable person can draw from this conundrum is that the result must be zero. Thus, we finally have for the time-dependent uncertainties:

$$(\triangle x)^2 = \frac{\hbar}{2m_e\omega}\cos^2\Omega t + \frac{1}{m_e^2\Omega^2}\frac{\hbar m_e\omega}{2}\sin^2\Omega t = \frac{\hbar}{2m_e\omega}\left(\cos^2\Omega t + \frac{\omega^2}{\Omega^2}\sin^2\Omega t\right)$$

$$(\triangle p)^2 = \frac{\hbar m_e\omega}{2}\cos^2\omega t + m_e^2\Omega^2\frac{\hbar}{2m_e\omega}\sin^2\omega t = \frac{\hbar m_e\omega}{2}\left(\cos^2\Omega t + \frac{\Omega^2}{\omega^2}\sin^2\Omega t\right),$$

and for their product

$$(\triangle x)^2\,(\triangle p)^2 = \frac{\hbar^2}{4}\left[\cos^4\Omega t + \sin^4\Omega t + \left(\frac{\omega^2}{\Omega^2} + \frac{\Omega^2}{\omega^2}\right)\cos^2\Omega t\sin^2\Omega t\right] =$$

$$\frac{\hbar^2}{4}\left[\cos^4\Omega t + \sin^4\Omega t + 2\cos^2\Omega t\sin^2\Omega t + \left(\frac{\omega^2}{\Omega^2} + \frac{\Omega^2}{\omega^2} - 2\right)\cos^2\Omega t\sin^2\Omega t\right] =$$

$$\frac{\hbar^2}{4}\left[1 + \left(\frac{\omega^2}{\Omega^2} + \frac{\Omega^2}{\omega^2} - 2\right)\cos^2\Omega t\sin^2\Omega t\right],$$

where in the second line I added and subtracted term $2\cos^2\Omega t\sin^2\Omega t$ and in the third line used identity $\cos^4\Omega t + \sin^4\Omega t + 2\cos^2\Omega t\sin^2\Omega t =$

$\left(\cos^2 \Omega t + \sin^2 \Omega t\right)^2 = 1$. Function $y + 1/y$, which appears in the final result, has a minimum at $y = 1$ ($\Omega = \omega$), at which point the product of the uncertainties becomes $\hbar^2/4$. For all other relations between the two frequencies, the product of the uncertainties exceeds this value in full agreement with the uncertainty principle. It is interesting to note that as the uncertainties oscillate, their product returns to its minimum value at times $t_n = \pi n/(2\Omega)$.

In the Schrödinger picture, the dynamics of quantum systems is described by the time dependence of the vectors representing quantum states. For the initial state given by Eq. 7.57, the time-dependent state can be presented as (see Eq. 4.15)

$$|\chi(t)\rangle = \sum_{n=0}^{\infty} c_n e^{-i\omega(n+1/2)} |n\rangle . \tag{7.65}$$

Computing the expectation value of the coordinate with this state and using again the representation of the coordinate operator in terms of lowering and raising operators, I have

$$\langle x \rangle = \sqrt{\frac{\hbar}{2m_e\omega}} \left[\sum_{n=0}^{\infty} \sum_{m=0}^{\infty} c_m^* c_n e^{i\omega(m-n)t} \langle m| \hat{a} |n\rangle + \right.$$

$$\left. \sum_{n=0}^{\infty} \sum_{m=0}^{\infty} c_m^* c_n e^{i\omega(m-n)t} \langle m| \hat{a}^\dagger |n\rangle \right] =$$

$$\sqrt{\frac{\hbar}{2m_e\omega}} \left[\sum_{n=0}^{\infty} \sum_{m=0}^{\infty} c_m^* c_n e^{i\omega(m-n)t} \sqrt{n}\delta_{m,n-1} + \right.$$

$$\left. \sum_{n=0}^{\infty} \sum_{m=0}^{\infty} c_m^* c_n e^{i\omega(m-n)t} \sqrt{n+1}\delta_{m,n+1} \right] =$$

$$\sqrt{\frac{\hbar}{2m_e\omega}} \left[e^{-i\omega t} \sum_{m=0}^{\infty} c_m^* c_{m+1} \sqrt{m+1} + e^{i\omega t} \sum_{m=0}^{\infty} c_{m+1}^* c_m \sqrt{m+1} \right] \tag{7.66}$$

in full agreement with Eq. 7.61 obtained using the Heisenberg representation. What is interesting about this result is that in the beginning of the computations, we had complex exponential functions with all frequencies $\omega (m-n)$. However, after the matrix elements of the lowering and raising operators have been taken into account, only terms with a single frequency ω survived. In the Heisenberg approach, frequencies $\omega (m-n)$ never appear because the properties of $\hat{a}$ and $\hat{a}^\dagger$ are incorporated from the very beginning at the level of the Heisenberg equations. Similar expressions can be easily derived for the expectation values of the momentum operator:

$$\langle p \rangle = i\sqrt{\frac{\hbar m_e\omega}{2}} \left[e^{i\omega t} \sum_{m=0}^{\infty} c_{m+1}^* c_m \sqrt{m+1} - e^{-i\omega t} \sum_{m=0}^{\infty} c_m^* c_{m+1} \sqrt{m+1} \right],$$

while generic expressions for the uncertainties of the coordinate and momentum operators in the Schrödinger picture are much more cumbersome and are more difficult to derive. Thus, I will illustrate the derivation of the uncertainties for the time-dependent states in the Schrödinger picture with the same example 18, which was previously solved in the Heisenberg picture.

Example 20 (Uncertainties of the Coordinate and Momentum of the Quantum Harmonic Oscillator in the Schrödinger Picture) Let me remind you that we are dealing with a harmonic oscillator prepared in a state

$$|\alpha_0\rangle = \frac{1}{\sqrt{2}} (|0\rangle + |1\rangle),$$

and we want to compute the uncertainties of the coordinate and momentum operators at an arbitrary time t using the Schrödinger picture.

Comparing the expression for the initial state with Eq. 7.66, expansion coefficients c_n in Eq. 7.65 can be identified as $c_0 = c_1 = 1/\sqrt{2}$ while all other coefficients vanish. Thus, the time-dependent state vector now becomes

$$|\chi(t)\rangle = \frac{1}{\sqrt{2}} \left[\exp\left(-\frac{1}{2}\omega t\right) |0\rangle + \exp\left(-\frac{3}{2}\omega t\right) |1\rangle \right].$$

The expectation values are immediately found from Eq. 7.66 to be as before

$$\langle x \rangle = \frac{1}{2} \sqrt{\frac{\hbar}{2m_e\omega}} \cos \omega t;$$

$$\langle p \rangle = -\frac{1}{2} \sqrt{\frac{\hbar\omega m_e}{2}} \sin \omega t.$$

To find the uncertainties, I need

$$\hat{x}^2 = \frac{\hbar}{2m_e\omega} \left(\hat{a}^2 + \left(\hat{a}^\dagger\right)^2 + \hat{a}\hat{a}^\dagger + \hat{a}^\dagger\hat{a} \right)$$

$$\hat{p}^2 = -\frac{\hbar\omega m_e}{2} \left(\hat{a}^2 + \left(\hat{a}^\dagger\right)^2 - \hat{a}\hat{a}^\dagger - \hat{a}^\dagger\hat{a} \right).$$

Using again the properties of the lowering and raising operators, I find

$$\hat{x}^2 |\chi(t)\rangle = \frac{\hbar}{2\sqrt{2}m_e\omega} \left(\sqrt{2}\exp\left(-\frac{1}{2}\omega t\right) |2\rangle + \exp\left(-\frac{1}{2}\omega t\right) |0\rangle + \right.$$

$$\left. \sqrt{6}\exp\left(-\frac{3}{2}\omega t\right) |3\rangle + 3\exp\left(-\frac{3}{2}\omega t\right) |1\rangle \right).$$

Now, premultiplying this result by $\langle \chi(t)|$ and using orthogonality of the eigenvectors, I find

$$\langle \hat{x}^2 \rangle = \frac{\hbar}{m_e \omega}$$

in complete agreement with the results obtained in the Heisenberg picture. I will leave computing of the result for the momentum operator to you.

7.2   Isotropic Three-Dimensional Harmonic Oscillator

Using the concept of normal coordinates, any three-dimensional (or even multi-particle) harmonic oscillator can be reduced to the collection of one-dimensional oscillators with total Hamiltonian being the sum of one-dimensional Hamiltonians. The spectrum of eigenvalues in this case is obtained by simply summing up the eigenvalues of each one-dimensional component, and the respective eigenvectors are obtained as direct product of one-dimensional eigenvectors. To illustrate this point, consider a Hamiltonian of the form

$$\hat{H} = \frac{\hat{p}_x^2}{2m_{ex}} + \frac{\hat{p}_y^2}{2m_{ey}} + \frac{\hat{p}_z^2}{2m_{ez}} + \frac{1}{2}\left(m_{ex}\omega_x^2\hat{x}^2 + m_{ey}\omega_y^2\hat{y}^2 + m_{ez}\omega_z^2\hat{z}^2\right) \qquad (7.67)$$

$$= \hat{H}_x + \hat{H}_y + \hat{H}_z.$$

I can define a state characterized by three quantum numbers $|n_x, n_y, n_z\rangle$ which can be considered as a "product" of the one-dimensional eigenvectors defined in the previous section $|n_x, n_y, n_z\rangle \equiv |n_x\rangle |n_y\rangle |n_z\rangle$, where the last notation does not presume any kind of actual "multiplication" but just serves as a reminder that the x-dependent part of the Hamiltonian 7.67 acts only on the $|n_x\rangle$ portion of the eigenvector, the $\hat{H}_y$ acts only on $|n_y\rangle$, and so on. Thus, as a result, I have

$$\left(\hat{H}_x + \hat{H}_y + \hat{H}_z\right)|n_x\rangle |n_y\rangle |n_z\rangle =$$

$$\left[\hbar\omega_x\left(n_x + \frac{1}{2}\right) + \hbar\omega_y\left(n_y + \frac{1}{2}\right) + \hbar\omega_z\left(n_z + \frac{1}{2}\right)\right]|n_x\rangle |n_y\rangle |n_z\rangle ,$$

where $n_{x,y,z}$ independently take integer values starting from zero. The position representation of the eigenvectors is obtained as

$$\varphi_{n_x,n_y,n_z}(x, y, z) = \langle x, y, z \,|n_x, n_y, n_z\rangle \equiv \langle x \,|n_x\rangle \langle y \,|n_y\rangle \langle z \,|n_z\rangle =$$

$$\varphi_{n_x}(x)\varphi_{n_y}(y)\varphi_{n_z}(z),$$

where each $\varphi_{n_i}(r_i)$ is given by Eq. 7.44.

In the most general case, when the parameters in $\hat{H}_x$, $\hat{H}_y$, and $\hat{H}_z$ are all different, we end up with distinct eigenvalues characterized by three independent integers. The energy of the ground state is characterized by $n_x = n_y = n_z = 0$ and is given by $E_{0,0,0} = \frac{1}{2}\hbar\left(\omega_x + \omega_y + \omega_z\right)$.

If, however, all masses and all three frequencies are equal to each other so that the Hamiltonian becomes

$$\hat{H} = \frac{\hat{p}_x^2 + \hat{p}_y^2 + \hat{p}_z^2}{2m_e} + \frac{1}{2}m_{ex}\omega^2\left(\hat{x}^2 + \hat{y}^2 + \hat{z}^2\right), \tag{7.68}$$

a new phenomenon emerges. The energy eigenvalues are now given by

$$E_{n_x,n_y,n_z} = \hbar\omega\left(\frac{3}{2} + n_x + n_y + n_z\right),$$

and it takes the same values for different eigenvectors as long as respective indexes obey condition $n = n_x + n_y + n_z$. In other words, the eigenvalues in this case become degenerate—several distinct vectors belong to the same eigenvalue. The number of degenerate eigenvectors is relatively easy to compute: for each n you can choose n_x to be anything between 0 and n, and once n_x is chosen, n_y can be anything between 0 and $n - n_x$, so there are $n - n_x + 1$ choices. Once n_x and n_y are determined, the remaining quantum number n_z becomes uniquely defined. Thus, the total number of choices of n_x and n_y for any given n can be found as

$$\sum_{n_x=0}^{n_x=n}(n - n_x + 1) = (n + 1)(n + 1) - n(n + 1)/2 = (n + 1)(n + 2)/2.$$

This degeneracy can be easily traced to the symmetry of the system, which has emerged once I made the parameters of the oscillator independent of the direction.

7.2.1   Isotropic Oscillator in Spherical Coordinates

Even though we already know the solution to the problem of an isotropic harmonic oscillator, it is instructive to reconsider it by working in the position representation and using the spherical coordinate system instead of the Cartesian one. The position representation of the Hamiltonian in this case becomes

$$\hat{H} = -\frac{\hbar^2}{2m_e}\nabla^2 + \frac{1}{2}m_e\omega^2 r^2 =$$

$$-\frac{\hbar^2}{2m_e r^2}\frac{\partial}{\partial r}\left(r^2\frac{\partial}{\partial r}\right) + \frac{\hat{L}^2}{2m_e r^2} + \frac{1}{2}m_e\omega^2 r^2, \tag{7.69}$$

where in the second line I used Eq. 5.62 representing Laplacian operator in terms of the radial coordinate r and operator $\hat{L}^2$. It is obvious that the Hamiltonian commutes with both $\hat{L}^2$ and $\hat{L}_z$ so that the eigenvectors of the Hamiltonian in the position representation can be written as

$$\psi_{n_r,l,m}(r,\theta,\varphi) = Y_l^m(\theta,\varphi) R_{n_r,l}(r). \tag{7.70}$$

Substituting Eq. 7.70 into the time-independent Schrödinger equation

$$\hat{H}\psi = E\psi$$

with Hamiltonian given by Eq. 7.69, you can derive for the radial function $R_{n_r,l}(r)$:

$$-\frac{\hbar^2}{2m_e r^2}\frac{d}{dr}\left(r^2\frac{\partial R_{n_r,l}}{\partial r}\right) + \frac{\hbar^2 l(l+1)}{2m_e r^2}R_{n_r,l} + \frac{1}{2}m_e\omega^2 r^2 R_{n_r,l} = E_{l,n_r}R_{n_r,l}. \tag{7.71}$$

It is convenient to introduce an auxiliary function $u_{n_r,l}(r) = rR_{n_r,l}$, which, when inserted into the radial equation above, turns it into

$$-\frac{\hbar^2}{2m_e}\frac{d^2 u_{n_r,l}}{dr^2} + \frac{\hbar^2 l(l+1)}{2m_e r^2}u_{n_r,l} + \frac{1}{2}m_e\omega^2 r^2 u_{n_r,l} = E_{l,n_r}u_{n_r,l}. \tag{7.72}$$

Equation 7.72 looks exactly like a one-dimensional Schrödinger equation with effective potential:

$$V_{eff} = \frac{\hbar^2 l(l+1)}{2m_e r^2} + \frac{1}{2}m_e\omega^2 r^2.$$

The plot of this potential (Fig. 7.3) shows that it possesses a minimum

$$V_{eff}^{min} = \hbar\omega\sqrt{l(l+1)}$$

Fig. 7.3 The schematic of the effective potential for the radial Schrödinger equation for isotropic 3-D harmonic oscillator in arbitrary units

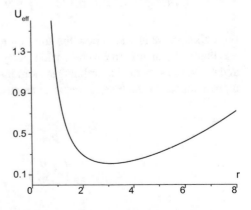

at

$$r^2_{min} = \frac{\hbar}{m_e\omega}\sqrt{l(l+1)}.$$

(Of course, you do not need to plot this function to know that it has a minimum—just compute the derivative and find its zero.) For any given l, the allowed values of energy obeying inequality $E > \hbar\omega\sqrt{l(l+1)}$ correspond to the classical bound motion; thus all energy levels in this effective potential are discrete (which, of course, is nobody's surprise, but still is a nice fact to know). States with $l = 0$ are described by the Schrödinger equation, which is the exact replica of the equation for the one-dimensional oscillator. You, however, should not rush to pull out of the drawer old dusty solutions of the one-dimensional problem (OK, not that old and dusty, but still). A huge difference with the purely one-dimensional case is the fact that the domain of the radial coordinate r is $[0, \infty]$, unlike the domain of a coordinate in the one-dimensional problem, which is $[-\infty, \infty]$. Consequently, the wave function $u_{n_r,l}$ must obey a boundary condition at $r = 0$. Given that the actual radial function $R_{n_r,l}$ must remain finite at $r = 0$, it is clear that $u_{n_r,l}(0)=0$. Now you can go ahead, brush the dust from the solutions of the one-dimensional harmonic oscillator problem, and see which fit this requirement. A bit of examination reveals that we have to throw out all even solutions with quantum numbers $0, 2, 4\cdots$ which do not satisfy the boundary condition at the origin. At the same time, all odd solutions, characterized by quantum numbers $1, 3, 5\cdots$, satisfy both the Schrödinger equation and the newly minted boundary condition at $r = 0$, so they (restricted to the positive values of the coordinate) do represent eigenvectors of the isotropic oscillator with zero angular momentum.

Solving this problem with $l > 0$ requires a bit more work. To make it somewhat easier to follow, I will begin by introducing a dimensionless radial coordinate $\varsigma = r/\xi$, where $\xi = \sqrt{h/m_e\omega}$ is the same length scale that was used in the one-dimensional problem. The Schrödinger equation rewritten in this variable becomes

$$-\frac{\hbar\omega}{2}\frac{d^2 u_{n_r,l}}{d\varsigma^2} + \frac{\hbar\omega l(l+1)}{2\varsigma^2}u_{n_r,l} + \frac{1}{2}\hbar\omega\varsigma^2 u_{n_r,l} = E_{l,n_r}u_{n_r,l}$$

$$\frac{d^2 u_{n_r,l}}{d\varsigma^2} - \frac{l(l+1)}{\varsigma^2}u_{n_r,l} - \varsigma^2 u_{n_r,l} + \epsilon_{l,n_r}u_{n_r,l} = 0 \qquad (7.73)$$

where I introduced dimensionless energy

$$\epsilon_{l,n_r} = 2E_{l,n_r}/\hbar\omega.$$

The resulting differential equation obviously does not have simple solutions expressible in terms of elementary functions. One of the approaches to solving it is to present a solution in the form of a power series $\sum c_j\varsigma^j$ and search for unknown coefficients c_j. In principle, knowing these coefficients is equivalent to knowing the entire function. Before attempting this approach, however, it would be wise to try to extract whatever information about the solution this equation might contain. For instance, you can ask about the solution's behavior at very small and very large values of ς. When $\varsigma \ll 1$, the main term in Eq. 7.73 is the angular momentum contribution to the effective potential. Neglecting all other terms, you end up with an equation

$$\frac{d^2 u_{n_r,l}}{d\varsigma^2} - \frac{l(l+1)}{\varsigma^2} u_{n_r,l} = 0, \tag{7.74}$$

which has a simple power solution

$$u_{n_r,l} = A\varsigma^{l+1}. \tag{7.75}$$

You are welcome to plug it back in Eq. 7.74 and verify it by yourself. For those who think that I used magical divination to arrive at this result, I have disappointing news: Eq. 7.74 belongs to a well-known class of so-called homogeneous equations. This means that if I multiply ς by an arbitrary constant factor λ, the equation does not change (check it), with a consequence that if function $u(\varsigma)$ is the solution, so is function $u(\lambda\varsigma)$. Such equations are solved by power functions $u \propto \varsigma^\varrho$, where power ϱ is found by plugging this function into the equation.

In the limit of large $\varsigma \gg 1$, the main contribution to Eq. 7.73 comes from the harmonic potential. We know from solving the one-dimensional problem that the respective wave functions contain an exponential term $\exp\left(-\tilde{x}^2/2\right)$ for x direction and similar terms for two other coordinates. When multiplying all these wave functions together to obtain a three-dimensional wave function, these exponential terms turn into $\exp\left(-\varsigma^2/2\right)$; thus it is natural to expect that the radial function $u_{n_r,l}$ will contain such a factor as well. To verify this assumption, I am going to substitute $\exp\left(-\varsigma^2/2\right)$ into Eq. 7.73 and see if it will satisfy the equation, at least in the limit $\varsigma \to \infty$. Neglecting all terms small compared to ς^2, I find

$$\frac{d^2 u}{d\varsigma^2} = -e^{-\varsigma^2/2} + \varsigma^2 e^{-\varsigma^2/2} \approx \varsigma^2 e^{-\varsigma^2/2}.$$

Substituting this result in Eq. 7.73, and neglecting all terms except of the harmonic potential, I find that this function is, indeed, an asymptotically accurate solution of this equation. I want you to really appreciate this result: in order to reproduce exponential decay of the wave function, which, by the way, almost ensures its normalizability, using a power series, we would have to keep track of all the infinite number of terms in it, which is quite difficult if not outright impossible. By pulling out this exponential term as well as the power law for small ς, you might entertain some hope that the remaining dependence on ς is simple enough to be dug out.

Thus, my next step is to present function $u_{n_r,l}(\varsigma)$ as

$$u_{l,n_r}(\varsigma) = A\varsigma^{l+1} \exp\left(-\varsigma^2/2\right) v_{l,n_r}(\varsigma) \tag{7.76}$$

and derive a differential equation for the remaining function $v_{n_r,l}(\varsigma)$. To this end, I first compute

$$\frac{d^2 u_{n_r,l}}{d\varsigma^2} = \frac{d}{d\varsigma}\Big[(l+1)\,\varsigma^l \exp\left(-\varsigma^2/2\right) v_{n_r,l}(\varsigma) -$$

$$\varsigma^{l+2} \exp\left(-\varsigma^2/2\right) v_{n_r,l}(\varsigma) + \varsigma^{l+1} \exp\left(-\varsigma^2/2\right) \frac{dv_{n_r,l}}{d\varsigma}\Big] =$$

$$l(l+1)\varsigma^{l-1} \exp\left(-\varsigma^2/2\right) v_{n_r,l}(\varsigma) - (l+1)\,\varsigma^{l+1} \exp\left(-\varsigma^2/2\right) v_{n_r,l}(\varsigma) +$$

$$(l+1)\,\varsigma^l \exp\left(-\varsigma^2/2\right) \frac{dv_{n_r,l}}{d\varsigma} - (l+2)\,\varsigma^{l+1} \exp\left(-\varsigma^2/2\right) v_{n_r,l}(\varsigma) +$$

$$\varsigma^{l+3} \exp\left(-\varsigma^2/2\right) v_{n_r,l}(\varsigma) - \varsigma^{l+2} \exp\left(-\varsigma^2/2\right) \frac{dv_{n_r,l}}{d\varsigma} +$$

$$(l+1)\,\varsigma^l \exp\left(-\varsigma^2/2\right) \frac{dv_{n_r,l}}{d\varsigma} - \varsigma^{l+2} \exp\left(-\varsigma^2/2\right) \frac{dv_{n_r,l}}{d\varsigma} +$$

$$\varsigma^{l+1} \exp\left(-\varsigma^2/2\right) \frac{d^2 v_{n_r,l}}{d\varsigma^2} =$$

$$\exp\left(-\varsigma^2/2\right) \varsigma^{l-1} v_{n_r}(\varsigma) \left(l(l+1) - \varsigma^2(2l+3) + \varsigma^4\right) +$$

$$\varsigma^l \exp\left(-\varsigma^2/2\right) \frac{dv_{n_r}}{d\varsigma} \left(2l+2-2\varsigma^2\right) + \varsigma^{l+1} \exp\left(-\varsigma^2/2\right) \frac{d^2 v_{n_r}}{d\varsigma^2}.$$

Frankly speaking, I did not have to torture you with these tedious calculations: such computational platforms as Mathematica or Maple work with symbolic expressions and can perform this computation faster and more reliably (and, yes, I did check my result against Mathematica's). Substituting this expression to Eq. 7.73, I get (and here you are on your own, or you can try computer algebra to reproduce this result)

$$\varsigma \frac{dv_{n_r,l}^2}{d\varsigma^2} + 2\left(l+1-\varsigma^2\right) \frac{dv_{n_r,l}}{d\varsigma} + \varsigma v_{n_r,l}(\varsigma)\left(\epsilon_{l,n_r} - 2l - 3\right) = 0. \qquad (7.77)$$

Now I can start solving this equation by presenting the unknown function $v_{n_r,l}(\varsigma)$ as a power series and trying to find the corresponding coefficients:

$$v_{n_r,l}(\varsigma) = \sum_{j=0}^{\infty} c_j \varsigma^j. \qquad (7.78)$$

The goal is to plug this expression into Eq. 7.77 and collect coefficients in front of equal powers of ς. First, I blindly substitute the series into Eq. 7.77 and separate all sums with different powers of ς:

$$\sum_{j=0}^{\infty} c_j j(j-1)\varsigma^{j-1} + 2(l+1)\sum_{j=0}^{\infty} c_j j\varsigma^{j-1} - 2\sum_{j=0}^{\infty} c_j j\varsigma^{j+1} +$$

$$(\epsilon_{l,n_r} - 2l - 3)\sum_{j=0}^{\infty} c_j \varsigma^{j+1} = 0.$$

Combining the first two and last two sums, I get

$$\sum_{j=0}^{\infty} j\,[j-1+2l+2]\,c_j\varsigma^{j-1} + \sum_{j=0}^{\infty} [\epsilon_{l,n_r} - 2l - 3 - 2j]\,c_j\varsigma^{j+1} = 0.$$

Next I notice that in the first sum, contributions from terms with $j = 0$ vanish, so that this sum starts with $j = 1$. I can reset the count of the summation index back to zero by introducing new index $k = j - 1$, so that this sum becomes

$$\sum_{k=0}^{\infty} c_{k+1}\,(k+1)\,(k+2l+2)\varsigma^k.$$

Renaming k back to j (this is a dummy index, so you can call it whatever you want, it does not care), we rewrite the previous equation as

$$\sum_{j=0}^{\infty} (j+1)\,[j+2l+2]\,c_{j+1}\varsigma^j + \sum_{j=0}^{\infty} [\epsilon_{l,n_r} - 2l - 3 - 2j]\,c_j\varsigma^{j+1} = 0.$$

The first sum in this expression begins with ς^0 term multiplied by coefficient c_1. The second sum, however, begins with linear in ς term and does not contain ς^0 at all. To satisfy the equation, coefficients in front of each power of ς must vanish independently of each other, so we have to set $c_1 = 0$. This makes the first sum again to start with $j = 1$. Utilizing the same trick as before, I am replacing j with $j+1$ while restarting count from new $j = 0$ again. The result is as follows:

$$\sum_{j=0}^{\infty} (j+2)\,[j+2l+3]\,c_{j+2}\varsigma^{j+1} + \sum_{j=0}^{\infty} [\epsilon_{l,n_r} - 2l - 3 - 2j]\,c_j\varsigma^{j+1} = 0.$$

Now I can, finally, combine the two sums and equate the resulting coefficient in front of ς^{j+1} to zero:

$$(j+2)\,[j+2l+3]\,c_{j+2} = [2l+3+2j - \epsilon_{l,n_r}]\,c_j$$

or

$$c_{j+2} = \frac{2l + 3 + 2j - \epsilon_{l,n_r}}{(j + 2) [j + 2l + 3]} c_j. \tag{7.79}$$

This is a so-called recursion relation, which allows computing all expansion coefficients recursively starting with the first one. It is important to note that Eq. 7.79 connects only coefficients with indexes of the same parity: all coefficients with even indexes are expressed in terms of c_0, and all coefficients with odd indexes are expressed in terms of c_1. But, wait, have not we determined a few lines back that $c_1 = 0$? Actually, we did determine that, and now, thanks to Eq. 7.79, I can establish that not only c_1 but all coefficients with odd indexes are zeroes. So, it looks like I achieved the announced goal—finding all coefficients in the power series expansion of $v_{n_r,l}$. Formally speaking, I did, indeed, but it is a bit too early to dance around the fire and celebrate. First, I still do not know what values of the dimensionless energy ϵ_{l,n_r} correspond to the respective eigenvectors, and, second, I have to verify that the found solution is, indeed, normalizable. The last issue is not trivial because we are dealing with an infinite series here, so there are always questions about its convergence and the behavior of the function it represents. As I shall demonstrate now, both these questions are connected and will be answered together.

Whether a function is normalizable or not is determined by its behavior for large values of its argument. I pulled out an exponentially decreasing factor from the solution hoping that it would be sufficient to guarantee normalization, but to be sure I need to consider the behavior of $v_{n_r,l}$ at $\varsigma \to \infty$. Any finite number of terms in the expansion 7.78 cannot overcome the exponentially decreasing factor $\exp\left(-\varsigma^2/2\right)$, so the anticipated danger can only come from the tail of the power series, i.e., from coefficients c_j with $j \to \infty$. In this limit the recursion relation 7.79 can be simplified to

$$c_{j+2} \approx \frac{2}{j} c_j, \tag{7.80}$$

which, when applied repeatedly, yields

$$c_{2j_0+2N} = \frac{2^{2N}}{2j_0(2j_0 + 2) \cdots (2j_0 + 2N)} c_{2j_0} = \frac{1}{j_0 (j_0 + 1) \cdots (j_0 + N)} c_{2j_0}.$$

When writing this expression, I explicitly took into account that there are only even indexes, which can be presented as $2j_0+2k$ with the total number of recursive factors being $2N$. Even though this expression is only valid for $j_0 \gg 1$, I can extend it to all values of j_0 because as I pointed out earlier, any finite number of terms in the power series would not affect its asymptotic behavior. That means that the large ς behavior of the series in question is the same as that of the series:

$$\sum_{j=0}^{\infty} \frac{\varsigma^{2j}}{j!} = e^{\varsigma^2}.$$

Even after combining this result with $\exp\left(-\varsigma^2/2\right)$ factor, which was pulled out earlier, I still end up with function $u_{n_r,l}(\varsigma)$ behaving as $\exp\left(\varsigma^2/2\right)$ at infinity. What a bummer! It is disappointing, but not really surprising: it is easy to check that $\exp\left(\varsigma^2/2\right)$ is the second possible asymptotic solution of Eq. 7.73, which I choose to discard because of its non-normalizable nature. Well, this is how it often is—you chase math out of the door, but it always comes back through the window to bite you. So, the question now is if there is anything I can do to save the normalizability of our solution. That light at the end of the tunnel will appear if you recall that any power series with a finite number of terms cannot overpower an exponentially decreasing function. Therefore, if I find a way to terminate the series at some finite number of terms, our conundrum will be resolved. To see how this is possible, let's take another look at the recursion relation, Eq. 7.79. What if at some value of j, which I will call $2n_r$ to emphasize its evenness, the numerator of this relation turns zero? If this were to happen, then coefficient $c_{j_{mx}+2}$ would vanish and vanquish all subsequent coefficients as well, so that instead of an infinite series, I will end up with a finite sum. This will surely guarantee the normalizability of the found solution. The condition for the numerator to vanish reads as

$$2l + 3 + 4n_r - \epsilon_{l,n_r} = 0$$

which is immediately recognizable as an equation for the dimensionless energy ϵ_{l,n_r}! While resolving the normalization problem, I just automatically solved finding the eigenvalue problem. Using

$$\epsilon_{l,n_r} = 3 + 2(l + 2n_r)$$

as well as the relation between ϵ_{l,n_r} and actual energy eigenvalues, I obtain

$$E_{l,n_r} = \hbar\omega\left(\frac{3}{2} + l + 2n_r\right).$$

Thus, for each l and n_r, you have an energy value and a respective wave function

$$u_{l,n_r}(\varsigma) = \varsigma^{l+1} \exp\left(-\varsigma^2/2\right) \sum_{j=0}^{2n_r} c_j \varsigma^j \tag{7.81}$$

where coefficients c_j are given by Eq. 7.79. To get a better feeling for this result, consider a few special examples.

1. $n_r = 0$. In this case the sum in Eq. 7.81 contains a single term c_0, so the non-normalized wave function becomes

$$u_{l,0}(\varsigma) = c_0\varsigma^{l+1} \exp\left(-\varsigma^2/2\right)$$

with respective energy value $E_{l,0} = \hbar\omega\left(\frac{3}{2} + l\right)$.

2. $n_r = 1$. Using Eq. 7.79 with $\epsilon_{l,n_r} = 3 + 2(l + 2)$, I find for c_2 (substituting $j = 0$ into Eq. 7.79):

$$c_2 = \frac{2l + 3 - (3 + 2l + 4)}{2\,[2l + 3]}c_0 = -\frac{2}{2l + 3}c_0$$

so that

$$u_{l,1}\left(\varsigma\right) = c_0\varsigma^{l+1}\exp\left(-\varsigma^2/2\right)\left(1 - \frac{2\varsigma^2}{2l + 3}\right).$$

Following this pattern you can compute the wave functions belonging to any eigenvalue. For higher energy eigenvalues, it would take more time and efforts, of course, but you can always give this task to a computer. Before finishing this section, I would like to note that the energy eigenvalues depend only on the sum $l + 2n_r$ rather than on each of these quantum numbers separately. It makes sense, therefore, to introduce a main quantum number $n = l + 2n_r$ and use it to characterize energy values:

$$E_n = \hbar\omega\left(\frac{3}{2} + n\right). \tag{7.82}$$

Then, the radial wave functions will be labeled by indexes l and n with a requirement $n - l = 2n_r \geq 0$, while the total wave function includes spherical harmonics and an additional index m. In actual physical variables, it becomes

$$\psi_{n,l,m} = \frac{1}{\xi}\left(\frac{r}{\xi}\right)^l\exp\left(-\frac{r^2}{2\xi^2}\right)\sum_{j=0}^{n-l}c_j\left(\frac{r}{\xi}\right)^j Y_l^m\left(\theta, \varphi\right) \tag{7.83}$$

where I reintroduced radial function $R_{n_r,l} = u_{n_r,l}/r$. This function is not normalized until the value of coefficient c_0 in its radial part is defined, but I am not going to bother you with that. Instead, I will compute the degree of degeneracy of an energy eigenvalue characterized by main number n, which is much more fun. Taking into account that for each l there are $2l + 1$ possible values of m, and that l runs all the way down from n in increments of 2 ($n - l$ must remain an even number), the total number of states with given n is

$$\sum\left(2l + 1\right) = (n + 1) + n(n + 1)/2 = (n + 1)\,(n + 2)\,/2,$$

where the summation over l is carried with increments of 2. It is a nice feeling to realize that this expression for degeneracy agrees with the one obtained using Cartesian coordinates.

The resulting expression for the wave function given by Eq. 7.83 is an alternative way to produce a position representation of the harmonic oscillator wave function and is quite remarkably different from the one obtained using Cartesian coordinates. One might wonder why it is at all possible to have such distinct ways to represent a same eigenvector. After all, isn't a representation, once chosen, supposed to provide a unique way to describe a quantum state? The matter of fact is that it is, indeed, so only if a corresponding eigenvalue is non-degenerate. In the degenerate case, one can form an infinite number of the linear combinations of the eigenvectors, and any one of them will realize the same representation of the corresponding state. In the case of isotropic harmonic oscillator, it means that the wave functions expressed in spherical coordinates can be presented as linear combinations of their Cartesian counterparts and vice versa.

7.3   Quantization of Electromagnetic Field and Harmonic Oscillators

7.3.1   Electromagnetic Field as a Harmonic Oscillator

Even though the idea of photons—the quanta of electromagnetic field—was one of the first quantum ideas introduced into the conscience of physicists by Einstein in 1905,[1] the full quantum description of electromagnetic field turned out to be a rather difficult problem. The first serious attempt in developing quantum electrodynamics was undertaken by Paul Dirac in his famous 1927 paper,[2] which was just the beginning of a long and difficult path walked by too many brilliant physicists to be mentioned in this book. Here are just a few names of those who made critical theoretical contributions to this field: German-American Hans Bethe, Japanese Sin-Itiro Tomonaga, and Americans Julian Schwinger, Richard Feynman, and Freeman Dyson. Quantum electrodynamics is a difficult subject addressed in multiple specialized books and is beyond the scope of this text. Nevertheless, I would love to scratch a bit from the surface of this field and demonstrate how ideas developed in the course of studying the harmonic oscillator emerge in new and unexpected places.

---

[1] The irony is that an explanation of photoelectric effect did not require the quantization of light despite what you might have read or heard. All experimental data could have been explained treating light classically while describing electrons in metals by the Schrödinger equation. Fortunately Einstein did not have the Schrödinger equation in 1905 and couldn't know that. The science does evolve in mysterious ways: Einstein's erroneous idea about the photoelectric effect inspired de Broglie and Schrödinger and brought about the Schrödinger equation, which could have been used to disprove the idea. Compton's effect, on the other hand, can indeed be considered as a proof of reality of photons.

[2] P.A.M. Dirac, The quantum theory of the emission and absorption of radiation. Proc. R. Soc. Lond. **114**, 243 (1927).

To this end, I propose considering a toy model of electromagnetic field, in which the field is described by single components of electric and magnetic fields:

$$E_x = a\mathcal{E}_0(t) \sin kz \tag{7.84}$$

$$B_y = -\frac{1}{c} a\mathcal{B}_0(t) \cos kz \tag{7.85}$$

where I introduced a normalization coefficient a to be defined later; extra factor $1/c$, where c is the speed of light in vacuum, in the formula for the magnetic field, ensures that amplitudes $\mathcal{E}_0$ and $\mathcal{B}_0$ have the same dimension (you might remember from the introductory course on electromagnetism relation $E = cB$ between electric and magnetic fields in a plane wave), and the negative sign is included for future convenience. The Maxwell equations for the electromagnetic field in this simplified case take the form

$$\frac{\partial E_x}{\partial z} = -\frac{\partial B_y}{\partial t}$$

$$\frac{\partial B_y}{\partial z} = -\frac{1}{c^2}\frac{\partial E_x}{\partial t}.$$

Plugging in the expressions for electric and magnetic fields given by Eqs. 7.84 and 7.85, you will find that the spatial dependence chosen for the fields in these equations is indeed consistent with the Maxwell equations, which will be reduced to the system of ordinary differential equations:

$$\frac{d\mathcal{B}_0}{dt} = \omega\mathcal{E}_0(t) \tag{7.86}$$

$$\frac{d\mathcal{E}_0}{dt} = -\omega\mathcal{B}_0(t). \tag{7.87}$$

Parameter ω appearing in these equations is defined as $\omega = ck$. It is easy to see that amplitudes of both electric and magnetic fields obey the same differential equation as a harmonic oscillator. For instance, differentiating the first of these equations with respect to time and using the second equation to replace the time derivative of the electric field, you will get

$$\frac{d^2\mathcal{B}_0}{dt^2} + \omega^2\mathcal{B}_0 = 0.$$

Similar equation can be derived for $\mathcal{E}_0$. You can also notice that Eqs. 7.86 and 7.87 have some resemblance to the Hamiltonian equations of classical mechanics, and this may make you wonder if they can be derived from some kind of a Hamiltonian. If you are asking why on earth would I want to re-derive these equations from a Hamiltonian, you were not paying attention to the first 130 pages of the book. Hamiltonian formalism allows us to introduce canonical pairs of variables, which we can turn into operators obeying canonical commutation relations; thus a Hamiltonian formulation is the key to turning classical theory of electromagnetic field into the quantum one.

How would one go about introducing a Hamiltonian for the electromagnetic fields? Naturally, one starts by remembering that Hamiltonian is the energy of the system and that the energy of the electromagnetic field is given by

$$\mathcal{H} = \int_V d^3r \left(\frac{1}{2} \varepsilon_0 E^2 + \frac{1}{2\mu_0} B^2 \right), \tag{7.88}$$

where integration is carried over the entire space occupied by the field. However, if you attempt to directly compute this integral using Eqs. 7.84 and 7.85 for electric and magnetic fields, you will encounter a problem: the integral is infinite. This happens because the field occupies the entire infinite space and does not decrease with distance. To fix the problem, I introduce a large but finite region of volume $V = L_z S_{xy}$, where L_z is the linear dimension of this region in z direction and S_{xy} is the area of the limiting plane perpendicular to it, and assume that the field vanishes outside of this region. This trick is very popular in physics, and you will encounter it in different circumstances later in the book. It can be justified by noting that the notion of a field occupying the entire space is by itself quite artificial with no relation to reality. It is also natural to assume that the properties of the field far away from the region of actual interest should not affect any observable phenomena, so that we can choose them to be as convenient for us as possible.

With this in mind, I can write the integral in Eq. 7.88 as

$$\mathcal{H} = a^2 S_{xy} \left[\frac{1}{2} \varepsilon_0 \mathcal{E}_0^2 \int_0^L dz \sin^2 kz + \frac{1}{2\mu_0} \mathcal{B}_0^2 \int_0^L dz \cos^2 kz \right] = $$
$$\frac{1}{4} a^2 \varepsilon_0 S_{xy} L \left[\mathcal{E}_0^2 + \mathcal{B}_0^2 \right],$$

where I assumed that k satisfies condition $kL = \pi n$, $n = 1, 2, \cdots$, making $\cos 2kz = 1$ at both the lower and upper integration limits so that the respective terms cancel out. Also, at the last step, I made a substitution $(\mu_0 \varepsilon_0)^{-1} = c^2$. You might, of course, object to the artificial discretization of the wave number and imposition of the arbitrary conditions on the values of the electric and magnetic fields at $z = L$. So, what can I say in my defense? First, in the limit $L \to \infty$, which I can make after everything is said and done, the discretization will disappear, and as you will see in a few short minutes, I will make the dependence on the volume which popped up in the last expression for the Hamiltonian, disappear as well. Second, I can invoke the same argument I just made when limiting the field to the finite volume: the behavior of the field in any finite region of space shall not be affected by its values at an infinitely remote plane. If you are still not convinced, I have my last line of defense: it works!

Now, I am ready to fix the normalization parameter a introduced in Eqs. 7.84 and 7.85. For the reasons which will become clear later, I will choose it to be

$$a = \sqrt{2\omega/(\varepsilon_0 V)}, \tag{7.89}$$

so that the final expression for the energy of the field becomes

$$\mathcal{H} = \frac{\omega}{2}\left(\mathcal{E}_0^2 + \mathcal{B}_0^2\right). \tag{7.90}$$

Did you notice that the dependence on the volume in the Hamiltonian is gone? This is fiction of course, because I have simply hidden it inside formulas for the fields, but in all expressions concerned with actual physical observables, it will vanish in all honesty.

Equation 7.90 looks very much like the Hamiltonian of a harmonic oscillator. The first term can be interpreted as kinetic energy with $\mathcal{E}_0$ playing the role of the canonical momentum and term $1/\omega$ replacing the mass, and the second term is an analog of the potential energy with $\mathcal{B}_0$ as a conjugated coordinate (note that the coefficient $m_e\omega^2/2$ in the harmonic oscillator potential energy is reduced to $\omega/2$ factor in Eq. 7.90 if you replace m_e with $1/\omega$). If you wonder why I chose the electric field to represent the momentum and the magnetic field to be the coordinate, and not vice versa, just compare Eqs. 7.86 and 7.87 with Hamiltonian equations 7.8 and 7.7, paying attention to the placement of the negative sign in these equations. You can easily see that the Hamiltonian equations reproduce Eqs. 7.86 and 7.87 justifying this identification. But do not be fooled. Identifying magnetic field with coordinate and electric field with momentum is, of course, a matter of convention resulting from the choice to place the negative sign in Eq. 7.85.

The Hamiltonian formulation of the classical Maxwell equations allows me now to introduce the quantum description of the fields. This is done by promoting $\mathcal{E}_0$ and $\mathcal{B}_0$ to operators with the standard canonical commutation relation:

$$\left[\hat{\mathcal{B}}_0, \hat{\mathcal{E}}_0\right] = i\hbar. \tag{7.91}$$

As a result, the classical Hamiltonian, Eq. 7.90, becomes a Hamiltonian operator:

$$\hat{H} = \frac{\omega}{2}\left(\hat{\mathcal{E}}_0^2 + \hat{\mathcal{B}}_0^2\right). \tag{7.92}$$

It is easy to see from Eq. 7.90 that both $\mathcal{E}_0$ and $\mathcal{B}_0$ have the dimension of $\sqrt{energy \times time}$ so that the dimension of the commutator on the left-hand side of Eq. 7.91 is $energy \times time$, which coincides with the dimension of Planck's constant, as it should. This result is not particularly surprising, of course, but it is always useful to check your dimensions once in a while just to make sure that your theory does not have any of the most basic problems. Using Eq. 7.91 together with Eqs. 7.84 and 7.85, I can compute the commutator of the non-zero components of the electric and magnetic fields, which, of course, are now also operators:

$$\left[\hat{B}_y, \hat{E}_x\right] = -\frac{i\hbar\omega}{\varepsilon_0 cV} \sin 2kz. \tag{7.93}$$

One immediate consequence of this result is the uncertainty relation for these components:

$$\Delta B_y \Delta E_x \geq \frac{\hbar\omega}{2\varepsilon_0 cV} \left|\sin 2kz\right|, \tag{7.94}$$

which shows that just like the coordinate and momentum, electric and magnetic fields cannot both be known with certainty in the same quantum state.

Canonical commutator, Eq. 7.91, also indicates that in the representation using eigenvectors of $\hat{B}_0$ as a basis, in which states are represented by wave functions dependent on the magnetic field amplitude $\hat{B}_0$, the representation of the electric field amplitude operator $\hat{\mathcal{E}}_0$ is

$$\hat{\mathcal{E}}_0 = -i\hbar \frac{\partial}{\partial B_0},$$

while the Hamiltonian takes the form

$$\hat{H} = \frac{\omega}{2} \left(-\hbar^2 \frac{\partial^2}{\partial B_0^2} + B_0^2\right).$$

Comparing this expression with the quantum Hamiltonian of the harmonic oscillator in the coordinate representation, you can see that they are mathematically identical if again you replace m_e with $1/\omega$. The wave functions representing eigenvectors of this Hamiltonian can be in this representation written down as

$$\varphi_n(B_0) = \frac{1}{\sqrt{2^n n! \xi_{em} \sqrt{\pi}}} \exp\left(-\frac{B_0^2}{2\xi_{em}^2}\right) \mathcal{H}_n\left(\frac{B_0}{\xi_{em}}\right) \tag{7.95}$$

where the characteristic scale of the quantum fluctuations of the magnetic field, ξ_{em}, is determined solely by Planck's constant $\xi_{em} = \sqrt{\hbar}$ (this result follows from Eq. 7.42 after the substitution $m_e = 1/\omega$). As with any wave function, $|\varphi_n(B_0)|^2$ determines the probability density function for the magnetic field amplitude.

While it is interesting to see how one can turn the coordinate representation of the harmonic oscillator into the magnetic field representation of the quantum electromagnetic theory, the practical value of this representation is quite limited. Much more important, from both theoretical and practical points of view, is the opportunity to introduce electromagnetic analogs of lowering and raising operators. In order to distinguish these operators from those used in the harmonic oscillator problem, I will use notation $\hat{b}$ and $\hat{b}^\dagger$ (do not confuse these operators with variables

b used in the description of the classical oscillator), where

$$\hat{b} = \sqrt{\frac{1}{2\hbar}}\hat{B}_0 + i\frac{\hat{\mathcal{E}}_0}{\sqrt{2\hbar}} \tag{7.96}$$

$$\hat{b}^\dagger = \sqrt{\frac{1}{2\hbar}}\hat{B}_0 - i\frac{\hat{\mathcal{E}}_0}{\sqrt{2\hbar}}. \tag{7.97}$$

Equations 7.96 and 7.97 are obtained from Eqs. 7.22 and 7.23 by setting $m_e\omega = 1$ and replacing $\hat{x}$ and $\hat{p}$ by $\hat{B}_0$ and $\hat{\mathcal{E}}_0$ correspondingly. Hamiltonian 7.92 expressed in terms of these operators acquires a familiar form:

$$\hat{H} = \hbar\omega\left(\hat{b}^\dagger\hat{b} + 1/2\right).$$

All commutators, which were computed in Sect. 7.1.1, remain exactly the same, so I can simply reproduce the results from that section: the energy eigenvalues of the electromagnetic field are given again by

$$E_n = \hbar\omega\left(n + \frac{1}{2}\right), \tag{7.98}$$

while eigenvectors can be constructed from the ground state $|0\rangle$ as

$$|n\rangle = \frac{1}{\sqrt{n!}}\left(\hat{b}^\dagger\right)^n|0\rangle. \tag{7.99}$$

Formally, both these results are exactly the same as in the case of the harmonic oscillator. However, the physical interpretation of the integer n in these expressions and, therefore, of both energy values and eigenvectors is completely different.

Indeed, in the case of a harmonic oscillator, we have a material particle, which can be placed in states with different energies, counted by the integer n. The electromagnetic field, on the other hand, once created, carries a certain amount of energy, and the same field cannot be made to have "more" energy. To produce a field with higher energy, you need to increase its amplitude, i.e., add "more" field. The discrete nature of allowed energy levels tells us that the energy of the field can only be increased in finite increments: to go from a state of electromagnetic field with energy E_n to the state with energy E_{n+1}, you have to add a discrete "quantum" of field with energy $\hbar\omega$. This discrete energy quantum is what was introduced by Einstein in 1905 as "das Lichtquantas." Replacing the term "quantum of light" with the term "photon,"[3] you can say that number n is the number of photons in a given state and that going from state $|n\rangle$ to state $|n + 1\rangle$ amounts to generating

---

[3]It is interesting that the term "photon" was used for the first time in an obscure paper by an American chemist Gilbert Lewis in 1926. His paper is forgotten, but the term he coined lives on.

or creating an extra photon, while transitioning to state $|n - 1\rangle$ means removing or annihilating a photon. To emphasize this point, operators $\hat{b}^\dagger$ and $\hat{b}$ are called in the context of quantum electromagnetic field theory "creation" and "annihilation" operators, respectively, rather than lowering and raising operators. The ground state $|0\rangle$ in this interpretation is the state with zero photons and is called, therefore, the vacuum state. A counterintuitive aspect of the vacuum state is that even though it is devoid of photons, it still has non-zero energy, which in our oversimplified model is just $\hbar\omega/2$. To one's mind it might appear as a nonsensical result: how can zero photons have non-zero energy? I hope it will not blow your mind away if I say that in a more complete theory, which takes into account multiple modes (waves with different wave vectors k) of electromagnetic field, the "vacuum" energy might become formally infinite. In order to wrap your mind around this weird result, consider the following.

The photon is not just "a quantum of electromagnetic field" as you might have read in popular books and introductory physics texts. The concept of a "photon" has a quite specific mathematically rigorous meaning: a single photon is an eigenvector of the electromagnetic Hamiltonian characterized by $n = 1$. Eigenvectors characterized by higher values of n describe n-photon states. The states described by eigenvectors of the Hamiltonian are not the states in which the electric or magnetic field has any definite value. Moreover, the commutation relation, Eq. 7.93, and following from it uncertainty relation 7.94 indicate that there are no states in which electric and magnetic fields both have definite values. Moreover, in the states with fixed photon numbers, the expectation values of electric and magnetic fields are zeroes just like the expectation values of coordinate and momentum operators of the mechanical harmonic oscillator. At the same time, the expectation values of the squares of the fields are not zeroes, and these are the quantities which determine the energy of the fields. These are what we call *vacuum fluctuations of electromagnetic field*, where vacuum has, again, a very specific meaning—it is not just emptiness or a void; it is a state with zero photons, which is not the same as a state with zero field.

The second issue which needs to be discussed in connection with vacuum energy is, again, the fact that a zero level of energy is always established arbitrarily. The vacuum energy, which we found, is counted from the (non-existent in quantum theory) state, in which both electric and magnetic fields are presumed to be zeroes. As long as the energy of the vacuum state does not change, while the phenomena we are interested in play out, we can set the vacuum energy to zero with no consequences for any physically significant results. To provide a counterexample to this statement, let me briefly describe a situation in which this assumption might not be true. If you consider the electromagnetic field between two conducting plates, the modes of the field and, therefore, its vacuum energy depend on the distance between the plates. This distance can be changed, in which case the vacuum energy also changes. Because of this capacity to change, it becomes relevant resulting in a tiny but observable attractive force acting between the plates known as the Casimir force. In most other situations, however, the vacuum energy is just a constant, whose value (finite or infinite) has no physical significance.

Thus, the eigenvectors of the electromagnetic Hamiltonian representing states with a definite number of photons, n, bear little resemblance to classical electromagnetic waves just like stationary states of the harmonic oscillator have no relation to the motion of the classical pendulum. At the same time, in Sect. 7.1.2, I demonstrated that a generic nonstationary state reproduces oscillations of the expectation values of coordinate and momentum resembling those of their classical counterparts. While this result is true for a generic initial state, and the behavior of the expectation values to a large extent does not depend on their details, not all initial states are created equal. However, to notice the difference between them, we have to go beyond the expectation values and consider the uncertainties of both coordinate and momentum or, in the electromagnetic context, of electric and magnetic fields. The fact that different initial states result in different behavior of uncertainties has already been demonstrated in the examples presented in Sect. 7.1.2. However, out of all the multitude of various initial states, there exists one, for which these uncertainties are minimized in a sense that their product has the smallest allowed by the uncertainty principle value. In the electromagnetic case it means that the sign $\geq$ in Eq. 7.94 is replaced with $=$. These states are called "coherent" states, and they are much more important in the electrodynamics rather than in mechanical context, so this is where I shall deal with them.

7.3.2   Coherent States of the Electromagnetic Field

The coherent states are defined as eigenvectors of the annihilation operator:

$$\hat{b} |\alpha\rangle = \alpha |\alpha\rangle . \tag{7.100}$$

Since the annihilation operator is not Hermitian, you should not expect the eigenvalues to be real, and we do not know yet if they are continuous or discrete. I can, however, try to find the representation of vectors $|\alpha\rangle$ in the basis of the eigenvectors $|n\rangle$ of the electromagnetic Hamiltonian:

$$|\alpha\rangle = \sum_{n=0}^{\infty} c_n |n\rangle , \tag{7.101}$$

where $c_n = \langle n | \alpha \rangle$. The Hermitian conjugation of Eq. 7.99 yields

$$\langle n| = \frac{1}{\sqrt{n!}} \langle 0| \left(\hat{b} \right)^n \tag{7.102}$$

so that I can find for the expansion coefficients

$$c_n = \frac{1}{\sqrt{n!}} \langle 0| \left(\hat{b}\right)^n |\alpha\rangle = \frac{\alpha^n}{\sqrt{n!}} \langle 0| \alpha\rangle .$$

The only unknown quantity here is $c_0 = \langle 0| \alpha\rangle$, which I find by requiring that $|\alpha\rangle$ is normalized, which means that $\sum_n |c_n|^2 = 1$. Applying this last condition, I have

$$|c_0|^2 \sum_{n=0}^{\infty} \frac{|\alpha|^{2n}}{n!} = |c_0|^2 \exp\left(|\alpha|^2\right) = 1$$

where I recalled that $\sum (x^n/n!)$ is a power series expansion for the exponential function of x. Thus, choosing c_0 to be real-valued, I have the following final expression for the expansion coefficients:

$$c_n = e^{-\frac{|\alpha|^2}{2}} \frac{\alpha^n}{\sqrt{n!}}. \tag{7.103}$$

Equation 7.103 together with Eq. 7.101 completely defines a coherent state with eigenvalue α. Since the derivation of the eigenvector did not produce any restrictions on α, it must be presumed to be a continuous complex-valued variable. The vector that I found describes a state which is the superposition of states with different numbers of photons and, respectively, with different energies. Respectively, the number of photons in this case is a random quantity with a probability distribution given by

$$p_n = |c_n|^2 = e^{-|\alpha|^2} \frac{|\alpha|^{2n}}{n!}. \tag{7.104}$$

Equation 7.104 describes a well-known probability distribution, called the Poisson distribution, which appears in a large number of physical and mathematical problems. This distribution describes the probability that n events will happen within some fixed interval (of time or of distances) provided that the probability of each event is independent of the occurrence of the others and all events are happening at a constant rate (probability per unit time or unit length or unit volume does not depend upon time or position). This distribution describes, for instance, the probability that n atoms will undergo radioactive decay within some time interval or the number of uniformly distributed non-interacting gas molecules that will be found occupying some volume in space. For more examples of the Poisson distribution, just google it. The entire Poisson distribution depends on a single parameter $|\alpha|^2$, whose physical meaning can be elucidated by computing the mean (or expectation value) of the number of photons $\bar{n}_\alpha$ in the state $|\alpha\rangle$:

$$\bar{n}_\alpha = \sum_{n=0}^{\infty} n p_n = e^{-|\alpha|^2} \sum_{n=0}^{\infty} n \frac{|\alpha|^{2n}}{n!} =$$

$$e^{-|\alpha|^2} \sum_{n=1}^{\infty} \frac{|\alpha|^{2n}}{(n-1)!} = e^{-|\alpha|^2} \sum_{k=0}^{\infty} \frac{|\alpha|^{2(k+1)}}{k!} =$$

$$e^{-|\alpha|^2} |\alpha|^2 \sum_{k=0}^{\infty} \frac{|\alpha|^{2k}}{k!} = e^{-|\alpha|^2} |\alpha|^2 e^{-|\alpha|} = |\alpha|^2,$$

where in the second line, I first took into account that the $n = 0$ term in the sum is multiplied by $n = 0$ and, therefore, does not contribute. Accordingly I started the sum with $n = 1$, after which I introduced a new index $k = n - 1$, which reset the counter back to zero. As a result, I gained an extra term $|\alpha|^2$, while the remaining sum became just an exponential function canceling out the normalization term $e^{-|\alpha|^2}$. This calculation shows that $|\alpha|^2$ has the meaning of the average number of photons in the state with eigenvalue α. It is also interesting to compute the uncertainty of the number of photons in this state $\Delta n = \sqrt{\left\langle (n - \bar{n}_\alpha)^2 \right\rangle} = \sqrt{\langle n^2 \rangle_\alpha - \bar{n}_\alpha^2}$. First, I compute $\langle n^2 \rangle_\alpha$:

$$\langle n^2 \rangle_\alpha = e^{-|\alpha|^2} \sum_{n=0}^{\infty} n^2 \frac{|\alpha|^{2n}}{n!} = e^{-|\alpha|^2} \sum_{n=1}^{\infty} n \frac{|\alpha|^{2n}}{(n-1)!} =$$

$$e^{-|\alpha|^2} \sum_{k=0}^{\infty} \frac{(k+1) |\alpha|^{2(k+1)}}{k!} = e^{-|\alpha|^2} |\alpha|^2 \sum_{k=0}^{\infty} \frac{|\alpha|^{2k}}{k!} +$$

$$e^{-|\alpha|^2} |\alpha|^2 \sum_{k=0}^{\infty} \frac{k |\alpha|^{2k}}{k!} = |\alpha|^2 + |\alpha|^4,$$

where I used the same trick with the sum as above, twice. Now I can find that $\Delta n = \sqrt{\bar{n}_\alpha}$. The relative uncertainty of the photon numbers $\Delta n / \bar{n}_\alpha = 1/\sqrt{\bar{n}_\alpha}$ and becomes progressively smaller as the average number of photons increases. The decrease of the quantum fluctuations signifies transition to classical behavior, and one can suppose, therefore, that in the limit $\bar{n}_\alpha \gg 1$, the electric and magnetic fields in this state will reproduce behavior typical for a classical electromagnetic wave. To verify this assumption, I will compute the expectation values and uncertainties of the electric and magnetic fields for this state as well as will consider their time dependence.

Reversing Eqs. 7.96 and 7.97, I find for the fields

$$\hat{\mathcal{B}}_0 = \sqrt{\frac{\hbar}{2}} \left(\hat{b} + \hat{b}^\dagger \right) \tag{7.105}$$

$$\hat{\mathcal{E}}_0 = i\sqrt{\frac{\hbar}{2}} \left(\hat{b}^\dagger - \hat{b} \right). \tag{7.106}$$

Taking squares of these expressions yields

$$\hat{\mathcal{B}}_0^2 = \frac{\hbar}{2}\left(\hat{b}^2 + \hat{b}^{\dagger 2} + \hat{b}\hat{b}^{\dagger} + \hat{b}^{\dagger}\hat{b}\right) = \frac{\hbar}{2}\left(\hat{b}^2 + \hat{b}^{\dagger 2} + 2\hat{b}^{\dagger}\hat{b} + 1\right) \qquad (7.107)$$

$$\hat{\mathcal{E}}_0^2 = -\frac{\hbar}{2}\left(\hat{b}^2 + \hat{b}^{\dagger 2} - \hat{b}\hat{b}^{\dagger} - \hat{b}^{\dagger}\hat{b}\right) = -\frac{\hbar}{2}\left(\hat{b}^2 + \hat{b}^{\dagger 2} - 2\hat{b}^{\dagger}\hat{b} - 1\right) \quad (7.108)$$

where I changed the order of operators in $\hat{b}\hat{b}^{\dagger}$ using the commutation relation $\left[\hat{b}, \hat{b}^{\dagger}\right] = 1$. Now I am ready to tackle both the expectation values and the uncertainties. The computation of the expectation values $\left\langle \hat{\mathcal{B}}_0 \right\rangle$ and $\left\langle \hat{\mathcal{E}}_0 \right\rangle$ is almost trivial: taking into account that $\langle \alpha | \hat{b} | \alpha \rangle = \alpha$ and $\langle \alpha | \hat{b}^{\dagger} | \alpha \rangle = \langle \alpha | \hat{b} | \alpha \rangle^* = \alpha^*$, I have

$$\left\langle \hat{\mathcal{B}}_0 \right\rangle = \sqrt{\frac{\hbar}{2}}\left(\alpha + \alpha^*\right) \qquad (7.109)$$

$$\left\langle \hat{\mathcal{E}}_0 \right\rangle = i\sqrt{\frac{\hbar}{2}}\left(\alpha^* - \alpha\right). \qquad (7.110)$$

The expectation values of the squares of the fields take just a bit more work: before computing $\langle \alpha | \hat{b}^{\dagger}\hat{b} | \alpha \rangle$, I first need to realize that the Hermitian conjugate of expression $\hat{b} | \alpha \rangle = \alpha$ is $\langle \alpha | \hat{b}^{\dagger} = \alpha^*$. With this little insight, the rest of the computation is as trivial as that for the expectation values. The result is

$$\left\langle \hat{\mathcal{B}}_0^2 \right\rangle = \frac{\hbar}{2}\left(\alpha + \alpha^*\right)^2 + \frac{\hbar}{2} \qquad (7.111)$$

$$\left\langle \hat{\mathcal{E}}_0^2 \right\rangle = \frac{\hbar}{2}\left(\alpha^* - \alpha\right)^2 + \frac{\hbar}{2}. \qquad (7.112)$$

Finally, the uncertainties of both fields are found to be independent of α and equal to

$$\Delta \hat{\mathcal{B}}_0 = \Delta \hat{\mathcal{E}}_0 = \sqrt{\frac{\hbar}{2}}$$

so that their product indeed is the smallest allowed by the uncertainty principle $\Delta \hat{\mathcal{B}}_0 \Delta \hat{\mathcal{E}}_0 = \hbar/2$. Relative uncertainties $\Delta \hat{\mathcal{B}}_0 / \left\langle \hat{\mathcal{B}}_0 \right\rangle$ diminish with the increase in $|\alpha| = \sqrt{\bar{n}_\alpha}$ and vanish in the limit $\bar{n}_\alpha \to \infty$, which obviously corresponds to the classical (no quantum fluctuations) limit. This result provides an additional reinforcement to the idea that the electromagnetic field in the coherent states is as close to a classical wave as possible.

Finally, I will consider how these quantities (the expectation values and uncertainties) change with time. The easiest way to do this is to use the Heisenberg picture, in which all dynamics is given by the time dependence of the annihilation operator, which as we know from the consideration of harmonic oscillator is very simple $\hat{b}_H(t) = \hat{b} \exp(-i\omega t)$, so that $\langle \alpha | \hat{b}_H | \alpha \rangle = \alpha \exp(-i\omega t)$. With this I immediately find for the field expectation values

$$\left\langle \hat{B}_0(t) \right\rangle = \sqrt{\frac{\hbar}{2}} \left(\alpha e^{-i\omega t} + \alpha^* e^{i\omega t} \right) \tag{7.113}$$

$$\left\langle \hat{\mathcal{E}}_0(t) \right\rangle = i\sqrt{\frac{\hbar}{2}} \left(\alpha^* e^{i\omega t} - \alpha e^{-i\omega t} \right) \tag{7.114}$$

and for their squares

$$\left\langle \hat{B}_0^2(t) \right\rangle = \frac{\hbar}{2} \left(\alpha e^{i\omega t} + \alpha^* e^{i\omega t} \right)^2 + \frac{\hbar}{2} \tag{7.115}$$

$$\left\langle \hat{\mathcal{E}}_0^2(t) \right\rangle = \frac{\hbar}{2} \left(\alpha^* e^{i\omega t} - \alpha e^{i\omega t} \right)^2 + \frac{\hbar}{2}. \tag{7.116}$$

It is remarkable that the uncertainties of the fields $\left\langle \hat{B}_0^2(t) \right\rangle - \left\langle \hat{B}_0(t) \right\rangle^2$, $\left\langle \hat{\mathcal{E}}_0^2(t) \right\rangle -$ $\left\langle \hat{\mathcal{E}}_0(t) \right\rangle^2$ remain time independent and satisfy the minimal form of the uncertainty principle at all times. While the harmonic time dependence of the expectation value is typical for almost any initial state, the uncovered behavior of the uncertainties is the special property of the coherent states and is what makes them so special. This also guarantees that the shape of the coherent superposition of the stationary states does not get distorted with time similar to what one would expect from a classical electromagnetic wave.

7.4 Problems

Problems for Sect. 7.1

Problem 83 Using Eq. 7.16 together with Eqs. 7.12 and 7.13, find the time dependence of the coordinate x and momentum p. Comparing the found result with Eq. 7.9, find the relation between parameters $b_0, b_0^\dagger$ and x_0, p_0.

Problem 84 Verify that Eqs. 7.14 and 7.15 are equivalent to the Hamiltonian equations for the regular coordinate and momentum by computing the time derivatives of the variables b and $b^\dagger$ using Eqs. 7.12 and 7.13 together with Eqs. 7.7 and 7.8.

Problem 85 Prove that $\hat{a} |n\rangle = \sqrt{n} |n-1\rangle$.

Problem 86 Suppose that a harmonic oscillator is at $t = 0$ in the state described by the following superposition:

$$|\alpha_0\rangle = a\left(\sqrt{2}\,|0\rangle + \sqrt{3}\,|1\rangle\right).$$

1. Normalize the state.
2. Find a vector $|\alpha(t)\rangle$ representing the state of the oscillator at an arbitrary time t.
3. Calculate the uncertainties of the coordinate and momentum operators in this state, and check that the uncertainty relation is fulfilled at all times.

Problem 87 Using the method of mathematical induction, prove that

$$\left(y - \frac{d}{dy}\right)^n \exp\left(-\frac{y^2}{2}\right) = (-1)^n \exp\left(-\frac{y^2}{2}\right) \frac{d^n \exp\left(-y^2\right)}{dy^n}$$

and derive Eq. 7.44 for the coordinate representation of an eigenvector of the Hamiltonian of the harmonic oscillator.

Problem 88 Using matrices $a_{mn} = \langle m|\,\hat{a}\,|n\rangle$; $a^{\dagger}_{mn} = \langle m|\,\hat{a}^{\dagger}\,|n\rangle$, demonstrate by direct matrix multiplication that

$$\left(\hat{a}^{\dagger}\hat{a}\right)_{mn} = \sum_k a^{\dagger}_{mk} a_{kn} = m\delta_{mn}.$$

Problem 89 Using the coordinate representation of the lowering operator $\hat{a}$, apply it to the coordinate representation of the $n = 3$ stationary state of the harmonic oscillator. Is the result normalized? If not, normalize it and compare the found normalization factor with Eq. 7.40.

Problem 90 Using lowering and raising operators, compute the expectation value of the kinetic, $\hat{K}$, and potential, $\hat{V}$, energies of a harmonic oscillator in an arbitrary stationary state $|n\rangle$. Check that

$$\left\langle \hat{K} \right\rangle = \left\langle \hat{V} \right\rangle.$$

This result is a particular case of the so-called virial theorem relating the expectation values of kinetic and potential energies of a particle in the potential described by $V = kx^p$. The general form of the theorem is $2\left\langle \hat{K} \right\rangle = p\left\langle \hat{V} \right\rangle$, which for $p = 2$ (harmonic oscillator) is reduced to the result of this problem.

Problem 91 Derive explicit expressions for Hermite polynomials with $n = 3, 4, 5$ (of course, you can always google it, but do it by yourselves—you can learn something), and demonstrate explicitly that they obey the orthogonality relation:

$$\int\limits_{-\infty}^{\infty} \exp\left(-x^2\right) H_m(x) H_n(x) dx = 0, \quad m \neq n.$$

Problem 92

1. Find eigenvectors $|\alpha\rangle$ of the lowering operator $\hat{a}$: $\hat{a}\,|\alpha\rangle = \alpha\,|\alpha\rangle$ in the coordinate representation. Normalize them.
2. Show that the raising operator $\hat{a}^{\dagger}$ does not have normalizable eigenvectors.

Problem 93 Compute a probability that a measurement of the coordinate will yield the value in the classically forbidden region for the oscillator prepared in each of the following stationary states: $|0\rangle$, $|1\rangle$, and $|3\rangle$. (Note that the boundary of classically allowed regions is different for each of these states.)

Problem 94 Consider an electron with mass m_e and charge $-e$ in a harmonic potential $\hat{V} = m_e\omega^2 x^2/2$ also subjected to a uniform electric field $\mathcal{E}$ in the positive x direction.

1. Write down the Hamiltonian for this system.
2. Using operator identities from Sect. 3.2.2, prove that

$$\exp\left(\frac{i\hat{p}_x d}{\hbar}\right)\hat{x}\exp\left(-\frac{i\hat{p}_x d}{\hbar}\right) = \hat{x} + d, \qquad (7.117)$$

where $\hat{x}$ and $\hat{p}_x$ are regular operators of the coordinate and the respective components of the momentum and d is a real number.
3. In Sect. 5.1.2 I already demonstrated using the example of a parity operator that if two vectors are related to each other as $|\beta\rangle = \hat{T}\,|\alpha\rangle$, while vectors $\left|\tilde{\beta}\right\rangle$ and $|\tilde{\alpha}\rangle$ are defined as $\left|\tilde{\beta}\right\rangle = \hat{U}\,|\beta\rangle$, $|\tilde{\alpha}\rangle = \hat{U}\,|\alpha\rangle$, one can show that $\left|\tilde{\beta}\right\rangle = \hat{T}'\,|\tilde{\alpha}\rangle$, where $\hat{T}' = \hat{U}\hat{T}\hat{U}^{-1}$. Use this relation together with Eq. 7.117 to reduce the Hamiltonian found in Part I of this problem to that of a harmonic oscillator without the electric field, and express the eigenvectors of the Hamiltonian with the field (perturbed Hamiltonian) in terms of the eigenvectors of the Hamiltonian without the field (unperturbed).
4. Write down the coordinate wave function representing the states of the perturbed Hamiltonian in terms of the wave functions representing the states of the unperturbed Hamiltonian. Comment on the results. Explain how it can be derived by manipulating the classical Hamiltonian before its quantization.
5. If the electron is in its ground state before the electric field is turned on, find the probability that the electron will be found in the ground state of the Hamiltonian with the electric field on. (Hint: You will need to use operator identities concerning with the exponential function of the sum of the operators discussed in Sect. 3.2.2 and the representation of the momentum operator in terms of raising and lowering operators. Remember: The exponential function of an operator is defined as a corresponding power series.)

Problems for Sect. 7.1.2

Problem 95 Using the Heisenberg representation, find the uncertainty of coordinate and momentum operators at an arbitrary time t for the state

$$|\alpha\rangle = \frac{1}{\sqrt{3}} \left(|1\rangle + |2\rangle + |3\rangle \right),$$

where $|n\rangle$ is nth stationary state of the harmonic oscillator. Verify that the uncertainty relation is fulfilled at all times.

Problem 96 Solve the previous problem using the Schrödinger representation.

Problem 97 Consider the system described in Problem 94, but work now in the Heisenberg picture.

1. Write down the Hamiltonian of the electron in the Heisenberg picture.
2. Write down the Heisenberg equations for lowering and raising operators and solve them.
3. Now, assume that the electric field was turned on at $t = 0$, when the electron was in the ground state $|0\rangle$ of the unperturbed Hamiltonian, and turned back off at $t = t_f$. In the Heisenberg picture, the state of the system does not change, so that all time evolution is described by the operators. Let us call the lowering and raising operators at $t = 0$ $\hat{a}_{in}$, $\hat{a}_{in}^{\dagger}$ (these are, obviously, the same operators that appear as initial conditions in the solutions of the Heisenberg equations found in Part I of the problem). These operators are just lowering and raising operators in the Schrödinger picture, so that the initial state obeys equation $\hat{a}_{in} |0\rangle = 0$. In the Heisenberg picture, raising and lowering operators change with time according to the expressions found in Part I. Considering these expressions at $t = t_f$, you will find $\hat{a}_f \equiv \hat{a}(t_f)$, and $\hat{a}_f^{\dagger} = \hat{a}^{\dagger}(t_f)$. Verify that these operators have the same commutation relation as their Schrödinger counterparts.
4. The time evolution of the Hamiltonian, which at all times has the form found in Part I, is completely described by the time dependence of lowering and raising operators. Using the expressions for $\hat{a}_f$ and $\hat{a}_f^{\dagger}$ found in the previous part of the problem, write down the Hamiltonian of the electron at times $t > t_f$ in terms of operators $\hat{a}_{in}$, $\hat{a}_{in}^{\dagger}$.
5. Using the found expression for the Hamiltonian, find the expectation value of energy in the given initial state.
6. The Hamiltonian of the electron at $t > t_f$ has the same form in terms of operators $\hat{a}_f$, $\hat{a}_f^{\dagger}$, as the Hamiltonian for $t < t_0$ has in terms of operators $\hat{a}_{in}$, $\hat{a}_{in}^{\dagger}$. Also, it has been shown in Part III that $\hat{a}_f$, $\hat{a}_f^{\dagger}$ have the same commutation relations as $\hat{a}_{in}$, $\hat{a}_{in}^{\dagger}$. This means that Hamiltonian $t > t_f$ has the same eigenvalues, and its eigenvectors satisfy the same relations:

$$\hat{a}_f |0\rangle_f = 0$$

$$|n\rangle_f = \frac{1}{\sqrt{n!}}\hat{a}_f^\dagger |0\rangle_f$$

where the first equation defines the new vacuum state $|0\rangle_f$ and the second equation defines new eigenvectors. Since operators $\hat{a}_f$, $\hat{a}_f^\dagger$ differ from $\hat{a}_{in}$, $\hat{a}_{in}^\dagger$, the new ground state and the new eigenvectors will be different from those of the initial Hamiltonian. Using the representation of $\hat{a}_f$ in terms of $\hat{a}_{in}$, find the probability that if the system started out in the ground state of the initial Hamiltonian, it will be found in the new ground state $|0\rangle_f$.

Problems for Sect. 7.2

Problem 98 Verify Eqs. 7.71 and 7.72.

Problem 99 Rewrite the Schrödinger equation for the stationary states of a 3-D isotropic harmonic oscillator in cylindrical coordinates, ρ, φ, z. Show that the wave function can be written down $\Psi_{n_1,n_2,m} = Z_{n_1}(z)R_{n_2}(\rho)\exp(im\varphi)$, and derive equations for functions $Z_{n_1}(z)$ and $R_{n_2}(\rho)$. The first of these equations will coincide with the Schrödinger equation for a one-dimensional harmonic oscillator, so you can use the results of Sect. 7.1.1 to determine this function and the corresponding contribution to energy, but the equation for $R_{n_2}(\rho)$ will have to be solved from scratch. Do it using the power series method developed in the text for the spherical coordinates.

Problem 100 You just saw that the wave functions of an isotropic oscillator can be presented using Cartesian, spherical, and cylindrical coordinates. While each of these functions, corresponding to the same degenerate energy value, has very different forms, since all of them represent the same eigenvectors belonging to the corresponding eigenvalue, you shall be able to present each of them as linear combinations of the others belonging to the same eigenvalue. Verify that this is indeed the case for states belonging to energy value $E = 5\hbar\omega/2$ by explicitly expressing wave functions written in Cartesian coordinates in terms of their spherical and cylindrical coordinate counterparts.

Problems for Sect. 7.3.2

Problem 101 Verify Eqs. 7.115 and 7.116 for time-dependent expectation values of the squares of electric and magnetic fields.

Problem 102 The flow of the energy of the electromagnetic field is described by the Poynting vector, which in SI units is given by

$$S = \frac{1}{\mu_0} E \times B.$$

In our toy model of the electromagnetic field, the Poynting vector becomes simply

$$S = \frac{1}{\mu_0} E_x B_y.$$

In quantum theory, the Poynting vector becomes an operator. Find the time-dependent expectation value and uncertainty of this operator in the coherent state.

Chapter 8
Hydrogen Atom

8.1 Transition to a One-Body Problem

Quantum mechanics of the atom of hydrogen, understood as a system consisting of a positively charged nucleus and a single negatively charged electron, is remarkable in many respects. This is one of the very few exactly solvable three-dimensional models with realistic interaction potential. As such it provides the foundation for much of our qualitative as well as quantitative understanding of optical properties of atoms at least as a first approximation for more complicated situations. A similar model also arises in the physics of semiconductors, where bound states of negative and positive charges form entities known as excitons, as well as in the situations involving a single conductance electron interacting with a charged impurity. Another curious property of this model is that the energy eigenvalues emerging from the exact solution of the Schrödinger equation coincide with energy levels predicted by the heuristic Bohr model based on a rather arbitrary combination of Newton's laws with a simple quantization rule for the angular momentum. While it might seem as a pure coincidence of a limited significance given that by now we have harnessed the full power of quantum theory and do not really need Bohr's quantization rules, one still might wonder by how much the development of quantum physics would have been delayed if it were not for this "coincidence."

I will begin exploration of this model with a brief reminder of how classical mechanics deals with the problem. There are two different aspects to it which need to be addressed. First, unlike all previous models considered so far, which involved a single particle, this is a two-body problem. Luckily for us, this problem only pretends to be two-body and can be easily reduced to two single-particle problems. This is how it is done in classical physics. The classical Hamiltonian of the problem has the following form:

$$H = \frac{p_1^2}{2m_p} + \frac{p_2^2}{2m_e} + V\left(|r_1 - r_2|\right), \tag{8.1}$$

© Springer International Publishing AG, part of Springer Nature 2018
L.I. Deych, *Advanced Undergraduate Quantum Mechanics*,
https://doi.org/10.1007/978-3-319-71550-6_8

where p_1, r_1 and p_2, r_2 are momentums and positions of the particles with corresponding masses m_e and m_p and $V(|r_1 - r_2|)$ is the Coulomb potential energy, which in the SI units can be written as

$$V(|r_1 - r_2|) = -\frac{1}{4\pi\varepsilon_r\varepsilon_0}\frac{Ze^2}{|r_1 - r_2|}.$$
(8.2)

Here e is the elementary charge, Z is the atomic number of the nucleus introduced to allow dealing with heavier hydrogen-like atoms such as atoms of alkali metals (or a charged impurity), and ε_r is the relative dielectric permittivity accounting for a possibility that the interacting particles are inside a dielectric medium. To separate this problem into two single-particle problems, I introduce new coordinates:

$$R = \frac{m_p r_1 + m_e r_2}{m_p + m_e},$$
(8.3)

$$r = r_1 - r_2.$$
(8.4)

I hope you have recognized in R a coordinate of the center of mass of two particles and in r their relative position vector. Now, I need to find the new momentums associated with these coordinates. For your sake I will avoid using formalism of canonical transformations in Hamiltonian mechanics and will begin with defining the kinetic energy in terms of respective velocities. Reversing Eqs. 8.3 and 8.4, I get

$$r_1 = R + \frac{m_e}{m_p + m_e}r,$$

$$r_2 = R - \frac{m_p}{m_p + m_e}r,$$

so that the kinetic energy can be found as

$$K = \frac{1}{2}m_p\left(\frac{dR}{dt} + \frac{m_e}{m_p + m_e}\frac{dr}{dt}\right)^2 + \frac{1}{2}m_e\left(\frac{dR}{dt} - \frac{m_p}{m_p + m_e}\frac{dr}{dt}\right)^2 =$$

$$\frac{1}{2}(m_p + m_e)\left(\frac{dR}{dt}\right)^2 + \frac{m_p m_e^2}{(m_p + m_e)^2}\left(\frac{dr}{dt}\right)^2 + \frac{m_e m_p^2}{(m_p + m_e)^2}\left(\frac{dr}{dt}\right)^2 =$$

$$\frac{1}{2}(m_p + m_e)\left(\frac{dR}{dt}\right)^2 + \frac{m_p m_e}{m_p + m_e}\left(\frac{dr}{dt}\right)^2.$$

Introducing two new masses—total mass of the system $M = m_p + m_e$ and reduced mass $\mu = m_p m_e/(m_p + m_e)$—I can define the momentum of the center of mass:

$$p_R = M\frac{dR}{dt},$$

and relative momentum

$$p_r = \mu \frac{dr}{dt},$$

so that the Hamiltonian, Eq. 8.1, can be rewritten as

$$H = \frac{p_R^2}{2M} + \frac{p_r^2}{2\mu} + V(r).$$

The corresponding Hamiltonian equations are separated into a pair of equations for the position and momentum of the center of mass:

$$\frac{dR}{dt} = \frac{p_R}{M}$$

$$\frac{dp_R}{dt} = 0,$$

and for the relative motion

$$\frac{dr}{dt} = \frac{p_r}{M}$$

$$\frac{dp_r}{dt} = -\frac{dV}{dr}.$$

The first pair of these equations describes a uniform motion of a free particle—the center of mass of the system—while the second pair describes the motion of a single particle in potential $V(r)$.

I have little doubts that variables r and p_r form a canonically conjugated pair, and so I can transition to the quantum description by promoting them to operators with standard commutation relation $[r_i, p_{rj}] = i\hbar\delta_{i,j}$. However, in order to be 100% sure and convince all the possible skeptics, I do need to verify this fact by computing Poisson brackets with these variables. To this end I need to express r and p_r in terms of initial coordinates and momentums. Expression for r is given by Eq. 8.4, so I only need to figure out p_r:

$$p_r = \frac{m_e m_p}{m_e + m_p}\left(\frac{dr_1}{dt} - \frac{dr_2}{dt}\right) = \frac{m_e}{m_e + m_p}p_1 - \frac{m_p}{m_e + m_p}p_2. \tag{8.5}$$

Let me focus for concreteness on x-components of the momentum and coordinate. Equation 3.4 for the Poisson bracket, where summation must include the sum over coordinates of both particles, yields

$$\{x, p_{rx}\} = \frac{\partial x}{\partial x_1}\frac{\partial p_{rx}}{\partial p_{1x}} + \frac{\partial x}{\partial x_2}\frac{\partial p_{rx}}{\partial p_{2x}} = \frac{m_e}{m_e + m_p} + \frac{m_p}{m_e + m_p} = 1$$

as expected. All other Poisson brackets also predictably produce necessary results, so you can start breathing again.

8.2 Eigenvalues and Eigenvectors

It is important that the portion of the Hamiltonian describing the motion of the center of mass, $\hat{H}_R = \hat{p}_R^2/2M$, is completely independent from the part responsible for the relative motion

$$\hat{H}_r = \frac{\hat{p}_r^2}{2\mu} + V(\hat{r}) \tag{8.6}$$

so that the eigenvectors of the total Hamiltonian $\hat{H} = \hat{H}_R + \hat{H}_r$ can be written down as $|\chi_R\rangle\,|\chi_r\rangle$, where the first vector is an eigenvector of $\hat{H}_R$ with eigenvalue E_R, while the second vector is the eigenvector of $\hat{H}_r$ with its own eigenvalue E_r. The eigenvalue of the total Hamiltonian is easily verified to be $E_R + E_r$ (when verifying this statement, remember that $\hat{H}_R$ acts only on $|\chi_R\rangle$, while $\hat{H}_r$ only affects $|\chi_r\rangle$). I am going to ignore the center of mass motion and will focus on the Hamiltonian $\hat{H}_r$, Eq. 8.6, with the Coulomb potential energy, Eq. 8.2. In what follows I will omit subindex r in the Hamiltonian.

What I am dealing with here is yet another example of a particle moving in a central potential, similar to the isotropic harmonic oscillator problem considered in Sect. 7.2. Just like in the case of a harmonic oscillator, Hamiltonian 8.6 commutes with angular momentum operators $\hat{L}^2$ and $\hat{L}_z$; thus its eigenvectors are also eigenvectors of the angular momentum. Working in the position representation and using spherical coordinates to represent the position, I can again write down for the wave function

$$\psi_{n,l,m}(r, \theta, \varphi) = Y_l^m(\theta, \varphi) R_{nl}(r)$$

where $Y_l^m(\theta, \varphi)$ are spherical harmonics—coordinate representation of the eigenvectors of angular momentum operators. The equation for the remaining radial function $R_{nl}(r)$ is derived in exactly the same way as in Sect. 7.2 and takes the form similar to Eq. 7.71:

$$-\frac{\hbar^2}{2\mu r^2}\frac{d}{dr}\left(r^2 \frac{\partial R_{n_r,l}}{\partial r}\right) + \frac{\hbar^2 l(l+1)}{2\mu r^2} R_{n_r,l} - \frac{1}{4\pi\varepsilon_r\varepsilon_0}\frac{Ze^2}{r} R_{n_r,l} = E_{l,n_r} R_{n_r,l} \tag{8.7}$$

with obvious replacements of $m_e \to \mu$ and quadratic harmonic oscillator potential for the Coulomb potential. Eigenvalues of energy $E_{l,n}$ are found by looking for normalizable solutions to this equation. My choice of the indexes to label the eigenvalues reflects the fact that the eigenvalues of the Hamiltonian with any central potential do not depend on m. Indeed, quantum number m is defined with respect to a particular choice of the polar axis Z, but since the energy of a system with a central potential cannot depend upon an arbitrary axis choice, it should not depend on this quantum number. Here is another example of how symmetry consideration helps to analyze the problem.

I will begin by reducing Eq. 8.7 to a dimensionless form as it is customary in this type of situations. What I need for this is a characteristic length scale, which in this problem, unlike the harmonic oscillator case, is not that obvious. But there is a trick which I can use to find it, and I am going to share it with you. Just by looking at the radial equation, I know that there are three main parameters: mass μ, charge e, and Planck's constant $\hbar$ in this problem, and I need to find their combination with the dimension of length. This is done by first writing down this combination in the most generic form as $\mu^\alpha \tilde{e}^\beta \hbar^\gamma$, where $\tilde{e} = e/\sqrt{4\pi\varepsilon_0\varepsilon_r}$ is the combination of charge, vacuum, and relative permittivity, ε_0 and ε_r correspondingly appearing in the Coulomb law in SI units, while α, β, and γ are unknown powers to be determined. In the next step, I will present the dimension of each factor in this expression in terms of the basic quantities: length, time, and mass. For instance, the dimension of $\tilde{e}$ can be found from the Coulomb law as $[\tilde{e}] = [F]^{1/2}[L]$, where $[F]$ stands for the dimension of force and $[L]$ stands for the dimension of length. The dimension of force in basic quantities is $[F] = [M][L][T]^{-2}$, where $[M]$ represents the dimension of mass and $[T]$ represents the dimension of time (think of Newton's second law). So, for the effective charge, I have $[\tilde{e}] = [M]^{1/2}[L]^{3/2}[T]^{-1}$. The dimension of Planck's constant can be determined from the Einstein–de Broglie relation between energy and frequency as $[\hbar] = [E][T] = [M][L]^2[T]^{-1}$, where in the second step I expressed the dimension of energy $[E]$ as $[E] = [F][L]$. Combining the results for the charge and Planck's constant, I find

$$\mu^\alpha \tilde{e}^\beta \hbar^\gamma = [M]^\alpha [M]^{\beta/2}[L]^{3\beta/2}[T]^{-\beta}[M]^\gamma[L]^{2\gamma}[T]^{-\gamma} =$$
$$[M]^{\alpha+\beta/2+\gamma}[L]^{3\beta/2+2\gamma}[T]^{-\beta-\gamma}.$$

If I want this expression to have the dimension of length $[L]$, I need to eliminate the excessive dimensions such as $[M]$ and $[T]$. Remembering that any quantity raised to the power of zero turns to unity and becomes dimensionless, I can eliminate $[M]$ and $[T]$ requiring that their corresponding powers vanish:

$$\alpha + \beta/2 + \gamma = 0$$
$$\beta + \gamma = 0.$$

Then all what is left to do is to make the power of L equal to unity:

$$3\beta/2 + 2\gamma = 1.$$

The result is the system of equations for unknown powers, solving which I find $\gamma = 2$, $\beta = -2$, and $\alpha = -1$, i.e., the characteristic length scale can be constructed using the parameters at our disposal as

$$a_B = \frac{4\pi\varepsilon_0\varepsilon_r\hbar^2}{e^2\mu}. \tag{8.8}$$

The found characteristic length is actually well known from Bohr's theory of atomic spectra and is called *Bohr radius*. It can be used to introduce a dimensionless coordinate $\varsigma = r/a_B$ and rewrite Eq. 8.7 as

$$-\frac{1}{\varsigma^2}\frac{d}{d\varsigma}\left(\varsigma^2\frac{dR_{n_r,l}}{d\varsigma}\right) + \frac{l(l+1)}{\varsigma^2}R_{n_r,l} - \frac{2Z}{\varsigma}R_{n_r,l} = \frac{2\left(4\pi\varepsilon_0\varepsilon_r\right)^2\hbar^2}{e^4\mu}E_{l,n_r}R_{n_r,l}.$$

(8.9)

You can verify (do it yourselves) that quantity

$$\tilde{E} = \frac{e^4\mu}{32\pi^2\varepsilon_0^2\varepsilon_r^2\hbar^2}$$

(8.10)

has the dimension of energy, so that I can present the right-hand side of this equation in terms of dimensionless energy parameter

$$\epsilon_{l,n} = E_{l,n}/\tilde{E}.$$

Finally, introducing auxiliary radial function $u_{n,l} = \varsigma R_{n,l}$ (the same as in the harmonic oscillator problem), I obtain the effective one-dimensional Schrödinger equation similar to Eq. 7.73:

$$-\frac{d^2u_{n_r,l}}{d\varsigma^2} + \frac{l(l+1)}{\varsigma^2}u_{n_r,l} - \frac{2Z}{\varsigma}u_{n_r,l} = \epsilon_{l,n_r}u_{n_r,l}.$$

(8.11)

The effective potential in Eq. 8.11 is positively infinite at small ς, but as ς increases, it, unlike the harmonic oscillator problem, becomes negative, reaches a minimum value of $-Z^2/[l(l+1)]$ at $\varsigma = l(l+1)/Z$, and remains negative while approaching zero for $\varsigma \to \infty$; see Fig. 8.1. Classical behavior in such a potential is bound for negative values of energy and unbound for positive energies. In the former case, we are dealing with a particle moving along a closed elliptical orbit, while in the

Fig. 8.1 Dimensionless effective potential as a function of dimensionless radial coordinate

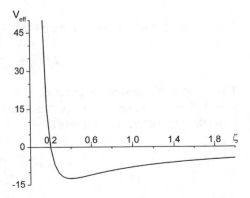

latter case, the situation is better described in terms of scattering of a particle by the potential. In quantum description, as usual, we shall expect states characterized by the discrete spectrum of eigenvalues for classically bound motion (negative energies) and states with continuous spectrum for positive energies. The wave functions representing states of continuum spectrum are well known but rather complex mathematically and are not used too frequently, so I shall avoid dealing with them for the sake of keeping everyone sane. The range of negative energies is much more important for understanding the physical processes in atoms and is more tractable. In terms of atomic physics, the states with negative energies correspond to intact atoms, where the electron is bound to its nucleus, and the probability that it will turn up infinitely far from the nucleus is zero. The states of continuous spectrum correspond to ionized atoms, where energy of the electron is too large for the nucleus to be able to "catch" it so that the electron can be found at an arbitrary large distances from the nucleus.

The process of finding the solution of Eq. 8.11 follows the same steps as solving a similar equation for the harmonic oscillator: find an asymptotic behavior at small and large ς, factor it out, and present the residual function as a power series. I, however, can simplify the form of the equation a bit more by replacing variable ς with a new variable $\rho = \varsigma \sqrt{-\epsilon_{l,n}}$ (remember $\epsilon_{l,n_r} < 0$!). Equation 8.11 now takes the following form:

$$\frac{d^2 u_{n_r,l}}{d\rho^2} - \frac{l(l+1)}{\rho^2} u_{n_r,l} + \frac{\kappa_{l,n_r}}{\rho} u_{n_r,l} - u_{n_r,l} = 0 \qquad (8.12)$$

where I introduced a new parameter

$$\kappa_{l,n_r} = \frac{2Z}{\sqrt{-\epsilon_{l,n_r}}}.$$

The asymptotic behavior at low ρ is determined by the contribution from the angular momentum and is the same as for the harmonic oscillator, $u_{n_r,l} \propto \rho^{l+1}$, but the large ρ limit is now determined by $u_{n_r,l}$ term. The resulting equation

$$\frac{d^2 u_{n_r,l}}{d\rho^2} = u_{n_r,l}$$

has two obvious solutions

$$u_{n_r,l} \propto \exp(\pm\rho)$$

of which I will only keep an exponentially decreasing one in hopes to end up with a normalizable solution. Thus, I am looking for the solution in the form

$$u_{n_r,l} = \rho^{l+1} \exp(-\rho) v_{n,l}(\rho), \qquad (8.13)$$

where a differential equation for the reduced function $v_{n,l}$ is derived by substituting Eq. 8.13 into Eq. 8.11. The rest of the procedure is quite similar to the one I outlined in the harmonic oscillator problem: present $v_{n,l}$ as a power series with respect to ρ, derive recursion relation for the coefficients of the expansion, verify that the asymptotic behavior of resulting power series yields a non-normalizable wave function, restore normalizability by requiring that the power series terminates after a final number of terms, and obtain an equation for the energy values consistent with the normalizability requirement. Leaving details of this analysis to the readers as an exercise (of course, you can always cheat by looking it up in a number of other textbooks, but you will gain so much more in terms of your technical prowess and self-respect by doing it yourselves!), I will present the result. The only way to ensure normalizability of the resulting wave functions is to require that parameter κ_{l,n_r} satisfies the following condition:

$$\kappa_{l,n_r} = 2(j_{max} + l + 1) \tag{8.14}$$

where j_{max} is the number of the largest non-zero coefficient in the power series expansion of function

$$v_{n,l}(\rho) = \sum_j c_j \rho^j$$

and takes arbitrary integer values starting from 0. However, since j_{max} appears in Eq. 8.14 only in combination with l, the actual allowed values of κ_{l,n_r} depend on a single parameter, called a *principal quantum number* n, which takes any integer values starting from $n = 1$. Independence of the energy eigenvalues of a hydrogen Hamiltonian of the angular momentum number l is a peculiarity of the Coulomb potential and reflects an additional symmetry present in this problem. In classical mechanics this symmetry manifests itself via the existence of a supplemental (to energy and angular momentum) conserving quantity called Laplace–Runge–Lenz vector

$$A = p \times L + \mu \frac{Ze^2}{4\pi \varepsilon_r \varepsilon_0} e_r$$

where e_r is a unit vector in the radial direction. In quantum theory this vector can be promoted to a Hermitian operator, but this procedure is not trivial because operators p and L do not commute. However, I am afraid that if I continue talking about the quantum version of Laplace–Runge–Lenz vector, I might open myself to a lawsuit for inflicting cruel and unusual punishment on the readers, so I will restrain myself. Those who are not afraid may look it up, but quantum treatment of Laplace–Runge–Lenz vector is not very common even in the wild prairies of the Internet.

Anyway, I can drop now the double-index notation in κ and dimensionless energy ϵ and classify the latter with a single index—principal quantum number n. Taking into account Eq. 8.14 and introducing

$$n = j_{max} + l + 1, \tag{8.15}$$

I find for the allowed energy values

$$E_n = \tilde{E}\epsilon_n = -\frac{Z^2}{n^2}\tilde{E} = -\frac{Z^2 e^4 \mu}{32\pi^2 \varepsilon_r^2 \varepsilon_0^2 \hbar^2}\frac{1}{n^2} = -\frac{E_g}{n^2} \qquad (8.16)$$

where I introduced a separate notation E_g for the ground state energy. It is also useful sometimes to have this expression written in terms of the Bohr radius a_B defined by Eq. 8.8:

$$E_n = -\frac{Z^2 e^2}{8\pi \varepsilon_r \varepsilon_0 a_B}\frac{1}{n^2}. \qquad (8.17)$$

For pure hydrogen atom in vacuum $Z = 1$, $\varepsilon_r = 1$, and taking into account that the ratio of the mass of a proton (the nucleus of a hydrogen atom is a single proton) to the mass of the electron is approximately $m_p/m_e \approx 1.8 \times 10^4$, I can replace the reduced mass μ with the electron mass. In this case, the numerical coefficient in front of $1/n^2$ contains only universal constants and can be computed once and for all. It defines the so-called Rydberg unit of energy 1 Ry, which in electron volts is approximately equal to 13.6 eV. This is one of those numbers which is actually worth remembering, just like a few first digits of number π. The physical meaning of this number has several interpretations. First of all, it is the ground state energy of the hydrogen atom, but taking into account that transition from the discrete energy levels to the continuous spectrum (ionization of atom) amounts to raising of the energy above zero, you can also interpret this value as the binding energy or ionization energy of a hydrogen atom—a work required to change the electron's energy from the ground state to zero. Also, this number fixes the scale of atomic energies in general. Transition to atoms heavier than hydrogen, which are characterized by larger atomic numbers, makes the ground state energy more negative increasing the binding energy of the atom, which of course totally makes sense (atoms with larger charge attract electrons stronger).

If you apply Eq. 8.16 to excitons in semiconductors, it will yield a very different energy scale. It happens for several reasons. First, masses of the interacting positive and negative charges forming an exciton are comparable in magnitude, so one does need to compute the reduced mass. Second, these masses are often by an order of magnitude smaller than the mass of the free electron, which results in significant decrease of the binding energy. This decrease is further enhanced by a relatively large dielectric constant of semiconductors ε_r. All these factors taken together result in a much larger ground state energy of excitons (remember the energy is negative!) with much smaller ionization or binding energy, which varies across different semiconductors and can take values from of the order of 10^{-3} to 10^{-2} eV.

Finally, I would like to point out the fact that the discrete energy levels of hydrogen-like atoms occupy the final spectral region between the ground state and zero. Despite this fact, the number of these levels is infinite, unlike, for instance, in the case of one-dimensional square potential well. This means that with increasing principal quantum number n, the separation between the adjacent levels becomes smaller and smaller, and at some point the discreteness of the energy becomes unrecognizable, even though the probability to observe the electron infinitely far from the nucleus is still zero. One can think about this phenomenon as approaching a classical limit, in which an electron's motion is finite, it is still bound to the nucleus, but quantum effects become negligibly small.

For each value of n, there are several combinations of j_{max} and l satisfying Eq. 8.15, which means that all energy levels, except of the ground state, are degenerate with several wave functions belonging to the same energy eigenvalue that differ from each other by values of l and m. The total degree of degeneracy is easy to compute taking into account that for each n, there are $n - 1$ values of l (l obeys obvious inequality $l < n$), and for each l, there are $2l + 1$ possible values of m. The total number of wave functions corresponding to the same value of energy is, therefore, given by

$$\sum_{l=0}^{n-1}(2l + 1) = 2\frac{(n - 1)n}{2} + n = n^2. \tag{8.18}$$

As expected, this formula yields a single state for $n = 1$ for which $j_{max} = 0$ and $l = m = 0$. For the energy level with $n = 2$, Eq. 8.18 predicts the existence of four states, which we easily recognize as one state $l = m = 0$ (in which case $j_{max} = 1$) and three more characterized by $l = 1$, $m = -1$; $l = 1$, $m = 0$; and $l = 1$, $m = 1$. For all of them, the maximum power of the polynomial function in the solution is $j_{max} = 0$. For some murky and not very important historical reasons, $l = 0$ states are called s-states, $l = 1$ are called p-states, $l = 2$ are d-states, and, finally, letter f is reserved for $l = 3$ states. The origin of these nomenclature comes from German words for various types of optical spectra associated with each of these states, but I am not going into this issue any further. Those who are interested are welcome to google it.

Replacing all dimensionless variables with physical radial coordinate $r = \rho n a_B / Z$, I find that the radial wave function $R_{n,l} = u_{n,l}/r$ with fixed values of n and l is a product of $(Zr/na_B)^l$, exponential function $\exp(-Zr/na_B)$, and a polynomial $v_{n,l}(Zr/na_B)$ of the order $n - l - 1$. The polynomials, which emerge in this problem, are well known in mathematical physics as associated Laguerre polynomials defined by two indexes as $L_{q-p}^p(x)$. The definition of these polynomials can be found in many textbooks as well as online, but for your convenience, I will provide it here as well. The definition is somewhat cumbersome and involves an additional polynomial called simply a Laguerre polynomial (no associate here):

$$L_q(x) = e^x \left(\frac{d}{dx}\right)^q \left(e^{-x}x^q\right). \tag{8.19}$$

To define the associated Laguerre polynomial, one needs to carry out some additional differentiation of the simply Laguerre polynomial:

$$L_{q-p}^p(x) = (-1)^p \left(\frac{d}{dx}\right)^p L_q(x). \tag{8.20}$$

It is quite obvious that index q in Eq. 8.19 specifies the degree of the respective polynomial (exponential functions obviously cancel each other after differentiation is performed). At the same time, index $q - p$ in Eq. 8.20 specifies the ultimate degree of the associated polynomial (differentiating p times a polynomial of degree q will reduce it exactly by this amount). So, in terms of these functions, the polynomial appearing in the hydrogen model can be written as $v_{n,l}(Zr/na_B) = L_{n-l-1}^{2l+1}(2Zr/na_B)$. The total normalized radial wave function $R_{n,l}(r)$ can be shown to be

$$R_{n,l}(r) = \sqrt{\left(\frac{2Z}{na_B}\right)^3 \frac{(n-l-1)!}{2n(n+l)!}} \exp\left(-\frac{Zr}{na_B}\right) \left(\frac{2Zr}{na_B}\right)^l L_{n-l-1}^{2l+1}\left(\frac{2Zr}{na_B}\right). \tag{8.21}$$

I surely hope you are impressed by the complexity of this expression and can appreciate the amount of labor that went into finding the normalization coefficient here, which you are given as a gift. I also have to warn you that different authors may use different definitions of the Laguerre polynomials, which affect the appearance of Eq. 8.21. More specifically, one might include an extra factor $1/(n+l)!$ either in the definition of the polynomial or in the normalization factor. Equation 8.21 is written according to the former convention, while if the latter is accepted, the term $(n+l)!$ must be replaced with $[(n+l)!]^3$. You might find both versions of the hydrogen wave function on the Internet or in the literature, and my choice was completely determined by the convention adapted by the popular computational platform MATHEMATICA ©, which I am using a lot to perform computations needed for this book. The total hydrogen wave function, which in the abstract notation can be presented as $|n, l, m\rangle$, is obtained by multiplying the radial function and the spherical harmonics $Y_{l,m}(\theta, \varphi)$:

$$\psi_{n,l,m}(r, \theta, \varphi) = R_{n,l}(r) Y_{l,m}(\theta, \varphi). \tag{8.22}$$

Different factors in Eqs. 8.22 and 8.21 are responsible for different physical effects; however, before giving them any useful interpretation, I have to remind you that the respective probability distribution density is given by

$$P(r, \theta) = |\psi_{n,l,m}(r, \theta, \varphi)|^2 r^2 \sin\theta =$$

$$\left(\frac{2Z}{na_B}\right)^3 \frac{(n-l-1)!}{2n[(n+1)!]} \exp\left(-\frac{2Zr}{na_B}\right) \left(\frac{2Zr}{na_B}\right)^{2l} r^2 \times$$

$$\left[L_{n-l-1}^{2l+1}\left(\frac{2Zr}{na_B}\right)\right]^2 \left[P_l^m(\cos\theta)\right]^2 \sin\theta \tag{8.23}$$

where I replaced spherical harmonics by the product of associated Legendre functions $P_l^m (\cos \theta)$ and $\exp (im\varphi)$ and took into account that the latter disappears after multiplication by the respective complex-conjugated expression. Additional factors r^2 and $\sin \theta$ are due to spherical volume element, which has the form of $dV = r^2 \sin \theta d\theta d\varphi dr$.

The exponential factor describes how fast the wave function decreases at infinity. The respective characteristic scale

$$r_{at} = \frac{na_B}{Z} \tag{8.24}$$

can be interpreted (quite loosely though) as a size of the atom, because the probability to find the electron at distance $r \gg r_{at}$ becomes exponentially small. In the case of the atom in the ground state ($n = 1$), it is easy to show (i.e., if you remember that a maximum of a function is given by a zero of its first derivative) that the distance $r = r_{at}$ corresponds to the maximum of the probability $P(r) = \int_0^\pi P(r, \theta)d\theta$. In the case of a hydrogen atom ($Z = 1$), $r_{at} = a_B$, and if this atom is in vacuum ($\varepsilon_r = 1$), the Bohr radius is determined by fundamental constants only. Replacing the reduced mass with the mass of the electron, you can find that the size of the hydrogen atom in the ground state is $a_B \approx 0.5 \times 10^{-10}$ m. This number sets up the atomic spatial scale just as 13.6 eV sets up the typical energy scale. In the case of excitons in semiconductors, the characteristic scale becomes much larger for the same reasons why the energy scale becomes smaller: large dielectric constant and smaller masses yield larger a_B; see Eq. 8.8. As a result the typical size of the exciton can be as large as 10^{-8} m, which is extremely important for semiconductor physics, as it allows significant simplification of the quantum description of excitons.

Radial distribution for higher lying states can have several maximums, so such a direct interpretation of r_{at} becomes impossible, but it still can be thought of as a cutoff distance, starting from which the probability for the electron to wander off dramatically decreases. It is interesting that this parameter increases with n, so excited atoms not only have more energy, but they are also larger in size. Figure 8.2 presents a number of radial functions for your perusal, which illustrates most of the properties discussed here.

The factors containing the power of the radial coordinate are responsible for electrons not falling on the nucleus—the probability that $r = 0$ is strictly zero. This probability for $r < r_{at}$ decreases with increasing angular momentum number l, which can be interpreted as a manifestation of the "centrifugal" force keeping rotating particles away from the center of rotation. The Laguerre polynomial factor is essentially responsible for the behavior of the radial wave function at the intermediate distances between zero and r_{at}: the degree of the respective polynomial determines how many zeroes the radial wave function has. Finally, the Legendre function $P_l^m (\cos \theta)$ is responsible for directionality of the probability distribution with respect to Z-axis. States with zero angular momentum are described by a completely isotropic wave function, which does not depend on direction at all. For

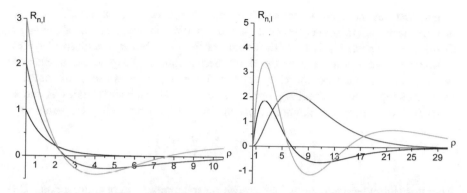

Fig. 8.2 A few radial functions with different values of principal and orbital numbers n and l. All graphs on the left figure correspond to $l = 0$ and increasing n between 1 and 3. The number of zeroes of the functions is equal to $n - l - 1$. The graphs in the right figure correspond to $n = 3$, $l = 1$; $n = 3$, $l = 2$; and $n = 4$, $l = 1$ (which one is which you can figure out yourselves by counting zeroes). The functions are not normalized for convenience of display

states with non-zero l, an important parameter is $l - m$, which yields the number of zeroes of the Legendre function and can also be used to determine the number of the respective maximums. The properties of the Legendre functions have been already discussed in Sect. 5.1.4, and the plots illustrating them were presented in Fig. 5.3, which you might want to consult to refresh your memory.

8.3 Virial and Feynman–Hellmann Theorems and Expectation Values of the Radial Coordinate in a Hydrogen Atom

I will finish the chapter by discussing one apparently very special and technical but at the same time practically very important problem of calculating the expectation values $\langle r^p \rangle$ of various powers of the radial coordinate r^p, where p can be any negative or positive integer, in the stationary states of a hydrogen atom. Formally, calculation of these expectation values involves evaluation of the integrals

$$\langle r^p \rangle = \int_0^\infty dr r^{p+2} \left[R_{nl}(r) \right]^2 dr \tag{8.25}$$

where $R_{nl}(r)$ has been defined in Eq. 8.21 and the extra 2 in r^{p+2} comes from the term r^2 in the probability distribution generated by the hydrogen-like wave function, Eq. 8.23. Direct calculation of the integral in Eq. 8.25 is a hopeless task given the complexity of the radial function, but it is possible to circumvent the problem by relying on the radial equation, Eq. 8.7, itself, rather than on the explicit form of its solution, Eq. 8.21.

But first, let me derive a remarkable relation between the expectation values of kinetic and potential energies of a quantum particle known as a *virial theorem*. Consider the expectation value of the operator $\hat{r} \cdot \hat{p}$ in an arbitrary quantum state and compute its time derivative using the Heisenberg picture of the quantum mechanics (the expectation values do not really depend on which picture is used, but working with time-dependent Heisenberg operators and time-independent states is more convenient than using the Ehrenfest theorem, Eq. 4.17, for Schrödinger operators):

$$\frac{d}{dt}\langle \hat{r} \cdot \hat{p} \rangle = \left\langle \frac{d\hat{r}}{dt} \cdot \hat{p} \right\rangle + \left\langle \hat{r} \cdot \frac{d\hat{p}}{dt} \right\rangle.$$

Applying Heisenberg equations for the position and momentum operators, Eqs. 4.28 and 4.29, to this expression, I obtain

$$\frac{d}{dt}\langle \hat{r} \cdot \hat{p} \rangle = \left\langle \frac{\hat{p}}{m} \cdot \hat{p} \right\rangle - \left\langle \hat{r} \cdot \nabla \hat{V} \right\rangle = 2\left\langle \hat{K} \right\rangle - \left\langle \hat{r} \cdot \nabla \hat{V} \right\rangle. \tag{8.26}$$

The left-hand side of Eq. 8.26 must vanish if the state used to compute the expectation value is an eigenvector of the Hamiltonian (a stationary state in the Schrödinger picture) because the expectation value of any operator in a stationary state is time-independent. This allows me to conclude that in the stationary states, the expectation values of kinetic and potential energies satisfy the relation

$$2\left\langle \hat{K} \right\rangle = \left\langle \hat{r} \cdot \nabla \hat{V} \right\rangle \tag{8.27}$$

known as virial theorem. In the case of the Coulomb potential of the hydrogen atom Hamiltonian, this theorem yields

$$2\left\langle \hat{K} \right\rangle = \frac{Ze^2}{4\pi \varepsilon_r \varepsilon_0} \left\langle \frac{1}{r} \right\rangle. \tag{8.28}$$

Since the expectation value of the Hamiltonian in its own stationary state is simply equal to the respective eigenvalue, I can write for the hydrogen-like Hamiltonian:

$$\left\langle \hat{H} \right\rangle = \left\langle \hat{K} \right\rangle - \frac{Ze^2}{4\pi \varepsilon_r \varepsilon_0} \left\langle \frac{1}{r} \right\rangle \Rightarrow$$

$$E_n = -\frac{Ze^2}{8\pi \varepsilon_r \varepsilon_0} \left\langle \frac{1}{r} \right\rangle$$

where I replaced the expectation value of the Hamiltonian with its eigenvalue for the n-th stationary state and used Eq. 8.28 to eliminate the expectation value of the kinetic energy. Finally, using Eq. 8.16 for E_n, I have

$$\frac{Z^2 e^4 \mu}{32\pi^2 \varepsilon_r^2 \varepsilon_0^2 \hbar^2} \frac{1}{n^2} = \frac{Ze^2}{8\pi \varepsilon_r \varepsilon_0} \left\langle \frac{1}{r} \right\rangle \Rightarrow$$

$$\left\langle \frac{1}{r} \right\rangle = \frac{Ze^2 \mu}{4\pi \varepsilon_r \varepsilon_0 \hbar^2} \frac{1}{n^2} = \frac{Z}{a_B n^2} \tag{8.29}$$

where in the last step I used Eq. 8.8 for Bohr radius a_B. The expectation values $\langle r^p \rangle$ for almost all other values of p can be derived using so-called Kramers' recursion relations, which I provide here without proof:

$$\frac{p+1}{n^2} \langle r^p \rangle - (2p+1) \frac{a_B}{Z} \langle r^{p-1} \rangle + \frac{pa_B^2}{4Z^2} \left[(2l+1)^2 - p^2 \right] \langle r^{p-2} \rangle = 0. \tag{8.30}$$

It is easy to see that I can indeed use Eqs. 8.29 and 8.30 to find $\langle r^p \rangle$ for any positive p, but Kramers' relations fail to yield $\langle r^{-2} \rangle$, because this term could arise if you set $p = 0$, but, unfortunately, the corresponding term vanishes because of the factor p in it. Therefore, I have to find an independent way of computing $\langle r^{-2} \rangle$. Luckily, there exists a cool theorem, which Richard Feynman derived while working on his undergraduate thesis, called Feynman–Hellmann theorem.[1] The derivation of this theorem is based on the obvious identity, which is valid for an arbitrary Hamiltonian and which I have already mentioned when deriving Eq. 8.29. To reiterate, the identity states that

$$E_n = \langle \psi_n | \hat{H} | \psi_n \rangle$$

if $|\psi_n\rangle$ are the eigenvectors of $\hat{H}$. Now assume that the Hamiltonian $\hat{H}$ depends on some parameter λ. It can be, for instance, a mass of a particle, or its charge, or something else. It is obvious then that the eigenvalues and the eigenvectors also depend on the same parameter. Differentiating this identity with respect to this parameter, you get

$$\frac{\partial E_n}{\partial \lambda} = \left\langle \frac{\partial \psi_n}{\partial \lambda} \middle| \hat{H} | \psi_n \right\rangle + \left\langle \psi_n | \hat{H} \middle| \frac{\partial \psi_n}{\partial \lambda} \right\rangle + \left\langle \psi_n \middle| \frac{\partial \hat{H}}{\partial \lambda} \middle| \psi_n \right\rangle .$$

The first two terms in this expression can be transformed as

$$\left\langle \frac{\partial \psi_n}{\partial \lambda} \middle| \hat{H} | \psi_n \right\rangle + \left\langle \psi_n | \hat{H} \middle| \frac{\partial \psi_n}{\partial \lambda} \right\rangle = E_n \left(\left\langle \frac{\partial \psi_n}{\partial \lambda} \middle| \psi_n \right\rangle + \left\langle \psi_n \middle| \frac{\partial \psi_n}{\partial \lambda} \right\rangle \right) =$$

$$E_n \frac{\partial \langle \psi_n | \psi_n \rangle}{\partial \lambda} = 0$$

---

[1]Hellmann derived this theorem 4 years before Feynman but published it in an obscure Russian journal, so it remained unknown until Feynman rediscovered it.

where I used the fact that all eigenvectors are normalized to unity, so that their norm, appearing in the last line of the above derivation, is just a constant. Thus, here is the statement of the Feynman–Hellmann theorem:

$$\frac{\partial E_n}{\partial \lambda} = \langle \psi_n | \frac{\partial \hat{H}}{\partial \lambda} | \psi_n \rangle . \tag{8.31}$$

This is a very simple, almost trivial result, and it is quite amazing that it can be used to solve rather complicated problems, such as finding the expectation value $\langle r^{-2} \rangle$ in the hydrogen atom problem. So, let's see how this is achieved. Going back to Eq. 8.7, you can recognize that this equation can be seen as an eigenvalue equation for Hamiltonian

$$\hat{H}_r = -\frac{\hbar^2}{2\mu r^2} \frac{d}{dr} \left(r^2 \frac{\partial}{\partial r} \right) + \frac{\hbar^2 l(l+1)}{2\mu r^2} - \frac{1}{4\pi \varepsilon_r \varepsilon_0} \frac{Ze^2}{r} \tag{8.32}$$

and that hydrogen energies are eigenvalues of this Hamiltonian. Therefore, I can apply the Feynman–Hellmann theorem to this Hamiltonian, choosing, for instance, the orbital quantum number l as a parameter λ. Differentiation of Eq. 8.32 with respect to l yields

$$\frac{\partial \hat{H}_r}{\partial l} = \frac{\hbar^2 (2l+1)}{2\mu r^2}.$$

In order to find derivative $\partial E_n / \partial l$, one needs to recall that the principal quantum number is related to the orbital number as $n = l + n_r + 1$ so that

$$\frac{\partial E_n}{\partial l} = \frac{\partial E_n}{\partial n} = \frac{Z^2 e^2}{4\pi \varepsilon_r \varepsilon_0 a_B} \frac{1}{n^3}.$$

Now, applying the Feynman–Hellmann theorem, I can write

$$\frac{\hbar^2 (2l+1)}{2\mu} \left\langle \frac{1}{r^2} \right\rangle = \frac{Z^2 e^2}{4\pi \varepsilon_r \varepsilon_0 a_B} \frac{1}{n^3}$$

where I used Eq. 8.17 for the energy. Rearranging this result and applying Eq. 8.8 for the Bohr radius, I obtain the final expression for $\langle r^{-2} \rangle$:

$$\left\langle \frac{1}{r^2} \right\rangle = \frac{Z^2 e^2 \mu}{2\pi \hbar^2 \varepsilon_r \varepsilon_0 a_B} \frac{1}{(2l+1) n^3} = \frac{2Z^2}{a_B^2} \frac{1}{(2l+1) n^3} . \tag{8.33}$$

Now, boys and girls, if what you have just witnessed is not a piece of pure magic with the Feynman–Hellmann theorem working as a magic wand, I do not know what else you would call it. And if you are not able to appreciate the awesomeness of this derivation, you probably shouldn't be studying quantum mechanics or physics at all for that matter. This result is also a key to finding, with the help of Kramers' relations, Eq. 8.30, of the expectation values $\langle r^p \rangle$ for any p. For instance, to find $\langle r^{-3} \rangle$, you just need to use Eq. 8.30 with $p = -1$:

$$\frac{a_B}{Z} \langle r^{-2} \rangle - \frac{a_B^2}{4Z^2} \left[(2l+1)^2 - 1 \right] \langle r^{-3} \rangle = 0 \Rightarrow$$

$$\left\langle \frac{1}{r^3} \right\rangle = \frac{4Z}{a_B} \frac{1}{(2l+1)^2 - 1} \left\langle \frac{1}{r^2} \right\rangle = \left(\frac{Z}{a_B} \right)^3 \frac{2}{l(l+1)(2l+1)n^3}. \tag{8.34}$$

If the sheer wonder at our ability to compute $\langle r^p \rangle$ without using the unseemly Laguerre polynomials is not a sufficient justification for you to vindicate spending some time doing these calculations, you will have to wait till Chap. 14, where I will put this result to actual use in understanding the fine structure of the spectra of hydrogen-like atoms.

8.4 Problems

Problem 103 Using Eqs. 8.4 and 8.5 together with canonical commutation relations for single-particle coordinates and momentums, derive the commutator between relative position vector r and corresponding momentum p_r to convince yourself that these variables, indeed, obey the canonical commutation relations.

Problem 104 Verify that Eq. 8.10 defines a quantity of the dimension of energy.

Problem 105

1. Derive Eq. 8.14 by applying the power series method to Eq. 8.12 and carrying out the procedure outlined in the text.
2. Find all radial functions with $n = 1$ and $n = 2$. Normalize them.

Problem 106 Using the definition of the associate Laguerre functions provided in the text, find explicit expressions for the radial functions corresponding to the states considered in the previous problem. Normalize them and make sure that the results are identical to those obtained previously.

Problem 107 An operator of the dipole moment is defined as $\hat{d} = e\hat{r}$ where e is the elementary charge and $\hat{r}$ is the position operator of the electron in the hydrogen atom. A dipole moment of a transition is defined as a matrix element of this operator between initial and final states of a system: $d_{nlm,n'l'm'} \equiv \langle nlm| \hat{d} |n'l'm' \rangle$. Evaluate this dipole moment for the transitions between ground state of the atom and all degenerate states characterized by $n = 2$.

Problem 108 Find the expectation values $\langle r \rangle$, $\langle 1/r \rangle$, and $\langle r^2 \rangle$ for a hydrogen atom in $|2, 1, m \rangle$ state.

Problem 109 Using the results of the previous problem and full 3-D Schrödinger equation with non-separated variables, find $\langle \hat{p}^2 \rangle$. Find a relation between the expectation values of the potential and kinetic energies.

Problem 110 A hydrogen atom is prepared in an initial state:

$$\phi\,(r,0) = \frac{1}{\sqrt{2}}\,(\psi_{2,1,1}\,(r,\theta,\varphi) + \psi_{1,0,0}\,(r,\theta,\varphi))\,.$$

Find the expectation value of the potential energy as a function of time.

Problem 111 Consider a hydrogen atom in a state described by the following wave function:

$$\phi(r) = R_{1,0}(r) + a\frac{z - \sqrt{2}x}{r}R_{2,1}(r)$$

where

$$R_{n,l}(r) = (r/na_B)^{l+1}\exp\left(-\frac{r}{na_B}\right)L_{n-l-1}^{2l+1}\,(2r/na_B)\,.$$

1. Rewrite this function in terms of normalized hydrogen wave functions.
2. Find the values of coefficient a that would make the entire function normalized.
3. If you measure $\hat{L}^2$ and $\hat{L}_z$, what values can you get and with what probabilities?
4. If you measure energy, which values are possible and what are their probabilities?
5. Find the probability that the measurement of the particle's position will find it in the direction specified by the polar angle θ $-44° < \theta < 46°$.
6. Find the probability that the measurement of the particle's position will find the particle at a distance $0.5a_B < r < a_B$ from the nucleus.

Chapter 9
Spin 1/2

9.1 Introduction: Why Spin?

The model of a pure spin 1/2, detached from all other degrees of freedom of a particle, is one of the simplest in quantum mechanics. Yet, it defies our intuition and resists developing that pleasant sensation of being able to relate a new concept to something that we think we already know (or at least are used to thinking about). We call this feeling "intuitive understanding," and it does play an important albeit mysterious role in our ability to use new concepts. The reason for this difficulty, of course, lies in the fact that spin is a purely quantum phenomenon with no reasonable way to model it on something that we know from classical physics. While the only known to me bulletproof remedy for this predicament is practice, I will try to somehow ease your pain by taking the time to develop the concept of spin and by providing empirical and theoretical arguments for its inevitability.

Experimentally spin manifests itself most directly via interaction between electrons and magnetic field and can be defined as an inherent property of electrons responsible for this interaction. This definition is akin to the definition of electric charge as a property responsible for electron's interaction with the electric field or of mass as a characteristic determining electron's acceleration under an action of a force. The substantial difference, of course, is that charge and mass are immutable scalar quantities, our views of which do not change when we transition from classical to quantum theories of nature. The concept of spin, on the other hand, is purely quantum and embodies two distinct types of entities. First is a Hermitian vector operator, characterized by two distinct eigenvalues and corresponding eigenvectors, which specify the possible experimental outcomes when one attempts to measure spin. Second are the spinors—particular type of vectors subjected to the action of the spin operator and representing various spin states; they control the probability of one or another outcome of the measurement.

© Springer International Publishing AG, part of Springer Nature 2018
L.I. Deych, *Advanced Undergraduate Quantum Mechanics*,
https://doi.org/10.1007/978-3-319-71550-6_9

To untangle the connections between spin, angular momentum, and magnetic interactions, let me begin with a simple example of a classical electron moving along a circular orbit of radius R with period T. Taken literally, this example does not make much sense, but it does produce surprisingly reasonable results, so it can be considered as a convenient and meaningful metaphor. So, imagine an observer placed at some point on the orbit and counting the number of times the electron passes by during some time $t \gg T$. The number of "sightings" of the electron, n, is related to the duration of the experiment t and the period T as $t = nT$. The total amount of charge that passes by the observer is obviously $q = ne = et/T$, where e is the elementary charge. The amount of charge passed across per unit time is what we call the electric current, which can be found as $I = q/t = et/(Tt) = e/T$. This crude trick replaced a circulating electron by a stationary electric current, which, of course, only makes sense if I spread the entire charge of the electron along its orbit by some kind of averaging procedure. But as I said, I am treating this model only as a metaphor. Accepting this metaphor, I can follow up by remembering that the interaction between a steady loop of current and a uniform magnetic field is described by the loop's magnetic dipole moment, μ defined as $\mu = IA\boldsymbol{n}$, where A is the area of the loop and $\boldsymbol{n}$ is the unit vector normal to the plane of the loop with direction determined by the right-hand rule (do you remember the right-hand rule?). In the case of the orbiting electron, the loop area is $A = \pi R^2$, so I have

$$\boldsymbol{\mu}_L = \frac{e}{T}\pi R^2 \boldsymbol{n} = \frac{ev}{2\pi R}\pi R^2 \boldsymbol{n} = \frac{em_e vR}{2m_e}\boldsymbol{n} = -\frac{e}{2m_e}\boldsymbol{L} \qquad (9.1)$$

where I (a) expressed period T in terms of the circumference $2\pi R$ and orbital velocity v: $T = 2\pi R/v$, (b) multiplied the numerator and the denominator of the resulting expression by electron's mass m_e, and (c) recognized that $m_e vR\boldsymbol{n}$ is a vector, which is equal in magnitude and opposite in direction to the orbital momentum of the electron $\boldsymbol{L}$. To figure out the "opposite" part of the last statement, recall that the magnetic moment is defined by the direction of the current—motion of the positive charges—while the charge of our orbiting electron is negative, and, therefore, it rotates in the direction opposite to the current. Equation 9.1 establishes the connection between the magnetic dipole moment of the electron and its orbital angular momentum.

The interaction between a classical magnetic dipole and a uniform magnetic field $\boldsymbol{B}$ can be described by a potential energy:

$$U_B = -\boldsymbol{\mu}_L \cdot \boldsymbol{B} = \frac{e}{2m_e}\boldsymbol{L} \cdot \boldsymbol{B}. \qquad (9.2)$$

According to this expression, the potential energy has a minimum when the magnetic dipole is oriented along the magnetic field and a maximum when they are oriented antiparallel to each other. For both these orientations, the torque on the dipole $\boldsymbol{\tau} = \boldsymbol{\mu}_L \times \boldsymbol{B}$ is zero, so these are two equilibrium positions, but while the former is the stable equilibrium, the latter is unstable. Equation 9.2 also establishes the connection between the potential energy U_B and electron's angular momentum L, which is quite useful for transitioning to quantum description. Quantization in this case consists merely in promoting the components of the angular momentum to the status of the operators. This newly born operator $\hat{U}_B$ can now be added to the Hamiltonian $\hat{H}_0$ describing the electron in the absence of the magnetic field to yield

$$\hat{H} = \hat{H}_0 + \frac{e}{2m_e} \mathbf{B} \cdot \hat{\mathbf{L}}. \tag{9.3}$$

$\hat{H}_0$, for instance, can describe an electron moving in some central potential $V(r)$ (the Coulomb potential would be a good example), and I will assume that its eigenvalues $E_{n,l}$ and eigenvectors $|n, l, m\rangle$ are known. The choice of notation here reflects the fact that the eigenvectors of the Hamiltonian with central potential must also be the eigenvectors of angular momentum operators $\hat{L}^2$ and $\hat{L}_z$ and that its eigenvalues do not depend on magnetic quantum number m.

It is quite easy to verify that if I choose the polar (Z)-axis of the coordinate system in the direction of the uniform magnetic field $\mathbf{B}$, eigenvectors $|n, l, m\rangle$ of $\hat{H}_0$ remain also eigenvectors of the total Hamiltonian given by Eq. 9.3. The corresponding eigenvalues are found as

$$\left(\hat{H}_0 + \frac{eB}{2m_e} L_z \right) |n, l, m\rangle = E_{n,l} |n, l, m\rangle + \frac{eB}{2m_e} \hbar m |n, l, m\rangle =$$

$$\left(E_{n,l} + \hbar \frac{eB}{2m_e} m \right) |n, l, m\rangle. \tag{9.4}$$

The combination of fundamental constant $e\hbar/2m_e$ has a dimension of magnetic dipole moment and is prominent enough to warrant giving it its own name. Bohr magneton μ_B is defined as

$$\mu_B = \frac{e\hbar}{2m_e} \tag{9.5}$$

so that the expression for the energy eigenvalues can be written down as

$$E_{n,l,m}^Z = E_{n,l} + m\mu_B B. \tag{9.6}$$

Term $m\mu_B B$ can be interpreted as the energy of interaction between the uniform magnetic field and a quantized magnetic moment with values which are multiples of μ_B. In this sense, the Bohr magneton can be thought of as a quantum of magnetic dipole moment. The most remarkable prediction of this simple computation is the m-dependence of the resulting energy levels, which is responsible for lifting the original $2l + 1$ degeneracy of the energy eigenvectors. Since magnetic field is the primary reason for this, it seems quite natural to give quantum number m the name of "magnetic" number.

Experimentally, this degeneracy lifting is observed via the Zeeman effect—splitting of the absorption or emission spectral lines in the presence of the magnetic field. I will discuss the relation between the absorption/emission of light and atomic energy levels in more detail in Part III of the book, but at this point, it is sufficient to recall old Bohr's postulates, one of which relates frequencies of the absorbed or emitted light to atomic energy levels:

$$\omega_{\alpha,\beta} = \frac{E_\alpha - E_\beta}{\hbar},$$

where α, β are composite indexes replacing groups of n, l, m for the sake of simplicity of notation. So, if you, say, observe a light emission due to the transition from the first excited state of hydrogen atom with $n = 2$ to the ground state, in the absence of magnetic field, you would see just one emission line formed by transitions between states $|2,0,0\rangle$, $|2,1,-1\rangle$, $|2,1,0\rangle$, and $|2,1,1\rangle$, all of which have the same energy, $E_2 = -\tilde{E}/4$, where $\tilde{E}$ was defined in Eq. 8.10. When the magnetic field is turned on, two of these states, $|2,1,-1\rangle$ and $|2,1,1\rangle$, acquire magnetic field-related corrections:

$$E_{2,-1} = -\tilde{E}/4 - \mu_B B$$

$$E_{2,1} = -\tilde{E}/4 + \mu_B B,$$

making their energy different from each other and $E_{2,0}$. As a result, instead of a single emission line with frequency $\omega = (E_2 - E_1)/\hbar = 3\tilde{E}/4\hbar$, an experimentalist would observe three lines at frequencies:

$$\omega_{-1} = 3\tilde{E}/4\hbar - \frac{\mu_B}{\hbar}B$$

$$\omega_0 = 3\tilde{E}/4\hbar$$

$$\omega_1 = 3\tilde{E}/4\hbar + \frac{\mu_B}{\hbar}B.$$

You should not think though that by deriving Eq. 9.4, I completely solved the Zeeman effect. The actual problem is much more complicated and involves addition of orbital and spin magnetic moments, as well as multi-electron effects, relativistic corrections, magnetic moment of nuclei, etc. What I did was just an illustration designed to make a particular point—the magnetic field lifting of the $2l + 1$ degeneracy of atomic levels gives rise to the odd number of closely positioned spectral lines. While for some atoms the odd number of lines is indeed observed, a large number of other observations manifest splitting into **even** number of lines. This phenomenon, called anomalous Zeeman effect, cannot be explained by interaction with orbital magnetic moment, because an even number of lines implies half-integer values of l. To explain this effect, we have to admit that in addition to "normal" orbital angular momentum, electrons also have another magnetic moment, which cannot be constructed from the coordinate and momentum operators and has to be, therefore, an intrinsic property of the electron not related to other regular (spatial–temporal) observables. The lowest number of observed split lines was equal to two. Equating $2l + 1$ to 2, you find that this splitting corresponds to $l = 1/2$. If you also recall that l is the maximum value of the magnetic number m, you might realize that m in this case can have only two values $m = \pm 1/2$.

A meticulous and mischievous reader might ask, of course, if it is absolutely necessary to derive a magnetic dipole momentum from an angular momentum. Can a magnetic momentum exist just by itself with no angular momentum attached to it? The answer to the first question is yes and to the second one is, obviously, no. To justify these answers, however, is not so easy, and the path toward realizing that electrons do possess an intrinsic angular momentum, which can be in one of two possible states, was a long one. Such physicists as Wolfgang Pauli (Austria–Switzerland–USA) and Arnold Sommerfeld (Germany) recognized very early that purely orbital state of electrons proposed in Bohr's model of atoms could not explain all experimental data, which consistently indicated that the actual number of states is double of what Bohr's model predicted. Pauli was writing about the "two-valuedness" of electrons in early 1925 as he needed it to explain the structure of atoms and formulate its famous Pauli exclusion principle. Later in 1925 two graduate students of Paul Ehrenfest from Leiden, the Netherlands, Goudsmit and Uhlenbeck, published a paper, in which they proposed that the required additional states come from intrinsic angular momentum of electrons due to their "spinning" on their own axis. They postulated that this new angular momentum of electrons, S, is related to its magnetic moment μ_s in a way similar to the relation between orbital momentum and orbital magnetic moment, but in order to fit experimental data, they had to multiply the Bohr magneton by 2:

$$\mu_s = -2 \times \frac{e}{2m_e}S = -2\frac{\mu_B}{\hbar}S. \tag{9.7}$$

The idea appeared so ridiculous to many serious physicists (such as Lorentz) that the students almost withdrew their paper, but luckily for them (and for physics), it was too late, and the paper was published. Eventually, it was recognized that while it was indeed wrong to think that a point-like particle such as electron can actually spin about its axis (estimates of the required spinning speed would put it well above the speed of light), so this classical mechanistic interpretation had to go, the idea of the intrinsic angular momentum, which "just is" as one of the attributes of electrons, survived, committing the names of Goudsmit and Uhlenbeck to the history of physics. Ironically, this was the highest achievement of their lives: they both made decent careers in physics, moving to the USA and securing respectable professorial positions, but they have never did anything as significant as their almost withdrawn student paper on spin.

There are other purely theoretical arguments for understanding spin as a different kind of the angular momentum, but this discussion is for a different time and a different book. At this point let me just mention that if we want to be able to add orbital angular momentum and spin angular momentum, which is absolutely necessary to explain a host of effects in atomic spectra, we must require that they both are described by objects of the same mathematical nature. This means that if the orbital momentum is described in quantum mechanics by three operator components $\hat{L}_x$, $\hat{L}_y$, and $\hat{L}_z$ of the angular momentum vector with commutation relations given by Eqs. 3.53–3.55, spin angular momentum must also be described by three operator components $\hat{S}_x$, $\hat{S}_y$, and $\hat{S}_z$ with the same commutation relations. Our calculations in Sect. 3.3.4 demonstrated that these commutation relations ensure that one of the operator components (usually it is chosen to be the z-component) and the operator of the square of the angular momentum $\hat{L}^2$(or $\hat{S}^2$) can have a common system of eigenvectors characterized by a set of two eigenvalues, $\hbar m_l$ for the z-component and $\hbar^2 l(l+1)$ for the square operator, where m_l can take values $m_l = -l, -l+1, \cdots, l-1, l$ and can be either integer or half-integer. The results of Sect. 5.1.4 indicated that orbital angular momentum can only be characterized by integer eigenvalues, but, as you can see, half-integer values are needed to deal with the spin angular momentum. It is amusing to think that nature tends to find use for everything, which appears in abstract mathematical theories! To distinguish between spin and orbital moments, I will replace notation l for the maximum eigenvalue of operator $\hat{L}_z$ with notation s for the maximum eigenvalue of $\hat{S}_z$. The lowest value that s can take is $1/2$, which means that there are only two possible eigenvalues of this operator, $-\hbar/2$ and $\hbar/2$. The eigenvalue of the operator $\hat{S}^2$ is $\hbar^2 s(s+1) = 3\hbar^2/4$, but it is the value of s that we have in mind when we are talking about electron having spin $1/2$. Thus, Pauli's two-valuedness of the electron comes here in the form of two eigenvectors and two eigenvalues of the z-component of the spin operator. The idea that the spin is an intrinsic and immutable property of electrons means that the $1/2$ value of quantum number s (or $3\hbar^2/4$ eigenvalue of operator $\hat{S}^2$) is as unchangeable as an electron's mass or charge, but at the same time, the electron can be in various distinct spin states described by eigenvectors of $\hat{S}_z$ or their arbitrary superposition.

9.2   Spin 1/2 Operators and Spinors

While spin $1/2$ operators are characterized by the same commutation relations as the operator of the orbital angular momentum, they act on vectors that live in a two-dimensional space of spin states or spinors. There are no reasons to panic at the sound of the unfamiliar word. The term spinor is used to describe a specific class of abstract vectors, which have all the same properties as any other vectors belonging to a Hilbert space, only much simpler because the dimensionality of the spinor space is just 2. One can introduce a ket spinor $|\chi\rangle$, its adjoint bra spinor $\langle\chi|$, and inner product of spinors $\langle\chi|\chi'\rangle$, which has the same property as any other inner products $\langle\chi|\chi'\rangle = (\langle\chi'|\chi\rangle)^*$. A basis in this space can be formed by two eigenvectors of operator $\hat{S}_z$, for which physicists use several, different in appearance, but otherwise equivalent notations. Two of the popular ways to designate these eigenvectors are $|1/2\rangle$ for the state belonging to the eigenvalue $\hbar/2$ and $|-1/2\rangle$ for its counterpart accompanying eigenvalue $-\hbar/2$. Alternatively, states with the positive eigenvalue are often called *spin-up* states with corresponding notation $|\uparrow\rangle$, while states with the negative eigenvalue are called *spin-down* states and are notated as $|\downarrow\rangle$. The main difference between spinors and vectors representing other states of quantum systems is that the spinors do not have the coordinate representation. They exist separately from the vector spaces formed by the eigenvectors of position or momentum operators or any other observables related to them. Spinors describe intrinsic properties of electrons, while vectors from other spaces represent their extrinsic spatial–temporal states.

This basis of the eigenvectors of operator $\hat{S}_z$ can be used to construct a particular representation of spinors and spin operators—as I demonstrated about 100 pages ago in Sect. 5.2.3. Generic spinors in this basis are represented by 2×1 column vectors:

$$|\chi\rangle = \begin{bmatrix} a \\ b \end{bmatrix} = a \begin{bmatrix} 1 \\ 0 \end{bmatrix} + b \begin{bmatrix} 0 \\ 1 \end{bmatrix}, \tag{9.8}$$

where $|\uparrow\rangle = \begin{bmatrix} 1 \\ 0 \end{bmatrix}$ represents the spin-up or $m = 1/2$ eigenvector, while $|\downarrow\rangle = \begin{bmatrix} 0 \\ 1 \end{bmatrix}$ represents the spin-down or $m = -1/2$ eigenvector. The representation of the respective bra vector is given by

$$\langle\chi| = [a^*\ b^*] = a^* [1\ 0] + b^* [0\ 1], \tag{9.9}$$

and the norm is

$$\langle\chi| \chi\rangle = a^*a + b^*b. \tag{9.10}$$

Normalized spinors obviously obey the condition

$$|a|^2 + |b|^2 = 1. \tag{9.11}$$

Spin operators $\hat{S}_x$, $\hat{S}_y$, and $\hat{S}_z$ defined with respect to a particular Cartesian coordinate system are represented in the basis of the eigenvectors of $\hat{S}_z$ by two-by-two matrices derived in Sect. 5.2.3:

$$\hat{S}_x = \frac{\hbar}{2} \begin{bmatrix} 0 & 1 \\ 1 & 0 \end{bmatrix} \tag{9.12}$$

$$\hat{S}_y = \frac{\hbar}{2} \begin{bmatrix} 0 & -i \\ i & 0 \end{bmatrix} \tag{9.13}$$

$$\hat{S}_z = \frac{\hbar}{2} \begin{bmatrix} 1 & 0 \\ 0 & -1 \end{bmatrix}. \tag{9.14}$$

Equations 9.12–9.14 are just a recapitulation of Eqs. 5.110 and 5.111 from Sect. 5.2.3, which I placed here for your convenience. Spin operators are often expressed in terms of so-called Pauli matrices $\hat{\sigma}_x$, $\hat{\sigma}_y$, and $\hat{\sigma}_z$ defined as

$$\hat{\sigma}_x = \begin{bmatrix} 0 & 1 \\ 1 & 0 \end{bmatrix} \tag{9.15}$$

$$\hat{\sigma}_y = \begin{bmatrix} 0 & -i \\ i & 0 \end{bmatrix} \tag{9.16}$$

$$\hat{\sigma}_z = \begin{bmatrix} 1 & 0 \\ 0 & -1 \end{bmatrix}. \tag{9.17}$$

These matrices have a number of important properties such as

$$\hat{\sigma}_x^2 = \hat{\sigma}_y^2 = \hat{\sigma}_z^2 = \hat{I}, \tag{9.18}$$

which means that they are simultaneously Hermitian and unitary, and

$$\hat{\sigma}_x\hat{\sigma}_y + \hat{\sigma}_y\hat{\sigma}_x = 0$$
$$\hat{\sigma}_x\hat{\sigma}_z + \hat{\sigma}_z\hat{\sigma}_x = 0 \tag{9.19}$$
$$\hat{\sigma}_z\hat{\sigma}_y + \hat{\sigma}_y\hat{\sigma}_z = 0,$$

which is often expressed as an anticommutativity property. Pauli matrices are used quite often in quantum mechanics, so it makes sense to acquaint yourselves with their properties. For instance, one can prove that the property expressed by Eq. 9.18 is valid for any matrix of the form $\sigma_n = \hat{\sigma} \cdot n$, where n is an arbitrary unit vector and $\hat{\sigma}$ is a vector with components given by Pauli matrices. Using the presentation of the unit vector in spherical coordinates

$$n_x = \sin\theta \cos\varphi$$
$$n_y = \sin\theta \sin\varphi \tag{9.20}$$
$$n_z = \cos\theta,$$

where θ and φ are polar and azimuthal angles defining the direction of n with respect to a particular system of Cartesian coordinate axis (see Fig. 9.1), you can derive for the matrix $\sigma_n = \sin\theta \cos\varphi\sigma_x + \sin\theta \sin\varphi\sigma_y + \cos\theta\sigma_z$:

$$\sigma_n = \begin{bmatrix} \cos\theta & \sin\theta e^{-i\varphi} \\ \sin\theta e^{i\varphi} & -\cos\theta \end{bmatrix}.$$

Fig. 9.1 Unit vector in the
Cartesian coordinate system

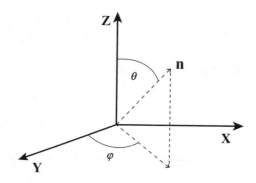

Squaring it will get you

$$\sigma_n^2 = \begin{bmatrix} \cos\theta & \sin\theta e^{-i\varphi} \\ \sin\theta e^{i\varphi} & -\cos\theta \end{bmatrix} \begin{bmatrix} \cos\theta & \sin\theta e^{-i\varphi} \\ \sin\theta e^{i\varphi} & -\cos\theta \end{bmatrix} =$$

$$\begin{bmatrix} \cos^2\theta + \sin^2\theta & \cos\theta\sin\theta e^{-i\varphi} - \cos\theta\sin\theta e^{-i\varphi} \\ \cos\theta\sin\theta e^{i\varphi} - \cos\theta\sin\theta e^{i\varphi} & \cos^2\theta + \sin^2\theta \end{bmatrix} =$$

$$\begin{bmatrix} 1 & 0 \\ 0 & 1 \end{bmatrix}.$$

This property makes the evaluation of various functions with Pauli matrices as arguments relatively easy. One of the popular examples is the exponential function $\exp(i\hat{\sigma} \cdot n)$, which you will enjoy computing when you get to the problem section of this chapter.

To help you become more comfortable with spin operators, I will now consider a few examples.

Example 21 (Measurement of the y-Component of the Spin) Assume that a single unmovable electron is placed in the state described by the spin-up eigenvector of operator $\hat{S}_z$. Using magnetic field directed along the Y-axis of the coordinate system, you are probing possible values of the y-component of the spin. What are these possible values and what are their probabilities?

Solution

As with any observable, possible results of its measurement are given by the eigenvalues of the respective operator. In this case this operator is $\hat{S}_y$ and you need to determine its eigenvalues. The answer is, of course, obvious ($+\hbar/2$ and $-\hbar/2$), but let's play the game and compute it. Besides, along the way you will determine the eigenvectors, which you need to answer the probability question. So, the eigenvector equation is

$$\frac{\hbar}{2}\begin{bmatrix} 0 & -i \\ i & 0 \end{bmatrix}\begin{bmatrix} a \\ b \end{bmatrix} = \frac{\hbar}{2}\lambda\begin{bmatrix} a \\ b \end{bmatrix},$$

which produces a set of two linear equations

$$-ib = \lambda a$$
$$ia = \lambda b. \tag{9.21}$$

The condition for the existence of nontrivial solutions given by zero of the determinant

$$\left\| \begin{matrix} \lambda & i \\ -i & \lambda \end{matrix} \right\|$$

becomes $\lambda^2 - 1 = 0$, yielding $\lambda_{1,2} = \pm 1$. Thus, recalling factor $\hbar/2$ that I prudently pulled out, you can conclude that the eigenvalues are, indeed, as predicted $\pm\hbar/2$. The first eigenvector is found by substituting $\lambda = 1$ to Eq. 9.21. This gives $a = -ib$, so that the respective eigenvector can be written as

$$|\hbar/2_y\rangle = b\begin{bmatrix} -i \\ 1 \end{bmatrix} = \frac{1}{\sqrt{2}}\begin{bmatrix} -i \\ 1 \end{bmatrix} \tag{9.22}$$

where at the last step I normalized it requiring that $2|b|^2 = 1$. Repeating this procedure with $\lambda = -1$, I find

$$|-\hbar/2_y\rangle = \frac{1}{\sqrt{2}}\begin{bmatrix} i \\ 1 \end{bmatrix}. \tag{9.23}$$

Taking into account that the initial state was $|\hbar/2_z\rangle = \begin{bmatrix} 1 \\ 0 \end{bmatrix}$, I find that the probabilities of the corresponding eigenvalues are

$$p_{\pm\hbar/2} = |\langle \pm\hbar/2_y| \hbar/2_z\rangle|^2 = \frac{1}{2}\left| [\pm i \ 1]\begin{bmatrix} 1 \\ 0 \end{bmatrix}\right|^2 = \frac{1}{2}.$$

Not a huge surprise, really.

Example 22 (Measurement of the Arbitrary Directed Planar Spin) What if we want to measure a component of the spin along a direction not necessarily aligned with one of the coordinate axes? Let me consider an example in which the measured component of the spin is in the Y–Z plane at an angle θ with the Z-axis and find possible outcomes and their probabilities assuming the same initial state as before.

Solution

I can define the specified direction by a unit vector with y-component $\sin \theta$ and z-component $\cos \theta$. Introducing unit vectors $\boldsymbol{e}_y$ and $\boldsymbol{e}_z$ along the respective axis, this vector can be conveniently presented as $\boldsymbol{n} = \boldsymbol{e}_y \sin \theta + \boldsymbol{e}_z \cos \theta$. The component of the spin in the direction of $\boldsymbol{n}$ is given by a dot product $\hat{S}_n = \hat{\boldsymbol{S}} \cdot \boldsymbol{n} = \hat{S}_y \sin \theta + \hat{S}_z \cos \theta$. Using the matrix representation of the spin operators in the basis of the eigenvectors of $\hat{S}_z$, Eqs. 9.12–9.14, I find for $\hat{S}_n$

$$\hat{S}_n = \frac{\hbar}{2} \sin \theta \begin{bmatrix} 0 & -i \\ i & 0 \end{bmatrix} + \frac{\hbar}{2} \cos \theta \begin{bmatrix} 1 & 0 \\ 0 & -1 \end{bmatrix} = \frac{\hbar}{2} \begin{bmatrix} \cos \theta & -i \sin \theta \\ i \sin \theta & -\cos \theta \end{bmatrix}.$$

The respective eigenvector equation becomes

$$\begin{bmatrix} \cos \theta & -i \sin \theta \\ i \sin \theta & -\cos \theta \end{bmatrix} \begin{bmatrix} a \\ b \end{bmatrix} = \lambda \begin{bmatrix} a \\ b \end{bmatrix},$$

and the equation for the eigenvalues takes the form

$$\begin{Vmatrix} \cos \theta - \lambda & -i \sin \theta \\ i \sin \theta & -\cos \theta - \lambda \end{Vmatrix} = -(\cos \theta - \lambda)(\cos \theta + \lambda) - \sin^2 \theta = \lambda^2 - 1 = 0.$$

I am not going to pretend that I am surprised that the eigenvalues are again $\pm \hbar/2$ as what else can they be?

Equations for the eigenvectors can be written as

1. $\lambda = 1$

$$a \cos \theta - ib \sin \theta = a$$

$$-ib \sin \theta = a(1 - \cos \theta)$$

$$-2ib \sin \frac{\theta}{2} \cos \frac{\theta}{2} = 2a \sin^2 \frac{\theta}{2}$$

$$-ib \cos \frac{\theta}{2} = a \sin \frac{\theta}{2}.$$

There are, of course, multiple choices of the coefficients in this equation, but I want to make the final form of the eigenvector as symmetric as possible, so I will choose $a = A \cos \frac{\theta}{2}$ and $b = iA \sin \frac{\theta}{2}$, which obviously satisfy the equation with an arbitrary A. The latter can be found from the normalization condition $|a|^2 + |b|^2 = 1$, which obviously gives $A = 1$. Now, I can write the first eigenvector as

$$|\hbar/2_n\rangle = \begin{bmatrix} \cos \frac{\theta}{2} \\ i \sin \frac{\theta}{2} \end{bmatrix} \tag{9.24}$$

2. $\lambda = -1$

$$a \cos \theta - ib \sin \theta = -a$$

$$ib \sin \theta = a (1 + \cos \theta)$$

$$2ib \sin \frac{\theta}{2} \cos \frac{\theta}{2} = 2a \cos^2 \frac{\theta}{2}$$

$$ib \sin \frac{\theta}{2} = a \cos \frac{\theta}{2}.$$

Using the same trick as previously, I find this eigenvector to be

$$|-\hbar/2_n\rangle = \begin{bmatrix} \sin \frac{\theta}{2} \\ -i \cos \frac{\theta}{2} \end{bmatrix}. \tag{9.25}$$

The direction described by $\theta = \pi/2$ corresponds to unit vector $\mathbf{n}$ pointing in the direction of the Y-axis, reducing this example to the previous one. Naturally, you would expect the eigenvector found here to reduce to the respective eigenvectors from the previous example. However, by substituting $\theta = \pi/2$ into Eqs. 9.24 and 9.25, you find that the resulting vectors do not coincide with Eqs. 9.22 and 9.23. Did I do something wrong here? Not really, because it is easy to notice that the difference between the two results is a mere factor of i, and we know that multiplication of an eigenvector by i or by any other complex number of the form $\exp(i\varphi)$, where φ is an arbitrary real number, does not change a quantum state and has no observable consequences. Finally, the probabilities that the measurements of the spin will produce one of the found eigenvalues are

$$p_{\hbar/2} = \left| \begin{bmatrix} \cos \frac{\theta}{2} & -i \sin \frac{\theta}{2} \end{bmatrix} \begin{bmatrix} 1 \\ 0 \end{bmatrix} \right|^2 = \cos^2 \frac{\theta}{2}$$

$$p_{-\hbar/2} = \left| \begin{bmatrix} \sin \frac{\theta}{2} & i \cos \frac{\theta}{2} \end{bmatrix} \begin{bmatrix} 1 \\ 0 \end{bmatrix} \right|^2 = \sin^2 \frac{\theta}{2}.$$

I can also use this result to find the expectation value of the operator $\hat{S}_n$ in the state $\begin{bmatrix} 1 \\ 0 \end{bmatrix}$. The probabilistic definition of the mean $\sum x_i p_i$, where x_i is the value of the variable and p_i is its probability, yields

$$\langle \hat{S}_n \rangle = (\hbar/2) \cos^2 \frac{\theta}{2} - (\hbar/2) \sin^2 \frac{\theta}{2} = (\hbar/2) \cos \theta,$$

which is exactly the value you should have expected from a classical vector oriented along the Z-axis when computing its component in the direction of $\mathbf{n}$. The same result is obtained by computing the expectation value using the operator definition:

$$\langle \hat{S}_n \rangle = \langle \uparrow_z | \hat{S}_n | \uparrow_z \rangle = \frac{\hbar}{2} \begin{bmatrix} 1 & 0 \end{bmatrix} \begin{bmatrix} \cos\theta & -i\sin\theta \\ i\sin\theta & -\cos\theta \end{bmatrix} \begin{bmatrix} 1 \\ 0 \end{bmatrix} =$$

$$\frac{\hbar}{2} \begin{bmatrix} 1 & 0 \end{bmatrix} \begin{bmatrix} \cos\theta \\ i\sin\theta \end{bmatrix} = \frac{\hbar}{2} \cos\theta.$$

Example 23 (Measuring of the Z-Component in an Arbitrary Spinor State) You can also ask a question, what if the spin was prepared in a state presented by one of the eigenvectors of $\hat{S}_n$, say, $|\hbar/2_n\rangle$ and we were measuring the z-component of the spin? What would be the probabilities of obtaining $\hbar/2$ or $-\hbar/2$ and the expectation value of $\hat{S}_z$ in this situation?

Solution

The corresponding probabilities are given by the following expressions:

$$\left| \begin{bmatrix} 1 & 0 \end{bmatrix} \begin{bmatrix} \cos\frac{\theta}{2} \\ i\sin\frac{\theta}{2} \end{bmatrix} \right|^2 = \cos^2\frac{\theta}{2}$$

$$\left| \begin{bmatrix} 0 & 1 \end{bmatrix} \begin{bmatrix} \cos\frac{\theta}{2} \\ i\sin\frac{\theta}{2}0 \end{bmatrix} \right|^2 = \sin^2\frac{\theta}{2}$$

yielding exactly the same results. Obviously the expectation value will also be the same.

These examples were designed to prepare you to answer an important but rather confusing question. The concept of spin is supposed to represent a vector quantity existing in our regular physical three-dimensional space. At the same time, the quantum objects used to describe spin, operators, and spinors have little relation to this space. While spin operators do have three components, they are not regular vectors, and the question about the "direction" of a vector operator does not make much sense. Spinors, representing spin states, are objects existing in an abstract two-dimensional space. Thus, the question is how these objects are connected with the physical space in which all our measurement apparatuses live. One might attempt to deflect this question by saying that after taking the expectation values of the spin operators for a given spin state, we will end up with a regular vector, which will provide us with the information about the spin and its direction. I can counter this by saying that this information is very limited. Indeed, I can also compute the uncertainty of each spin component, which will also give me a regular vector. The problem is that in the most generic situation, the vector obtained from the expectation values and the vector obtained from the uncertainties do not have to have the same direction, making it difficult to come up with a reasonable interpretation of these results. One way to avoid this ambiguity is to focus on eigenvectors, in which case expectation values provide the complete description of the situation. You only need to figure out the connection between the spatial direction, spin operators and corresponding eigenvectors.

One way to answer this question is to do what we just did in the previous example: introduce a component of the spin operator in the direction of interest, find its eigenvectors, and analyze their connection to this direction. But I want to add a bit more intrigue to the issue and will use a different approach. Let me ask you this: what is the best way to write down a generic spinor? Equation 9.8, which does it by introducing two complex parameters, a and b, is too general and does not contain all the information available about even the most generic spin states. Indeed, two complex numbers contain four independent real parameters, which can be brought out explicitly by writing a and b in the exponential form: $a = |a| \exp{(i\phi_a)}$ and $b = |b| \exp{(i\phi_b)}$. I can do better than that and reduce the number of parameters to just two without making the spinor any less generic.

First, I am going to use the freedom of choice of the overall phase of the spinor. To this end I will multiply both a and b by $\exp{[-i(\phi_b + \varphi_a)/2]}$, bringing the spinor in the following form:

$$|\chi\rangle = \begin{bmatrix} |a| \exp{(-i\varphi/2)} \\ |b| \exp{(i\varphi/2)} \end{bmatrix},$$

where $\varphi = \phi_a - \phi_b$, and there are only three parameters left to worry about. Obviously, this is not the only way to eliminate one of the phases, but this one presents the spinor in a rather symmetric form, and just like all physicists, I have a sweet spot for symmetry. Besides, frankly speaking, I know where I want to go and just taking you along for the ride. The normalization imposes additional condition on these parameters, telling me that I can use it to eliminate another one of them reducing the total number to just two. After a few seconds of staring at Eq. 9.11, it can descend upon you that this equation looks similar to the fundamental trigonometric identity $\cos^2 x + \sin^2 x = 1$ and that you can automatically satisfy the normalization condition by choosing $|a| = \cos{(\theta/2)}$ and $|b| = \sin{(\theta/2)}$, expressing both $|a|$ and $|b|$ in terms of a single parameter $\theta/2$. If you are asking why $\theta/2$ and not just θ, you will have the answer in a few minutes, just keep reading. Now, as promised, I have the expression for the generic normalized spinor:

$$|\chi_1\rangle = \begin{bmatrix} \cos{(\theta/2)} \exp{(-i\varphi/2)} \\ \sin{(\theta/2)} \exp{(i\varphi/2)} \end{bmatrix} \tag{9.26}$$

with only two parameters, θ and φ. The choice I made for $|a|$ and $|b|$ is not unique, and I can generate another spinor by assigning $|a| = \sin{(\theta/2)}$ and $|b| = -\cos{(\theta/2)}$:

$$|\chi_2\rangle = \begin{bmatrix} \sin{(\theta/2)} \exp{(-i\varphi/2)} \\ -\cos{(\theta/2)} \exp{(i\varphi/2)} \end{bmatrix}. \tag{9.27}$$

It is easy to verify by computing $\langle \chi_1 | \chi_2 \rangle$ that these spinors are orthogonal (of course, I designed them with this particular goal in mind), and by generating matrices

$$|\chi_1\rangle\langle\chi_1| = \begin{bmatrix} \cos^2(\theta/2) & \cos(\theta/2)\sin(\theta/2)\exp(-i\varphi) \\ \cos(\theta/2)\sin(\theta/2)\exp(i\varphi) & \sin^2(\theta/2) \end{bmatrix}$$

and

$$|\chi_2\rangle\langle\chi_2| = \begin{bmatrix} \sin^2(\theta/2) & -\cos(\theta/2)\sin(\theta/2)\exp(-i\varphi) \\ -\cos(\theta/2)\sin(\theta/2)\exp(i\varphi) & \cos^2(\theta/2) \end{bmatrix},$$

you can also check that

$$|\chi_1\rangle\langle\chi_1| + |\chi_2\rangle\langle\chi_2| = \hat{I},$$

indicating that these two spinors form a complete set. (When trying to reproduce these calculations, do not forget complex conjugation when converting kets into respective bra vectors.)

Thus, with little efforts, I have constructed a complete set of two generic mutually orthogonal spinors characterized by parameters, which can be interpreted as angles, and this must mean something. The found representation of spinors establishes a one-to-one relationship between two-dimensional space of spin states and points on the surface of a regular three-dimensional sphere of unit radius (see Fig. 9.2). The points at the north and south poles of the sphere, characterized by $\theta = 0$ and $\theta = \pi$, describe the eigenvectors of $\hat{S}_z$ operators $|\uparrow\rangle$ and $|\downarrow\rangle$, respectively (angle φ is not defined for these points, but it is not a problem because respective factors $\exp(-i\varphi/2)$ become in these cases simply insignificant phase factors). It is also easy to notice that the antipodal points lying on the opposite ends of an arbitrarily oriented diameter of the sphere correspond to two mutually perpendicular spin states. Indeed, spherical coordinates of the antipodal points are related to each other as $\theta_2 = \pi - \theta_1$, $\varphi_2 = \varphi_1 + \pi$. Substituting these expressions into Eq. 9.26, you will immediately obtain the spinor presented in Eq. 9.27. While performing this operation, you can appreciate the wisdom of using half-angles $\theta/2$ and $\varphi/2$ in these expressions.

In order to further figure out the physical meaning of the mapping between spinors and directions in regular 3-D space, consider the same operator, $\hat{S}_n = \hat{\mathbf{S}} \cdot \mathbf{n}$, which I discussed in the preceding example, but with a unit vector $\mathbf{n}$ defining a generic direction characterized by the same angles θ, φ as in Fig. 9.2. This is the same vector which I introduced in connection with Pauli matrices, Eq. 9.20, so that the operator $\hat{S}_n$ becomes

$$\hat{S}_n = \frac{\hbar}{2}\hat{\sigma}_n = \frac{\hbar}{2}\begin{bmatrix} \cos\theta & \sin\theta e^{-i\varphi} \\ \sin\theta e^{i\varphi} & -\cos\theta \end{bmatrix}.$$

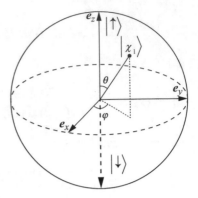

Now, let me apply this operator to the spinors presented in Eq. 9.26:

$$\frac{\hbar}{2} \begin{bmatrix} \cos\theta & \sin\theta e^{-i\varphi} \\ \sin\theta e^{i\varphi} & -\cos\theta \end{bmatrix} \begin{bmatrix} \cos(\theta/2)\exp(-i\varphi/2) \\ \sin(\theta/2)\exp(i\varphi/2) \end{bmatrix} =$$

$$\frac{\hbar}{2} \begin{bmatrix} \cos\theta\cos(\theta/2)\exp(-i\varphi/2) + \sin\theta\sin(\theta/2)\exp(-i\varphi/2) \\ \sin\theta\cos(\theta/2)\exp(i\varphi/2) - \cos\theta\sin(\theta/2)\exp(i\varphi/2) \end{bmatrix} =$$

$$\frac{\hbar}{2} \begin{bmatrix} \cos(\theta/2)\exp(-i\varphi/2)\left(\cos\theta + 2\sin^2(\theta/2)\right) \\ \sin(\theta/2)\exp(i\varphi/2)\left(2\cos^2(\theta/2) - \cos\theta\right) \end{bmatrix} =$$

$$\frac{\hbar}{2} \begin{bmatrix} \cos(\theta/2)\exp(-i\varphi/2)\left(2\cos^2(\theta/2) - 1 + 2\sin^2(\theta/2)\right) \\ \sin(\theta/2)\exp(i\varphi/2)\left(2\cos^2(\theta/2) - 2\cos^2(\theta/2) + 1\right) \end{bmatrix} =$$

$$\frac{\hbar}{2} \begin{bmatrix} \cos(\theta/2)\exp(-i\varphi/2) \\ \sin(\theta/2)\exp(i\varphi/2) \end{bmatrix}.$$

Isn't that nice? A generic spinor with arbitrarily introduced parameters θ and φ turned out to be an eigenvector of an operator representing the component of the spin in the direction defined by these parameters. It probably would not come as a particularly great surprise now that the second eigenvector I conjured up is also an eigenvector of the same operator but corresponding to the second eigenvalue, namely, $-\hbar/2$. (Check it out as an exercise. And by the way, did you notice that in the course of this computation, I used a couple of trigonometric identities such as $\cos x = 2\cos^2 x/2 - 1$ and $\sin x = 2\sin x/2\cos x/2$?) This exercise allows us to give more substance to an already established connection between spinors and directions in physical space: each spinor parametrized as in Eq. 9.26 or 9.27 is an eigenvector of a component of the spin in the direction specified by parameters θ and φ interpreted as spherical coordinates of the corresponding unit vector lying on the surface of the Bloch sphere. The measurement of the spin in this direction will yield definite results corresponding to the respective eigenvalue, so it can be interpreted as the direction of the spin for this particular spin state. It also makes sense that antipodal points on the Bloch sphere represent eigenvectors belonging to opposite eigenvalues of $\hat{S}_n$. Finally, by now, I hope you have the answer to the question why I used half-angles in the definition of the spinors.

9.3 Dynamic of Spin in a Uniform Magnetic Field

A bound (for instance, by attraction to a nucleus) electron in a uniform magnetic field, the system used earlier to introduce the Zeeman effect, is also the simplest somewhat realistic physical model allowing one to study quantum dynamics of a pure spin. Assuming that the interaction between the spin and the magnetic field does not affect the orbital state of the electron, one can ignore energy associated with the latter and omit the atomic part of the Hamiltonian (remember, energy only matters when it changes, and if it does not, we can always make it equal to zero). The Hamiltonian of this system is obtained by dropping $\hat{H}_0$ term from Eq. 9.3 and replacing orbital angular momentum $\hat{L}$ with $2\hat{S}$, where factor 2 takes into account the empirically established modification of the connection between spin angular and magnetic momenta, Eq. 9.7. The resulting Hamiltonian takes the form

$$\hat{H} = 2\frac{\mu_B}{\hbar}\hat{S} \cdot \boldsymbol{B}. \qquad (9.28)$$

Note that magnetic field $\boldsymbol{B}$ is not an operator because it describes a classical magnetic field created by a source whose physics is outside of our consideration. Since the field is uniform, it makes sense to use its direction as one of the axes of the coordinate system, which I have to specify in order to be able to carry out subsequent calculations. It is customary to choose axis Z as the one which is codirected with the magnetic field, in which case Hamiltonian 9.28 significantly simplifies

$$\hat{H} = 2\frac{\mu_B B}{\hbar}\hat{S}_z. \qquad (9.29)$$

In this section I will discuss the dynamics of spin described by this Hamiltonian using both Schrödinger and Heisenberg pictures of quantum mechanics.

9.3.1 Schrödinger Picture

In the Schrödinger picture, we always begin by establishing the eigenvalues and eigenvectors of the Hamiltonian. It is obvious that the eigenvectors of the Hamiltonian given by Eq. 9.29 coincide with those of operator $\hat{S}_z$, which I will denote here as $|\uparrow\rangle$ with eigenvalue $\hbar/2$ (spin-up) and $|\downarrow\rangle$ with eigenvalue $-\hbar/2$

(spin-down). The respective eigenvalues of the Hamiltonian are quite obvious and
are

$$E_\uparrow = \mu_B B$$

$$E_\downarrow = -\mu_B B. \tag{9.30}$$

A solution of the time-dependent Schrödinger equation for an arbitrary time-
dependent spinor $|\chi(t)\rangle$

$$-i\hbar \frac{\partial |\chi(t)\rangle}{\partial t} = 2\frac{\mu_B B}{\hbar} \hat{S}_z |\chi(t)\rangle$$

can be presented as a linear combination of the stationary states of the Hamiltonian

$$|\chi(t)\rangle = a \exp\left(i\frac{\mu_B B}{\hbar} t\right) |\uparrow\rangle + b \exp\left(-i\frac{\mu_B B}{\hbar} t\right) |\downarrow\rangle \tag{9.31}$$

where coefficients a and b are determined by the initial state of the spin

$$|\chi(0)\rangle = a |\uparrow\rangle + b |\downarrow\rangle. \tag{9.32}$$

Equation 9.31 essentially solves the problem of the dynamics of a single spin in a
uniform magnetic field. It, however, does little to develop our intuition about the
physical phenomena, which this solution describes. In a typical experimental situation, one is rarely dealing with a single spin. Most frequently, an experimentalist
would measure a signal from an ensemble of many spins, and if we can neglect any
kind of interaction between them, as well as assume that all spins are in the same
initial state,[1] the experimental results can be described by finding the expectation
values of the spin operators. So, let me compute these expectation values for the
state described by Eq. 9.31.

To this end I will use the representation of a generic spinor in the form of Eq. 9.26
and rewrite coefficients a and b as

$$a = \cos(\theta/2) \exp(-i\varphi/2)$$

$$b = \sin(\theta/2) \exp(i\varphi/2). \tag{9.33}$$

Substituting these expressions for a and b into Eq. 9.31 and using regular represen-
tation of basis spinors $|\uparrow\rangle$ and $|\downarrow\rangle$, you can find

---

[1]The assumption about the same initial state is the most difficult to realize experimentally and can
be justified only at zero temperature.

$$|\chi(t)\rangle = \begin{bmatrix} \cos{(\theta/2)} \exp{(-i\varphi/2)} \exp{\left(i\frac{\mu_B B}{\hbar}t\right)} \\ \sin{(\theta/2)} \exp{(i\varphi/2)} \exp{\left(-i\frac{\mu_B B}{\hbar}t\right)} \end{bmatrix}. \tag{9.34}$$

It is easiest to compute the expectation value of $\hat{S}_z$:

$$\left\langle \hat{S}_z \right\rangle = \frac{\hbar}{2} \left(|a|^2 - |b|^2 \right) = \frac{\hbar}{2} \cos{\theta}. \tag{9.35}$$

I derived this expression taking advantage of the fact that $|\uparrow\rangle$ and $|\downarrow\rangle$ in Eq. 9.31 are eigenvectors of $\hat{S}_z$, and, therefore, coefficients in front of them (their absolute values squared, of course) determine the probabilities of the respective eigenvalues. To find the expectation values of two other components, I will have to do a little bit more work computing

$$\left\langle \hat{S}_{x,y} \right\rangle = \langle \chi(t)| \hat{S}_{x,y} |\chi(t)\rangle.$$

I begin with the x-component and first compute the right half of this expression $\hat{S}_x |\chi(t)\rangle$:

$$\hat{S}_x |\chi(t)\rangle = \frac{\hbar}{2} \begin{bmatrix} 0 & 1 \\ 1 & 0 \end{bmatrix} \begin{bmatrix} \cos{(\theta/2)} \exp{(-i\varphi/2)} \exp{\left(i\frac{\mu_B B}{\hbar}t\right)} \\ \sin{(\theta/2)} \exp{(i\varphi/2)} \exp{\left(-i\frac{\mu_B B}{\hbar}t\right)} \end{bmatrix} =$$

$$\frac{\hbar}{2} \begin{bmatrix} \sin{(\theta/2)} \exp{(i\varphi/2)} \exp{\left(-i\frac{\mu_B B}{\hbar}t\right)} \\ \cos{(\theta/2)} \exp{(-i\varphi/2)} \exp{\left(i\frac{\mu_B B}{\hbar}t\right)} \end{bmatrix}.$$

By the way, have you noticed how operator $\hat{S}_x$ flips the components of the spinor? Anyway, to complete this computation, I find the inner product of this ket with the bra version of spinor in Eq. 9.31:

$$\left\langle \hat{S}_x \right\rangle = \frac{\hbar}{2} \left[\cos{(\theta/2)} \sin{(\theta/2)} \exp{(i\varphi)} \exp{\left(-i\frac{2\mu_B B}{\hbar}t\right)} \right.$$

$$\left. + \cos{(\theta/2)} \sin{(\theta/2)} \exp{(-i\varphi)} \exp{\left(i\frac{2\mu_B B}{\hbar}t\right)} \right] -$$

$$\frac{\hbar}{2} \sin{\theta} \cos{\left(\frac{2\mu_B B}{\hbar}t - \varphi\right)}.$$

Similar calculations with the y-component operator yield

$$\left\langle \hat{S}_y \right\rangle = \frac{\hbar}{2} \sin{\theta} \sin{\left(\frac{2\mu_B B}{\hbar}t - \varphi\right)}.$$

Let's collect all these results together to get the better picture:

$$\langle \hat{S}_z \rangle = \frac{\hbar}{2} \cos \theta$$

$$\langle \hat{S}_x \rangle = \frac{\hbar}{2} \sin \theta \cos \left(\frac{2\mu_B B}{\hbar} t - \varphi \right)$$

$$\langle \hat{S}_y \rangle = \frac{\hbar}{2} \sin \theta \sin \left(\frac{2\mu_B B}{\hbar} t - \varphi \right). \tag{9.36}$$

Here is what we have: a vector of length $\hbar/2$ remains at all times at angle θ with respect to the magnetic field, but its projection on the X–Y plane of the coordinate system (which is perpendicular to the magnetic field) rotates with frequency $\omega_L = 2\mu_B B/\hbar = eB/m_e$, where I substituted Eq. 9.5 for the Bohr magneton. A remarkable fact about this result is the disappearance of Planck's constant from the final expression for frequency, which signals that this phenomenon must exist in classical physics as well, and it, indeed, does. Equation 9.36 describes a very well-known effect—Larmor precession—which is observed every time when a magnetic moment (of any nature) interacts with a uniform magnetic field. However, the frequency of the precession might be different for different magnetic moments because of its dependence on the so-called gyromagnetic ratio defined as a coefficient of proportionality between the magnetic dipole moment and the angular momentum. For the orbital angular momentum, this ratio is $-e/(2m_e)$ as given by Eq. 9.1, while for the spin, it is two times larger, resulting in twice as big precession frequency.

9.3.2   Heisenberg Picture

To describe spin precession in the Heisenberg picture, I have to solve Heisenberg equations 4.24 for spin operators. To simplify notations I will omit subindex H, which I used to distinguish Schrödinger from Heisenberg operators. However, it is important to note that the angular momentum commutation relations, Eqs. 3.53–3.55, remain the same for both pictures, provided that we take Heisenberg operators at the same time. If you do not see how to verify this statement, imagine sandwiching both sides of the commutation relation between operators $\exp \left(i\hat{H}t/\hbar \right)$ and $\exp \left(-i\hat{H}t/\hbar \right)$ and also inserting the products of these operators (which is equal to unity, by the way) between the products of any two operators in the commutator.

Thus, using the necessary commutation relation, I obtain the following equations:

$$\frac{d\hat{S}_z}{dt} = -\frac{i}{\hbar} \omega_L \left[\hat{S}_z, \hat{S}_z \right] = 0 \tag{9.37}$$

$$\frac{d\hat{S}_x}{dt} = -\frac{i}{\hbar}\omega_L\left[\hat{S}_x, \hat{S}_z\right] = -\omega_L\hat{S}_y \tag{9.38}$$

$$\frac{d\hat{S}_y}{dt} = -\frac{i}{\hbar}\omega_L\left[\hat{S}_y, \hat{S}_z\right] = \omega_L\hat{S}_x, \tag{9.39}$$

where I introduced the Larmor frequency defined in the previous section to the Hamiltonian. Differentiating Eqs. 9.38 and 9.39 with respect to time, I can separate them into two independent differential equations of the second order:

$$\frac{d^2\hat{S}_x}{dt^2} = -\omega_L^2\hat{S}_x$$

$$\frac{d^2\hat{S}_y}{dt^2} = -\omega_L^2\hat{S}_y$$

with obvious solutions

$$\hat{S}_x(t) = \hat{A}_x \sin\omega_L t + \hat{B}_x \cos\omega_L t$$

$$\hat{S}_y(t) = \hat{A}_y \sin\omega_L t + \hat{B}_y \cos\omega_L t.$$

Unknown constant operators $\hat{A}_{x,y}$ and $\hat{B}_{x,y}$ are determined by the initial conditions for the spin operators and their derivatives:

$$\hat{B}_x = \hat{S}_x(0); \ \hat{A}_x = \frac{1}{\omega_L}\frac{d\hat{S}_x}{dt} = -\hat{S}_y(0)$$

$$\hat{B}_y = \hat{S}_y(0); \ \hat{A}_y = \frac{1}{\omega_L}\frac{d\hat{S}_y}{dt} = \hat{S}_x(0)$$

where $\hat{S}_{x,y}(0)$ coincide with the Schrödinger spin operators. Thus, I have

$$\hat{S}_x(t) = -\hat{S}_y(0)\sin\omega_L t + \hat{S}_x(0)\cos\omega_L t \tag{9.40}$$

$$\hat{S}_y(t) = \hat{S}_x(0)\sin\omega_L t + \hat{S}_y(0)\cos\omega_L t. \tag{9.41}$$

All that is left to do is to compute the expectation values of the Schrödinger spin operators in the initial state given by Eqs. 9.32 and 9.33. However, I do not have to repeat these calculations as we can just read them off Eq. 9.36 at $t - 0$. This yields

$$\left\langle \hat{S}_x(t) \right\rangle = \frac{\hbar}{2}\sin\theta\cos\varphi\cos\omega_L t + \frac{\hbar}{2}\sin\theta\sin\varphi\sin\omega_L t = \frac{\hbar}{2}\sin\theta\cos(\omega_L t - \varphi)$$

$$\left\langle \hat{S}_y(t) \right\rangle = \frac{\hbar}{2}\sin\theta\cos\varphi\sin\omega_L t - \frac{\hbar}{2}\sin\theta\sin\varphi\cos\omega_L t = \frac{\hbar}{2}\sin\theta\sin(\omega_L t - \varphi)$$

in complete agreement with the results obtained from the Schrödinger picture.

9.4 Spin of a Two-Electron System

9.4.1 Space of Two-Particle States

I will complete the discussion of the spin by considering a system of two electrons.
The goal of this exercise is to figure out if it makes sense to talk about a total spin of
this system understood as a some kind of the sum of two individual spins $\hat{S}_1 + \hat{S}_2$. In
classical physics that would have been a trivial question—of course, we can define
the total angular momentum of several particles—just add them up remembering
that they are vectors. You can even derive a total angular momentum conservation
law valid in the absence of external torques, just like you can derive a total linear
momentum conservation law if the system of particles is not exposed to external
forces. In quantum mechanics, when individual spins are presented by operators
acting in different spaces containing spin states of each particle, the answer to this
question is more complex. It is still affirmative: yes, it is possible to define the total
spin of a system of two (or more particles) by introducing a new operator, which
can be formally defined as

$$\hat{S}^{(tp)} = \hat{S}^{(1)} + \hat{S}^{(2)}, \tag{9.42}$$

where the upper index is an abbreviation of "two-particle." However, so far Eq. 9.42
is a purely formal expression, in which even the meaning of the sign "+" is not
clear. What I need to do now is to figure out the properties of $\hat{S}^{(tp)}$ and their relation
to the properties of $\hat{S}^{(1)}$ and $\hat{S}^{(2)}$, which is not a trivial task.

Operators are defined by their action on vectors, and, since vectors live in a
certain vector space, the first step in defining an operator is to understand the space
where vectors, on which the operator acts, live. Operators $\hat{S}_1$ and $\hat{S}_2$ operate on
vectors that live in different and unrelated spaces: one acts on spin states of one
particle and the other on the states of a completely different particle. I can, however,
combine these spaces to form a new extended space, which would include spin states
of both particles. To define such a space, all what I need is to define a basis in it,
and then any other vector can be presented as a linear combination of the vectors of
basis. The space containing spin states of each individual particle is defined by two
basis vectors for each particle. These states are eigenvectors of operators $\hat{S}_z^{(1)}$ and $\hat{S}_z^{(2)}$
(obviously defined in the same coordinate system) and can be depicted symbolically
in a few equivalent ways discussed in previous sections. Here I will use spin-up
and spin-down notations indicated by the vertical arrows $|\uparrow_{1,2}\rangle$ or $|\downarrow_{1,2}\rangle$, where
subindexes 1 and 2 simply indicate a particle whose states these kets represent.
In a system of two particles, there exist four different combinations of their spin
states: both spins up, both spins down, the first spins up, the second spins down,
and vice versa. You can create a notation for these states either by putting two state
signifiers inside a single ket, like that $|\uparrow_1, \downarrow_2\rangle$, or by sticking together two kets like
this: $|\uparrow_1\rangle |\downarrow_2\rangle$. The difference between the two notations is superfluous, and either

one can be used freely, while the second notation with two separate kets is slightly more convenient when one needs to write down matrix elements corresponding to operators acting on different particles. Thus, I will present the four basis vectors in a new four-dimensional space containing states of a two-spin system as

$$|1\rangle \equiv |\uparrow_1\rangle |\uparrow_2\rangle, \ |2\rangle \equiv |\uparrow_1\rangle |\downarrow_2\rangle, \ |3\rangle \equiv |\downarrow_1\rangle |\uparrow_2\rangle \ |4\rangle \equiv |\downarrow_1\rangle |\downarrow_2\rangle. \tag{9.43}$$

Conversion from kets to bras occurs following all the standard rules of Hermitian conjugation applied to the states of both particles.

A larger space formed from two smaller spaces in the described manner is called in mathematics a *tensor product* of spaces. It has all the standard algebraic properties of a linear vector space discussed in Sect. 2.2, and I only need to add the distributive properties involving vectors belonging to different components of the tensor product:

$$(|e_1\rangle + |e_2\rangle) |v_1\rangle \equiv |e_1\rangle |v_1\rangle + |e_2\rangle |v_1\rangle$$

$$|e_1\rangle (|v_1\rangle + |v_2\rangle) \equiv |e_1\rangle |v_1\rangle + |e_1\rangle |v_2\rangle. \tag{9.44}$$

The inner product in the tensor space is defined as

$$(\langle e_1| \langle v_1|) (|e_2\rangle |v_2\rangle) = \langle e_1| e_2\rangle \langle v_1| v_2\rangle, \tag{9.45}$$

and it is quite obvious that this definition preserves the main property of the inner product, namely, that $\langle \beta |\alpha\rangle = \langle \alpha |\beta\rangle^*$. In the case of the two-spin system, you can, for instance, find using the notation from Eq. 9.43:

$$\langle 1| 1\rangle = \langle \uparrow_1| \uparrow_1\rangle \langle \uparrow_2| \uparrow_2\rangle = 1$$

where it is presumed that vectors $|\uparrow_{1,2}\rangle$ are normalized. You can also find that the inner products involving different vectors from the basis vanish such as

$$\langle 1| 2\rangle = \langle \uparrow_1| \uparrow_1\rangle \langle \uparrow_2| \downarrow_2\rangle = 0$$

$$\langle 1| 3\rangle = \langle \uparrow_1| \downarrow_1\rangle \langle \uparrow_2| \uparrow_2\rangle = 0.$$

In reality you have already encountered the tensor product of spaces earlier in this book, even though I never used the name. One example was the construction of states of a three-dimensional harmonic oscillator from the states of the one-dimensional oscillators.

To illustrate calculations of the inner product between vectors belonging to such a tensor product space, consider the following example.

Example 24 (Working with Vectors from a Tensor Product Space) Compute the norms of the following vectors as well as the inner product $\langle \beta |\alpha\rangle$:

$$|\alpha\rangle = (3i\,|\!\uparrow_1\rangle + 4\,|\!\downarrow_1\rangle)\,(2\,|\!\uparrow_2\rangle - i\,|\!\downarrow_2\rangle)$$
$$|\beta\rangle = (2\,|\!\uparrow_1\rangle - i\,|\!\downarrow_1\rangle)\,(2\,|\!\uparrow_2\rangle - 3\,|\!\downarrow_2\rangle)\,.$$

Solution

Since all the vectors in adjacent parenthetic expressions are kets and belong to different spaces, it is clear that I am dealing here with the tensor product of two spaces. Distribution properties, expressed by Eq. 9.44, allow me to convert these expressions into

$$|\alpha\rangle = 6i\,|\!\uparrow_1\rangle\,|\!\uparrow_2\rangle - 4i\,|\!\downarrow_1\rangle\,|\!\downarrow_2\rangle + 3\,|\!\uparrow_1\rangle\,|\!\downarrow_2\rangle + 8\,|\!\downarrow_1\rangle\,|\!\uparrow_2\rangle$$
$$|\beta\rangle = 4\,|\!\uparrow_1\rangle\,|\!\uparrow_2\rangle + 3i\,|\!\downarrow_1\rangle\,|\!\downarrow_2\rangle - 6\,|\!\uparrow_1\rangle\,|\!\downarrow_2\rangle - 2i\,|\!\downarrow_1\rangle\,|\!\uparrow_2\rangle\,.$$

Note that the order in which vectors belonging to different spaces are stacked together is completely irrelevant. Using the normalized and orthogonal basis introduced in Eq. 9.43, I can rewrite this expression as

$$|\alpha\rangle = 6i\,|1\rangle - 4i\,|4\rangle + 3\,|2\rangle + 8\,|3\rangle$$
$$|\beta\rangle = 4\,|1\rangle + 3i\,|4\rangle - 6\,|2\rangle - 2i\,|3\rangle\,.$$

Now I can compute the norms and the inner product following the standard procedure, which yields

$$\|\alpha\| = \sqrt{36 + 16 + 9 + 64} = \sqrt{125}$$
$$\|\beta\| = \sqrt{16 + 9 + 36 + 4} = \sqrt{65}$$
$$\langle\beta\,|\alpha\rangle = 4 \times 6i - 4i\,(-3i) - 3 \times 6 + 8 \times (2i) = -30 + 40i.$$

Finally, I need to introduce the rule describing how spin operators act on the vectors in the tensor product space. The rule is actually very simple: the operators affect only the states of their own particles. To illustrate this rule, consider the following example.

Example 25 (Operator Action in Tensor Spaces) For the state $|\alpha\rangle$ from the previous example, compute

(a)

$$\left(\hat{S}_z^{(1)} + \hat{S}_z^{(2)}\right)|\alpha\rangle$$

(b)

$$\left(\hat{S}_+^{(1)} + \hat{S}_+^{(2)}\right)|\alpha\rangle$$

Solution

(a)

$$\left(\hat{S}_z^{(1)} + \hat{S}_z^{(2)}\right)\left(6i\,|\!\uparrow_1\rangle\,|\!\uparrow_2\rangle - 4i\,|\!\downarrow_1\rangle\,|\!\downarrow_2\rangle + 3\,|\!\uparrow_1\rangle\,|\!\downarrow_2\rangle + 8\,|\!\downarrow_1\rangle\,|\!\uparrow_2\rangle\right) =$$

$$6i\hat{S}_z^{(1)}\,|\!\uparrow_1\rangle\,|\!\uparrow_2\rangle + 6i\,|\!\uparrow_1\rangle\,\hat{S}_z^{(2)}\,|\!\uparrow_2\rangle - 4i\hat{S}_z^{(1)}\,|\!\downarrow_1\rangle\,|\!\downarrow_2\rangle - 4i\,|\!\downarrow_1\rangle\,\hat{S}_z^{(2)}\,|\!\downarrow_2\rangle +$$

$$3\hat{S}_z^{(1)}\,|\!\uparrow_1\rangle\,|\!\downarrow_2\rangle + 3\,|\!\uparrow_1\rangle\,\hat{S}_z^{(2)}\,|\!\downarrow_2\rangle + 8\hat{S}_z^{(1)}\,|\!\downarrow_1\rangle\,|\!\uparrow_2\rangle + 8\,|\!\downarrow_1\rangle\,\hat{S}_z^{(2)}\,|\!\uparrow_2\rangle =$$

$$3i\hbar\,|\!\uparrow_1\rangle\,|\!\uparrow_2\rangle + 3i\hbar\,|\!\uparrow_1\rangle\,|\!\uparrow_2\rangle + 2i\hbar\,|\!\downarrow_1\rangle\,|\!\downarrow_2\rangle + 2i\hbar\,|\!\downarrow_1\rangle\,|\!\downarrow_2\rangle +$$

$$\frac{3\hbar}{2}\,|\!\uparrow_1\rangle\,|\!\downarrow_2\rangle - \frac{3\hbar}{2}\,|\!\uparrow_1\rangle\,|\!\downarrow_2\rangle - 4\hbar\,|\!\downarrow_1\rangle\,|\!\uparrow_2\rangle + 4\hbar\,|\!\downarrow_1\rangle\,|\!\uparrow_2\rangle =$$

$$6i\hbar\,|\!\uparrow_1\rangle\,|\!\uparrow_2\rangle + 4i\hbar\,|\!\downarrow_1\rangle\,|\!\downarrow_2\rangle$$

(b)

$$\left(\hat{S}_+^{(1)} + \hat{S}_+^{(2)}\right)\left(6i\,|\!\uparrow_1\rangle\,|\!\uparrow_2\rangle - 4i\,|\!\downarrow_1\rangle\,|\!\downarrow_2\rangle + 3\,|\!\uparrow_1\rangle\,|\!\downarrow_2\rangle + 8\,|\!\downarrow_1\rangle\,|\!\uparrow_2\rangle\right) =$$

$$6i\hat{S}_+^{(1)}\,|\!\uparrow_1\rangle\,|\!\uparrow_2\rangle + 6i\,|\!\uparrow_1\rangle\,\hat{S}_+^{(2)}\,|\!\uparrow_2\rangle - 4i\hat{S}_+^{(1)}\,|\!\downarrow_1\rangle\,|\!\downarrow_2\rangle - 4i\,|\!\downarrow_1\rangle\,\hat{S}_+^{(2)}\,|\!\downarrow_2\rangle +$$

$$3\hat{S}_+^{(1)}\,|\!\uparrow_1\rangle\,|\!\downarrow_2\rangle + 3\,|\!\uparrow_1\rangle\,\hat{S}_+^{(2)}\,|\!\downarrow_2\rangle + 8\hat{S}_+^{(1)}\,|\!\downarrow_1\rangle\,|\!\uparrow_2\rangle + 8\,|\!\downarrow_1\rangle\,\hat{S}_+^{(2)}\,|\!\uparrow_2\rangle =$$

$$-4i\hbar\,|\!\uparrow_1\rangle\,|\!\downarrow_2\rangle - 4i\hbar\,|\!\downarrow_1\rangle\,|\!\uparrow_2\rangle + 3\hbar\,|\!\uparrow_1\rangle\,|\!\uparrow_2\rangle + 8\hbar\,|\!\uparrow_1\rangle\,|\!\uparrow_2\rangle =$$

$$11\hbar\,|\!\uparrow_1\rangle\,|\!\uparrow_2\rangle - 4i\hbar\left(|\!\uparrow_1\rangle\,|\!\downarrow_2\rangle + |\!\downarrow_1\rangle\,|\!\uparrow_2\rangle\right)$$

9.4.2 Operator of the Total Spin

The concept of the tensor product gives an exact mathematical meaning to Eq. 9.42 and the "plus" sign in it as illustrated by the previous example. Indeed, if each operator appearing on the right-hand side of this equation is defined to act in a common space of two-particle states, then the plus sign generates the regular operator sum as defined in earlier chapters of this book.

Now I can tackle the main question: what are the eigenvalues and eigenvectors of the components of the total spin operator defined by Eq. 9.42, and of its square $\left(\hat{S}^{tp}\right)^2 = \left(\hat{S}^{(1)} + \hat{S}^{(2)}\right)^2$? When discussing any systems of operators, the first question you must be concerned with is the commutation relations between these operators. The first commutator that needs to be dealt with is between operators $\hat{S}^{(1)}$ and $\hat{S}^{(2)}$, and it is quite obvious that any two components of these operators commute, i.e.,

$$\left[\hat{S}_i^{(1)}, \hat{S}_j^{(2)}\right] = 0$$

for all i, j taking values x, y, and z. Indeed, since $\hat{S}_i^{(1)}$ only affects the states of the particle 1, and $\hat{S}_i^{(2)}$ acts only on the states of particle 2, an order in which these operators are applied is irrelevant. Now it is quite easy to establish that all commutation relations for the components of operator $\hat{\mathbf{S}}^{tp}$ and its square are exactly the same as for any operator of angular momentum. This justifies the claim that there exists a system of vectors, $|S, M\rangle$, which are common eigenvectors of one of the components of $\hat{\mathbf{S}}^{tp}$, usually chosen to be $\hat{S}_z^{tp}$, and of the operator $\left(\hat{\mathbf{S}}^{tp}\right)^2$ characterized by two numbers M and S such that

$$\hat{S}_z^{tp} |S, M\rangle = \hbar M |S, M\rangle$$

$$\left(\hat{\mathbf{S}}^{tp}\right)^2 |S, M\rangle = \hbar^2 S \left(S + 1\right) |S, M\rangle.$$

It also can be claimed that $|M| \leq S$ and that these numbers take integer of half-integer values. What is missing at this point is the information about the actual values that S can take and its relation to eigenvalues of the spin operators of the individual particles. Also, one would like to know about the connection between eigenvectors of $\hat{S}_z^{tp}$ and $\left(\hat{\mathbf{S}}^{tp}\right)^2$ and their single-particle counterparts. To answer these questions, I am going to generate matrix representations of operators $\hat{S}_z^{tp}$ and $\left(\hat{\mathbf{S}}^{tp}\right)^2$, using the basis vectors defined in Eq. 9.43. I will start with operator $\hat{S}_z^{tp}$. The application of this operator to the basis vectors yields (see examples above)

$$\hat{S}_z^{tp} |\uparrow_1\rangle |\uparrow_2\rangle = \hbar |\uparrow_1\rangle |\uparrow_2\rangle \tag{9.46}$$

$$\hat{S}_z^{tp} |\uparrow_1\rangle |\downarrow_2\rangle = 0 \tag{9.47}$$

$$\hat{S}_z^{tp} |\downarrow_1\rangle |\uparrow_2\rangle = 0 \tag{9.48}$$

$$\hat{S}_z^{tp} |\downarrow_1\rangle |\downarrow_2\rangle = -\hbar |\downarrow_1\rangle |\downarrow_2\rangle. \tag{9.49}$$

These results indicate that the basis vectors defined in Eq. 9.43 are also eigenvectors of the operator $\hat{S}_z^{tp}$ with eigenvalues $\pm\hbar$ and a double-degenerate eigenvalue 0. Thus, the matrix of this operator in this basis is a diagonal 4×4 matrix:

$$\hat{S}_z^{tp} = \hbar \begin{bmatrix} 1 & 0 & 0 & 0 \\ 0 & 0 & 0 & 0 \\ 0 & 0 & 0 & 0 \\ 0 & 0 & 0 & -1 \end{bmatrix}$$

where I have positioned the matrix elements in accord with the numeration of eigenvectors introduced in Eq. 9.43. For instance, the right-hand side of Eq. 9.46, where operator $\hat{S}_z^{tp}$ acts on the first of the basis vectors, represents the first column of the matrix, which contains a single non-zero element, the right-hand side of Eq. 9.47 yields the second column, where all elements are zeroes, and so on and so forth.

Operator $\left(\hat{S}^{tp}\right)^2$ requires more work. First, let me rewrite it in terms of the particle's operators $\hat{S}^{(1)}$ and $\hat{S}^{(2)}$:

$$\left(\hat{S}^{tp}\right)^2 = \left(\hat{S}^{(1)} + \hat{S}^{(2)}\right)^2 = \left(\hat{S}^{(1)}\right)^2 + \left(\hat{S}^{(2)}\right)^2 + 2\hat{S}^{(1)} \cdot \hat{S}^{(2)} =$$

$$\left(\hat{S}^{(1)}\right)^2 + \left(\hat{S}^{(2)}\right)^2 + 2\left(\hat{S}_x^{(1)}\hat{S}_x^{(2)} + \hat{S}_y^{(1)}\hat{S}_y^{(2)} + \hat{S}_z^{(1)}\hat{S}_z^{(2)}\right) =$$

$$\left(\hat{S}^{(1)}\right)^2 + \left(\hat{S}^{(2)}\right)^2 + 2\hat{S}_z^{(1)}\hat{S}_z^{(2)} +$$

$$2\left[\frac{\hat{S}_+^{(1)} + \hat{S}_-^{(1)}}{2}\frac{\hat{S}_+^{(2)} + \hat{S}_-^{(2)}}{2} + \frac{\hat{S}_+^{(1)} - \hat{S}_-^{(1)}}{2i}\frac{\hat{S}_+^{(2)} - \hat{S}_-^{(2)}}{2i}\right] =$$

$$\left(\hat{S}^{(1)}\right)^2 + \left(\hat{S}^{(2)}\right)^2 + 2\hat{S}_z^{(1)}\hat{S}_z^{(2)} + \hat{S}_+^{(1)}\hat{S}_-^{(2)} + \hat{S}_-^{(1)}\hat{S}_+^{(2)}$$

where I replaced the x- and y-components of the spin operator by ladder operators defined in Eqs. 3.59 and 3.60 adapted for spin operators. The last expression is perfectly suited for generating the matrix of $\left(\hat{S}^{tp}\right)^2$. Applying this operator to each of the basis vectors, I can again simply read out the columns of this matrix:

$$\left(\hat{S}^{tp}\right)^2 |1\rangle =$$

$$\left[\left(\hat{S}^{(1)}\right)^2 + \left(\hat{S}^{(2)}\right)^2 + 2\hat{S}_z^{(1)}\hat{S}_z^{(2)} + \hat{S}_+^{(1)}\hat{S}_-^{(2)} + \hat{S}_-^{(1)}\hat{S}_+^{(2)}\right] |\uparrow_1\rangle |\uparrow_2\rangle = \quad (9.50)$$

$$\frac{3}{4}\hbar^2 |\uparrow_1\rangle |\uparrow_2\rangle + \frac{3}{4}\hbar^2 |\uparrow_1\rangle |\uparrow_2\rangle + \frac{1}{2}\hbar^2 |\uparrow_1\rangle |\uparrow_2\rangle = 2\hbar^2 |\uparrow_1\rangle |\uparrow_2\rangle \equiv 2\hbar^2 |1\rangle .$$

The ladder operators do not contribute to the final result because the raising operator applied to the spin-up vector yields zero. All other terms in this expression follow from the standard properties of the spin operators. Continue

$$\left(\hat{S}^{tp}\right)^2 |2\rangle =$$

$$\left[\left(\hat{S}^{(1)}\right)^2 + \left(\hat{S}^{(2)}\right)^2 + 2\hat{S}_z^{(1)}\hat{S}_z^{(2)} + \hat{S}_+^{(1)}\hat{S}_-^{(2)} + \hat{S}_-^{(1)}\hat{S}_+^{(2)}\right] |\uparrow_1\rangle |\downarrow_2\rangle =$$

$$\frac{3}{4}\hbar^2 |\uparrow_1\rangle |\downarrow_2\rangle + \frac{3}{4}\hbar^2 |\uparrow_1\rangle |\downarrow_2\rangle - \frac{1}{2}\hbar^2 |\uparrow_1\rangle |\downarrow_2\rangle + \hbar^2 |\downarrow_1\rangle |\uparrow_2\rangle = \quad (9.51)$$

$$\hbar^2 |\uparrow_1\rangle |\downarrow_2\rangle + \hbar^2 |\downarrow_1\rangle |\uparrow_2\rangle \equiv \hbar^2 |2\rangle + \hbar^2 |3\rangle$$

where the ladder operators in term $\hat{S}_-^{(1)}\hat{S}_+^{(2)}$ are responsible for the non-zero contribution and where the spin of each particle becomes upside down. And again

$$\left(\hat{s}^{tp}\right)^{2}|3\rangle =$$

$$\left[\left(\hat{s}^{(1)}\right)^{2}+\left(\hat{s}^{(2)}\right)^{2}+2\hat{S}_{z}^{(1)}\hat{S}_{z}^{(2)}+\hat{S}_{+}^{(1)}\hat{S}_{-}^{(2)}+\hat{S}_{-}^{(1)}\hat{S}_{+}^{(2)}\right]|\downarrow_{1}\rangle\,|\uparrow_{2}\rangle =$$

$$\frac{3}{4}\hbar^{2}|\downarrow_{1}\rangle\,|\uparrow_{2}\rangle+\frac{3}{4}\hbar^{2}|\downarrow_{1}\rangle\,|\uparrow_{2}\rangle-\frac{1}{2}\hbar^{2}|\downarrow_{1}\rangle\,|\uparrow_{2}\rangle+\hbar^{2}|\uparrow_{1}\rangle\,|\downarrow_{2}\rangle = \qquad (9.52)$$

$$\hbar^{2}|\uparrow_{1}\rangle\,|\downarrow_{2}\rangle+\hbar^{2}|\downarrow_{1}\rangle\,|\uparrow_{2}\rangle \equiv \hbar^{2}|3\rangle+\hbar^{2}|2\rangle$$

where the inversion of the spins in the last term is due to operators $\hat{S}_{+}^{(1)}\hat{S}_{-}^{(2)}$. Finally

$$\left(\hat{s}^{tp}\right)^{2}|4\rangle =$$

$$\left[\left(\hat{s}^{(1)}\right)^{2}+\left(\hat{s}^{(2)}\right)^{2}+2\hat{S}_{z}^{(1)}\hat{S}_{z}^{(2)}+\hat{S}_{+}^{(1)}\hat{S}_{-}^{(2)}+\hat{S}_{-}^{(1)}\hat{S}_{+}^{(2)}\right]|\downarrow_{1}\rangle\,|\downarrow_{2}\rangle = \qquad (9.53)$$

$$\frac{3}{4}\hbar^{2}|\downarrow_{1}\rangle\,|\downarrow_{2}\rangle+\frac{3}{4}\hbar^{2}|\downarrow_{1}\rangle\,|\downarrow_{2}\rangle+\frac{1}{2}\hbar^{2}|\downarrow_{1}\rangle\,|\downarrow_{2}\rangle = 2\hbar^{2}|\downarrow_{1}\rangle\,|\downarrow_{2}\rangle \equiv 2\hbar^{2}|4\rangle.$$

Reading out columns 1 through 4 from Eqs. 9.50 to 9.53 correspondingly, I generate the desired matrix:

$$\left(\hat{s}^{tp}\right)_{ij}^{2}=\hbar^{2}\begin{bmatrix}2&0&0&0\\0&1&1&0\\0&1&1&0\\0&0&0&2\end{bmatrix}.$$

What is left now is to find its eigenvalues and eigenvectors, i.e., to solve the eigenvalue problem:

$$\hbar^{2}\begin{bmatrix}2&0&0&0\\0&1&1&0\\0&1&1&0\\0&0&0&2\end{bmatrix}\begin{bmatrix}a_{1}\\a_{2}\\a_{3}\\a_{4}\end{bmatrix}=\lambda\hbar^{2}\begin{bmatrix}a_{1}\\a_{2}\\a_{3}\\a_{4}\end{bmatrix}.$$

Two eigenvalues can be found just by looking at Eqs. 9.50 and 9.53, which indicate that vectors $|1\rangle$ and $|4\rangle$ are eigenvectors of this matrix with eigenvalues $2\hbar^{2}$. This circumstance is reflected in the structure of the matrix, where the first and fourth rows as well as the first and fourth columns contain single non-zero elements. Such matrices are known as block-diagonal, and what makes them special is that each block can be considered independently of the others and treated accordingly. For instance, the equations for a_{1} and a_{4} will not contain any other coefficients, while equations for elements a_{2} and a_{3} will only contain these two elements. Since I already know that solutions with $a_{1}=1$, $a_{2,3,4}=0$ and $a_{4}=1$, $a_{1,2,3}=0$ are eigenvectors corresponding to $\lambda = 2\hbar^2$, I only need to deal with the remaining two coefficients a_2 and a_3 satisfying equations

$$a_2 + a_3 = \lambda a_2$$
$$a_2 + a_3 = \lambda a_3.$$

It immediately follows from this system that either $a_2 = a_3$ or $\lambda = 0$. In the former case, I have

$$\lambda = 2,$$

while the latter one gives me

$$a_2 = -a_3.$$

Thus, I end up once again with eigenvalue $2\hbar^2$, but now it belongs to the eigenvector

$$\frac{1}{\sqrt{2}} (|2\rangle + |3\rangle) = \frac{1}{\sqrt{2}} (|\uparrow_1\rangle \, |\downarrow_2\rangle + |\downarrow_1\rangle \, |\uparrow_2\rangle)$$

where I set $a_2 = a_3 = 1/\sqrt{2}$ to make this vector normalized. I also got a new eigenvalue equal to zero with eigenvector

$$\frac{1}{\sqrt{2}} (|2\rangle - |3\rangle) = \frac{1}{\sqrt{2}} (|\uparrow_1\rangle \, |\downarrow_2\rangle - |\downarrow_1\rangle \, |\uparrow_2\rangle).$$

Recalling that eigenvalues of the $\left(\hat{S}^{tp}\right)^2$ must have the form $\hbar^2 S(S+1)$, I can immediately deduce that eigenvalue $2\hbar^2$ corresponds to $S = 1$, while eigenvalue zero, obviously, corresponds to $S = 0$.

It is time to put all these results together. Here is what I have: a triple degenerate eigenvalue characterized by spin $S = 1$ and three eigenvectors

$$|1, 1\rangle = |\uparrow_1\rangle \, |\uparrow_2\rangle$$

$$|1, 0\rangle = \frac{1}{\sqrt{2}} (|\uparrow_1\rangle \, |\downarrow_2\rangle + |\downarrow_1\rangle \, |\uparrow_2\rangle) \tag{9.54}$$

$$|1, -1\rangle = |\downarrow_1\rangle \, |\downarrow_2\rangle$$

and a single non-degenerate eigenvalue corresponding to $S = 0$ with eigenvector

$$|0, 0\rangle = \frac{1}{\sqrt{2}} (|\uparrow_1\rangle \, |\downarrow_2\rangle - |\downarrow_1\rangle \, |\uparrow_2\rangle) \tag{9.55}$$

attached to it. Notations used for these eigenvectors follow the traditional scheme $|S, M\rangle$ and reflect the facts that all three eigenvectors in Eq. 9.54 are simultaneously eigenvectors of operator $\hat{S}_z^{tp}$ with corresponding quantum numbers $M = 1, M = 0$, and $M = -1$, while a single eigenvector in Eq. 9.55 is also an eigenvector of $\hat{S}_z^{tp}$ corresponding to $M = 0$. You might want to pay attention to the fact that both superposition eigenvectors $|2, 0\rangle$ and $|0, 0\rangle$ are linear combinations of the eigenvectors of $\hat{S}_z^{tp}$ established in Eqs. 9.47 and 9.48 belonging to a double-degenerate eigenvalue 0 of $\hat{S}_z^{tp}$, which reflects the general notion that linear combinations of degenerate eigenvectors are also eigenvectors belonging to the same eigenvalue. The particular combinations appearing in Eqs. 9.54 and 9.55 ensure that these vectors are simultaneously eigenvectors of the operator $\left(\hat{S}^{tp}\right)^2$. The results presented in these equations also reflect the general property of the angular momentum operators: the value of quantum number S determines the maximum and minimum allowed values of the second quantum number M and, respectively, the total number $2S + 1$ of eigenvectors belonging to the given eigenvalue of $\left(\hat{S}^{tp}\right)^2$. Indeed, for $S = 1$, we have three vectors with M ranging from -1 to 1, while for $S = 0$, there exists a single vector with $M = 0$. This situation is often described by saying that the system of two-spin $1/2$ particles can be in two states characterized by the total spin equal to one or zero. The former is called a *triplet* state reflecting the existence of three distinct states with the same S and different magnetic numbers M, and the latter is called a *singlet* for obvious enough reasons. People also often say that in the triplet state, the spins of the particles are parallel to each other, while in the singlet state, they are antiparallel, but this is highly misleading. Even leaving aside the obvious quantum mechanical fact that the direction of spin in quantum mechanics is not defined because only one component of the vector can have a definite value in a given state, parallel or antiparallel can refer only to the sign of the z-component of the spin determined by the value of M. As we have just seen, this number can be equal to zero, reflecting the "antiparallel" orientation of the particle's spins, when the particles are either in the $S = 1$ or $S = 0$ state. Therefore, more accurate verbal description of the situation (if you really need one) may sound like this: in the triplet spin states, the particle's spins can be either parallel or antiparallel, while in the singlet state, they can only be antiparallel.

To complete this discussion, let me direct your attention to another interesting difference between triplet and singlet states. The former are symmetric with respect to the *exchange* of the particles, while the latter are antisymmetric. What it means is that if you replace particle indexes 1 and 2 in Eqs. 9.54 and 9.55 (exchange the particles one and two), the states described by the former equation do not change, while the singlet state described by the latter equation changes its sign. The operation of the particle's exchange reflects the classical idea that you can somehow distinguish between the particles marking them as one and two and then swap them by placing particle one in the state of particle two and vice versa. In quantum mechanics two electrons are not really distinguishable, and, therefore, the swapping operation shouldn't change the properties of the system. This topic will be discussed in much more detail in Chap. 11 devoted to quantum mechanics of many identical particles. Here I just want to mention, giving you a brief preview of what is coming, that the symmetry and antisymmetry of the spin states of the two-particle system are a reflection of quantum indistinguishability of electrons.

9.5   Operator of Total Angular Momentum

9.5.1   Combining Orbital and Spin Degrees of Freedom

When discussing the model of spin $1/2$ or addition of two such spins, I intentionally ignored the fact that the spin is "attached" to a particle, which can be involved in all kinds of crazy things such as being a part of an atom or rushing through a piece of metal delivering an electron current. At the same time, such phenomena as resonant tunneling or hydrogen atom in the previous chapters were treated with utter ignorance of the fact that in addition to "regular" observables, such as position or momentum, electrons also carry around their spin, which is as unalienable as their mass or charge. Now the time has come to design a formalism allowing to treat spin and orbital properties[2] of the electrons (and other particles with spin) together.

First of all, one needs to recognize that the spinors and orbital vectors are completely different animals and inhabit different habitats. For instance, while you can represent eigenvectors of momentum and angular momentum in the same, say, position representation or express them in terms of each other, it is impossible to construct a position representation for the eigenvectors of the spin operators or present momentum eigenvectors as a linear combination of spinors. Accordingly, operators acting on orbital vectors do not affect spinors, and spin operators are indifferent to vectors representing orbital states. One of the trivial consequences of this is, of course, that orbital and spin operators always commute. Giving these statements a bit of a thought, you can notice a certain similarity with the just discussed two-spin problem, where we also had to deal with vectors belonging to two unrelated spaces and being acted upon only by their "native" operators. That situation was handled by combining spinors representing spin states of different particles into a common space formed as a tensor product of the spaces of each individual spin. Similarly, spin and orbital spaces of a single particle can also be combined into a tensor product space by stacking together all combinations of the basis vectors from both spaces. Assuming that the orbital space is described by some discrete basis $\left| q_k^{(1)}, q_m^{(2)}, \cdots q_p^{(N_{max})} \right\rangle$ based on a set of mutually consistent observables, a typical basis vector in a compound tensor product space can be made to look something like this:

$$\left| q_k^{(1)}, q_m^{(2)}, \cdots q_p^{(N_{max})} \right\rangle \left| m_{s_i} \right\rangle , \tag{9.56}$$

---

[2] By orbital properties I understand all those properties of the particle that can be described using quantum states related to position or momentum operators or a combination thereof. In what follows I will call these states and vectors representing them orbital states or orbital vectors.

where $|m_{s_i}\rangle$ is a basis spinor. Since there are only two of those, the dimension of the combined space is two times the dimensionality of the orbital space. Indeed, attaching the spin state to each orbital basis vector $\left| q_k^{(1)}, q_m^{(2)}, \cdots q_p^{(N_{max})} \right\rangle$, I am generating two new basis vectors:

$$\left| q_k^{(1)}, q_m^{(2)}, \cdots q_p^{(N_{max})} \right\rangle |1/2\rangle$$

and

$$\left| q_k^{(1)}, q_m^{(2)}, \cdots q_p^{(N_{max})} \right\rangle |-1/2\rangle,$$

or, if you prefer,

$$\left| q_k^{(1)}, q_m^{(2)}, \cdots q_p^{(N_{max})} \right\rangle |\uparrow\rangle$$

and

$$\left| q_k^{(1)}, q_m^{(2)}, \cdots q_p^{(N_{max})} \right\rangle |\downarrow\rangle.$$

Sometimes the indicator of a spin state is put inside a single ket or bra vector together with the signifiers of all other observables:

$$\left| q_k^{(1)}, q_m^{(2)}, \cdots q_p^{(N_{max})} \right\rangle |m_{s_i}\rangle \equiv \left| q_k^{(1)}, q_m^{(2)}, \cdots q_p^{(N_{max})}, m_{s_i} \right\rangle, \tag{9.57}$$

but this notation hides the critical difference between the spin and orbital observables and makes some calculations less intuitive, so I would prefer using the notation of Eq. 9.56 most of the time. Nevertheless, sometimes it might be appropriate to use the simplified notation of Eq. 9.57, and if you notice me doing it, do not start throwing stones—this is just a notation, chosen based on convenience and a moment's expedience.

An arbitrary vector $|\chi\rangle$ residing in the tensor product space can be presented as

$$|\chi\rangle = \sum_{km,\cdots p} a_{km\cdots p;\uparrow} \left| q_k^{(1)}, q_m^{(2)}, \cdots q_p^{(N_{max})} \right\rangle |\uparrow\rangle +$$

$$\sum_{km,\cdots p} a_{km\cdots p;\downarrow} \left| q_k^{(1)}, q_m^{(2)}, \cdots q_p^{(N_{max})} \right\rangle |\downarrow\rangle. \tag{9.58}$$

Expansion coefficients $a_{km\cdots p;\uparrow}$ now define the probability $\left| a_{km\cdots p;\uparrow} \right|^2$ that the measurement of the mutually consistent observables including a component of the spin will yield values $k.m \cdots p$ for regular observables and $\hbar/2$ for the spin's component. The set of coefficients $a_{km\cdots p;\downarrow}$ defines the probability $\left| a_{km\cdots p;\downarrow} \right|^2$ that the observation will produce the same values of all the orbital observables and value $-\hbar/2$ for the spin. The sum of probabilities

$$p_{km\cdots p} = \left| a_{km\cdots p;\uparrow} \right|^2 + \left| a_{km\cdots p;\downarrow} \right|^2$$

yields the probability to observe the given values of the observables provided that the spin is not measured, while the sums

$$p_\uparrow = \sum_{km,\cdots p} \left| a_{km\cdots p;\uparrow} \right|^2$$

or

$$p_\downarrow = \sum_{km,\cdots p} \left| a_{km\cdots p;\downarrow} \right|^2$$

generate probabilities of getting values of the spin component $\hbar/2$ or $-\hbar/2$, respectively, regardless of the values of other observables. Finally, the normalization condition for the expansion coefficients must now include the summation over all available variables:

$$\sum_{km,\cdots p} \left[\left| a_{km\cdots p;\uparrow} \right|^2 + \left| a_{km\cdots p;\downarrow} \right|^2 \right] = 1. \tag{9.59}$$

Equations 9.56–9.58 are written under the assumption that the basis in the orbital space is discrete. However, they can be easily adapted to representations in a continuous basis by replacing all the sums with integrals and probabilities with corresponding probability densities. For instance, in the basis of the position eigenvectors $|r\rangle$, Eqs. 9.56 and 9.58 become $|r\rangle |m_s\rangle$ and

$$|\chi\rangle = \int d^3 r \psi_\uparrow(r) |r\rangle |\uparrow\rangle + \int d^3 r \psi_\downarrow(r) |r\rangle |\downarrow\rangle . \tag{9.60}$$

$|\psi_{m_s}(r)|^2$ now gives the position probability density for the corresponding spin state $|m_s\rangle$, $|\psi_\uparrow(r)|^2 + |\psi_\downarrow(r)|^2$ yields the same, but when the spin state is not important, and $\int d^3 r |\psi_{m_s}(r)|^2$ generates the probability of finding the particle in the spin state $|m_s\rangle$. The normalization Eq. 9.59 now becomes

$$\int d^3 r \left[|\psi_\uparrow(r)|^2 + |\psi_\downarrow(r)|^2 \right] = 1. \tag{9.61}$$

One can generate particular representations for the generic vectors $|\chi\rangle$ by choosing specific bases for the orbital and spinor components of the states. One of the most popular choices is to use the position representation for the orbital vectors and eigenvectors of operator $\hat{S}_z$ for the spinor component. The respective representation is generated by premultiplying $\left|q_k^{(1)}, q_m^{(2)}, \cdots q_p^{(N_{max})}\right\rangle$ by $\langle r |$, which yields

$$\psi_{q_k^{(1)}, q_m^{(2)}, \cdots q_p^{(N_{max})}}(r) = \left\langle r \left| q_k^{(1)}, q_m^{(2)}, \cdots q_p^{(N_{max})} \right\rangle \right. ,$$

and by replacing $|m_{s_i}\rangle$ with a corresponding two-component column $\begin{bmatrix} 1 \\ 0 \end{bmatrix}$ for the spin-up (or $+1/2$) state and $\begin{bmatrix} 0 \\ 1 \end{bmatrix}$ for the spin-down (or $-1/2$) state. Since the coordinate representation for the orbital states is almost always used in conjunction with the representation of spinors in the basis of the eigenvectors of $\hat{S}_z$ operator, I will call this form the coordinate–spinor representation. Then the combined spin–orbital state takes the form

$$\psi_{q_k^{(1)}, q_m^{(2)}, \cdots q_p^{(N_{max})}}(r) \begin{bmatrix} 1 \\ 0 \end{bmatrix}$$

or

$$\psi_{q_k^{(1)}, q_m^{(2)}, \cdots q_p^{(N_{max})}}(r) \begin{bmatrix} 0 \\ 1 \end{bmatrix} ,$$

depending on the chosen spin state. The generic state vector represented by Eq. 9.58 in this representation becomes (I will keep the same notation for the abstract vector and its coordinate–spinor representation to avoid introducing new symbols, when it is not really necessary and should not cause any confusion)

$$|\chi\rangle = \sum_{km, \cdots p} a_{km \cdots p; \uparrow} \psi_{q_k^{(1)}, q_m^{(2)}, \cdots q_p^{(N_{max})}}(r) \begin{bmatrix} 1 \\ 0 \end{bmatrix} +$$

$$\sum_{km, \cdots p} a_{km \cdots p; \downarrow} \psi_{q_k^{(1)}, q_m^{(2)}, \cdots q_p^{(N_{max})}}(r) \begin{bmatrix} 0 \\ 1 \end{bmatrix} =$$

$$\Psi_\uparrow(r, t) \begin{bmatrix} 1 \\ 0 \end{bmatrix} + \Psi_\downarrow(r, t) \begin{bmatrix} 0 \\ 1 \end{bmatrix} = \begin{bmatrix} \Psi_\uparrow(r, t) \\ \Psi_\downarrow(r, t) \end{bmatrix} , \qquad (9.62)$$

where

$$\Psi_\uparrow(r, t) = \sum_{km, \cdots p} a_{km \cdots p; \uparrow}(t) \psi_{q_k^{(1)}, q_m^{(2)}, \cdots q_p^{(N_{max})}}(r)$$

$$\Psi_\downarrow(r, t) = \sum_{km, \cdots p} a_{km \cdots p; \downarrow}(t) \psi_{q_k^{(1)}, q_m^{(2)}, \cdots q_p^{(N_{max})}}(r) \qquad (9.63)$$

are the orbital wave functions corresponding to spin-up and spin-down states correspondingly. These functions appear in Eq. 9.63 as linear combinations of the initial basis vectors transformed in their position representations. Obviously, $\Psi_\uparrow(r, t)$ and $\Psi_\downarrow(r, t)$ in these expressions are the same functions, which appear in Eq. 9.60 presenting expansion of an abstract vector $|\chi\rangle$ in the basis of position eigenvectors.

Any combination of orbital and spin operators act on vectors defined by Eq. 9.58 or 9.62 following a simple rule: orbital operators act on orbital component of the vector, and spin operators affect only its spin component. To illustrate this point, consider the following example.

Example 26 (Using Operators of Orbital and Spin Angular Momentum.) Consider the following vector representing a state of an electron in a hydrogen atom:

$$|\alpha\rangle = \frac{2}{3}|2, 1, -1\rangle |\uparrow\rangle + \frac{1}{3}|1, 0, 0\rangle |\downarrow\rangle - \frac{1}{3}|2, 0, 0\rangle |\uparrow\rangle + \frac{1}{\sqrt{3}}|2, 1, 1\rangle |\downarrow\rangle,$$

where the orbital portion of the state follows the standard notation $|n, l, m\rangle$. Compute the following expressions:

1. $\langle\alpha| \hat{H} |\alpha\rangle$, where $\hat{H}$ is the Hamiltonian of a hydrogen atom, Eq. 8.6.
2. $\left(\hat{L}_+\hat{S}_- + \hat{L}_-\hat{S}_+\right) |\alpha\rangle$.
3. $\left(\hat{L}_z + \hat{S}_z\right) |\alpha\rangle$.
4. Write down vector $|\alpha\rangle$ in the coordinate–spinor representation.

Solution

1. I begin by computing

$$\hat{H} |\alpha\rangle = -\frac{2}{3}\frac{E_1}{4}|2, 1, -1\rangle |\uparrow\rangle - \frac{1}{3}E_1 |1, 0, 0\rangle |\downarrow\rangle + \frac{1}{3}\frac{E_1}{4}|2, 0, 0\rangle |\uparrow\rangle$$
$$-\frac{1}{\sqrt{3}}\frac{E_1}{4}|2, 1, 1\rangle |\downarrow\rangle,$$

where $-E_1$ is the hydrogen ground state energy. Now I find

$$\langle\alpha| \hat{H} |\alpha\rangle = -\frac{E_1}{9} - \frac{E_1}{9} - \frac{E_1}{36} - \frac{E_1}{12} = -\frac{E_1}{3},$$

where I took into account that all terms in the expression above remain mutually orthogonal, so that all cross-product terms in the inner product vanish. The spin components of the state are not affected by the Hamiltonian because it does not contain any spin operators.

2.

$$\left(\hat{L}_+\hat{S}_- + \hat{L}_-\hat{S}_+\right)|\alpha\rangle = \frac{2}{3}\sqrt{2}\hbar^2\,|2,1,0\rangle\,|\downarrow\rangle + \frac{1}{\sqrt{3}}\sqrt{2}\hbar^2\,|2,1,0\rangle\,|\uparrow\rangle =$$

$$\sqrt{\frac{2}{3}}\hbar^2\,|2,1,0\rangle\left(\frac{2}{\sqrt{3}}|\downarrow\rangle + |\uparrow\rangle\right),$$

where I applied orbital and spin ladder operators separately to corresponding orbital and spin portions of the vectors using correspondingly Eqs. 3.75, 3.76, 5.104, and 5.106. In particular I found that

$$\hat{L}_+\hat{S}_-\,|2,1,0\rangle\,|\downarrow\rangle = \hat{L}_+\,|2,1,0\rangle\,\hat{S}_-\,|\downarrow\rangle = 0$$

as well as that

$$\hat{L}_-\hat{S}_+\,|2,1,0\rangle\,|\uparrow\rangle = \hat{L}_-\,|2,1,0\rangle\,\hat{S}_+\,|\uparrow\rangle = 0.$$

3.

$$\left(\hat{L}_z + \hat{S}_z\right)|\alpha\rangle = -\hbar\frac{2}{3}\,|2,1,-1\rangle\,|\uparrow\rangle + \frac{2}{3}\frac{\hbar}{2}\,|2,1,-1\rangle\,|\uparrow\rangle -$$

$$\frac{1}{3}\frac{\hbar}{2}\,|1,0,0\rangle\,|\downarrow\rangle - \frac{1}{3}\frac{\hbar}{2}\,|2,0,0\rangle\,|\uparrow\rangle + \frac{1}{\sqrt{3}}\hbar\,|2,1,1\rangle\,|\downarrow\rangle - \frac{1}{\sqrt{3}}\frac{\hbar}{2}\,|2,1,1\rangle\,|\downarrow\rangle =$$

$$\frac{\hbar}{2}\left[-\frac{2}{3}\,|2,1,-1\rangle\,|\uparrow\rangle - \frac{1}{3}\,|1,0,0\rangle\,|\downarrow\rangle - \frac{1}{3}\,|2,0,0\rangle\,|\uparrow\rangle + \frac{1}{\sqrt{3}}\,|2,1,1\rangle\,|\downarrow\rangle\right]$$

4. A coordinate–spinor representation of vector $|\alpha\rangle$ looks like this:

$$\begin{bmatrix} \frac{2}{3}R_{21}(r)Y_1^{-1}(\theta,\varphi) - \frac{1}{3\sqrt{4\pi}}R_{20}(r) \\ \frac{1}{3\sqrt{4\pi}}R_{10}(r) + \frac{1}{\sqrt{3}}R_{21}(r)Y_1^1(\theta,\varphi) \end{bmatrix}.$$

If $\Psi_\uparrow(r,t)$ and $\Psi_\downarrow(r,t)$ can be written down as

$$\Psi_\uparrow(r,t) = a_1(t)\psi(r,t); \quad \Psi_\downarrow(r,t) = a_2(t)\psi(r,t), \tag{9.64}$$

Eq. 9.62 becomes

$$|\chi\rangle = \psi(r,t)\begin{bmatrix} a_1(t) \\ a_2(t) \end{bmatrix}. \tag{9.65}$$

resulting in the separation of spin and orbital components of the state. The spin and orbital properties of the particle in such a state are completely independent of each other, and changing one of them wouldn't affect the other. In a more generic case, when $\Psi_\uparrow(r)$ and $\Psi_\downarrow(r)$ are two different functions, the orbital state of the particle depends on its spin state and vice versa. This interdependence is called "spin–orbit coupling" and is responsible for many important phenomena. Some of them are old, known for a century, while others have been discovered only recently. For instance, spin–orbit interaction is responsible for the fine structure of atomic spectra (an old phenomenon known from the earlier days of quantum mechanics), but it also gave birth to the entire new "hot" research area in contemporary semiconductor physics known as *spintronics*. Researchers working in this field seek to control the flow of electrons using their spin as a steering wheel and also to control the orientation of an electron's spin by affecting its electric current. I will talk more about spin–orbit coupling and its effect on atomic spectra in Chap. 14, but for the spintronics effects, you will have to consult a more specialized book.

While the abstract form of the Schrödinger equation

$$i\hbar \frac{\partial |\chi\rangle}{\partial t} = \hat{H}|\chi\rangle$$

stays the same even when the spin and orbital degrees of freedom are combined, its position representation, which is frequently used for practical calculations, needs to be modified. Indeed, in the representation described by Eq. 9.62, a state of a particle is described by two wave functions corresponding to two different spin states. Respectively, a single Schrödinger equation becomes a system of two equations, whose form depends on the interactions included in the Hamiltonian. To find the explicit form of these equations, you will need to convert operator $\hat{H}$ into the combined position–spinor representation. This can be done independently for the orbital and spin portions of the Hamiltonian with the result, which can be presented in the form

$$\hat{H} \to \hat{H}_{m_s,m_s'}(r) \equiv \langle m_s| \hat{H}(r) |m_s'\rangle$$

where m_s, m_s' take values 1 or 2 corresponding, respectively, to $m_s = 1/2$ and $m_s = -1/2$. Thus, the Hamiltonian in the presence of the spin becomes a 2×2 matrix, and its action on the state presented in the form of Eq. 9.62 involves (in addition to what it normally does to orbital vectors) the multiplication of a matrix and a spinor. In the most trivial case, when the Hamiltonian does not contain any spin operators and does not act, therefore, on spin states, this matrix becomes

$$\hat{H}_{m_s,m_s'}(r) \equiv \langle m_s| \hat{H}(r) |m_s'\rangle = \hat{H}(r) \langle m_s |m_s'\rangle = \hat{H}(r) \delta_{m_s,m_s'}$$

so that the Schrödinger equations for both wave function components $\Psi_\uparrow(r)$ and $\Psi_\downarrow(r)$ are identical. In this case the total state of the system is described by the vector of the form given by Eq. 9.65, in which the coefficients a_1 and a_2 of the spinor component can be chosen arbitrarily. Physically this means that in the absence of the spin-related terms in the Hamiltonian, the spin state of the particle does not change with time and is determined by the initial conditions.

Now let me consider a less trivial case, when the Hamiltonian includes a stand-alone spin operator, something like what we dealt with in Sect. 9.3:

$$\hat{H} = \hat{H}_{orb} + 2\frac{\mu_B B}{\hbar}\hat{S}_z. \tag{9.66}$$

Here $\hat{H}_{orb}$ is a spin-independent portion of the Hamiltonian, and the second term, as you know, describes the interaction of the spin with uniform magnetic field B directed along the Z-axis. In the matrix form, this Hamiltonian becomes

$$\hat{H}_{m_s,m'_s} = \hat{H}_{orb}\delta_{m_s,m'_s} + \mu_B B\,(\hat{\sigma}_z)_{m_s,m'_s} \tag{9.67}$$

where I used the representation of the spin operators in terms of the corresponding Pauli matrices introduced in Eqs. 9.15–9.17. The explicit matrix form of the stationary Schrödinger equation becomes

$$\begin{bmatrix} \hat{H}_{orb} & 0 \\ 0 & \hat{H}_{orb} \end{bmatrix}\begin{bmatrix} \Psi_\uparrow(r) \\ \Psi_\downarrow(r) \end{bmatrix} + \mu_B B\begin{bmatrix} 1 & 0 \\ 0 & -1 \end{bmatrix}\begin{bmatrix} \Psi_\uparrow(r) \\ \Psi_\downarrow(r) \end{bmatrix} = E\begin{bmatrix} \Psi_\uparrow(r) \\ \Psi_\downarrow(r) \end{bmatrix}$$

and translates into two independent equations:

$$\hat{H}_{orb}\Psi_\uparrow(r) + \mu_B B\Psi_\uparrow(r) = E\Psi_\uparrow(r) \tag{9.68}$$

$$\hat{H}_{orb}\Psi_\downarrow(r) - \mu_B B\Psi_\downarrow(r) = E\Psi_\downarrow(r). \tag{9.69}$$

This independence signifies the absence of any spin–orbit coupling in this system: the functions $\Psi_\uparrow(r)$ and $\Psi_\downarrow(r)$ can be chosen in the form of Eq. 9.64 where $\psi(r)$ is a solution of the orbital equation $\hat{H}_{orb}\psi(r) = E_{orb}\psi(r)$. With this, Eqs. 9.68 and 9.69 can be converted into equations

$$a_1\,(E - E_{orb} - \mu_B B) = 0$$

$$a_2\,(E - E_{orb} + \mu_B B) = 0,$$

yielding two eigenvalues $E^{(1)} = E_{orb} + \mu_B B$ and $E^{(2)} = E_{orb} - \mu_B B$, with two respective eigenvectors $a_1^{(1)} = 1$, $a_2^{(1)} = 0$ and $a_1^{(2)} = 0$, $a_2^{(2)} = 1$. Choosing the zero level of energy at E_{orb} and disregarding the orbital part of the resulting spinors

$$|\eta^{(1)}\rangle = \psi(r)\begin{bmatrix} 1 \\ 0 \end{bmatrix}$$

$$|\eta^{(2)}\rangle = \psi(r)\begin{bmatrix} 0 \\ 1 \end{bmatrix}$$

which does not affect any of the phenomena associated with the action of the magnetic field on electron's spin, you end up with eigenvalues

$$E^{(1,2)} = \pm\mu_B B$$

and eigenvectors

$$|\eta^{(1)}\rangle = \begin{bmatrix} 1 \\ 0 \end{bmatrix}; \ |\eta^{(2)}\rangle = \begin{bmatrix} 0 \\ 1 \end{bmatrix}$$

identical to those found for a single spin in the magnetic field in Sect. 9.3. This example demonstrates that the "pure" spin approach, which ignores orbital components of the total state of a particle, is justified as long as the presence of the spin does not change its orbital state, i.e., in the absence of the spin–orbit interaction.

9.5.2   Total Angular Momentum: Eigenvalues and Eigenvectors

In Example 26 in the preceding section, you learned that working in the tensor product of spin and orbital spaces, you can operate with expressions combining orbital and spin operators such as $\hat{L}_z + \hat{S}_z$. This is a z-component of a vector operator

$$\hat{\boldsymbol{J}} = \hat{\boldsymbol{L}} + \hat{\boldsymbol{S}} \tag{9.70}$$

called the operator of a total angular momentum, which plays an important role in the general structure of quantum mechanics as well as in a variety of its applications. For instance, this operator is crucial for understanding the energy levels of hydrogen atom in the presence of spin–orbit coupling and magnetic field; I will introduce you to these topics in Chap. 14. Here my objective is to elucidate the general properties of this operator, which appears to be a logical conclusion to the discussion started in the previous section.

I begin by stating that components of vector $\hat{\boldsymbol{J}}$ obey the same commutation relations as those of its constituent vectors $\hat{\boldsymbol{L}}$ and $\hat{\boldsymbol{S}}$. This statement is easy to verify, taking into account that orbital and spin operators commute. For instance, you can check that

$$\hat{J}_x\hat{J}_y - \hat{J}_y\hat{J}_x = \hat{L}_x\hat{L}_y - \hat{L}_y\hat{L}_x + \hat{S}_x\hat{S}_y - \hat{S}_y\hat{S}_x = i\hbar\hat{L}_z + i\hbar\hat{S}_z = i\hbar\hat{J}_z \tag{9.71}$$

where I canceled terms like $\hat{L}_x\hat{S}_y - \hat{S}_y\hat{L}_x = 0$. Once the commutation relations for the components of $\hat{\boldsymbol{J}}$ are established, one can immediately claim that all components of $\hat{\boldsymbol{J}}$ commute with operator $\hat{\boldsymbol{J}}^2$, which can be written down as

$$\hat{\boldsymbol{J}}^2 = \hat{\boldsymbol{L}}^2 + \hat{\boldsymbol{S}}^2 + 2\hat{\boldsymbol{L}} \cdot \hat{\boldsymbol{S}}. \tag{9.72}$$

Indeed, the proof of the similar statement for orbital angular momentum carried out in Sect. 3.3.2 was based exclusively on the inter-component commutation relations and is, therefore, automatically expanded to all operators with the same commutation relations. If you go back to Sect. 3.3.4, you will recall that the derivation of the eigenvalues of the orbital angular momentum operators carried out there also relied exclusively on the commutation relations. Therefore, you can immediately claim, without fear of retribution or embarrassment, that operators $\hat{J}^2$ and $\hat{J}_z$ possess a common system of eigenvectors, characterized by two numbers j and m_J, satisfying inequality $-j \leq m_J \leq j$, taking either integers or half-integer values, and which generate eigenvalues of these operators according to

$$\hat{J}^2 |j, m_J\rangle = \hbar^2 j(j+1) |j, m_J\rangle \tag{9.73}$$

$$\hat{J}_z |j, m_J\rangle = \hbar m_J |j, m_J\rangle . \tag{9.74}$$

However, it would be wrong for you to think that Eqs. 9.72 and 9.74 are the end of the story. While these equations do give you some information about the eigenvalues and eigenvectors of $\hat{J}^2$ and $\hat{J}_z$, this information is quite limited and does not allow you, for instance, to generate representations of these vectors in any basis except of their own or to help you evaluate the results of the application of various combinations of orbital and spin angular momentum operators to these states. To be able to do all this, you need to answer more rather tough questions such as (a) what is a relation between numbers j, m_J on the one hand and numbers l, s, m, and m_s on the other, and (b) how are vectors $|j, m_J\rangle$ connected with vectors $|l, m_l\rangle$ and $|m_s\rangle$? Finding answers to these questions requires substantial additional efforts, so that Eqs. 9.73 and 9.74 are not the end but just the beginning of the journey.

And as a first step, I would note an additional property of the operators $\hat{J}^2$ and $\hat{J}_z$, which they possess by the virtue of being the sum of orbital and spin operators: they both commute with operators $\hat{L}^2$ and $\hat{S}^2$. Proof of this statement is quite straightforward and is based on Eq. 9.72 as well as on the fact that both $\hat{L}^2$ and $\hat{S}^2$ commute with all their components (well, $\hat{S}^2$ for spin $1/2$ is proportional to a unity matrix and, therefore, commutes with everything). This means that operators $\hat{J}^2, \hat{J}_z, \hat{L}^2$, and $\hat{S}^2$ have a common set of eigenvectors so that numbers j and m_J do not provide a full description of these vectors. To have these vectors fully characterized, one needs to throw number l into the mix replacing $|j, m_J\rangle$ with $|j, l, m_J\rangle$ and adding equation

$$\hat{L}^2 |j, l, m_J\rangle = \hbar^2 l(l+1) |j, l, m_J\rangle \tag{9.75}$$

to Eqs. 9.73 and 9.74. Strictly speaking, I would need to include here a spin number s as well, but since I am going to limit this discussion to only spin $1/2$ particles, this number never changes so that its inclusion would just superfluously increase the clumsiness of the notations.

A relation between vectors $|j, l, m_J\rangle$ and individual eigenvectors of the orbital and spin operators can be established by using the latter as a basis in the combined spin–orbital space defined in Sect. 9.5.1 as a tensor product of the orbital and spinor spaces. Specializing a generic Eq. 9.58 to the particular case, when the basis in the orbital space is presented by vectors $|l, m\rangle$, I can write for an arbitrary member $|\chi\rangle$ of the tensor product space:

$$|\chi\rangle = \sum_{l', m, m_s} C^{l'}_{m, m_s} |l', m\rangle |m_s\rangle . \tag{9.76}$$

However, when applying this expansion to the particular case of vectors $|j, l, m_J\rangle$, I need to take into account that these vectors are eigenvectors of $\hat{l}^2$, i.e., that they must obey Eq. 9.75:

$$\hat{L}^2 \sum_{l', m, m_s} C^{l'}_{m, m_s} |l', m\rangle |m_s\rangle = \sum_{l', m, m_s} C^{l'}_{m, m_s} \hat{L}^2 |l', m\rangle |m_s\rangle =$$

$$\hbar^2 \sum_{l', m, m_s} C^{l'}_{m, m_s} l' (l' + 1) |l', m\rangle |m_s\rangle = \hbar^2 l(l + 1) |l, m\rangle |m_s\rangle .$$

Because of the orthogonality of the basis vectors $|l, m\rangle |m_s\rangle$, the only way to satisfy the equality in the last line is to make sure that $l' = l$ is the only term in the sum. This is achieved by setting $C^{l'}_{m, m_s} = C^{l}_{m, m_s} \delta_{l, l'}$ and thereby vanquishing the summation over l'. In a less formal way, you can argue that for the vector defined by Eq. 9.76 to be an eigenvector of $\hat{L}^2$, it cannot be a combination of vectors with different values of l. Thus, I can conclude that a representation of $|j, l, m_J\rangle$ in the basis of $|l, m\rangle |m_s\rangle$ must have the following form:

$$|j, l, m_J\rangle = \sum_{m, m_s} C^{l, j}_{m, m_s, m_J} |l, m\rangle |m_s\rangle \tag{9.77}$$

where I also added upper index j and lower index m_J to the expansion coefficients to make it explicit that the expansion is for eigenvectors of operators $\hat{J}^2$ and $\hat{J}_z$ characterized by quantum numbers j and m_J.

The task now is to find coefficients $C^{l, j}_{m, m_s m_J}$, which are a particular case of so-called Clebsch–Gordan coefficients.[3] To this end I will first apply operator $\hat{J}_z$ to the left-hand side of Eq. 9.77 and operator $\hat{L}_z + \hat{S}_z$ to its right-hand side. Using Eq. 9.74 on the left-hand side and similar properties of orbital and spin angular momentum operators on the right-hand side, I obtain

---

[3]Clebsch–Gordan coefficients allow to present eigenvectors of an operator $\hat{J}_1 + \hat{J}_2$ in terms of eigenvectors of generic angular momentum operators $\hat{J}_1$ and $\hat{J}_2$.

$$m_J \, |j, l, m_J\rangle = \sum_{m, m_s} C^{l,j}_{m, m_s, m_J} \, (m + m_s) \, |l, m\rangle \, |m_s\rangle \Rightarrow$$

$$m_J \sum_{m, m_s} C^{l,j}_{m, m_s, m_J} \, |l, m\rangle \, |m_s\rangle = \sum_{m, m_s} C^{l,j}_{m, m_s, m_J} \, (m + m_s) \, |l, m\rangle \, |m_s\rangle \Rightarrow$$

$$\sum_{m, m_s} C^{l,j}_{m, m_s, m_J} \, (m_J - m - m_s) \, |l, m\rangle \, |m_s\rangle = 0.$$

For the equation in the last line to be true, one of two things should happen: either $m_J = m + m_s$ or $C^{l,j}_{m, m_s, m_J} = 0$. This means that the Clebsch–Gordan coefficients vanish unless $m = m_J - m_s$ so that they can be presented as

$$C^{l,j}_{m, m_s, m_J} = C^{l,j}_{m_s, m_J} \delta_{m, m_J - m_s}.$$

Substituting this result into Eq. 9.77, I can eliminate the summation over m and obtain a simplified form of this expansion:

$$|j, l, m_J\rangle = \sum_{m_s} C^{l,j}_{m_s, m_J} \, |l, m\rangle \, |m_s\rangle =$$

$$C^{l,j}_{1/2_s, m_J} \left|l, m_J - \frac{1}{2}\right\rangle |{\uparrow}\rangle + C^{l,j}_{-1/2_s, m_J} \left|l, m_J + \frac{1}{2}\right\rangle |{\downarrow}\rangle \tag{9.78}$$

where the last line explicitly accounts for the fact that the spin number m_s only takes two values $1/2$ and $-1/2$. Equation 9.77 contains all the information about Clebsch–Gordan coefficients that I could extract from operator $\hat{J}_z$ (which is not that much), but hopefully I can learn more from operator $\hat{J}^2$.

The idea is the same: apply $\hat{J}^2$ to the left-hand side of Eq. 9.78, its reincarnation in the form $\hat{L}^2 + \hat{S}^2 + 2\hat{L} \cdot \hat{S}$ to this equation's right-hand side, and find conditions that the two sides of the equation agree. The first step is just a recapitulation of Eq. 9.73:

$$\hat{J}^2 \, |j, l, m_J\rangle = \hbar^2 j(j+1) \, |j, l, m_J\rangle =$$

$$\hbar^2 j(j+1) \left[C^{l,j}_{1/2_s, m_J} \left|l, m_J - \frac{1}{2}\right\rangle |{\uparrow}\rangle + C^{l,j}_{-1/2_s, m_J} \left|l, m_J + \frac{1}{2}\right\rangle |{\downarrow}\rangle \right] \tag{9.79}$$

but the second one results in rather long expressions, which couldn't even fit to a single page. Therefore, I will deal with different terms in $\hat{L}^2 + \hat{S}^2 + 2\hat{L} \cdot \hat{S}$ separately. First I will do $\hat{L}^2 + \hat{S}^2$, which is the easiest to handle:

$$\left(\hat{L}^2 + \hat{S}^2\right)\left[C^{l,j}_{1/2_s, m_J}\left|l, m_J - \frac{1}{2}\right\rangle|\uparrow\rangle + C^{l,j}_{-1/2_s, m_J}\left|l, m_J + \frac{1}{2}\right\rangle|\downarrow\rangle\right] =$$

$$\hbar^2 l(l+1)\left[C^{l,j}_{1/2_s, m_J}\left|l, m_J - \frac{1}{2}\right\rangle|\uparrow\rangle + C^{l,j}_{-1/2_s, m_J}\left|l, m_J + \frac{1}{2}\right\rangle|\downarrow\rangle\right] +$$

$$\frac{3}{4}\hbar^2\left[C^{l,j}_{1/2_s, m_J}\left|l, m_J - \frac{1}{2}\right\rangle|\uparrow\rangle + C^{l,j}_{-1/2_s, m_J}\left|l, m_J + \frac{1}{2}\right\rangle|\downarrow\rangle\right] +$$

$$\hbar^2\left(l(l+1) + \frac{3}{4}\right)\left[C^{l,j}_{1/2_s, m_J}\left|l, m_J - \frac{1}{2}\right\rangle|\uparrow\rangle + C^{l,j}_{-1/2_s, m_J}\left|l, m_J + \frac{1}{2}\right\rangle|\downarrow\rangle\right].$$

$$(9.80)$$

To evaluate the remaining $\hat{L} \cdot \hat{S}$ term, I first give it a makeover using ladder operators $\hat{L}_\pm$ and $\hat{S}_\pm$:

$$\hat{L} \cdot \hat{S} = \hat{L}_x\hat{S}_x + \hat{L}_y\hat{S}_y + \hat{L}_z\hat{S}_z =$$

$$\hat{L}_z\hat{S}_z + \frac{1}{2}\left(\hat{L}_+ + \hat{L}_-\right)\frac{1}{2}\left(\hat{S}_+ + \hat{S}_-\right) +$$

$$\frac{1}{2i}\left(\hat{L}_+ - \hat{L}_-\right)\frac{1}{2i}\left(\hat{S}_+ - \hat{S}_-\right) =$$

$$\hat{L}_z\hat{S}_z + \frac{1}{2}\left(\hat{L}_-\hat{S}_+ + \hat{L}_+\hat{S}_-\right) \qquad (9.81)$$

where I used Eqs. 3.59 and 3.60 for orbital and Eqs. 5.109 and 5.108 for spin opera-tors. Using the fact that $\left|l, m_J - \frac{1}{2}\right\rangle|\uparrow\rangle$ and $\left|l, m_J + \frac{1}{2}\right\rangle|\downarrow\rangle$ are eigenvectors of $\hat{L}_z$ and $\hat{S}_z$ with eigenvalues $\hbar\left(m_J - 1/2\right), \hbar/2$ and $\hbar\left(m_J + 1/2\right), -\hbar/2$ correspondingly, I get for the first term in the last line of Eq. 9.81:

$$\hat{L}_z\hat{S}_z\left[C^{l,j}_{1/2_s, m_J}\left|l, m_J - \frac{1}{2}\right\rangle|\uparrow\rangle + C^{l,j}_{-1/2_s, m_J}\left|l, m_J + \frac{1}{2}\right\rangle|\downarrow\rangle\right] =$$

$$\frac{\hbar^2}{2}\left(m_J - \frac{1}{2}\right)C^{l,j}_{1/2_s, m_J}\left|l, m_J - \frac{1}{2}\right\rangle|\uparrow\rangle -$$

$$\frac{\hbar^2}{2}\left(m_J + \frac{1}{2}\right)C^{l,j}_{-1/2_s, m_J}\left|l, m_J + \frac{1}{2}\right\rangle|\downarrow\rangle. \qquad (9.82)$$

To compute the contribution from $\hat{L}_-\hat{S}_+$ and $\hat{L}_+\hat{S}_-$, you need to recall that $\hat{S}_+|\uparrow\rangle = 0$, $\hat{S}_-|\downarrow\rangle = 0$, $\hat{S}_+|\downarrow\rangle = \hbar|\uparrow\rangle$, $\hat{S}_-|\uparrow\rangle = \hbar|\downarrow\rangle$ (These formulas originally appeared in Sect. 5.2.3, Eqs. 5.104 and 5.106, but I am reposting them here for your convenience.) You will also need to go back to Eqs. 3.75 and 3.76 to figure out the part related to operators $\hat{L}_\pm$. Refreshing this way your memory of the ladder operators, you can get

$$\hat{L}_-\hat{S}_+\left[C^{l,j}_{1/2_s,m_J}\left|l,m_J-\frac{1}{2}\right\rangle|\uparrow\rangle + C^{l,j}_{-1/2_s,m_J}\left|l,m_J+\frac{1}{2}\right\rangle|\downarrow\rangle\right] =$$

$$\hbar^2\sqrt{l(l+1)-\left(m_J+\frac{1}{2}\right)\left(m_J-\frac{1}{2}\right)}C^{l,j}_{-1/2_s,m_J}\left|l,m_J-\frac{1}{2}\right\rangle|\uparrow\rangle \qquad (9.83)$$

and

$$\hat{L}_+\hat{S}_-\left[C^{l,j}_{1/2_s,m_J}\left|l,m_J-\frac{1}{2}\right\rangle|\uparrow\rangle + C^{l,j}_{-1/2_s,m_J}\left|l,m_J+\frac{1}{2}\right\rangle|\downarrow\rangle\right] =$$

$$\hbar^2\sqrt{l(l+1)-\left(m_J+\frac{1}{2}\right)\left(m_J-\frac{1}{2}\right)}C^{l,j}_{1/2_s,m_J}\left|l,m_J+\frac{1}{2}\right\rangle|\downarrow\rangle. \qquad (9.84)$$

Finally, you just need to bring together all Eqs. 9.80–9.84 and apply some simple algebra (just group together the like terms) to cross the goal line:

$$\left(\hat{L}^2+\hat{S}^2+2\hat{L}\cdot\hat{S}\right)\left[C^{l,j}_{1/2_s,m_J}\left|l,m_J-\frac{1}{2}\right\rangle|\uparrow\rangle + C^{l,j}_{-1/2_s,m_J}\left|l,m_J+\frac{1}{2}\right\rangle|\downarrow\rangle\right] =$$

$$\hbar^2\left[C^{l,j}_{1/2_s,m_J}\left(l(l+1)+m_J+\frac{1}{4}\right)+\right.$$

$$\left.\sqrt{l(l+1)-\left(m_J+\frac{1}{2}\right)\left(m_J-\frac{1}{2}\right)}C^{l,j}_{-1/2_s,m_J}\right]\left|l,m_J-\frac{1}{2}\right\rangle|\uparrow\rangle +$$

$$\hbar^2\left[C^{l,j}_{-1/2_s,m_J}\left(l(l+1)-m_J+\frac{1}{4}\right)+\right.$$

$$\left.\sqrt{l(l+1)-\left(m_J+\frac{1}{2}\right)\left(m_J-\frac{1}{2}\right)}C^{l,j}_{1/2_s,m_J}\right]\left|l,m_J+\frac{1}{2}\right\rangle|\downarrow\rangle.$$

Comparing this against Eq. 9.79 and equating coefficients in front of each of the vectors, you will end up with the following system of equations for coefficients $C^{l,j}_{1/2_s,m_J}$ and $C^{l,j}_{-1/2_s,m_J}$:

$$\left(l(l+1)-j(j+1)+m_J+\frac{1}{4}\right)C^{l,j}_{1/2_s,m_J}+$$

$$\sqrt{l(l+1)-\left(m_J+\frac{1}{2}\right)\left(m_J-\frac{1}{2}\right)}C^{l,j}_{-1/2_s,m_J} = 0 \qquad (9.85)$$

$$\sqrt{l(l+1) - \left(m_J + \frac{1}{2}\right)\left(m_J - \frac{1}{2}\right)} C^{l,j}_{1/2s,m_J} +$$

$$\left(l(l+1) - j(j+1) + \frac{1}{4} - m_J\right) C^{l,j}_{-1/2s,m_J} = 0. \qquad (9.86)$$

And once again you are looking for non-zero solutions of a homogeneous system of linear equations, and once again you need to find zeroes of the determinant formed by its coefficients:

$$\left\| \begin{matrix} l(l+1) - j(j+1) + m_J + \frac{1}{4}; & \sqrt{l(l+1) - \left(m_J + \frac{1}{2}\right)\left(m_J - \frac{1}{2}\right)} \\ \sqrt{l(l+1) - \left(m_J + \frac{1}{2}\right)\left(m_J - \frac{1}{2}\right)}; & l(l+1) - j(j+1) + \frac{1}{4} - m_J \end{matrix} \right\| = 0.$$

Evaluation of the determinate yields

$$\left(l(l+1) - j(j+1) + \frac{1}{4} + m_J\right)\left(l(l+1) - j(j+1) + \frac{1}{4} - m_J\right) -$$

$$l(l+1) + \left(m_J + \frac{1}{2}\right)\left(m_J - \frac{1}{2}\right) =$$

$$\left(l(l+1) - j(j+1) + \frac{1}{4}\right)^2 - l(l+1) - \frac{1}{4} =$$

$$\left[\left(l + \frac{1}{2}\right)^2 - j(j+1)\right]^2 - \left(l + \frac{1}{2}\right)^2$$

where I used easily verified identity

$$l(l+1) + \frac{1}{4} \equiv \left(l + \frac{1}{2}\right)^2. \qquad (9.87)$$

Now it is quite easy to find that equation

$$\left[\left(l + \frac{1}{2}\right)^2 - j(j+1)\right]^2 - \left(l + \frac{1}{2}\right)^2 = 0$$

is satisfied for

$$j(j+1) = \left(l + \frac{1}{2}\right)\left(l + \frac{3}{2}\right)$$

or

$$j(j+1) = \left(l + \frac{1}{2}\right)\left(l - \frac{1}{2}\right).$$

The only physically meaningful solutions of these equations are

$$j_1 = l + \frac{1}{2} \tag{9.88}$$

and

$$j_2 = l - \frac{1}{2}. \tag{9.89}$$

(Two other solutions $-l - 3/2$ and $-l - 1/2$ are negative and must be ignored.) The obtained result means that for any value of the orbital quantum number l, operator $\hat{J}^2$ has two possible eigenvalues $\hbar^2 j_1 (j_1 + 1)$ and $\hbar^2 j_2 (j_2 + 1)$ with j_1 and j_2 defined above. For each value of j, there are $2j + 1$ values of m_J, $m_J = -j, -j + 1 \cdots j - 1, j$ so that the total number of states $|j, l, m_J\rangle$ (for a given l) is $2(l + 1/2) + 1 + 2((l - 1/2) + 1) = 2(2l + 1)$, which is exactly the same as the number of states $|l, m\rangle |m_s\rangle$. One important conclusion from this arithmetic is that orthogonal and linearly independent states $|j, l, m_J\rangle$ and other orthogonal and independent states $|l, m\rangle |m_s\rangle$ represent two alternative bases in the same vector space: vectors of the former basis are defined by the states in which the measurement of the total angular momentum and its component would yield determinate results, and vectors of the latter basis correspond to the states in which orbital and spin momenta separately would have definite values.

Now I can go back to Eqs. 9.85 and 9.86 and find the Clebsch–Gordan coefficients that establish a connection between vectors $|j, l, m_J\rangle$ and vectors $|l, m\rangle |m_s\rangle$, signaling the close end of this journey. Substituting the found values for j_1 and j_2 to Eqs.9.85 and 9.86, I find the two sets of the coefficients:

$$C^{l,j_1}_{-1/2 s, m_J} = \frac{l + \frac{1}{2} - m_J}{\sqrt{l(l + 1) - m_J^2 + \frac{1}{4}}} C^{l,j_1}_{1/2 s, m_J} = \sqrt{\frac{l + \frac{1}{2} - m_J}{l + \frac{1}{2} + m_J}} C^{l,j_1}_{1/2 s, m_J} \tag{9.90}$$

$$C^{l,j_2}_{1/2 s, m_J} = -\frac{l + \frac{1}{2} - m_J}{\sqrt{l(l + 1) - m_J^2 + \frac{1}{4}}} C^{l,j_2}_{-1/2 s, m_J} = -\sqrt{\frac{l + \frac{1}{2} - m_J}{l + \frac{1}{2} + m_J}} C^{l,j_2}_{-1/2 s, m_J} \tag{9.91}$$

where I again used Eq. 9.87. As usual, Eqs. 9.85 and 9.86 yield only the ratio of the coefficients, and in order to find the coefficients themselves, the normalization requirement, complemented by the convention that the Clebsch–Gordan coefficients remain real, needs to be invoked. Substituting Eqs. 9.90 and 9.91 into the normalization condition

$$\left| C^{l,j}_{-1/2 s, m_J} \right|^2 + \left| C^{l,j}_{1/2 s, m_J} \right|^2 = 1,$$

I find after some trivial algebra

$$
C^{l,j_1}_{1/2_s,m_J} = \sqrt{\frac{l + \frac{1}{2} + m_J}{2l + 1}}; \ C^{l,j_1}_{-1/2_s,m_J} = \sqrt{\frac{l + \frac{1}{2} - m_J}{2l + 1}}
$$

$$
C^{l,j_2}_{1/2_s,m_J} = \sqrt{\frac{l + \frac{1}{2} - m_J}{2l + 1}}; \ C^{l,j_2}_{-1/2_s,m_J} = -\sqrt{\frac{l + \frac{1}{2} + m_J}{2l + 1}}. \tag{9.92}
$$

Now you just plug Eq. 9.92 into Eq. 9.78 to derive the final expressions for the two eigenvectors of operator $\hat{J}^2$ characterized by quantum numbers j_1 and j_2 in terms of linear combination of the orbital and spin angular momentum eigenvectors:

$$
|l + 1/2, l, m_J\rangle = \frac{1}{\sqrt{2l + 1}} \left[\sqrt{l + m_J + 1/2} \left| l, m_J - \frac{1}{2} \right\rangle |\uparrow\rangle + \right.
$$
$$
\left. \sqrt{l - m_J + 1/2} \left| l, m_J + \frac{1}{2} \right\rangle |\downarrow\rangle \right] \tag{9.93}
$$

$$
|l - 1/2, l, m_J\rangle = \frac{1}{\sqrt{2l + 1}} \left[\sqrt{l - m_J + 1/2} \left| l, m_J - \frac{1}{2} \right\rangle |\uparrow\rangle - \right.
$$
$$
\left. \sqrt{l + m_J + 1/2} \left| l, m_J + \frac{1}{2} \right\rangle |\downarrow\rangle \right]. \tag{9.94}
$$

It is quite easy to verify that vectors $|l + 1/2, l, m_J\rangle$ and $|l - 1/2, l, m_J\rangle$ are normalized and orthogonal, as they shall be. One can interpret this result by saying that if an electron is prepared in a state with determinate values of total angular momentum $\hbar^2 j(j + 1)$, one of its components, $\hbar m_J$, and total orbital momentum $\hbar^2 l(l + 1)$, the values of the corresponding components of its orbital momentum $\hbar m$ and spin $\hbar m_s$ remain uncertain. An attempt to measure them will produce the combination $m = m_J - 1/2, m_s = 1/2$ with probabilities

$$
P_{m_J - 1/2, 1/2} = \begin{cases} \frac{l + m_J + 1/2}{2l + 1} & , j = l + 1/2 \\ \frac{l - m_J + 1/2}{2l + 1} & j = l - 1/2 \end{cases} \tag{9.95}
$$

or combination $m = m_J + 1/2, m_s = -1/2$ with probabilities

$$
P_{m_J + 1/2, -1/2} = \begin{cases} \frac{l - m_J + 1/2}{2l + 1} & , j = l + 1/2 \\ \frac{l + m_J + 1/2}{2l + 1} & j = l - 1/2 \end{cases}. \tag{9.96}
$$

To help you feel better about these results, let me illustrate the application of Eqs. 9.95 and 9.96 by a few examples.

Example 27 (Measuring Spin and Orbital Angular Momentums in the State with the Definite Value of the Total Angular Momentum) Assume that an electron is in a state with a given orbital momentum l, total angular momentum $j = l - 1/2$, and its z-component $m_J = l - 3/2$ and that you have a magic instrument allowing you to measure the z-components of electron's orbital momentum and its spin. What are the possible outcomes of such a measurement and their probabilities?

Solution

Value $m_J = l - 3/2$ can be obtained in two different ways—when $m_s = 1/2$ and $m = l - 2$ or $m_s = -1/2$ and $m = l - 1$. The probability of the first outcome is (second line in Eq. 9.95)

$$p_{l-2,1/2} = \frac{l - (l - 3/2) + 1/2}{2l + 1} = \frac{2}{2l + 1},$$

and the probability of the second outcome (second line in Eq. 9.96) is

$$p_{l-1,-1/2} = \frac{l + (l - 3/2) + 1/2}{2l + 1} = \frac{2l - 1}{2l + 1}.$$

Obviously the sum of the two probabilities is equal to one, and for large values of l, the second outcome is significantly more probable.

Example 28 (More on Measurement of Spin and Orbital Momentums) Let me modify the previous example by assuming that the value of the total angular momentum is not known, but it is known that the electron can be in either state of the total angular momentum with equal probability. How will the answer to the previous example change in this case?

Solution

Now you have to take into account that both possible outcomes discussed in the previous example can come either from the state with $j = l + 1/2$ or the state with $j = l - 1/2$. Respectively, the total probability of the outcomes becomes

$$p_{l-2,1/2} = \frac{1}{2}\frac{l - (l - 3/2) + 1/2}{2l + 1} + \frac{1}{2}\frac{l + (l - 3/2) + 1/2}{2l + 1} = \frac{l + 1/2}{2l + 1} = \frac{1}{2}$$

$$p_{l-1,-1/2} = \frac{1}{2}\frac{l + (l - 3/2) + 1/2}{2l + 1} + \frac{1}{2}\frac{l - (l - 3/2) + 1/2}{2l + 1} = \frac{1}{2}.$$

Even though, generally speaking, either m_J or m and m_s cannot be known with certainty in the same state, there exist two states in which all three of these quantum numbers have definite values. These are the states with the largest $m_J = l + 1/2$ and smallest $m_J = -l - 1/2$ values of m_J, for which one of the Clebsch–Gordan coefficients vanishes, while the other one turns to unity, reducing Eq. 9.93 to

$$|l + 1/2, l, l + 1/2\rangle = |l, l\rangle |\uparrow\rangle \, ; \, |l + 1/2, l, -l - 1/2\rangle = |l, -l\rangle |\downarrow\rangle .$$

You can easily understand this fact by noting that $m_J = l + 1/2$ or $m_J = -l - 1/2$ can be obtained only by a single combination of m and m_s: $m_J = l + 1/2$ corresponds to the choice $m = l$ and $m_s = 1/2$, while $m_J = -l - 1/2$ can only be generated by $m = -l$, $m_s = -1/2$.

Equations 9.88 and 9.89 together with Eqs. 9.93 and 9.94 provide answers to all the questions posed in the beginning of this subsection: you now know the relation between total, orbital, and spin angular momentum quantum numbers as well as between corresponding eigenvectors. In particular, Eqs. 9.93 and 9.94 allow generating any representation for $|j, l, m_J\rangle$, using corresponding representations for vectors $|l, m\rangle$ and $|m_s\rangle$, as well as define the action of any combination of orbital and spin operators on these vectors. To illustrate this point, I will write down the coordinate–spinor representation of $|l + 1/2, l, m_J\rangle$ using Eq. 9.93 and the corresponding representations for $|l, m\rangle$ and $|m_s\rangle$:

$$|l + 1/2, l, m_J\rangle = \frac{1}{\sqrt{2l+1}} \left[\begin{array}{c} \sqrt{l + m_J + 1/2}\, Y_l^{m_J - 1/2}(\theta, \varphi) \\ \sqrt{l - m_J + 1/2}\, Y_l^{m_J + 1/2}(\theta, \varphi) \end{array} \right].$$

I can now use this to compute, e.g., $\hat{L}_+ \hat{S}_- |l + 1/2, l, m_J\rangle$. Taking the matrix representation for $\hat{S}_-$ from Eq. 5.107 and recalling that any orbital operator in the spinor representation is multiplied by a unity matrix, I can rewrite this expression as

$$\hat{L}_+ \hat{S}_- |l + 1/2, l, m_J\rangle = \frac{\hbar}{\sqrt{2l+1}} \left[\begin{array}{cc} \hat{L}_+ & 0 \\ 0 & \hat{L}_+ \end{array} \right] \left[\begin{array}{cc} 0 & 0 \\ 1 & 0 \end{array} \right] \left[\begin{array}{c} \sqrt{l + m_J + 1/2}\, Y_l^{m_J - 1/2}(\theta, \varphi) \\ \sqrt{l - m_J + 1/2}\, Y_l^{m_J + 1/2}(\theta, \varphi) \end{array} \right] =$$

$$\frac{\hbar \sqrt{l + m_J + 1/2}}{\sqrt{2l+1}} \left[\begin{array}{cc} \hat{L}_+ & 0 \\ 0 & \hat{L}_+ \end{array} \right] \left[\begin{array}{c} 0 \\ Y_l^{m_J - 1/2}(\theta, \varphi) \end{array} \right] = \frac{\hbar \sqrt{l + m_J + 1/2}}{\sqrt{2l+1}} \left[\begin{array}{c} 0 \\ \hat{L}_+ Y_l^{m_J - 1/2}(\theta, \varphi) \end{array} \right] -$$

$$\frac{\hbar \sqrt{l + m_J + 1/2}}{\sqrt{2l+1}} \sqrt{l(l+1) - (m_J - 1/2)(m_J + 1/2)} \left[\begin{array}{c} 0 \\ Y_l^{m_J + 1/2}(\theta, \varphi) \end{array} \right] =$$

$$\hbar (l + m_J + 1/2) \sqrt{\frac{l - m_J + 1/2}{2l + 1}} \left[\begin{array}{c} 0 \\ Y_l^{m_J + 1/2}(\theta, \varphi) \end{array} \right].$$

9.6 Problems

Section 9.2

Problem 112 Write down a spinor corresponding to the point on the Bloch sphere with coordinates $\theta = \pi/4$, $\varphi = 3\pi/2$.

Problem 113 The impossibility of half-integer values of the angular momentum for orbital angular momentum operators expressed in terms of coordinate and momentum operators can be demonstrated by considering the following example. Imagine that there exists a state of the orbital angular momentum with $l = 1/2$. Then in the coordinate representation, these states would be represented by two functions $f_{1/2}(\theta, \varphi)$ and $f_{-1/2}(\theta, \varphi)$ corresponding to the values of the magnetic quantum number $m = 1/2$ and $m = -1/2$, respectively. These functions must obey the following set of equations:

$$\hat{L}_+ f_{1/2}(\theta, \varphi) = 0; \ \hat{L}_- f_{-1/2}(\theta, \varphi) = 0$$

$$\hat{L}_+ f_{-1/2}(\theta, \varphi) = f_{1/2}(\theta, \varphi); \ \hat{L}_- f_{+1/2}(\theta, \varphi) = f_{-1/2}(\theta, \varphi).$$

Using the coordinate representation of the ladder operators, show that these equations are mutually inconsistent.

Problem 114 An electron is in spin state described by (non-normalized) spinor:

$$|\chi\rangle = \begin{bmatrix} 2i - 3 \\ 4 \end{bmatrix}.$$

1. Normalize this spinor.
2. If you measure the z-component of the spin, what are the probabilities of various outcomes?
3. What is the expectation value of the z-component of the spin in this state?
4. Answer the same questions for x- and y-components.

Problem 115

1. Consider a spin in state

$$\begin{bmatrix} 1 \\ 0 \end{bmatrix}.$$

You measure the component of the spin in the direction of the unit vector n characterized by angles θ, φ of the spherical coordinate system. What is a probability of obtaining value $-\hbar/2$ as an outcome of this measurement?
2. Imagine that you conduct two measurements in a quick succession: first you carry out the measurement described in the previous part of the problem, and right after that, you measure the y-component of the spin. Find the probability of getting $\hbar/2$ as an outcome of the last measurement. (Hint: Do not forget to consider all possible paths that could lead to this outcome.)

Problem 116 Consider a particle with spin 1/2 in a state in which a component of the spin in a specified direction is equal to $\hbar/2$. Choose a coordinate system with the Z-axis along this direction and some arbitrary positions for X- and Y-axes in the perpendicular plane. Now imagine that you measure a component of the spin in a direction making angle 30° with the Z-axis and lying in the XZ plane. Find the probabilities of the various outcomes of this measurement.

Section 9.3

Problem 117 Derive the expression for the expectation value of the y-component of the spin in the state specified by Eq. 9.34.

Problem 118 Consider a spin in the initial state characterized by angles $\theta = \pi/6$ and $\varphi = \pi/3$ of the Bloch sphere. At time $t = 0$, the magnetic field B directed along the polar axes of the spherical coordinate system is turned on and remains on for $t = \pi/(2\omega_L)$ seconds. After the field is off, an experimentalist measures the z-component of the spin. What is the probability that the measurement yields $\hbar/2$? $-\hbar/2$? Answer the same questions if it is the x-component of the spin that is being measured.

Problem 119 In the last problem to Chap. 5, you found matrices $\hat{S}_x$, $\hat{S}_y$, and $\hat{S}_z$ for a particle with spin 3/2. Assume that an interaction of this particle with its surrounding is described by Hamiltonian:

$$\hat{H} = \frac{\varepsilon_0}{\hbar^2} \left(\hat{S}_x^2 - \hat{S}_y^2 \right) - \frac{\varepsilon_1}{\hbar^2} \hat{S}_z^2.$$

1. Find the stationary states of this Hamiltonian.
2. Assuming that the initial state of the particle is given by a generic spinor of the form

$$|\chi_0\rangle = \begin{bmatrix} 1 \\ 0 \\ 0 \\ 0 \end{bmatrix},$$

   find the spin state of the particle at time t.
3. Calculate the time-dependent expectation values of all three components of the spin operator.

Problem 120 Consider a spin 1/2 particle in a time-dependent magnetic field, which rotates with angular velocity Ω in the X–Y plane:

$$B = iB_0 \cos \Omega t + jB_0 \sin \Omega t,$$

where i and j are unit vectors in the directions of X and Y coordinate axes, respectively. Derive the Heisenberg equations for the spin operators and solve them. Note, since the Hamiltonian of this system is time-dependent, you cannot claim the same form for the Hamiltonian in Schrödinger and Heisenberg pictures based upon the notion that the time-evolution operator $\hat{U}$ commutes with the Hamiltonian (it does not because it does not have the form of $\exp\left(-i\hat{H}t/\hbar\right)$, which is only valid for time-independent Hamiltonians). Nevertheless, since the time-dependent factor in the Hamiltonian does not contain operators, you can still show that the Heisenberg form of the Hamiltonian, which in the Schrödinger picture has the form

$$\hat{H} = 2\frac{\mu_B}{\hbar}\hat{S}\cdot B,$$

has exactly the same form in the Heisenberg picture if the Schrödinger spin operator is replaced with its time-dependent Heisenberg operator.

1. Convince yourself that this is, indeed, the case.
2. Derive the Heisenberg equations for all three components of the spin operators.
3. Solve these equations and find the time dependence of the spin operators. (Hint: You might want to introduce new time-dependent operators defined as

$$\hat{P} = \hat{S}_x \cos \Omega t + \hat{S}_y \sin \Omega t$$

$$\hat{Q} = \hat{S}_y \cos \Omega t - \hat{S}_x \sin \Omega t$$

and derive equations for them.)

Section 9.4

Problem 121 Normalize the following vector belonging to the tensor product of two spaces:

$$|\psi\rangle = 2i\left|e_1^{(1)}\right\rangle\left(\left|e_1^{(2)}\right\rangle - 3i\left|e_2^{(2)}\right\rangle\right) + \left(2\left|e_1^{(1)}\right\rangle - 3\left|e_2^{(1)}\right\rangle\right)\left|e_2^{(2)}\right\rangle,$$

assuming that vectors $\left|e_{1,2}^{(1)}\right\rangle$ and $\left|e_{1,2}^{(2)}\right\rangle$ are normalized and mutually orthogonal.

Problem 122 Compute commutators $\left[\hat{S}_i^{(tp)}, \hat{S}_j^{(tp)}\right]$ for all $i \neq j$ and $\left[\hat{S}_i^{(tp)}, \left(\hat{S}^{(tp)}\right)^2\right]$, where i,j take values x, y, and z.

Problem 123 Assuming that vectors $\left|e_{1,2}^{(1)}\right\rangle$ and $\left|e_{1,2}^{(2)}\right\rangle$ in Problem 121 correspond to spin-up and spin-down states of two particles as defined by operators $\hat{S}_z^{(1,2)}$ correspondingly, compute

$$\langle\psi|\hat{S}^{(1)}\cdot\hat{S}^{(2)}|\psi\rangle,$$

where vector $|\psi\rangle$ is also defined in Problem 121.

Problem 124 Derive Eqs. 9.46 through 9.49.

Problem 125 Consider a system of two interacting spins described by Hamiltonian:

$$\hat{H} = \frac{2\mu_B}{\hbar}\hat{S}^{(1)}B + \frac{2\mu_B}{\hbar}\hat{S}^{(2)}B + J\hat{S}^{(1)} \cdot \hat{S}^{(2)}.$$

Find the eigenvalues and eigenvectors of this Hamiltonian. Do it in two different ways: first, use eigenvectors of individual $\hat{S}_z^{(1,2)}$ operators as a basis, and second, use eigenvectors of the operators of the total spin. Find the ground state of the system for different relations between the magnetic field and parameter J. Consider cases $J > 0$ and $J < 0$.

For Sect. 9.5.1

Problem 126 Using the approach presented in Sect. 9.4, consider addition of the operators of the orbital angular momentum and spin, limiting your consideration to the orbital states with $l = 1$.

1. Construct the matrix of the operator $\hat{J}^2$, where $\hat{J} = \hat{L} + \hat{S}$, in the basis of eigenvectors of operators $\hat{L}^2$, $\hat{L}_z$, and $\hat{S}_z$, taking into account only those eigenvectors which belong to the orbital quantum number $l = 1$. (Hint: Your basis will consist of 6 vectors, so that you are looking for a 6×6 matrix.)
2. Diagonalize the matrix and confirm that eigenvectors of $\hat{J}^2$ are characterized by quantum numbers $j = 1/2$ and $j = 3/2$.
3. Find the eigenvectors of $\hat{J}^2$ in this basis.

Problem 127

1. Write down an expression for a spinor describing the equal superposition of states, in which an electron in the ground state of an infinite one-dimensional potential is also in a spin-up state, while an electron in the first excited state of this potential is also in the spin-down state. The potential confines the electron's motion in x direction, while spin-up and spin-down states correspond to the z-component of the spin.
2. Imagine that you have measured a component of the spin in the x direction and obtained value $\hbar/2$. Find the probability distribution of the electron's coordinate right after this measurement.

Problem 128 A one-dimensional harmonic oscillator is placed in a state

$$|\alpha\rangle = \frac{1}{\sqrt{2}} \left[|0\rangle\,|\uparrow\rangle + |1\rangle\,|\downarrow\rangle \right],$$

where spin-up and spin-down states are defined with respect to the z-component of the spin operator and kets $|0\rangle$ and $|1\rangle$ correspond to the ground state and the first excited state of a harmonic oscillator. At time $t = 0$ an experimentalist turns on a uniform magnetic field in the z direction. Find the state of the system at a later time t, and compute the expectation values of oscillator's coordinate and momentum. (Hint: You can use Eqs. 9.68 and 9.69 with the orbital part of the Hamiltonian taken to be that of a harmonic oscillator.)

For Sect. 9.5.2

Problem 129 Compute the expectation value of all components of the operator

$$\hat{\boldsymbol{J}} = \hat{\boldsymbol{L}} + \hat{\boldsymbol{S}}$$

as well as of operator $\hat{\boldsymbol{J}}^2$ in state

$$|\chi\rangle = \frac{1}{\sqrt{14}} \left[Y_l^{l-2}(\theta, \varphi) \left| \frac{1}{2} \right\rangle - 2 Y_l^l (\theta, \varphi) \left| -\frac{1}{2} \right\rangle + 3i Y_l^2 (\theta, \varphi) \left| \frac{1}{2} \right\rangle \right].$$

Problem 130 Derive Eq. 9.92.

Problem 131 Consider an electron in a state with $l = 2$, $j = 3/2$, and $m_J = 0$. If one measures the z-components of the electron orbital momentum and spin, what are the possible values and their probabilities?

Problem 132 Let me reverse the previous problem: assume that the electron is in the state with $l = 2$, $m = 1$, and $m_s = -1/2$. What are the possible values of j and their probabilities?

Problem 133 Consider an electron in the following state (in the coordinate representation):

$$|\alpha\rangle = \frac{2}{\sqrt{10}} Y_1^1 (\theta, \varphi) \left| \frac{1}{2} \right\rangle + \frac{1}{\sqrt{10}} Y_2^0 (\theta, \varphi) \left| -\frac{1}{2} \right\rangle + \frac{1}{\sqrt{10}} Y_1^{-1} (\theta, \varphi) \left| \frac{1}{2} \right\rangle$$

$$+ \frac{2}{\sqrt{10}} Y_2^1 (\theta, \varphi) \left| -\frac{1}{2} \right\rangle.$$

1. If one measures $\hat{\boldsymbol{J}}^2$ and $\hat{J}_z$, what values can one expect to observe and what are their probabilities?
2. Present this vector as a linear combination of appropriate vectors $|j, l, m_J\rangle$.

Section 9.5.2

Problem 134 Compute commutators $\left[\hat{J}_y, \hat{J}_z\right]$ and $\left[\hat{J}_x, \hat{J}_z\right]$, and demonstrate that they have a standard for the angular momentum operators form.

Problem 135 Write down the position–spinor representation of vector $|l - 1/2, l, m_J\rangle$, and compute $\hat{L}_-\hat{S}^+ |l - 1/2, l, m_J\rangle$ using this representation.

Chapter 10
Two-Level System in a Periodic External Field

I have already mentioned somewhere in the beginning of this book that while vectors representing states of realistic physical systems generally belong to an infinite-dimensional vector space, we can always (well, almost, always) justify limiting our consideration to a subspace of states with a reasonably small dimension. The smallest nontrivial subspace containing states that can be assumed to be isolated from the rest of the space is two-dimensional. One relatively clean example of such a subspace is formed by two-dimensional spinors in the situations when one can neglect interactions between spins of different particles as well as by the spin–orbital interaction. An approximately isolated two-dimensional subspace can also be found in systems described by Hamiltonians with discrete spectrum, if this spectrum is strongly non-equidistant, i.e., the energy intervals between adjacent energy levels $\Delta_i = E_{i+1} - E_i$ are different for different pairs of levels. Two-level models are very popular in various areas of physics because, on one hand, they are remarkably simple, while on the other hand, they capture essential properties of many real physical systems ranging from atoms to semiconductors.

The most popular (and useful) version of this model involves an interaction of a two-level system with a periodic time-dependent external "potential." This can be an electric dipole potential describing interaction of an atomic electron with electric field or magnetic "potential" describing interaction of an electron spin with time-dependent magnetic field. Since I am not going to go into concrete details of a physical system, which this model is supposed to represent, I will introduce it by assuming that its Hamiltonian is a sum of a time-independent "unperturbed" part $\hat{H}_0$ and the time-dependent "perturbation" $\hat{V}(t)$. I will also assume that $\hat{H}_0$ has only two linearly independent and orthogonal eigenvectors, which I will designate as $|1\rangle$ and $|2\rangle$, and two corresponding eigenvalues $E_1^{(0)}$ and $E_2^{(0)}$, which may be degenerate.

It is easy to see now that $\hat{H}_0$ can be written as

$$\hat{H}_0 = E_1^{(0)} |1\rangle \langle 1| + E_2^{(0)} |2\rangle \langle 2|. \tag{10.1}$$

© Springer International Publishing AG, part of Springer Nature 2018
L.I. Deych, *Advanced Undergraduate Quantum Mechanics*,
https://doi.org/10.1007/978-3-319-71550-6_10

Indeed, taking into account the orthogonality and normalization of $|1\rangle$ and $|2\rangle$, you can find

$$\hat{H}_0 |1\rangle = E_1^{(0)} |1\rangle \langle 1| 1\rangle + E_2^{(0)} |2\rangle \langle 2| 1\rangle = E_1^{(0)} |1\rangle \,,$$

and

$$\hat{H}_0 |2\rangle = E_1^{(0)} |1\rangle \langle 1| 2\rangle + E_2^{(0)} |2\rangle \langle 2| 2\rangle = E_2^{(0)} |2\rangle \,,$$

confirming that the Hamiltonian given by Eq. 10.1 does, indeed, have the properties prescribed to it. It is obvious that in the basis of these eigenvectors, $\hat{H}_0$ is presented by a diagonal matrix with eigenvalues along the main diagonal. In the most general form, the interaction term can be written down as

$$\hat{V} = V_{11} |1\rangle \langle 1| + V_{22} |2\rangle \langle 2| + V_{12} |1\rangle \langle 2| + V_{21} |2\rangle \langle 1| \,.$$

The diagonal elements in this expression, $V_{ii}(t) = \langle i| \hat{V} |i\rangle$, often vanish, thanks to the symmetry of the system. Indeed, if the initial Hamiltonian is symmetric with respect to inversion, its eigenvectors have definite parity—they are either odd or even. If, in addition, the interaction Hamiltonian is odd (which is quite common—for instance, the electric–dipole interaction is proportional to $\hat{r} \cdot \mathcal{E}$, where $\mathcal{E}$ is the electric field, and position operator changes sign upon inversion), the diagonal elements of the interaction term must vanish (details of the arguments can be found in Sect. 7.1). Also, the requirement that the operator must be Hermitian demands that $V_{21} = V_{12}^*$.

10.1   Two-Level System with a Time-Independent Interaction: Avoided Level Crossing

I begin by considering the properties of the two-level model with a time-independent interaction term, so that the complete Hamiltonian of the system becomes

$$\hat{H} = E_1^{(0)} |1\rangle \langle 1| + E_2^{(0)} |2\rangle \langle 2| + V_{12} |1\rangle \langle 2| + V_{12}^* |2\rangle \langle 1| \qquad (10.2)$$

where V_{ij} are in general complex constant parameters. Since this is a time-independent Hamiltonian, it makes sense to explore its eigenvectors and eigenvalues using vectors $|1\rangle$ and $|2\rangle$ as a basis. The Hamiltonian in this representation becomes a 2×2 matrix so that the eigenvector equation can be written in the matrix form

$$\begin{bmatrix} E_1^{(0)} & V_{12} \\ V_{12}^* & E_2^{(0)} \end{bmatrix} \begin{bmatrix} a_1 \\ a_2 \end{bmatrix} = E \begin{bmatrix} a_1 \\ a_2 \end{bmatrix}, \qquad (10.3)$$

and the corresponding equation for the eigenvalues becomes

$$\begin{Vmatrix} E_1^{(0)} - E & V_{12} \\ V_{12}^* & E_2^{(0)} - E \end{Vmatrix} = 0.$$

Evaluation of the determinant turns it into a simple quadratic equation:

$$E^2 - E\left(E_1^{(0)} + E_2^{(0)}\right) + E_1^{(0)}E_2^{(0)} - |V_{12}|^2 = 0$$

with two solutions (I provided a lot of detailed derivations in this book, but I am not going to show how to solve quadratic equations!)

$$E_1 = \frac{1}{2}\left(E_1^{(0)} + E_2^{(0)}\right) + \frac{1}{2}\sqrt{\left(E_1^{(0)} - E_2^{(0)}\right)^2 + 4\,|V_{12}|^2} \tag{10.4}$$

$$E_2 = \frac{1}{2}\left(E_1^{(0)} + E_2^{(0)}\right) - \frac{1}{2}\sqrt{\left(E_1^{(0)} - E_2^{(0)}\right)^2 + 4\,|V_{12}|^2}. \tag{10.5}$$

Substituting the first of these solutions into

$$\left(E_1^{(0)} - E\right)a_1 + V_{12}a_2 = 0$$

(the first of the equations encoded in the matrix form in Eq. 10.3), I find the ratio of the coefficients representing the first eigenvector of the Hamiltonian:

$$\frac{a_1^{(1)}}{a_2^{(1)}} = -2\frac{V_{12}}{E_1^{(0)} - E_2^{(0)} - \sqrt{\left(E_1^{(0)} - E_2^{(0)}\right)^2 + 4\,|V_{12}|^2}}. \tag{10.6}$$

Repeating this calculation with the second eigenvalue, I find the ratio of the coefficients for the second eigenvector:

$$\frac{a_1^{(2)}}{a_2^{(2)}} = -2\frac{V_{12}}{E_1^{(0)} - E_2^{(0)} + \sqrt{\left(E_1^{(0)} - E_2^{(0)}\right)^2 + 4\,|V_{12}|^2}}. \tag{10.7}$$

The normalization coefficients for these eigenvectors are too cumbersome and are not too informative, so I will leave the eigenvectors non-normalized. Both of them can be written as a superposition of vectors $|1\rangle$ and $|2\rangle$ with coefficients $a_{1,2}^{(1,2)}$ defined by Eqs. 10.6 and 10.7:

$$|E_{1,2}\rangle = a_1^{(1,2)}\,|1\rangle + a_2^{(1,2)}\,|2\rangle \tag{10.8}$$

where I used eigenvalues to label the corresponding eigenvectors.

The ratios of the coefficients in this superposition determine relative contributions of each of the original states into $|E_{1,2}\rangle$. These ratios depend on the relation between the inter-level spectral distance $\left|E_1^{(0)} - E_2^{(0)}\right|$ and the interaction matrix element $|V_{12}|$. If the former is much larger than the latter, I can expand the denominators of Eqs. 10.6 and 10.7 as

$$\sqrt{\left(E_1^{(0)} - E_2^{(0)}\right)^2 + 4|V_{12}|^2} \approx E_1^{(0)} - E_2^{(0)} + \frac{2|V_{12}|^2}{E_1^{(0)} - E_2^{(0)}}$$

where it is assumed for concreteness that $E_1^{(0)} > E_2^{(0)}$. Then Eqs. 10.6 and 10.7 yield

$$\frac{a_1^{(1)}}{a_2^{(1)}} \approx \frac{V_{12}\left(E_1^{(0)} - E_2^{(0)}\right)}{|V_{12}|^2} \gg 1$$

$$\frac{a_1^{(2)}}{a_2^{(2)}} \approx -\frac{V_{12}}{E_1^{(0)} - E_2^{(0)}} \ll 1.$$

Thus, the contributions of the state presented by vector $|2\rangle$ into the eigenvector $|E_1\rangle$ and of state $|1\rangle$ into the eigenvector $|E_2\rangle$ are very small. Not surprisingly, the energy E_1 in this limit is close to $E_1^{(0)}$, and E_2 is close to $E_2^{(0)}$ (check it out, please). These results justify the assumption lying in the foundation of the two-level model: contributions from energetically remote states can, indeed, be neglected. It also provides a quantitative condition for validity of this approximation: $\left|E_n^{(0)} - E_m^{(0)}\right| \gg |V_{nm}|$, where n, m are the labels for energy levels and the corresponding states.

It is easy to verify that if I reversed inequality $E_1^{(0)} > E_2^{(0)}$ and assumed instead that $E_1^{(0)} < E_2^{(0)}$, the role of vectors $|1\rangle$ and $|2\rangle$ would have interchanged: the main contribution to state $|E_1\rangle$ would have come from initial vector $|2\rangle$, and state $|E_2\rangle$ would have been mostly determined by $|1\rangle$. This flipping between the initial vectors is due to trivial but often overlooked property of the square root, $\sqrt{x^2} = |x|$, which is x when x is positive and $-x$ when it is negative. In one of the exercises, you are asked to verify this flipping phenomenon.

In the opposite limit $\left|E_1^{(0)} - E_2^{(0)}\right| \ll |V_{12}|$, the radical in Eqs. 10.6 and 10.7 can be approximated as

$$\sqrt{\left(E_1^{(0)} - E_2^{(0)}\right)^2 + 4|V_{12}|^2} \approx 2|V_{12}| \tag{10.9}$$

which is valid with accuracy up to terms of the order of $\left(E_1^{(0)} - E_2^{(0)}\right)^2 / |V_{12}|^2 \ll 1$. The ratios of the coefficients in this case become

$$\frac{a_1^{(1)}}{a_2^{(1)}} = -2\frac{V_{12}}{E_1^{(0)} - E_2^{(0)} - 2\,|V_{12}|} \approx e^{i\delta_V}\left(1 + \frac{E_1^{(0)} - E_2^{(0)}}{2\,|V_{12}|}\right)$$

$$\frac{a_1^{(2)}}{a_2^{(2)}} = -2\frac{V_{12}}{E_1^{(0)} - E_2^{(0)} + 2\,|V_{12}|} \approx -e^{i\delta_V}\left(1 - \frac{E_1^{(0)} - E_2^{(0)}}{2\,|V_{12}|}\right)$$

where I introduced the phase of the matrix element $V_{12} = |V_{12}|\exp(i\delta_V)$ and used approximation for $(1 + x)^{-1} \approx 1 - x$. Note that the correction to the main terms ($\pm \exp[i\delta_V]$) in both expressions is linear in $\left(E_1^{(0)} - E_2^{(0)}\right)/|V_{12}|$, which justifies approximation for the radical used in Eq. 10.9 (neglected quadratic terms are smaller than the linear ones kept in the expressions for the coefficients). The contributions of the initial eigenvectors in this limit are almost equal to each other in magnitude while differing in their phase by π (do I need to remind you that $-1 = \exp(i\pi)$?). Approximate expressions for the energy eigenvalues, Eqs. 10.6 and 10.7 in this limit, become (again neglecting quadratic terms in $\left(E_1^{(0)} - E_2^{(0)}\right)/|V_{12}|$)

$$E_1 = \frac{1}{2}\left(E_1^{(0)} + E_2^{(0)}\right) + |V_{12}| \tag{10.10}$$

$$E_2 = \frac{1}{2}\left(E_1^{(0)} + E_2^{(0)}\right) - |V_{12}|. \tag{10.11}$$

What is significant about this result is that even when the difference between initial energy levels is very small compared to the matrix element of the interaction, the difference between the actual eigenvalues is $|V_{12}|$ and is not small at all.

Experimentalists love the two-level models because they are simple (all what you need to know is how to solve quadratic equations), and they are tempted to use it as often as they can in disparate fields of physics. Theoreticians, of course, hate this model with as much fervor because if all of the physics could have been explained by a two-level model, all theoreticians would have lost their jobs. Luckily, this is not the case.

The physics described by this model becomes particularly interesting (and important) if the initial Hamiltonian $\hat{H}_0$ depends on some parameters, which can be controlled experimentally in such a way that the sign of the difference $E_1^{(0)} - E_2^{(0)}$ can be continuously altered. In this case, at certain value of this parameter, the two initial energy levels become degenerate, and if one plots dependence of $E_1^{(0)}$ and $E_2^{(0)}$ as functions of this parameter, the corresponding curves would cross at some point. This is an example of an accidental degeneracy, which is not related to any symmetry and occurs only at particular values of a system's parameters. Still, it happens in a number of physical systems and is of great interest because it affects how the system reacts to various stimuli. If, however, one plots the dependence of the actual eigenvalues as functions of the same parameter, the curves would not

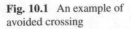

Fig. 10.1 An example of
avoided crossing

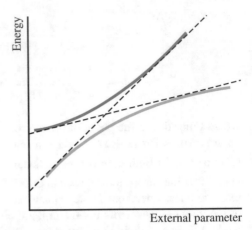

cross each other as is obvious from Eqs. 10.10 and 10.11. The curves representing
this dependence will now look like the ones shown in Fig. 10.1. You can see that
the curves do not cross each other anymore giving this phenomenon the name of
avoided level crossing.

This is a remarkable phenomenon, which is not easily appreciated. Let me try to
help you to understand what is so special about these two curves not crossing each
other. Let's begin far on the left from the point of the degeneracy, where $E_1^{(0)} > E_2^{(0)}$.
We ascertained that in this case the lower curve describes the energy of a state,
which is mostly $|2\rangle$, while the state whose energy belongs to the upper curve is
mostly $|1\rangle$. At the point of avoided crossing, the eigenvectors describing the state of
the system consist of both $|1\rangle$ and $|2\rangle$ in equal proportions. Now let's keep moving
along the lower curve, which means that we are turning the dial and experimentally
gradually changing our control parameter. After we will have passed the point of
avoided crossing, the relation between initial energy levels has changed: now we
have $E_1^{(0)} < E_2^{(0)}$. Now, the main contribution to the superposition represented by
the points on the lower curve comes from the state $|1\rangle$,[1] and if I move the system
far enough from the avoided crossing point, I will have a state mostly consisting
of the state $|1\rangle$. Now think about it: we started with the state of the system being
predominantly $|2\rangle$, and by continuously changing our parameter, we transformed
this state in the one which is now predominantly state $|1\rangle$. This is better than any
Hogwarts style transformation wizardry simply because it is not a magic and not an
illusion—just honest to earth quantum mechanics!

---

[1]Recall a comment I made at the end of the discussion of the limit $\left| E_1^{(0)} - E_2^{(0)} \right| \gg |V_{12}|$.

10.2 Two-Level System in a Harmonic Electric Field: Rabi Oscillations

Now let me switch gears and allow the perturbation operator $\hat{V}$ to become a function of time. More specifically, I will assume that perturbation matrix elements V_{12} and V_{21} have the following form:

$$V_{21}(t) = V_{12}(t) = \mathcal{E} \cos \Omega t,$$

where $\mathcal{E}$ is real. This form of the perturbation describes, for instance, a dipole interaction between a two-level system and a harmonic electric field and appears in many realistic situations. The Hamiltonian of the system in this case reads

$$\hat{H} = E_1^{(0)} |1\rangle \langle 1| + E_2^{(0)} |2\rangle \langle 2| + \mathcal{E} \cos \Omega t \left(|1\rangle \langle 2| + |2\rangle \langle 1|\right). \qquad (10.12)$$

This is the first time you are dealing with an explicitly time-dependent Hamiltonian in the Schrödinger picture, and this requires certain adjustments in the way of thinking about the problem. First of all, you have to accept the fact that you cannot present solutions in the form of $\exp\left(-iEt/\hbar\right)|\psi\rangle$, with $|\psi\rangle$ being an eigenvector of the Hamiltonian. Equation $\hat{H}|\psi\rangle = E|\psi\rangle$ with time-dependent Hamiltonian and time-independent $|\psi\rangle$ does not make sense anymore. In other words, the stationary states do not exist in the case of time-dependent Hamiltonians, and we need, therefore, a new way of solving the time-dependent Schrödinger equation. No one can forbid you, however, to use eigenvectors of any time-independent Hamiltonian as a basis, because basis is a basis regardless of the properties of the Hamiltonian. The choice of the basis is determined solely by the reason of convenience, and it is especially convenient in this case to use eigenvectors of $\hat{H}_0$ presented by vectors $|1\rangle$ and $|2\rangle$. Thus, let me present the unknown time-dependent state vector $|\psi(t)\rangle$ as a linear combination of these vectors:

$$|\psi(t)\rangle = a_1(t) \exp\left(-\frac{iE_1^{(0)}t}{\hbar}\right)|1\rangle + a_2(t) \exp\left(-\frac{iE_2^{(0)}t}{\hbar}\right)|2\rangle \qquad (10.13)$$

with some unknown coefficients $a_{1,2}$. This expression reminds very much Eq. 4.15 for a general solution with a time-independent Hamiltonian but with two significant differences—first, the basis used in Eq. 10.13 is not formed by eigenvectors of the total Hamiltonian $\hat{H}$, which does not have eigenvectors (at least not in a regular sense of the word), and second, the expansion coefficients now are unknown functions of time, while their counterparts in Eq. 4.15 were constants. You might wonder at this point if I am allowed to separate the exponential factors characteristic of the time dependence of the stationary states. A simple answer is: "Why not?" As long as I allow for the yet undetermined time dependence of the residual coefficients, I can factor out any time-dependent function I want. It will affect the equations, which these coefficients obey, but not the final result. The most meticulous of you might also ask that even if it is allowed to pull out these factors, why bother doing it? This is a more valid question, which deserves a more detailed answer. Let me begin by saying that I did not have to do it: the earth would not stop in its tracks if I did not, and we would still solve the problem. However, by doing so, I reflect a somewhat deeper understanding of two distinct sources of the time dependence of the vector states. One is a trivial dependence given by these exponential factors, which would have existed even if the Hamiltonian did not depend on time. These exponential factors have nothing to do with the time dependence of the Hamiltonian. Factoring them out right away, I ensure that the remaining time dependence of the coefficients reflects only genuine nontrivial dynamics. As an extra bonus, I hope that by doing so, I will arrive at equations that are easier to analyze.

Substitution of Eq. 10.13 to the left-hand side of the Schrödinger equation $i\hbar d \,|\psi\rangle \,/dt$ yields

$$
i\hbar \frac{d \,|\psi\rangle}{dt} = E_1^{(0)} a_1(t) \exp \left(-\frac{iE_1^{(0)}t}{\hbar} \right) |1\rangle + i\hbar \frac{da_1(t)}{dt} \exp \left(-\frac{iE_1^{(0)}t}{\hbar} \right) |1\rangle +
$$

$$
E_2^{(0)} a_2(t) \exp \left(\frac{iE_2^{(0)}t}{\hbar} \right) |2\rangle + i\hbar \frac{da_2(t)}{dt} \exp \left(-\frac{iE_2^{(0)}t}{\hbar} \right) |2\rangle . \qquad (10.14)
$$

The right-hand side of this equation, $\hat{H} \,|\psi\rangle$, with $\hat{H}$ defined by Eq 10.12 and $|\psi\rangle$ by Eq. 10.13 becomes

$$
\hat{H} \,|\psi\rangle = E_1^{(0)} a_1(t) \exp \left(-\frac{iE_1^{(0)}t}{\hbar} \right) |1\rangle + E_2^{(0)} a_2(t) \exp \left(\frac{iE_2^{(0)}t}{\hbar} \right) |2\rangle +
$$

$$
\mathcal{E} a_2(t) \cos \Omega t \exp \left(-\frac{iE_2^{(0)}t}{\hbar} \right) |1\rangle + \mathcal{E} a_1(t) \cos \Omega t \exp \left(-\frac{iE_1^{(0)}t}{\hbar} \right) |2\rangle \qquad (10.15)
$$

where I took into account the orthogonality of the basis states. Equating coefficients in front of vectors $|1\rangle$ and $|2\rangle$ on the left- and right-hand sides of the Schrödinger equation (Eqs. 10.14 and 10.15 correspondingly) results in differential equations for the time-dependent coefficients $a_{1,2}(t)$:

$$
i\hbar \frac{da_1(t)}{dt} = \mathcal{E} a_2(t) \cos \Omega t \exp \left(\frac{i\left[E_1^{(0)} - E_2^{(0)} \right]t}{\hbar} \right) \qquad (10.16)
$$

$$
i\hbar \frac{da_2(t)}{dt} = \mathcal{E} a_1(t) \cos \Omega t \exp \left(-\frac{i\left[E_1^{(0)} - E_2^{(0)} \right]t}{\hbar} \right) . \qquad (10.17)
$$

Factors $\exp\left(\pm i\left[E_1^{(0)} - E_2^{(0)}\right] t/\hbar\right)$ on the right-hand side in these equations appeared as a result of eliminating the corresponding exponential factors $\exp\left(-iE_{1,2}^{(0)}t/\hbar\right)$ from their left-hand sides. Note that energy eigenvalues appear in these equations only in the form of their difference, which is just another manifestation of the already mentioned fact that the absolute values of the energy levels are irrelevant. To simplify the notations, let me introduce a so-called transition frequency:

$$\omega_{12} = \frac{E_1^{(0)} - E_2^{(0)}}{\hbar} \tag{10.18}$$

where I again for concreteness assumed that $E_1^{(0)} - E_2^{(0)} > 0$. Introducing this notation and replacing $\cos \Omega t$ by the sum of the respective exponential functions, I can rewrite Eqs. 10.16 and 10.17 in the following form:

$$i\hbar \frac{da_1(t)}{dt} = \frac{1}{2}\mathcal{E}a_2(t)\left(\exp\left[i\left(\omega_{12} - \Omega\right)t\right] + \exp\left[i\left(\omega_{12} + \Omega\right)t\right]\right) \tag{10.19}$$

$$i\hbar \frac{da_2(t)}{dt} = \frac{1}{2}\mathcal{E}a_1(t)\left(\exp\left[-i\left(\omega_{12} - \Omega\right)t\right] + \exp\left[-i\left(\omega_{12} + \Omega\right)t\right]\right) \tag{10.20}$$

Equations 10.19 and 10.20 cannot be solved analytically. However, the most interesting phenomena described by these equations occur when $\omega_{12} - \Omega \ll \omega_{12} + \Omega$, in which case I can introduce an effective approximation capturing the most important properties of the model (obviously, something will be left out, and there might be situations when this something becomes important, but I am going to pretend that such situations do not concern me at all). In order to formulate this approximation, it is convenient to introduce a parameter $\Delta = \omega_{12} - \Omega$ called *frequency detuning*. In the case of the small detuning, the two exponential terms in Eqs. 10.19 and 10.20 change with time on significantly different time scales. Terms containing $\omega_{12} - \Omega$ oscillate with a much larger period (much slower) as compared to the terms containing $\omega_{12} + \Omega$, which exhibit comparatively fast oscillations.

In order to understand why fast oscillations are not effective in influencing the behavior of the system, imagine a regular pendulum acted upon by a force, which changes its direction faster than the pendulum manages to react to it (it is called inertia, in case you forgot, and it takes some time for any quantity to change by any appreciable amount). What will happen to the pendulum in this case? Right before it has any chance to move in the initial direction of the force, the force will have already changed and push the pendulum in the opposite direction. This is a very frustrating situation, so the pendulum will just stay where it is. This effect in a scientific jargon is called self-averaging—the force changes so much faster than the reaction time of the pendulum that it effectively averages itself out to zero. Taking advantage of this self-averaging effect, I will drop the fast-changing terms in Eqs. 10.19 and 10.20, turning them into

$$ i\hbar\frac{da_1(t)}{dt} = \frac{1}{2}\mathcal{E}a_2(t)\exp(i\Delta t) \tag{10.21} $$

$$ i\hbar\frac{da_2(t)}{dt} = \frac{1}{2}\mathcal{E}a_1(t)\exp(-i\Delta t). \tag{10.22} $$

Differentiating the first of these equations with respect to time, I get

$$ i\hbar\frac{da_1^2(t)}{dt^2} = \frac{1}{2}\mathcal{E}\frac{da_2(t)}{dt}\exp(i\Delta t) + \frac{1}{2}i\Delta\mathcal{E}a_2(t)\exp(i\Delta t). $$

Now, taking da_2/dt from Eq. 10.22 while expressing $a_2(t)$ in terms of da_1/dt using Eq. 10.21, I am getting rid of coefficient a_2 and derive an equation containing only a_1:

$$ \frac{da_1^2(t)}{dt^2} - i\Delta\frac{da_1(t)}{dt} + \frac{1}{4\hbar^2}\mathcal{E}^2 a_1(t) = 0. \tag{10.23} $$

Did you notice how the time-dependent exponents in Eq. 10.23 magically disappeared turning it into a regular linear differential equation of the second order with constant coefficients? You might notice that this is the same equation which describes (among other things) a motion of a damped harmonic oscillator with damping represented by a term with the first time derivative. This might appear a bit troublesome, because the motion of a damped harmonic oscillator is characterized by exponential decay of the respective quantities with time, and this is not the behavior which we would like our quantum state to have. However, before going into a panic mode, look at the equation a bit more carefully, and then you might notice that "the damping" coefficient (whatever appears in front of da_1/dt) is purely imaginary, so no real damping takes place, and you can breathe easier.

Damping or no damping, I know that equations of the type of Eq. 10.23 are solved by an exponential function, which I choose in the form of $\exp(i\omega t)$. Substitution of this function into Eq. 10.23 yields an equation for the yet unknown parameter ω:

$$ \omega^2 - \Delta\omega - \frac{1}{4}\Omega_R^2 = 0, $$

where I introduced a new quantity of the dimension of frequency

$$ \Omega_R = \frac{\mathcal{E}}{\hbar}, \tag{10.24} $$

which plays an important role in the phenomena we are about to uncover. The quadratic equation for ω has two solutions:

$$ \omega_\pm = \frac{1}{2}\Delta \pm \frac{1}{2}\sqrt{\Delta^2 + \Omega_R^2} \tag{10.25} $$

(both of which are, by the way, real) so that the general solution to Eq. 10.23 takes the form

$$a_1 = A \exp(i\omega_+ t) + B \exp(i\omega_- t).$$ (10.26)

Expression for the second coefficient, a_2, is found using Eq. 10.21:

$$a_2 = \frac{2i\hbar}{\mathcal{E}} \exp(-i\Delta t) \frac{da_1(t)}{dt} =$$

$$-\frac{2}{\Omega_R} \exp(-i\Delta t) [A\omega_+ \exp(i\omega_+ t) + B\omega_- \exp(i\omega_- t)].$$

Combining exponential functions in this equation, you might notice the emergence of two frequencies, $\omega_+ - \Delta$ and $\omega_- - \Delta$, which can be evaluated into

$$\omega_+ - \Delta = -\frac{1}{2}\Delta + \frac{1}{2}\sqrt{\Delta^2 + \Omega_R^2} = -\omega_-$$

$$\omega_- - \Delta = -\frac{1}{2}\Delta - \frac{1}{2}\sqrt{\Delta^2 + \Omega_R^2} = -\omega_+$$

allowing you to write an expression for a_2 as

$$a_2 = -\frac{2}{\Omega_R} [A\omega_+ \exp(-i\omega_- t) + B\omega_- \exp(-i\omega_+ t)].$$ (10.27)

Amplitudes A and B in Eqs. 10.26 and 10.27 are yet undetermined; to find them I have to specify initial conditions for Eqs. 10.21 and 10.22, the issue which I have not even mentioned yet. At the same time, you are perfectly aware that any problem involving a time evolution is not complete without initial conditions, which in quantum mechanics mean a state of the system at some instant of time defined as $t = 0$.

It is usually assumed in this type of problems that one can "turn on" the time-dependent interaction at some instant determined by the will of the experimentalist, and in many cases it does make sense. For instance, the time-dependent term in Hamiltonian 10.12 can represent a laser beam, which you can, indeed, turn on and off at will. In this case one can prepare the system to be in a specific state before the laser is turned on and study how this state will evolve due to the interaction with the laser radiation. It is simplest to prepare the system in the lowest energy stationary state, and so this is what I will choose as the initial condition:

$$|\psi(0)\rangle = |2\rangle.$$

Taking into account Eq. 10.13, I can translate it into the following initial conditions for the dynamic variables a_1 and a_2:

$$a_1(0) = 0 \tag{10.28}$$

$$a_2(0) = 1. \tag{10.29}$$

Substituting $t = 0$ into Eqs. 10.26 and 10.27 and using Eqs. 10.28 and 10.29, I derive the following equations for amplitudes A and B:

$$A + B = 0;$$

$$-\frac{2}{\Omega_R}[A\omega_+ + B\omega_-] = 1,$$

which are easily solved to yield

$$A = -B = -\frac{\Omega_R}{2(\omega_+ - \omega_-)}.$$

It is easy to see using Eq. 10.25 that

$$\omega_+ - \omega_- = \sqrt{\Delta^2 + \Omega_R^2},$$

so that the amplitudes take on the value

$$A = -B = -\frac{\Omega_R}{2\sqrt{\Delta^2 + \Omega_R^2}}.$$

Having found A and B, I can write down the final solutions for the time-dependent coefficients $a_{1,2}(t)$:

$$a_1 = \frac{\Omega_R}{2\sqrt{\Delta^2 + \Omega_R^2}}[\exp(i\omega_- t) - \exp(i\omega_+ t)] \tag{10.30}$$

$$a_2 = \frac{1}{\sqrt{\Delta^2 + \Omega_R^2}}[\omega_- \exp(-i\omega_+ t) - \omega_+ \exp(-i\omega_- t)]. \tag{10.31}$$

These equations formally solve the problem I set out for you to solve: you now know the time-dependent state of the two-level system described by Hamiltonian 10.12 at any instant of time. But I wouldn't blame you if you still have this annoying gnawing feeling of not being quite satisfied, probably because you are not quite sure what to do with this solution and what kind of useful physical information you can dig out from it. Indeed, the standard interpretation of coefficients in expressions similar to Eq. 10.13 as probability amplitudes, whose squared absolute values yield the probability of obtaining a corresponding value of an observable whose eigenvectors are used as a basis, wouldn't work here. The problem is that we are using the basis provided by eigenvectors of the Hamiltonian of a system, which does not exist anymore, so that this traditional interpretation does not make much sense.

One way to make sense out of Eqs. 10.26 and 10.27 is to recognize that in a typical experiment, the time-dependent interaction does not last forever—it starts at some instant, which you can designate as $t = 0$, and it usually ends at some time $t = t_f$ (for instance, when a graduate student running the experiment gets tired, turns the laser off, and goes on a date). So, after the time-dependent part of the Hamiltonian vanishes, you are back to the standard situation, but the system is now in a superposition state defined by the values of the coefficients $a_{1,2}$ at the time, when the laser got switched off. Now, you can quickly take the measurement of the energy and interpret the results in terms of probabilities of getting one of two values: $E_1^{(0)}$ or $E_2^{(0)}$. The probability $p\left(E_1^{(0)}\right)$ that the measurement would yield $E_1^{(0)}$ is given as usual by $|a_1|^2$, which according to Eq. 10.30 is

$$p\left(E_1^{(0)}\right) = \frac{\Omega_R^2}{4\left(\Delta^2 + \Omega_R^2\right)}\left(2 - \exp\left[i\left(\omega_+ - \omega_-\right)t_f\right] - \exp\left[-i\left(\omega_+ - \omega_-\right)t_f\right]\right) =$$

$$\frac{\Omega_R^2}{2\left(\Delta^2 + \Omega_R^2\right)}\left[1 - \cos\left(\omega_+ - \omega_-\right)t_f\right] = \frac{\Omega_R^2}{\left(\Delta^2 + \Omega_R^2\right)}\sin^2\frac{\omega_+ - \omega_-}{2}t_f =$$

$$\frac{\Omega_R^2}{\left(\Delta^2 + \Omega_R^2\right)}\sin^2\frac{\sqrt{\Delta^2 + \Omega_R^2}}{2}t_f. \tag{10.32}$$

The probability that this measurement would yield value $E_2^{(0)}$ could have been computed in exactly the same manner, and I will give you a chance to do it, as an exercise, but here I will be smart and take advantage of the fact that $p\left(E_1^{(0)}\right) + p\left(E_2^{(0)}\right) = 1$, so that without much ado, I can present you with

$$p\left(E_2^{(0)}\right) = 1 - \frac{\Omega_R^2}{\left(\Delta^2 + \Omega_R^2\right)}\sin^2\frac{\sqrt{\Delta^2 + \Omega_R^2}}{2}t_f =$$

$$\frac{\Delta^2}{\left(\Delta^2 + \Omega_R^2\right)}\sin^2\frac{\sqrt{\Delta^2 + \Omega_R^2}}{2}t_f + \cos^2\frac{\sqrt{\Delta^2 + \Omega_R^2}}{2}t_f. \tag{10.33}$$

Equations 10.32 and 10.33 create a clear physical picture of what is happening with our system. The first thing to note is the periodic oscillations of the probabilities with time with frequency $\Omega_{GR} = \sqrt{\Delta^2 + \Omega_R^2}$ called generalized Rabi frequency (note that the factor $1/2$ in the arguments of the cos and sin functions in these equations is the result of transition from $\cos x$ to the functions of $x/2$ and is, therefore, not included into the definition of the frequency). There exist special times $t_{f_n} =$

Fig. 10.2 Oscillations of $p\left(E_1^{(0)}\right)$ for three values of the detuning: $\triangle = 0$, $\triangle/\Omega_R = 0.5$, and $\triangle/\Omega_R = 1.5$

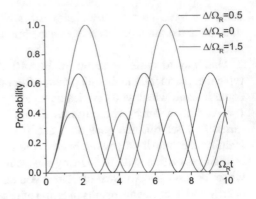

$\pi n/\Omega_{GR}$, where n is an integer, when the probability that the system will be found in the higher energy state is zero, and there are times $t_{f_n} = \pi n/\Omega_{GR} + \pi/2$ when this probability acquires its maximum value $\Omega_R^2/\left(\triangle^2 + \Omega_R^2\right)$. For probability $p\left(E_2^{(0)}\right)$, the situation is reversed—the probability reaches its value of unity at certain times $t_{f_n} = \pi n/\Omega_{GR}$, but its minimum value occurring at $t_{f_n} = \pi n/\Omega_{GR} + \pi/2$ is not zero, but is equal to $\triangle^2/\left(\triangle^2 + \Omega_R^2\right)$. Figure 10.2 depicts these oscillations of probability known as Rabi oscillations. The period of these oscillations as well as maximum and minimum values of the corresponding probabilities depend on the detuning parameter $\triangle$ controlled by the experimentalists. For large detuning $\triangle \gg \Omega_R$, the frequency of oscillations is determined mostly by $\triangle$, but their swing (a difference between largest and smallest values) diminishes. For instance, the largest value of $p\left(E_1^{(0)}\right)$ becomes of the order of $\Omega_R^2/\triangle^2 \ll 1$, while the smallest value of $p\left(E_2^{(0)}\right)$ is in this case very close to unity: $1 - \Omega_R^2/\triangle^2$. For both probabilities there are not much oscillations to speak of. A more interesting situation arises in the case of small detuning, with the special case of zero detuning being of most interest. The frequency of Rabi oscillations in this case becomes smallest and is equal to Ω_R, which is called Rabi frequency, and the probabilities swing between exact zero and exact unity becoming the most pronounced.

If you are interested how one can observe Rabi oscillations, here is an example of how it can be done. Imagine that you subject an ensemble of two-level systems to a strong time-periodic electric field with small detuning and turn it off at different times. The timing of the switching-off will determine the probability of a two-level system to be in the higher energy state. The fraction of the systems in the ensemble in this state is proportional to the corresponding probability. The systems will eventually undergo transition to the lower energy level and emit light. The intensity of the emitted light will be proportional to the number of systems in the upper state and will change periodically with the switching-off time. In real experiments there is no actual need to turn the electric field on and off all the time because spontaneous transitions of the system from upper to lower energy states happen even with the electric field on, and when this transition happens, the system is kicked off its normal dynamic so hard that it forgets everything about what was happening to it before that, so that the whole process starts anew. These kicks serve effectively as switches for the electric field. Oscillations in this case can be observed as functions of Rabi frequency controlled by the strength of the applied electric field. It is important that Rabi oscillations can be observed only if their period is shorter than the time interval between the "kicks." To fulfill this condition, the applied electric field must be strong enough to yield oscillations with a sufficiently high frequency.

10.3 Problems

Problems for Sect. 10.1

Problem 136 Find the approximate expression for energy levels of a two-level system with a time-independent perturbation in the limit $|V_{12}| \ll |E_1 - E_2|$ for two cases: $E_1 > E_2$ and $E_1 < E_2$.

Problem 137 Assume that the perturbation part of the Hamiltonian is given by $\hat{V} = \hat{z}\mathcal{E}$, where $\mathcal{E}$ is the electric field and $\hat{z}$ is the respective coordinate operator. Assume also that the wave functions of the states included in the Hamiltonian are described (in the coordinate representation) by wave functions defined on the one-dimensional interval $-\infty < z < \infty$:

$$\langle z | E_1 \rangle = \frac{1}{\sqrt{a_B}} \exp\left(-|z|/a_B\right)$$

$$\langle z | E_2 \rangle = \sqrt{\frac{2}{a_B^3}}\, z \exp\left(-|z|/a_B\right),$$

where a_B is the Bohr radius for an electron in the hydrogen atom, and that the electric field is given in terms of the binding energy of the hydrogen atom W_b and electron's charge e as $\mathcal{E} = W_b/ea_B$. The unperturbed energy levels are given as

$$E_1 = W_b(1 + u)$$

$$E_2 = W_b(1 - u),$$

where u is a dimensionless parameter that can be changed between values -2 and 2.

1. Find the perturbation matrix V_{ij}.
2. Find the eigenvalues of the full Hamiltonian, and plot them as a function of the u.
3. Also find the eigenvectors of the full Hamiltonian, and plot the ratio of the relative weights of the initial vectors $|E_{1,2}\rangle$ to the both found eigenvectors as functions of u.
4. Also, consider phases of the ratio of the coefficients c_1/c_2 for both eigenvectors, and plot their dependence on parameter u.

5. In all plots pay special attention to the region around $u = 0$, and describe how the behavior of the eigenvalues of the perturbed Hamiltonian differs from the corresponding behavior of the unperturbed energies.
6. Describe the behavior of the absolute values and the phases of c_1/c_2 in the vicinity of $u = 0$.

Problems for Sect. 10.2

Problem 138 Find the probability that the measurement of the energy will yield value E_2 directly from the coefficient a_2 in Eq. 10.33, and verify that the expression for this probability given in the text is correct.

Problem 139 Find the time dependence of the probabilities $p(E_{1,2})$ assuming that at time $t = 0$ the system was in the state $|1\rangle$.

Chapter 11
Non-interacting Many-Particle Systems

11.1 Identical Particles in the Quantum World: Bosons and Fermions

Quantum mechanical properties of a single particle are an important starting point for studying quantum mechanics, but in real experimental and practical situations, you will rarely deal with just a single particle. Most frequently you encounter systems consisting of many (from two to infinity) interacting particles. The main difficulty in dealing with many-particle systems comes from a significantly increased dimensionality of space, where all possible states of such systems reside. In Sect. 9.4 you saw that the states of the system of two spins belong to a four-dimensional spinor space. It is not too difficult to see that the states of a system consisting of N spins would need a 2^N-dimensional space to fit them all. Indeed, adding each new spin $1/2$ particle with two new spin states, you double the number of basis vectors in the respective tensor product, and even the system of as few as ten particles inhabits a space requiring 1024 basis vectors. More generally, imagine that you have a particle which can be in one of M mutually exclusive states, represented obviously by M mutually orthogonal vectors (I will call them single-particle states), which can be used as a basis in this single-particle M-dimensional space. You can generate a tensor product of single-particle spaces by stacking together M basis vectors from each single-particle space. Naively you might think that the dimension of the resulting space will be M^N, but it is not always so. The reality is more interesting, and to get the dimensionality of many-particle states correctly, you need to dig deeper into the concept of identity of quantum particles.

In the classical world, we know that all electrons are the same—the same charge and the same mass—but if necessary we can still distinguish between them saying that this is an electron with such and such initial coordinates and initial velocity, and therefore, it follows this particular trajectory. A second electron, which is exactly the same as the first one, but starting out with different initial conditions, follows its own trajectory. And even if these two electrons interact, scatter off each other, we can still

© Springer International Publishing AG, part of Springer Nature 2018

L.I. Deych, *Advanced Undergraduate Quantum Mechanics*,

https://doi.org/10.1007/978-3-319-71550-6_11

Fig. 11.1 Two distinguishable classical electrons interact with each other and follow their own distinguishable trajectory. We can easily say which electron follows which trajectory

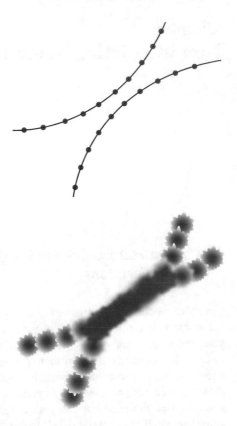

Fig. 11.2 Propagating clouds of probabilities representing the particles. In the interaction region, the clouds overlap, and the individuality of the particles is lost (Warning: it is dangerous to take this cartoon too seriously!)

say which electron is which by following their trajectories (see Fig. 11.1). Thus, we say that classical electrons, even though they are identical, are still distinguishable.

The situation changes when you are talking about quantum particles. In essentially the same setup—two particles approach each other from opposite directions, interact, and move each in its new direction—the situation becomes completely different. Now instead of two well-localized particles with perfectly defined trajectories, you are dealing with moving amorphous clouds of probabilities, and when they approach each other and overlap, all you can measure is the probability to find one or two particles within a certain region of space, but you have no means to tell which of the observed particles is which (Fig. 11.2). In quantum mechanics the individuality of particles is completely lost—they are not just identical, but they are indistinguishable.

Now the questions arise: how to formally describe this indistinguishability, and what are the observable consequences of this property? To begin, let me formally assign numbers 1 and 2 to the two particles and assume that particle 1 is in the state described by vector $|\alpha^{(1)}\rangle$, where α indicates a particular quantum state and 1 assigns this state to the first particle, and the second particle is in the state $|\beta^{(2)}\rangle$. The space of the two-particle states can be generated by the tensor product of the single-particle states with a two-vector basis:

$$\left|\psi_1^{(tp)}\right\rangle = \left|\alpha^{(1)}\right\rangle \left|\beta^{(2)}\right\rangle \tag{11.1}$$

$$\left|\psi_2^{(tp)}\right\rangle = \left|\alpha^{(2)}\right\rangle \left|\beta^{(1)}\right\rangle, \tag{11.2}$$

where the second vector is obtained from the first by replacing particle 1 with particle 2. This operation can be formally described by a special "exchange" operator $\hat{P}(1,2)$ whose job is to interchange indexes of the particles assigned to each state:

$$\left|\alpha^{(2)}\right\rangle \left|\beta^{(1)}\right\rangle = \hat{P}(1,2)\left|\alpha^{(1)}\right\rangle \left|\beta^{(2)}\right\rangle.$$

When applied twice, this operator obviously leaves the initial vector intact (two exchanges $1 \to 2$, $2 \to 1$ are equivalent to no exchange at all), meaning that $\hat{P}^2(1,2) = \hat{I}$. An immediate consequence of this identity is that eigenvalues of this operator are either equal to 1 or -1.

Using the exchange operator, the concept of indistinguishability can be formulated in a precise and formal way. Consider an arbitrary state of two particles represented by vector $|\psi(1,2)\rangle$. If particles 1 and 2 are truly indistinguishable, then vector $|\psi(2,1)\rangle = \hat{P}(1,2)|\psi(1,2)\rangle$ and initial vector $|\psi(1,2)\rangle$ must represent the same state, which means that they can differ from each other only by a constant factor. The formal representation of the last statement looks like this:

$$\hat{P}(1,2)|\psi(1,2)\rangle = |\psi(2,1)\rangle = \lambda |\psi(1,2)\rangle, \tag{11.3}$$

which makes it clear that if $|\psi(1,2)\rangle$ represents a state of indistinguishable particles, it must be an eigenvector of $\hat{P}(1,2)$. The remarkable thing about this conclusion is that there are only two types of such eigenvectors—those corresponding to eigenvalue 1 and those that belong to eigenvalue -1, i.e., any vector describing a state of indistinguishable particles must belong to one of two classes: symmetric (even) with respect to the exchange of the particles when

$$\hat{P}(1,2)|\psi(1,2)\rangle = |\psi(1,2)\rangle \tag{11.4}$$

or antisymmetric (odd) if

$$\hat{P}(1,2)|\psi(1,2)\rangle = -|\psi(1,2)\rangle. \tag{11.5}$$

Moreover, Hamiltonians of indistinguishable particles obviously do not change when the particles are exchanged (otherwise they wouldn't be indistinguishable), which means that the exchange operator and the Hamiltonian commute:

$$\left[\hat{H}, \hat{P}(1,2)\right] = 0 \tag{11.6}$$

(if it is not clear where it comes from, check the discussion around the parity operator in Sect. 5.1, where similar issues were raised). In the context of the exchange operator, Eq. 11.6 signifies two things. First is that the Hamiltonian and the exchange operator are compatible and share a common system of eigenvectors. In other words, eigenvectors of a Hamiltonian of two indistinguishable particles must be either symmetric or antisymmetric. Second, if you treat the exchange operator as a representative of a specific observable that takes only two values depending on the symmetry of the state, Eq. 11.6 indicates that the expectation value of this observable does not change with time (see Sect. 4.1.3 and the discussion around Eq. 4.17 there). Accordingly, Eq. 11.6 ensures that if a two-particle system starts out in a symmetric or antisymmetric state, it will remain in this state forever.

While it is useful to know that all states of indistinguishable particles must belong to one of the two symmetry classes and that a system put in a symmetry class at some instant of time will stay in this class forever, we still do not know how to relate the symmetry of the states of the particular system of particles to their other properties: does it depend on the particle's charges, masses, and potential they are moving on, or can it be somehow created deliberately through a clever measurement process? I personally find the answer to all these questions, which I am about to reveal to you, quite amazing: the symmetry of any state of indistinguishable particles cannot be "chosen" or changed; the particles are born with predestined fate to be only in states with one or another symmetry predetermined by their **spin**. It turns out that particles with half-integer spin can exist only in antisymmetric states, while particles with integer spins can be only in symmetric states. This statement is called a spin-statistics theorem and is, in my view, one of the most amazing fundamental results, which follows purely mathematically from the requirement that quantum mechanics agrees with the relativity theory. Just stop to think about it: quantum mechanics deals with phenomena occurring at very small spatial, temporal, mass, and energy scales, while relativity theory explains the behavior of nature at very large velocities. Apparently, quantum mechanics and relativity overlap when the interaction between light and matter is involved and, at high energies, when particle-antiparticle phenomena become important. However, the theoretical requirements of the self-consistency of the theory, one of which is the spin-statistics theorem, are felt well outside of these overlap areas and penetrate all of quantum mechanics from atomic energy structure to electric, magnetic, and optical properties of solids and fluids. Two better known phenomena made possible by the spin-statistics connection are superfluidity and superconductivity. The proof of this theorem relies heavily on quantum field theory, and you will sleep better at night by just accepting it as one of the axioms of quantum mechanics.

The spin-statistics theorem was proved by Wolfgang Pauli in 1939 but published only in 1940. The footnote to Pauli's paper in *Physical Review* states that the paper is a part of the report prepared for the Solvay Congress 1939, which did not take place because of the war in Europe. By the time of that publication, Pauli had moved from Zurich to Princeton, because Switzerland rejected his request for Swiss citizenship on the ground of him becoming a German citizen after Hitler annexed his native Austria. Anyway, to finish with this theorem, I only need to mention that the particles with half-integer spins are called *fermions* (in honor of Enrico Fermi, an Italian physicist, who had to leave Italy after Mussolini came to power; he moved to the USA, where he created the world's first nuclear reactor and played a crucial role in the Manhattan Project), while particles with whole spins are called *bosons* after Indian physicist Satyendra Nath Bose, who worked on the system of indistinguishable photons as early as in 1924. It is interesting that after an initial attempt to publish his paper on this topic failed, Bose sent it to Einstein asking Einstein's opinion and assistance with publication. Einstein translated the paper into German and published in the leading German physics journal of the time *Zeitschrift für Physik* (under Bose's name, of course).

Before getting back to the business of doing practical quantum mechanics with many-particle systems, a few additional words about fermions and bosons might be useful. Among elementary particles constituting regular matter, fermions are most abundant: electrons, protons, and neutrons—the main building blocks of atoms, molecules, and the rest of the material world are all spin 1/2 fermions. The only elementary boson you would encounter in regular setting would be a photon— a quantum of the electromagnetic field—and not a regular material particle. This can be taken as a general rule—as long as we are talking about elementary particles, the matter is represented by fermions, while the interaction fields, and other objects, which would classically be presented as waves in quantum mechanics, become bosons. Other examples of bosons you can find are quantized elastic waves (phonons) or quantized magnetic waves (magnons).

However, the concept of fermions and bosons can be extended to composite particles as long as the processes they are taking part in do not change their internal structure. The most famous examples of such composite particles are electron Cooper pairs (Cooperons, named after American physicist Leon Cooper who discovered them in 1956), responsible for superconductivity phenomenon, and He^4 nuclei, which in addition to two mandatory protons contain two neutrons, making the total number of particles in the nucleus equal to 4. In both these examples, we are dealing with composite bosons. Indeed, a pair of electrons, as you already know, can be in the state with total spin either 0 or 1, i.e., the spin of the pair is in either case integer. In the case of He^4 nucleus, there are four spin 1/2 particles, and by diving them into two pairs, you can also see that the total spin of this system can again only be an integer. Another interesting example of composite bosons is an exciton in semiconductors, which consists of an electron and a hole,[1] both with spin 1/2. The extent to which the inner structure in all these examples can be neglected and the particles can be treated as bosons depends on the amount of energy required to disintegrate them into their constituent parts. For Cooperons this energy is quite small—of the order of 10^{-3} eV—which explains

---

[1] Energy levels in a semiconductor are organized in bands separated by large gaps. The band, all energy levels of which are filled with electrons, is called a valence band, and the closest to its empty band is a conduction band. When an electron gets excited from the valence band, the conduction band acquires an electron, and a valence band losses an electron, which leaves in its stead a positively charged hole. Here you have an electron-hole pair behaving as real positively and negatively charged spin 1/2 particles.

why they can only survive at very low (below $10\,K$) temperatures; exciton binding energies vary over a rather large range between several millielectronvolts and several hundred millielectronvolts, depending upon the material, and they, therefore, survive at temperatures between $10\,K$ and the room temperature ($300\,K$). He^4 nucleus, of course, is the most stable of all the composite particles discussed: it takes the whole 28.3 mega-electronvolts to take it apart.

After this short and hopefully entertaining detour, it is time to get back to business of figuring out how to implement the requirements of the spin-statistics theorem in practical calculations. A generic vector representing a two-particle state and expressed as a linear combination of basis vectors 11.1 and 11.2

$$|\psi(1,2)\rangle = a_1 \left|\alpha^{(1)}\right\rangle\left|\beta^{(2)}\right\rangle + a_2 \left|\alpha^{(2)}\right\rangle\left|\beta^{(1)}\right\rangle$$

with arbitrary coefficients $a_{1,2}$ does not obey the required symmetry condition. However, after a few minutes of contemplation and silent staring at this expression, you will probably see that you can satisfy the symmetry requirements of Eq. 11.4 by choosing $a_1 = a_2$, while Eq. 11.5 can be made happy with the choice $a_1 = -a_2$. (If you are not that big on contemplation, just switch the particles in the expression for $|\psi(1,2)\rangle$, and write down the conditions of Eq. 11.4 or 11.5 explicitly.) If in addition to symmetry you want your two-particle states to be also normalized, you can choose for fermions

$$|\psi_f(1,2)\rangle = \frac{1}{\sqrt{2}}\left[\left|\alpha^{(1)}\right\rangle\left|\beta^{(2)}\right\rangle - \left|\alpha^{(2)}\right\rangle\left|\beta^{(1)}\right\rangle\right] \tag{11.7}$$

and for bosons

$$\left|\psi_b^{(1)}(1,2)\right\rangle = \frac{1}{\sqrt{2}}\left[\left|\alpha^{(1)}\right\rangle\left|\beta^{(2)}\right\rangle + \left|\alpha^{(2)}\right\rangle\left|\beta^{(1)}\right\rangle\right]. \tag{11.8}$$

While Eq. 11.7 exhausts all possible two-particle states for fermions, in the case of bosons, two more states, in which different particles occupy the same single-particle state, can be constructed:

$$\left|\psi_b^{(2)}(1,2)\right\rangle = \left|\alpha^{(1)}\right\rangle\left|\alpha^{(2)}\right\rangle \tag{11.9}$$

$$\left|\psi_b^{(3)}(1,2)\right\rangle = \left|\beta^{(1)}\right\rangle\left|\beta^{(2)}\right\rangle. \tag{11.10}$$

An attempt to arrange a similar state for fermions fails because you cannot have an antisymmetric expression with two identical states—they simply cancel each other giving you a zero. In other words, it is impossible to have a two-particle state of fermions, in which each fermion is in the same single-particle state. This is essentially an expression of famous Pauli's exclusion principle, which Pauli formulated in 1925 trying to explain why atoms with even number of electrons are more chemically stable than atoms with odd electron numbers. He realized that this can be explained requiring that there can only be one electron per single-electron state. If one takes into account only orbital quantum numbers such as principal number n, orbital number $l < n$, and magnetic number $|m| \leq l$ (see Chap. 8), the total number of available states is equal to n^2, which does not have to be even. So, Pauli postulated the existence of yet another quantum quantity, which can only take two different values, making the total amount of quantum numbers characterizing a state of an electron in atom equal to 4 and the total number of available states $2n^2$. The initial formulation of this principle was concerned only with electrons and was stated approximately like this: no two electrons in a many-electron atom can have the same values of four quantum numbers. Despite the success of this principle in explaining the periodic table, Pauli remained unsatisfied for two principal reasons: (a) he had no idea which physical quantity the fourth quantum number represents, and (b) he was not able to derive his principle from more fundamental postulates of quantum mechanics. The first of his concerns was resolved with the emergence of the idea of spin (see Sect. 9.1), but it took him 14 long years to finally prove the spin-statistics theorem, of which his exclusion principle is a simple corollary.

Before continuing I would like to clear up one terminological problem. When dealing with many-particle systems, the word "state" might have different meanings when used in different contexts. On one hand, I will talk about states characterizing the actual many-particle system; Eqs. 11.7 through 11.10 give examples of such states for the two-particle system. On the other hand, I use single-particle states, such as $\left|\alpha^{(1)}\right\rangle$ or $\left|\beta^{(2)}\right\rangle$, to construct the many-particle states $\left|\psi_f(1,2)\right\rangle$ or $\left|\psi_b^{(i)}(1,2)\right\rangle$. So, in order to avoid misunderstandings and misconceptions, let's agree that the term "state" from now on will always refer to an actual state of a many-particle system, while single-particle states from this point forward will be called single-particle *orbitals*. Understood literally orbitals are usually used to describe single-electron states of atomic electrons, but I will take the liberty to expand this term to any single-electron state. Getting this out of the way, I now want to direct your attention to the following fact. In the system of two fermions with only two available orbitals, we ended up with just a single two-particle state. At the same time, in the case of the same number of bosons and the same number of orbitals, there are three linearly independent orthogonal two-particle states, and if we were to forget about symmetry requirements (as we would if dealing with distinguishable particles), we would have ended up with a four-dimensional space of two-particle states just like in the two-spin problem from Sect. 9.4. You can see now that the dimensionality of the space containing many-particle states severely depends on the symmetry requirements, and the naive prediction for this dimension to be M^N turned to be only correct for distinguishable particles.

11.2   Constructing a Basis in a Many-Fermion Space

While identical bosons are responsible for some fascinating phenomena such as superfluidity and superconductivity, the systems of many fermions are much more ubiquitous in the practical applications of quantum theory, and, therefore, I will mostly focus on them from now on. As always, the first thing to understand is the structure of the space in which vectors representing the states of interest live. This includes finding its dimension and constructing a basis. The problem of finding the dimension of a many-particle space is an exercise in combinatorics—the science of counting the number of different combinations of various objects. In the case of fermions, the problem is formulated quite simply: given N objects (particles) and M boxes (orbitals), you need to compute in how many different ways you can fill the boxes assuming that each box can hold only one particle, and an order in which the particles are distributed among the boxes is not important. Once you find one distribution of the particles among the boxes, it becomes a seed for one many-particle state. The state itself is found by permuting the particles among the boxes, adding a negative sign for each permutation and summing up the results. To understand the situation, better begin with a simplest case: $M = N$. When the number of particles is equal to the number of orbitals, you do not have much of a choice: you just have to put one particle in each box, and then do the permutations— you end up with a single antisymmetric state. As an example, consider three particles that can be in one of three available orbitals $\left| \alpha_i^{(s)} \right\rangle$, where the lower index enumerates the orbitals and the upper index refers to the particles. Assume that you put the first particle in the first box, the second particle in the second, and the third one in the third, generating the following combination of the orbitals: $\left| \alpha_1^{(1)} \right\rangle \left| \alpha_2^{(2)} \right\rangle \left| \alpha_3^{(3)} \right\rangle$. Now, let me switch particles 1 and 2, generating combination $- \left| \alpha_1^{(2)} \right\rangle \left| \alpha_2^{(1)} \right\rangle \left| \alpha_3^{(3)} \right\rangle$. If I switch the particles again, say, particles 1 and 3, I will get the new combination $\left| \alpha_1^{(2)} \right\rangle \left| \alpha_2^{(3)} \right\rangle \left| \alpha_3^{(1)} \right\rangle$. Note that the negative sign has disappeared because each new permutation brings about a change of sign. Making all 6 (3!) permutations, you will end up with a single three-particle state:

$$\left| \alpha_1^{(1)} \right\rangle \left| \alpha_2^{(2)} \right\rangle \left| \alpha_3^{(3)} \right\rangle - \left| \alpha_1^{(2)} \right\rangle \left| \alpha_2^{(1)} \right\rangle \left| \alpha_3^{(3)} \right\rangle + \left| \alpha_1^{(3)} \right\rangle \left| \alpha_2^{(1)} \right\rangle \left| \alpha_3^{(2)} \right\rangle -$$

$$\left| \alpha_1^{(3)} \right\rangle \left| \alpha_2^{(2)} \right\rangle \left| \alpha_3^{(1)} \right\rangle + \left| \alpha_1^{(2)} \right\rangle \left| \alpha_2^{(3)} \right\rangle \left| \alpha_3^{(1)} \right\rangle - \left| \alpha_1^{(1)} \right\rangle \left| \alpha_2^{(3)} \right\rangle \left| \alpha_3^{(2)} \right\rangle . \qquad (11.11)$$

In agreement with the permutation rules described above, all terms in Eq. 11.11 with negative signs in front of them can be obtained from the first term by exchanging just one pair of particles, while the terms with the positive sign are obtained by permutation of two particles. It makes sense, of course, because, as I said before, an exchange of any two fermions is complemented by a change of sign, in which case an exchange of two pairs of fermions is equivalent to changing the sign twice: $+ \rightarrow - \rightarrow +$, which is, of course, the same argument as I made when deriving this expression. Finally, if you are wondering how to choose the first, seeding, term, the answer is simple: it does not matter and you can start with any of them. The only difference, which you might notice, is that all negative terms could become positive and vice versa, which amounts to a simple overall negative sign in front of the whole expression, and this makes no physical difference whatsoever.

Now, since, for every selection of the number of boxes equal to the number of the particles, you end up with just a single many-particle state, the total number of states is simply equal to the number of ways you can select N boxes out of M. This is a classical combinatorial problem with a well-known solution given by the number of combinations for N objects chosen out of M. Thus, the number of distinct linearly independent and orthogonal N-fermion states based on M available single-fermion orbitals (the dimensionality $D(N, M)$ of the corresponding space) is

$$D(N, M) = \binom{M}{N} = \frac{M!}{N!(M-N)!}. \tag{11.12}$$

You can verify this general results with a few simple examples. Let's say that now you want to build a space of three-fermion states using five available single-fermion orbitals. According to Eq. 11.12 this space possesses $5!/(3!2!) = 10$ basis vectors. Using the same notation $\left|\alpha_i^{(s)}\right\rangle$ as before, but now allowing index i to run from 1 to 5, you can generate the following ten seed vectors, in which each particle is assigned to a different orbital:

$$\left|\alpha_1^{(1)}\right\rangle\left|\alpha_2^{(2)}\right\rangle\left|\alpha_3^{(3)}\right\rangle, \ \left|\alpha_1^{(1)}\right\rangle\left|\alpha_2^{(2)}\right\rangle\left|\alpha_4^{(3)}\right\rangle, \ \left|\alpha_1^{(1)}\right\rangle\left|\alpha_2^{(2)}\right\rangle\left|\alpha_5^{(3)}\right\rangle$$

$$\left|\alpha_1^{(1)}\right\rangle\left|\alpha_3^{(2)}\right\rangle\left|\alpha_4^{(3)}\right\rangle, \ \left|\alpha_1^{(1)}\right\rangle\left|\alpha_3^{(2)}\right\rangle\left|\alpha_5^{(3)}\right\rangle, \ \left|\alpha_1^{(1)}\right\rangle\left|\alpha_4^{(2)}\right\rangle\left|\alpha_5^{(3)}\right\rangle$$

$$\left|\alpha_2^{(1)}\right\rangle\left|\alpha_3^{(2)}\right\rangle\left|\alpha_4^{(3)}\right\rangle\left|\alpha_2^{(1)}\right\rangle\left|\alpha_3^{(2)}\right\rangle\left|\alpha_5^{(3)}\right\rangle, \ \left|\alpha_2^{(1)}\right\rangle\left|\alpha_4^{(2)}\right\rangle\left|\alpha_5^{(3)}\right\rangle,$$

$$\left|\alpha_3^{(1)}\right\rangle\left|\alpha_4^{(2)}\right\rangle\left|\alpha_5^{(3)}\right\rangle.$$

Each of these seeds yields a single antisymmetric state in an exactly the same way as in the previous example.

A bit of gazing at Eq. 11.11 might reveal you an ultimate truth about the structure of this expression: a sum of products of various distinct combinations of nine elements $\left|\alpha_i^{(j)}\right\rangle$ where each index takes three different values is grouped in three with alternating positive and negative signs. Some digging in your associative memory will bring to the surface that this is nothing but a determinant of a matrix whose rows are the three participating orbitals with different particles assigned to each row:

$$|\alpha_1, \alpha_2, \alpha_3\rangle = \begin{Vmatrix} \left|\alpha_1^{(1)}\right\rangle & \left|\alpha_1^{(2)}\right\rangle & \left|\alpha_1^{(3)}\right\rangle \\ \left|\alpha_2^{(1)}\right\rangle & \left|\alpha_2^{(2)}\right\rangle & \left|\alpha_2^{(3)}\right\rangle \\ \left|\alpha_3^{(1)}\right\rangle & \left|\alpha_3^{(2)}\right\rangle & \left|\alpha_3^{(3)}\right\rangle \end{Vmatrix} \tag{11.13}$$

where on the left of this equation I introduced a notation $|\alpha_1, \alpha_2, \alpha_3\rangle$ which contains all the information you need to know about the state presented on the right, namely, that this three-fermion state is formed by distributing three particles among orbitals $|\alpha_1\rangle$, $|\alpha_2\rangle$, and $|\alpha_3\rangle$. The right-hand side of this expression gives you a good mnemonic rule about how to combine these three orbitals into an antisymmetric three-fermion state. Arranging the orbitals into determinants makes the antisymmetry of the corresponding state obvious: the exchange of particles becomes mathematically equivalent to the interchange of the columns of the determinant, and this operation is well known to reverse its sign.

The idea to arrange orbitals into determinants in order to construct automatically antisymmetric many-fermion states was first used independently by Heisenberg and Dirac in their 1926 papers and expressed in a more formal way by John C. Slater, an American physicist, in 1929, and for this reason these determinants bear his name. That was a time when American physicists had to travel for postdoctoral positions to Europe, and not the other way around, so after getting his Ph.D. from Harvard, Slater moved to Cambridge and then to Copenhagen before coming back to the USA and joining the Physics Department at Harvard as a faculty member.

A word of caution: the fermion states in the form of the Slater determinant are not necessarily the eigenvectors of a many-particle Hamiltonian, which, in general, can be presented in the form

$$\hat{H}^{(N)} = \sum_{i=1}^{N} \hat{H}_i + \frac{1}{2} \sum_{i=1}^{N} \sum_{j}^{N} \hat{V}_{i,j} \qquad (11.14)$$

where the first term is the sum of the single-particle Hamiltonians for each particle, which includes operators of the particle's kinetic energy and might include a term describing the interaction of each particle with some external object, e.g., electric field, while the second term describes the interaction between the particles, most frequently the Coulomb repulsion between negatively charged electrons. The factor $1/2$ in front of the second term takes into account that the double summation over i and j counts the interaction between each pair of particles twice: once as $\hat{V}_{i,j}$ and the second time as $\hat{V}_{j,i}$. The principal difference between these two terms is that while each $\hat{H}_i$ acts only on the orbitals of "its own" particle, the interaction term acts on the orbitals of two particles. As a result, any simple tensor product of single-particle orbitals is an eigenvector of the first term of the many-particle Hamiltonian, but not of the entire Hamiltonian. Consider, for instance, the three-particle state from the previous example. Picking up just one term from Eq. 11.11, I can write

$$\left(\hat{H}_1 + \hat{H}_2 + \hat{H}_3\right) \left|\alpha_1^{(1)}\right\rangle \left|\alpha_2^{(2)}\right\rangle \left|\alpha_3^{(3)}\right\rangle =$$

$$\left|\alpha_2^{(2)}\right\rangle \left|\alpha_3^{(3)}\right\rangle \hat{H}_1 \left|\alpha_1^{(1)}\right\rangle + \left|\alpha_1^{(1)}\right\rangle \left|\alpha_3^{(3)}\right\rangle \hat{H}_2 \left|\alpha_2^{(2)}\right\rangle +$$

$$\left|\alpha_1^{(1)}\right\rangle \left|\alpha_2^{(2)}\right\rangle \hat{H}_3 \left|\alpha_3^{(3)}\right\rangle = \qquad (11.15)$$

$$(E_1 + E_2 + E_3) \left|\alpha_1^{(1)}\right\rangle \left|\alpha_2^{(2)}\right\rangle \left|\alpha_3^{(3)}\right\rangle.$$

Since all other terms in Eq. 11.11 feature the same three orbitals, it is obvious that all of them are eigenvectors of this Hamiltonian with the same eigenvalue, so that the entire antisymmetric three-particle state given by the Slater determinant, Eq. 11.13, is also its eigenvector. It is also clear that for any Slater determinant state, the eigenvalue of the non-interacting Hamiltonian is always a sum of the single-particle energies of the orbitals used to construct the determinant. If, however, one adds the interaction term to the picture, the situation changes as none of the single-particle orbitals can be eigenvectors of $\hat{V}_{i,j}$, which acts on states of two particles, so that the Slater determinants are no longer stationary states of many-fermion system. This does not mean, of course, that they are useless—they form a convenient basis in the space of many-particle states, which ensures that all states represented in this basis are antisymmetric. This brings me back to Eq. 11.12, defining the dimension of this space and highlighting the main difficulty of dealing with interacting many-particle systems—the space containing the corresponding states is just too large.

Consider, for instance, an atom of carbon, with its six electrons. You can start building the basis for the six-electron space starting with lowest energy orbitals and continuing until you have enough basis vectors. The two lowest energy orbitals correspond to principal quantum number $n = 1$, orbital and magnetic numbers equal to zero, and two spin numbers $\pm 1/2$: $|1, 0, 0, 1/2\rangle$ and $|1, 0, 0, -1/2\rangle$. This is definitely not enough for six electrons, so you need to go to orbitals with $n = 2$, of which there are 8: $|2, 0, 0, 1/2\rangle$, $|2, 0, 0, -1/2\rangle$, $|2, 1, -1, 1/2\rangle$, $|2, 1, -1, -1/2\rangle$, $|2, 1, 0, 1/2\rangle$, $|2, 1, 0, -1/2\rangle$, $|2, 1, 1, 1/2\rangle$, and $|2, 1, 1, -1/2\rangle$, where the notation follows the regular scheme $|n, l, m, m_s\rangle$ (I combined the spin number m_s with orbital quantum numbers for the sake of simplifying the notation). If I limit the space to just these ten orbitals (and it is not the fact that orbitals with $n = 3$ should not be included), the total number of basis vectors in this space will be $10!/(6!4!) = 210$. It means that using the Slater determinants as a basis in this space, I will end up with the Hamiltonian of the system represented by a 210×210 matrix. Allowing the electrons to occupy additional $n = 3$ orbitals, all 18 of them, will bring the dimensionality of the six-electron space to 376,740. I hope these examples give you a clear picture of how difficult problems with many interacting particles can be and explain why people were busy inventing a great variety of different approximate ways of dealing with them. Very often, the idea behind these methods is to replace the Hamiltonian in Eq. 11.14 by an effective Hamiltonian without an interaction term. The effects of the interaction in such approaches are always hidden in "new" single-particle Hamiltonians retaining some information about the interaction with other particles. A more detailed exposition of this issue is way beyond the scope of this book and can be found in many texts on atomic physics and quantum chemistry.

Before continuing to the next section, let me consider a few examples involving non-interacting indistinguishable particles so that you could get a better feel for the quantum mechanical indistinguishability.

Example 29 (Non-interacting Particles in a Potential Well.) Consider a system of three non-interacting particles in an infinite one-dimensional potential well. Assuming that the particles are (a) distinguishable spinless atoms of equal mass

m_a, (b) electrons, and (c) indistinguishable spinless bosons, find three lowest energy eigenvalues of this system, and write down the corresponding wave functions (spinors when necessary).

Solution

(a) In the case of three distinguishable atoms, no symmetry requirements can be imposed on the three-particle wave function, so the ground state energy corresponds to a state in which all three atoms are in the same single-particle ground state orbital:

$$\psi_1^{(3)}(z_1, z_2, z_3) = \sqrt{\left(\frac{2}{L}\right)^3} \sin\frac{\pi z_1}{L} \sin\frac{\pi z_2}{L} \sin\frac{\pi z_3}{L} \tag{11.16}$$

with corresponding energy

$$E_{1,1,1} = \frac{3\hbar^2\pi^2}{2L_z^2 m_a}. \tag{11.17}$$

The second energy level would correspond to moving one of the atoms to the second single-particle orbital, so that I have for the three degenerate three-particle states

$$\psi_{2,1}^{(3)}(z_1, z_2, z_3) = \sqrt{\left(\frac{2}{L}\right)^3} \sin\frac{2\pi z_1}{L} \sin\frac{\pi z_2}{L} \sin\frac{\pi z_3}{L}$$

$$\psi_{2,2}^{(3)}(z_1, z_2, z_3) = \sqrt{\left(\frac{2}{L}\right)^3} \sin\frac{\pi z_1}{L} \sin\frac{2\pi z_2}{L} \sin\frac{\pi z_3}{L} \tag{11.18}$$

$$\psi_{2,3}^{(3)}(z_1, z_2, z_3) = \sqrt{\left(\frac{2}{L}\right)^3} \sin\frac{\pi z_1}{L} \sin\frac{\pi z_2}{L} \sin\frac{2\pi z_3}{L}$$

with the corresponding energy

$$E_{2,1,1} = \frac{6\hbar^2\pi^2}{2L_z^2 m_a}. \tag{11.19}$$

Finally, the next lowest energy will correspond to two particles moved to the second single-particle level with the wave functions and triple-degenerate energy level given by

$$\psi_{3,1}^{(3)}(z_1, z_2, z_3) = \sqrt{\left(\frac{2}{L}\right)^3} \sin\frac{2\pi z_1}{L} \sin\frac{2\pi z_2}{L} \sin\frac{\pi z_3}{L}$$

$$\psi_{3,2}^{(3)}(z_1, z_2, z_3) = \sqrt{\left(\frac{2}{L}\right)^3} \sin\frac{\pi z_1}{L} \sin\frac{2\pi z_2}{L} \sin\frac{2\pi z_3}{L} \qquad (11.20)$$

$$\psi_{3,3}^{(3)}(z_1, z_2, z_3) = \sqrt{\left(\frac{2}{L}\right)^3} \sin\frac{2\pi z_1}{L} \sin\frac{\pi z_2}{L} \sin\frac{2\pi z_3}{L}$$

$$E_{2,2,1} = \frac{9\hbar^2\pi^2}{2L_z^2 m_a}. \qquad (11.21)$$

(b) Electrons are indistinguishable fermions, so their many-particle states must be antisymmetric. The single-particle orbitals are spinors, formed as a tensor product of the eigenvectors of the infinite potential well and of the spin operator $\hat{S}_z$. For convenience, I will begin by writing down the single-particle orbitals in the symbolic form $|n, m_s\rangle$, where n corresponds to an energy level in the infinite well and m_s is a spin magnetic number. To construct the vector representing the ground state of the three-electron system, I need to include three different orbitals with the lowest single-particle energies. Obviously these are $|1, \uparrow\rangle$, $|1, \downarrow\rangle$, $|2, m_s\rangle$. The choice of the spin state in the third orbital is arbitrary, so that there are two different ground states with the same energy. The respective Slater determinant becomes

$$|1, 1, 2\rangle = \begin{Vmatrix} |1, \uparrow\rangle_1 & |1, \uparrow\rangle_2 & |1, \uparrow\rangle_3 \\ |1, \downarrow\rangle_1 & |1, \downarrow\rangle_2 & |1, \downarrow\rangle_3 \\ |2, \uparrow\rangle_1 & |2, \uparrow\rangle_2 & |2, \uparrow\rangle_3 \end{Vmatrix}$$

where the lower subindex enumerates electrons, and I chose for concreteness the spin-up state for the spin portion of the third orbital. Notation $|1, 1, 2\rangle$ for the three-electron state was chosen in the form, which reflects the eigenvectors of the infinite potential, "occupied"[2] by electrons in this state. Expanding the determinant and pulling out the spin number into a separate ket, I have

$$|1, 1, 2\rangle = |1\rangle_1 |\uparrow\rangle_1 |1\rangle_2 |\downarrow\rangle_2 |2\rangle_3 |\uparrow\rangle_3 + |1\rangle_1 |\downarrow\rangle_1 |1\rangle_2 |\uparrow\rangle_2 |2\rangle_3 |\uparrow\rangle_3 +$$
$$|1\rangle_1 |\uparrow\rangle_1 |2\rangle_2 |\uparrow\rangle_2 |1\rangle_3 |\downarrow\rangle_3 - |2\rangle_1 |\uparrow\rangle_1 |1\rangle_2 |\downarrow\rangle_2 |1\rangle_3 |\uparrow\rangle_3 - |1\rangle_1 |\downarrow\rangle_1 |1\rangle_2 |\uparrow\rangle_2$$
$$|2\rangle_3 |\uparrow\rangle_3 - |1\rangle_1 |\uparrow\rangle_1 |2\rangle_2 |\uparrow\rangle_2 |1\rangle_3 |\downarrow\rangle_3.$$

---

[2]"Occupied" in this context means that a given orbital participates in the formation of a given many-particle state.

Bringing back the position representation of the eigenvectors of the well, the last result can be written down as

$$
|1, 1, 2\rangle^{(1)} = \sqrt{\left(\frac{2}{L}\right)^3} \times
$$

$$
\left\{ \begin{bmatrix} \sin\frac{\pi z_1}{L} \\ 0 \end{bmatrix} \begin{bmatrix} 0 \\ \sin\frac{\pi z_2}{L} \end{bmatrix} \begin{bmatrix} \sin\frac{2\pi z_3}{L} \\ 0 \end{bmatrix} + \begin{bmatrix} 0 \\ \sin\frac{\pi z_1}{L} \end{bmatrix} \begin{bmatrix} \sin\frac{\pi z_2}{L} \\ 0 \end{bmatrix} \begin{bmatrix} \sin\frac{2\pi z_3}{L} \\ 0 \end{bmatrix} + \right.
$$

$$
\begin{bmatrix} 0 \\ \sin\frac{\pi z_1}{L} \end{bmatrix} \begin{bmatrix} \sin\frac{2\pi z_2}{L} \\ 0 \end{bmatrix} \begin{bmatrix} 0 \\ \sin\frac{\pi z_3}{L} \end{bmatrix} - \begin{bmatrix} \sin\frac{2\pi z_1}{L} \\ 0 \end{bmatrix} \begin{bmatrix} 0 \\ \sin\frac{\pi z_2}{L} \end{bmatrix} \begin{bmatrix} \sin\frac{\pi z_3}{L} \\ 0 \end{bmatrix} -
$$

$$
\left. \begin{bmatrix} 0 \\ \sin\frac{\pi z_1}{L} \end{bmatrix} \begin{bmatrix} \sin\frac{\pi z_2}{L} \\ 0 \end{bmatrix} \begin{bmatrix} \sin\frac{2\pi z_3}{L} \\ 0 \end{bmatrix} - \begin{bmatrix} \sin\frac{\pi z_1}{L} \\ 0 \end{bmatrix} \begin{bmatrix} \sin\frac{2\pi z_2}{L} \\ 0 \end{bmatrix} \begin{bmatrix} 0 \\ \sin\frac{\pi z_3}{L} \end{bmatrix} \right\} . \qquad (11.22)
$$

To get a bit more comfortable with this expression, let's apply operator

$$
\hat{H} = \hat{H}^{(1)} + \hat{H}^{(2)} + \hat{H}^{(3)},
$$

where $\hat{H}^{(i)}$ is a single-electron infinite potential well Hamiltonian, which in the spinor representation is proportional to a unit matrix:

$$
\hat{H}|1, 1, 2\rangle = \sqrt{\left(\frac{2}{L}\right)^3} \times
$$

$$
\left\{ \begin{bmatrix} \hat{H}^{(1)}\sin\frac{\pi z_1}{L} \\ 0 \end{bmatrix} \begin{bmatrix} 0 \\ \sin\frac{\pi z_2}{L} \end{bmatrix} \begin{bmatrix} \sin\frac{2\pi z_3}{L} \\ 0 \end{bmatrix} + \begin{bmatrix} 0 \\ \hat{H}^{(1)}\sin\frac{\pi z_1}{L} \end{bmatrix} \begin{bmatrix} \sin\frac{\pi z_2}{L} \\ 0 \end{bmatrix} \begin{bmatrix} \sin\frac{2\pi z_3}{L} \\ 0 \end{bmatrix} + \right.
$$

$$
\begin{bmatrix} 0 \\ \hat{H}^{(1)}\sin\frac{\pi z_1}{L} \end{bmatrix} \begin{bmatrix} \sin\frac{2\pi z_2}{L} \\ 0 \end{bmatrix} \begin{bmatrix} 0 \\ \sin\frac{\pi z_3}{L} \end{bmatrix} - \begin{bmatrix} \hat{H}^{(1)}\sin\frac{2\pi z_1}{L} \\ 0 \end{bmatrix} \begin{bmatrix} 0 \\ \sin\frac{\pi z_2}{L} \end{bmatrix} \begin{bmatrix} \sin\frac{\pi z_3}{L} \\ 0 \end{bmatrix} -
$$

$$
\begin{bmatrix} 0 \\ \hat{H}^{(1)}\sin\frac{\pi z_1}{L} \end{bmatrix} \begin{bmatrix} \sin\frac{\pi z_2}{L} \\ 0 \end{bmatrix} \begin{bmatrix} \sin\frac{2\pi z_3}{L} \\ 0 \end{bmatrix} - \begin{bmatrix} \hat{H}^{(1)}\sin\frac{\pi z_1}{L} \\ 0 \end{bmatrix} \begin{bmatrix} \sin\frac{2\pi z_2}{L} \\ 0 \end{bmatrix} \begin{bmatrix} 0 \\ \sin\frac{\pi z_3}{L} \end{bmatrix} +
$$

$$
\begin{bmatrix} \sin\frac{\pi z_1}{L} \\ 0 \end{bmatrix} \begin{bmatrix} 0 \\ \hat{H}^{(2)}\sin\frac{\pi z_2}{L} \end{bmatrix} \begin{bmatrix} \sin\frac{2\pi z_3}{L} \\ 0 \end{bmatrix} + \begin{bmatrix} 0 \\ \sin\frac{\pi z_1}{L} \end{bmatrix} \begin{bmatrix} \hat{H}^{(2)}\sin\frac{\pi z_2}{L} \\ 0 \end{bmatrix} \begin{bmatrix} \sin\frac{2\pi z_3}{L} \\ 0 \end{bmatrix} +
$$

$$
\begin{bmatrix} 0 \\ \sin\frac{\pi z_1}{L} \end{bmatrix} \begin{bmatrix} \hat{H}^{(2)}\sin\frac{2\pi z_2}{L} \\ 0 \end{bmatrix} \begin{bmatrix} 0 \\ \sin\frac{\pi z_3}{L} \end{bmatrix} - \begin{bmatrix} \sin\frac{2\pi z_1}{L} \\ 0 \end{bmatrix} \begin{bmatrix} 0 \\ \hat{H}^{(2)}\sin\frac{\pi z_2}{L} \end{bmatrix} \begin{bmatrix} \sin\frac{\pi z_3}{L} \\ 0 \end{bmatrix} -
$$

$$
\begin{bmatrix} 0 \\ \sin\frac{\pi z_1}{L} \end{bmatrix} \begin{bmatrix} \hat{H}^{(2)}\sin\frac{\pi z_2}{L} \\ 0 \end{bmatrix} \begin{bmatrix} \sin\frac{2\pi z_3}{L} \\ 0 \end{bmatrix} - \begin{bmatrix} \sin\frac{\pi z_1}{L} \\ 0 \end{bmatrix} \begin{bmatrix} \hat{H}^{(2)}\sin\frac{2\pi z_2}{L} \\ 0 \end{bmatrix} \begin{bmatrix} 0 \\ \sin\frac{\pi z_3}{L} \end{bmatrix} +
$$

$$
\begin{bmatrix} \sin\frac{\pi z_1}{L} \\ 0 \end{bmatrix} \begin{bmatrix} 0 \\ \sin\frac{\pi z_2}{L} \end{bmatrix} \begin{bmatrix} \hat{H}^{(3)}\sin\frac{2\pi z_3}{L} \\ 0 \end{bmatrix} + \begin{bmatrix} 0 \\ \sin\frac{\pi z_1}{L} \end{bmatrix} \begin{bmatrix} \sin\frac{\pi z_2}{L} \\ 0 \end{bmatrix} \begin{bmatrix} \hat{H}^{(3)}\sin\frac{2\pi z_3}{L} \\ 0 \end{bmatrix} +
$$

$$\left[\begin{matrix}0\\ \sin\frac{\pi z_1}{L}\end{matrix}\right]\left[\begin{matrix}\sin\frac{2\pi z_2}{L}\\ 0\end{matrix}\right]\left[\begin{matrix}0\\ \hat{H}^{(3)}\sin\frac{\pi z_3}{L}\end{matrix}\right] - \left[\begin{matrix}\sin\frac{2\pi z_1}{L}\\ 0\end{matrix}\right]\left[\begin{matrix}0\\ \sin\frac{\pi z_2}{L}\end{matrix}\right]\left[\begin{matrix}\hat{H}^{(3)}\sin\frac{\pi z_3}{L}\\ 0\end{matrix}\right] -$$

$$\left[\begin{matrix}0\\ \sin\frac{\pi z_1}{L}\end{matrix}\right]\left[\begin{matrix}\sin\frac{\pi z_2}{L}\\ 0\end{matrix}\right]\left[\begin{matrix}\hat{H}^{(3)}\sin\frac{2\pi z_3}{L}\\ 0\end{matrix}\right] - \left[\begin{matrix}\sin\frac{\pi z_1}{L}\\ 0\end{matrix}\right]\left[\begin{matrix}\sin\frac{2\pi z_2}{L}\\ 0\end{matrix}\right]\left[\begin{matrix}0\\ \hat{H}^{(3)}\sin\frac{\pi z_3}{L}\end{matrix}\right]\Bigg\}.$$

I understand that this expression looks awfully intimidating (or just awful), but I still want you to gather your wits and go through it line by line, and let the force be with you. The first thing that you shall notice is that every single-particle Hamiltonian affects only those orbitals that contain its own particle. Now remembering that each of the orbitals is an eigenvector of the corresponding Hamiltonian, you can rewrite the above expression as

$$\hat{H}\,|1,1,2\rangle = \sqrt{\left(\frac{2}{L}\right)^3}\times$$

$$\left\{E_1\left[\begin{matrix}\sin\frac{\pi z_1}{L}\\ 0\end{matrix}\right]\left[\begin{matrix}0\\ \sin\frac{\pi z_2}{L}\end{matrix}\right]\left[\begin{matrix}\sin\frac{2\pi z_3}{L}\\ 0\end{matrix}\right] + E_1\left[\begin{matrix}0\\ \sin\frac{\pi z_1}{L}\end{matrix}\right]\left[\begin{matrix}\sin\frac{\pi z_2}{L}\\ 0\end{matrix}\right]\left[\begin{matrix}\sin\frac{2\pi z_3}{L}\\ 0\end{matrix}\right] +$$

$$E_1\left[\begin{matrix}0\\ \sin\frac{\pi z_1}{L}\end{matrix}\right]\left[\begin{matrix}\sin\frac{2\pi z_2}{L}\\ 0\end{matrix}\right]\left[\begin{matrix}0\\ \sin\frac{\pi z_3}{L}\end{matrix}\right] - E_2\left[\begin{matrix}\sin\frac{2\pi z_1}{L}\\ 0\end{matrix}\right]\left[\begin{matrix}0\\ \sin\frac{\pi z_2}{L}\end{matrix}\right]\left[\begin{matrix}\sin\frac{\pi z_3}{L}\\ 0\end{matrix}\right] -$$

$$E_1\left[\begin{matrix}0\\ \sin\frac{\pi z_1}{L}\end{matrix}\right]\left[\begin{matrix}\sin\frac{\pi z_2}{L}\\ 0\end{matrix}\right]\left[\begin{matrix}\sin\frac{2\pi z_3}{L}\\ 0\end{matrix}\right] - E_1\left[\begin{matrix}\sin\frac{\pi z_1}{L}\\ 0\end{matrix}\right]\left[\begin{matrix}\sin\frac{2\pi z_2}{L}\\ 0\end{matrix}\right]\left[\begin{matrix}0\\ \sin\frac{\pi z_3}{L}\end{matrix}\right] +$$

$$E_1\left[\begin{matrix}\sin\frac{\pi z_1}{L}\\ 0\end{matrix}\right]\left[\begin{matrix}0\\ \sin\frac{\pi z_2}{L}\end{matrix}\right]\left[\begin{matrix}\sin\frac{2\pi z_3}{L}\\ 0\end{matrix}\right] + E_1\left[\begin{matrix}0\\ \sin\frac{\pi z_1}{L}\end{matrix}\right]\left[\begin{matrix}\sin\frac{\pi z_2}{L}\\ 0\end{matrix}\right]\left[\begin{matrix}\sin\frac{2\pi z_3}{L}\\ 0\end{matrix}\right] +$$

$$E_2\left[\begin{matrix}0\\ \sin\frac{\pi z_1}{L}\end{matrix}\right]\left[\begin{matrix}\sin\frac{2\pi z_2}{L}\\ 0\end{matrix}\right]\left[\begin{matrix}0\\ \sin\frac{\pi z_3}{L}\end{matrix}\right] - E_1\left[\begin{matrix}\sin\frac{2\pi z_1}{L}\\ 0\end{matrix}\right]\left[\begin{matrix}0\\ \sin\frac{\pi z_2}{L}\end{matrix}\right]\left[\begin{matrix}\sin\frac{\pi z_3}{L}\\ 0\end{matrix}\right] -$$

$$E_1\left[\begin{matrix}0\\ \sin\frac{\pi z_1}{L}\end{matrix}\right]\left[\begin{matrix}\sin\frac{\pi z_2}{L}\\ 0\end{matrix}\right]\left[\begin{matrix}\sin\frac{2\pi z_3}{L}\\ 0\end{matrix}\right] - E_2\left[\begin{matrix}\sin\frac{\pi z_1}{L}\\ 0\end{matrix}\right]\left[\begin{matrix}\sin\frac{2\pi z_2}{L}\\ 0\end{matrix}\right]\left[\begin{matrix}0\\ \sin\frac{\pi z_3}{L}\end{matrix}\right] +$$

$$E_2\left[\begin{matrix}\sin\frac{\pi z_1}{L}\\ 0\end{matrix}\right]\left[\begin{matrix}0\\ \sin\frac{\pi z_2}{L}\end{matrix}\right]\left[\begin{matrix}\sin\frac{2\pi z_3}{L}\\ 0\end{matrix}\right] + E_2\left[\begin{matrix}0\\ \sin\frac{\pi z_1}{L}\end{matrix}\right]\left[\begin{matrix}\sin\frac{\pi z_2}{L}\\ 0\end{matrix}\right]\left[\begin{matrix}\sin\frac{2\pi z_3}{L}\\ 0\end{matrix}\right] +$$

$$E_1\left[\begin{matrix}0\\ \sin\frac{\pi z_1}{L}\end{matrix}\right]\left[\begin{matrix}\sin\frac{2\pi z_2}{L}\\ 0\end{matrix}\right]\left[\begin{matrix}0\\ \sin\frac{\pi z_3}{L}\end{matrix}\right] - E_1\left[\begin{matrix}\sin\frac{2\pi z_1}{L}\\ 0\end{matrix}\right]\left[\begin{matrix}0\\ \sin\frac{\pi z_2}{L}\end{matrix}\right]\left[\begin{matrix}\sin\frac{\pi z_3}{L}\\ 0\end{matrix}\right] -$$

$$E_2\left[\begin{matrix}0\\ \sin\frac{\pi z_1}{L}\end{matrix}\right]\left[\begin{matrix}\sin\frac{\pi z_2}{L}\\ 0\end{matrix}\right]\left[\begin{matrix}\sin\frac{2\pi z_3}{L}\\ 0\end{matrix}\right] - E_1\left[\begin{matrix}\sin\frac{\pi z_1}{L}\\ 0\end{matrix}\right]\left[\begin{matrix}\sin\frac{2\pi z_2}{L}\\ 0\end{matrix}\right]\left[\begin{matrix}0\\ \sin\frac{\pi z_3}{L}\end{matrix}\right]\right\},$$

where $E_{1,2}$ are eigenvalues of energy corresponding to eigenvectors $|1\rangle$, $|2\rangle$ of the infinite potential well. Combining the like terms (terms with the same combination of single-particle orbitals), you will find

$$\hat{H}\,|1,1,2\rangle = (2E_1 + E_2)\,|1,1,2\rangle\,.$$

The second eigenvector belonging to this eigenvalue can be generated by changing the spin state paired with the orbital state $|2\rangle$ from spin-up to spin-down, which yields

$$|1,1,2\rangle^{(2)} = \sqrt{\left(\frac{2}{L}\right)^3} \times$$

$$\left\{ \begin{bmatrix} \sin\frac{\pi z_1}{L} \\ 0 \end{bmatrix} \begin{bmatrix} 0 \\ \sin\frac{\pi z_2}{L} \end{bmatrix} \begin{bmatrix} 0 \\ \sin\frac{2\pi z_3}{L} \end{bmatrix} + \begin{bmatrix} 0 \\ \sin\frac{\pi z_1}{L} \end{bmatrix} \begin{bmatrix} \sin\frac{\pi z_2}{L} \\ 0 \end{bmatrix} \begin{bmatrix} 0 \\ \sin\frac{2\pi z_3}{L} \end{bmatrix} + \right.$$

$$\begin{bmatrix} 0 \\ \sin\frac{\pi z_1}{L} \end{bmatrix} \begin{bmatrix} 0 \\ \sin\frac{2\pi z_2}{L} \end{bmatrix} \begin{bmatrix} 0 \\ \sin\frac{\pi z_3}{L} \end{bmatrix} - \begin{bmatrix} 0 \\ \sin\frac{2\pi z_1}{L} \end{bmatrix} \begin{bmatrix} 0 \\ \sin\frac{\pi z_2}{L} \end{bmatrix} \begin{bmatrix} \sin\frac{\pi z_3}{L} \\ 0 \end{bmatrix} - $$

$$\left. \begin{bmatrix} 0 \\ \sin\frac{\pi z_1}{L} \end{bmatrix} \begin{bmatrix} \sin\frac{\pi z_2}{L} \\ 0 \end{bmatrix} \begin{bmatrix} 0 \\ \sin\frac{2\pi z_3}{L} \end{bmatrix} - \begin{bmatrix} \sin\frac{\pi z_1}{L} \\ 0 \end{bmatrix} \begin{bmatrix} 0 \\ \sin\frac{2\pi z_2}{L} \end{bmatrix} \begin{bmatrix} 0 \\ \sin\frac{\pi z_3}{L} \end{bmatrix} \right\}\,.$$

To get the next energy level and the corresponding eigenvector, I just need to move one of the particles to the orbital $|2\rangle\,|m_s\rangle$, which means that the Slater determinant is now formed by orbitals $|1,m_s\rangle$, $|2,\downarrow\rangle$, $|2,\uparrow\rangle$ with the arbitrary value of the spin state in the single-particle ground state. Using for concreteness the spin-up value in $|1,m_s\rangle$, I can write

$$|1,2,2\rangle^{(1)} = \sqrt{\left(\frac{2}{L}\right)^3} \times$$

$$\left\{ \begin{bmatrix} \sin\frac{\pi z_1}{L} \\ 0 \end{bmatrix} \begin{bmatrix} \sin\frac{2\pi z_2}{L} \\ 0 \end{bmatrix} \begin{bmatrix} 0 \\ \sin\frac{2\pi z_3}{L} \end{bmatrix} + \begin{bmatrix} \sin\frac{2\pi z_1}{L} \\ 0 \end{bmatrix} \begin{bmatrix} 0 \\ \sin\frac{2\pi z_2}{L} \end{bmatrix} \begin{bmatrix} \sin\frac{\pi z_3}{L} \\ 0 \end{bmatrix} + \right.$$

$$\begin{bmatrix} 0 \\ \sin\frac{2\pi z_1}{L} \end{bmatrix} \begin{bmatrix} \sin\frac{\pi z_2}{L} \\ 0 \end{bmatrix} \begin{bmatrix} \sin\frac{2\pi z_3}{L} \\ 0 \end{bmatrix} - \begin{bmatrix} 0 \\ \sin\frac{2\pi z_1}{L} \end{bmatrix} \begin{bmatrix} \sin\frac{2\pi z_2}{L} \\ 0 \end{bmatrix} \begin{bmatrix} \sin\frac{\pi z_3}{L} \\ 0 \end{bmatrix} - $$

$$\left. \begin{bmatrix} \sin\frac{2\pi z_1}{L} \\ 0 \end{bmatrix} \begin{bmatrix} \sin\frac{\pi z_2}{L} \\ 0 \end{bmatrix} \begin{bmatrix} 0 \\ \sin\frac{2\pi z_3}{L} \end{bmatrix} - \begin{bmatrix} \sin\frac{\pi z_1}{L} \\ 0 \end{bmatrix} \begin{bmatrix} 0 \\ \sin\frac{2\pi z_2}{L} \end{bmatrix} \begin{bmatrix} \sin\frac{2\pi z_3}{L} \\ 0 \end{bmatrix} \right\}\,.$$

The energy corresponding to this state is $E_{2,2,1} = E_1 + 2E_2$ and coincides with Eq. 11.20 for energy of the second excited in the system of the distinguishable particles. Finally, to generate the next lowest energy level, one has to keep two orbitals corresponding to the second excited level of the well with different values of the spin number, and then the only choice for the third orbital would be to use one of two $|3\rangle\,|m_s\rangle$ orbitals, which result in two degenerate eigenvectors, one of which is shown below:

$$|2,2,3\rangle^{(2)} = \sqrt{\left(\frac{2}{L}\right)^3} \times$$

$$\left\{ \begin{bmatrix} \sin\frac{3\pi z_1}{L} \\ 0 \end{bmatrix} \begin{bmatrix} \sin\frac{2\pi z_2}{L} \\ 0 \end{bmatrix} \begin{bmatrix} 0 \\ \sin\frac{2\pi z_3}{L} \end{bmatrix} + \begin{bmatrix} \sin\frac{2\pi z_1}{L} \\ 0 \end{bmatrix} \begin{bmatrix} 0 \\ \sin\frac{2\pi z_2}{L} \end{bmatrix} \begin{bmatrix} \sin\frac{3\pi z_3}{L} \\ 0 \end{bmatrix} + \right.$$

$$\begin{bmatrix} 0 \\ \sin\frac{2\pi z_1}{L} \end{bmatrix} \begin{bmatrix} \sin\frac{3\pi z_2}{L} \\ 0 \end{bmatrix} \begin{bmatrix} \sin\frac{2\pi z_3}{L} \\ 0 \end{bmatrix} - \begin{bmatrix} 0 \\ \sin\frac{2\pi z_1}{L} \end{bmatrix} \begin{bmatrix} \sin\frac{2\pi z_2}{L} \\ 0 \end{bmatrix} \begin{bmatrix} \sin\frac{3\pi z_3}{L} \\ 0 \end{bmatrix} -$$

$$\left. \begin{bmatrix} \sin\frac{2\pi z_1}{L} \\ 0 \end{bmatrix} \begin{bmatrix} \sin\frac{3\pi z_2}{L} \\ 0 \end{bmatrix} \begin{bmatrix} 0 \\ \sin\frac{2\pi z_3}{L} \end{bmatrix} - \begin{bmatrix} \sin\frac{3\pi z_1}{L} \\ 0 \end{bmatrix} \begin{bmatrix} 0 \\ \sin\frac{2\pi z_2}{L} \end{bmatrix} \begin{bmatrix} \sin\frac{2\pi z_3}{L} \\ 0 \end{bmatrix} \right\}.$$

(I derived this expression by simply replacing $\sin\frac{\pi z_i}{L}$ everywhere with $\sin\frac{3\pi z_i}{L}$.)
The respective energy value is given by

$$E_{2,2,3} = E_3 + 2E_2 = \frac{17\hbar^2\pi^2}{2L_z^2 m_e}.$$

(c) Now, let me deal with the system of three identical spinless bosons. The symmetry requirement for the three-boson system allows using all identical orbitals (the resulting state is automatically symmetric); thus, the ground state can be built of a single orbital $|1\rangle$ and turns out to be the same as in the case of distinguishable particles and with the same energy value (Eq. 11.17). A difference from distinguishable particles arises when transitioning to excited states. Now, to satisfy the symmetry requirements, I have to turn three degenerate states of Eqs. 11.18 and 11.20 with energies given by Eqs. 11.19 and 11.21 into single non-degenerate states:

$$\psi_{2,1,1}^{(3)}(z_1, z_2, z_3) = \sqrt{\left(\frac{2}{L}\right)^3}\left[\sin\frac{2\pi z_1}{L} \sin\frac{\pi z_2}{L} \sin\frac{\pi z_3}{L} + \right.$$

$$\left. \sin\frac{\pi z_1}{L} \sin\frac{2\pi z_2}{L} \sin\frac{\pi z_3}{L} + \sin\frac{\pi z_1}{L} \sin\frac{\pi z_2}{L} \sin\frac{2\pi z_3}{L} \right]$$

$$\psi_{2,2,1}^{(3)}(z_1, z_2, z_3) = \sqrt{\left(\frac{2}{L}\right)^3}\left[\sin\frac{2\pi z_1}{L} \sin\frac{2\pi z_2}{L} \sin\frac{\pi z_3}{L} + \right.$$

$$\left. \sin\frac{\pi z_1}{L} \sin\frac{2\pi z_2}{L} \sin\frac{2\pi z_3}{L} + \sin\frac{2\pi z_1}{L} \sin\frac{\pi z_2}{L} \sin\frac{2\pi z_3}{L} \right].$$

11.3 Pauli Principle and Periodic Table of Elements: Electronic Structure of Atoms

While we are not equipped to deal with systems of large numbers of interacting particles, you can still appreciate how Pauli's idea of exclusion principle helped understand the periodicity in the properties of the atoms. In order to follow the arguments, you need to keep in mind two important points. First, when discussing the chemical properties of atoms, people are interested foremost in the many-particle ground state, i.e., a state of many electrons, which would have the lowest possible energy. Second, since the Pauli principle forbids states in which two electrons occupy the same orbital, you have to build many-particle states using at least as many orbitals as many particles are in your system, starting with ground state orbitals and adding new orbitals in a way, which would minimize an unavoidable increase of the sum of single-particle energies of all involved electrons. This last point implicitly assumes that the lowest energy of non-interacting particles would remain the lowest energy even if the interaction is taken into account. This assumption is not always true, but the discussion of this issue is beyond the scope of this book. Anyway, having these two points in mind, let's consider what happens with states of electrons as we are moving along the periodic table. Helium occupies the second place in the first row and is known as an inert gas, meaning that it is very stable and is not eager to participate in chemical reactions or form chemical bonds. It has two electrons, and therefore you need only two orbitals, which can have the same value of the principal number $n = 1$ to construct a two-electron state:

$$|1, 0, 0, 1/2\rangle_1 \, |1, 0, 0, -1/2\rangle_2 - |1, 0, 0, 1/2\rangle_2 \, |1, 0, 0, -1/2\rangle_1 \, .$$

These two orbitals exhaust all available states with the same principal number. In chemical language, we can say the electrons in helium atom belong to a complete or closed shell. Going to the next atom, lithium Li, you will notice that it has very different chemical properties—lithium is an active alkali metal, which readily participates in a variety of chemical reactions and forms a number of different compounds gladly offering one of its electrons for chemical bonding. Three lithium electrons need more than two orbitals to form a three-electron state, so you must start dealing with orbitals characterized by principal number $n = 2$. There are *eight* of them, but only one is really required to form the lowest energy three-electron state, and as a result seven of those orbitals remain, using physicist's jargon, "unoccupied." Once you go along the second row of the periodic table, the number of electrons increases to four in the case of beryllium, five for boron, six for carbon, seven for nitrogen, eight for oxygen, nine for fluorine, and finally ten for neon. With an increasing number of electrons, you must add additional orbitals to be able to create corresponding many-electron states, so that the number of "unoccupied" orbitals decreases. As the number of available, unused orbitals is getting smaller, the chemical activity of the corresponding substances diminishes, until you reach another inert gas neon. To construct a many-electron state for neon

Table 11.1 Elements of the second row of the periodic table and electronic configurations of their ground states in terms of single-electron orbitals and the term symbols

Element	Configuration	Term symbol
Li_3	$1s^2 2s^1$	$^2S_{1/2}$
Be_4	$1s^2 2s^2$	$^1S_0$
B_5	$1s^2 2s^2 2p^1$	$^2P_{1/2}$
C_6	$1s^2 2s^2 2p^2$	$^3P_0$
N_7	$1s^2 2s^2 2p^3$	$^4S_{3/2}$
O_8	$1s^2 2s^2 2p^4$	$^3P_2$
F_9	$1s^2 2s^2 2p^5$	$^2P_{3/2}$
Ne_{10}	$1s^2 2s^2 2p^6$	$^1S_0$

with ten electrons, you have to use all ten available orbitals with $n = 1$ and $n = 2$. Consequently, the electron structure of neon is again characterized as a closed shell configuration. A popular way to visualize this process of filling up the available orbitals consists in assigning numbers $1, 2, \cdots$ to the principal quantum number n, and letters s, p, f, and d to orbitals with orbital angular momentum number l equal to 0, 1, 2, and 3, respectively. The configuration of helium in this notation, primarily used in atomic physics and quantum chemistry, would be $1s^2$, where the first number stays for the principal number, and the upper index indicates the number of electrons available for assignment to orbitals with $l = 0$. The electronic structure of elements in the second row of the periodic table discussed above is shown in Table 11.1.

You can see from this table that $l = 0$ orbitals are added first to the list of available single-electron states, and only after that additional six orbitals with $l = 1$ and different values of m and m_s are thrown in. The supposition here is that single-electron states with $l = 0$ would contribute less energy than the $l = 1$ states[3]. Therefore, these two orbitals must be incorporated into the basis first. The assumption that the orbitals with larger n and larger l would contribute more energy, and, therefore, the corresponding orbitals must be added only after the orbitals with lower values of these numbers are filled, is not always correct, and for some elements orbitals with lower l and higher n contribute less energy than orbitals with higher l and lower n. This happens, for instance, with orbital $4s$, which contributes less energy than the orbital $3d$, but there are no simple hand-waving arguments that could explain or predict this behavior. Anyway, going now to the third row of the periodic table, you again start with the new set of orbitals characterized by $n = 3$, plenty of which are available for 11 electrons in the first element, another alkali metal, sodium. I think you get the gist of how it is working, but on the other hand, you shall be aware that this line of arguments is still a gross oversimplification, and periodic table of elements is not that periodic in some instances, and there are lots of elements that do not fit this simple model of closed shells.

Single-electron orbitals $|n, l, m, m_s\rangle$ based on eigenvectors of operators of orbital and spin angular momenta are not the only way to characterize the ground states

---

[3]I have to remind you that while the hydrogen energy levels are degenerate with respect to l, for other atoms this is not true because of the interaction with other electrons.

of atoms. An alternative approach is based on using eigenvectors of total orbital angular momentum $\hat{L}^{(tot)} = \sum_i \hat{L}^{(i)}$ (sum of the orbital momenta of all electrons), total spin of all electrons $\hat{S}^{(tot)} = \sum_i \hat{S}^{(i)}$, and grand total momentum $\hat{J} = \hat{L}^{(tot)} + \hat{S}^{(tot)}$.

Properties of the sum of two arbitrary angular momentum operators, $\hat{J}^{(1)}$ and $\hat{J}^{(2)}$, can be figured out by generalizing the results for the sum of two spins or the spin $1/2$ and the angular momentum presented in Chap. 9. The eigenvectors of the operator $\left(\hat{J}^{(1)} + \hat{J}^{(2)}\right)^2$ are characterized by quantum number j, which can take values

$$|j_1 - j_2| \le j \le j_1 + j_2, \tag{11.23}$$

where j_1 and j_2 refer to eigenvalues of $\left(\hat{J}^{(1)}\right)^2$ and $\left(\hat{J}^{(2)}\right)^2$, respectively. For each j, eigenvalues of $\hat{J}_z^{(1)} + \hat{J}_z^{(2)}$ are characterized by magnetic numbers M_j obeying usual inequality $|M_j| \le j$ and related to individual magnetic numbers m_{j_1} and m_{j_2} of $\hat{J}_z^{(1)}$ and $\hat{J}_z^{(2)}$ correspondingly as

$$M_j = m_{j_1} + m_{j_2}. \tag{11.24}$$

While Eq. 11.24 can be easily derived, proving Eq. 11.23 is a bit more than you can chew at this stage, but you may at least verify that it agrees with the cases considered in Chap. 9: for two $1/2$ spins, Eq. 11.23 gives two values for j: $j = 1, 0$ in agreement with Eqs. 9.54 and 9.55, and for the sum of the orbital momentum and the $1/2$ spin, Eq. 11.23 yields $j = l \pm 1/2$ again in agreement with Sect. 9.5.2.

The transition from the description of many-fermion states in terms of single-particle orbitals to the basis formed by eigenvectors of total orbital momentum, total spin, and grand total angular momentum raises an important issue of separate symmetry properties of many-particle orbital and spin states. Consider again for simplicity two fermions that can individually be in orbital states $|\psi_1\rangle$ and $|\psi_2\rangle$ and spin states $|\uparrow\rangle$ and $|\downarrow\rangle$. In the description, where spin and orbital states are lumped together in a one single-particle orbital (this is what I did writing equations such as Eq. 11.11 or 11.13), I would have introduced four single-electron orbitals $|\alpha_i\rangle$:

$$|\alpha_1\rangle \equiv |\psi_1, \uparrow\rangle; \ |\alpha_2\rangle \equiv |\psi_2, \uparrow\rangle; \ |\alpha_3\rangle \equiv |\psi_1, \downarrow\rangle; \ |\alpha_4\rangle \equiv |\psi_2, \downarrow\rangle$$

and used them as a basis in a $4!/(2!2!) = $ six-dimensional two-fermion space. If, however, I preferred to use eigenvectors of the total spin of the two particles as a basis in the spin sector of the total spin–orbital two-particle space, separating thereby the orbital and spin states, I would have to make sure that both the former and the latter components separately possess a definite parity. Four eigenvectors of the total spin of two spin $1/2$ particles, indeed, contain a symmetric triplet $|1, M_S\rangle$ of states with total $S^{(tot)} = 1$ (see Eq. 9.54) and one antisymmetric singlet state (Eq. 9.55) with total $S^{(tot)} = 0$. Thus, if I take these states as the spin components of the total basis of two-particle fermion states, then the symmetry of the spin component will dictate the symmetry of the orbital portion. Indeed, to make the entire two-fermion state antisymmetric, the orbital components paired with any of the symmetric two-spin state $|1, M_S\rangle$ must itself be antisymmetric. Two available orbital states can only yield a single antisymmetric combination resulting in three basis vectors characterized by the value of total spin $S^{(tot)} = 1$:

$$\frac{1}{\sqrt{2}} \left[\left| \psi_1^{(1)} \right\rangle \left| \psi_2^{(2)} \right\rangle - \left| \psi_2^{(1)} \right\rangle \left| \psi_1^{(2)} \right\rangle \right] |1, -1\rangle$$

$$\frac{1}{\sqrt{2}} \left[\left| \psi_1^{(1)} \right\rangle \left| \psi_2^{(2)} \right\rangle - \left| \psi_2^{(1)} \right\rangle \left| \psi_1^{(2)} \right\rangle \right] |1, 0\rangle \qquad (11.25)$$

$$\frac{1}{\sqrt{2}} \left[\left| \psi_1^{(1)} \right\rangle \left| \psi_2^{(2)} \right\rangle - \left| \psi_2^{(1)} \right\rangle \left| \psi_1^{(2)} \right\rangle \right] |1, 1\rangle ,$$

where $1/\sqrt{2}$ factor ensures the normalization of the vector representing the orbital portion of the state. The remaining total spin eigenvector corresponding to $S = 0$ is an antisymmetric singlet $|0, 0\rangle$. Consequently, the corresponding orbital part of the two-particle state must be symmetric resulting in three additional possible states:

$$\left| \psi_1^{(1)} \right\rangle \left| \psi_1^{(2)} \right\rangle |0, 0\rangle$$

$$\left| \psi_2^{(1)} \right\rangle \left| \psi_2^{(2)} \right\rangle |0, 0\rangle \qquad (11.26)$$

$$\frac{1}{\sqrt{2}} \left[\left| \psi_1^{(1)} \right\rangle \left| \psi_2^{(2)} \right\rangle + \left| \psi_2^{(1)} \right\rangle \left| \psi_1^{(2)} \right\rangle \right] |0, 0\rangle .$$

You may notice that the first two of these states are formed by identical orbitals. This is not forbidden by the Pauli principle because the spin state of two electrons in this case is antisymmetric. This situation is often described by saying that the two electrons in the same orbital state have "opposite" spins, which is not exactly accurate. Indeed, "opposite" can refer only to the possible values of the z-component of spin, but those can have opposite values in the singlet state as well as in the triplet state with $M_s = 0$. Thus, it is more accurate to describe this situation as a total spin zero or a singlet state. Combining three spin-antisymmetric states, Eq. 11.26, with three antisymmetric-orbital states, Eq. 11.25, you find that the total number of basis vectors in this representation is the same (six) as in the single-particle orbital basis, confirming that this is just an alternative basis in the same vector space.

A more realistic example of a basis based on the separation of many-fermion spin and orbital states would include two particles and at least three single-particle orbital states corresponding to $l = 1, m = -1, 0, 1$. The total orbital angular momentum of two electrons in this case can take three values: $L = 0, 1, 2$ with the total number of corresponding states being $1 + 3 + 5 = 9$ with various values of magnetic number M. To figure out the symmetry of these states, you would need to present them as a linear combination of single-particle states using the Clebsch–Gordan coefficients, similar to what I did in Sect. 9.5.2:

$$|L, l_1, l_2, M\rangle = \sum_{m_1, m_2} C^{L, l_1, l_2}_{M, m_1, m_2} |l_1 m_1\rangle |l_2, m_2\rangle \delta_{m_2, M - m_1}, \tag{11.27}$$

where Kronecker's delta makes sure that Eq. 11.24 is respected. The particle's exchange symmetry of the states presented by $|L, l_1, l_2, M\rangle$ is determined by the transformation rule of the Clebsch–Gordan coefficients with respect to the transposition of indexes l_1, m_1 and l_2, m_2, which you will have to accept without proof:

$$C^{L, l_1, l_2}_{M, m_1, m_2} = (-1)^{L - l_1 - l_2} C^{L, l_2, l_1}_{M, m_2, m_1}. \tag{11.28}$$

Indeed, applying the exchange operator $\hat{P}(1, 2)$ to Eq. 11.27, you will see that its action on the right-hand side of the equation consists in the interchange of indexes l_1 and l_2 in the Clebsch–Gordan coefficients:

$$\hat{P}(1, 2) |L, l_1, l_2, M\rangle = \sum_{m_1, m_2} C^{L, l_2, l_1}_{M, m_2, m_1} |l_1 m_1\rangle |l_2, m_2\rangle \delta_{m_1, M - m_2} =$$

$$(-1)^{L - l_1 - l_2} \sum_{m_1, m_2} C^{L, l_1, l_2}_{M, m_1, m_2} |l_1 m_1\rangle |l_2, m_2\rangle \delta_{m_2, M - m_1}$$

$$= (-1)^{L - l_1 - l_2} |L, l_1, l_2, M\rangle.$$

In the second line of this expression, I used the transposition property of $C^{L, l_1, l_2}_{M, m_1, m_2}$, Eq. 11.28. With this it becomes quite evident that state $|L, l_1, l_2, M\rangle$ is symmetric with respect to the exchange of particles if $L - l_1 - l_2$ is even and is antisymmetric if $L - l_1 - l_2$ is odd. In the example when $l_1 = l_2 = 1$, which I am trying to figure out now, this rule yields that the states with $L = 2$ and $L = 0$ are symmetric and the state with $L = 1$ is antisymmetric. Correspondingly, the latter must be paired with a triplet spin state, while the former two must go together with the zero spin state. Since the total number of single-electron orbitals in this case is 6, the expected number of two-particle antisymmetric basis vectors is $6!/(4!2!) = 15$, and if you insist I can list all of them below (I will use a simplified notation $|L, M\rangle$ omitting l_1 and l_2):

$$|2, M\rangle |0, 0\rangle$$

$$|1, M\rangle |1, M_s\rangle \tag{11.29}$$

$$|0, 0\rangle |0, 0\rangle.$$

The first line in this expression contains five vectors with $-2 \leq M \leq 2$, the second line represents $3 \times 3 = 9$ vectors with both M and M_s taking three values each, and finally, the last line supplies the last 15th vector to the basis.

Finally, to complete the picture, I can rewrite these vectors in terms of the grand total momentum $\hat{\boldsymbol{J}}$. The first five vectors from the expression above obviously correspond to $j = 2$, so one can easily replace this line with vectors $|2, M_J\rangle$, where the first number now corresponds to the value of j. The nine vectors from the second line correspond to three values of j: $j = 2, 1, 0$. While this situation appears terribly similar to the case of $l_1 = 1$ and $l_2 = 1$ states considered previously, the significant difference is that vectors $|1, M\rangle |1, M_s\rangle$ are no longer associated with just one or another particle, so that Eq. 11.28 has no relation to the symmetry properties of the resulting states $|j, 1, 1, M_J\rangle$ with respect to the exchange of the particles. All these states are as asymmetric under operator $\hat{P}(1, 2)$ as states $|1, M\rangle |1, M_s\rangle$. The last line in Eq. 11.29 obviously corresponds to a single state with zero grand total angular momentum, which simply coincides with $|0, 0\rangle |0, 0\rangle$. In summary, the antisymmetric basis in terms of eigenvectors of operators $\hat{J}^2$, $\left(\hat{L}^{(tot)}\right)^2$, $\left(\hat{S}^{(tot)}\right)^2$, and $\hat{J}_z$ is formed by vectors $|j, L, S, M_J\rangle$:

$$|2, 2, 0, M_J\rangle, |2, 1, 1, M_J\rangle, |1, 1, 1, M_J\rangle, |0, 1, 1, 0\rangle, |0, 0, 0, 0\rangle. \qquad (11.30)$$

It is easy to check that this basis also consists of $5+5+3+1+1 = 15$ vectors. They can be expressed as linear combinations of eigenvectors of the total orbital and total spin momenta (Eq. 11.29) with the help of Eq. 11.27 and the same Clebsch–Gordan coefficients, which can always be found on the Internet. Just to illustrate this point, let me do it for the grand total eigenvector $|1, 1, 1, 0\rangle$ using one of the tables of the Clebsch–Gordan coefficients that Google dug out for me in the depth of the World Wide Web:

$$|2, 1, 1, 0\rangle = \sqrt{\frac{1}{6}} |1, 1\rangle |1, -1\rangle + \sqrt{\frac{1}{6}} |1, -1\rangle |1, 1\rangle + \sqrt{\frac{2}{3}} |1, 0\rangle |1, 0\rangle.$$

Values of the total orbital, spin, and grand total momentum are often used to designate the electronic structure of atoms, instead of single-electron orbitals, in the form of the so-called term symbol:

$$^{2S+1}L_J. \qquad (11.31)$$

Here the center symbol designates the value of the total orbital momentum using the same correspondence between numerical values and letters: S, P, D, F for $0, 1, 2, 3$ correspondingly similar to the single-electron orbital case but with capital rather than lowercase letters. The right subscript shows the value of the grand total momentum, and the left superscript shows the multiplicity of the respective energy configuration with respect to the total spin magnetic number M_s. For instance, using this notation, the states $|1, 1, 1, M_J\rangle$ can be described as $^3P_1$, while states $|2, 2, 0, M_J\rangle$ become $^1D_2$.

The example of two electrons and three available single-particle orbital states is more realistic than the one with only two such states, but it is still a far cry

from what people have to deal when analyzing real atoms. The system of only
two electrons corresponds to helium atom, and one only needs one orbital state
with $l_{1,2} = 0$ to construct an antisymmetric two-electron ground state. In terms
of eigenvectors of total angular momentum and total spin, this state corresponds to
$L = 0, S = 0: |0,0\rangle |0,0\rangle$, where the orbital component is symmetric (both electrons
are in the same orbital state), and the spin component is antisymmetric (spins are
in an antisymmetric singlet state). The term symbol for this state is obviously $^1S_0$.
Going from helium to lithium, you already have to deal with three electrons, with the
corresponding structure in terms of single-electron orbitals shown in the first line of
Table 11.1. To figure out the values of the total orbital, spin, and grand total momenta
for this element, you can start with the one established for helium atom and add an
additional electron to it assuming that it does not disturb the existing configuration
of the two electrons in the closed shell. Since we know that this electron goes to
the orbital with $l = 0$, the total orbital momentum remains zero, and the total spin
becomes $1/2$ (you add a single spin to a state with $S = 0$, so what else can you
get?), so the grand total moment becomes $J = 0 + 1/2 = 1/2$, so that the term
symbol for Li becomes the same as for hydrogen $^2S_{1/2}$ emphasizing the periodic
property of the electronic properties of the elements. For the same reason, the term
symbol for the next element, beryllium, is exactly the same as the one we derived
for helium (see Table 11.1). To figure out the term symbol for boron, ignore the two
electrons in the first closed shell, which do not contribute anything to the total orbital
or spin momenta, and focus on the three electrons in the second shell. For these three
electrons, you have available two orbitals with the same orbital state, $l_1 = l_2 = 0$,
and opposite spin states and an extra orbital with $l_3 = 1$ and $s_3 = 1/2$. The total
orbital and spin momenta in this case can only be equal to $L = 1$ and $S = 1/2$,
while the grand total momentum can be either $J_1 = 1/2$ or $J_2 = 3/2$. Thus, boron
can be in one of two configurations $^2P_{1/2}$ or $^2P_{3/2}$, but so far we have no means
of figuring out which of these two configurations have a lower energy. To answer
this question, we can ask help from German physicist Friedrich Hermann Hund,
who formulated a set of empiric rules determining which term symbol describes the
electron configurations in atoms with lowest energy. These rules can be formulated
as follows:

1. For a given configuration, a term with the largest total spin has the lowest energy.
2. Among the terms with the same multiplicity, a term with the largest total orbital
   momentum has the lowest energy.
3. For the terms with the same total spin and total orbital momentum, the value of
   the grand total momentum corresponding to the lowest energy is determined by
   the filling of the outermost shell. If the outermost shell is half-filled or less than
   half-filled, then the term with the lowest value of the grand total momentum has
   the lowest energy, but if the outermost shell is more than half-filled, the term with
   the largest value of the grand total momentum has the lowest energy.

In the case of boron, you have to go straight to the third of Hund's rules because
the first two do not disambiguate between the corresponding terms. Checking
Table 11.1, you can see that the outermost shell for boron is the one characterized by

principal number $n = 2$, and the total number of single-particle orbitals on this shell is 8. Since boron has three electrons on this shell, the shell is less than half-filled, so that the third Hund's rule tells you that the ground state configuration of boron is $^2P_{1/2}$.

The case of carbon is even more interesting. Ignoring again two electrons with $L = 0$ and $S = 0$, I focus on two p-electrons with $l_1 = l_2 = 1$. Speaking of total orbital momentum and total spin, you can identify the following possible values for L and S: $L = 0, 1, 2$ and $S = 0, 1$. However, one needs to remember that the overall state, including its spin and orbital components, must be antisymmetric, so that not all combinations of L and S are possible. For instance, you already know that $L = 2$ orbitals are all symmetric; therefore, they can only coexist with spin singlet $S = 0$. The corresponding grand total momentum is $J = 2$, so that the respective term is $^1D_2$. The state with total orbital momentum $L = 1$ is antisymmetric and, therefore, demands a symmetric triplet spin state $S = 1$. This combination of orbital and spin momenta can generate grand total momentum $J = 2, 1, 0$, so that we have the following terms: $^3P_2, {}^3P_1, {}^3P_0$. Finally, symmetric $L = 0$ state must be coupled with the spin singlet giving rise to term $^1S_0$. In summary, I identified five possible terms consistent with the antisymmetry requirement: $^1D_2, {}^3P_2, {}^3P_1, {}^3P_0$, and $^1S_0$. Using the first two Hund's rules, you can limit the choice of the ground state configuration to the P states, and since the number of electrons in C atom on the outer shell is only 4, it is half-filled, and the third Hund's rule yields that the ground state configuration for carbon is $^3P_0$. Figuring out term symbols for elements where there are more than two electrons in the incomplete subshell (orbitals with the same value of the single-particle orbital momentum), such as nitrogen (three electrons on the p-subshell), is more complex, so I give you the term symbols for the rest of the elements in the second row of the periodic table in Table 11.1 without proof for you to contemplate.

11.4 Exchange Energy and Other Exchange Effects

11.4.1 Exchange Interaction

Some of the examples discussed in the previous section have already demonstrated a weird interconnectedness between spin and orbital components of many-particle states, which has nothing to do with any kind of real spin–orbital interaction. Recall, for instance, Eqs. 11.25 and 11.26 for two-fermion states: the triplet spin state in Eq. 11.25 requires asymmetric orbital state, while the singlet spin state in Eq. 11.26 asks for the orbital states to be symmetric. In the absence of interaction between electrons, all three $S = 1$ states are degenerate and belong to the energy eigenvalue $E_1 + E_2$, where $E_{1,2}$ are eigenvalues of the single-particle Hamiltonian corresponding to the "occupied" orbital states. At the same time, $S = 0$ states correspond to three different energies $2E_1$, $2E_2$, and $E_1 + E_2$, depending upon the orbital components used in their construction. The $E_1 + E_2$ energy level is, therefore, fourfold degenerate, with the corresponding eigenvectors formed by symmetric and antisymmetric combinations of the same two orbital functions $|\psi_1\rangle$ and $|\psi_2\rangle$. It is important to emphasize that three of these degenerate states correspond to the total spin of the system $S = 1$, and the fourth one possesses total spin $S = 0$. An interaction between electrons, however, might lift the degeneracy, making the energy of a two-electron system dependent on its spin state even in the absence of any actual spin-dependent interactions. This is yet another fascinating evidence of the weirdness of the quantum world.

I will demonstrate this phenomenon using a simple spin-independent Coulomb interaction potential:

$$\hat{V}(1, 2) = \frac{e^2}{4\pi\varepsilon_0 |\hat{r}_1 - \hat{r}_2|}$$

which describes the repulsion between two electrons in the atom of helium and is added to an attractive potential responsible for the interaction between the electrons and the nucleus. While a mathematically rigorous solution of a quantum three-body problem is too complicated for us to handle, what I can do is to compute the expectation value of the potential $\hat{V}(1, 2)$ using eigenvectors of the non-interacting electrons. As you will find out later in Chap. 13, such an expectation value gives you an approximation for the interaction-induced correction to the eigenvalues of the Hamiltonian.

Let me begin with the two-fermion state described by the vector presented in Eq. 11.25, which is characterized by an antisymmetric orbital component. The interaction potential does not contain any spin-related operators, allowing me to ignore the spin component of this state (it will simply yield $\langle 1, M_s| 1, M_s\rangle = 1$) and write the expectation value as follows:

$$\left\langle \hat{V}(1, 2) \right\rangle =$$

$$\frac{1}{2}\left[\left\langle\psi_2^{(2)}\right|\left\langle\psi_1^{(1)}\right| - \left\langle\psi_1^{(2)}\right|\left\langle\psi_2^{(1)}\right|\right]\hat{V}(1, 2)\left[\left|\psi_1^{(1)}\right\rangle\left|\psi_2^{(2)}\right\rangle - \left|\psi_2^{(1)}\right\rangle\left|\psi_1^{(2)}\right\rangle\right] =$$

$$\frac{1}{2}\left[\left\langle\psi_2^{(2)}\right|\left\langle\psi_1^{(1)}\right|\hat{V}(1, 2)\left|\psi_1^{(1)}\right\rangle\left|\psi_2^{(2)}\right\rangle + \left\langle\psi_1^{(2)}\right|\left\langle\psi_2^{(1)}\right|\hat{V}(1, 2)\left|\psi_2^{(1)}\right\rangle\left|\psi_1^{(2)}\right\rangle\right] - \tag{11.32}$$

$$\frac{1}{2}\left[\left\langle\psi_1^{(2)}\right|\left\langle\psi_2^{(1)}\right|\hat{V}(1, 2)\left|\psi_1^{(1)}\right\rangle\left|\psi_2^{(2)}\right\rangle + \left\langle\psi_2^{(2)}\right|\left\langle\psi_1^{(1)}\right|\hat{V}(1, 2)\left|\psi_2^{(1)}\right\rangle\left|\psi_1^{(2)}\right\rangle\right]. \tag{11.33}$$

If you carefully compare the terms in the third and fourth lines of the expression above, you will notice a striking difference between them. In both terms in the third line (Eq. 11.32), the ket and bra vectors, describing any of the two particles, represent the same state ($\left|\psi_1^{(1)}\right\rangle$ and $\left\langle\psi_1^{(1)}\right|$, $\left|\psi_2^{(2)}\right\rangle$ and $\left\langle\psi_2^{(2)}\right|$), while the ket and bra vectors of the same particle in the fourth line (Eq. 11.33) correspond to different states ($\left|\psi_1^{(1)}\right\rangle$ and $\left\langle\psi_2^{(1)}\right|$, $\left|\psi_2^{(2)}\right\rangle$ and $\left\langle\psi_1^{(2)}\right|$). In other words, the terms in the line labeled as Eq. 11.32 look like regular single-particle expectation values, while the terms in the next line look like non-diagonal matrix elements computed between different states for each particle. You can also notice that the two terms in Eq. 11.32 can be transformed into the other by exchange operator $\hat{P}(1, 2)$. Since the particles are identical, no matrix elements must change as a result of the transposition, which means that these terms are equal to each other. If you, however, apply the exchange operator to the terms in Eq. 11.33, you will generate expressions, where ket and bra vectors are reversed, meaning that these terms are complex conjugates of each other. Finally, you can easily see that the expression in Eq. 11.32 would have exactly the same form even if the particles in question were distinguishable, while Eq. 11.33 results from the antisymmetrization requirements imposed on the two-electron state.

Taking all this into account, the interaction expectation value can be presented as

$$\left\langle \hat{V}(1, 2) \right\rangle = V_C + V_{exc}, \tag{11.34}$$

where V_C is defined as

$$V_C = \left\langle \psi_2^{(2)} \middle| \left\langle \psi_1^{(1)} \middle| \hat{V}(1, 2) \middle| \psi_1^{(1)} \right\rangle \middle| \psi_2^{(2)} \right\rangle$$

and V_{exc} as

$$V_{exc} = -Re\left[\left\langle \psi_1^{(2)} \middle| \left\langle \psi_2^{(1)} \middle| \hat{V}(1, 2) \middle| \psi_1^{(1)} \right\rangle \middle| \psi_2^{(2)} \right\rangle \right].$$

Using the position representation for the orbital states, the expression for V_C can be written down in the explicit form

$$V_C = \frac{e^2}{4\pi\varepsilon_0} \int d^3 r_1 \int d^3 r_2 \frac{|\psi_1(r_1)|^2 |\psi_2(r_2)|^2}{|r_1 - r_2|}, \tag{11.35}$$

which makes all statements made about V_C rather obvious. If you agree to identify $e|\psi(r)|^2$ with the charge density, you can interpret Eq. 11.35 as a classical energy of the Coulomb interaction between two continuously distributed charges with densities $e|\psi_1(r)|^2$ and $e|\psi_2(r)|^2$.

The expression for V_{exc} in the position representation takes the form

$$V_{exc} = -\frac{e^2}{4\pi\varepsilon_0} Re\left[\int d^3 r_1 \int d^3 r_2 \frac{\psi_1^*(r_2)\psi_2^*(r_1)\psi_1(r_1)\psi_2(r_2)}{|r_1 - r_2|} \right], \tag{11.36}$$

which does not have any classical interpretation. This contribution to the energy is called *exchange energy*, and its origin can be directly traced to the antisymmetrization requirement. The expectation value computed with the symmetric orbital state would have the same form as in Eq. 11.34, with one but important difference—a different sign in front of the exchange energy term. Thus, previously degenerate states are now split by the interaction of the amount equal to $2V_{exc}$ on the basis of their spin states. Just think about it—in the absence of any special spin–orbit interaction term in the Hamiltonian, the energies of the two-electron states composed of the same single-particle orbitals depend on their spin state! This is a purely quantum effect, one of the manifestations of the oddity of quantum mechanics, which has profound experimental and technological implications. However, first of all, I want you to get some feeling about the actual magnitude of this effect; for this reason, I am going to compute the Coulomb and exchange energies for a simple example of a two-electron state of the helium atom.

For concreteness (and to simplify calculations), I will presume that the orbitals participating in the construction of the two-electron state are $|1, 0, 0\rangle$ and $|2, 0, 0\rangle$, where I used the notation for the states from Chap. 8. In the position representation, the corresponding wave functions are $\psi_1(r_1) = R_{10}(r_1)/\sqrt{4\pi}$ and $\psi_2(r_2) = R_{20}(r_2)/\sqrt{4\pi}$, where R_{10} and R_{20} are hydrogen radial wave functions, and the factor $1/\sqrt{4\pi}$ is what is left of the spherical harmonics with zero orbital momentum. When integrating Eq. 11.35 with respect to r_1, I can choose the Z-axis of the spherical coordinate system in the direction of r_2, in which case the denominator in this equation can be written down as

$$|r_1 - r_2| = \sqrt{r_1^2 + r_2^2 - 2r_1r_2 \cos \theta_1}.$$

The integral over r_1 now becomes

$$I(r_2) = \frac{32}{4\pi a_B^3} \int\limits_0^\infty dr_1 \int\limits_0^\pi d\theta_1 \int\limits_0^{2\pi} d\varphi_1 r_1^2 \sin \theta_1 \frac{e^{-4r_1/a_B}}{\sqrt{r_1^2 + r_2^2 - 2r_1r_2 \cos \theta_1}} =$$

$$\frac{64}{a_B^3} \int\limits_0^\infty dr_1 r_1^2 e^{-4r_1/a_B} \int\limits_{-1}^1 dx \frac{1}{\sqrt{r_1^2 + r_2^2 - 2r_1r_2 x}},$$

where I substituted

$$R_{10} = 2 (2/a_B)^{3/2} \exp (-2r/a_B)$$

(remember that $Z = 2$ for He). Integral over x yields

$$\int\limits_{-1}^1 dx \frac{1}{\sqrt{r_1^2 + r_2^2 - 2r_1r_2 x}} = \frac{1}{2r_1r_2} \int\limits_{-2r_1r_2}^{2r_1r_2} \frac{dz}{\sqrt{r_1^2 + r_2^2 + z}} =$$

$$\frac{1}{r_1r_2} \left[\sqrt{r_1^2 + r_2^2 + 2r_1r_2} - \sqrt{r_1^2 + r_2^2 - 2r_1r_2} \right] = \frac{r_1 + r_2 - |r_1 - r_2|}{r_1r_2}. \qquad (11.37)$$

Evaluating this expression separately for $r_1 > r_2$ and $r_1 < r_2$, I find for $I(r_2)$

$$I(r_2) = \frac{128}{a_B^3 r_2} \int_0^{r_2} dr_1 r_1^2 e^{-4r_1/a_B} + \frac{128}{a_B^3} \int_{r_2}^{\infty} dr_1 r_1 e^{-4r_1/a_B} =$$

$$\frac{4}{r_2} \left[1 - \left(1 + \frac{2r_2}{a_B} \right) e^{-4r_2/a_B} \right]. \qquad (11.38)$$

Now, using

$$R_{20} = 2 \left(\frac{1}{a_B} \right)^{3/2} \left(1 - \frac{r}{a_B} \right) \exp \left(-\frac{r}{a_B} \right).$$

I get for V_C

$$V_C = \frac{e^2}{4\pi\varepsilon_0} \frac{8}{4\pi a_B^3} \int_0^{\infty} dr_2 \int_0^{\pi} d\theta_2 \int_0^{2\pi} d\varphi_2 \sin\theta_2 r_2^2 I(r_2) \left(1 - \frac{r_2}{a_B} \right)^2 \exp\left(-\frac{2r_2}{a_B} \right) =$$

$$\frac{e^2}{4\pi\varepsilon_0} \frac{32}{a_B^3} \int_0^{\infty} dr_2 r_2 \left[1 - \left(1 + \frac{2r_2}{a_B} \right) e^{-4r_2/a_B} \right] \left(1 - \frac{r_2}{a_B} \right)^2 \exp\left(-\frac{2r_2}{a_B} \right) =$$

$$\frac{272}{81} \frac{e^2}{4\pi\varepsilon_0 a_B} = 3.35 Ry \cong 46.34 \, \text{eV}$$

where I used Eq. 8.17 with Z set to unity and notation $Ry = 13.8 \, \text{eV}$ for hydrogen's ground state (in vacuum).

Now, I will compute the exchange energy correction. Keeping the same notation $I(r_2)$ for the first integral with respect to r_1, I can present it, using expressions for the radial functions provided above, as

$$I_2(r_2) = \frac{8\sqrt{2}}{4\pi a_B^3} \int_0^{\infty} dr_1 \int_0^{\pi} d\theta_1 \int_0^{2\pi} d\varphi_1 r_1^2 \sin\theta_1 \frac{\left(1 - \frac{r_1}{a_B} \right) \exp\left(-\frac{3r_1}{a_B} \right)}{\sqrt{r_1^2 + r_2^2 - 2r_1 r_2 \cos\theta_1}} =$$

$$\frac{4\sqrt{2}}{a_B^3} \int_0^{\infty} dr_1 r_1^2 \exp\left(-\frac{3r_1}{a_B} \right) \left(1 - \frac{r_1}{a_B} \right) \int_{-1}^{1} dx \frac{1}{\sqrt{r_1^2 + r_2^2 - 2r_1 r_2 x}}.$$

Equation 11.37 for the angular integral and Mathematica © for the remaining radial integrals yield

$$I(r_2) = \frac{8\sqrt{2}}{a_B^3 r_2} \int\limits_0^{r_2} dr_1 r_1^2 \exp\left(-\frac{3r_1}{a_B}\right)\left(1 - \frac{r_1}{a_B}\right) +$$

$$\frac{8\sqrt{2}}{a_B^3} \int\limits_{r_2}^{\infty} dr_1 r_1 \exp\left(-\frac{3r_1}{a_B}\right)\left(1 - \frac{r_1}{a_B}\right) = \frac{8\sqrt{2}}{27 a_B}\left(1 + \frac{3r_2}{a_B}\right) e^{-3r_2/a_B}.$$

Plugging it into Eq. 11.36 and dropping the real value sign because all the functions in the integral are real, I have

$$V_{exc} = -\frac{e^2}{4\pi\varepsilon_0} \frac{8\sqrt{2}}{27 a_B} 4 \left(\frac{2}{a_B}\right)^{3/2}\left(\frac{1}{a_B}\right)^{3/2} \times$$

$$\int\limits_0^{\infty} dr_2 r_2^2 \exp\left(-\frac{2r_2}{a_B}\right)\left(1 - \frac{r_2}{a_B}\right) \exp\left(-\frac{r_2}{a_B}\right)\left(1 + \frac{3r_2}{a_B}\right) \exp\left(-\frac{3r_2}{a_B}\right) =$$

$$\frac{128}{27} \frac{e^2}{4\pi\varepsilon_0 a_B^4} \int\limits_0^{\infty} dr_2 r_2^2 \exp\left(-\frac{6r_2}{a_B}\right)\left(1 - \frac{r_2}{a_B}\right)\left(1 + \frac{3r_2}{a_B}\right) \approx -1.21\,\text{eV}.$$

Thus, this calculation showed that a state with $S = 1$ has a smaller energy than a state with $S = 0$ by $2 \times 1.21 = 2.42\,\text{eV}$. However, the sign of the integral in the exchange term depends on the fine details of wave functions representing single-electron orbitals and is not predestined. If single-particle orbitals of electrons were different, describing, for instance, electrons with non-zero orbital momentum in outer shells of heavier elements, or electrons in metals, the situation might have been reversed, and the singlet spin state could have a lower energy than a triplet.

This difference between energies of symmetric and antisymmetric spin states gives rise to something known as an exchange interaction between spins and plays an extremely important role in magnetic properties of materials. In particular, this "interaction," which is simply a result of the fermion nature of electrons, is responsible for the formation of ordered spin arrangements responsible for such phenomena as ferromagnetism or antiferromagnetism.

Ferromagnets—materials with permanent magnetization—have been known since the earliest days of human civilization, but the origin of their magnetic properties remained a mystery for a very long time. André-Marie Ampère (a French physicist who lived between 1775 and 1836 and made seminal contributions to electromagnetism) proposed that magnetization is the result of the alignment of all dipole magnetic moments formed by circular electron currents of each atom in the same direction. This alignment, he believed, was due to the magnetostatic interaction between the dipoles, which made the energy of the system lowest when all dipoles point in the same direction. Unfortunately, calculations showed that the magnetostatic interaction is so weak that thermal fluctuations would destroy the ferromagnetic order even at temperatures as small as a few Kelvins. The energy of the spin exchange interaction is much bigger (if you think that 2 eV is a small energy, you will be delighted to know that it corresponds to a temperature of more than 20,000 K). The temperature at which iron loses its ferromagnetic properties due to thermal agitation is about 1043 K, which corresponds to the exchange energy of only 80 meV, which is the real culprit behind the ordering of spin magnetic moments. In the classical picture, ordering would mean that all magnetic moments are aligned in the same direction, but describing this phenomenon quantum mechanically, we would say that N spins S are aligned if they are in the symmetric state with total spin equal to NS. For such a state to correspond to the ground state of the system of N spins, the exchange energy must favor (be lower for) symmetric states over the antisymmetric states. An antisymmetric state classically could be described as an array of magnetic moments, each pair of which is aligned in the opposite directions, while quantum mechanically we would say that each pair of spins in the array is in the spin zero state. Materials with such an arrangement of magnetic moments are known as antiferromagnetics, and for the antiferromagnetic state to be a ground state, the exchange energy must change its sign compared to the ferromagnetic case. The complete theory of magnetic order in solids is rather complicated, so you should not think that this brief glimpse into this area gives you any kind of even remotely complete picture, but beyond all these complexities, there is a main underlying physical mechanism—the exchange energy.

11.4.2 Exchange Correlations

The symmetry requirement on the many-particle states of indistinguishable particles affects not only their interaction energy but also their spatial positions. To illustrate this point, I will compute the expectation value of the distance between two electrons defined as

$$\left\langle (r_1 - r_2)^2 \right\rangle = \left\langle r_1^2 \right\rangle + \left\langle r_2^2 \right\rangle - 2 \left\langle r_1 r_2 \right\rangle. \tag{11.39}$$

This time around, I will assume that the two electrons belong to two different hydrogen-like atoms separated by a distance R small enough for the wave functions describing states of each electron to have significant spatial overlap. I will also assume that the electrons are in the same atomic orbital $|n, l, m\rangle$, but since each of these orbitals belongs to two different atoms, they represent different states, even if they are described by the same set of quantum numbers. To distinguish between these orbitals, I will add another parameter to the set of quantum numbers characterizing the position of the nucleus: $|n, l, m, R\rangle$. If the atoms are separated by a significant distance, you can quite clearly ascribe an electron to an atom it belongs. However, when the distance between the nuclei becomes comparable with the characteristic size of the electron's wave function, this identification is no longer possible, and you have to introduce two orbitals for each electron, $|n, l, m, R_1\rangle_i$

and $|n, l, m, R_2\rangle_i$, where lower index i outside of the ket symbol takes values 1 or 2 signifying one or another electron. This two-electron system can be again in a singlet or triplet spin state demanding symmetric or antisymmetric two-electron orbital state:

$$|\pm\rangle = \frac{1}{\sqrt{2}} \left[|n, l, m, R_1\rangle_1 \, |n, l, m, R_2\rangle_2 \pm |n, l, m, R_1\rangle_2 \, |n, l, m, R_2\rangle_1 \right]. \qquad (11.40)$$

The first two terms in Eq. 11.39 are determined by single-particle orbitals:

$$\langle \pm | \, r_1^2 \, | \pm \rangle = \frac{1}{2} \Big[{}_1 \langle n, l, m, R_1 | \, r_1^2 \, | n, l, m, R_1 \rangle_1 + {}_1 \langle n, l, m, R_2 | \, r_1^2 \, | n, l, m, R_2 \rangle_1 \pm$$

$${}_1 \langle n, l, m, R_1 | \, r_1^2 \, | n, l, m, R_2 \rangle_1 \times {}_2 \langle n, l, m, R_1 | \, n, l, m, R_2 \rangle_2 \pm$$

$${}_1 \langle n, l, m, R_2 | \, r_1^2 \, | n, l, m, R_1 \rangle_1 \times {}_2 \langle n, l, m, R_2 | \, n, l, m, R_1 \rangle_2 \Big].$$

When writing this expression, I took into account that the orbitals belonging to the same atom are normalized, ${}_2 \langle n, l, m, R_2 | \, n, l, m, R_2 \rangle_2 = 1$, but orbitals belonging to different atoms are not necessarily orthogonal: ${}_2 \langle n, l, m, R_1 | \, n, l, m, R_2 \rangle_2 \neq 0$. Similar expression for $\langle \pm | \, r_2^2 \, | \pm \rangle$ is

$$\langle \pm | \, r_2^2 \, | \pm \rangle = \frac{1}{2} \Big[{}_2 \langle n, l, m, R_1 | \, r_2^2 \, | n, l, m, R_1 \rangle_2 + {}_2 \langle n, l, m, R_2 | \, r_2^2 \, | n, l, m, R_2 \rangle_2 \pm$$

$${}_2 \langle n, l, m, R_1 | \, r_2^2 \, | n, l, m, R_2 \rangle_2 \times {}_1 \langle n, l, m, R_1 | \, n, l, m, R_2 \rangle_1 \pm$$

$${}_2 \langle n, l, m, R_2 | \, r_2^2 \, | n, l, m, R_1 \rangle_2 \times {}_1 \langle n, l, m, R_2 | \, n, l, m, R_1 \rangle_1 \Big].$$

Since both atoms are assumed to be identical, the following must be true:

$${}_1 \langle n, l, m, R_1 | \, r_1^2 \, | n, l, m, R_1 \rangle_1 = {}_2 \langle n, l, m, R_2 | \, r_2^2 \, | n, l, m, R_2 \rangle_2 \equiv a^2$$

$${}_1 \langle n, l, m, R_2 | \, r_1^2 \, | n, l, m, R_2 \rangle_1 = {}_2 \langle n, l, m, R_1 | \, r_2^2 \, | n, l, m, R_1 \rangle_2 \equiv b^2$$

$${}_1 \langle n, l, m, R_1 | \, r_1^2 \, | n, l, m, R_2 \rangle_1 = {}_2 \langle n, l, m, R_2 | \, r_2^2 \, | n, l, m, R_1 \rangle_2 \equiv u$$

$${}_2 \langle n, l, m, R_1 | \, n, l, m, R_2 \rangle_2 = {}_1 \langle n, l, m, R_2 | \, n, l, m, R_1 \rangle_1 \equiv v.$$

All these relations can be obtained by noticing that the system remains unchanged if you replace $R_1 \rightarrow R_2$ and simultaneously change electron indexes 1 and 2. Taking into account these relations and corresponding simplified notations, I can write for $\langle \pm | \, r_{1,2}^2 \, | \pm \rangle$:

$$\langle \pm | \, r_1^2 \, | \pm \rangle = \langle \pm | \, r_2^2 \, | \pm \rangle = \frac{1}{2} \left[a^2 + b^2 \pm \left(uv + u * v^* \right) \right].$$

The next step is to evaluate $\langle r_1 r_2 \rangle$:

$$\langle \pm | r_1 r_2 | \pm \rangle =$$

$$\frac{1}{2} \left[{}_1 \langle n, l, m, R_1 | r_1 | n, l, m, R_1 \rangle_1 \times {}_2 \langle n, l, m, R_2 | r_2 | n, l, m, R_2 \rangle_2 + \right. \tag{11.41}$$

$${}_1 \langle n, l, m, R_2 | r_1 | n, l, m, R_2 \rangle_1 \times {}_2 \langle n, l, m, R_1 | r_2 | n, l, m, R_1 \rangle_2 \pm$$

$${}_1 \langle n, l, m, R_1 | r_1 | n, l, m, R_2 \rangle_1 \times {}_2 \langle n, l, m, R_1 | r_2 | n, l, m, R_2 \rangle_2 \pm$$

$$\left. {}_1 \langle n, l, m, R_2 | r_1 | n, l, m, R_1 \rangle_1 \times {}_2 \langle n, l, m, R_2 | r_2 | n, l, m, R_1 \rangle_2 \right] . \tag{11.42}$$

The evaluation of these expressions requires a more explicit determination of the point with respect to which electron position vectors are defined. Assuming for concreteness that the origin of the coordinate system is at the nucleus of atom 1, I can immediately note that the symmetry with respect to inversion kills first two terms in Eq. 11.42 since $_1 \langle n, l, m, R_1 | r_1 | n, l, m, R_1 \rangle_1 = {}_2 \langle n, l, m, R_1 | r_2 | n, l, m, R_1 \rangle_2 = 0$. The remaining two terms survive and can be written as

$$\langle \pm | r_1 r_2 | \pm \rangle = \pm |d|^2$$

where I introduced vector d defined as follows:

$$_1 \langle n, l, m, R_1 | r_1 | n, l, m, R_2 \rangle_1 = {}_2 \langle n, l, m, R_2 | r_2 | n, l, m, R_1 \rangle_2 \equiv d.$$

Finally, combining all the obtained results together, I can write

$$\left\langle (r_1 - r_2)^2 \right\rangle = a^2 + b^2 \pm \left(uv + u * v^* - 2 |d|^2 \right). \tag{11.43}$$

While the actual computation of matrix elements appearing in Eq. 11.43 is rather difficult and will not be attempted here, you can still learn something from this exercise. Its main lesson is that the spin state of the electrons affects how close the electrons of the two atoms can be. Assuming for concreteness that the expression in the parentheses in Eq. 11.43 is negative, which is favored by the term $|d|^2$ (the actual sign depends on the single-electron orbitals), one can conclude that the antisymmetric spin state promoting symmetric orbital state (+ sign in $\pm$) results in electrons being closer together, than in the case of the symmetric spin state. This is an interesting quantum mechanical effect: electrons appear to be "pushed" closer toward each other or further away from each other depending on their spin state even though there is no actual physical force doing the "pushing." This phenomenon plays an important role in chemical bonding between atoms, because electrons, when "pushed" toward each other, pull their nuclei along making the formation of a stable bi-atomic molecule more likely.

11.5 Fermi Energy

The behavior of systems consisting of many identical particles (and by many here I mean really huge, something like Avogadro's number) is studied by a special field of physics called quantum statistics. Even a sketchy review of this field would take us well outside the scope of this book, but there is one problem involving a very large number of fermions, which we can handle. The issue in question is the structure of the ground state and its energy for the system on $N \gg 1$ non-interacting free electrons (an ideal electron gas) confined within a box of volume V. Each electron is a free particle characterized by a momentum p, corresponding single-particle energy $E_p = p^2/2m_e$, and a single particle wave function (in the position representation) $\psi_p(r) = A_p \exp(ip \cdot r/\hbar)$, where A_p is a normalization parameter, which was chosen in Sect. 5.1.1 to be $1/\sqrt{2\pi\hbar}$ to generate a delta-function normalized wave function. Here it is more convenient to choose an alternative normalization, which would explicitly include volume V occupied by the electrons. To achieve this, I will impose so-called periodic boundary conditions:

$$\psi_p(r + L) = \psi_p(r), \tag{11.44}$$

where L is a vector with components L_x, L_y, L_z such that $L_x L_y L_z = V$. This boundary condition is the most popular choice in solid-state physics, and if you are wondering about its physical meaning and any kind of relation to reality, it does not really have any. The logic of using it is based upon two ideas. First, it is more convenient than, say, particle-in-the-box boundary conditions $\psi_p(L) = 0$, implying that the electrons are confined in an infinite potential well, because it keeps the wave functions in the complex exponential form rather than forcing them to become much less convenient real-valued sin functions. Second, it is believed that as long as we are not interested in specific surface-related phenomena, the behavior of the wave functions at the boundary of a solid shall not have any impact on its bulk properties. I have used a similar idea when computing the total energy of electromagnetic field in Sect. 7.3.1.

This boundary condition imposes restrictions on the allowed values of the electron's momentum:

$$\exp(ip \cdot (r + L)/\hbar) = \exp(ip \cdot r/\hbar) \Rightarrow$$

$$\exp(ip \cdot L/\hbar) = 1 \Rightarrow \frac{p_i \cdot L_i}{\hbar} = 2\pi n_i, \tag{11.45}$$

where $i = x, y, z$ and $n_i = \pm1, \pm2, \pm3 \cdots$. In addition to making the spectrum of the momentum operator discrete, the periodic boundary condition also allows an alternative normalization of the wave function:

$$|A_p|^2 \int\limits_{-L_x/2}^{L_x/2} \int\limits_{-L_y/2}^{L_y/2} \int\limits_{-L_z/2}^{L_z/2} e^{-i\frac{\mathbf{p}\cdot\mathbf{r}}{\hbar}} e^{i\frac{\mathbf{p}\cdot\mathbf{r}}{\hbar}} dxdydz = 1$$

which yields $A_p = 1/\sqrt{V}$. The system of normalized and orthogonal single-electron wave function takes the form

$$\psi_{n_1,n_2,n_3}(r) = \frac{1}{\sqrt{V}} \exp\left[i\left(\frac{2\pi}{L_x}n_1 x + \frac{2\pi}{L_y}n_2 y + \frac{2\pi}{L_z}n_3 z\right)\right],$$

while the single-electron energies form a discrete spectrum defined by

$$\epsilon_{n_1 n_2 n_3} = \frac{(2\pi\hbar)^2}{2m_e}\left(\frac{n_1^2}{L_x^2} + \frac{n_2^2}{L_y^2} + \frac{n_3^2}{L_z^2}\right). \tag{11.46}$$

This wave function generates two single-electron orbitals characterized by two different values of the spin magnetic number m_s, which is perfectly suitable for generating many-particle states of the N-electron system. The ground state of this system is given by the Slater determinant formed by N single-electron orbitals with the lowest possible single-particle energies ranging from $\epsilon_{1,1,1}$ to some maximum value ϵ_F corresponding to the last orbital making it into the determinant. Thus, all single-particle orbitals of electrons are divided into two groups: those that are included (occupied) into the Slater determinant for the ground state and those that are not (empty or vacant). The occupied and empty orbitals are separated by energy ϵ_F known as the Fermi energy. The Fermi energy is an important characteristic of an electron gas, which obviously depends on the number of electrons N and determines much of its ground state properties. Thus, let's spend some time trying to figure it out.

In principle, finding ϵ_F is quite straightforward: one needs to find the total number of orbitals $M(\epsilon)$ with energies less than ϵ. Then the Fermi energy is found from equation

$$M(\epsilon_F) = N. \tag{11.47}$$

However, counting the orbitals and finding $M(\epsilon)$ are not quite trivial because energy values defined by Eq. 11.46 are highly degenerate, and what is even worse is that there is no known analytical formula for the degree of the degeneracy as a function of energy. The problem, however, can be solved in the limit when $N \to \infty$ and $V \to \infty$ so that the concentration of electrons N/V remains constant. In this limit the discrete granular structure of the energy spectrum becomes negligible (the spectrum in this case is called quasi-continuous), and the function $M(\epsilon)$ can be determined.

You might think that I am nuts because I first introduce finite V to make the spectrum discrete and then go to the limit $V \to \infty$ to make it continuous again. The thing is that if I had begun with the infinite volume and continuous spectrum, the only information I would have had about the number of states is that it is infinite (the number of states of continuous spectrum is infinite for any finite interval of energies), which does not help me at all. What I am getting out of this roundabout approach is the knowledge about *how* the number of states turns infinite when the volume goes to infinity, and as you will see, this is exactly what we need to find the Fermi energy.

In order to find $M(\epsilon)$, it is convenient to visualize the states that need to be counted. This can be done by presenting each single-electron orbital graphically as points with coordinates n_1, n_2, n_3 in a three-dimensional space defined by a regular Cartesian coordinate system with axes X, Y, and Z. Each point here represents two orbitals with different values of the spin magnetic number. Surrounding each point by little cubes with sides equal to unity, I can cover the entire three-dimensional space containing the electron's orbitals. Since each cube has a unit volume, the total volume covered by the cubes is equal to the number of points within the covered region. Since each point represents two orbitals with opposite spins, the number of all orbitals in this region is twice the number of points.

For simplicity let me assume that all $L_z = L_y = L_z \equiv L$, which allows me to rewrite Eq. 11.46 in the form

$$n^2(\epsilon) = n_1^2 + n_2^2 + n_3^2 \tag{11.48}$$

where I introduced

$$n^2(\epsilon) = \frac{2m_e\epsilon_{n_1n_2n_3}L^2}{(2\pi\hbar)^2}. \tag{11.49}$$

Equation 11.48 defines a sphere in the space of electron orbitals with radius $n \propto L\sqrt{\epsilon}$. If you allow non-integer values for numbers $n_{1,2,3}$, you could say that each point on the surface of the sphere corresponds to states with the same energy ϵ (such surface is called isoenergetic). All points in the interior of the surface correspond to states with energies less than ϵ, while all points in the exterior represent states with energies larger than ϵ. Now, the number of orbitals encompassed by the surface is, as I just explained, simply equal to the volume of the corresponding region multiplied by two to account for two values of spin. Thus, I can write for the number of states with energies less than ϵ[4]:

$$M(\epsilon) = 2\frac{4}{3}\pi\left(\frac{L}{2\pi\hbar}\right)^3 (2m_e\epsilon)^{3/2} = V\frac{(2m_e\epsilon)^{3/2}}{3\pi^2\hbar^3}. \tag{11.50}$$

---

[4]If instead of periodic boundary conditions you would use the particle-in-the-box boundary conditions requiring that the wave function vanishes at the boundary of the region $L_x \times L_y \times L_z$, you would have ended up with $p_i = \pi n_i/L_i$, where n_i now can only take positive values because wave functions $\sin(\pi n_1 x/L_x)\sin(\pi n_2 y/L_y)\sin(\pi n_3 z/L_z)$ with positive and negative values of n_i represent the same function, while function $\exp i\left(\frac{2\pi}{L_x}n_1 x + \frac{2\pi}{L_y}n_2 y + \frac{2\pi}{L_z}n_3 z\right)$ with positive and negative indexes represents two linearly independent states. As a result, Eq. 11.50 when used in this case would have an extra factor $1/8$ reflecting the fact that only $1/8$ of a sphere correspond to points with all positive coordinates.

Fig. 11.3 Two-dimensional version of a state counting procedure described in the texts: squares replace cubes, a circle represents a sphere, but the points are still the states specified by two instead of three integer numbers. The 2-D version is easier to process visually, but illustrates all the important points

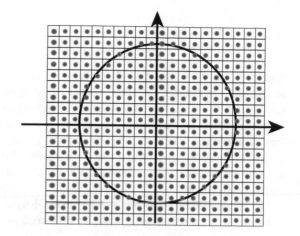

The problem with this calculation is, of course, that the points on the surface do not necessarily correspond to integer values of n_1, n_2, and n_3 so that this surface cuts through the little cubes in its immediate vicinity (see Fig. 11.3 representing a two-dimensional version of the described construction). As a result, some states with energies in the thin layer surrounding the spherical surface cannot be counted accurately. The number of such states is obviously proportional to the area of the enclosing sphere, which is $\propto L^2$, while the number of states counted correctly is $\propto L^3$, so that the relative error of the outlined procedure behaves as $1/L$ and approaches zero as L goes to infinity. Thus, Eq. 11.50 can be considered to be asymptotically correct in the limit $L \to \infty$. Now you can see the value of this procedure with the initial discretization followed by passing to the quasi continuous limit. It allowed me to establish the exact dependence of M as a function of volume V expressed by Eq. 11.50. Now I can easily find the Fermi energy by substituting Eq. 11.50 into Eq. 11.47:

$$V\frac{(2m_e\epsilon_F)^{3/2}}{3\pi^2\hbar^3} = N \Rightarrow$$

$$\epsilon_F = \frac{\hbar^2}{2m_e}\left(\frac{3\pi^2 N}{V}\right)^{2/3}. \tag{11.51}$$

The most important feature of Eq. 11.51 is that the number of electrons N and the volume they occupy V, the two quantities which are supposed to go to infinity, appear in this equation only in the form of the ratio N/V which we shall obviously keep constant when passing to the limit $V \to \infty$, $N \to \infty$. The ratio N/V specifies the number of electrons per unit volume and is also known as the electron concentration. This is one of the most important characteristics of the electron gas.

It is important to understand that the Fermi energy is the single-electron energy of the last "occupied" single-particle orbital and is not the energy of the many-electron ground state. To find the latter I need to add energies of all occupied single-electron orbitals:

$$E = 2 \sum_{n_1,n_2,n_3}^{N_{max}} \epsilon_{n_1,n_2,n_3}, \qquad (11.52)$$

where N_{max} is the collection of indexes n_1, n_2, n_3 corresponding to the last occupied state and the factor of 2 accounts for the spin variable. I will compute this sum again in the limit $V \to \infty$, $N \to \infty$ (which, by the way, is called thermodynamic limit), and while doing so I will show you a trick of converting discrete sums in integrals. This operation is, again, possible because in the thermodynamic limit the discrete spectrum becomes quasi-continuous, and the arguments I am going to employ here are essentially the same as the one used to compute the Fermi energy but with a slightly different flavor.

So I begin. When an orbital index n_i changes by one, the change of the respective component of the momentum p_i can be presented as

$$\Delta p_i = \frac{2\pi\hbar}{L_i} \Delta n_i,$$

where $\Delta n_i = 1$. With this relation in mind, I can rewrite Eq. 11.52 as

$$E = 2 \sum_{n_1,n_2,n_3}^{N_{max}} \epsilon_{n_1,n_2,n_3} \Delta n_1 \Delta n_2 \Delta n_3 = 2 \frac{L^3}{(2\pi\hbar)^3} \sum_{p_x,p_y,p_z}^{N_{max}} \epsilon_{p_x,p_y,p_z} \Delta p_x \Delta p_y \Delta p_z$$

where I again set $L_z = L_y = L_z \equiv L$ (remember that $\Delta n_i = 1$, so by including factors $\Delta n_1 \Delta n_2 \Delta n_3$ into the original sum, I did not really change anything). Now, when $L \to \infty$, Δn_i remains equal to unity, but $\Delta p_i \to 0$, so that the corresponding sum in the preceding expression turns into an integral:

$$E = 2 \frac{V}{(2\pi\hbar)^3} \int_{-p_x^F}^{p_x^F} \int_{-p_y^F}^{p_y^F} \int_{-p_z^F}^{p_z^F} dp_x dp_y dp_z \epsilon\left(p_x, p_y, p_z\right) \qquad (11.53)$$

where I changed the notation for the energy to emphasize that momentum is now not a discrete index, but a continuous variable. Since the single-particle energy $\epsilon\left(p_x, p_y, p_z\right)$ depends only upon p^2, it makes sense to compute the integral in Eq. 11.53 using the representation of the momentum vector in spherical coordinates. Replacing the Cartesian volume element $dp_x dp_y dp_z$ with its spherical counterpart $p^2 \sin\theta d\theta d\varphi dp$, where θ and φ are polar and azimuthal angles characterizing the direction of vector $\boldsymbol{p}$, I can rewrite Eq. 11.53 as

$$E = 2 \frac{V}{(2\pi\hbar)^3} \int_0^{p_F} \int_0^{\pi} \int_0^{2\pi} \epsilon\left(p\right) p^2 \sin\theta d\theta d\varphi dp,$$

where p_F is the magnitude of the momentum corresponding to the Fermi energy ϵ_F. I proceed replacing the integration variable p with another variable ϵ according to relation $p = \sqrt{2m_e\epsilon}$:

$$E = 2\frac{V(2m_e)^{3/2}}{2(2\pi\hbar)^3}4\pi\int_0^{\epsilon_F}\epsilon\sqrt{\epsilon}d\epsilon. \tag{11.54}$$

Before computing this integral, let me point out that it can be rewritten in the following form:

$$E = \int_0^{\epsilon_F}\epsilon g(\epsilon)\,d\epsilon \tag{11.55}$$

where I introduced quantity

$$g(\epsilon) = \frac{V(2m_e)^{3/2}\sqrt{\epsilon}}{2\pi^2\hbar^3} \tag{11.56}$$

called *density of states*. This quantity, which means the number of states per unit energy interval, is an important characteristic of any many-particle system. Actually it is so important to give me an incentive to deviate from the original goal of computing the integral in Eq. 11.54 and spend more time talking about it.

To convince you that $g(\epsilon)\,d\epsilon$ can, indeed, be interpreted as a number of states with energies within the interval $[\epsilon, \epsilon + d\epsilon]$, I will simply compute this quantity directly using the same state counting technique, which I used to find the Fermi energy. However, this time around I am interested in a number of states within a spherical shell with inner radius $n(\epsilon)$ and outer radius $n(\epsilon + d\epsilon)$:

$$n(\epsilon + d\epsilon) = n(\epsilon) + d\epsilon\,(dn/d\epsilon) = \frac{L}{2\pi\hbar}\sqrt{2m_e\epsilon} + \frac{L}{4\pi\hbar}\sqrt{\frac{2m_e}{\epsilon}}d\epsilon$$

where I used Eq. 11.49 for $n(\epsilon)$. The volume occupied by this shell is

$$\Delta V = 4\pi n^2 dn = 4\pi n^2\frac{dn}{d\epsilon}d\epsilon -$$

$$4\pi\frac{L^2 2m_e\epsilon}{4\pi^2\hbar^2}\frac{L}{4\pi\hbar}\sqrt{\frac{2m_e}{\epsilon}}d\epsilon = \frac{V(2m_e)^{3/2}\sqrt{\epsilon}}{4\pi^2\hbar^3}d\epsilon.$$

Using again the fact that the volume allocated to a single point in Fig. 11.3 is equal to one and that there are two single-electron orbitals per point, the total number of states within this spherical layer is

$$\frac{V(2m_e)^{3/2}\sqrt{\epsilon}}{2\pi^2\hbar^3}d\epsilon$$

which according to Eq. 11.56 is exactly $g(\epsilon)\,d\epsilon$.

Now I can go back to Eq. 11.55 and complete the computation of the integral, which is quite straightforward and yields

$$E = \frac{V(2m_e)^{3/2}}{2\pi^2\hbar^3}\int_0^{\epsilon_F}\epsilon\sqrt{\epsilon}\,d\epsilon = \frac{V(2m_e)^{3/2}}{5\pi^2\hbar^3}\epsilon_F^{5/2} =$$

$$\frac{3}{5}N\left(\frac{V(2m_e)^{3/2}}{3\pi^2 N(\hbar^2)^{3/2}}\right)\epsilon_F^{5/2}.$$

In the last line, I rearranged the expression for the energy to make it clear (with the help of Eq. 11.51) that the expression in the parentheses is $\epsilon_F^{-3/2}$ and that the ground state energy of the non-interacting free electron gas can be written down as

$$E = \frac{3}{5}N\epsilon_F. \tag{11.57}$$

This expression can also be rewritten in another no less illuminating form. Substituting Eq. 11.51 into Eq. 11.57, I can present energy E of the gas as a function of volume:

$$E = \frac{3\hbar^2(3\pi^2)^{2/3}}{10m_e}\frac{N^{5/3}}{V^{2/3}}.$$

The fact that this energy depends on the volume draws out an important point: if you try to expand (or contract) the volume occupied by the gas, its energy changes, which means that someone has to do some work to affect this change. Recalling a simple formula from introductory thermodynamics class $dW = PdV$, where W is work and P is pressure exerted by the gas on the walls of containing vessel, and taking into account that for the fixed number of particles, energy E depends only on volume V (no temperature), you can relate dW to $-dE$ and determine the pressure exerted by the non-interacting electrons on the walls of the container as

$$P = -\frac{dE}{dV} = \frac{\hbar^2(3\pi^2)^{2/3}}{5m_e}\frac{N^{5/3}}{V^{5/3}}.$$

Thus, even in the ground state (which, by the way, from the thermodynamic point of view corresponds to zero temperature), an electron gas exerts a pressure on the surrounding medium which depends on the concentration of the electrons. The coolest thing about this result is that unlike the case of a classical ideal gas, this pressure has nothing to do with the thermal motion of the electrons because they are in the ground state, which is equivalent to their temperature being equal to zero. This pressure is a purely quantum effect solely due to the indistinguishability of the electrons and their fermion nature.

11.6 Problems

Problems for Sect. 11.1

Problem 140 Consider the following configuration of single-particle orbitals for a system of four identical fermions:

$$\left|\alpha_1^{(1)}\right\rangle \left|\alpha_2^{(2)}\right\rangle \left|\alpha_3^{(3)}\right\rangle \left|\alpha_4^{(4)}\right\rangle .$$

Applying exchange operator $\hat{P}(i,j)$ to all pairs of particles in this configuration, generate all possible transpositions of the particles and determine the correct signs in front of them. Write down the correct antisymmetric four-fermion state involving these single-particle orbitals.

Problem 141 Consider the system of two bosons that can be in one of four single-particle orbitals. List all possible two-boson states adhering to the symmetrization requirements.

Problem 142 Consider two non-interacting electrons in a one-dimensional harmonic oscillator potential characterized by classical frequency ω.

1. Consider single-electron orbitals $|\alpha_{n,m_s}\rangle = |n\rangle |m_s\rangle$ where $|n\rangle$ is an eigenvector of the harmonic oscillator and $|m_s\rangle$ is a spinor describing one of two possible eigenvectors of operator $\hat{S}_z$. Using orbitals $|\alpha_{n,m_s}\rangle$, write down the Slater determinant for the two-electron ground state, and find the corresponding ground state energy.
2. Do the same for the first excited state(s) of this system.
3. Write the two-electron states found in Parts I and II in the position-spinor representation.
4. Now use the eigenvectors of the total spin of the two particles to construct the two-particle ground and first excited states. Find the relations between the two-particle states found here with those found in Parts I and II.
5. Compute the expectation value $\left\langle (z_1 - z_2)^2 \right\rangle$ where $z_{1,2}$ are coordinates of the two electrons in the states determined above.

Problem 143 Repeat Problem 142 for two non-interacting bosons.

Problems for Sect. 11.3

Problem 144 Consider an atom of nitrogen, which has three electrons in $l = 1$ states.

1. Using single-particle orbitals with $l = 1$ and different values of orbital and spin magnetic numbers, construct all possible Slater determinants representing possible three-electron states.
2. Applying operators $\hat{S}_z^{(1)} + \hat{S}_z^{(2)} + \hat{S}_z^{(3)}$ to all found three-particle states, figure out the possible values of the total spin in these states.

Problems for Sect. 11.4

Problem 145 Consider two identical non-interacting particles both in the ground states of their respective harmonic oscillator potentials. Particle 1 is in the potential $V_1 = \frac{1}{2}m\omega^2 x_1^2$, while particle 2 is in the potential $V_2 = \frac{1}{2}m\omega^2 (x_2 - d)^2$.

1. Assuming that particles are spin $1/2$ fermions in a singlet spin state, write down the orbital portion of the two-particle state and compute the expectation value of the two-particle Hamiltonian

$$\hat{H} = \frac{\hat{p}_1^2}{2m_e} + \frac{\hat{p}_2^2}{2m_e} + \frac{1}{2}m\omega^2 x_1^2 + \frac{1}{2}m\omega^2 (x_2 - d)^2$$

   in this state.
2. Repeat the calculations assuming that the particles are in the state with total spin $S = 1$.
3. The energy you found in Parts I and II depends upon the distance d between the equilibrium points of the potential. Classically such a dependence would mean that there is a force associated with this energy and describing repulsive or attractive interaction between the two particles. In the case under consideration, there is no real interaction, and what you have found is a purely quantum effect due to symmetry requirements on the two-particle states. Still, you can describe the result in terms of the effective "force" of interaction between the particles. Find this force for both singlet and triplet spin states, and specify its character (attractive or repulsive).

Problem 146 Consider two electrons confined in a one-dimensional infinite potential well of width d and interacting with each other via potential $V_{int} = -E_0 (z_1 - z_2)^2$ where E_0 is a real positive constant and $z_{1,2}$ are coordinates of the electrons.

1. Construct the ground state two-electron wave function assuming that electrons are (a) in a singlet spin state and (b) in a triplet spin state.

2. Compute the expectation value of the interaction potential in each of these states.
3. With interaction term included, which spin configuration would have smaller energy?

Problem for Sect. 11.5

Problem 147 Consider an ideal gas of N electrons confined in a three-dimensional harmonic oscillator potential:

$$\hat{V} = \frac{1}{2}m_e\omega^2\left(x^2 + y^2 + z^2\right) \equiv \frac{1}{2}m_e\omega^2 r^2.$$

Find the Fermi energy of this system and the total energy of the many-electron ground state. Hint: The degeneracy degrees of the single-particle energy levels in this case can be easily found analytically, so no transition to quasi-continuous spectrum and from summation to integration is necessary.

Part III
Quantum Phenomena and Methods

In this part of the book, I will introduce you into the wonderful world of actual experimentally observable quantum mechanical phenomena. Theoretical description of each of these phenomena will require developing special technical methods, which I will present as we go along. So, let the journey begin.

Chapter 12
Resonant Tunneling

12.1 Transfer-Matrix Approach in One-Dimensional Quantum Mechanics

12.1.1 Transfer Matrix: General Formulation

In Sects. 6.2 and 6.3 of Chap. 6, I introduced one-dimensional quantum mechanical models, in which potential energy of a particle was described by a simplest piecewise constant function (or its extreme case—a delta-function), defining a single potential well or barrier. A natural extension of this model is a potential energy profile corresponding to several wells and/or barriers (or several delta-functions). In principle, one can approach the multi-barrier problem in the same way as a single well/barrier situation: divide the entire range of the coordinate into regions of constant potential energy, and use the continuity conditions for the wave function and its derivative to "stitch" the solutions from different regions. However, it is easier said than done. Each new discontinuity point adds two new unknown coefficients and correspondingly two equations. If in the case of a single barrier you had to deal with the system of four equations, a dual-barrier problem would require solving the system of eight equations, and soon even writing those equations down becomes a serious burden, and I do not even want to think about having to solve them.

Luckily, there is a better way of dealing with the ever-increasing number of the boundary conditions in problems with multiple jumps of the potential energy. In this section I will show you a convenient method of arranging the unknown amplitudes of the wave functions and relating them to each other across the point of the discontinuity.

© Springer International Publishing AG, part of Springer Nature 2018
L.I. Deych, *Advanced Undergraduate Quantum Mechanics*,
https://doi.org/10.1007/978-3-319-71550-6_12

Let's move forward by going back to the simplest problem of a step potential with a single discontinuity:

$$V(z) = \begin{cases} V_0 & z < 0 \\ V_1 & z > 0, \end{cases} \tag{12.1}$$

where I assigned the coordinate $z = 0$ (the origin of the coordinate axes) to the point where the potential makes its jump. If I were to ask a good diligent student of quantum mechanics to write down a wave function of a particle with energy E exceeding both V_0 and V_1, I would have most likely been presented with the following expression:

$$\psi(z) = \begin{cases} A_0 \exp(ik_0 z) + B_0 \exp(-ik_0 z) & z < 0 \\ A_1 \exp(ik_1 z) & z > 0, \end{cases} \tag{12.2}$$

where

$$k_0 = \frac{\sqrt{2m_e(E - V_0)}}{\hbar}$$

$$k_1 = \frac{\sqrt{2m_e(E - V_1)}}{\hbar}.$$

This wave function would have been perfectly fine if all what I were after was just the single step-potential problem. In this section, however, I have further-reaching goals, so I need to generalize this expression allowing for a possibility to have a wave function component corresponding to the particles propagating in the negative z direction for $z > 0$ as well as for $z < 0$. If you wonder where these backward propagating particles could come from, just imagine that there might be another discontinuity in the potential somewhere down the line, at a positive value of z, which would create a flux of reflected particles propagating in the negative z direction at $z > 0$. To take this possibility into account, I will replace Eq. 12.2 with a wave function of a more general form

$$\psi(z) = \begin{cases} A_0 \exp(ik_0 z) + B_0 \exp(-ik_0 z) & z < 0 \\ A_1 \exp(ik_1 z) + B_1 \exp(-ik_1 z) & z > 0. \end{cases} \tag{12.3}$$

The continuity of the wave function and its derivative at $z = 0$ then yields

$$A_0 + B_0 = A_1 + B_1 \tag{12.4}$$

$$k_0(A_0 - B_0) = k_1(A_1 - B_1). \tag{12.5}$$

Quite similarly to what I already did in Sect. 6.2.1, I can rewrite these equations as

$$A_1 = \frac{1}{2}\left(1 + \frac{k_0}{k_1}\right) A_0 + \frac{1}{2}\left(1 - \frac{k_0}{k_1}\right) B_0 \tag{12.6}$$

$$B_1 = \frac{1}{2}\left(1 - \frac{k_0}{k_1}\right) A_0 + \frac{1}{2}\left(1 + \frac{k_0}{k_1}\right) B_0. \tag{12.7}$$

However, for the next step, I prepared for you something new. After spending some time staring at these two equations, you might divine that they can be presented in a matrix form if amplitudes $A_{1,0}$ and $B_{1,0}$ are arranged into a two-dimensional column vector, while the coefficients in front of A_0 and B_0 are arranged into a 2×2 matrix:

$$\begin{bmatrix} A_1 \\ B_1 \end{bmatrix} = \begin{bmatrix} \frac{1}{2}\left(1 + \frac{k_0}{k_1}\right) & \frac{1}{2}\left(1 - \frac{k_0}{k_1}\right) \\ \frac{1}{2}\left(1 - \frac{k_0}{k_1}\right) & \frac{1}{2}\left(1 + \frac{k_0}{k_1}\right) \end{bmatrix} \begin{bmatrix} A_0 \\ B_0 \end{bmatrix}. \tag{12.8}$$

Go ahead, perform matrix multiplication in Eq. 12.8, and convince yourself that the result is, indeed, the system of Eqs. 12.6 and 12.7. If we agree to always use an amplitude of the forward propagating component of the wave function (whatever appears in front of $\exp(ik_i z)$) as the first element in the two-dimensional column and the amplitude of the backward propagating component (the one appearing in front of $\exp(-ik_i z)$) as the second element, I can introduce notation $v_{0,1}$ for the respective columns, $D^{(1,0)}$ for the matrix

$$D^{(1,0)} = \begin{bmatrix} \frac{k_1 + k_0}{2k_1} & \frac{k_1 - k_0}{2k_1} \\ \frac{k_1 - k_0}{2k_1} & \frac{k_1 + k_0}{2k_1} \end{bmatrix}, \tag{12.9}$$

and rewrite Eq. 12.8 as a compact matrix equation:

$$v_1 = D^{(1,0)} v_0. \tag{12.10}$$

Upper indexes in the notation for this matrix are supposed to be read from right to left and symbolize a transition across a boundary between potentials V_0 and V_1.

I will not be surprised if at this point you feel a bit disappointed and thinking: "so what, dude? This is just a fancy way of presenting what we already know." But be patient: patience is a virtue and is usually rewarded. The real utility of the matrix notation becomes apparent only when you have to deal with potentials featuring multiple discontinuities. So, let's get to it and assume that at some point with coordinate $z = z_1$, the potential experiences another jump changing abruptly from V_1 to V_2. If asked to write the expression for the wave function in the regions between $z = 0$ and $z = z_1$ and for $z > z_1$, you would have probably written

$$\psi(z) = \begin{cases} A_1 \exp(ik_1 z) + B_1 \exp(-ik_1 z) & 0 < z < z_1 \\ A_2 \exp(ik_2 z) + B_2 \exp(-ik_2 z) & z > z_1 \end{cases} \tag{12.11}$$

which is, of course, a perfectly reasonable and correct expression. However, if you tried to write down the continuity equations at $z = z_1$ using this wave function and present them in a matrix form, you would have ended up with a matrix containing exponential factors like $\exp(\pm ik_{1,2}z_1)$ and which would not look at all like simple matrix $D^{(1,0)}$ from Eq. 12.9. I can try to make the situation more attractive by rewriting the expression for the wave function in a form, in which arguments of the exponential functions vanish at $z = z_1$:

$$\psi(z) = \begin{cases} A_1^{(L)} \exp[ik_1(z - z_1)] + B_1^{(L)} \exp[-ik_1(z - z_1)] & 0 < z < z_1 \\ A_1^{(R)} \exp[ik_2(z - z_1)] + B_1^{(R)} \exp[ik_2(z - z_1)] & z > z_1. \end{cases} \tag{12.12}$$

This amounts to redefining amplitudes appearing in front of the respective exponents as you will see for yourselves when doing Problem 2 in the exercise section for this chapter. Please note the change in the notations: instead of distinguishing the amplitudes by their lower indexes (1 or 2), I introduced upper indexes L and R, indicating that these coefficients describe the wave function immediately to the left or to the right of the discontinuity point, correspondingly. At the same time, the lower indexes in all coefficients are now set to be 1, implying that we are dealing with the discontinuity at point $z = z_1$. In terms of these new coefficients, the stitching conditions take a form

$$A_1^{(R)} = \frac{1}{2}\left(1 + \frac{k_1}{k_2}\right)A_1^{(L)} + \frac{1}{2}\left(1 - \frac{k_1}{k_2}\right)B_1^{(L)} \tag{12.13}$$

$$B_1^{(R)} = \frac{1}{2}\left(1 - \frac{k_1}{k_2}\right)A_1^{(L)} + \frac{1}{2}\left(1 + \frac{k_1}{k_2}\right)B_1^{(L)}, \tag{12.14}$$

which, with obvious substitutions $k_0 \to k_1$ and $k_1 \to k_2$, become identical to Eqs. 12.6 and 12.7. These equations can again be written in the matrix form

$$v_1^{(R)} = D^{(2,1)} v_1^{(L)}, \tag{12.15}$$

where $v_1^{(R)}$ is formed by coefficients $A_1^{(R)}$ and $B_1^{(R)}$, $v_1^{(L)}$—by coefficients $A_1^{(L)}$ and $B_1^{(L)}$, while matrix $D^{(2,1)}$ is defined as

$$D^{(2,1)} = \begin{bmatrix} \frac{k_2+k_1}{2k_2} & \frac{k_2-k_1}{2k_2} \\ \frac{k_2-k_1}{2k_2} & \frac{k_2+k_1}{2k_2} \end{bmatrix}. \tag{12.16}$$

You might have noticed by now the common features of the matrices $D^{(1,0)}$ and $D^{(2,1)}$: (a) they both describe the transition across a boundary between two values of the potential (V_0 to V_1 for the former and V_1 to V_2 for the latter), (b) they both connect the pairs of coefficients characterizing the wave functions immediately on the left of the potential jump with those specifying the wave function immediately on the right of the jump, and, finally, (c) they have a similar structure, recognizing which enables you to write down a matrix connecting the wave function amplitudes across a generic discontinuity as

$$
\boldsymbol{D}^{(i+1,i)} = \begin{bmatrix} \frac{k_{i+1}+k_i}{2k_{i+1}} & \frac{k_{i+1}-k_i}{2k_{i+1}} \\ \frac{k_{i+1}-k_i}{2k_{i+1}} & \frac{k_{i+1}+k_i}{2k_{i+1}} \end{bmatrix},
\tag{12.17}
$$

where

$$
k_i = \frac{\sqrt{2m_e\,(E - V_i)}}{\hbar}
$$

is determined by the potential to the left of the discontinuity and

$$
k_{i+1} = \frac{\sqrt{2m_e\,(E - V_{i+1})}}{\hbar}
$$

corresponds to the value of the potential to the right of it. It is also not too difficult to rewrite Eqs. 12.3 and 12.12 in a situation, when a jump of potential occurs at an arbitrary point $z = z_i$:

$$
\psi(z) = \begin{cases} A_i^{(L)} \exp[ik_i\,(z - z_i)] + B_i^{(L)} \exp[-ik_i\,(z - z_i)] & z_{i-1} < z < z_i \\ A_i^{(R)} \exp[ik_{i+1}\,(z - z_i)] + B_i^{(R)} \exp[ik_{i+1}\,(z - z_i)] & z > z_i. \end{cases}
\tag{12.18}
$$

Correspondingly, Eq. 12.15 becomes

$$
\boldsymbol{v}_i^{(R)} = \boldsymbol{D}^{(i+1,i)} \boldsymbol{v}_i^{(L)},
\tag{12.19}
$$

where $\boldsymbol{v}_i^{(R)}$ contains $A_i^{(R)}$ and $B_i^{(R)}$, while $\boldsymbol{v}_i^{(L)}$ contains $A_i^{(L)}$ and $B_i^{(L)}$.

I hope that by now I have managed to convince you that using the suggested matrix notations does have its benefits, but I also suspect that some of you might become somewhat skeptical about the generality of this approach. You might be thinking that all these formulas that I so confidently presented here can only be valid for energies exceeding all potentials V_i and that this fact strongly limits the utility of the method. If this did occur to you, accept my commendation for paying attention, but reality is not as bad as it appears. Let's see what happens if one of V_i turns out to be larger than E. Obviously, in this case the respective k_i becomes imaginary and can be written as

$$
k_i = \frac{\sqrt{2m_e\,(E - V_i)}}{\hbar} = i\frac{\sqrt{2m_e\,(V_i - E)}}{\hbar} \equiv i\kappa_i,
\tag{12.20}
$$

where I introduced a new real-valued parameter

$$\kappa_i = \frac{\sqrt{2m_e \left(V_i - E\right)}}{\hbar}.$$

(12.21)

The corresponding wave function at $z < z_i$ becomes

$$\psi(z) = A_i^{(L)} \exp\left[-\kappa_i \left(z - z_i\right)\right] + B_i^{(L)} \exp\left[\kappa_i \left(z - z_i\right)\right].$$

The continuity condition of the wave function at $z = z_i$ remains the same as Eq. 12.4:

$$A_i^{(L)} + B_i^{(L)} = A_i^{(R)} + B_i^{(R)},$$

while the continuity of the derivative of the wave function yields this instead of Eq. 12.5

$$-\kappa_i \left(A_i^{(L)} - B_i^{(L)}\right) = ik_{i+1} \left(A_i^{(R)} - B_i^{(R)}\right),$$

where I assumed for the sake of argument that $E > V_{i+1}$. Combining these two equations, I get, instead of Eqs. 12.6 and 12.7,

$$A_i^{(R)} = \frac{1}{2}A_i^{(L)} \left(1 - \frac{\kappa_i}{ik_{i+1}}\right) + \frac{1}{2}B_i^{(L)} \left(1 + \frac{\kappa_i}{ik_{i+1}}\right)$$

$$B_i^{(R)} = \frac{1}{2}A_i^{(L)} \left(1 + \frac{\kappa_i}{ik_{i+1}}\right) + \frac{1}{2}B_i^{(L)} \left(1 - \frac{\kappa_i}{ik_{i+1}}\right),$$

which can be written again in the form of Eq. 12.19 with a new matrix

$$\tilde{D}^{(i+1,i)} = \begin{bmatrix} \frac{ik_{i+1}-\kappa_i}{2ik_{i+1}} & \frac{ik_{i+1}+\kappa_i}{2ik_{i+1}} \\ \frac{ik_{i+1}+\kappa_i}{2ik_{i+1}} & \frac{ik_{i+1}-\kappa_i}{2ik_{i+1}} \end{bmatrix} = \begin{bmatrix} \frac{k_{i+1}+i\kappa_i}{2k_{i+1}} & \frac{k_{i+1}-i\kappa_i}{2k_{i+1}} \\ \frac{k_{i+1}-i\kappa_i}{2k_{i+1}} & \frac{k_{i+1}+i\kappa_i}{2k_{i+1}} \end{bmatrix}.$$

Comparing $\tilde{D}^{(i+1,i)}$ with $D^{(i+1,i)}$ in Eq. 12.17, you can immediately see that the latter can be obtained from the former with the simple substitution defined in Eq. 12.20. Therefore, you do not really have to worry about the relation between energy and the respective value of the potential: Eq. 12.17 works in all cases, and if k_i turns out to be imaginary, you just need to replace it with $i\kappa_i$ as prescribed by Eq. 12.20 (or just let the computer to do it for you). Consequently, matrix $\tilde{D}^{(i+1,i)}$ turns out to be perfectly unnecessary and will not be used any more, but there is one circumstance which you must pay close attention to. Special significance of Eq. 12.20 is that when k_i turns imaginary, it forces it to have a positive imaginary part (square root obviously allows for either positive or negative signs). As a result, the exponential factor at the amplitude designated as A_i acquires a negative positive argument, while the exponential factor multiplied by amplitude B_i gets a real positive argument. You need to take this into account when designating corresponding amplitudes as A or B and placing them in the first or the second row of your column vector. (Obviously, it is not the actual symbols used to designate the amplitudes that are important, but their places in the column vector.)

I hope that your head is not spinning yet, but as a prophylactic measure, let me summarize what we have achieved so far. We are considering a particle moving in a piecewise constant potential, which has interruptions of continuity at a number of points with coordinates $z = z_i$ (the first discontinuity occurs at $z_0 = 0$). When crossing z_i, the potential jumps from V_i to V_{i+1}. In the vicinity of each discontinuity point, the wave function is presented by Eq. 12.18, organized in such a way that coefficients with upper index L determine amplitudes of the right- and left-propagating components of the wave function on the left of the discontinuity and coefficients with upper index R determine the same amplitudes on the right of z_i. The connection between these pairs of coefficients is described by the matrix equation as presented by Eq. 12.19.

To help you get a better feeling of why this matrix representation is useful, let me put together the matrix equations for a few successive discontinuity points:

$$v_2^{(R)} = D^{(3,2)} v_2^{(L)}; \ v_1^{(R)} = D^{(2,1)} v_1^{(L)}, \ v_0^{(R)} = D^{(2,1)} v_0^{(L)} \tag{12.22}$$

The structure of these equations indicates that it might be possible to relate column vector $v_2^{(R)}$ to $v_0^{(L)}$ by consecutive matrix multiplication if we had matrices relating $v_2^{(L)}$ to $v_1^{(R)}$, $v_1^{(L)}$ to $v_0^{(R)}$, and, in general, $v_i^{(L)}$ to $v_{i-1}^{(R)}$. To find these matrices, I have to take you back to Eq. 12.18, where you shall notice that the pairs of coefficients $A_{i-1}^{(R)}, B_{i-1}^{(R)}$ and $A_i^{(L)}, B_i^{(L)}$ describe the wave function defined on the same interval $z_i < z < z_{i+1}$. Accordingly, the following must be true:

$$A_i^{(L)} \exp \left[ik_i \left(z - z_i\right)\right] + B_i^{(L)} \exp \left[-ik_i \left(z - z_i\right)\right] =$$

$$A_{i-1}^{(R)} \exp \left[ik_i \left(z - z_{i-1}\right)\right] + B_{i-1}^{(R)} \exp \left[-ik_i \left(z - z_{i-1}\right)\right]$$

which is satisfied if

$$A_i^{(L)} \exp \left[ik_i \left(z - z_i\right)\right] = A_{i-1}^{(R)} \exp \left[ik_i \left(z - z_{i-1}\right)\right]$$

and

$$B_i^{(L)} \exp \left[-ik_i \left(z - z_i\right)\right] = B_{i-1}^{(R)} \exp \left[-ik_i \left(z - z_{i-1}\right)\right]$$

Canceling the common factor $\exp \left(ik_i z\right)$, you find

$$A_i^{(L)} = \exp \left[ik_i \left(z_i - z_{i-1}\right)\right] A_{i-1}^{(R)} \tag{12.23}$$

$$B_i^{(L)} = \exp \left[-ik_i \left(z_i - z_{i-1}\right)\right] B_{i-1}^{(R)} \tag{12.24}$$

which can be presented in the matrix form as

$$
\begin{bmatrix} A_i^{(L)} \\ B_i^{(L)} \end{bmatrix} = \begin{bmatrix} \exp\left[ik_i\left(z_i - z_{i-1}\right)\right] & 0 \\ 0 & \exp\left[-ik_i\left(z_i - z_{i-1}\right)\right] \end{bmatrix} \begin{bmatrix} A_{i-1}^{(R)} \\ B_{i-1}^{(R)} \end{bmatrix}. \tag{12.25}
$$

Introducing the diagonal matrix

$$
\boldsymbol{M}^{(i)} = \begin{bmatrix} \exp\left[ik_i\left(z_i - z_{i-1}\right)\right] & 0 \\ 0 & \exp\left[-ik_i\left(z_i - z_{i-1}\right)\right] \end{bmatrix} \tag{12.26}
$$

I can give Eq. 12.25 the form

$$
\boldsymbol{v}_i^{(L)} = \boldsymbol{M}^{(i)} \boldsymbol{v}_{i-1}^{(R)}, \tag{12.27}
$$

which you can recognize as the missing relation between $\boldsymbol{v}_i^{(L)}$ and $\boldsymbol{v}_{i-1}^{(R)}$. Note that the upper index in $\boldsymbol{M}^{(i)}$ signifies that it corresponds to the region of coordinates $z_{i-1} < z < z_i$, where the potential is equal to V_i. It is important to note that Eq. 12.26 can be used even if k_i turns out to be imaginary. All you will need to do in this case is to replace k_i with $i\kappa_i$ according to Eqs. 12.20 and 12.21. Now, complimenting Eq. 12.22 with the missing links, you get

$$
\boldsymbol{v}_2^{(R)} = \boldsymbol{D}^{(3,2)} \boldsymbol{v}_2^{(L)}; \quad \boldsymbol{v}_2^{(L)} = \boldsymbol{M}^{(2)} \boldsymbol{v}_1^{(R)}; \quad \boldsymbol{v}_1^{(R)} = \boldsymbol{D}^{(2,1)} \boldsymbol{v}_1^{(L)}, \tag{12.28}
$$

$$
\boldsymbol{v}_1^{(L)} = \boldsymbol{M}^{(1)} \boldsymbol{v}_0^{(R)}; \quad \boldsymbol{v}_0^{(R)} = \boldsymbol{D}^{(2,1)} \boldsymbol{v}_0^{(L)},
$$

which, after combining all successive matrix relations, yields

$$
\boldsymbol{v}_2^{(R)} = \boldsymbol{D}^{(3,2)} \boldsymbol{M}^{(2)} \boldsymbol{D}^{(2,1)} \boldsymbol{M}^{(1)} \boldsymbol{D}^{(1,0)} \boldsymbol{v}_0^{(L)}. \tag{12.29}
$$

This result illuminates the power of the method, which is presented here: the amplitudes of the wave function after the particles have encountered three discontinuity points of the potential are expressed in terms of the amplitudes specifying the wave function in the region before the first discontinuity via a simple matrix relation, $\boldsymbol{v}_2^{(R)} = \boldsymbol{T}^{(3)} \boldsymbol{v}_0^{(L)}$, where matrix $\boldsymbol{T}^{(3)}$, called the transfer matrix, is the product of five matrices of two different kinds:

$$
\boldsymbol{T}^{(3)} = \boldsymbol{D}^{(3,2)} \boldsymbol{M}^{(2)} \boldsymbol{D}^{(2,1)} \boldsymbol{M}^{(1)} \boldsymbol{D}^{(1,0)}.
$$

Matrices $\boldsymbol{D}^{(i+1,i)}$ can be called interface matrices as they describe transformation of the wave function amplitudes due to crossing of an interface between two distinct values of the potential, and you can use the name "free propagation matrices" for $\boldsymbol{M}^{(i)}$ because they describe the evolution of the wave function due to free propagation of the particle between two discontinuities. Equation 12.29 has a simple physical interpretation if you read it from right to left: a particle begins

Fig. 12.1 A potential profile corresponding to Eq. 12.29

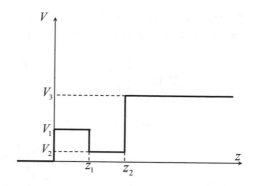

with a wave function characterized by column vector v_0. It encounters the first discontinuity at $z = z_1$, and upon crossing it the wave function coefficients undergo transformation prescribed by matrix $D^{(1,0)}$. After that the wave function evolves as it were for a free particle in potential V_1—this evolution is described by the propagation matrix $M^{(1)}$. The crossing of the boundary between V_1 and V_2 regions is represented by the interface matrix $D^{(2,1)}$ and so on and so forth. One of the possible potential profiles that could have been described by Eq. 12.29 is shown in Fig. 12.1.

Equation 12.29 is trivially generalized to the case of an arbitrary number, N, of the discontinuities, located at points z_i, $i = 0, 1, 2 \cdots N - 1$ with $z_0 = 0$:

$$v_{N-1}^{(R)} = T^{(N)} v_0^{(L)} \tag{12.30}$$

with a corresponding transfer matrix defined as

$$T^{(N)} = D^{(N,N-1)} M^{(N-1)} \cdots D^{(2,1)} M^{(1)} D^{(1,0)}. \tag{12.31}$$

Once the transfer matrix is known, you can use it to obtain all the information about wave functions (and corresponding energy eigenvalues when appropriate) of the particle in the corresponding potential both in the continuous and discrete segments of the energy spectrum. The next section in this chapter discusses how this can be done.

12.1.2 Application of Transfer-Matrix Formalism to Generic Scattering and Bound State Problems

12.1.2.1 Generic Scattering Problem via the Transfer Matrix

Having defined a generic transfer matrix $T^{(N)}$, I can now solve a typical scattering problem similar to the one discussed in Sect. 6.2.1. Setting it up amounts to specifying the wave function of the particle at $z < 0$ (before the particle encounters the first break of the continuity) and at $z > z_{N-1}$ (after the particle passes through the last discontinuity point). The scattering wave function introduced in Sect. 6.2.1

$$\psi(z) = \begin{cases} \exp(ik_0 z) + r \exp(-ik_0 z), & z < 0 \\ t \exp(ik_N z) & z > z_{N-1} \end{cases} \qquad (12.32)$$

is in the transfer-matrix formalism described by column vectors v_0 and v_N:

$$v_0^{(L)} = \begin{bmatrix} 1 \\ r \end{bmatrix}; \quad v_{N-1}^{(R)} = \begin{bmatrix} t \\ 0 \end{bmatrix} \qquad (12.33)$$

Presenting the generic T-matrix by its (presumably known) elements

$$T^{(N)} = \begin{bmatrix} t_{11} & t_{12} \\ t_{21} & t_{22} \end{bmatrix}$$

I can rewrite Eq. 12.30 in the expanded form as

$$\begin{bmatrix} t \\ 0 \end{bmatrix} = \begin{bmatrix} t_{11} & t_{12} \\ t_{21} & t_{22} \end{bmatrix} \begin{bmatrix} 1 \\ r \end{bmatrix}.$$

This translates into the system of linear equations:

$$t = t_{11} + r t_{12}$$
$$0 = t_{21} + r t_{22}.$$

From the second of these equations, I immediately have

$$r = -\frac{t_{21}}{t_{22}}, \qquad (12.34)$$

and substituting this result into the first one, I find

$$t = t_{11} - \frac{t_{12} t_{21}}{t_{22}} = \frac{det\left(T^{(N)}\right)}{t_{22}}. \qquad (12.35)$$

Here $det\left(T^{(N)}\right) \equiv t_{11} t_{22} - t_{12} t_{21}$ is the determinant of the T-matrix $T^{(N)}$, which, believe it or not, can actually be quite easily computed for the most general transfer matrix defined in Eq. 12.31.

To do so you must, first, recall that the determinant of the product of the matrices is equal to the product of the determinants of the individual factors:

$$det\left(T^{(N)}\right) = det\left(D^{(N,N-1)}\right) det\left(M^{(N-1)}\right)\cdots\times$$
$$det\left(D^{(2,1)}\right) det\left(M^{(1)}\right) det\left(D^{(1,0)}\right). \tag{12.36}$$

It is easy to see that $det\left(M^{(i)}\right) = 1$ for any i, so all these factors can be omitted from Eq. 12.36 yielding

$$det\left(T^{(N)}\right) = det\left(D^{(N,N-1)}\right) det\left(D^{(N-1,N-2)}\right)\cdots\times$$
$$det\left(D^{(2,1)}\right) det\left(D^{(1,0)}\right). \tag{12.37}$$

Now all I need is to compute the determinant of the generic matrix $D^{(i+1,i)}$. Using Eq. 12.17, I find

$$det\left(D^{(i+1,i)}\right) = \left(\frac{k_{i+1}+k_i}{2k_{i+1}}\right)^2 - \left(\frac{k_{i+1}-k_i}{2k_{i+1}}\right)^2 = \frac{k_i}{k_{i+1}},$$

which leads to the following expression for $det\left(T^{(N)}\right)$:

$$det\left(T^{(N)}\right) = \frac{k_{N-1}}{k_N}\frac{k_{N-2}}{k_{N-1}}\cdots\frac{k_1}{k_2}\frac{k_0}{k_1} = \frac{k_0}{k_N} \tag{12.38}$$

Isn't it amazing how all k_i in the intermediate regions got canceled, so that the determinant depends only upon the wave numbers (real or imaginary) in the first and the last region of the constant potential. Using this result in Eq. 12.35, I can find a simplified expression for the transmission amplitude

$$t = \frac{k_0}{k_N}\frac{1}{t_{22}} \tag{12.39}$$

which becomes even simpler if the potential for $z < 0$ and for $z > z_{N-1}$ is the same. In this case the determinant of the transfer matrix becomes equal to unity and $t = 1/t_{22}$. Having found r and t, I can restore the wave function in the entire range of the coordinate by consequently applying interface and propagation matrices constituting the total transfer matrix $T^{(N)}$.

With help of Eq. 6.53 from Sect. 6.2.1, I can also find the corresponding reflection and transmission probabilities:

$$R = |r|^2 = \left|\frac{t_{21}}{t_{22}}\right|^2$$

$$T = \frac{k_N^2}{k_0^2}|t|^2 = \left|\frac{1}{t_{22}}\right|^2$$

Fig. 12.2 An example of a potential with discrete spectrum

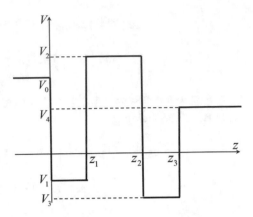

where I used Eq. 12.39 for t. Since reflection and transmission probabilities must obey the condition $R + T = 1$, it imposes the following general condition on the elements of the transfer matrix:

$$|t_{22}|^2 - |t_{12}|^2 = 1.$$

12.1.2.2 Finding Bound States with the Transfer Matrix

Now let me show how transfer-matrix method can be used to find energies of the bound states, if they are allowed by the potential. Consider, for instance, a potential shown in Fig. 12.2. After eyeballing this figure for a few moments and recalling that discrete energy levels correspond to classically bound motion, you shall be able to conclude that states with energies in the interval $V_3 < E < V_4$ must belong to the discrete spectrum. An important general point to make here is that the discrete spectrum in such a Hamiltonian exists at energies, which are smaller than the limiting values of the potential at $z \to \pm\infty$ and larger than the potential's smallest value, provided that these conditions are not self-contradictory. For such values of energy, the solutions of the Schrödinger equation for $z < 0$ and $z > z_{N-1}$ (classically forbidden regions) take the form of real exponential functions, so that instead of Eq. 12.32, I have

$$\psi(z) = \begin{cases} B_0 \exp(\kappa_0 z) & z < 0 \\ A_N \exp(-\kappa_N z) & z > z_{N-1}, \end{cases} \tag{12.40}$$

where

$$\kappa_0 = \frac{\sqrt{2m(V_0 - E)}}{\hbar}$$

$$\kappa_N = \frac{\sqrt{2m(V_N - E)}}{\hbar}$$

Before continuing I have to reiterate a point that I already made earlier in this section. Equation 12.40 was obtained by making transition from parameters k_0 and k_N, which become imaginary for the chosen values of energy, to real parameters κ_0 and κ_N with the help of Eq. 12.20. This procedure turns exponential functions $\exp(\pm ikz)$ into $\exp(\mp\kappa z)$. Accordingly, in order to preserve the structure of my transfer matrices, I have to designate amplitude coefficients in front of $\exp(\kappa_i z)$ as B_i and coefficients in front of $\exp(-\kappa_i z)$ as A_i. Finally, I feel obliged to remind you that I discarded exponentially growing terms in Eq. 12.40 in order to preserve normalizability of the wave function. Thus, now, initial vectors v_0 and v_N, instead of Eq. 12.33, take the form

$$v_0^{(L)} = \begin{bmatrix} 0 \\ B_0 \end{bmatrix}; \quad v_{N\ 1}^{(R)} = \begin{bmatrix} A_N \\ 0 \end{bmatrix}$$

The resulting transfer-matrix equation in this case becomes

$$\begin{bmatrix} A_N \\ 0 \end{bmatrix} = \begin{bmatrix} t_{11} & t_{12} \\ t_{21} & t_{22} \end{bmatrix} \begin{bmatrix} 0 \\ B_0 \end{bmatrix}$$

which yields

$$A_N = t_{12}B_0$$

$$0 = t_{22}B_0$$

The last of these equations produces an equation for the allowed energy values, since it can only be fulfilled for nonvanishing B_0 and A_0 if

$$t_{22}(E) = 0 \tag{12.41}$$

The first of these equations express A_N in terms of the remaining undetermined coefficient B_0, which can be fixed by the normalization requirement.

12.1.3   Application of the Transfer Matrix to a Symmetrical Potential Well

To illustrate the transfer-matrix method, I will now apply it to a problem, which we have already solved in Sect. 6.2.1—the states of a particle in a symmetric rectangular potential well. To facilitate application of the transfer-matrix approach, I will describe this potential by function

$$V(z) = \begin{cases} V_b & z < 0 \\ V_w & 0 < z < d \\ V_b & z > d, \end{cases} \qquad (12.42)$$

which differs from the one used in Sect. 6.2.1 by the choice of the origin of the coordinate axis for z. This potential has two discontinuity points: it changes from V_b to V_w at $z_0 = 0$ and then, again, from V_w to V_b at $z_1 = d$, where it is assumed that $V_b > V_w$. Correspondingly, I need to introduce two interface matrices: $\boldsymbol{D}^{(1,0)}$ as defined in Eq. 12.9 with $k_0 = \sqrt{2m_e(E - V_b)}$ and $k_1 = \sqrt{2m_e(E - V_w)}$ and $\boldsymbol{D}^{(2,1)}$ defined in Eq. 12.16 with $k_2 = k_0$.

12.1.3.1 Scattering States

Scattering states (continuous spectrum) of this potential correspond to energies $E > V_b$, in which case parameters k_0 and k_1 are regular real-valued wave numbers. Inserting the free propagation matrix $\boldsymbol{M}^{(1)}$ from Eq. 12.26 between $\boldsymbol{D}^{(2,1)}$ and $\boldsymbol{D}^{(1,0)}$ according to Eq. 12.31 and taking into account that $z_0 = 0$ and $z_1 = d$, I obtain the total T-matrix

$$\boldsymbol{T}^{(2)} = \boldsymbol{D}^{(2,1)}\boldsymbol{M}^{(1)}\boldsymbol{D}^{(1,0)} =$$

$$\begin{bmatrix} \frac{k_0+k_1}{2k_0} & \frac{k_0-k_1}{2k_0} \\ \frac{k_0-k_1}{2k_0} & \frac{k_0+k_1}{2k_0} \end{bmatrix} \begin{bmatrix} \exp(ik_1 d) & 0 \\ 0 & \exp(-ik_1 d) \end{bmatrix} \begin{bmatrix} \frac{k_1+k_0}{2k_1} & \frac{k_1-k_0}{2k_1} \\ \frac{k_1-k_0}{2k_1} & \frac{k_1+k_0}{2k_1} \end{bmatrix} =$$

$$\begin{bmatrix} \frac{k_0+k_1}{2k_0}\exp(ik_1 d) & \frac{k_0-k_1}{2k_0}\exp(-ik_1 d) \\ \frac{k_0-k_1}{2k_0}\exp(ik_1 d) & \frac{k_0+k_1}{2k_0}\exp(-ik_1 d) \end{bmatrix} \begin{bmatrix} \frac{k_1+k_0}{2k_1} & \frac{k_1-k_0}{2k_1} \\ \frac{k_1-k_0}{2k_1} & \frac{k_1+k_0}{2k_1} \end{bmatrix} =$$

$$\begin{bmatrix} \frac{(k_0+k_1)^2\exp(ik_1 d)-(k_0-k_1)^2\exp(-ik_1 d)}{4k_0 k_1} & \frac{(k_1^2-k_0^2)[\exp(ik_1 d)-\exp(-ik_1 d)]}{4k_0 k_1} \\ -\frac{(k_1^2-k_0^2)[\exp(ik_1 d)-\exp(-ik_1 d)]}{4k_0 k_1} & \frac{(k_0+k_1)^2\exp(-ik_1 d)-(k_0-k_1)^2\exp(ik_1 d)}{4k_0 k_1} \end{bmatrix} =$$

$$\frac{1}{2k_0 k_1} \times$$

$$\begin{bmatrix} i\left(k_0^2+k_1^2\right)\sin k_1 d + 2k_0 k_1 \cos k_1 d & i\left(k_1^2-k_0^2\right)\sin k_1 d \\ -i\left(k_1^2-k_0^2\right)\sin k_1 d & -i\left(k_0^2+k_1^2\right)\sin k_1 d + 2k_0 k_1 \cos k_1 d \end{bmatrix}.$$
$$(12.43)$$

Substitution of the corresponding elements of the T-matrix from the last expression into Eqs. 12.34 and 12.39 yields the amplitude reflection and transmission coefficients:

$$r = \frac{i\left(k_1^2 - k_0^2\right)\sin k_1 d}{-i\left(k_0^2 + k_1^2\right)\sin k_1 d + 2k_0 k_1 \cos k_1 d} = \tag{12.44}$$

$$\frac{\left(k_0^2 - k_1^2\right)\sin k_1 d}{\left(k_0^2 + k_1^2\right)\sin k_1 d + 2ik_0 k_1 \cos k_1 d}.$$

$$t = \frac{2k_0 k_1}{-i\left(k_0^2 + k_1^2\right)\sin k_1 d + 2k_0 k_1 \cos k_1 d} = \tag{12.45}$$

$$\frac{2ik_0 k_1}{\left(k_0^2 + k_1^2\right)\sin k_1 d + 2ik_0 k_1 \cos k_1 d}.$$

where at the last steps, the numerators and denominators of the expressions for r and t were multiplied by i. The resulting expressions coincide with Eqs. 6.58 and 12.35 of Sect. 6.2, which, of course, is not surprising. Having found the reflection and transmission amplitudes, I can easily restore the entire wave function. Indeed, substitution of Eqs. 12.45 and 12.44 into Eq. 12.32 yields the wave function for $z < 0$ and $z > d$. Next, using Eq. 12.10 with v_0 in the form

$$v_0 = \begin{bmatrix} 1 \\ r \end{bmatrix}$$

I find coefficients $A_0^{(R)}$ and $B_0^{(R)}$:

$$\begin{bmatrix} A_0^{(R)} \\ B_0^{(R)} \end{bmatrix} = \begin{bmatrix} \frac{k_1+k_0}{2k_1} & \frac{k_1-k_0}{2k_1} \\ \frac{k_1-k_0}{2k_1} & \frac{k_1+k_0}{2k_1} \end{bmatrix} \begin{bmatrix} 1 \\ r \end{bmatrix} \Rightarrow$$

$$A_0^{(R)} = \frac{k_1 + k_0 + r(k_1 - k_0)}{2k_1} \tag{12.46}$$

$$B_0^{(R)} = \frac{k_1 - k_0 + r(k_1 + k_0)}{2k_1}, \tag{12.47}$$

which generate the wave function in the region $0 < z < d$:

$$\psi(z) = \frac{k_1(1+r) + k_0(1-r)}{2k_1} e^{ik_1 z} + \frac{k_1(1+r) - k_0(1-r)}{2k_1} e^{-ik_1 z}.$$

I will leave it as an exercise to demonstrate that coefficients $A_0^{(R)}$ and $B_0^{(R)}$ in Eqs. 12.46 and 12.47 coincide with coefficients A_2 and B_2 in Eqs. 6.60 and 6.61 in Sect. 6.2. Rewriting the expression for the wave function as

$$\psi(z) = \frac{k_1(1+r) + k_0(1-r)}{2k_1} e^{ik_1 d} e^{ik_1(z-d)} +$$

$$\frac{k_1(1+r) - k_0(1-r)}{2k_1} e^{-ik_1 d} e^{-ik_1(z-d)},$$

where I simply multiplied each term by $\exp(ik_1d)\exp(-ik_1d) \equiv 1$, you can identify coefficients $A_1^{(L)}$ and $B_1^{(L)}$ as

$$A_1^{(L)} = \frac{k_1 + k_0 + r\,(k_1 - k_0)}{2k_1} e^{ik_1d}$$

$$B_1^{(L)} = \frac{k_1 - k_0 + r\,(k_1 + k_0)}{2k_1} e^{-ik_1d}.$$

The same expressions for $A_1^{(L)}$ and $B_1^{(L)}$ can obviously be found by multiplying diagonal matrix $\boldsymbol{M}^{(1)}$ by $\boldsymbol{v}_0^{(R)}$ formed by coefficients $A_0^{(R)}$ and $B_0^{(R)}$. Finally, in order to convince the skeptics that the outlined procedure is self-consistent, you can try to apply the interface matrix $\boldsymbol{D}^{(1,2)}$ to $A_1^{(L)}$ and $B_1^{(L)}$:

$$\begin{bmatrix} A_2^{(R)} \\ B_2^{(R)} \end{bmatrix} = \begin{bmatrix} \frac{k_0+k_1}{2k_0} & \frac{k_0-k_1}{2k_0} \\ \frac{k_0-k_1}{2k_0} & \frac{k_0+k_1}{2k_0} \end{bmatrix} \begin{bmatrix} \frac{k_1+k_0+r(k_1-k_0)}{2k_1} e^{ik_1d} \\ \frac{k_1-k_0+r(k_1+k_0)}{2k_1} e^{-ik_1d} \end{bmatrix} \tag{12.48}$$

yielding for $A_2^{(R)}$

$$A_2^{(R)} = \frac{(k_0 + k_1)^2}{4k_0k_1} e^{ik_1d} + r\frac{k_1^2 - k_0^2}{4k_0k_1} e^{ik_1d} -$$

$$\frac{(k_0 - k_1)^2}{4k_0k_1} e^{-ik_1d} - r\frac{k_1^2 - k_0^2}{4k_0k_1} e^{-ik_1d} =$$

$$e^{ik_1d}\left(\frac{k_0(1-r)}{4k_1} + \frac{k_1(1+r)}{4k_0} + \frac{1}{2}\right) -$$

$$e^{-ik_1d}\left(\frac{k_0(1-r)}{4k_1} + \frac{k_1(1+r)}{4k_0} - \frac{1}{2}\right).$$

To continue I have to use the reflection coefficient r given by Eq. 12.44. Evaluating parts of the expression for $A_2^{(R)}$ separately, I find

$$\frac{k_0(1-r)}{4k_1} = \frac{k_0}{4k_1}\left[1 - \frac{(k_0^2 - k_1^2)\sin k_1d}{(k_0^2 + k_1^2)\sin k_1d + 2ik_0k_1\cos k_1d}\right] =$$

$$\frac{k_0}{2}\frac{k_1\sin k_1d + ik_0\cos k_1d}{(k_0^2 + k_1^2)\sin k_1d + 2ik_0k_1\cos k_1d}$$

$$\frac{k_1(1+r)}{4k_0} = \frac{k_1}{4k_0}\left[1 + \frac{(k_0^2 - k_1^2)\sin k_1d}{(k_0^2 + k_1^2)\sin k_1d + 2ik_0k_1\cos k_1d}\right] =$$

$$\frac{k_1}{2}\frac{k_0\sin k_1d + ik_1\cos k_1d}{(k_0^2 + k_1^2)\sin k_1d + 2ik_0k_1\cos k_1d}.$$

Lastly,

$$\frac{k_0(1-r)}{4k_1} + \frac{k_1(1+r)}{4k_0} + \frac{1}{2} =$$

$$\frac{1}{2}\left[\frac{2k_0k_1\sin k_1 d + i\left(k_0^2 + k_1^2\right)\cos k_1 d}{\left(k_0^2 + k_1^2\right)\sin k_1 d + 2ik_0k_1\cos k_1 d} + 1\right] -$$

$$\frac{i}{2}\frac{(k_0 + k_1)^2 e^{-ik_1 d}}{\left(k_0^2 + k_1^2\right)\sin k_1 d + 2ik_0k_1\cos k_1 d},$$

where at the last step, I replaced $\sin k_1 d + i\cos k_1 d$ with $i\exp(-ik_1 d)$. Similarly,

$$\frac{k_0(1-r)}{4k_1} + \frac{k_1(1+r)}{4k_0} - \frac{1}{2} =$$

$$\frac{1}{2}\left[\frac{2k_0k_1\sin k_1 d + i\left(k_0^2 + k_1^2\right)\cos k_1 d}{\left(k_0^2 + k_1^2\right)\sin k_1 d + 2ik_0k_1\cos k_1 d} - 1\right] =$$

$$\frac{i}{2}\frac{(k_0 - k_1)^2 e^{ik_1 d}}{\left(k_0^2 + k_1^2\right)\sin k_1 d + 2ik_0k_1\cos k_1 d}.$$

Combining all these results, I finally get $A_2^{(R)}$:

$$A_2^{(R)} = \frac{i}{2}\frac{(k_0 + k_1)^2 - (k_0 - k_1)^2}{\left(k_0^2 + k_1^2\right)\sin k_1 d + 2ik_0k_1\cos k_1 d} =$$

$$\frac{2ik_0k_1}{\left(k_0^2 + k_1^2\right)\sin k_1 d + 2ik_0k_1\cos k_1 d}. \tag{12.49}$$

Catching my breath after this marathon calculations (OK—half marathon), I am eager to compare Eq. 12.49 with Eq. 12.45 for the transmission amplitude. With a sigh of relief, I find that they, indeed, coincide. I will leave it as an exercise to demonstrate that $B_2^{(R)}$ vanishes as it should.

12.1.3.2 Bound States

Now I will illustrate application of the transfer-matrix approach to bound states of the square potential well described by the same by Eq. 12.42. Discrete spectrum of this potential is expected to exist in the interval of energies defined as $V_w < E < V_b$. The transfer matrix given in Eq. 12.43 can be adapted to this case by replacing wave number k_0 with $i\kappa_0$, where κ_0 in this context is defined as

$$\kappa_0 = \frac{\sqrt{2m(V_b - E)}}{\hbar}$$

This procedure yields

$$T = \frac{1}{2\kappa_0 k_1} \times$$

$$\begin{bmatrix} \left(-\kappa_0^2 + k_1^2\right) \sin k_1 d + 2\kappa_0 k_1 \cos k_1 d & \left(k_1^2 + \kappa_0^2\right) \sin k_1 d \\ -\left(k_1^2 + \kappa_0^2\right) \sin k_1 d & -\left(-\kappa_0^2 + k_1^2\right) \sin k_1 d + 2\kappa_0 k_1 \cos k_1 d \end{bmatrix}$$

and Eq. 12.41 for the bound state energies takes the following form:

$$2\kappa_0 k_1 \cos k_1 d = \left(-\kappa_0^2 + k_1^2\right) \sin k_1 d$$

or

$$\tan (k_1 d) = \frac{2\kappa_0 k_1}{-\kappa_0^2 + k_1^2} \tag{12.50}$$

At the first glance, this result does not agree with the one I derived in Sect. 6.2.1, where states were segregated according to their parity with different equations for the energy levels of the even and odd states. Equation 12.50, on the other hand, is a single equation, and the parity of the states has never been even mentioned. If, however, you pause to think about it, you will see that the differences between results obtained here and in Sect. 6.2.1 are purely superficial.

First of all, you need to notice that the coordinates used here and in Sect. 6.2.1 have different origins. Placing the origin of the coordinate at the center of the well made the inversion symmetry of the potential with respect to its center reflected in its coordinate dependence. Consequently, we were able to classify states by their parity. This immediate benefit of the symmetry is lost once the origin of the coordinate is displaced from the center of the well. This, of course, did not change the underlying symmetry of the potential (it has nothing to do with such artificial things as our choice of the coordinate system), but it masked it. The wave functions written in the coordinate system centered at the edge of the potential well do not have a definite parity with respect to point $z = 0$, and it is not surprising that my derivation of the eigenvalue equation naturally yielded a single equation for all energy eigenvalues. However, it is not too difficult to demonstrate that our single Equation 12.50 is in reality equivalent to two equations of Sect. 6.2.1, but it does take some extra efforts.

First, you shall notice that trigonometric functions in Eqs. 6.39 and 6.42 are expressed in terms of $kd/2$, while Eq. 12.50 contains $\tan (k_1 d)$. Thus, it makes sense to try to express $\tan (k_1 d)$ in terms of $k_1 d/2$ using a well-known identity

$$\tan (k_1 d) = \frac{2 \tan (k_1 d/2)}{1 - \tan^2 (k_1 d/2)},$$

which yields

$$\frac{\tan (k_1 d/2)}{1 - \tan^2 (k_1 d/2)} = \frac{\kappa_0 k_1}{-\kappa_0^2 + k_1^2}.$$

To simplify algebra, it is useful to temporarily introduce notations $x = \tan (k_1 d/2)$, $\sigma = \left(-\kappa_0^2 + k_1^2\right)/\kappa_0 k_1$, and rewrite the preceding equation as a quadratic equation for x:

$$x^2 + x\sigma - 1 = 0$$

This equation has two solutions:

$$x_{1,2} = -\frac{1}{2}\sigma \pm \frac{1}{2}\sqrt{\sigma^2 + 4}$$

Computing $\sigma^2 + 4$ you will easily find that

$$\sigma^2 + 4 = \frac{k_1^4 + \kappa_0^4 - 2k_1^2\kappa_0^2}{k_1^2\kappa_0^2} + 4 = \frac{\left(\kappa_0^2 + k_1^2\right)^2}{k_1^2\kappa_0^2}$$

which yield the following for x_1 and x_2:

$$x_1 = -\frac{-\kappa_0^2 + k_1^2}{2\kappa_0 k_1} - \frac{\kappa_0^2 + k_1^2}{2k_1\kappa_0} = -\frac{k_1}{\kappa_0}$$

$$x_2 = -\frac{-\kappa_0^2 + k_1^2}{2\kappa_0 k_1} + \frac{\kappa_0^2 + k_1^2}{2k_1\kappa_0} = \frac{\kappa_0}{k1}$$

Recalling what x stands for, you can see that one equation 12.50 is now replaced by two equations:

$$\tan (k_1 d/2) = -\frac{k_1}{\kappa_0} \tag{12.51}$$

$$\tan (k_1 d/2) = \frac{\kappa_0}{k1}. \tag{12.52}$$

which are exactly the eigenvalue equations for odd and even wave functions derived in Sect. 6.2.1. Isn't it beautiful, really?

Having figured out the situation with the eigenvalues, I can take care of the eigenvectors. The ratio of the wave function amplitudes A_2/B_0 is given by

$$\frac{A_2}{B_0} = t_{12} = \frac{\left(k_1^2 + \kappa_0^2\right)\sin k_1 d}{2\kappa_0 k_1}, \tag{12.53}$$

while amplitudes of the wave functions in the region $0 < z < d$ are found from

$$\begin{bmatrix} A_1 \\ B_1 \end{bmatrix} = D^{(1,0)} \begin{bmatrix} 0 \\ B_0 \end{bmatrix}.$$

Matrix $D^{(1,0)}$ is adapted to the case under consideration by the same substitution $k_0 \rightarrow i\kappa_0$ as before:

$$D^{(1,0)} = \begin{bmatrix} \frac{k_1 + i\kappa_0}{2k_1} & \frac{k_1 - i\kappa_0}{2k_1} \\ \frac{k_1 - i\kappa_0}{2k_1} & \frac{k_1 + i\kappa_0}{2k_1} \end{bmatrix}.$$

Using this matrix, you easily find

$$A_1 = \frac{k_1 - i\kappa_0}{2k_1} B_0$$

$$B_1 = \frac{k_1 + i\kappa_0}{2k_1} B_0,$$

which yields the following expression for the wave function inside the well:

$$\psi(z) = B_0 \left(\frac{k_1 - i\kappa_0}{2k_1} \exp(ik_1 z) + \frac{k_1 + i\kappa_0}{2k_1} \exp(-ik_1 z) \right) =$$

$$B_0 \left(\cos k_1 z + \frac{\kappa_0}{k_1} \sin k_1 z \right).$$

You can replace the ratio κ_0/k_1 in this expression with $\tan(k_1 d/2)$ or with $-\cot(k_1 d/2)$ according to Eqs. 12.51 and 12.52 and obtain the following expressions for the wave function representing two different types of states:

$$\psi(z) = \begin{cases} \frac{B_0}{\cos(k_1 d/2)} \cos[k_1 (z - d/2)], & \kappa_0/k_1 = \tan(k_1 d/2) \\ -\frac{B_0}{\sin(k_1 d/2)} \sin[k_1 (z - d/2)], & \kappa_0/k_1 = -\cot(k_1 d/2) \end{cases}$$

It is quite obvious now that the found wave functions are even and odd with respect to variable $\tilde{z} = z - d/2$, which is merely a coordinate defined in the coordinate system with the origin at the center of the well, just like in Sect. 6.2.1. One can also show that Eq. 12.53 is reduced to $A_2 = \pm B_0$ for two different types of the wave function, again in agreement with the results of Sect. 6.2.1. This proof I will leave to you as an exercise.

12.2 Resonant Tunneling

In this section I will apply the transfer-matrix method to describe an interesting and practically important phenomenon of resonant tunneling. This phenomenon arises when one considers quantum states of a particle in a potential, which consists of two (or more) potential barriers separated by a potential well. An example of such a potential is shown in Fig. 12.3. I am interested here in the states corresponding to under-barrier values of energies E: $0 < E < V$. It was established in Sect. 6.2.1 that in the case of a single barrier whose width d satisfies inequality $d\kappa \gg 1$, where $\kappa = \sqrt{2m_e(V-E)}/\hbar$, such states are characterized by an exponentially small transmission probability $T \propto \exp(-\kappa d)$, which is responsible for the effect of quantum tunneling—a particle incident on the barrier has a non-zero probability to "tunnel" through it and continue its free propagation on the other side of the barrier. You might wonder if adding a second barrier will result in any new and interesting effects. A common sense based on "classical" probability theory suggests that in the presence of the second barrier, the total transmission probability will simply be a product of transmission coefficients for each of the barriers $T \propto T_1 T_2 \propto \exp(-\kappa_1 d_1 - \kappa_2 d_2)$, further reducing the probability that the particle tunnels through the barriers. However, as it often happens, the reality is more complex (and sometimes more intriguing) than our initial intuited insight. So, let's see if our intuition leads us astray in this case.

To simplify algebra, I will assume that both barriers have the same width d and height V and that they are separated by a region of zero potential of length w. This potential profile is characterized by four discontinuity points with coordinates

$$x_0 = 0; \; x_1 = d; \; x_2 = d + w; \; x_3 = 2d + w. \tag{12.54}$$

Accordingly, the propagation of a particle through this potential is described by four interface matrices, $D^{(1,0)}$, $D^{(2,1)}$, $D^{(3,2)}$, and $D^{(4,3)}$, and three free propagating matrices $M^{(1)}$, $M^{(2)}$, and $M^{(3)}$. Matrices $D^{(1,0)}$ and $D^{(2,1)}$ are obviously identical to matrices $D^{(3,2)}$ and $D^{(4,3)}$, correspondingly, and can be obtained from those appearing in the first line of Eq. 12.43 by replacing $k_0 \rightarrow k = \sqrt{2m_e E}/\hbar$ and $k_1 \rightarrow i\kappa = i\sqrt{2m_e(V-E)}/\hbar$:

$$D^{(1,0)} = D^{(3,2)} = \begin{bmatrix} \frac{i\kappa+k}{2i\kappa} & \frac{i\kappa-k}{2i\kappa} \\ \frac{i\kappa-k}{2i\kappa} & \frac{i\kappa+k}{2i\kappa} \end{bmatrix}; \tag{12.55}$$

Fig. 12.3 Double-barrier potential

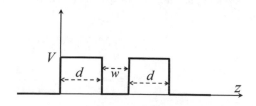

$$D^{(2,1)} = D^{(4,3)} = \begin{bmatrix} \frac{k+i\kappa}{2k} & \frac{k-i\kappa}{2k} \\ \frac{k-i\kappa}{2k} & \frac{k+i\kappa}{2k} \end{bmatrix}. \tag{12.56}$$

For matrices $M^{(1)}$, $M^{(2)}$, and $M^{(3)}$, I can write, using general definition, Eq. 12.26 and expressions for the corresponding coordinates given in Eq. 12.54:

$$M^{(1)} = M^{(3)} = \begin{bmatrix} \exp(-\kappa d) & 0 \\ 0 & \exp(\kappa d) \end{bmatrix} \tag{12.57}$$

$$M^{(2)} = \begin{bmatrix} \exp(ikw) & 0 \\ 0 & \exp(-ikw) \end{bmatrix}. \tag{12.58}$$

The total transfer matrix T then becomes

$$T^{(4)} = D^{(4,3)}M^{(3)}D^{(3,2)}M^{(2)}D^{(2,1)}M^{(1)}D^{(1,0)} =$$

$$D^{(2,1)}M^{(1)}D^{(1,0)}M^{(2)}D^{(2,1)}M^{(1)}D^{(1,0)} \equiv T^{(2)}M^{(2)}T^{(2)}, \tag{12.59}$$

where $T^{(2)}$ is the transfer matrix describing the single barrier. I do not have to calculate this matrix from scratch. Instead, I can again replace k_0 with k and k_1 with $i\kappa$ in Eq. 12.43:

$$T^{(2)} = \frac{1}{2kk_1} \times$$

$$\begin{bmatrix} i\left(k^2 + k_1^2\right)\sin k_1 d + 2kk_1\cos k_1 d & i\left(k_1^2 - k^2\right)\sin k_1 d \\ -i\left(k_1^2 - k\right)\sin k_1 d & -i\left(k^2 + k_1^2\right)\sin k_1 d + 2kk_1\cos k_1 d \end{bmatrix} \rightarrow$$

$$\frac{1}{2ik\kappa}\begin{bmatrix} i\left(k^2 - \kappa^2\right)\sin(i\kappa d) + 2ik\kappa\cos(i\kappa d) & -i\left(\kappa^2 + k^2\right)\sin(i\kappa d) \\ i\left(\kappa^2 + k^2\right)\sin(i\kappa d) & -i\left(k^2 - \kappa^2\right)\sin(i\kappa d) + 2ik\kappa\cos(i\kappa d) \end{bmatrix} =$$

$$\frac{1}{2ik\kappa}\begin{bmatrix} -\left(k^2 - \kappa^2\right)\sinh(\kappa d) + 2ik\kappa\cosh(\kappa d) & \left(\kappa^2 + k^2\right)\sinh(\kappa d) \\ -\left(\kappa^2 + k^2\right)\sinh(\kappa d) & \left(k^2 - \kappa^2\right)\sinh(\kappa d) + 2ik\kappa\cosh(\kappa d) \end{bmatrix} \tag{12.60}$$

At the last step of this derivation, I used identities connecting trigonometric and hyperbolic functions: $\sin(iz) = i\sinh z$ and $\cos(iz) = \cosh z$. The elements of this matrix determine amplitude reflection and transmission coefficients for a single-barrier potential, r_1 and t_1 correspondingly, as established by Eqs. 12.34 and 12.39:

$$t_1 = \frac{2ik\kappa}{\left(k^2 - \kappa^2\right)\sinh(\kappa d) + 2ik\kappa\cosh(\kappa d)} \tag{12.61}$$

$$r_1 = -\frac{\left(\kappa^2 + k^2\right)\sinh(\kappa d)}{\left(k^2 - \kappa^2\right)\sinh(\kappa d) + 2ik\kappa\cosh(\kappa d)} \tag{12.62}$$

Equations 12.61 and 12.62, obviously, can be derived from Eqs. 12.44 and 12.45 for the single-well problem with the same replacements of k_0 and k_1 used to obtain the T-matrix itself.

In order to simplify further computations and also to provide an easier way to relate the properties of the double-barrier structure to those of its single-barrier components, I am going use Eqs. 12.34 and 12.39 to rewrite the transfer matrix in terms of the amplitude reflection and transmission coefficients, r_1 and t_1:

$$T^{(2)}_{22} = \frac{1}{t_1}; \quad T^{(2)}_{21} = -\frac{r_1}{t_1}.$$

Using the explicit form of the matrix $T^{(2)}$, Eq. 12.60, you can determine that $T^{(2)}_{11} = \left(T^{(2)}_{22}\right)^*$ and $T^{(2)}_{12} = \left(T^{(2)}_{21}\right)^*$, so that the entire $T^{(2)}$ can be written down as

$$T^{(2)} = \begin{bmatrix} 1/t_1^* & -r_1^*/t_1^* \\ -r_1/t_1 & 1/t_1 \end{bmatrix}.$$

Multiplying this by $M^{(2)}$ from Eq. 12.58, I get

$$T^{(2)}M^{(2)} = \begin{bmatrix} \exp{(ikw)}/t_1^* & -\exp{(-ikw)}\,r_1^*/t_1^* \\ -\exp{(ikw)}\,r_1/t_1 & \exp{(-ikw)}/t_1 \end{bmatrix} \begin{bmatrix} 1/t_1^* & r_1^*/t_1^* \\ -r_1/t_1 & 1/t_1 \end{bmatrix},$$

and, finally, multiplying this matrix by $T^{(2)}$ (from the left), I find the total double-barrier T-matrix $T^{(4)}$:

$$T^{(4)} = \begin{bmatrix} \frac{\exp(ikw)}{\left(t_1^*\right)^2} + \frac{\exp(-ikw)|r_1|^2}{|t_1|^2} & \frac{\exp(ikw)r_1^*}{\left(t_1^*\right)^2} - \frac{\exp(-ikw)r_1^*}{|t_1|^2} \\ -\frac{\exp(ikw)r_1}{|t_1|^2} - \frac{\exp(-ikw)r_1}{t_1^2} & -\frac{\exp(ikw)|r_1|^2}{|t_1|^2} + \frac{\exp(-ikw)}{t_1^2} \end{bmatrix}$$

$$= \frac{1}{|t_1|^2} \begin{bmatrix} \frac{t_1\exp(ikw)}{t_1^*} + |r_1|^2\exp{(-ikw)} & \frac{t_1\exp(ikw)r_1^*}{t_1^*} - \exp{(-ikw)}\,r_1^* \\ -r_1\exp{(ikw)} - \frac{t_1^*\exp(-ikw)r_1}{t_1} & -|r_1|^2\exp{(ikw)} + \frac{t_1^*\exp(-ikw)}{t_1} \end{bmatrix}.$$

This expression can be simplified by introducing

$$t_1 = |t|\exp{(i\varphi_t)}$$

$$r_1 = |r|\exp{(i\psi_r)},$$

which yields

$$T^{(4)} = \frac{1}{|t_1|^2} \times$$

$$\begin{bmatrix} e^{i(kw+2\varphi_t)} + |r_1|^2\,e^{-ikw} & r_1^*\left[e^{i(kw+2\varphi_t)} - e^{-ikw}\right] \\ -r_1\left[e^{-i(kw+2\varphi_t)} + e^{ikw}\right] & -|r_1|^2\,e^{ikw} + e^{-i(kw+2\varphi_t)} \end{bmatrix} =$$

$$\frac{1}{|t_1|^2} \times$$

$$\begin{bmatrix} e^{i\varphi_t}\left[e^{i(kw+\varphi_t)} + |r_1|^2\, e^{-i(kw+\varphi_t)}\right] & 2ir_1^* e^{i\varphi_t} \sin(kw+\varphi_t) \\ -2r_1 e^{-i\varphi_t} \cos(kw+\varphi_t) & e^{-i\varphi_t}\left[-|r_1|^2\, e^{i(kw+\varphi_t)} + e^{-i(kw+\varphi_t)}\right] \end{bmatrix}.$$

$$(12.63)$$

At the last step I factored out $\exp(i\varphi_t)$ to make residual expressions more symmetrical with respect to the phases of the remaining exponential functions and used Euler's identities $\cos x = (\exp(ix) + \exp(-ix))/2$ and $\sin x = (\exp(ix) - \exp(-ix))/2i$. Now you can simply read out the expressions for the total amplitude reflection and transmission coefficients:

$$t_{db} = \frac{|t_1|^2 \exp(i\varphi_t)}{-|r_1|^2 \exp(ikw + i\varphi_t) + \exp(-ikw - i\varphi_t)} \tag{12.64}$$

$$r_{db} = \frac{r_1^* \exp(2i\varphi_t)\left[\exp(ikw + i\varphi_t) - \exp(-ikw - i\varphi_t)\right]}{-|r_1|^2 \exp(ikw + i\varphi_t) + \exp(-ikw - i\varphi_t)}, \tag{12.65}$$

where subindex db stands for the *double barrier*.

I will begin the analysis of the obtained expression with the transmission probability $T_{db} = |t_{db}|^2$:

$$T_{db} = \frac{|t_1|^4}{\left|\left(1 - |r_1|^2\right)\cos(kw + \varphi_t) - i\left(1 + |r_1|^2\right)\sin(kw + \varphi_t)\right|^2}$$

At this point it is useful to recall that transmission and reflection probabilities obey the probability conservation condition $|t_1|^2 + |r_1|^2 = 1$, which allows to rewrite the expression for T_{db} in the simplified form

$$T_{db} = \frac{|t_1|^4}{|t_1|^4 \cos^2(kw + \varphi_t) + \left(1 + |r_1|^2\right)^2 \sin^2(kw + \varphi_t)}. \tag{12.66}$$

The corresponding expression for the reflection probability becomes

$$R_{db} = \frac{4\,|r_1|^2 \sin^2(kw + \varphi_t)}{|t_1|^4 \cos^2(kw + \varphi_t) + \left(1 + |r_1|^2\right)^2 \sin^2(kw + \varphi_t)} \tag{12.67}$$

Before going any further, it is always useful to check that the results obtained obey
the probability conservation condition $R_{db} + T_{db} = 1$. To prove that this is indeed
true, you just need to demonstrate that

$$|t_1|^4 + 4\,|r_1|^2 \sin^2{(kw + \varphi_t)} = |t_1|^4 \cos^2{(kw + \varphi_t)} + \left(1 + |r_1|^2\right)^2 \sin^2{(kw + \varphi_t)}.$$

You might probably find an easier way to prove this identity, but this is how I did it:

$$|t_1|^4 + 4\,|r_1|^2 \sin^2{(kw + \varphi_t)} =$$

$$|t_1|^4 \left(\cos^2{(kw + \varphi_t)} + \sin^2{(kw + \varphi_t)}\right) + 4\,|r_1|^2 \sin^2{(kw + \varphi_t)} =$$

$$|t_1|^4 \cos^2{(kw + \varphi_t)} + \left(4\,|r_1|^2 + |t_1|^4\right) \sin^2{(kw + \varphi_t)} =$$

$$|t_1|^4 \cos^2{(kw + \varphi_t)} + \left(4\,|r_1|^2 + \left(1 - |r_1|^2\right)^2\right) \sin^2{(kw + \varphi_t)} =$$

$$|t_1|^4 \cos^2{(kw + \varphi_t)} + \left(1 + |r_1|^2\right)^2 \sin^2{(kw + \varphi_t)}. \qquad (12.68)$$

Having verified that my calculations are not obviously wrong, I can proceed with
their analysis. If you remember that the naive expectation, which I described in the
beginning of this section, was that adding a second barrier would result in a total
transmission being just a product of the transmission probabilities through each
barrier, which in our case of identical barriers would mean $T_{db} = |t_1|^4$. Looking
at Eq. 12.66, you can indeed notice the factor $|t_1|^4$ in its numerator, but you will
also see that this factor is accompanied by a denominator, which is responsible
for breaking our naive expectations. What this denominator does, it selects special
energies, namely, the ones obeying the condition

$$\varsigma\,(E) = k(E)w + \varphi_t(E) = \pi n, \ n = 1, 2, 3 \cdots, \qquad (12.69)$$

which turns $\sin{(kw + \varphi_t)}$ in Eqs. 12.66 and 12.67 to zero. For energies satisfying
Eq. 12.69, the reflection coefficient vanishes and the transmission coefficient turns
to unity. So much for the second barrier suppressing the transmission probability!
In reality, the presence of the second barrier somehow magically helps the quantum
particle to penetrate both barriers without any reflection, albeit only at special
energies. This phenomenon is called *resonant tunneling*, and it is a wonderful
manifestation of importance of quantum superposition of states or, as one could
say, of the wave nature of quantum particles. Energy values at which the resonant
tunneling takes place are called *tunneling resonances*.

To analyze this effect in more details, it is useful to rearrange terms in the
denominator of Eq. 12.66. Using identity in Eq. 12.68, I can rewrite the expression
for the transmission probability T_2 as

$$T_{db} = \frac{|t_1|^4}{|t_1|^4 + 4|r_1|^2 \sin^2(kw + \varphi_t)} =$$

$$\frac{1}{1 + \frac{4|r_1|^2}{|t_1|^4} \sin^2(kw + \varphi_t)}. \tag{12.70}$$

This expression makes it even more obvious that every time when the energy of the particle obeys the resonance condition, Eq. 12.69, the transmission turns to unity, but it also reveals the role of the parameter:

$$\Gamma = \frac{|t_1|^2}{|r_1|}. \tag{12.71}$$

Indeed, let me find the values of the energy for which the transmission drops to the half of its maximum value, i.e., becomes equal to $1/2$. Quite obviously, this happens whenever

$$\frac{4}{\Gamma^2} \sin^2 \varsigma = 1 \iff |\sin \varsigma| = \Gamma/2. \tag{12.72}$$

In the case of the thick individual barriers, when the effect of the resonant transmission is most drastic, the single-barrier transmission $|t_1|$ is small, while the reflection $|r_1|$ is almost unity. In this case Eq. 12.71 can be approximated as follows:

$$\Gamma = \frac{|t_1|^2}{\sqrt{1 - |t_1|^2}} \approx \frac{|t_1|^2}{1 - |t_1|^2/2} \approx |t_1|^2 \left(1 + |t_1|^2/2\right) \approx |t_1|^2, \tag{12.73}$$

where I neglected terms smaller than $|t_1|^2$. This approximation shows that Γ is as small as $|t_1|^2$ meaning that according to Eq. 12.72, the value of the phase $\varsigma(E)$ at the energy values corresponding to $T_{db} = 1/2$ only weakly deviates from the resonant value E_n with $\varsigma(E_n) = \pi n$. Accordingly, $\varsigma(E)$ can be presented as $\varsigma(E) = \pi n + \delta\varsigma_n$ where $\delta\varsigma_n \ll 1$, allowing to simplify Eq. 12.72 as

$$|\sin(\delta\varsigma_n)| \approx |\delta\varsigma_n| = \Gamma/2. \tag{12.74}$$

Thus, parameter $\Gamma/2$ determines the magnitude of the deviation of the phase $\varsigma(E)$ from its resonant value required to bring down the transmission coefficient by half. The smaller the Γ, the smaller is such deviation, which means, in other words, that smaller Γ results in steeper decrease of transmission when particle's energy shifts away from the resonance. Deviation of the phase can be translated into the respective deviation of energy by presenting

$$\varsigma(E) \approx \varsigma(E_n) + \frac{d\varsigma(E)}{dE} \delta E$$

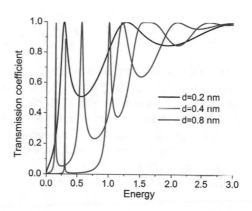

Fig. 12.4 Double-barrier transmission of an electron as a function of energy for the structure with barrier height 1 eV, the distance between the barriers $w = 1.2$ nm, and three barrier widths: blue line corresponds to $d = 0.8$ nm, red to $d = 0.4$ nm, and black to $d = 0.2$ nm. Energy is given in dimensionless units of $2m_e E w^2 / \hbar^2$

The deviation of the phase equal to $\Gamma/2$ corresponds to the deviation of energy equal to

$$\frac{\Gamma_E}{2} = \left[\frac{d\varsigma(E)}{dE} \right]^{-1} \frac{\Gamma}{2} \tag{12.75}$$

If one plots transmission as a function of particle energy, the resonant values will appear as peaks of the transmission, while parameter Γ_E will determine the width of these peaks. More accurately $\Gamma_E/2$ is called the half-width at half-maximum (HWHM). The origin of "half-maximum" in this term is obvious, and half-width refers to the fact that Eq. 12.74 has two solutions $\pm\Gamma/2$, and the total width of the resonance at half-maximum is $(E_n + \Gamma_E/2) - (E_n - \Gamma_E/2) = \Gamma_E$. Widening of the barriers results in decreasing Γ, which can be qualitatively described as narrowing of the resonances. You can observe this phenomenon in Fig. 12.4, presenting transmission as a function of energy for several barrier widths. You can also see that the resonances broaden with increasing energy. This is the result of the energy dependence of the elements of the single-barrier transfer matrix and, correspondingly, of the parameters Γ and the derivative of the phase $d\varsigma/dE$. The explicit expression for this derivative can be found from Eq. 12.61 for the amplitude transmission coefficient, but the result is rather cumbersome and can be left out.

This figure reveals that parameter Γ also determines how small the transmission becomes between the maximums and, therefore, how prominent the resonances are. In order to see where this effect comes from, it is useful to rewrite Eq. 12.70 for transmission as

$$T_{db} = \frac{(\Gamma/2)^2}{(\Gamma/2)^2 + \sin^2(kw + \varphi_t)} \tag{12.76}$$

One can see now that the minimum of transmission, which occurs whenever $\sin(kw + \varphi_t)$ reaches its largest value of unity, is

$$T_{db}^{(min)} = \frac{\Gamma^2}{\Gamma^2 + 1} \approx \Gamma^2$$

where I assumed at the last step that $\Gamma \ll 1$, i.e., it increases with increasing Γ. You may also notice that the position of the resonances is different for different barrier thicknesses. This result seems to be contrary to Eq. 12.69, which shows explicitly only the dependence of the resonant energies on the distance between the barriers, w. The observed effect of the dependence of the resonances on d emphasizes the role of the phase factor φ_t, which does depend on the thickness of the barriers, but is often overlooked.

In the vicinity of the resonance $\varsigma_n = \pi n$, one can expand the $\sin(\varsigma)$ as

$$\sin(\varsigma - \varsigma_n + \pi n) = (-1)^n \sin(\varsigma - \varsigma_n) \approx (-1)^n (\varsigma - \varsigma_n) \approx (-1)^n (d\varsigma/dE)(E - E_n).$$

With this approximation Eq. 12.76 for transmission can be presented in the vicinity of the resonance as

$$T_{db} = \frac{(\Gamma_E/2)^2}{(\Gamma_E/2)^2 + (E - E_n)^2} \tag{12.77}$$

Resonance behavior of this type occurs frequently in various areas of physics and is called a Breit–Wigner resonance, while Eq. 12.77 bears the name of a Breit–Wigner formula.[1]

The treatment of the resonant tunneling, which I have developed, is remarkably independent on the details of the shapes of the barriers constituting the double-barrier structure. As long as the boundaries of the barriers are clearly defined so that I can write down a single-barrier transfer matrix $T^{(2)}$ and the distance between the barriers, w, I can use the results of this section. Do not get me wrong—the parameters of $T^{(2)}$, of course, depend on the details of the barrier's shape, but what I want to say is that $T^{(2)}$ can be computed independently of the double-barrier problem once and for all, numerically if needed, and then used in the analysis of the resonant tunneling.

So, I hope you are convinced by now that the resonant tunneling is a remarkable phenomenon, which can be relatively simply described in terms of the reflection and

---

[1]Gregory Breit was an American physicist, known for his work in high energy physics and involvement at the earlier stages of the Manhattan project. Eugene Wigner was a Hungarian-American theoretical physicist, winner of the half of 1963 Nobel Prize "for his contributions to the theory of the atomic nucleus and the elementary particles, particularly through the discovery and application of fundamental symmetry principles." In 1939 he participated in a faithful Einstein-Szilard meeting resulting in a letter to President Roosevelt prompting him to initiate work on development of atomic bombs. You might find this comment of his particularly intriguing, "It was not possible to formulate the laws of quantum mechanics in a fully consistent way without reference to consciousness," which he made in one of his essays published in collection "Symmetries and Reflections – Scientific Essays (1995)."

transmission coefficients of a single barrier. Still, you might feel certain dissatisfaction because all these calculations do not really *explain* how passing through two thick barriers instead of one can improve the probability of transmission, leave alone make it equal to one. They also do not clarify the role of the quantum superposition, which, I claimed, played a crucial role in this phenomenon. There are several distinct ways to develop a more qualitative, intuitive understanding of this situation. First is naturally based on thinking about quantum mechanical properties of the particle in terms of waves, their superposition and interference. To see how these ideas play out, consider an expression for the amplitude reflection coefficient r_{db}, which determines the relative contribution of the backward propagating wave in the wave function representing the state of the particle in the region $z < 0$:

$$\psi(z) - \exp(ikz) + r_{db} \exp(-ikz).$$

A careful look at Eq. 12.65 reveals that this expression describes a superposition of two waves, both propagating backward, but with different phases. The origin of these contributions is the multiple reflections of the waves representing the particle's state between the boundaries of both barriers (this is why the second barrier is crucial for this effect to occur). The only terms contributing to the phase difference between them are $\exp(ikw + i\varphi_t)$ and $\exp(-ikw - i\varphi_t + i\pi)$, where the extra $i\pi$ in the argument of the exponent takes care of the negative sign appearing in front of this expression in Eq. 12.65. The phase difference between these contributions to the reflected (backward propagating) component of the wave function is $\Delta\phi = 2kw + 2\varphi_t + \pi$, and if we want to suppress reflection by destructive interference, we must require that $\Delta\phi = \pi + 2\pi n$, which results in exactly the condition for the transmission resonance $kw + \varphi_t = \pi n$.

It is also instructive to take a look at the spatial dependence of the probability density $|\psi(z)|^2$ for resonant and off-resonant values of energy. The analytical expression for this quantity is quite cumbersome, especially off the resonance, so I will spare you from having to suffer through its derivation, presenting instead only the corresponding graphs obtained for the same values of the parameters as in Fig. 12.4 for off- and on-resonance values of the particle's energy. The first two graphs in Fig. 12.5 correspond to the value of energy smaller and larger than the energy of the first tunneling resonance. In both cases you can observe oscillations of the probability in the region $z < 0$ due to interference between incident and reflected waves. You should also notice that the relative probability to find the particle in front of the barrier at the maximums of the interference pattern significantly exceeds the probability to find the particle between the barriers or behind them (the right boundary of the second barrier can be clearly identified from the graphs by the absence of any interference pattern in the transmitted wave) for energies both below and above the resonance. The situation, however, changes completely at the resonance (the last graph in the figure). The most remarkable feature of this graph is a pronounced increase of the likelihood that the particle is located between the barriers. If we are dealing with a beam of many electrons incident on the structure, this effect will result in an accumulation of electrons between the barriers making

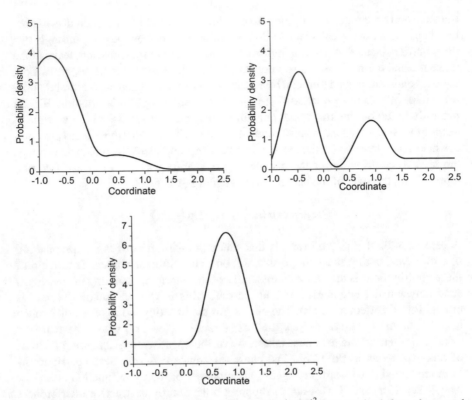

Fig. 12.5 Spatial dependence of the probability density $|\psi(z)|^2$ for energies below, above, and equal to the energy of the first tunneling resonance. Parameters of the double-barrier structure are the same as in Fig. 12.4 with the barrier width $d = 0.4\,\text{nm}$

this region strongly negatively charged. Electric field associated with this strong charge will repel incoming electrons making it more difficult for additional electrons to penetrate the barriers. This effect, called Coulomb blockade, can be noticed as increase in the number of reflected electrons as we increase the density of electrons in the beam. For very small distance between the barriers, the effect of Coulomb blockade can be so strong that even a single electron is capable of preventing other electrons from entering the structure. Thanks to this phenomenon, physicists and engineers gain ability to count individual electrons and develop single-electron devices.

It is important to notice that the resonance probability distribution featured in Fig. 12.5 corresponds to the smallest of the resonance energies, which satisfies Eq. 12.69 with $n = 1$. Now I want you to take a look at the probability distributions corresponding to resonance energies satisfying Eq. 12.69 with $n = 2$ and $n = 3$ presented in Fig. 12.6.

Ignore for the second that the functions depicted in Figs. 12.5 and 12.6 do not vanish at infinity, and compare them to those shown in Fig. 6.6 , which present the

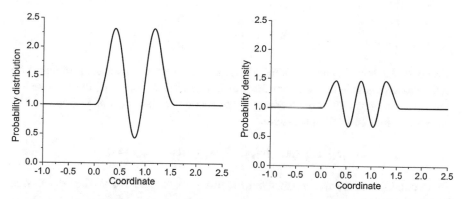

Fig. 12.6 Spatial dependence of the probability density $|\psi(z)|^2$ at the resonance energies of the second and third order $(n = 2, 3)$

wave functions corresponding to the first three bound energy levels in a rectangular potential. Taking into consideration the obvious difference stemming from the fact that the graphs in Fig. 6.6 are those of the real-valued wave functions, while Figs. 12.5 and 12.6 depict $|\psi(z)|^2$, you cannot help noticing the eerie resemblance between the two sets of graphs. You might also notice that the resonance condition, Eq. 12.69, resembles an equation for the energy eigenvalues of the bound states. Actually, in the limit $d \rightarrow \infty$, this equation must exactly reproduce Eq. 12.50 with obvious replacements $d \rightarrow w$ and $V_w \rightarrow 0$, and it would be interesting to demonstrate that. Will you dare to try? Do not get deceived by the term kw in Eq. 12.69, which might make you think about the bound states of an infinite potential well. The finite nature of the potential barriers arising in the limit $d \rightarrow \infty$ is hidden in the phase term φ_t, which shall play the main role when recasting Eqs. 12.69 into the form of Eq. 12.50.

Anyway, this similarity between resonance wave functions and those of the bound states offers an alternative interpretation of the phenomenon of the resonant tunneling. Imagine that you start with a potential, in which the barriers are infinitely thick, so that you can place a particle in one of the stationary states of the respective potential well. Then, using a magic wand, you reduce the thickness of the barriers to a finite value. What will happen to the particle in this situation? Using what we learned about the tunneling effect, you can intuit that the particle will "tunnel out" of the potential well and escape to the infinity. In formal mathematical language, this can be rephrased by saying that the boundary condition for the corresponding Schrödinger equation at $z \rightarrow \pm\infty$ must take a form of a wave propagating to the right $(\exp(ikz))$ for $z \rightarrow \infty$ and a wave propagating to the left $(\exp(-ikz))$ for $z \rightarrow -\infty$. These boundary conditions differ from the ones we used when deriving the transmission and reflection coefficients by the absence of the wave $\exp(ikz)$ incident on the potential from negative infinity. Correspondingly, the wave function at $z < 0$ and $z > 2d + w$ is now presented by the column vectors

$$v_0 = \begin{bmatrix} 0 \\ r \end{bmatrix} ; \ v_4 = \begin{bmatrix} t \\ 0 \end{bmatrix}$$

similar to the bound state problem. Also, similar to the bound state problem, you will have to conclude that the transfer-matrix equation $T^{(4)} v_0 = v_4$ in this case has non-zero solutions only if the element in the second row and second column of $T^{(4)}$ vanishes. Equation 12.63 then yields

$$- |r_1|^2 \exp{(ikw + i\varphi_t)} + \exp{(-ikw - i\varphi_t)} = 0$$

This equation can be transformed into a more convenient for further discussion form:

$$\exp{(2ikw + 2i\varphi_t)} \equiv \exp{(2i\varsigma)} = \frac{1}{|r_1|^2} \tag{12.78}$$

where I brought back the same notation for the phase $\varsigma = kw + \varphi$ used when discussing tunneling resonances. Trying to solve this equation, for instance, by graphing its left-hand and right-hand sides, you will immediately realize that this equation does not have real-valued solutions (it's left-hand side is complex-valued, while the right-hand side is always real). More accurately, I shall say that this equation might have real solutions only if $|r_1|$ is equal to unity for all frequencies, which is equivalent to the requirement that the thickness of the barriers d becomes infinite. If this is the case, Eq. 12.78 can be satisfied if $2\varsigma = 2\pi n$, which is, of course, just Eq. 12.69, and I hope that by now you have already demonstrated that this equation is equivalent to Eq. 12.50 in the limit of the infinitely thick barriers. We, however, are interested in the situation when the barriers are thick but finite, so that $|r_1|^2$ is less than one, but not by much. Using the probability conservation equation, $|t_1|^2 + |r_1|^2 = 1$, I can rewrite Eq. 12.78 as

$$\exp{(2i\varsigma)} = \frac{1}{1 - |t_1|^2} \tag{12.79}$$

and using condition $|t_1|^2 \ll 1$, approximate it as

$$\exp{(2i\varsigma)} \approx 1 + |t_1|^2 \tag{12.80}$$

where I used well-known approximation

$$(1 + x)^\alpha \approx 1 + \alpha x$$

which is just the first two terms in the power series expansion of function $(1 + x)^\alpha$ with $\alpha = -1$. Expecting that the solution to this equation deviates only slightly from $\varsigma_n = \pi n$, I will present ς as $\varsigma = \pi n + \chi$, where $\chi \ll 1$. The exponential left-hand side of Eq. 12.80 in this case becomes

$$\exp\left(2i\pi n + 2i\chi\right) = e^{2i\chi} \approx 1 + 2i\chi$$

and substituting it into Eq. 12.80, I find that χ is a purely imaginary quantity equal to

$$\chi = -\frac{1}{2}i\,|t_1|^2$$

Thus, the wave number satisfying Eq. 12.80 acquires an imaginary part defined by equation

$$kw + \varphi_t = \pi n - \frac{1}{2}i\,|t_1|^2 \approx \pi n - \frac{1}{2}i\Gamma \tag{12.81}$$

where I used Eq. 12.73 to replace $|t_1|^2$ with Γ.

Gazing for some time at Eq. 12.81, you will realize that something not quite kosher is happening here. When starting the calculations, we postulated that the wave function at infinity is described by propagating waves with real wave numbers. Well, it turns out that it is not possible to keep this assumption and satisfy all other boundary conditions. If you now substitute Eq. 12.81 into the $\exp\left(\pm ikz\right)$, it will turn into $\exp\left[\pm i\left(\pi n - \varphi_t\right)z/w\right]\exp\left(\pm \Gamma z/2\right)$, which explodes exponentially for both $z < 0$ and $z > 2d + w$. Quite obviously this is not an acceptable wave function as it cannot be normalized neither in the regular nor in δ-function sense. So, does it mean that all our efforts for the last hour, hour and a half (I guess this is how long it would take you to get through this segment of the book, but, believe me, it took me much, much longer to write it), were in vain? Well, not quite, of course, why would I bother you with this if it were. What I want to do now to save my face is to take the phase $\varsigma\left(E\right)$ in Eq. 12.81 and expand it as a function of energy around the point E_n, where E_n is a resonant frequency obeying equation $\varsigma\left(E_n\right) = \pi n$. It will give me

$$\varsigma\left(E\right) \approx \pi n + \left(d\varsigma/dE\right)\left(E - E_n\right),$$

which I will substitute to Eq. 12.81

$$\pi n + \left(d\varsigma/dE\right)\left(E - E_n\right) = \pi n - \frac{1}{2}i\Gamma \Rightarrow$$

$$E = E_n - \frac{1}{2}i\Gamma\left[\frac{d\varsigma}{dE}\right]^{-1} = E_n - \frac{1}{2}i\Gamma_E \tag{12.82}$$

where I used Eq. 12.75 to introduce energy HWHM parameter Γ_E. Quite clearly, Eq. 12.81 cannot be satisfied with real values of energy, so that the found solutions cannot be eigenvalues of a Hermitian operator, (which must be real), and of course, they are not. The problem, which we have been trying to solve, lost its Hermitian nature once it was allowed for the wave function not to vanish at infinity. So, if the found solutions are not "true" energy eigenvalues, what are they? Can we prescribe them at least some physical meaning? Well, just by looking at Eq. 12.82, you can notice that its real part coincides with the energy of the tunneling resonances, while its imaginary part is equal to the HWHM parameter of those resonances. Plugging this equation into the time-dependent portion of the wave function (which has been ignored so far), $\exp(-iEt/\hbar)$, will get you

$$\psi(t) \propto e^{-iE_n t/\hbar} e^{-\frac{1}{2}\Gamma_E t/\hbar} \qquad (12.83)$$

The respective probability distribution, which normally wouldn't depend on time, is now exponentially decreasing:

$$P = |\psi|^2 \propto \exp(-\Gamma_E t/\hbar) \qquad (12.84)$$

with a characteristic time scale $\tau_E = \hbar/\Gamma_E$. And Eq. 12.84, actually, admits quite a natural physical interpretation. To see this, you need to recall the very initial assumption that I made starting discussing this approach to the resonant tunneling. The question I posed at that time was: What would happen to a particle placed in a stationary state of a potential wall with infinitely thick potential barriers if the width of the barriers would become large but finite? Physical intuition told us that a particle in this case would be able to tunnel out of the well through the barriers, which means that the probability to locate the particle inside the well would diminish with time. We can understand Eq. 12.84 as a formal description of this decay of probability due to tunneling. Thus, even though the wave functions, which I calculated, do not appear to have much physical or mathematical meaning, the complex eigenvalues given by Eq. 12.82 contain an important physically relevant information: its real part yields the energy of the tunneling resonances, while its imaginary part describes both the resonance width, Γ_E, and the time of the decay of the probability due to tunneling τ.

These complex eigenvalues are called *quasi-energies*, while respective states are known as "quasi-stationary states," "quasi-modes," or "resonance states." The term quasi-stationary implies that a particle placed in such a state would not stay there forever and would tunnel out during some time; the time τ_E can be understood as an average lifetime of such states. It is remarkable that the product $\Gamma_E \tau_E$ is simply equal to Planck's constant $\hbar$, making relationship between the width of the resonance and the lifetime of the respective quasi-stationary state similar to the uncertainty relation between, say, coordinate and momentum operators.

More accurate treatment of such time-dependent tunneling requires solving the time-dependent Schrödinger equation with an appropriate initial condition, but even such, less than rigorous, but intuitively appealing approach, can be used to infer important information about behavior of the particle in potentials similar to the double-barrier potential considered here. This approach, together with the concept of quasi-stationary states, was first introduced by George Gamow, an influential Russian-American physicist, born in Odessa (Russian Empire, presently Ukraine), educated in Soviet Union, and defected to the West in 1933 as Stalin's purges began to intensify. (One of his closest university friends, Matvei Bronstein, was executed

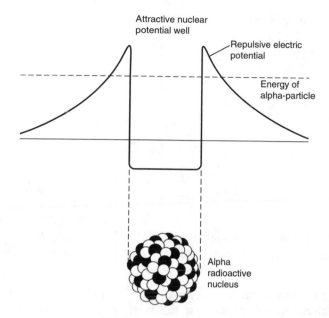

Fig. 12.7 A schematic of a potential barrier experienced by an alpha-particle inside of a nucleus

by Soviet authorities in 1938 on trumped-up treason charges.) Gamow introduced these states (sometimes called Gamow states) while developing the theory of *alpha-particle* radioactivity. His idea was that the alpha-particles contained inside of the nucleus of a radioactive atom experience a potential in the form of a well followed by a thick, but finite, barrier (see Fig. 12.7).

Radioactive decay in Gamow's theory was understood as a slow tunneling of α-particles out of the nucleus. Gamow's approach can actually be modified to give it more mathematical rigor and to turn the wave functions representing the quasi-states into physically and mathematically meaningful objects. However, the modern variations of the concept of quasi-states is a topic lying far outside of the scope of this book, so let me just finish this chapter now.

12.3 Problems

Problem 148 Verify that the matrix equation, Eq. 12.8, is, indeed, equivalent to the system of equations, Eqs. 12.6 and 12.7.

Problem 149

1. Write down boundary conditions for the wave function and its derivative presented in Eq. 12.11.

Fig. 12.8 Potential described
in Problem 7

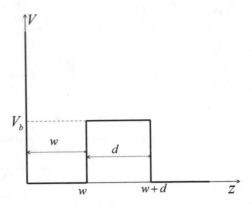

2. Find relations between amplitudes $A_1^{(L,R)}$, $B_1^{(L,R)}$ and $A_{1,2}$, $B_{1,2}$, which would convert the boundary conditions found in Part I of the problem into Eqs. 12.13 and 12.14.

Problem 150 Demonstrate that coefficients $A_0^{(R)}$ and $B_0^{(R)}$ in Eqs. 12.46 and 12.47 coincide with coefficients A_2 and B_2 in Eqs. 6.60 and 6.61 in Sect. 6.2.

Problem 151 Show that coefficient $B_2^{(R)}$ in Eq. 12.48 vanishes.

Problem 152 Prove that Eq. 12.53 is, indeed, reduced to $A_2 = \pm B_0$ for energies satisfying dispersion equations 12.51 and 12.52.

Problem 153 Use the transfer-matrix method to find the equation for the bound state in the asymmetric potential well:

$$V(z) = \begin{cases} V_1 & z < 0 \\ V_w & 0 < z < d \\ V_2 & z > d \end{cases}$$

where $V_2 > V_1$.

Problem 154 Consider an electron moving in a potential described as (see Fig. 12.8)

$$V(x) = \begin{cases} \infty & x < 0 \\ 0 & 0 < x < w \\ V_b & w < x < w + d \\ 0 & x > w + d \end{cases}$$

You are interested in the properties of the electron in this potential for energies $0 < E < V_b$. It is clear that for the particle incident on this potential from the left, which is the only direction it can be incident from, the probability of reflection is always equal to one, simply because the wave function at $z < 0$ vanishes and there can be no transmitted particles. So, it appears that the effects of tunneling resonance discussed in Sect. 12.2 have no place in this potential. At the same time, in the limit $d \to \infty$, this potential allows for at least one bound state localized mostly within the region of the potential well. When the barrier width d becomes finite, this bound stationary state begins leaking outside due to the tunneling effect, just like we discussed in the section on the resonant tunneling. Accordingly, we must expect this potential to possess quasi-stationary states, but it is not clear how they are related to the reflective properties of the potential. I suggest that you try to figure it out.

1. First, find the amplitude reflection coefficient assuming that to the right of the potential, there are both incident and reflected waves, so that the column vector representing the wave function for $z > w + d$ would look like

$$v_R = \begin{bmatrix} r \\ 1 \end{bmatrix}$$

(As always, the first element is occupied by the amplitude in front of the wave propagating to the right, but in the case under consideration, this wave represents reflected particles.) The wave function for $0 < z < w$ must turn zero at $z = 0$, which is achieved by the function of the form

$$\psi(z) = Ae^{ikz} - Ae^{-ikz} = Ae^{ikw}e^{ik(z-w)} - Ae^{-ikw}e^{-ik(z-w)}$$

so that the wave function to the very left of the potential discontinuity at $z = w$ is presented by the column

$$v_1 = \begin{bmatrix} Ae^{ikw} \\ -Ae^{-ikw} \end{bmatrix}$$

The wave function for $z > w$ can be found using standard combination of the interface and free propagation matrices (you will need two of the former and one of the latter). Complete the transfer-matrix calculations and find parameter r. Look for any traces of possible resonant behavior paying special attention to the phase of r.

2. Now repeat these calculations assuming that there are no incident particles so that the wave function for $z > d + w$ is represented by a column

$$v_R = \begin{bmatrix} r \\ 0 \end{bmatrix}$$

Derive equation for the complex quasi-energies of the respective quasi-stationary states. Assuming that the imaginary part of the quasi-energies is small, separate this equation into an equation for the real and imaginary parts, just like we did in Sect. 12.2. Compare the results with those of the preceding calculations.

Chapter 13
Perturbation Theory for Stationary States: Stark Effect and Polarizability of Atoms

Only few models in quantum mechanics allow for an exact analytical solution. Most of the problems, which are relevant to the real-world situations and are important for understanding the fundamental nature of things or for applications, can only be solved using one or another type of approximation. In this chapter I will introduce a method designed for finding approximate solutions for eigenvalues and corresponding eigenvectors of a time-independent Hamiltonian with a discrete spectrum. This method works for Hamiltonians that can be written as a sum of two parts: the main or unperturbed Hamiltonian $\hat{H}_0$, whose eigenvalues, $E_s^{(0)}$, and eigenvectors, $|s\rangle$, are presumed to be known, and a perturbation $\epsilon\hat{V}$. Parameter ϵ that I pulled out of $\hat{V}$ has a formal meaning of the strength of the perturbation, but this can be understood literally only in the sense that the perturbation vanishes when $\epsilon = 0$. The actual parameter determining the strength of the perturbation emerges only *post factum*, after the problem is solved. I will mainly use ϵ as a technical bookkeeping device (you will know what it means when you see it) and set it equal to unity at the end. The index s appearing in the notation for the eigenvalues and the eigenvectors can be a composite index, consisting of several subindexes. For instance, if $\hat{H}_0$ is the Hamiltonian of a hydrogen-like atom, then s contains principal, orbital, and magnetic numbers n, l, m. It is also presumed that the perturbation $\hat{V}$ can be considered small in some yet undefined sense so that the eigenvalues and eigenvectors of the total Hamiltonian

$$\hat{H} = \hat{H}_0 + \epsilon\hat{V} \tag{13.1}$$

do not deviate too much from the $E_s^{(0)}$ and $|s\rangle$ correspondingly. Consequently, one might hope that they can be found approximately using the eigenvalues and eigenvectors of $\hat{H}_0$ as a starting point.

The development of the method is quite different for non-degenerate unperturbed eigenvalues and the degenerate ones. You can see where the difference is coming from by pondering over the following. Whatever the approximate expression I will

© Springer International Publishing AG, part of Springer Nature 2018
L.I. Deych, *Advanced Undergraduate Quantum Mechanics*,
https://doi.org/10.1007/978-3-319-71550-6_13

derive for the eigenvalues and eigenvectors, they must reduce to $E_s^{(0)}$ and $\left|s^{(0)}\right\rangle$ as I set $\epsilon = 0$. If $E_s^{(0)}$ is a non-degenerate eigenvalue so that $|s\rangle$ is the only eigenvector belonging to it, this process does not raise any issues. If, however, $E_s^{(0)}$ is degenerate, meaning that there are several unperturbed orthonormal eigenvectors $\left|s_i^{(0)}\right\rangle$ belonging to it, together with infinitely many linear combinations thereof, an outcome of the transition $\epsilon \to 0$ in this case becomes a mystery. You will learn eventually that as improbable as it might sound, it is the perturbation operator $\hat{V}$ that "decides" the outcome of this transition even as its own "strength" goes to zero. You can consider these somewhat vague remarks as a teaser designed to spur your curiosity. All this will become (hopefully) much clearer once we get down to it. At this point, my goal is simply to justify the importance of separate consideration of degenerate and non-degenerate unperturbed eigenvalues.

13.1   Non-degenerate Perturbation Theory

The non-degenerate case is more straightforward, so this is what I am going to begin with. The idea is to present the unknown eigenvalues E_s and eigenvectors $|s\rangle$ as power series of the form

$$E_s = E_s^{(0)} + \epsilon E_s^{(1)} + \epsilon^2 E_s^{(2)} + \epsilon^3 E_s^{(3)} + \cdots \tag{13.2}$$

$$|s\rangle = \left|s^{(0)}\right\rangle + \epsilon \left|s^{(1)}\right\rangle + \epsilon^2 \left|s^{(2)}\right\rangle + \epsilon^3 \left|s^{(3)}\right\rangle + \cdots \tag{13.3}$$

and plug them into the stationary Schrödinger equation:

$$\left(\hat{H}_0 + \epsilon \hat{V}\right) |s\rangle = E_s |s\rangle . \tag{13.4}$$

This procedure yields

$$\hat{H}_0 \left|s^{(0)}\right\rangle + \epsilon \hat{H}_0 \left|s^{(1)}\right\rangle + \epsilon^2 \hat{H}_0 \left|s^{(2)}\right\rangle + \cdots +$$
$$\epsilon \hat{V} \left|s^{(0)}\right\rangle + \epsilon^2 \hat{V} \left|s^{(1)}\right\rangle + \epsilon^3 \hat{V} \left|s^{(2)}\right\rangle + \cdots =$$
$$E_s^{(0)} \left|s^{(0)}\right\rangle + \epsilon E_s^{(0)} \left|s^{(1)}\right\rangle + \epsilon E_s^{(1)} \left|s^{(0)}\right\rangle + \epsilon^2 E_s^{(0)} \left|s^{(2)}\right\rangle + \tag{13.5}$$
$$\epsilon^2 E_s^{(2)} \left|s^{(0)}\right\rangle + \epsilon^2 E_s^{(1)} \left|s^{(1)}\right\rangle + \cdots .$$

For this expression to be true for an arbitrary value of the perturbation parameter ϵ, it is necessary that the terms with the same power of ϵ on the left-hand side and on the right-hand side of this equation were individually equal to each other:

$$\epsilon^0 : \hat{H}_0 \left| s^{(0)} \right\rangle = E_s^{(0)} \left| s^{(0)} \right\rangle, \tag{13.6}$$

$$\epsilon^1 : \hat{H}_0 \left| s^{(1)} \right\rangle + \hat{V} \left| s^{(0)} \right\rangle = E_s^{(0)} \left| s^{(1)} \right\rangle + E_s^{(1)} \left| s^{(0)} \right\rangle, \tag{13.7}$$

$$\epsilon^2 : \hat{H}_0 \left| s^{(2)} \right\rangle + \hat{V} \left| s^{(1)} \right\rangle = E_s^{(0)} \left| s^{(2)} \right\rangle + E_s^{(2)} \left| s^{(0)} \right\rangle + E_s^{(1)} \left| s^{(1)} \right\rangle. \tag{13.8}$$

Now you can see what I meant in saying that ϵ will only be used for bookkeeping purposes: I use it to identify different approximation orders, and once it is done, it can be set to unity.

Equation 13.6 is just the eigenvalue equation for the unperturbed Hamiltonian, which, as I presumed, is fulfilled by E_s^0 and $\left| s^{(0)} \right\rangle$. Corrections to the eigenvalue and the eigenvector proportional to ϵ (I will call them the first-order corrections) should supposedly be found from Eq. 13.7. You might wonder if it is possible to find both these unknown quantities from a single equation. Well, let's see.

First, I multiply Eq. 13.7 by $\left\langle s^{(0)} \right|$ from the left:

$$\left\langle s^{(0)} \right| \hat{H}_0 \left| s^{(1)} \right\rangle + \left\langle s^{(0)} \right| \hat{V} \left| s^{(0)} \right\rangle = E_s^{(0)} \left\langle s^{(0)} \left| s^{(1)} \right\rangle + E_s^{(1)} \left\langle s^{(0)} \left| s^{(0)} \right\rangle \right..$$

Taking into account that $\left\langle s^{(0)} \left| s^{(0)} \right\rangle = 1 \right.$ (normalization) and that $\left\langle s^{(0)} \right| \hat{H}_0 = E_s^{(0)} \left\langle s^{(0)} \right|$ (Hermitian property of the Hamiltonian), I transform this expression into

$$E_s^{(0)} \langle s^{(0)} | s^{(1)} \rangle + \left\langle s^{(0)} \right| \hat{V} \left| s^{(0)} \right\rangle = E_s^{(0)} \langle s^{(0)} | s^{(1)} \rangle + E_s^{(1)},$$

which gives me the first-order correction to the energy eigenvalue

$$E_s^{(1)} = \left\langle s^{(0)} \right| \hat{V} \left| s^{(0)} \right\rangle. \tag{13.9}$$

So far so good—I got the correction to the energy, but how about the correction to the eigenvector, $\left| s^{(1)} \right\rangle$, which got canceled out? But fret you not—the cancelation of the $\left| s^{(1)} \right\rangle$ is not a bug, but a feature, which allowed me to isolate and determine the energy correction. In order to obtain $\left| s^{(1)} \right\rangle$, I need to do something else, something which would eliminate the $E_s^{(1)}$ term from Eq. 13.7. One way to achieve this is to premultiply the equation by an eigenvector of the unperturbed Hamiltonian, different from $\left\langle s^{(0)} \right|$, say, $\left\langle q^{(0)} \right|$, where $q = 1, 2, 3, s-1, s+1, \cdots$. Indeed, in this case the term with $E_s^{(1)}$ vanishes because of the orthogonality condition: $\langle q^{(0)} | s^{(0)} \rangle = 0$, for $q \neq s$. The remaining expression in this case becomes

$$\left\langle q^{(0)} \right| \hat{H}_0 \left| s^{(1)} \right\rangle + \left\langle q^{(0)} \right| \hat{V} \left| s^{(0)} \right\rangle = E_s^{(0)} \langle q^{(0)} | s^{(1)} \rangle \Rightarrow$$
$$E_q^{(0)} \langle q^{(0)} | s^{(1)} \rangle + V_{qs} = E_s^{(0)} \langle q^{(0)} | s^{(1)} \rangle,$$

where I again used $\left\langle q^{(0)} \right| \hat{H}_0 = E_q^{(0)} \left\langle q^{(0)} \right|$ and introduced the matrix element $V_{qs} = \left\langle q^{(0)} \right| \hat{V} \left| s^{(0)} \right\rangle$ (note that the order of indexes in V_{qs} follows their order in $\left\langle q^{(0)} \right| \hat{V} \left| s^{(0)} \right\rangle$ from left to right). The result is an equation for $\langle q^{(0)} | s^{(1)} \rangle$, which yields

$$\langle q^{(0)} \, | s^{(1)} \rangle = \frac{V_{qs}}{E_s^{(0)} - E_q^{(0)}}. \tag{13.10}$$

This quantity is a component of the unknown vector $|s^{(1)}\rangle$ in the direction of the vector $|q^{(0)}\rangle$ and can be used to find the entire vector $|s^{(1)}\rangle$ as follows. Since $|q^{(0)}\rangle$ are eigenvectors of a Hermitian operator and, therefore, form a basis, I can expand $|s^{(1)}\rangle$ in this basis as

$$|s^{(1)}\rangle = \sum_q |q^{(0)}\rangle \langle q^{(0)} | \, s^{(1)}\rangle =$$

$$\sum_{q \neq s} |q^{(0)}\rangle \langle q^{(0)} | \, s^{(1)}\rangle + \langle s^{(0)} | \, s^{(1)}\rangle |s^{(0)}\rangle,$$

where in the sum in the second line I separated out the term with $q = s$. With $\langle q^{(0)} \, | s^{(1)} \rangle$ found, I am just one step away from finding the entire vector $|s^{(1)}\rangle$: all I need is the value of $\langle s^{(0)} | \, s^{(1)}\rangle$, which so far remains unknown. The help comes from a familiar place—the normalization condition. Consider the found eigenvector with accuracy up to the first order in ϵ:

$$|s\rangle = |s^{(0)}\rangle + \epsilon \langle s^{(0)} | \, s^{(1)}\rangle |s^{(0)}\rangle + \epsilon \sum_{q \neq s} \frac{V_{qs}}{E_s^{(0)} - E_q^{(0)}} |q^{(0)}\rangle,$$

and compute its norm $\langle s| \, s\rangle$:

$$\langle s| \, s\rangle = \langle s^{(0)} | \, s^{(0)}\rangle + \epsilon \langle s^{(0)} | \, s^{(1)}\rangle \langle s^{(0)} | \, s^{(0)}\rangle +$$

$$\epsilon \sum_{q \neq s} \frac{V_{qs}}{E_s^{(0)} - E_q^{(0)}} \langle s^{(0)} \, | q^{(0)}\rangle + \epsilon \sum_{q \neq s} \frac{V_{qs}^*}{E_s^{(0)} - E_q^{(0)}} \langle q^0) | s^0\rangle + O(\epsilon^2).$$

I cannot include in this expression the terms of the second order in ϵ^2 or higher because other terms of the same order were omitted from the initial expression for $|s\rangle$. The second line in this expression is actually equal to zero because of the orthogonality condition $\langle s^{(0)} \, | q^{(0)}\rangle = \langle q^0) | s^0\rangle = 0$, so all that is left is

$$\langle s| \, s\rangle = 1 + \epsilon \langle s^{(0)} | \, s^{(1)}\rangle,$$

where I took into account that $\langle s^{(0)} | \, s^{(0)}\rangle = 1$. Thus, if I want the norm of $|s\rangle$ to be equal to unity, I have to set $\langle s^{(0)} | \, s^{(1)}\rangle = 0$. This is the last piece I needed to find the first-order correction to the eigenvector, which I can now write down as

$$|s\rangle = |s^{(0)}\rangle + \epsilon \sum_{q \neq s} \frac{V_{qs}}{E_s^{(0)} - E_q^{(0)}} |q^{(0)}\rangle. \tag{13.11}$$

So, I am done with the first-order corrections to the eigenvalues and eigenvectors, and now I can turn to finding the corrections proportional to ϵ^2, which are often called second-order corrections. Going back to Eq. 13.8 and performing the same magic trick of premultiplying it by $\langle s^{(0)}|$, I find

$$\langle s^{(0)}|\, \hat{H}_0\, |s^{(2)}\rangle + \langle s^{(0)}|\, \hat{V}\, |s^{(1)}\rangle =$$
$$E_s^{(0)} \langle s^{(0)}\, |s^{(2)}\rangle + E_s^{(2)} \langle s^{(0)}\, |s^{(0)}\rangle + E_s^{(1)} \langle s^{(0)}\, |s^{(1)}\rangle.$$

The first term on the left evaluates to $E_s^{(0)} \langle s^{(0)}\, |s^{(2)}\rangle$ and cancels the first term on the right, while the remaining expression, remembering that $\langle s^{(0)}\, |s^{(0)}\rangle = 1$ and $\langle s^{(0)}\, |s^{(1)}\rangle = 0$, becomes

$$E_s^{(2)} = \langle s^{(0)}|\, \hat{V}\, |s^{(1)}\rangle = \sum_{q \neq s} \frac{V_{sq} V_{qs}}{E_s^{(0)} - E_q^{(0)}} = \sum_{q \neq s} \frac{|V_{sq}|^2}{E_s^{(0)} - E_q^{(0)}}, \qquad (13.12)$$

where I again introduced a matrix element $V_{sq} = \langle s^{(0)}|\, \hat{V}\, |q^{(0)}\rangle$ and took into account that $V_{qs} = V_{sq}^*$.

Premultiplying Eq. 13.8 by $\langle q^{(0)}|$, I obtain

$$\langle q^{(0)}|\, \hat{H}_0\, |s^{(2)}\rangle + \langle q^{(0)}|\, \hat{V}\, |s^{(1)}\rangle =$$
$$E_s^{(0)} \langle q^{(0)}\, |s^{(2)}\rangle + E_s^{(2)} \langle q^{(0)}\, |s^{(0)}\rangle + E_s^{(1)} \langle q^{(0)}\, |s^{(1)}\rangle.$$

Using again the Hermitian property of the Hamiltonian to compute the first term in the first line and orthogonality of the zero-order eigenvectors to eliminate the middle term in the second line, I turn this expression into

$$E_q^{(0)} \langle q^{(0)}\, |s^{(2)}\rangle + \langle q^{(0)}|\, \hat{V}\, |s^{(1)}\rangle = E_s^{(0)} \langle q^{(0)}\, |s^{(2)}\rangle + E_s^{(1)} \langle q^{(0)}\, |s^{(1)}\rangle.$$

Now, using Eq. 13.9 as well as Eqs. 13.10 and 13.11, I can convert it into the following:

$$\left(E_q^{(0)} - E_s^{(0)}\right) \langle q^{(0)}\, |s^{(2)}\rangle - \langle s^{(0)}|\, \hat{V}\, |s^{(0)}\rangle \frac{V_{qs}}{E_s^{(0)} - E_q^{(0)}} - \sum_{p \neq s} \frac{V_{qp} V_{ps}}{E_s^{(0)} - E_p^{(0)}} \Rightarrow$$
$$\langle q^{(0)}\, |s^{(2)}\rangle = - \frac{V_{ss} V_{qs}}{\left(E_s^{(0)} - E_q^{(0)}\right)^2} + \sum_{p \neq s} \frac{V_{qp} V_{ps}}{\left(E_s^{(0)} - E_p^{(0)}\right) \left(E_s^{(0)} - E_q^{(0)}\right)}.$$

Respectively, the second-order correction to the eigenvector becomes

$$\left|s^{(2)}\right\rangle = -\sum_{q\neq s} \frac{V_{ss}V_{qs}}{\left(E_s^{(0)} - E_q^{(0)}\right)^2}\left|q^{(0)}\right\rangle + \sum_{q\neq s}\sum_{p\neq s} \frac{V_{qp}V_{ps}}{\left(E_s^{(0)} - E_p^{(0)}\right)\left(E_s^{(0)} - E_q^{(0)}\right)}\left|q^{(0)}\right\rangle,$$

(13.13)

where I again set $\left\langle s^{(0)}\left|s^{(2)}\right.\right\rangle$ to zero as a normalization condition. Equations 13.12 and 13.13 complete my program of derivation of the lowest-order corrections to the non-degenerate eigenvalues and the eigenvectors of Hamiltonian $\hat{H}$. Finding the third- and higher-order corrections becomes progressively more cumbersome and is rarely necessary.

Ideally, from the point of view of minimizing one's efforts, it would be preferable to get all the answers just from the first-order terms. Often, however, as we already discussed in Chap. 10 on the two-level model, and Sect. 7.1.2 on quantum harmonic oscillator, the diagonal elements of the perturbation part of the Hamiltonian, those that determine the first-order corrections to the eigenvalues, vanish. This happens, for instance, when the unperturbed Hamiltonian $\hat{H}_0$ is invariant with respect to the parity operator, so that its eigenvectors can be classified as being even or odd. If, in addition, the perturbation operator $\hat{V}$ is odd (changes sign upon application of the inversion operator), then $\left\langle s^{(0)}\left|\hat{V}\right|s^{(0)}\right\rangle$ vanishes and takes along the first-order correction to the energy. If the first-order correction to the energy turns zero, you do not have a choice as to rely on the second-order correction. If the second-order terms are not sufficient as well, it usually means (there are exceptions, of course) that the perturbation approach is not suitable for the problem at hand.

13.1.1 Quadratic Stark Effect

The perturbation theory plays a crucial role in understanding the responses of a quantum system to external influences such as electric or magnetic fields. Leaving the effects due to a magnetic field for a separate chapter, in this section I will focus on the interaction between an atom and an electric field, $\mathcal{E}$, which, in many instances, can be assumed to be spatially uniform. I will also assume that this electric field is static, i.e., does not depend on time. Then, what I need to do as the first matter of business is to find the electric field-induced corrections to the energy spectrum and corresponding eigenvectors of the unperturbed Hamiltonian. When this is done, I can move on to analyze how these changes manifest themselves in observable phenomena.

As a practical example, I will consider the effects of the static electric field on the ground state of a hydrogen atom (Chap. 8), which is the only non-degenerate energy level of hydrogen and can be, therefore, studied using the developed method. So, assuming that $\hat{H}^{(0)}$ in Eq. 13.1 describes a hydrogen-like system considered in Chap. 8, I can replace the abstract zero-order eigenvectors $\left|s^{(0)}\right\rangle$ of the previous section with their more concrete version $|nlm\rangle$ and energy eigenvalues with $E_n^{(0)} = -E_g/n^2$. The ground state is described by eigenvector $|100\rangle$ and energy $-E_g$, the expression for which can be found in Chap. 8. Indexes s and q are now replaced by three indexes, n, m, and l, and summation over q involves the summation over all three indexes subject to regular restrictions on their values determined in Chap. 8. The operator of perturbation $\hat{V}$ for a uniform electric field has a simple form, which you already encountered in this book several times, for instance, in Chap. 10:

$$\hat{V} = -e\mathcal{E}\hat{z}. \tag{13.14}$$

Here I chose the Z-axis of the coordinate system used to define the components of the operators of the angular momentum along the direction of the electric field. The first-order correction to the ground state vanishes as was explained above because the perturbation potential is odd with respect to inversion. Those who are skeptical about symmetry-based arguments can verify this statement directly working, for instance, in the position representation

$$\langle 100|\,\hat{V}\,|100\rangle = -e\mathcal{E}\frac{1}{4\pi}\int_0^\pi\int_0^{2\pi}\int_0^\infty d\theta d\varphi dr \sin\theta r^2\,[R_{10}(r)]^2\,r\cos\theta,$$

$R_{10}(r)$ is the radial component of the wave function representing the ground state of the hydrogen-like system, and I converted z into spherical coordinates. Now you only need to compute the integral $\int_0^\pi d\theta\sin\theta\cos\theta$ over the polar angle θ to convince yourself (which shouldn't be too difficult) that it vanishes.

With this issue clarified, let's take on a more difficult problem of finding the second-order correction to the ground state energy using Eq. 13.12. The most important thing to realize when using this equation is that the sum over q is now three sums: over n, l, and m. The sum over n starts with $n = 2$, because the term $n = 1$ is excluded from the summation by the condition $q \neq s$, which in our case translates to $n \neq 1$, the sum over l runs from 0 to $n - 1$, and the sum over m covers values from $-l$ to l:

$$E_1^{(2)} = \sum_{n=?}^\infty\sum_{l=0}^{n-1}\sum_{m=-l}^l \frac{|V_{100,nlm}|^2}{E_1^{(0)} - E_n^{(0)}} =$$

$$e^2\mathcal{E}^2\sum_{n=2}^\infty\sum_{l=0}^{n-1}\sum_{m=-l}^l \frac{|z_{100,nlm}|^2}{E_1^{(0)} - E_n^{(0)}}. \tag{13.15}$$

It should be noted that Eq. 13.15 does not tell the entire story since the spectrum of a hydrogen atom contains also a continuous segment, which, strictly speaking, needs to be included. Moreover, since the potential due to a constant external field grows progressively more negative with the growing value of coordinate z, at some point the total potential felt by the electron will become less negative than a given negative energy of a bound state, providing for a possibility for the electron to tunnel out of the nucleus's potential (at which point the atom becomes ionized). This turns the stationary state of the electron into quasi-stationary, similar to the situation considered in Sect. 12.2. All these complications, however, can be ignored for a weak enough field because (a) for states from continuous spectrum, the energy denominators in Eq. 13.12 become so large that the contribution from the corresponding terms can be safely neglected, and (b) even though the bound states become formally quasi-stationary, if the field is not too strong, their lifetime is long enough to treat them as normal stationary states.

Now my task is to compute the matrix elements $z_{100,nlm}$, which, using the hydrogen wave functions from Chap. 8, can be written down as

$$z_{100,nlm} = \frac{1}{\sqrt{4\pi}} \int_0^\pi \int_0^{2\pi} \int_0^\infty d\theta d\varphi dr \sin\theta r^3 \cos\theta Y_l^m(\theta,\varphi) R_{n,l}(r) R_{1,0}(r). \quad (13.16)$$

To evaluate the angular portion of the integral

$$\int_0^\pi \int_0^{2\pi} d\theta d\varphi \sin\theta \cos\theta Y_l^m(\theta,\varphi)$$

let me first notice that this integral vanishes for all values of the magnetic number m, with exception of $m = 0$. To see this just recall that the spherical harmonics $Y_l^m(\theta,\varphi)$ contain the factor $\exp(im\varphi)$, which in this integral is the only factor containing the azimuthal angle φ. Integration of $\exp(im\varphi)$ over the entire range of φ between 0 and 2π yields zero unless $m = 0$, when the value of the integral becomes 2π. Having disposed of the integration with respect to φ, I am left with the integral over the polar angle

$$\int_0^\pi \int_0^{2\pi} d\theta d\varphi \sin\theta \cos\theta Y_l^m(\theta,\varphi) = 2\pi\sqrt{\frac{2l+1}{4\pi}}\delta_{m,0} \int_0^\pi d\theta \sin\theta \cos\theta P_l(\theta) =$$

$$\sqrt{\pi(2l+1)}\delta_{m,0} \int_{-1}^1 x P_l(x) dx,$$

where I replaced the spherical harmonic Y_l^0 with the regular Legendre polynomials and made a substitution of variables $x = \cos\theta$. To evaluate the remaining integral, I recall that $x = P_1(x)$ so that the last expression can be rewritten as

$$\sqrt{\pi(2l+1)}\delta_{m,0} \int_{-1}^1 P_1(x) P_l(x) dx.$$

All what is left now is to invoke orthogonality of the Legendre polynomials and use Eq. 5.71 from Sect. 5.1.4 adapted to the case $m = 0$:

$$\int_{-1}^{1} P_{l_1}(x)P_l(x)dx = \frac{2}{2l+1}\delta_{ll_1}.$$

Applying this formula to the case under consideration ($l_1 = 1$), I obtain the final result for the angular part of the matrix element:

$$\int_{0}^{\pi}\int_{0}^{2\pi} d\theta d\varphi \sin\theta \cos\theta Y_l^m(\theta,\varphi) =$$

$$\sqrt{\pi(2l+1)}\frac{2}{2l+1}\delta_{m,0}\delta_{l1} = \frac{2\sqrt{\pi}}{\sqrt{3}}\delta_{m,0}\delta_{l1}.$$

Substituting this result into Eq. 13.16, I get

$$z_{100,nlm} = \frac{1}{\sqrt{3}}\delta_{m,0}\delta_{l1}\int_{0}^{\infty} dr r^3 R_{n,1}(r)R_{1,0}(r).$$

Replacing the integration variable r with its dimensionless counterpart $x = Zr/(a_B)$, where all notations are taken from Chap. 8, the expression for the matrix element can be recast as

$$z_{100,nlm} = \frac{1}{\sqrt{3}}\delta_{m,0}\delta_{l1}\left(\frac{a_B}{Z}\right)^4\int_{0}^{\infty} dx x^3 R_{n,1}(x)R_{1,0}(x).$$

The radial wave functions can be read of Eq. 8.21 for the hydrogen wave functions. In terms of dimensionless variable x, the ground state ($n = 1, l = 0$) function takes the form

$$R_{1,0}(r) = 2\sqrt{\left(\frac{Z}{a_B}\right)^3}\exp(\ x)$$

(recall that the Laguerre polynomial $L_0^1(2x) \equiv 1$), while $R_{n,1}(x)$ becomes

$$R_{n,1}(x) = \sqrt{\left(\frac{2Z}{na_B}\right)^3\frac{(n-2)!}{2n(n+1)!}}\left(\frac{2x}{n}\right)\exp\left(-\frac{x}{n}\right)L_{n-2}^3\left(\frac{2x}{n}\right).$$

Substituting these expressions into the formula for the matrix element, I end up with

$$z_{100,nlm} = \frac{1}{\sqrt{3}}\delta_{m,0}\delta_{l1}\left(\frac{a_B}{Z}\right)^4 2\sqrt{\left(\frac{Z}{a_B}\right)^3}\sqrt{\left(\frac{2Z}{na_B}\right)^3 \frac{(n-2)!}{2n(n+1)!}}\frac{2}{n}\times$$

$$\int_0^\infty dx x^3 x e^{-x} e^{-x/n} L_{n-2}^3\left(\frac{2x}{n}\right) dx =$$

$$\delta_{m,0}\delta_{l1}\frac{8}{\sqrt{3}}\frac{a_B}{Zn^3}\sqrt{\frac{(n-2)!}{n(n+1)!}}\int_0^\infty dx x^4 e^{-x(1+1/n)} L_{n-2}^3\left(\frac{2x}{n}\right) dx =$$

$$\delta_{m,0}\delta_{l1}\frac{a_B}{Z}f(n), \qquad (13.17)$$

where I introduced function $f(n)$, which depends only on the principal quantum number n:

$$f(n) = \frac{8}{\sqrt{3}}\frac{1}{n^3}\sqrt{\frac{(n-2)!}{(n+1)!}}\int_0^\infty dx x^4 e^{-x(1+1/n)} L_{n-2}^3\left(\frac{2x}{n}\right) dx. \qquad (13.18)$$

Now the second-order correction to the ground state energy can be written down as

$$E_1^{(2)} = -\frac{8\pi\varepsilon_r\varepsilon_0}{Z^4}\mathcal{E}^2 a_B^3 \sum_{n=2}^\infty \frac{n^2}{n^2-1}f^2(n), \qquad (13.19)$$

where I replaced $E_n^{(0)}$ with their actual values $-E_g/n^2$ and used explicit expression for E_g in terms of Bohr radius using Eq. 8.17. In principle, function $f(n)$ can be found analytically for an arbitrary n, but the result is not worth the effort since we will end up with a nasty looking sum over n, which at any rate we wouldn't be able to find exactly. Thus, instead, I will evaluate $f(n)$ only for $n = 2, 3, 4, 5$ and use the results to compute $E_s^{(2)}$ approximately, including only these terms into the sum. To help you digest and reproduce these computations, I am providing some intermediate results in Table 13.1.

Table 13.1 Data for calculating the quadratic Stark effect

n	$L_{n-2}^3(2x)$	$f(n)$
2	1	0.7449
3	$4-2x$	0.2983
4	$2(5-5x+5x^2)$	0.1759
5	$20-30x+12x^2-\frac{4}{3}x^3$	0.1205

Using these data, you can now easily evaluate Eq. 13.19 to yield

$$E_1^{(2)} = -\frac{8\pi\varepsilon_r\varepsilon_0}{Z^4}\mathcal{E}^2 a_B^3\,(0.7399 + 0.1001 + 0.0330 + 0.01512 + \cdots) \approx$$

(13.20)

$$-1.78\frac{4\pi\varepsilon_r\varepsilon_0}{Z^4}\mathcal{E}^2 a_B^3.$$

Adding more terms to the sum does not affect the numerical coefficient too much: going from four terms to 200 changes this factor by 2%. It is interesting to note that the ground state energy of hydrogen in the electric field can be found exactly, without resorting to the perturbation theory. This theory is too complicated to discuss in this book, but no one can forbid me to use its result for comparison. The exact solution produces factor 2.5 instead of 1.78, which is a 20% difference. I would say that this is a pretty decent approximation given the difference in the amount of efforts required to derive the approximate and exact results.

Equation 13.20 can also be recast in another illuminating form. By multiplying the numerator and denominator of this equation by e^2, you can notice that the resulting expression contains the ground state energy E_g in its denominator. Making this fact explicit, you can rewrite Eq. 13.20 in the form

$$E_1^{(2)} = -\frac{0.89}{Z^2}\frac{e^2 a_B^2 \mathcal{E}^2}{E_g}$$

(13.21)

where the numerator has a clear physical meaning: $ea_B\mathcal{E}$ is the change of the potential energy of the electron in the field $\mathcal{E}$ over the distance equal to the "size" of the atom expressed by the Bohr radius a_B. This expression also makes it much easier to get a feeling for the numerical magnitude of the Stark effect, since I know that the Bohr radius (assuming that we are dealing with an actual hydrogen atom in vacuum) $a_B = 5.29 \times 10^{-11}$ m, and using it as a typical value for the electric field $\mathcal{E} = 10^6$ V/m, I find for $ea_B\mathcal{E}$ in electron volts $ea_B\mathcal{E} \approx 5 \times 10^{-5}$ eV. Recalling that the ground state energy of hydrogen in vacuum is 13.6 eV, I can estimate the quadratic Stark effect correction to the energy as (ignoring numerical coefficients of the order of unity) being of the order of 10^{-10} eV. This change in energy levels is observed by measuring the electric field-induced shift of the absorption or emission lines in the hydrogen spectrum. The energy shift of the order of 10^{-10} eV translates into a frequency shift of about 10^5 Hz. This shift of spectral lines is what is known as the Stark effect, and because in the case considered in this section the shift is quadratic in the field, it is qualified as quadratic Stark effect. The effect was discovered in 1913 by German physicist Johannes Stark who was awarded for this discovery the 1919 Nobel Prize in Physics. Stark was an active supporter of the Nazi regime and was closely involved in *Deutsche Physik* movement, whose goal was to cleanse German science from foreign mostly Jewish influence. It was he who described Heisenberg as a White Jew after Heisenberg publicly defended Einstein's relativity theory. He was probably the only one famous physicist who after the war was sentenced to a prison term for collaboration with Hitler's regime.

13.1.2 Atom's Polarizability

As you just saw, the modification of the energy eigenvalues in the presence of the electric field manifests itself as a Stark effect—shift of the spectral lines in the absorption or emission spectra of an atom. In this section I will focus on an effect resulting from the modification of the wave functions.

The wave functions, $\psi(r)$, as you all know, characterize the probability density for the electron's coordinate $P(r) = |\psi(r)|^2$, so that the related quantity $\rho(r) = e|\psi(r)|^2$ can be interpreted as a charge density. The easiest way to understand why this is so is to imagine that you are dealing with an ensemble of N non-interacting electrons (please indulge me here, pretending that such electrons do exist). Then $NP(r)d^3r$ determines the number of the electrons within the volume element d^3r, and since each of them carries a charge e, the charge within this volume would be $eNP(r)d^3r$. Accordingly, the charge density (charge per unit volume) would be $NP(r)$. Now go back to the case of a single electron ($N = 1$), and you have your charge density $\rho(r)$ as defined just a few lines above. If you want, you can imagine an atomic electron as being a continuously distributed charged cloud rather than as a point-like object.

In the absence of the electric field, the wave functions of the electron have a definite parity as was discussed in the previous section. The charge density $\rho(r)$, in this case, is always an even function. One of the important consequences of this is that the average position vector of the electron $\int r|\psi(r)|^2 d^3r$ is zero. Quantity

$$\langle d \rangle = -e \int r|\psi(r)|^2 d^3r$$

can be interpreted as an expectation value of a dipole operator $\hat{d} = -e\hat{r}$, which is also, obviously, zero (negative sign in the definition of the dipole moment reflects the negativity of the electron's charge, e). I can say that the hydrogen atom does not have a permanent (independent of the field) dipole moment. Now, try to imagine what would happen if the atom is placed in the electric field. Classically speaking, the field will exert a force on the electron shifting it to a new equilibrium position. When describing this effect in terms of the electron cloud, you can imagine that the cloud is displaced so that its center does not coincide with the position of the nucleus (see Fig. 13.1). As a result the expectation value of the electron's dipole moment takes on a non-zero value, dependent on the field. This process is called polarization, and a dipole whose dipole moment appears only in the presence of an external field is called polarizable dipole. My task in this section is to describe this effect quantitatively using the first-order corrections to the electron's wave function given by Eq. 13.10.

Formally speaking, what I need to do is to compute the expectation value of the dipole moment operator using the eigenvector perturbed by the field. The computation is pretty straightforward, so let me just get on with it:

Fig. 13.1 Spatial distribution of the electron charge density in the presence of the electric field. The large ellipse represents the electron cloud, while the small black circle is the nucleus. The center of the cloud is shifted with respect to the nucleus because of the electric field represented by green field lines

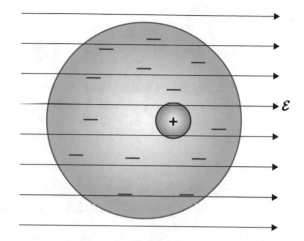

$$\langle s | \hat{\boldsymbol{d}} | s \rangle = \left[\langle s^{(0)} | + \epsilon \sum_{q \neq s} \frac{V_{qs}^*}{E_s^{(0)} - E_q^{(0)}} \langle q^{(0)} | \right] \hat{\boldsymbol{d}}$$

$$\left[|s^{(0)}\rangle + \epsilon \sum_{q \neq s} \frac{V_{qs}}{E_s^{(0)} - E_q^{(0)}} |q^{(0)}\rangle \right].$$

(I hope you have not yet forgotten that in transitioning from ket vectors to bra vectors, you are supposed to complex conjugate all complex quantities.) But, let me continue:

$$\langle s | \hat{\boldsymbol{d}} | s \rangle = \epsilon \sum_{n=2}^{\infty} \sum_{l=0}^{n-1} \sum_{m=-l}^{l} \frac{V_{nlm,100}}{E_1^{(0)} - E_n^{(0)}} \langle 100 | \hat{\boldsymbol{d}} | nlm \rangle +$$

$$\epsilon \sum_{n=2}^{\infty} \sum_{l=0}^{n-1} \sum_{m=-l}^{l} \frac{V_{nlm,100}^*}{E_1^{(0)} - E_n^{(0)}} \langle nlm | \hat{\boldsymbol{d}} | 100 \rangle =$$

$$2\epsilon \sum_{n=2}^{\infty} \sum_{l=0}^{n-1} \sum_{m=-l}^{l} \frac{Re \, [V_{nlm,100} \boldsymbol{d}_{100,nlm}]}{E_1^{(0)} - E_n^{(0)}}.$$

Here is what I have done here. First, I remembered that $\langle s^{(0)} | \hat{\boldsymbol{d}} | s^{(0)} \rangle = 0$ as we just discussed, then I dropped the terms proportional to ϵ^2 because I have no right to keep them. Finally, I used the defining property of the Hermitian operators $\langle q^{(0)} | \hat{\boldsymbol{d}} | s^{(0)} \rangle = \langle s^{(0)} | \hat{\boldsymbol{d}} | q^{(0)} \rangle^*$, which allowed me to replace

$$V_{qs} \langle s^{(0)} | \hat{\boldsymbol{d}} | q^{(0)} \rangle + V_{qs}^* \langle q^{(0)} | \hat{\boldsymbol{d}} | s^{(0)} \rangle$$

with

$$V_{qs}\left\langle s^{(0)}\left|\hat{d}\right|q^{(0)}\right\rangle + V_{qs}^{*}\left\langle s^{(0)}\left|\hat{d}\right|q^{(0)}\right\rangle^{*} = 2Re\left[V_{qs}\left\langle s^{(0)}\left|\hat{d}\right|q^{(0)}\right\rangle\right].$$

oh, yes, I also introduced a dipole matrix element $d_{sq} \equiv \left\langle s^{(0)}\left|\hat{d}\right|q^{(0)}\right\rangle$ and replaced the abridged indexes s and q with full notation $1, 0, 0$ and n, l, m reflecting the nature of the states in question. Now, let me recall that $\hat{d} = -e\hat{r}$, and, therefore, $d_{100,nlm} = -er_{100,nlm}$, while $V_{nlm,100s} = -e\mathcal{E}z_{nlm,100s}$, so that, for the dipole expectation value, I have

$$\langle s|\hat{d}|s\rangle = 2\epsilon e^{2}\mathcal{E}\sum_{n=2}^{\infty}\sum_{l=0}^{n-1}\sum_{m=-l}^{l}\frac{Re\left[z_{nlm,100}r_{100,nlm}\right]}{E_{1}^{(0)} - E_{n}^{(0)}}. \tag{13.22}$$

The matrix element $z_{nlm,100}$ has been calculated in the previous section, so let's focus on $r_{100,nlm}$. First of all, you need to recognize that this is a vector, and, therefore, you are dealing here with three matrix elements: $x_{100,nlm}$, $y_{100,nlm}$, and $z_{100,nlm}$, where x, y, and z are coordinates defined in the same coordinate system that I used to describe the Stark effect (Z-axis is along the electric field). The z-component of this matrix element, $z_{100,nlm}$, is obviously $z_{100,nlm} = z_{nlm,100}^{*}$, and since the latter is real, they both are given by Eq. 13.17. The arguments that led me to conclude that $z_{nlm,100}$ is zero unless $m = 0$ do not apply to $x_{100,nlm}$ and $y_{100,nlm}$ because both x and y coordinates, when expressed in a spherical coordinate system, contain the azimuthal angle φ: $x = r\sin\theta\cos\varphi$, $y = r\sin\theta\sin\varphi$. Instead of dealing with $x_{100,nlm}$ and $y_{100,nlm}$ separately, I find it more convenient to deal with $x + iy$, which in spherical coordinates becomes simply $r\sin\theta\exp(i\varphi)$, and the respective matrix element can now be presented as

$$x_{100,nlm} + iy_{100,nlm} =$$

$$\frac{1}{\sqrt{4\pi}}\int_{0}^{\pi}\int_{0}^{2\pi}\int_{0}^{\infty}d\theta d\varphi dr \sin\theta r^{3}\sin\theta e^{i\varphi}Y_{l}^{m}(\theta,\varphi)R_{n,l}(r)R_{1,0}(r). \tag{13.23}$$

As it has become clear from the discussion of $z_{100,nlm}$, for the integral over φ not to vanish, factor $\exp(i\varphi)$ in Eq. 13.23 must be canceled by an appropriate factor from spherical harmonics $Y_{l}^{m}(\theta,\varphi)$. Such a cancelation is only possible if index m in the spherical harmonic is $m = -1$, indicating that the entire matrix element is different from zero only for $m = -1$. This means that our job here is done: we can forget about both $x_{100,nlm}$ and $y_{100,nlm}$ components of the matrix element $\langle s|\hat{d}|s\rangle$. Indeed, Eq. 13.22 contains the product of $z_{100,nlm}$ with $x_{100,nlm}$ or $y_{100,nlm}$, or $z_{100,nlm}$, which does not vanish only if both factors in the product are not zero. We know that $z_{100,nlm}$ is zero for all $m \neq 0$, while $x_{100,nlm}$ and $y_{100,nlm}$ are not zero only if $m = -1$. No amount of Hogwarts magic can make these two conditions be fulfilled simultaneously, assuring that d_{z} is the only non-zero component of the dipole expectation value

$$\langle s| \hat{d}_z |s\rangle = 2\epsilon e^2 \mathcal{E} \sum_{n=2}^{\infty} \sum_{l=0}^{n-1} \sum_{m=-l}^{l} \frac{Re\,[z_{nlm,100}z_{100,nlm}]}{E_1^{(0)} - E_{nlm}^{(0)}} =$$

$$2\epsilon e^2 \mathcal{E} \sum_{n=2}^{\infty} \sum_{l=0}^{n-1} \sum_{m=-l}^{l} \frac{|z_{nlm,100}|^2}{E_1^{(0)} - E_n^{(0)}} \tag{13.24}$$

where, in the last line, I dropped the sign of taking the real part because $z_{nlm,100}z_{100,nlm} \equiv |z_{nlm,100}|^2$ is a real quantity. The fact that only z-component of the dipole moment survives the process of taking the expectation value shouldn't surprise anyone, of course. In the absence of "special" directions, the direction of the external field is the only one which the induced dipole moment of the atom can reasonably be expected to have.

Equation 13.24 demonstrates that the induced dipole moment is codirected with the external field and is proportional to it. This linear dependence between the dipole moment and the field allows to introduce an important quantity, called polarizability, which is defined as a coefficient of proportionality between the dipole moment and the field:

$$\alpha_p = 2e^2 \sum_{n=2}^{\infty} \sum_{l=0}^{n-1} \sum_{m=-l}^{l} \frac{|z_{nlm,100}|^2}{E_1^{(0)} - E_n^{(0)}}. \tag{13.25}$$

The expectation value of the dipole operator, $\langle d \rangle$, can now be written simply as

$$\langle d \rangle = \alpha_p \mathcal{E}. \tag{13.26}$$

The significance of Eq. 13.26 goes beyond this particular example: the linear relation between the induced dipole moment and the external field remains valid in a wide variety of cases, well outside of the particular model considered here.

Now, prepare to get surprised. Go a few pages back, find the expression for the second-order correction to the electron's energy, and compare it with Eq. 13.24. To make your life even easier, I will copy this expression here for your convenience:

$$E_1^{(2)} = e^2 \mathcal{E}^2 \sum_{n=2}^{\infty} \sum_{l=0}^{n-1} \sum_{m=-l}^{l} \frac{|z_{100,nlm}|^2}{E_1^{(0)} - E_n^{(0)}}.$$

Do you see that the expression for the correction to the energy level in the presence of the field, which can be interpreted as a change of electron's energy due to the field, can be written down as

$$E_1^{(2)} = \frac{1}{2}\alpha_p \mathcal{E}^2?. \tag{13.27}$$

What is remarkable about this expression is that it represents the energy of a classical polarizable dipole. Here is how it can be derived using nothing but classical electrostatics. You probably remember that the energy of a dipole d in a uniform electric field E is $U_d = d \cdot \mathcal{E}$. This formula, however, was derived for a permanent dipole, namely, the dipole whose dipole moment does not depend on the external field. In the case under consideration, d is dependent of the field, and this expression has to be modified. To take into account that the dipole moment changes with the field, I will build the dipole moment by small increments $\delta d = \alpha_p \delta E$. If the respective change of the field $\delta \mathcal{E}$ from a given value $\mathcal{E}$ is infinitesimally small, one can neglect the corresponding change of the dipole moment and use for the corresponding increment in the energy the formula

$$\delta U_d = \delta d \cdot \mathcal{E} = \alpha_p \mathcal{E} \delta \mathcal{E} = \alpha_p \mathcal{E} \delta \mathcal{E},$$

where in the last step I assumed that $\mathcal{E}$ and δE have the same direction, so that I can replace both vectors with their magnitudes. To find the total energy of the dipole induced by field $\mathcal{E}$, I now need to add together all these small increments starting from the ones corresponding to zero field and finishing when the desirable value of the field is reached. I probably did not have to use that many words to describe a standard operation of integration and could write right away

$$U = \alpha_p \int_0^{\mathcal{E}} \mathcal{E} \delta \mathcal{E} = \frac{1}{2} \alpha_p \mathcal{E}^2,$$

which is exactly Eq. 13.27. Probably, this is just my personal quirk, but at some point in my life, I wondered how come that the quantum expression for the energy of the dipole $\hat{d} \cdot \mathcal{E}$ which goes into the Hamiltonian and is valid even for induced dipoles does not have the factor $1/2$, while the classical version of the same expression for polarizable dipoles is $d \cdot \mathcal{E}/2$. How can this be reconciled with the correspondence principle, for instance? After learning how to compute the polarizability of atoms, I understood that the classical expression must be compared with the expectation value of the dipole moment, while the quantum formula is an operator expression. Once you go from the operator to the expectation value, the classical and quantum expression for the energy of the polarizable dipole start agreeing.

13.2   Degenerate Perturbation Theory and Applications

13.2.1   General Formulation

If we are interested in finding the corrections to a degenerate energy eigenvalue, the approach presented in Sect. 13.1 must be modified. In the introduction to this chapter, I have already outlined a conceptual need for a modification. Namely, I pointed out that in the degenerate case, the starting point of the perturbation expansion (the zero-order approximation) is not clearly defined. There is also another, more technical argument justifying the demand for a modified approach. If you look at Eq. 13.10 for the first-order corrections to the eigenvector or Eq. 13.12 for the second-order corrections to the eigenvalue, you will notice the energy denominator $E_s^{(0)} - E_q^{(0)}$ and the restriction $q \neq s$ on the summation. This restriction assures that the energy denominator does not vanish resulting in a blowout of the respective term. While this restriction works very well in the non-degenerate case, if $E_s^{(0)}$ is degenerate, the condition that $q \neq s$ does not guarantee that $E_s^{(0)} \neq E_q^{(0)}$ because there can be eigenvectors differing in some of their quantum numbers but having the same energy. The new perturbation theory, in addition to resolving the ambiguity about the zero-order term, must also ensure the absence of the exploding terms in the perturbation expansion.

I will begin by refining the notation for the eigenvectors of the unperturbed Hamiltonian belonging to degenerate eigenvalues. To this end I shall add an extra index reflecting degeneracy turning $|s^{(0)}\rangle$ into $|s^{(0)}, \lambda\rangle$. Index s, as before, will enumerate different eigenvalues, while the second index λ, which can combine several indexes, will distinguish between eigenvectors belonging to the same eigenvalue. In the case of the hydrogen atom, for instance, q would be the principal quantum number n, while λ would correspond to two indexes l and m.

Since I am not sure which of the eigenvectors belonging to a degenerate eigenvalues shall be used as a zero-order term in the perturbation expansion, I will begin with an arbitrary linear combination of those hoping that somehow I will be able to figure out what the "correct" combination is. Accordingly, in the derivation outlined in Eqs. 13.2–13.8, I will replace $|s^{(0)}\rangle$ by

$$|\chi_s\rangle = \sum_\lambda a_\lambda |s^{(0)}, \lambda\rangle, \qquad (13.28)$$

so that Eqs. 13.6 and 13.7 become

$$\hat{H}_0 \sum_\lambda a_\lambda |s^{(0)}, \lambda\rangle = E_s^{(0)} \sum_\lambda a_\lambda |s^{(0)}, \lambda\rangle \qquad (13.29)$$

$$\hat{H}_0 |s^{(1)}\rangle + \hat{V} \sum_\lambda a_\lambda |s^{(0)}, \lambda\rangle = E_s^{(0)} |s^{(1)}\rangle + E_s^{(1)} \sum_\lambda a_\lambda |s^{(0)}, \lambda\rangle. \qquad (13.30)$$

Equation 13.29 is fulfilled automatically by virtue of $|s^{(0)}, \lambda\rangle$ being eigenvectors of the Hamiltonian $\hat{H}^{(0)}$, so I only need to deal with Eq. 13.30. I shall follow the same procedure as in the non-degenerate case and first premultiply this equation by $\sum_\mu a_\mu^* \langle s^{(0)}, \mu|$:

$$\sum_{\mu} a_{\mu}^{*} \left\langle s^{(0)}, \mu \right| \hat{H}_0 \left| s^{(1)} \right\rangle + \sum_{\lambda} \sum_{\mu} a_{\lambda} a_{\mu}^{*} \left\langle s^{(0)}, \mu \right| \hat{V} \left| s^{(0)}, \lambda \right\rangle = \tag{13.31}$$

$$E_s^{(0)} \sum_{\mu} a_{\mu}^{*} \left\langle s^{(0)}, \mu \right| \left| s^{(1)} \right\rangle + E_s^{(1)} \sum_{\lambda} \sum_{\mu} a_{\lambda} a_{\mu}^{*} \left\langle s^{(0)}, \mu \right| \left| s^{(0)}, \lambda \right\rangle.$$

Recalling that $\left\langle s^{(0)}, \mu \right| \hat{H}^{(0)} = E_s^{(0)} \left\langle s^{(0)}, \mu \right|$, you can again, just like in the non-degenerate case, cancel the first term on the left-hand side of Eq. 13.31 with the first term on the right-hand side. Taking into account also that the eigenvectors, even those belonging to a degenerate eigenvalue, are orthogonal, you can rewrite Eq. 13.31 as

$$\sum_{\lambda} \sum_{\mu} a_{\lambda} a_{\mu}^{*} \left\langle s^{(0)}, \mu \right| \hat{V} \left| s^{(0)}, \lambda \right\rangle = E_s^{(1)} \sum_{\mu} a_{\mu} a_{\mu}^{*} \Rightarrow$$

$$\sum_{\mu} a_{\mu}^{*} \left(\sum_{\lambda} V_{\mu,\lambda}^{(s)} a_{\lambda} - E_s^{(1)} a_{\mu} \right) = 0 \Longrightarrow$$

$$\sum_{\lambda} V_{\mu,\lambda}^{(s)} a_{\lambda} = E_s^{(1)} a_{\mu} \tag{13.32}$$

where I introduced a perturbation matrix $V_{\lambda,\mu}^{(s)} = \left\langle s^{(0)}, \mu \right| \hat{V} \left| s^{(0)}, \lambda \right\rangle$ constructed using vectors $\left| s^{(0)}, \lambda \right\rangle$ **belonging to the same degenerate subspace defined by the eigenvalue** $E_s^{(0)}$. Equation 13.32 replaces Eq. 13.9 for degenerate eigenvalues. After prolonged staring and maybe some meditation, you will recognize that this is your old acquaintance: an eigenvalue equation, written this time for matrix $V_{\lambda,\mu}^{(s)}$. The eigenvalues are found, as usual, as zeroes of the determinant

$$\left\| V_{\lambda,\mu}^{(s)} - E_s^{(1)} \delta_{\lambda,\mu} \right\| = 0, \tag{13.33}$$

and substituting each of the solution to this equation (sometimes called *secular equation*) back to Eq. 13.32, you will find sets of corresponding coefficients a_{λ} engendering the "correct" zero-order eigenvectors.

To help you see a more clear picture of what you are dealing with here, I will now consider, as an example, a simplest possible case of a double-degenerate eigenvalue. In this case the indexes λ, μ take only two values so that $V_{\lambda,\mu}^{(s)}$ becomes a 2×2 matrix. The secular Eq. 13.33, accordingly, takes the following form:

$$\left\| \begin{matrix} V_{1,1}^{(s)} - E_s^{(1)} & V_{1,2}^{(s)} \\ V_{1,2}^{(s)*} & V_{2,2}^{(s)} - E_s^{(1)} \end{matrix} \right\| = \left(V_{1,1}^{(s)} - E_s^{(1)} \right) \left(V_{2,2}^{(s)} - E_s^{(1)} \right) - \left| V_{1,2}^{(s)} \right|^2 = 0.$$

This quadratic equation has, obviously, two easily found distinct solutions:

$$E_{s_{1,2}}^{(1)} = \frac{1}{2}\left(V_{1,1}^{(s)} + V_{2,2}^{(s)}\right) \pm \frac{1}{2}\sqrt{\left(V_{1,1}^{(s)} + V_{2,2}^{(s)}\right)^2 + 4\left|V_{1,2}^{(s)}\right|^2}. \tag{13.34}$$

If the diagonal elements of the perturbation matrix vanish, as you already know might happen quite often, Eq. 13.34 simplifies:

$$E_{s_{1,2}}^{(1)} = \pm\left|V_{1,2}^{(s)}\right|. \tag{13.35}$$

The main result revealed by Eqs. 13.34 and 13.35 is that a generic perturbation lifts the degeneracy of the initial eigenvalue $E_s^{(0)}$ and splits it into two new distinct eigenvalues

$$E_{s_{1,2}} = E_s^{(0)} + \left[\frac{1}{2}\left(V_{1,1}^{(s)} + V_{2,2}^{(s)}\right) \pm \frac{1}{2}\sqrt{\left(V_{1,1}^{(s)} + V_{2,2}^{(s)}\right)^2 + 4\left|V_{1,2}^{(s)}\right|^2}\right],$$

where I set $\epsilon = 1$. Even though Eq. 13.33 has been formally derived in the first order of the perturbation theory, do not let this circumstance deceive you: its dependence on the perturbation matrix elements is not linear. Actually, it is not even analytic as the square root in it cannot be expanded into any kind of a power series around the point where the perturbation turns zero. The dependence of eigenvalues on the perturbation is not linear even in the case of simplified Eq. 13.35, regardless of how "linear" this expression might look like: an absolute value of a complex number involves the square root as well.

Having found the corrections to energy, I can go back to Eq. 13.32 and find the corresponding coefficients a_λ. In the case of double-degenerate eigenvalues, I shall be looking for two sets of the coefficients—one for each of the corrections to the energy. To distinguish between them, I will add the upper indexes 1 and 2 to a_λ, so that the notation becomes $a_\lambda^{(1,2)}$. To make algebra less cumbersome so that you could focus on what is really important, I will compute $a_\lambda^{(1,2)}$ only for the case $V_{1,1}^{(s)} = V_{2,2}^{(s)} = 0$. Equation 13.32 for a double-degenerate eigenvalue with zero diagonal elements of the perturbation matrix takes the form

$$V_{1,2}^{(s)}a_2 = E_s^{(1)}a_1$$

$$V_{2,1}^{(s)}a_1 = E_s^{(1)}a_2.$$

Substituting $E_{s_1}^{(1)} = \left|V_{1,2}^{(s)}\right|$ into the above equations, I obtain

$$V_{1,2}^{(s)}a_2^{(1)} = \left|V_{1,2}^{(s)}\right|a_1^{(1)}$$

$$V_{2,1}^{(s)}a_1^{(1)} = \left|V_{1,2}^{(s)}\right|a_2^{(1)}.$$

Presenting the perturbation matrix element as $V_{1,2}^{(s)} = \left| V_{1,2}^{(s)} \right| \exp(i\phi_V)$ and recalling that $V_{2,1}^{(s)} = V_{1,2}^{(s)*}$, you can see that both equations above can be reduced to the same form

$$\frac{a_1^{(1)}}{a_2^{(1)}} = e^{i\phi_V}. \tag{13.36}$$

The second solution of the secular equation $E_{s2}^{(1)} = -\left| V_{1,2}^{(s)} \right|$ yields, instead of Eq. 13.36,

$$\frac{a_1^{(2)}}{a_2^{(2)}} = -e^{i\phi_V}. \tag{13.37}$$

To simplify the situation even further, let me assume that the matrix elements $V_{1,2}^{(s)}$ are real and positive ($\phi_V = 0$), in which case the two sets of coefficients become simply $a_1^{(1)} = a_2^{(1)}$ and $a_1^{(2)} = -a_2^{(2)}$. As it is typical for the eigenvalue problems, one of the coefficients remains undefined and is found, again, using the normalization condition $\left| a_1^{(1,2)} \right|^2 + \left| a_2^{(1,2)} \right|^2 = 1$. This yields

$$a_1^{(1)} = a_2^{(1)} = 1/\sqrt{2}$$

and

$$a_1^{(2)} = -a_2^{(2)} = 1/\sqrt{2}.$$

Now Eq. 13.28 yields for me two linearly independent orthogonal and normalized "correct" zero-order eigenvectors:

$$\left| \chi_s^{(1)} \right\rangle = \frac{1}{\sqrt{2}} \left(|s, 1\rangle^{(0)} + |s, 2\rangle^{(0)} \right) \tag{13.38}$$

$$\left| \chi_s^{(2)} \right\rangle = \frac{1}{\sqrt{2}} \left(|s, 1\rangle^{(0)} - |s, 2\rangle^{(0)} \right). \tag{13.39}$$

It might appear that no traces of the perturbation matrix are left in Eqs. 13.38 and 13.39, but this appearance is deceptive. While it is true that the magnitude of the perturbation matrix elements has vanished, you must remember that I made a very specific assumption about their phases. A different choice of the phase would have resulted in different eigenvector combinations as you will see yourself when working through the problems section of this chapter. Now the meaning of the statement that the perturbation affects the zero-order eigenvectors even though its magnitude goes to zero becomes clear: the transition to the zero perturbation limit occurs in such a way that keeps the phase of the perturbation matrix elements intact.

Finally, I would like to make an almost trivial but nevertheless tremendously important point: the off-diagonal matrix elements of the perturbation operator $\hat{V}$ built with "correct" zero-order eigenvectors, Eqs. 13.38 and 13.39, vanish. Indeed, by direct computation you may see that

$$\langle \chi_s^{(1)} | \hat{V} | \chi_s^{(2)} \rangle = \frac{1}{2} \left({}^{(0)} \langle s,1| + {}^{(0)} \langle s,2| \right) \hat{V} \left(|s,1\rangle^{(0)} - |s,2\rangle^{(0)} \right) =$$

$$\frac{1}{2} \left(V_{2,1}^{(s)} - V_{1,2}^{(s)} \right) = 0$$

where I took into account assumptions that went into Eqs. 13.38 and 13.39 ($V_{\lambda,\lambda}^{(s)} = 0$, $V_{1,2}^{(s)} = V_{2,1}^{(s)}$). Of course, there is no surprise here—after all I defined $\left| \chi_s^{(1,2)} \right\rangle$ as eigenvectors of $\hat{V}$, and the matrix of any operator in the basis of its eigenvector is diagonal. Still, it is helpful (for intuition development) to see the concrete realization of this abstract statement in this particular context. In a more general situation, I can state that vectors

$$\left| \chi_s^{(\lambda)} \right\rangle = \sum_{\lambda'} a_{\lambda'}^{(\lambda)} \left| s^{(0)}, \lambda' \right\rangle, \tag{13.40}$$

where coefficients $a_{\lambda'}^{(\lambda)}$, satisfying Eq. 13.32 for a given eigenvalue λ, diagonalize the perturbation operator. This means that

$$\left\langle \chi_s^{(\lambda)} \right| \hat{V} \left| \chi_s^{(\mu)} \right\rangle = E_s^{(1)} \delta_{\lambda,\mu}. \tag{13.41}$$

There is also another, rather abstract, but nonetheless important side to this story. Eigenvectors belonging to a degenerate eigenvalue form what mathematicians would call a subspace in a larger space spanned by all eigenvectors of the Hamiltonian. It means that any linear combination of the degenerate eigenvectors is again an eigenvector belonging to the same eigenvalue (see also Sect. 3.2.3). The degree of degeneracy determines the dimension of this subspace. For instance, in the example of a double-degenerate eigenvalue, we ended up with a two-dimensional subspace of the eigenvectors defined by the basis of two linearly independent orthogonal vectors $\left| \chi_s^{(1)} \right\rangle$ and $\left| \chi_s^{(2)} \right\rangle$. In the matrix representation of the total Hamiltonian $\hat{H}$, one can separate out a submatrix of smaller dimension, defined on a subspace of eigenvectors belonging to the same eigenvalue. If one chooses as a basis in this subspace vectors defined by Eq. 13.40, then the resulting submatrix $H_{\lambda,\mu}^{(s)} = \left\langle \chi_s^{(\lambda)} \right| \hat{H} \left| \chi_s^{(\mu)} \right\rangle$ will have only diagonal elements:

$$H_{\lambda,\mu}^{(s)} = E_s^{(0)} \delta_{\lambda,\mu} + E_s^{(1)} \delta_{\lambda,\mu}. \tag{13.42}$$

This conclusion is directly based on Eq. 13.41 and on the understanding that any mutually orthogonal combinations of the degenerate eigenvectors $|s, \lambda\rangle^{(0)}$ are eigenvectors of $\hat{H}^{(0)}$ and, therefore, turn it into a diagonal matrix. This result can be interpreted by saying that vectors defined by Eq. 13.40 are eigenvectors not only of the perturbation operator but of the entire Hamiltonian, if it is considered only in the subspace formed by degenerate eigenvectors. Another way to express the same idea is to say that finding the first-order corrections to degenerate energy eigenvalues is equivalent to *exact* diagonalization of the entire Hamiltonian performed on the subspace of the degenerate eigenvectors. I will illustrate these points with a simple example.

Example 30 (Diagonalization of a Hamiltonian on a Degenerate Subspace) Consider a Hamiltonian defined by matrix

$$\hat{H} = \begin{bmatrix} 0 & \epsilon & 2i\epsilon \\ \epsilon & 1 & \epsilon \\ -2i\epsilon & \epsilon & 1 \end{bmatrix}.$$

Presenting it in the form $\hat{H}_0 + \hat{V}$, where $\hat{H}_0$ is the diagonal matrix, find the "correct" zero-order eigenvectors belonging to a double-degenerate eigenvalue 1 and corresponding first-order eigenvalues.

Solution

The unperturbed Hamiltonian $\hat{H}_0$ in this example is

$$\hat{H}_0 = \begin{bmatrix} 0 & 0 & 0 \\ 0 & 1 & 0 \\ 0 & 0 & 1 \end{bmatrix},$$

while the perturbation matrix can be written down as

$$\hat{V} = \epsilon \begin{bmatrix} 0 & 1 & 2i \\ 1 & 0 & 1 \\ -2i & 1 & 0 \end{bmatrix}.$$

It is obvious that the unperturbed Hamiltonian has two eigenvalues: 0 and 1, with the latter being double degenerate. Corresponding eigenvectors are

$$|1, 1\rangle = \begin{bmatrix} 0 \\ 1 \\ 0 \end{bmatrix}, |1, 2\rangle = \begin{bmatrix} 0 \\ 0 \\ 1 \end{bmatrix}.$$

Now, using these two vectors, I need to build matrix $V_{\lambda,\mu}^{(1)}$, where the upper index points to the eigenvalue that I am dealing with.

$$V_{1,1}^{(1)} = \langle 1, 1| \hat{V} |1, 1\rangle =$$

$$\epsilon \begin{bmatrix} 0 & 1 & 0 \end{bmatrix} \begin{bmatrix} 0 & 1 & 2i \\ 1 & 0 & 1 \\ -2i & 1 & 0 \end{bmatrix} \begin{bmatrix} 0 \\ 1 \\ 0 \end{bmatrix} = \begin{bmatrix} 0 & 1 & 0 \end{bmatrix} \begin{bmatrix} 1 \\ 0 \\ 1 \end{bmatrix} = 0,$$

$$V_{1,2}^{(1)} = \langle 1, 1| \hat{V} |1, 2\rangle =$$

$$\epsilon \begin{bmatrix} 0 & 1 & 0 \end{bmatrix} \begin{bmatrix} 0 & 1 & 2i \\ 1 & 0 & 1 \\ -2i & 1 & 0 \end{bmatrix} \begin{bmatrix} 0 \\ 0 \\ 1 \end{bmatrix} = \epsilon \begin{bmatrix} 0 & 1 & 0 \end{bmatrix} \begin{bmatrix} 2i \\ 1 \\ 0 \end{bmatrix} = \epsilon,$$

$$V_{2,2}^{(1)} = \langle 1, 2| \hat{V} |1, 2\rangle =$$

$$\epsilon \begin{bmatrix} 0 & 0 & 1 \end{bmatrix} \begin{bmatrix} 0 & 1 & 2i \\ 1 & 0 & 1 \\ -2i & 1 & 0 \end{bmatrix} \begin{bmatrix} 0 \\ 0 \\ 1 \end{bmatrix} = \epsilon \begin{bmatrix} 0 & 0 & 1 \end{bmatrix} \begin{bmatrix} 2i \\ 1 \\ 0 \end{bmatrix} = 0.$$

Thus, the perturbation matrix in the subspace belonging to the degenerate eigenvalue takes the following form:

$$V_{\lambda,\mu}^{(0)} = \epsilon \begin{bmatrix} 0 & 1 \\ 1 & 0 \end{bmatrix}.$$

Eigenvalues $v_{1,2}$ of this matrix are found from equation

$$v^2 - \epsilon^2 = 0,$$

which yields $v_{1,2} = \pm\epsilon$. The respective eigenvectors are found from equations

$$\epsilon \begin{bmatrix} 0 & 1 \\ 1 & 0 \end{bmatrix} \begin{bmatrix} a_1 \\ a_2 \end{bmatrix} = \pm\epsilon \begin{bmatrix} a_1 \\ a_2 \end{bmatrix} \Rightarrow$$

$$a_2 = \pm a_1$$

so that in the normalized form they can be written down as

$$\left| \eta_0^{(1)} \right\rangle = \frac{1}{\sqrt{2}} \begin{bmatrix} 1 \\ 1 \end{bmatrix}; \ \left| \eta_0^{(2)} \right\rangle = \frac{1}{\sqrt{2}} \begin{bmatrix} 1 \\ -1 \end{bmatrix}.$$

I will complete this exercise by generating a 2×2 matrix using eigenvectors belonging to the degenerate eigenvalue out of the entire Hamiltonian, just to make the connection with abstract arguments in the preceding paragraph. Here is how it goes:

$$H_{1,1}^{(1)} = \langle 1,1| \hat{H} |1,1 \rangle =$$

$$\begin{bmatrix} 0 & 1 & 0 \end{bmatrix} \begin{bmatrix} 0 & \epsilon & 2i\epsilon \\ \epsilon & 1 & \epsilon \\ -2i\epsilon & \epsilon & 1 \end{bmatrix} \begin{bmatrix} 0 \\ 1 \\ 0 \end{bmatrix} = \begin{bmatrix} 0 & 1 & 0 \end{bmatrix} \begin{bmatrix} \epsilon \\ 1 \\ \epsilon \end{bmatrix} = 1,$$

$$H_{1,2}^{(1)} = \langle 1,1| \hat{H} |1,2 \rangle =$$

$$\begin{bmatrix} 0 & 1 & 0 \end{bmatrix} \begin{bmatrix} 0 & \epsilon & 2i\epsilon \\ \epsilon & 1 & \epsilon \\ -2i\epsilon & \epsilon & 1 \end{bmatrix} \begin{bmatrix} 0 \\ 0 \\ 1 \end{bmatrix} = \begin{bmatrix} 0 & 1 & 0 \end{bmatrix} \begin{bmatrix} 2i\epsilon \\ \epsilon \\ 1 \end{bmatrix} = \epsilon,$$

$$H_{2,2}^{(1)} = \langle 1,2| \hat{H} |1,2 \rangle =$$

$$\begin{bmatrix} 0 & 0 & 1 \end{bmatrix} \begin{bmatrix} 0 & \epsilon & 2i\epsilon \\ \epsilon & 1 & \epsilon \\ -2i\epsilon & \epsilon & 1 \end{bmatrix} \begin{bmatrix} 0 \\ 0 \\ 1 \end{bmatrix} = \begin{bmatrix} 0 & 0 & 1 \end{bmatrix} \begin{bmatrix} 2i\epsilon \\ \epsilon \\ 1 \end{bmatrix} = 1.$$

Putting all these elements together, I end up with a matrix

$$H^{(1)} = \begin{bmatrix} 1 & \epsilon \\ \epsilon & 1 \end{bmatrix},$$

which as you can see is, indeed, a submatrix of the total Hamiltonian formed by cutting out the block formed by elements in the second and third rows and columns. If I rewrite this matrix in a new basis formed by the "correct" eigenvectors (see Eq. 5.97), I will have

$$H^{(1)} = \frac{1}{2} \begin{bmatrix} 1 & 1 \\ 1 & -1 \end{bmatrix} \begin{bmatrix} 1 & \epsilon \\ \epsilon & 1 \end{bmatrix} \begin{bmatrix} 1 & 1 \\ 1 & -1 \end{bmatrix} =$$

$$\frac{1}{2} \begin{bmatrix} 1 & 1 \\ 1 & -1 \end{bmatrix} \begin{bmatrix} 1+\epsilon & 1-\epsilon \\ 1+\epsilon & -1+\epsilon \end{bmatrix} =$$

$$\begin{bmatrix} 1+\epsilon & 0 \\ 0 & 1-\epsilon \end{bmatrix} = \begin{bmatrix} 1 & 0 \\ 0 & 1 \end{bmatrix} + \begin{bmatrix} \epsilon & 0 \\ 0 & -\epsilon \end{bmatrix}$$

in agreement with Eq. 13.42. Here I used the rule for transformation of an operator from one basis to another as discussed in Sect. 5.2.2.

I will complete the discussion of the secular equation by noting that solving Eq. 13.33 amounts to finding the roots of a polynomial of N-th order, where N is the degree of degeneracy of the corresponding eigenvalue. The fundamental theorem of algebra states that the polynomial of N-th order has N solutions, but does not exclude a possibility that some solutions can occur several times. The fact that Eq. 13.33 is an eigenvalue equation of a Hermitian operator ensures that all roots of the corresponding polynomial are real. If all these solutions differ from each other, we say that the perturbation completely lifts the degeneracy, but there might be situations in which the removal of the degeneracy is only partial, and some of the new eigenvalues still remain degenerate.

Now I am ready to proceed and start looking for corrections to the degenerate eigenvectors, for which I have to go back to Eq. 13.30 and rewrite it in terms of one of the newly found zero-order eigenvectors $\left|\chi_s^{(\lambda)}\right\rangle$ corresponding to a particular eigenvalue λ of the perturbation matrix:

$$\hat{H}_0 \left|s^{(1)}, \lambda\right\rangle + \hat{V}\left|\chi_s^{(\lambda)}\right\rangle = E_s^{(0)}\left|s^{(1)}, \lambda\right\rangle + E_s^{(1)}\left|\chi_s^{(\lambda)}\right\rangle.$$

If I now premultiply this equation by $\left\langle\chi_s^{(\lambda)}\right|$, I will get, taking into account Eq. 13.41, $E_s^{(1)} = V_{\lambda\lambda}^{(s)}$ as expected, and if I premultiply it by $\left\langle\chi_s^{(\mu)}\right|$, where $\mu \neq \lambda$, I will get the identity

$$\left\langle\chi_s^{(\mu)}\middle| s^{(1)}, \lambda\right\rangle E_s^{(0)} = \left\langle\chi_s^{(\mu)}\middle| s^{(1)}, \lambda\right\rangle E_s^{(0)}$$

(remember that eigenvalues of $\hat{H}_0$ do not depend on μ, and $V_{\mu,\lambda} = 0$). Now, finally, let me premultiply it by $\left\langle\chi_q^{(\mu)}\right|$, which are eigenvectors of the perturbation operator $\hat{V}$ in the subspace defined by a different (possibly degenerate) eigenvalue $q \neq s$:

$$\left\langle\chi_q^{(\mu)}\middle| s^{(1)}, \lambda\right\rangle E_q^{(0)} + V_{q\mu,s0\lambda} = \left\langle q^{(0)}\middle| s^{(1)}, \lambda\right\rangle E_s^{(0)} \Rightarrow$$

$$\left\langle\chi_q^{(\mu)}\middle| s^{(1)}, \lambda\right\rangle = \frac{V_{q\mu,s\lambda}}{E_s^{(0)} - E_q^{(0)}},$$

where $V_{q\mu,s\lambda} = \left\langle\chi_q^{(\mu)}\middle|\hat{V}\middle|\chi_s^{(\lambda)}\right\rangle$ is the perturbation matrix element computed with the zero-order eigenvectors obeying Eq. 13.40. Expanding $\left|s^{(1)}, \lambda\right\rangle$ in the basis of the "correct" zero-order eigenvectors, I will find

$$\left|s^{(1)}, \lambda\right\rangle = \sum_{\mu} \sum_{q} \left|\chi_q^{\mu}\right\rangle \left\langle \chi_q^{(\mu)}\right| s^{(1)}, \lambda\rangle =$$

$$\sum_{\mu} \sum_{q \neq s} \left|\chi_q^{\mu}\right\rangle \left\langle \chi_q^{(\mu)}\right| s^{(1)}, \lambda\rangle + \sum_{\mu} \left|\chi_s^{\mu}\right\rangle \left\langle \chi_s^{(\mu)}\right| s^{(1)}, \lambda\rangle =$$

$$\sum_{\mu} \sum_{q \neq s} \frac{V_{q\mu, s\lambda}}{E_s^{(0)} - E_q^{(0)}} \left|\chi_q^{\mu}\right\rangle + \sum_{\mu} a_{\mu, \lambda}^{(s)} \left|\chi_s^{\mu}\right\rangle.$$

Parameters $a_{\mu, \lambda}^{(s)} = \left\langle \chi_s^{(\mu)}\right| s^{(1)}, \lambda\rangle$ in this expression are not yet defined (the fact routinely swept under the rug in way too many textbooks and online notes on this topic). A similar situation also occurred in the non-degenerate case, but then we had to deal with only one undefined parameter, and now their number is equal to the degree of degeneracy. The total eigenvector with accuracy up to the linear order in ϵ is

$$|s, \lambda\rangle = \left|\chi_s^{\lambda}\right\rangle \left(1 + a_{\lambda, \lambda}^{(s)}\right) + \sum_{\mu \neq \lambda} a_{\mu, \lambda}^{(s)} \left|\chi_s^{\mu}\right\rangle + \sum_{\mu} \sum_{q \neq s} \frac{V_{q\mu, s\lambda}}{E_s^{(0)} - E_q^{(0)}} \left|\chi_q^{\mu}\right\rangle. \tag{13.43}$$

Normalization condition for this vector yields, just like in the non-degenerate case:

$$\langle s, \lambda| s, \lambda\rangle = \left|1 + a_{\lambda, \lambda}^{(s)}\right| = 1$$

(all coefficients $a_{\mu, \lambda}^{(s)}$ vanish because vectors $\left|\chi_s^{\mu}\right\rangle$ and $\left|\chi_s^{\lambda}\right\rangle$ with $\mu \neq \lambda$ are orthogonal), from which you can safely conclude that $a_{\lambda, \lambda}^{(s)} = 0$. But what about all other still undefined coefficients $a_{\mu, \lambda}^{(s)}$? The normalization condition turned out to be useless in this regard, but I have one more trick up my sleeve. To avoid unnecessary complications, I will assume that the perturbation completely lifts the initial degeneracy of the eigenvalue so that all $E_s^{(1)}$ are different. All eigenvectors of Hermitian operators belonging to different eigenvalues must be mutually orthogonal. So, let me consider the inner product $\left\langle \chi_s^{(\mu)} \mid \chi_s^{\lambda}\right\rangle$. Taking into account all available orthogonality conditions, I find

$$\langle s, \nu| s, \lambda\rangle = \sum_{\mu \neq \lambda} a_{\mu, \lambda}^{(s)} \left\langle \chi_s^{(\nu)}\right| \chi_s^{\mu}\rangle = \sum_{\mu \neq \lambda} a_{\mu, \lambda}^{(s)} \delta_{\nu, \mu} = a_{\nu, \lambda}^{(s)} = 0.$$

Now, when I have found all coefficients in Eq. 13.43, I can write down the final expression for $|s, \lambda\rangle$:

$$|s, \lambda\rangle = \left|\chi_s^{\lambda}\right\rangle + \sum_{\mu} \sum_{q \neq s} \frac{V_{q\mu, s\lambda}}{E_s^{(0)} - E_q^{(0)}} \left|\chi_q^{\mu}\right\rangle \tag{13.44}$$

which is cited in most textbooks but often under a false pretense. It is important to understand that condition $q \neq s$ excludes all the terms with the same eigenvalue of the energy, so no issues with a denominator of this expression going to zero never arise. Often cited fact that the perturbation matrix elements $V_{q\mu,s0\lambda}$ turn zero for $q = s$ and $\mu \neq \lambda$, thanks to the special choice of the zero-order eigenvectors, while factually true, has very little to do with the justification of Eq. 13.44, contrary to what you might have read.

13.2.2 Linear Stark Effect

While the Stark effect observed for the non-degenerate ground state of the hydrogen atom is quadratic in field, the response of degenerate energy levels of a hydrogen atom to a uniform electric field can actually be linear. To illustrate the origin of this effect, I will consider the first-order corrections to the degenerate energy level of a hydrogen atom characterized by the principal quantum number $n > 1$ and eigenvectors $|n, l, m\rangle$. The subspace of the degenerate eigenvectors consists of all eigenvectors with fixed principal number n but varying orbital and magnetic numbers $l < n$ and $-l < m < l$. The total number of such degenerate states and, therefore, the dimensionality of the subspace are, as you know, n^2. The perturbation operator is given again by Eq. 13.14, so that the perturbation matrix, limited to the degenerate subspace, is

$$V_{nl_1m_1,nl_2m_2} = -e\mathcal{E} \langle nl_1m_1| \hat{z} |nl_2m_2\rangle .$$

Using the arguments similar to the ones I used when analyzing matrix elements for the non-degenerate case, I can show that this matrix is diagonal with respect to the magnetic number m:

$$V_{nl_1m_1,nl_2m_2} = -e\mathcal{E} \langle nl_1m_1| \hat{z} |nl_2m_1\rangle \, \delta_{m_1m_2} .$$

(In case you forgot, the argument goes like this: in the position representation, ket $|nl_2m_2\rangle$ contains a factor $\exp(im_2\varphi)$, and bra $\langle nl_1m_1|$ contributes $\exp(-im_1\varphi)$. Since there are no other factors dependent on the azimuthal angle, integration of $\exp[i(m_2 - m_1)\varphi]$ over the entire range of φ from 0 to 2π yields zero unless $m_2 = m_1$.) Another general property of the matrix element $\langle nl_1m_1| \hat{z} |nl_2m_1\rangle$ can be established by looking at its behavior with respect to the inversion operator. As I have already mentioned more than once, if a Hamiltonian is invariant with respect to inversion, the matrix elements computed with its eigenvectors must be even to have a non-zero value. Now, the matrix element $\langle nl_1m_1| \hat{z} |nl_2m_1\rangle$ consists of three components: an operator $\hat{z}$, which is odd (changes sign) upon inversion, ket vector $|nl_2m_1\rangle$, and bra vector $\langle nl_1m_1|$ that transforms upon inversion according to

$$\hat{P} |nl_2m_1\rangle = (-1)^{l_2} |nl_2m_1\rangle ; \quad \langle nl_1m_1| \hat{P} = (-1)^{l_1} \langle nl_1m_1| .$$

The transformation rule for the entire matrix element therefore becomes

$$\hat{P} \langle nl_1m_1| \hat{z} |nl_2m_1\rangle = -(-1)^{l_1+l_2} \langle nl_1m_1| \hat{z} |nl_2m_1\rangle.$$

You can clearly see now that for the matrix element to remain invariant under inversion and, therefore, to have a non-zero value, orbital numbers l_1 and l_2 must have opposite parities: if one is even, the other one must be odd. Obviously, all matrix elements with $l_1 = l_2$ vanish. Actually, it can be shown that the restriction on the values of orbital numbers l_1 and l_2 is even more stringent: the matrix element $\langle nl_1m_1| \hat{z} |nl_2m_1\rangle$ is different from zero only if $|l_1 - l_2| = 1$.

Having figured out the general properties of the perturbation matrix elements, I can now turn to finding the first-order corrections to the energy eigenvalues. Unfortunately, I cannot get too far in the general case of arbitrary n because what I would have to do is to search for roots on the polynomial of the order n^2, and there are no methods of doing this analytically for an arbitrary n. Besides, I do not really need it. To illustrate the effects of the electric field on a degenerate energy eigenvalue, it is sufficient to consider an eigenvalue with the lowest nontrivial degree of degeneracy, which is the eigenvalue with $n = 2$. There are four states belonging to the same energy value $E_2^{(0)} = -E_g/4$: $|2, 0, 0\rangle$, $|2, 1, -1\rangle$, $|2, 1, 0\rangle$, and $|2, 1, 1\rangle$ to which I will assign numbers $1, 2, 3,$ and 4 correspondingly, so that I can label the perturbation matrix with only two indexes, $V_{ij}^{(2)}$, where i and j take values from one to four, and as a reminder, the upper index refers to the principal quantum number $n = 2$. For instance, the matrix element $V_{1,2}^{(2)}$ corresponds to $\langle 2, 0, 0| \hat{V} |2, 1, -1\rangle$, $V_{1,3}^{(2)}$ corresponds to $\langle 2, 0, 0| \hat{V} |2, 1, 0\rangle$, and so on and so forth. The task of solving Eq. 13.33 for a 4×4 matrix might also appear to be too formidable; after all it involves finding roots of the polynomial of the fourth order, for which, in general, there are no analytic formulas. But if you do not succumb to an immediate panic attack, you may find out that in this particular case, nature prepares for us a nice surprise. Using the general properties of the matrix element established in the previous paragraph ($m_1 = m_2$, $|l_1 - l_2| = 1$), you can easily find that the only non-zero matrix elements of the perturbation matrix are

$$V_{1,3}^{(2)} = V_{3,1}^{(2)*} = -e\mathcal{E} \langle 200| \hat{z} |210\rangle. \tag{13.45}$$

All other matrix elements involving $l = 0$ state $|2, 0, 0\rangle$ vanish because of $m_1 = m_2$ selection rule, while all matrix elements between states with $l = 1$ vanish because of the rule $|l_1 - l_2| = 1$. This is a huge simplification of our task because the secular equation for the eigenvalue corrections now takes the form

$$\begin{Vmatrix} -E_2^{(1)} & 0 & V_{1,3}^{(2)} & 0 \\ 0 & -E_2^{(1)} & 0 & 0 \\ V_{1,3}^{(2)*} & 0 & -E_2^{(1)} & 0 \\ 0 & 0 & 0 & -E_2^{(1)} \end{Vmatrix} =$$

$$-E_2^{(1)} \begin{Vmatrix} -E_2^{(1)} & 0 & V_{1,3}^{(2)} \\ 0 & -E_2^{(1)} & 0 \\ V_{1,3}^{(2)*} & 0 & -E_2^{(1)} \end{Vmatrix} =$$

$$\left[E_2^{(1)}\right]^2 \begin{Vmatrix} -E_2^{(1)} & V_{1,3}^{(2)} \\ V_{1,3}^{(2)*} & -E_2^{(1)} \end{Vmatrix} =$$

$$\left[E_2^{(1)}\right]^2 \left(\left[E_2^{(1)}\right]^2 - \left|V_{1,3}^{(2)}\right|^2 \right) = 0.$$

Here I computed the 4×4 determinant using cofactor expansion first along the last row of the initial determinant, which contains a single non-zero term, and then along the second row of the remaining 3×3 determinant. The resulting equation has four solutions, as expected, two of which coincide and are equal to zero. The remaining two solutions are $E_{2;3,4}^{(1)} = \pm \left|V_{1,3}^{(2)}\right|$. What it means is that now, instead of a single initial fourfold degenerate eigenvalue, we have three distinct eigenvalues, one of which is double degenerate and is equal to the zero-order value:

$$E_{2;1,2} = E_2^{(0)},$$

$$E_{2;3,4} = E_2^{(0)} \pm \left|V_{1,3}^{(2)}\right|.$$

To complete this calculation, I would need to evaluate the matrix element $V_{1,3}^{(2)}$. The most straightforward path toward this goal is to use the position representation for the hydrogen eigenvectors, which I, for your convenience, present below (saving you a few minutes you would have to spend googling these formulas on your own):

$$|200\rangle = \frac{1}{\sqrt{\pi}} \left(\frac{Z}{2a_B}\right)^{3/2} \left(1 - \frac{Zr}{2a_B}\right) \exp\left(-\frac{Zr}{2a_B}\right)$$

$$|210\rangle = \frac{1}{2}\sqrt{\frac{1}{\pi}} \left(\frac{Z}{2a_B}\right)^{3/2} \left(\frac{Zr}{a_B}\right) \exp\left(-\frac{Zr}{2a_B}\right) \cos\theta.$$

Substituting these expressions in Eq. 13.45, I get

$$V_{1,3}^{(2)} = -e\mathcal{E} \left(\frac{Z}{2a_B}\right)^3 \int_0^\infty dr \left(1 - \frac{Zr}{2a_B}\right) \left(\frac{Zr^4}{a_B}\right) \exp\left(-\frac{Zr}{a_B}\right) \int_0^\pi d\theta \sin\theta \cos^2\theta,$$

where I expressed z in spherical coordinates as $r\cos\theta$; introduced spherical volume elements $r^2 \sin\theta dr d\theta d\varphi$; carried out trivial integration over φ, which yielded 2π; and separated integrals with respect to angular and radial variables. The angular integral is computed easily by a standard substitution of variables $x = \cos\theta$ and $dx = -\sin\theta d\theta$ and yields

$$\int_{-1}^{1} dx x^2 = \frac{2}{3}.$$

All what is left now is to do the radial integral:

$$V_{1,3}^{(2)} = -e\mathcal{E}\frac{2}{3}\left(\frac{Z}{2a_B}\right)^3 \int_0^\infty dr \left(1 - \frac{Zr}{2a_B}\right)\left(\frac{Zr^4}{a_B}\right)\exp\left(-\frac{Zr}{a_B}\right).$$

To turn the quite straightforward while tedious and boring job of computing the remaining integral into something a bit more interesting and fun, I challenge you to extract as much physical information from this expression without actually doing the integration. The trick is (and this is the kind of trick, which professional physicists learn to do as a pure reflexive action) to replace the integration variable r with something dimensionless. In the integral in question, such a substitution is almost obvious: $y = Zr/a_B$, after which the integral becomes

$$V_{1,3}^{(2)} = \frac{1}{12}e\mathcal{E}\frac{a_B}{Z}\int_0^\infty dx \left(\frac{x}{2} - 1\right)x^4 \exp\left(-x\right)$$

revealing that the matrix element is proportional to the drop of the potential energy of the electron in the external field over the size of the atom exemplified by Bohr radius a_B/Z. The remaining integral contributes only a numerical factor, which I will compute using computational platform Mathematica © with the result

$$V_{1,3}^{(2)} = 3e\mathcal{E}\frac{a_B}{Z}.$$

Now I can give the explicit expressions for the energy eigenvalues modified by the electric field:

$$E_{2;1,2} = -E_g/4$$

$$E_{2;3,4} = -E_g/4 \pm 3e\mathcal{E}\frac{a_B}{Z}. \tag{13.46}$$

The Stark effect in the case of degenerate energy eigenvalues differs from its non-degenerate counterpart in two important aspects. First, the corrections to the

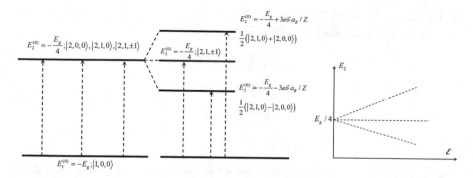

Fig. 13.2 Splitting of the $n = 2$ degenerate energy level of a hydrogen atom

eigenvalues are now linear in the field (hence the name—linear Stark effect). Second, instead of a simple shift of the energy, two new energy eigenvalues emerge above and below of the initial level. Experimentally this effect is seen as splitting of a single emission or absorption spectral line corresponding to a transition between $n = 1$ ground state and all $n = 2$ states into three distinct lines, the distance between which grows linearly with the increasing electric field. This phenomenon is illustrated in Fig. 13.2, where the left panel shows the electric field-induced changes in the energy spectrum of the atom together with possible quantum transitions responsible for the absorption of light, while the right panel illustrates the field-induced linear divergence of the emergent energy eigenvalues.

The next stop of our linear Stark effect train is the "correct" zero-order eigenvectors found by solving Eq. 13.32, which, taking into account the structure of the perturbation matrix, can be written as

$$
\begin{bmatrix}
0 & 0 & V_{1,3}^{(2)} & 0 \\
0 & 0 & 0 & 0 \\
V_{1,3}^{(2)} & 0 & 0 & 0 \\
0 & 0 & 0 & 0
\end{bmatrix}
\begin{bmatrix}
a_1 \\
a_2 \\
a_3 \\
a_4
\end{bmatrix}
= E_2^{(1)}
\begin{bmatrix}
a_1 \\
a_2 \\
a_3 \\
a_4
\end{bmatrix}.
$$

As far as the value of the matrix element $V_{1,3}^{(2)}$ is concerned, all what I need to know about it at this point is that it is real. The resulting matrix equation yields the following system of equations:

$$
V_{1,3}^{(2)} a_3 = E_2^{(1)} a_1
$$

$$
0 = E_2^{(1)} a_2
$$

$$
V_{1,3}^{(2)*} a_1 = E_2^{(1)} a_3 \tag{13.47}
$$

$$
0 = E_2^{(1)} a_4,
$$

which need to be solved for each of the found eigenvalues. Let's first dispose of the trivial case $E_2^{(1)} = 0$, which reduces this system to

$$V_{1,3}^{(2)} a_3 = 0$$

$$0 = 0$$

$$V_{1,3}^{(2)*} a_1 = 0$$

$$0 = 0.$$

The first and the third equations obviously yield $a_3 = a_1 = 0$, while the second and the fourth equations at first glance do not tell much—they are just trivial identities. What they are actually saying, however, is that coefficients a_2 and a_4 remain undefined and can be chosen at will. It is also important to remember that 0 is a degenerate eigenvalue, so that we have to choose two pairs of these coefficients, and we want to do in such a way that the resulting eigenvectors would be orthogonal to each other. Actually, this requirement is quite easy to fulfill by choosing for one pair $a_2^{(1)} = 1$, $a_4^{(1)} = 0$ and for the other $a_2^{(2)} = 0$, $a_4^{(2)} = 1$. With this choice we end up with two trivial and easily anticipated answers (check out the enumeration of the eigenvectors I introduced before these calculations were started!):

$$|2\rangle^{(1)} = |2, 1, -1\rangle \, ; \, |2\rangle^{(2)} = |2, 1, 1\rangle \, . \tag{13.48}$$

It is not surprising that eigenvalues unaffected by the perturbation are accompanied by eigenvectors from the initial zero-order set.

Substituting the third eigenvalue $E_{2;3}^{(1)} = \left| V_{1,3}^{(2)} \right| = V_{1,3}^{(2)}$ into Eq. 13.47, I have

$$V_{1,3}^{(2)} a_3 = V_{1,3}^{(2)} a_1$$

$$0 = V_{1,3}^{(2)} a_2$$

$$V_{1,3}^{(2)} a_1 = V_{1,3}^{(2)} a_3 \tag{13.49}$$

$$0 = V_{1,3}^{(2)} a_4 \, .$$

These equations yield $a_2^{(3)} = a_4^{(3)} = 0$ and $a_3 = a_1$. The respective eigenvector becomes

$$|2\rangle^{(3)} = \frac{1}{\sqrt{2}} \left(|2, 0, 0\rangle + |2, 1, 0\rangle \right) . \tag{13.50}$$

where I set $a_3 = a_1 = 1/\sqrt{2}$ for normalization purposes. I will let you have a bit of fun with the last eigenvalue $E_{2;4}^{(1)} = -V_{1,3}^{(2)}$ and provide only the final answer for the last eigenvector

$$|2\rangle^{(4)} = \frac{1}{\sqrt{2}} \left(|2, 0, 0\rangle - |2, 1, 0\rangle \right) . \tag{13.51}$$

13.3 Problems

Problems to Sect. 13.1

Problem 155 An electron in a one-dimensional infinite potential well

$$V(z) = \begin{cases} 0 & 0 \leq z \leq d \\ \infty & z < 0, z > d \end{cases}$$

is also subjected to an electric field also in the positive z direction.

1. Write down analytically and sketch the total potential energy of the electron in the presence of the field.
2. Find first the non-zero corrections to the electron's energy eigenvalues due to the field. For the ground state of the electron, evaluate the three first terms in the appropriate sum and give a numerical estimate for the corrections assuming that $d = 100\,\text{nm}$ and the electric field is $\mathcal{E} = 10^6$ V/m.
3. Obtain the first-order correction to the ground state wave function of the electron.
4. Determine the expectation value of the electron's coordinate in the presence of the field, and compare it to the expectation value without the field. Give a qualitative explanation of the result.
5. Where in the well the probability to find the electron is the largest?

Problem 156 Using the stationary perturbation theory, find first the nonvanishing corrections to the energy eigenvalues and eigenvectors of an electron moving in a one-dimensional harmonic oscillator potential $V_0 = m_e \omega^2 z^2 / 2$ due to perturbation potentials of two different kinds:

1. $V(z) = \eta z^3$
2. $V(z) = \varsigma z^4$

Do not use the position representation to compute the matrix elements, instead use the ladder operator formalism.

Problem 157 Consider an electron interacting with a positively charged ion via a harmonic oscillator potential $V_0 = m_e \omega^2 z^2 / 2$, where the origin of the coordinate system is chosen at the position of a positively charged particle assumed unmovable.

1. Find the corrections to the electron's energy eigenvalues due to a uniform electric field in the positive z direction.
2. Find the polarizability of the electron, assuming it is in the ground state.

Do not use the position representation to compute the matrix elements, instead use the ladder operator formalism.

Problem 158 An electron in an infinite potential well of width d (see Problem 1) is also subjected to a perturbation potential $V_p = \alpha\delta(x - d/2)$. Find the corrections to the energy eigenvalue up to the second order of the perturbation theory and the corrections to the eigenvectors up to the first order.

Problem 159 The potential energy of an electron is given by the following expression:

$$V(r) = -\frac{e}{4\pi\varepsilon_0 r} + A\frac{z^2}{r^3} + B\frac{x^2 + y^2}{r^3},$$

where the last two terms can be considered as a perturbation. Using non-degenerate perturbation theory, find the first-order corrections to the energy and the wave function of the ground state of the electron in a unperturbed Coulomb potential.

Problem 160 Consider a particle in the infinite potential well

$$V_0(z) = \begin{cases} 0 & -d/2 \leq z \leq d/2 \\ \infty & z < -d/2, z > d/2 \end{cases}$$

also subjected to a perturbation of the following form:

$$V_1(z) = \begin{cases} \varsigma\frac{2|z|}{d} & -d/2 \leq z \leq d/2 \\ 0 & z < -d/2, z > d/2. \end{cases}$$

Derive the expressions for the first non-zero corrections to the ground state energy in the infinite potential well and its corresponding wave function.

Problem 161 Consider a three-dimensional isotropic harmonic oscillator subjected to the following perturbation potential:

$$\hat{V} = k(xy + xz + zy).$$

Find first the nonvanishing correction to the ground state energy of the oscillator and the first-order correction to its wave function. Hint: It is easier to do this problem in the Cartesian coordinates.

Problems to Sect. 13.2

Problem 162 Find the "correct" zero-order wave functions for a double-degenerate energy level if the matrix elements of perturbation potential are

$$V_{12} = |V|\,(1 + i).$$

Problem 163 Consider a system described by the following Hamiltonian matrix:

$$
\hat{H} = \begin{bmatrix}
-1 & i\epsilon & 2\epsilon & \epsilon\,(1-i) \\
-i\epsilon & -1 & -\epsilon & \epsilon \\
2\epsilon & -\epsilon & 2 & 2i\epsilon \\
\epsilon\,(1+i) & \epsilon & -2i\epsilon & 2
\end{bmatrix},
$$

where ϵ is a small real parameter, $\epsilon \ll 1$.

1. Present this Hamiltonian in the form $\hat{H}_0 + \hat{V}$ where $\hat{H}_0$ is a diagonal matrix.
2. You shall see that $\hat{H}_0$ has two pairs of double-degenerate eigenvalues. Using degenerate perturbation theory, find the first-order corrections to both eigenvalues and the "correct" zero-order eigenvectors.
3. Find the first-order corrections to the eigenvectors and the second-order corrections to the eigenvalues.

Problem 164 Using degenerate perturbation theory, find the corrections to the energy of $n = 2$ hydrogen energy level due to perturbation of the form

$$
V = \lambda z x.
$$

Problem 165 Using the degenerate perturbation theory, find the first-order corrections to the energy and "correct" zero-order eigenvectors for a first excited state of the system described in Problem 161.

Chapter 14
Fine Structure of the Hydrogen Spectra and Zeeman Effect

14.1 Spin–Orbit Interaction and Fine Structure of the Energy Spectrum of Hydrogen

14.1.1 Spin–Orbit Contribution to the Hamiltonian

You might still have a vague recollection of me mentioning the spin–orbit coupling in Sect. 9.5.1, where I introduced the tensor product of spin and orbital spaces as a means to construct vectors representing both orbital and spin components of a quantum state (if you do not remember that, you would do yourself a favor by going back and rereading that part of the book). More specifically, the issue of spin–orbit coupling came up in the discussion of generic vectors in the tensor product space, which could be presented as a superposition of basis vectors, in which different spin states are paired with different orbital components. Such states can be called spin–orbit coupled because the orbital properties of a system in such a state can be changed by affecting its spin and vice versa. However, practically, such states can only be realized in systems with actual spin–orbit interaction contributing a special term containing a combination of spin and orbital operators to their energy and, correspondingly, quantum Hamiltonian. This interaction is quite common. It appears in many ordinary systems, such as atoms or semiconductors, and is responsible for a number of important phenomena. In atoms it gives rise to the spectral features known in the early days of quantum mechanics, while in semiconductors it brings about the relatively recently discovered effects allowing, for instance, to use the spin to control electron spatial flow. Combining spin and orbital phenomena in such nontrivial situations is never a simple task, even if merely because it doubles the number of equations that must be solved. At the same time, the phenomena resulting from the spin–orbit interaction are way too important to be simply ignored and shall be discussed even if you only start getting comfortable with intricacies of the quantum description of the world. Therefore, in this section, I am giving you

© Springer International Publishing AG, part of Springer Nature 2018
L.I. Deych, *Advanced Undergraduate Quantum Mechanics*,
https://doi.org/10.1007/978-3-319-71550-6_14

a chance to learn about some aspects of the spin–orbit interaction in a relatively non-threatening environment by considering, again, a simple model of a single electron in a hydrogen-like atom.

Before going into the details of the physical mechanism responsible for the spin–orbit interaction in this system, let me try to guess its general operator structure using simple symmetry arguments. I will begin by making quite a trivial remark that the energy term, whatever it might be, must be a scalar. This scalar, however, must include the spin which is described by vector, $\hat{S}$, and another vector operator (or a combination thereof) related to the orbital spatial–temporal behavior of the electron. There are only three vectors that could foot the bill: the position vector $\hat{r}$, the momentum vector $\hat{p}$, and the orbital momentum vector $\hat{L}$. All these three operators are indeed vectors, but not all vectors are created equal. Momentum and position are, as you already know, odd operators, while the orbital momentum is even. This distinction is true for quantum operators as well as for classical vectors: for this reason, the angular momentum and other vectors having the same property with respect to inversion are sometimes called "pseudo-vectors." Other examples of classical "pseudo-vectors" are magnetic moments and magnetic field; in general anything that can be produced from normal vectors or related to them via the operation of vector (cross) product is a pseudo-vector.

Since spin operators do not have a classical analog, it is impossible to use this criterion to determine how they would behave with respect to inversion. Luckily, there are other ways to make this determination. One can, for instance, look at the potential energy associated with the spin magnetic moment, Eq. 9.28. Since we know that the potential energy is invariant with respect to inversion (or any other transformation of coordinates for that matter), it immediately follows that the spin operator must have the same parity as the magnetic field, which is even. Reversing this argument, I can say that in order to produce an expression invariant with respect to inversion, one must combine the spin operator with another even operator, which excludes position and momentum operators, leaving the angular momentum as the only viable alternative.

Thus, I hope these arguments have convinced you that the spin–orbit term in the Hamiltonian must involve the dot product of spin and orbital momentum operators:

$$\hat{H}_{so} = \lambda \hat{L} \cdot \hat{S}, \tag{14.1}$$

where a proportionality constant λ cannot be determined without going into details of the actual physical mechanism underlying the spin–orbit interaction, and this is what I am going to do now. The origin of the spin–orbit coupling in atoms is best understood by looking at the electron–nucleus interaction from the point of view of a moving electron. In the electron's reference frame, the nucleus is seen as a positive electric charge Ze (Z is the atomic number of the respective element) orbiting the electron (which is at the center of the orbit) moving with the orbital speed of the electron, but in the opposite direction. From electrodynamics you must know that a moving electric charge creates a magnetic field B. Ignoring for now the fact that the charge is moving along a circular trajectory (I will get back to this point later), I can use the solution of the Maxwell equations for a charge moving at constant speed valid in the weakly relativistic limit $v \ll c$[1]:

$$B = -\frac{1}{c^2} v \times E,$$ (14.2)

where v is the velocity of the electron and E is the electric field of the charge. The negative sign in Eq. 14.2 reflects the fact that the velocity of the proton, which is supposed to appear in this expression, is opposite to that of the electron. In the limit $v \ll c$, the electric field is given by a standard Coulomb expression:

$$E = \frac{1}{4\pi\varepsilon_0\varepsilon_r} \frac{Ze}{r^2} e_r,$$ (14.3)

where e_r is the unit vector in the radial direction toward the electron and all other notations correspond to Chap. 8. Substitution of Eq. 14.3 into Eq. 14.2 yields for the magnetic field

$$B = -\frac{1}{4\pi\varepsilon_0\varepsilon_r c^2} \frac{Ze}{r^2} [v \times e_r] = -\frac{1}{4\pi\varepsilon_0\varepsilon_r c^2} \frac{Zem_e}{m_e r^3} [v \times r]$$

$$= \frac{1}{4\pi\varepsilon_0\varepsilon_r} \frac{Ze}{m_e c^2 r^3} L,$$

where I replaced unit vector e_r with the position vector of the electron relative to proton $e_r = r/r$, multiplied the numerator and the denominator by electron mass m_e, changed the order of multiplication in the cross product $[v \times r]$, and finally replaced $m_e [r \times v]$ with the orbital momentum L. The direction of L in this derivation determines the direction of the magnetic field, B. This fact can be also confirmed using the right-hand rule as it is clear from Fig. 14.1.

Finally, recalling Eq. 9.7 for the magnetic moment of the electron as well as the expression for the potential energy of the magnetic dipole in the magnetic field ($U = -\mu_s \cdot B$), you can find for the spin–orbit interaction energy:

$$\hat{H}_{so} = \frac{1}{4\pi\varepsilon_0\varepsilon_r} \frac{Ze^2}{m_e^2 c^2 R^3} \hat{L} \cdot \hat{S}.$$ (14.4)

The presented derivation of Eq. 14.4 suffers from one significant shortcoming: it is based on Eq. 14.2 derived for a particle moving at constant velocity, while the motion of the electron in an atom is nothing but uniform. For the same reason, an electron is hardly an inertial reference frame, which makes the transition to it, used in the derivation, also quite suspicious. The problems with this formula, however,

---

[1] You might need to refresh your memory of classical electrodynamics at this point using the Internet or one of the available undergraduate textbooks on electrodynamics.

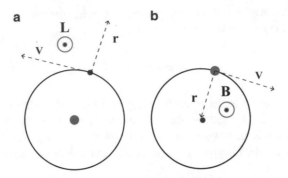

Fig. 14.1 (**a**) A hydrogen atom in the reference frame of the nucleus. An electron is orbiting the nucleus in the counterclockwise direction; vectors *v* and *r* represent velocity and position vector of the electron relative to the proton, so that its angular momentum *L* points out of the page. (**b**) A hydrogen atom in the reference frame of electron. The nucleus is traversing an orbit with an electron in its center in the clockwise direction creating a magnetic field at the electron's location also directed toward the viewer. Vectors *v* and *r* show directions of the differential current element *Idl* in the Biot–Savart law and position of the point of the observation of the magnetic field relative to the local current element. Magnetic field *B* is also directed out of the page as prescribed by the cross product *Idl* × *e_r*

were first noticed by experimentalists who found that theoretical predictions of the hydrogen spectra based on Eq. 14.4 deviate from the experimental results and that the theory can be brought into better agreement with measurements if one divides the expression in Eq. 14.4 by two. This discrepancy remained a mystery until a British physicist Llewellyn Hilleth Thomas[2] in 1926 re-derived the expression for the spin–orbit energy by carefully applying Lorentz transformations of the special theory of relativity when transitioning to the electron's reference frame. The key to the correct solution was to take into account the rotation of the reference frame attached to the electron as it traverses its trajectory around the nucleus. It is really quite amazing that these calculations yielded a spin–orbit energy differing from Eq. 14.4 exactly by factor $1/2$ as required to bring the theory in an agreement with the experiments. The kinematic effect due to the rotating reference frame, which actually has nothing to do with quantum mechanics and arises every time when one has to deal with rotating reference frames, is called *Thomas precession*, and the corresponding factor 1/2 is called *Thomas's half*. Taking Thomas's half into account yields the final expression for the spin–orbit contribution to the Hamiltonian

$$\hat{H}_{so} = \frac{1}{8\pi\varepsilon_0\varepsilon_r} \frac{Ze^2}{m_e^2 c^2 r^3} \hat{L} \cdot \hat{S}. \tag{14.5}$$

---

[2]Llewellyn Hilleth Thomas was a British physicist who eventually moved to the USA, where he held a professorial position in Ohio State University, was a member of the Watson Scientific Computing Laboratory at Columbia University, and was the IBM First Fellow in the Watson Research Center. His last position was at North Carolina State University.

This expression agrees with our symmetry based on Eq. 14.1, where the undefined parameter λ can now be identified as

$$\lambda(r) = \frac{1}{8\pi\varepsilon_0\varepsilon_r} \frac{Ze^2}{m_e^2 c^2 r^3} \tag{14.6}$$

and turns out to be a function of the radial coordinate of the electron.

14.1.2 Schrödinger Equation with Spin–Orbit Term

Having found the contribution of the spin–orbit interaction to the electron's Hamiltonian, I can attempt to write down the Schrödinger equation in the coordinate–spinor representation. Using representation of the spin matrices derived in Sect. 5.2.3, I can present operator product $\hat{\boldsymbol{L}} \cdot \hat{\boldsymbol{S}}$ as

$$\hat{\boldsymbol{L}} \cdot \hat{\boldsymbol{S}} = \hat{L}_x \hat{S}_x + \hat{L}_y \hat{S}_y + \hat{L}_z \hat{S}_z =$$
$$\frac{\hbar}{2} \begin{bmatrix} \hat{L}_z & \hat{L}_x - i\hat{L}_y \\ \hat{L}_x + i\hat{L}_y & -\hat{L}_z \end{bmatrix} = \frac{\hbar}{2} \begin{bmatrix} \hat{L}_z & \hat{L}_- \\ \hat{L}_+ & -\hat{L}_z \end{bmatrix}, \tag{14.7}$$

where in the last step I replaced x and y components of the angular momentum operator with the ladder operators $\hat{L}_\pm$ introduced in Eqs. 3.59 and 3.60. Now, the Schrödinger equation for the components $\Psi_\uparrow(r)$ and $\Psi_\downarrow(r)$ of the spinor $|\chi\rangle$, Eq. 9.62, can be written down as

$$\hat{H}_{orb}\Psi_\uparrow(r) + \lambda\frac{\hbar}{2}\left[\hat{L}_z\Psi_\uparrow(r) + \hat{L}_-\Psi_\downarrow(r)\right] = E\Psi_\uparrow(r), \tag{14.8}$$

$$\hat{H}_{orb}\Psi_\downarrow(r) + \lambda\frac{\hbar}{2}\left[\hat{L}_+\Psi_\uparrow(r) - \hat{L}_z\Psi_\downarrow(r)\right] = E\Psi_\downarrow(r). \tag{14.9}$$

Thanks to the spin–orbit term, Eqs. 14.8 and 14.9 become interdependent, which means that the spin and orbital states no longer can be chosen independently of each other.

Since the spin–orbit coupling parameter λ depends on the radial coordinate, it is probably hopeless to try solving these equations exactly, but I still want to see how far I can go trying, even if just to satisfy my curiosity. On a more serious note, I do believe that playing a bit with these equations might provide a better understanding of the correlation between the orbital and spin states generated by the spin–orbit coupling.

My first step would be naturally to try to separate the radial and angular variables. Since both the $\hat{H}_{orb}$ and spin–orbit terms in Eqs. 14.8 and 14.9 contain operators of the angular momentum, I can try to look for the solution in the form $\Psi_\uparrow(r) = R_\uparrow(r)Y_l^m(\theta, \varphi)$, where the first factor is yet an unknown radial function and $Y_l^m(\theta, \varphi)$

is the familiar spherical harmonic. However, inspecting the term containing $\Psi_\downarrow(r)$ in Eq. 14.8, I notice that it appears in the combination with the lowering ladder operator $\hat{L}_-$. This means that if I choose $\Psi_\downarrow(r)$ with the angular dependence given by the same spherical harmonics $Y_l^m(\theta, \varphi)$, I will be in trouble because $\hat{L}_- Y_l^m(\theta, \varphi) \propto Y_l^{m-1}(\theta, \varphi)$, and I would not be able to satisfy the equation (the angular parts of the wave functions would not cancel). However, if I do not panic, it might occur to me that by choosing $\Psi_\downarrow(r)$ as $\Psi_\downarrow(r) = R_\downarrow(r) Y_l^{m+1}(\theta, \varphi)$, I can actually save the day because $\hat{L}_- Y_l^{m+1}(\theta, \varphi) \propto Y_l^m(\theta, \varphi)$ so that the angular dependence in all terms of the equation will be given by $Y_l^m(\theta, \varphi)$, which I will be able to cancel leaving only the unknown radial components of the wave functions. For this to work, however, the same choices for $\Psi_\downarrow(r)$ and $\Psi_\uparrow(r)$ must do the same trick for Eq. 14.9. A bit of inspection will convince you that it actually does: $\hat{L}_+ Y_l^m(\theta, \varphi) \propto Y_l^{m+1}(\theta, \varphi)$, which coincides with the angular dependence of all other terms containing $\Psi_\downarrow(r)$.

To give these arguments a more polished and technical form, let me first recall from Chap. 8, Eq. 8.7 that

$$\hat{H}_{orb}\left[R(r)Y_l^m\right] = \left[-\frac{\hbar^2}{2\mu r^2}\frac{d}{dr}\left(r^2\frac{\partial R}{\partial r}\right) + \frac{\hbar^2 l(l+1)}{2\mu r^2}R - \frac{1}{4\pi\varepsilon_r\varepsilon_0}\frac{Ze^2}{r}R\right]Y_l^m$$

(see Chap. 8 for all notations). I will rewrite this expression introducing the radial orbital operator $\hat{H}_r$ as

$$\hat{H}_{orb}\left[R(r)Y_l^m\right] = Y_l^m \hat{H}_r R(r),$$

where

$$\hat{H}_r = -\frac{\hbar^2}{2\mu r^2}\frac{d}{dr}\left(r^2\frac{\partial}{\partial r}\right) + \frac{\hbar^2 l(l+1)}{2\mu r^2} - \frac{1}{4\pi\varepsilon_r\varepsilon_0}\frac{Ze^2}{r}.$$

I also need to recall that

$$\hat{L}_z Y_l^m(\theta, \varphi) = \hbar m Y_l^m(\theta, \varphi);$$

$$\hat{L}_- Y_l^{m+1}(\theta, \varphi) = \hbar\sqrt{l(l+1) - m(m+1)}Y_l^m(\theta, \varphi);$$

$$\hat{L}_+ Y_l^m(\theta, \varphi) = \hbar\sqrt{l(l+1) - m(m+1)}Y_l^{m+1}(\theta, \varphi).$$

Then Eq. 14.8 takes the form

$$Y_l^m \hat{H}_r R_\uparrow(r) + \frac{\hbar^2}{2}\lambda m R_\uparrow(r) Y_l^m(\theta, \varphi) +$$

$$\frac{\hbar^2}{2}\lambda\sqrt{l(l+1) - m(m+1)}R_\downarrow(r)Y_l^m(\theta, \varphi) = ER_\uparrow(r)Y_l^m(\theta, \varphi).$$

The similar procedure for Eq. 14.9 yields

$$Y_l^{m+1}\hat{H}_r R_\downarrow(r) - \frac{\hbar^2}{2}\lambda\,(m+1)\,R_\downarrow(r)Y_l^{m+1}(\theta,\varphi)+$$

$$\frac{\hbar^2}{2}\lambda\sqrt{l(l+1)-m(m+1)}R_\uparrow(r)Y_l^{m+1}(\theta,\varphi) = ER_\downarrow(r)Y_l^{m+1}(\theta,\varphi).$$

Now you can see that the spherical harmonics can, indeed, be canceled in both equations so that I am left with a system of purely radial equations

$$\left(\hat{H}_r + \frac{\hbar^2}{2}\lambda m\right)R_{n,l,m,\uparrow}(r)+$$

$$\frac{\hbar^2}{2}\lambda\sqrt{l(l+1)-m(m+1)}R_{n,l,m+1,\downarrow}(r) = ER_{n,l,m,\uparrow}(r), \qquad (14.10)$$

$$\left(\hat{H}_r - \frac{\hbar^2}{2}\lambda(m+1)\right)R_{n,l,m+1,\downarrow}(r)+$$

$$\frac{\hbar^2}{2}\lambda\sqrt{l(l+1)-m(m+1)}R_{n,l,m,\uparrow}(r) = ER_{n,l,m+1,\downarrow}(r), \qquad (14.11)$$

where I added additional indexes to the radial function to emphasize that it depends upon the four quantum numbers: the radial number n, which will emerge once the eigenvalues of energy are found from these equations, the orbital and magnetic numbers l and m, and finally the spin number, which in this case is designated by an arrow pointing up or down.

I am afraid that is it   I reached the end of this path as I am not going to attempt solving Eqs. 14.10 and 14.11. Now let's take a breath and try to understand if I can learn anything from this exercise. Leaving aside the issue of finding the radial functions, I do know now that the state of the atom in the presence of the spin–orbit interaction is given by a spinor of the following form:

$$|\chi\rangle = \begin{bmatrix} R_{n,l,m,\uparrow}Y_l^m(\theta,\varphi) \\ R_{n,l,m+1,\downarrow}(r)Y_l^{m+1}(\theta,\varphi) \end{bmatrix}, \qquad (14.12)$$

which can also be presented as a linear combination of basis spinors representing eigenvectors of the spin operator $\hat{S}_z$

$$|\chi\rangle = R_{n,l,m,\uparrow}(r)Y_l^m(\theta,\varphi)\,|\uparrow\rangle + R_{n,l,m+1,\downarrow}(r)Y_l^{m+1}(\theta,\varphi)\,|\downarrow\rangle. \qquad (14.13)$$

Even though I do not know the radial functions in this expression, it still provides me with some interesting information. For instance, it is clear from Eq. 14.13 that $|\chi\rangle$ is an eigenvector of the operator $\hat{\boldsymbol{L}}^2$:

$$
\hat{\boldsymbol{L}}^2 |\chi\rangle = R_{n,l,m,\uparrow}\hat{\boldsymbol{L}}^2 Y_l^m(\theta,\varphi)|\uparrow\rangle + R_{n,l,m+1,\downarrow}(r)\hat{\boldsymbol{L}}^2 Y_l^{m+1}(\theta,\varphi)|\downarrow\rangle =
$$
$$
\hbar^2 l(l+1)\left[R_{n,l,m,\uparrow}Y_l^m(\theta,\varphi)|\uparrow\rangle + R_{n,l,m+1,\downarrow}(r)Y_l^{m+1}(\theta,\varphi)\right]|\downarrow\rangle,
$$

where I took into account that the $\hat{\boldsymbol{L}}^2$ operator does not affect the radial functions or spin eigenvectors. At the same time, it is also clear that $|\chi\rangle$ is not an eigenvector of operators $\hat{L}_z$ or $\hat{S}_z$ (I will leave it to you to verify this fact on your own). But there is more. Since $|\chi\rangle$ combines spin and orbital states, it makes complete sense to see how this state will be affected by the operators of the total angular momentum $\hat{\boldsymbol{J}}$ introduced in Sect. 9.5.2, beginning with the operator $\hat{J}_z = \hat{L}_z + \hat{S}_z$. Remembering that $\hat{L}_z$ acts only on the orbital components of the state while $\hat{S}_z$ affects only its spinor components, I obtain

$$
\left(\hat{L}_z + \hat{S}_z\right)\left[R_{n,l,m,\uparrow}Y_l^m(\theta,\varphi)|\uparrow\rangle + R_{n,l,m+1,\downarrow}(r)Y_l^{m+1}(\theta,\varphi)|\downarrow\rangle\right] =
$$
$$
\hbar\left(m+\frac{1}{2}\right)R_{n,l,m,\uparrow}Y_l^m(\theta,\varphi)|\uparrow\rangle +
$$
$$
\hbar\left(m+1-\frac{1}{2}\right)R_{n,l,m+1,\downarrow}(r)Y_l^{m+1}(\theta,\varphi)|\downarrow\rangle =
$$
$$
\hbar\left(m+\frac{1}{2}\right)\left[R_{n,l,m,\uparrow}Y_l^m(\theta,\varphi)|\uparrow\rangle + R_{n,l,m+1,\downarrow}(r)Y_l^{m+1}(\theta,\varphi)|\downarrow\rangle\right].
$$

Rewriting this result in a more concise form

$$
\hat{J}_z |\chi\rangle = \hbar\left(m+\frac{1}{2}\right)|\chi\rangle
$$

makes it plainly obvious that $|\chi\rangle$ is an eigenvector of $\hat{J}_z$ with $m_J = m + 1/2$ (see Eq. 9.74). At this point it is only natural to check out the relationship between $|\chi\rangle$ and the operator $\hat{\boldsymbol{J}}^2$. On a hunch, I will compare Eq. 14.13 with Eqs. 9.93 and 9.94 representing eigenvectors of $\hat{\boldsymbol{J}}^2$, but to make the comparison easier, I will first rewrite Eq. 14.13 in an abstract form replacing index m with $m_J - 1/2$:

$$
|\chi\rangle = R_{n,l,m_J-1/2,\uparrow}(r)|l, m_J - 1/2\rangle|\uparrow\rangle +
$$
$$
R_{n,l,m_J+1/2,\downarrow}(r)|l, m_J + 1/2\rangle|\downarrow\rangle. \tag{14.14}
$$

Now it is plainly obvious that Eq. 14.14 has the same structure as eigenvectors of $\hat{\boldsymbol{J}}^2$ $|j_1, l, m_J\rangle$ and $|j_2, l, m_J\rangle$, where j_1 and j_2 are defined in Eqs. 9.88 and 9.89, and if I define the radial functions as

$$R_{nlm,\uparrow} = R_{nlm_J}^{j_1} \frac{1}{\sqrt{2l+1}} \sqrt{l + m_J + 1/2} \tag{14.15}$$

$$R_{nlm+1,\downarrow} = R_{nlm_J}^{j_1} \frac{1}{\sqrt{2l+1}} \sqrt{l - m_J + 1/2} \tag{14.16}$$

or as

$$R_{nlm,\uparrow} = R_{nlm_J}^{j_2} \frac{1}{\sqrt{2l+1}} \sqrt{l - m_J + 1/2} \tag{14.17}$$

$$R_{nlm,\downarrow} = -R_{nlm_J}^{j_2} \frac{1}{\sqrt{2l+1}} \sqrt{l + m_J + 1/2}, \tag{14.18}$$

I can rewrite Eq. 14.14 either as

$$|\chi_1\rangle = R_{nlm_J}^{j_1} |j_1, l, m_J\rangle$$

for $j_1 = l + 1/2$ or as

$$|\chi_2\rangle = R_{nlm_J}^{j_2} |j_2, l, m_J\rangle$$

for $j_2 = l - 1/2$. Thereby I am explicitly demonstrating that with these choices of radial functions, vectors $|\chi_1\rangle$ and $|\chi_2\rangle$ become eigenvectors of the operator $\hat{\boldsymbol{J}}^2$.

If you are wondering why this fact is such a big deal, just take a new look at Eq. 9.72 depicting the structure of the operator $\hat{\boldsymbol{J}}^2$, and you will see that it contains the same term $\hat{\boldsymbol{L}} \cdot \hat{\boldsymbol{S}}$ as the spin–orbit coupling contribution to the Hamiltonian. It does not take much now to demonstrate that $\hat{\boldsymbol{J}}^2$ commutes with the entire Hamiltonian, including the spin–orbit coupling term, and, therefore, common eigenvectors of both $\hat{\boldsymbol{J}}^2$ and $\hat{J}_z$ are also eigenvectors of the Hamiltonian. All what is left now is to substitute Eqs. 14.15 and 14.16 or Eqs. 14.17 and 14.18 into Eqs. 14.10 and 14.11, replacing along the way m with $m_J - 1/2$ and $m + 1$ with $m_J + 1/2$. First I deal with Eq. 14.10:

$$\left(\hat{H}_r + \frac{\hbar^2}{2} \lambda \left(m_J - \frac{1}{2} \right) \right) \frac{1}{\sqrt{2l+1}} \sqrt{l + m_J + 1/2} R_{nlm_J}^{j_1} +$$

$$\frac{\hbar^2}{2} \lambda \sqrt{l(l+1) - \left(m_J^2 - \frac{1}{4} \right)} \frac{1}{\sqrt{2l+1}} \sqrt{l - m_J + 1/2} R_{nlm_J}^{j_1} =$$

$$E_{n,j_1,l} R_{nlm_J}^{j_1} \frac{1}{\sqrt{2l+1}} \sqrt{l + m_J + 1/2},$$

where I added indexes to the energy eigenvalue E anticipating quantum numbers which this eigenvalue might depend upon. Recalling identity 9.87 to simplify $\sqrt{l(l+1) - (m_J^2 - \frac{1}{4})} = \sqrt{(l + 1/2 + m_J)(l + 1/2 - m_J)}$, I determine that $\sqrt{l + m_J + 1/2}/\sqrt{2l+1}$ is a common factor which can be canceled and that the remaining expression takes the following form:

$$\left(\hat{H}_r + \frac{\hbar^2}{2}\lambda\left(m_J - \frac{1}{2}\right) + \frac{\hbar^2}{2}\lambda\left(l - m_J + 1/2\right)\right)R_{nlm_J}^{j_1} = E_{n,j,l}R_{nlm_J}^{j_1} \Rightarrow$$

$$\left(\hat{H}_r + \frac{\hbar^2}{2}\lambda l\right)R_{nlm_J}^{j_1} = E_{n,j_1,l}R_{nlm_J}^{j_1}. \qquad (14.19)$$

This last equation surely looks nice and pretty, but before celebrating, we all would be well advised to make sure that Eq. 14.11 will be reduced to the same form after all these substitutions, because if it did not, I would have had two different equations for the same function, and this wouldn't be too good, right? So, here I go:

$$\left(\hat{H}_r - \frac{\hbar^2}{2}\lambda\left(m_J + \frac{1}{2}\right)\right)\frac{1}{\sqrt{2l+1}}\sqrt{l - m_J + 1/2}R_{nlm_J}^{j_1} +$$

$$\frac{\hbar^2}{2}\lambda\sqrt{l(l+1) - m(m+1)}\frac{1}{\sqrt{2l+1}}\sqrt{l + m_J + 1/2}R_{nlm_J}^{j_1} =$$

$$E_{n,j,l}\frac{1}{\sqrt{2l+1}}\sqrt{l - m_J + 1/2}R_{nlm_J}^{j_1}.$$

Using the same trick as above, I will now find that $\sqrt{l + m_J + 1/2}/\sqrt{2l+1}$ is the common factor, and after canceling it, I will end up with equation

$$\left(\hat{H}_r - \frac{\hbar^2}{2}\lambda\left(m_J + \frac{1}{2}\right) + \frac{\hbar^2}{2}\lambda\left(l + m_J + \frac{1}{2}\right)\right)R_{nlm_J}^{j_1} = E_{n,j_1,l}R_{nlm_J}^{j_1} \Rightarrow$$

$$\left(\hat{H}_r + \frac{\hbar^2}{2}\lambda l\right)R_{nlm_J}^{j_1} = E_{n,j_1,l}R_{nlm_J}^{j_1}$$

which is exactly the same as Eq. 14.19. Now we can celebrate—Eq. 14.19 is, indeed, a correct equation for the radial wave function corresponding to $j_1 = l + 1/2$. Repeating all these steps with Eqs. 14.17 and 14.18, you can obtain the radial equation for $R_{nlm_J}^{j_2}$ in the form

$$\left(\hat{H}_r - \frac{\hbar^2}{2}\lambda(l+1)\right)R_{nlm_J}^{j_2} = E_{n,j_2,l}R_{nlm_J}^{j_2}. \qquad (14.20)$$

The spin–orbit contribution to Eqs. 14.19 and 14.20 is different for different values of j, which justifies adding this number as an index in the notation for energy $E_{n,j,l}$.

Of course, nothing of this shall come as a big surprise; it is just an additional verification of the fact that eigenvectors of $\hat{J}^2$ and $\hat{J}_z$ are, indeed, the eigenvectors of the Hamiltonian.

Trigger warning: what I am going to show you now might get some of you really upset. All this work, which we did deriving Eqs. 14.19 and 14.20, was not really necessary as all these results can be derived faster and with less effort. So, if you feel the righteous indignation about me torturing you unnecessarily, I do have something to say in my defense. Now you have a much clearer understanding of the nuts and bolts of the spin–orbit interaction and its effect on the eigenvectors of the Hamiltonian than you would have had otherwise, and a few extra hours of me writing and you reading is a fair price to pay for it.

So, now let me show the easier way, and I will begin by writing down a complete Hamiltonian with the spin–orbit interaction in the form

$$\hat{H} = -\frac{\hbar^2}{2\mu r^2}\frac{\partial}{\partial r}\left(r^2\frac{\partial}{\partial r}\right) + \frac{\hat{L}^2}{2\mu r^2} - \frac{Ze^2}{4\pi\varepsilon_0\varepsilon_r r} + \lambda\hat{L}\cdot\hat{S}$$

where the kinetic energy is presented as a sum of radial and orbital momentum terms, and I use the same notations as in Chap. 8. Next I replace $\hat{L}\cdot\hat{S}$ in the spin–orbit interaction with

$$\hat{L}\cdot\hat{S} = \left(\hat{J}^2 - \hat{L}^2 - \hat{S}^2\right)/2$$

using Eq. 9.72, so that the Hamiltonian becomes

$$\hat{H} = -\frac{\hbar^2}{2\mu r^2}\frac{\partial}{\partial r}\left(r^2\frac{\partial}{\partial r}\right) + \frac{\hat{L}^2}{2\mu r^2} - \frac{Ze^2}{4\pi\varepsilon_0\varepsilon_r r} + \frac{1}{2}\lambda\left(\hat{J}^2 - \hat{L}^2 - \hat{S}^2\right). \tag{14.21}$$

Now acting by Hamiltonian 14.21 on

$$|\chi\rangle = R_{j,l}\,|j,l,m_J\rangle, \tag{14.22}$$

I get

$$\left[-\frac{\hbar^2}{2\mu r^2}\frac{\partial}{\partial r}\left(r^2\frac{\partial}{\partial r}\right) + \frac{\hat{L}^2}{2\mu r^2} - \frac{Ze^2}{4\pi\varepsilon_0\varepsilon_r r} + \frac{1}{2}\lambda\left(\hat{J}^2 - \hat{L}^2 - \hat{S}^2\right)\right] R_{j,l}(r)\,|j,l,m_J\rangle =$$

$$\left[-\frac{\hbar^2}{2\mu r^2}\frac{\partial}{\partial r}\left(r^2\frac{\partial}{\partial r}\right) + \frac{\hbar^2 l(l+1)}{2\mu r^2} - \frac{Ze^2}{4\pi\varepsilon_0\varepsilon_r r} + \frac{\hbar^2}{2}\lambda\right.$$

$$\left.\left(j(j+1) - l(l+1) - \frac{3}{4}\right)\right] R_{j,l}(r)\,|j,l,m_J\rangle.$$

Respectively, the radial part of the Schrödinger equation becomes

$$\left[-\frac{\hbar^2}{2\mu r^2}\frac{\partial}{\partial r}\left(r^2\frac{\partial}{\partial r}\right)+\frac{\hbar^2 l(l+1)}{2\mu r^2}-\frac{Ze^2}{4\pi\varepsilon_0\varepsilon_r r}+\frac{\hbar^2}{2}\right.$$

$$\left.\lambda\left(j(j+1)-l(l+1)-\frac{3}{4}\right)\right]R_{j,l}(r)=E_{n,j,l}R_{j,l}(r). \qquad (14.23)$$

The first three terms in this equation corresponds to $\hat{H}_r$ in Eqs. 14.19 and 14.20, and the last term is the spin–orbit contribution. Now, it is just a matter of simple algebra to convince oneself that $j(j+1)-l(l+1)-\frac{3}{4}$ is equal to l for $j=l+1/2$ and to $-l-1$ for $j=l-1/2$, making Eq. 14.23 equivalent to Eqs. 14.19 and 14.20.

14.1.3   Fine Structure of the Hydrogen Spectrum

To determine how spin–orbit interaction affects energy eigenvalues of the electron in the hydrogen atom, one would need to solve Eq. 14.23. Unfortunately, since spin–orbit coupling parameter λ depends on the radial coordinate $1/r^3$, which is different from both $1/r$ term in the Coulomb potential and $1/r^2$ term in the orbital angular momentum term, the exact analytical solution of this equation is out of reach, so the perturbation treatment discussed in Chap. 13 needs to be invoked with the spin–orbit term being treated as the perturbation. The first-order correction to the hydrogen energy levels is given by the expectation value of the perturbation term calculated with unperturbed radial hydrogen wave functions $R_{l,n}^{(0)}(r)$ described in Chap. 8:

$$E_{n,j,l}=E_n^{(0)}+\frac{1}{16\pi\varepsilon_0\varepsilon_r}\frac{Z\hbar^2 e^2}{\mu^2 c^2}\left(j(j+1)-l(l+1)-\frac{3}{4}\right)\left\langle\frac{1}{r^3}\right\rangle,$$

where λ was replaced with its explicit expression from Eq. 14.6. Using Eq. 8.34 for the expectation value

$$\left\langle\frac{1}{r^3}\right\rangle=\int_0^\infty dr\frac{1}{r}\left[R_{l,n}^{(0)}(r)\right]^2,$$

which was derived in Sect. 8.3, I find the first-order spin–orbit corrections to the energy levels to be

$$E_{n,j,l}=E_n^{(0)}+\frac{1}{8\pi\varepsilon_0\varepsilon_r}\frac{Z^4\hbar^2 e^2}{\mu^2 c^2 a_B^3}\frac{j(j+1)-l(l+1)-\frac{3}{4}}{l(l+1)(2l+1)n^3}=$$

$$E_n^{(0)}\left(1-\frac{Z^2\hbar^2}{\mu^2 c^2 a_B^2}\frac{j(j+1)-l(l+1)-\frac{3}{4}}{l(l+1)(2l+1)n}\right). \qquad (14.24)$$

Equation 8.34 for $\langle r^{-3} \rangle$ fails for $l = 0$, but it is not a big concern because the spin–orbit interaction vanishes at $l = 0$ anyway (spin has no one to interact with when the orbital momentum disappears; formally it becomes clear by substituting $j = 1/2$ into Eq. 14.23). Accordingly, Eq. 14.24 should be applied only to the states with $l > 0$.

The dimensionless constant $Z\hbar/(\mu c a_B)$ in Eq. 14.24 can be presented in several equivalent forms. First, replacing the Bohr radius with its expression from Eq. 8.8, this constant can be rewritten as

$$\frac{Z\hbar}{\mu c a_B} = \frac{Ze^2}{4\pi\varepsilon_0\varepsilon_r c\hbar} = \frac{Z}{\varepsilon_r}\alpha,$$

where I combined all fundamental parameters into the so-called *fine-structure constant* (ε_r and Z are material parameters and are not fundamental)

$$\alpha = \frac{e^2}{4\pi\varepsilon_0 c\hbar} \approx \frac{1}{137.036}. \tag{14.25}$$

This is one of the most famous constants in physics which plays a particularly important role in quantum electrodynamics appearing as a dimensionless strength of interaction between electrons and photons. One of the possible interpretations of this constant involves the ratio of the potential energy of the system of two electrons at a distance x from each other and the energy of a photon with wavelength $\lambda = 2\pi x$:

$$\alpha = \left(\frac{e^2}{4\pi\varepsilon_0 x} \right) / \left(\frac{\hbar c}{x} \right).$$

Introducing this fine-structure constant, Eq. 14.24 can be written down as

$$E_{n,j,l} = E_n^{(0)} \left(1 - \frac{Z^2}{\varepsilon_r^2}\alpha^2 \frac{j(j+1) - l(l+1) - \frac{3}{4}}{l(l+1)(2l+1)n} \right). \tag{14.26}$$

Alternatively, one can rewrite the same constant in terms of the zero-order hydrogen energies $E_n^{(0)}$

$$\left(\frac{Z\hbar}{\mu c a_B} \right)^2 = \frac{Z^2 e^4}{16\pi^2\varepsilon_0^2\varepsilon_r^2 c^2\hbar^2} = -\frac{2E_n^{(0)}}{\mu c^2}n^2$$

where I used the expression for $E_n^{(0)}$ from Eq. 8.16. Now the spin–orbit correction to the energy $\Delta E_{njl}^{(so)}$ can be written down as

$$\Delta E_{njl}^{(so)} = \frac{2\left[E_n^{(0)} \right]^2}{\mu c^2} n \frac{j(j+1) - l(l+1) - \frac{3}{4}}{l(l+1)(2l+1)}. \tag{14.27}$$

Unlike Eq. 14.26, the just-derived expression defines the spin–orbit correction to the energy in terms of the ratio of the unperturbed energy $E_n^{(0)}$ and the electron's rest energy μc^2. Both expressions for the energy predict that the spin–orbit interaction lifts the degeneracy of the hydrogen energy levels with respect to the orbital quantum number l, and, in addition, each energy level of hydrogen with $l \neq 0$ splits in two levels with different values of the total angular momentum $j_{1,2} = l \pm 1/2$.

At this point a question should pop up in the head of an attentive reader: why is it OK to derive Eq. 14.24 using the non-degenerate perturbation theory when everyone knows that all hydrogen energy levels (except of the ground state, of course) degenerate? This is a valid question, but the reason why the non-degenerate perturbation works here is quite simple: the perturbation matrix $\langle n,j,l,m_J | \lambda \hat{\boldsymbol{L}} \cdot \hat{\boldsymbol{S}} | n,j_1,l_1,m_{J_1} \rangle$ is diagonal in the basis of degenerate eigenvectors of the unperturbed Hamiltonian formed by all vectors $|n,j,l,m_J\rangle$ with a fixed value of the principal quantum number n. In other words, by choosing the eigenvectors of the operators $\hat{\boldsymbol{L}}^2$, $\hat{\boldsymbol{J}}^2$, and $\hat{J}_z$ as the zero-order eigenvectors, I automatically chose the "correct" basis diagonalizing the perturbation operator.

While we can be rightfully proud of being able to derive Eq. 14.26, it is a bit too early to pop champagne corks. Unfortunately, this result does not describe the experimental data correctly signaling that something in our analysis is missing. To figure out what it might be, note the presence of the speed of light in the fine-structure constant which hints that the spin–orbit interaction might have something to do with relativistic effects. After all, in the extreme nonrelativistic limit $c \to \infty$, the spin–orbit correction does vanish. Then the question you should be asking is if there are other relativistic effects which we might have missed and which can affect the hydrogen spectrum. The answer to this question is, indeed, affirmative: there are two more relativistic effects which generate contributions to the energy of the same order of magnitude as the spin–orbit interaction and, therefore, need to be included into consideration. One of these effects is due to the relativistic correction to the electron's kinetic energy, and the other is the so-called Darwin[3] term, whose origin is not that easy to explain without invoking a complete relativistic theory of electrons based on Dirac's equation. However, this term contributes only to the energy of $l = 0$ states, and as luck would have it, it is reproduced correctly by a simple sum of the spin–orbit and relativistic kinetic energy contributions to the energy.

The kinetic energy of a relativistic particle, according to Einstein's theory of special relativity, is related to the particle momentum p as

$$K_{rel} = \sqrt{p^2 c^2 + m_e^2 c^4} = m_e c^2 \sqrt{1 + \frac{p^2}{m_e^2 c^2}}.$$

---

[3]Charles Galton Darwin was an English physicist, another grandson of Charles Darwin, the author of the evolution theory. He became a director of the National Physics Laboratory in 1938 and remained in this position through the World War II participating in the Manhattan Project, where he was responsible for coordinating American, British, and Canadian efforts.

Since we are interested in small relativistic corrections only, I can expand this expression in a power series with respect to $(p/m_e c)^2$ and keep just the three first terms in the series

$$K_{rel} \approx m_e c^2 + \frac{p^2}{2m_e} - \frac{p^4}{8m_e^3 c^2}.$$ (14.28)

The first and the second terms in this equation are the rest energy of the electron (which is just a constant and can be ignored) and the nonrelativistic kinetic energy. The object of our interest is the third term in this expression whose effect on the hydrogen energy levels I intend to explore.

I cannot use the radial equation 14.23 for this purpose because it does not include the $\hat{p}^4$ term, and if I try to separate it into the radial and angular parts, the result would be disastrously cumbersome (just imagine having to square the first two terms in Eq. 14.21). Fortunately, I do not have to do it and can work instead with the Hamiltonian

$$\hat{H} = \frac{\hat{p}^2}{2\mu} - \frac{\hat{p}^4}{8\mu^3 c^2} - \frac{Ze^2}{4\pi\varepsilon_0 \varepsilon_r r} + \frac{1}{2}\lambda\left(\hat{J}^2 - \hat{L}^2 - \hat{S}^2\right),$$ (14.29)

where the kinetic energy operator remains intact, and in the relativistic correction term, I replaced electron mass m_e with the reduced mass μ to preserve the consistency of the notation. The first thing which is important to realize is that vectors $|jlm_J\rangle$ are eigenvectors of the operator $\hat{p}^4$, which is obvious because an eigenvector of any operator $\hat{A}$ is automatically an eigenvector of an operator $\hat{A}^2$, and[4] we have already seen that $|jlm_J\rangle$ is an eigenvector of $\hat{p}^2$. This means that the perturbation matrix built on the basis of these vectors is diagonal, and I can again use the non-degenerate perturbation theory to find the first-order corrections to the energy. The second important point that needs to be made is that the first order of the perturbation theory is linear with respect to the perturbation operator. As a result the correction to the energy due to the sum of two perturbation operators is equal to the sum of the corresponding energy corrections due to each of the perturbation separately. Accordingly, the modification of the hydrogen energy levels due to the relativistic term in Hamiltonian 14.29 can be written down as

$$\Delta E_{njl}^{(rel)} = -\frac{1}{8\mu^3 c^2}\langle n, j, l, m_J|\hat{p}^4|n, j, l, m\rangle,$$ (14.30)

---

[4]If it is not obvious for you, here is the proof: assume that $\hat{A}|q\rangle = a_q|q\rangle$. Then $\hat{A}\hat{A}|q\rangle = a_q\hat{A}|q\rangle = a_q^2|q\rangle$.

where $|n, j, l, m\rangle = R_{n,l}^{(0)} |j, l, m\rangle$ are complete zero-order hydrogen wave functions written in the basis of the vectors $|j, l, m\rangle$. In order to compute the matrix element in Eq. 14.30, I will present the operator $\hat{p}^4$ in it as

$$\Delta E_{njl}^{(rel)} = -\frac{1}{2\mu c^2} \langle n, j, l, m_J| \frac{\hat{p}^2}{2\mu} \frac{\hat{p}^2}{2\mu} |n, j, l, m\rangle \tag{14.31}$$

and will consider the action of the emerging kinetic energy operators to their respective ket (on the right) and bra (on the left) vectors. The expression

$$\frac{\hat{p}^2}{2\mu} |n, j, l, m\rangle$$

is evaluated by converting the time-independent Schrödinger equation

$$\left(\frac{\hat{p}^2}{2\mu} - \frac{Ze^2}{4\pi\varepsilon_0\varepsilon_r r} \right) |n, j, l, m\rangle = E_n^{(0)} |n, j, l, m\rangle$$

into

$$\frac{\hat{p}^2}{2\mu} |n, j, l, m\rangle = \left(E_n^{(0)} + \frac{Ze^2}{4\pi\varepsilon_0\varepsilon_r r} \right) |n, j, l, m\rangle .$$

Hermitian conjugation of this result yields

$$\langle n, j, l, m| \frac{\hat{p}^2}{2\mu} = \left(E_n^{(0)} + \frac{Ze^2}{4\pi\varepsilon_0\varepsilon_r r} \right) \langle n, j, l, m| .$$

Substituting these expressions into Eq. 14.31, I get

$$\Delta E_{njl}^{(rel)} = -\frac{1}{2\mu c^2} \langle n, j, l, m_J| \left(E_n^{(0)} + \frac{Ze^2}{4\pi\varepsilon_0\varepsilon_r r} \right)^2 |n, j, l, m\rangle =$$

$$-\frac{\left[E_n^{(0)} \right]^2}{2\mu c^2} - \frac{E_n^{(0)}}{\mu c^2} \frac{Ze^2}{4\pi\varepsilon_0\varepsilon_r} \langle n, j, l, m_J| \frac{1}{r} |n, j, l, m\rangle$$

$$-\frac{1}{2\mu c^2} \frac{Z^2 e^4}{16\pi^2\varepsilon_0^2\varepsilon_r^2} \langle n, j, l, m_J| \frac{1}{r^2} |n, j, l, m\rangle .$$

Both expectation values $\langle r^{-1} \rangle$ and $\langle r^{-2} \rangle$ appearing in this expression were computed in Sect. 8.3, and substitution of the corresponding expressions from Eqs. 8.29 and 8.33 yields

$$\Delta E_{njl}^{(rel)} = -\frac{\left[E_n^{(0)}\right]^2}{2\mu c^2} - \frac{E_n^{(0)}}{\mu c^2}\frac{Ze^2}{4\pi\varepsilon_0\varepsilon_r}\frac{Z}{a_B n^2} -$$

$$\frac{1}{2\mu c^2}\frac{Z^2 e^4}{16\pi^2\varepsilon_0^2\varepsilon_r^2}\frac{2Z^2}{a_B^2}\frac{1}{(2l+1)n^3} = -\frac{\left[E_n^{(0)}\right]^2}{2\mu c^2} \times$$

$$\left(1 + \frac{Z^2 e^2}{2\pi\varepsilon_0\varepsilon_r a_B n^2 E_n^{(0)}} + \frac{Z^4 e^4}{8\pi^2\varepsilon_0^2\varepsilon_r^2 a_B^2 \left[E_n^{(0)}\right]^2 (2l+1)n^3}\right) =$$

$$- \frac{\left[E_n^{(0)}\right]^2}{2\mu c^2}\left(\frac{8n}{2l+1} - 3\right) \equiv \alpha^2 \frac{Z^2}{\varepsilon_r^2}E_n^{(0)}\frac{1}{4n^2}\left(\frac{8n}{2l+1} - 3\right), \quad (14.32)$$

where I simplified the resulting expressions using Eqs. 8.8 and 8.17 for the Bohr radius and unperturbed energy of the hydrogen atom, respectively. Combining Eqs. 14.27 and 14.32, I obtain the total first-order correction to the energy levels of an electron in the hydrogen atom:

$$\Delta E_{njl} = -\frac{\left[E_n^{(0)}\right]^2}{\mu c^2}\left[\frac{2n}{2l+1}\left(2 - \frac{j(j+1) - l(l+1) - \frac{3}{4}}{l(l+1)}\right) - \frac{3}{2}\right].$$

Substituting $j = l+1/2$ or $j = l-1/2$, you can convince yourself that this expression can be simplified into

$$\Delta E_{njl}^{(fs)} = -\frac{\left[E_n^{(0)}\right]^2}{\mu c^2}\left[\frac{2n}{j+\frac{1}{2}} - \frac{3}{2}\right] \quad (14.33)$$

for any value of j. Surprisingly, but even for $l = 0$ $(j = 1/2)$), this expression gives the correct answer including the abovementioned Darwin term, which I did not even bother to consider. According to this result, energy levels of the electron in a hydrogen-like atom acquire dependence on the total angular momentum via the quantum number j but remain independent of the orbital quantum number l. For instance, this result predicts that states of the electron with $j = 1/2$ originating from orbitals with $l = 0$ and $l = 1$ would have the same energy, while states with $j = 1/2, l = 1$ and $j = 3/2, l = 1$ would have distinct energies. This difference between energy levels characterized by the same principal number n and different values of j is called the fine structure of the hydrogen atom.

While agreement between theoretical results predicted by Eq. 14.33 and the experiment is relatively good, more careful measurements reveal additional energy levels of hydrogen not predicted by this equation. The origin of these even more closely located energies, called hyperfine structure, can be traced to interaction between spins of electron and the nucleus. A consideration of this effect is outside of the scope of this book. Another interesting point related to the spectrum of hydrogen is concerned with the abovementioned predictions that both $j = 1/2, l = 0$ and $j = 1/2, l = 1$ have the same energy. In reality the latter state was found to have a slightly lower energy than the former. This difference between these energies is called the Lamb shift, and its discovery by Willis Lamb[5] and his graduate student Robert Retherford in 1947 was a triumph of experimental physics, for which Lamb was awarded the Nobel Prize in Physics in 1955. Lamb's shift is a stunning example of a phenomenon which combines a small magnitude with oversized significance for physics—explanation of its origin gave birth to modern quantum electrodynamics with its overreaching concepts of renormalization influencing desparate fields from phase transitions to the black holes. The calculations of the Lamb shift were first carried out by Hans Bethe in 1947,[6] whose paper on the Lamb shift was only two pages long and became one of the influential theoretical papers of the after-World War II period. Alas, you will have to wait till a more serious graduate level course in quantum electrodynamics to be able to appreciate the beauty of the theory explaining the Lamb shift.

14.2   Zeeman Effect

Changes in atomic spectra in the presence of a strong magnetic field were first observed in 1896 by Dutch physicist Pieter Zeeman, whose experiments were based on earlier theoretical work of Zeeman's compatriot Hendrik Antoon Lorentz (who also derived the famous Lorentz transformations used by Einstein in his relativity theory). Both Zeeman and Lorentz were awarded in 1902 the Nobel Prize in Physics

---

[5]Willis Lamb was an American experimental physicist who made significant contribution to quantum electrodynamics and the field of quantum measurements. He holds professorial positions at the University of Oxford and Yale, Columbia and Stanford Universities, and the University of Arizona.

[6]Hans Bethe was a German-born physicist who immigrated to the USA in 1935 (only 2 years later than Einstein) and became a professor at Cornell University where he worked till his death in 2005. He won the 1967 Nobel Prize in Physics for his work on the theory explaining the formation of chemical elements due to nuclear reaction within stars. During the war, he headed theoretical efforts within the Manhattan Project and played a critical role in calculating the critical mass of the weapons. After the war he was active in efforts to outlaw testing of nuclear weapons convincing Kennedy and Nixon administrations to sign the Partial Nuclear Test Ban Treaty (1963) and the Anti-Ballistic Missile Treaty (1972). He also made important contribution in solid-state physics.

for this discovery signifying its importance. While initially seen as a broadening of a spectral line, it was later found to be a splitting of an initial line into as many as 15 additional lines. This phenomenon is rightfully known as the Zeeman effect, and I have already mentioned it in Chap. 9 trying to provide experimental justifications for an electron's spin. The Zeeman effect proved to play an important role in many areas of physics and astrophysics, in addition to quantum mechanics. It is not surprising, therefore, that I am, as many other authors of quantum mechanics textbooks, eager to devote a significant amount of time to developing a proper quantum description of this phenomenon. Since the spin–orbit interaction plays an important role in this treatment, now seems like a suitable time to do so.

When describing the contribution of interaction with a magnetic field to the atom's Hamiltonian, one needs to take into account that an electron in a generic quantum state possesses two types of magnetic dipole moments: one, $\boldsymbol{\mu}_L$, is related to its orbital angular momentum $\boldsymbol{L}$, and the other, $\boldsymbol{\mu}_s$, is due to the electron's spin. While the corresponding contributions to the Hamiltonian are similar for both magnetic moments

$$\hat{H}_Z = -\left(\boldsymbol{\mu}_L + \boldsymbol{\mu}_s\right) \cdot \boldsymbol{B},$$

the theoretical description of the effect is complicated by the fact that gyromagnetic ratios, connecting the magnetic moment with the corresponding angular momentum, are different for the orbital angular moment and spin. As it was explained in Sect. 9.1, the former is two times larger than the latter (see Eqs. 9.1 and 9.7) so that the expression for the magnetic energy contribution to the Hamiltonian can be written down as

$$\hat{H}_Z = \frac{e}{2m_e}\left(\hat{L} + 2\hat{S}\right) \cdot \boldsymbol{B} =$$

$$\frac{eB}{2m_e}\left(\hat{L}_z + 2\hat{S}_z\right) \tag{14.34}$$

where I choose the axis Z of my coordinate system along the magnetic field $\boldsymbol{B}$. If the gyromagnetic ratios for orbital and spin momentums were the same, the magnetic energy would depend only on a z-component of the total angular momentum $\hat{J}_z$ so that the Zeeman energy and the entire Hamiltonian would commute with the operators $\hat{J}^2$ and $\hat{J}_z$ making the analysis as simple as that for the spin–orbit interaction. Unfortunately (or fortunately—makes life more interesting), this is not the case, and either orbital or spin momentums appear in the Zeeman energy separately. For instance, you can replace $\hat{L}_z$ with $\hat{J}_z - \hat{S}_z$ and end up with the following expression:

$$\hat{H}_Z = \frac{eB}{2m_e}\left(\hat{J}_z + \hat{S}_z\right). \tag{14.35}$$

Neither vectors $|l, m\rangle\,|m_s\rangle$ nor $|j, l, m_J\rangle$ are eigenvectors of the Hamiltonian $\hat{H}_Z$, and, therefore, I cannot simply invoke the first-order non-degenerate perturbation theory to find the corresponding corrections to the energy levels.

The situation can be simplified in two extreme limits: of very weak or very strong magnetic field. "Weakness" or "strength" of the field is meant here in comparison with the spin–orbit interaction. In the weak magnetic field, the fine structure of the spectrum due to the spin–orbit and relativistic corrections remains a predominant feature, while effects due to the magnetic field can be considered as a small perturbation. The strong magnetic field means that the Zeeman splitting, which we discussed in Sect. 9.1, is the main effect, while the spin–orbit and relativistic contributions play role of the small perturbation. I will first consider the small field limit.

14.2.1   Zeeman Effect in the Weak Magnetic Field

The weakness of the magnetic field means that the Hamiltonian of Eq. 14.35 can be treated as a perturbation term, while the atomic Hamiltonian with relativistic and spin–orbit contributions define the system of zero-order energy eigenvalues and eigenvectors. This point needs to be emphasized here because the zero-order eigenvalues determine the degeneracy of the unperturbed spectrum and, therefore, determine the type of the perturbation theory one has to use. As you learned in Sect. 13.2, the most important quantity in this regard is the perturbation matrix built on the basis of the degenerate eigenvectors. If the fine structure defines the zero-order eigenvalues given by Eq. 14.33, then I have to deal with the subspace of vectors $|n, j, l, m_J\rangle$ with fixed numbers n and j. The first term in Hamiltonian 14.35 is obviously diagonal in the basis of these states:

$$\langle n, j, l, m_J | \hat{J}_z | n, j, l', m_J' \rangle = \hbar m_J \delta_{l,l'} \delta_{m_J, m_J'}. \tag{14.36}$$

Thus, the main attention must be paid to the second term. In order to compute the matrix element $\langle n, j, l, m_J | \hat{S}_z | n, j, l', m_J' \rangle$, I will invoke Eqs. 9.93 and 9.94 expressing vectors $|n, j, l, m_J\rangle$ as linear combinations of vectors $|n, l, m\rangle |m_s\rangle$ (I added the radial index n to $|n, l, m\rangle$ in order to include the radial function, but it does not make any difference since all perturbation operators do not depend on the radial coordinate). Then for $j = l + 1/2$, I have

$$\hat{S}_z |n, j, l', m_J'\rangle = \frac{1}{\sqrt{2l'+1}} \left[\sqrt{l' + m_J' + 1/2} \left| n, l', m_J' - \frac{1}{2} \right\rangle \hat{S}_z |\uparrow\rangle + \right.$$

$$\left. \sqrt{l' - m_J' + 1/2} \left| n, l', m_J' + \frac{1}{2} \right\rangle \hat{S}_z |\downarrow\rangle \right] =$$

$$\frac{\hbar}{2} \frac{1}{\sqrt{2l'+1}} \left[\sqrt{l' + m_J' + 1/2} \left| n, l', m_J' - \frac{1}{2} \right\rangle |\uparrow\rangle - \right.$$

$$\left. \sqrt{l' - m_J' + 1/2} \left| n, l', m_J' + \frac{1}{2} \right\rangle |\downarrow\rangle \right]$$

and

$$\langle n,j,l,m_J| \hat{S}_z |n,j,l,m_J'\rangle = \frac{\hbar}{2}\frac{1}{2l+1}[l+m_J+1/2-l+m_J-1/2]\delta_{m_J m_J'}$$

$$= \hbar\frac{m_J}{2l+1}\delta_{l,l'}\delta_{m_J m_J'},$$

where I took into account that spinors describing spin-up and spin-down states are orthogonal and so are orbital states $|n,l,m\rangle$ with different values of m and l. Similarly for $j = l - 1/2$,

$$\hat{S}_z |n,j,l',m_J'\rangle =$$

$$\frac{1}{\sqrt{2l'+1}}\left[\sqrt{l'-m_J'+1/2}\left|n,l',m_J'-\frac{1}{2}\right\rangle\hat{S}_z |\uparrow\rangle -\right.$$

$$\left.\sqrt{l'+m+1/2}\left|n,l',m_J'+\frac{1}{2}\right\rangle\hat{S}_z |\downarrow\rangle\right] =$$

$$\frac{\hbar}{2}\frac{1}{\sqrt{2l'+1}}\left[\sqrt{l'-m_J'+1/2}\left|n,l',m_J'-\frac{1}{2}\right\rangle|\uparrow\rangle +\right.$$

$$\left.\sqrt{l'+m_J'+1/2}\left|n,l',m_J'+\frac{1}{2}\right\rangle|\downarrow\rangle\right]$$

$$\langle n,j,l,m_J| \hat{S}_z |n,j,l',m_J'\rangle = \frac{\hbar}{2}\frac{1}{2l+1}[l-m_J+1/2-l-m_J-1/2]\delta_{m_J m_J'} =$$

$$-\hbar\frac{m_J}{2l+1}\delta_{l,l'}\delta_{m_J m_J'}.$$

So, what I see here is that even though vectors $|n,j,l,m_J\rangle$ are not eigenvectors of $\hat{S}_z$, the matrix $\langle n,j,l',m_J| \hat{S}_z |n,j,l,m_J'\rangle$ is still diagonal with respect to indexes l and m_J. The significance of this fact is that now I do not need to diagonalize the perturbation matrix and can simply use the non-degenerate perturbation theory to find the magnetic field corrections to the fine-structure spectrum. The respective energy correction $\Delta E_{njl,m_J}$ is given by

$$\Delta E_{njl,m_J} = \frac{eB}{2m_e}\langle n,j,l,m_J| \left(\hat{J}_z + \hat{S}_z\right)|n,j,l,m_J\rangle =$$

$$\frac{eB}{2m_e}\hbar m_J\left(1\pm\frac{1}{2l+1}\right), \qquad (14.37)$$

where the plus sign corresponds to $j = l + 1/2$ and the minus sign refers to $j = l-1/2$. Both expressions appearing in Eq. 14.37 can be rewritten as a single formula with the arbitrary quantum number j as

$$\Delta E_{njl,m_J} = \frac{eB}{2m_e} \hbar m_J \left[1 + \frac{j(j+1) - l(l+1) + 3/4}{2j(j+1)} \right]. \tag{14.38}$$

If you are wondering how I managed to derive Eq. 14.38 from Eq. 14.37, I have to admit that in reality, I did nothing of the sort. To derive Eq. 14.38, one would have to use more general and sophisticated methods, which would be out of place in this book, but you can easily convince yourself that Eqs. 14.38 and 14.37 are indeed equivalent to each other. It should be noted that the magnetic field removes degeneracy of the energy levels forming the fine structure of the hydrogen spectrum not only with respect to the magnetic quantum number m_J but also with respect to the orbital index l. The expression in the square brackets in Eq. 14.38 is often called a *Lande g-factor*, $g_{J,l}$:

$$g_{J,l} = 1 + \frac{j(j+1) - l(l+1) + 3/4}{2j(j+1)}. \tag{14.39}$$

Equation 14.38 must be combined with Eq. 14.33 to get the full expression for the hydrogen spectra in the presence of a weak magnetic field:

$$E_{n,j,l,m_J} = E_n^{(0)} \left[1 - \frac{E_n^{(0)}}{\mu c^2} \left(\frac{2n}{j+\frac{1}{2}} - \frac{3}{2} \right) \right] + g_{J,l} \mu_B B m_J,$$

where I reintroduced the Bohr magneton μ_B defined in Eq. 9.5. The magnetic field contribution presented in such a form remains very much like the naive formula, Eq. 9.6 for the orbital Zeeman effect derived in Sect. 9.1. You can see that the entire effect of the spin, and spin–orbit coupling in this limit is reduced to the Lande g-factor, $g_{J,l}$.

14.2.2   Strong Magnetic Field

In the limit of the strong magnetic field, the Zeeman contribution to the atomic Hamiltonian is more important than the spin–orbit and relativistic corrections and must be included, therefore, into the zero-order unperturbed Hamiltonian. The resulting Hamiltonian is similar to the one considered in Sect. 9.1—the only difference is the presence of the spin contribution ignored in Eq. 9.3. However, the presence of the spin term does not change the main property of this Hamiltonian— it still commutes with operators $\hat{L}^2$, $\hat{L}_z$, and $\hat{S}_z$. Accordingly, vectors $|n, l, m\rangle |m_s\rangle$ are still eigenvectors of the Hamiltonian with the Zeeman term, as respective eigenvalues are, similarly, to Eq. 9.4:

$$\left[\hat{H}_{at} + \frac{eB}{2m_e}\left(\hat{L}_z + 2\hat{S}_z\right)\right]|n, l, m\rangle |m_s\rangle = E_n^{(0)} + \mu_B B \left(m + 2m_s\right).$$

The Zeeman contribution lifts the degeneracy of the hydrogen energy eigenvalues with respect to the magnetic and spin quantum numbers m and m_s, but the energy still does not depend on the orbital number l. The subspace of the degenerate eigenvectors is now formed by vectors $|n, l, m\rangle |m_s\rangle$, where all quantum numbers, except of l, are fixed. The spin–orbit coupling and relativistic correction are now being treated as a perturbation. The latter, which is proportional to $\hat{p}^4$, is obviously diagonal in this basis because vectors $|l, m\rangle$ (note the absence of n in the latter signifying that the radial function is not included at this point) are the eigenvectors of this operator. Accordingly, the contribution of this term to the energy remains to be given by Eq. 14.32 as in the absence of the magnetic field.

To deal with the spin–orbit term, I shall consider matrix

$$\langle m_s| \langle n, l, m| \left(\hat{L}_x\hat{S}_x + \hat{L}_y\hat{S}_y + \hat{L}_z\hat{S}_z\right) |n, l', m\rangle |m_s\rangle =$$

$$\langle n, l, m| \hat{L}_x |n, l', m\rangle \langle m_s| \hat{S}_x |m_s\rangle + \langle n, l, m| \hat{L}_y |n, l', m\rangle \langle m_s| \hat{S}_y |m_s\rangle +$$

$$\langle n, l, m| \hat{L}_z |n, l', m\rangle \langle m_s| \hat{S}_z |m_s\rangle .$$

Since expectation values of operators $\hat{S}_{x,y}$ in states presented by eigenvectors of $\hat{S}_z$, which appears in the expression above, are zeroes, this expression is reduced to

$$\left(\langle m_s| \langle n, l, m| \hat{L}_x\hat{S}_x + \hat{L}_y\hat{S}_y + \hat{L}_z\hat{S}_z\right) |n, l', m\rangle |m_s\rangle =$$

$$\langle n, l, m| \hat{L}_z |n, l', m\rangle \langle m_s| \hat{S}_z |m_s\rangle$$

regardless of the value of the orbital number l. The remaining expression is easy to evaluate if you remember that again, regardless of l, $|n, l, m\rangle$ is an eigenvector of $\hat{L}_z$:

$$\langle m_s| \langle n, l, m| \hat{L} \cdot \hat{S} |n, l', m\rangle |m_s\rangle = \hbar^2 m m_s \delta_{l,l'}. \tag{14.40}$$

Equation 14.40 demonstrates that the perturbation matrix defined on the degenerate subspace of vectors $|n, l, m\rangle |m_s\rangle$ is diagonal permitting me to use the non-degenerate perturbation theory to find the spin–orbit corrections to the energy in the first order of the perturbation theory. It yields

$$E_{nlmm_s} = E_n^{(0)} + \mu_B B \left(m + 2m_s\right) + m m_s \frac{1}{8\pi\varepsilon_0\varepsilon_r} \frac{Ze^2\hbar^2}{\mu^2c^2} \left\langle\frac{1}{r^3}\right\rangle =$$

$$E_n^{(0)} + \mu_B B \left(m + 2m_s\right) + m m_s \frac{1}{8\pi\varepsilon_0\varepsilon_r a_B^3} \frac{Z^4e^2\hbar^2}{\mu^2c^2} \frac{2}{l(l+1)(2l+1)n^3} =$$

$$E_n^{(0)} \left(1 - 2m m_s \frac{Z^2\alpha^2}{\varepsilon_r^2} \frac{1}{l(l+1)(2l+1)n}\right) + \mu_B B \left(m + 2m_s\right),$$

where I again introduced the fine-structure constant α. Including the relativistic correction from Eq. 14.32, I get

$$E_{nlmm_s} = E_n^{(0)} \left[1 - \frac{2Z^2\alpha^2}{\varepsilon_r^2 n} \left(\frac{mm_s - l(l+1)}{l(l+1)(2l+1)} + \frac{3}{8n} \right) \right] +$$
$$\mu_B B (m + 2m_s) . \tag{14.41}$$

The main difference with the limit of the weak field considered in Sect. 14.2.1 is that now the energy eigenvalues are determined by orbital and spin quantum numbers l, m, m_s rather than by the total angular momentum numbers j and m_J. In a hand-waving way, one can say that the strong magnetic field "breaks" the coupling between the orbital and spin angular momenta forcing them to precess independently around the direction of the field; while in the case of the weak field, the spin and orbital momenta remain coupled and precess together with the total angular momentum J.

14.2.3 Intermediate Magnetic Field

Between the two extremes of very weak and very strong magnetic field, there lies a terrain of moderate fields, which is the most difficult for exploration. In this regime the spin–orbit, relativistic, and Zeeman contributions to the Hamiltonian are of the same order and must be all treated as a perturbation. The zero-order Hamiltonian in this case is just an atomic Hamiltonian with zero-order eigenvalues $E_n^{(0)}$ given by a standard Bohr formula, Eq. 8.16. The degenerate subspace in this case is defined solely by a principal quantum number n. The dimension of this subspace is $2n^2$ (n^2 without the spin component), and the basis in this subspace can be formed either by vectors $|n, l, m\rangle |m_s\rangle$ with all allowed for a given n values of l, m, and m_s or by vectors $|n, j, l, m_J\rangle$ with again all allowed values of j, l, and m_J. Neither of these bases diagonalizes all three perturbation operators in the degenerate subspace: the Zeeman term is diagonal only in the basis $|n, l, m\rangle |m_s\rangle$, the spin–orbit Hamiltonian—only in the basis $|n, j, l, m_J\rangle$, and only the relativistic correction is diagonal in both bases.

A thoughtful reader at this point might say: "Wait a minute! In the weak field regime we used vectors $|n, j, l, m_J\rangle$ which were not the eigenvectors of the Zeeman term, but the Zeeman Hamiltonian turned out to be diagonal in this basis nonetheless. What has changed upon transition from the weak field to the moderate field regimes?" The answer to this question lies in understanding the structure of the degenerate subspace, which is different in the weak and moderate cases. If you remember, in the weak field case, the zero-order energy depended on the total angular momentum number j, which restricted the degenerate subspace to vectors with the same j and different l and m_J. And if you keep j the same, the Zeeman energy indeed becomes diagonal with respect to the remaining indexes l, m_J. The situation changes when the zero-order energy remains degenerate with respect to j, throwing eigenvectors with different j into the game. It is with respect to these vectors that the Zeeman Hamiltonian becomes nondiagonal complicating my and your life substantially.

Such a significant increase in the dimensionality of the degenerate subspace makes finding the first-order corrections to the energy analytically for an arbitrary n impossible, so I will have to limit my consideration to the simplest nontrivial case of the energy eigenvalue characterized by the principal quantum number $n = 2$ (case $n = 1$ is trivial since it only allows $l = 0$ and has only trivial degeneracy with respect to the spin index m_s). It is, of course, not as satisfactory as being able to derive general expressions for arbitrary values of all quantum numbers, but as they say, a bird in the hand is worth two in the bush, so let's get what we can.

There are eight degenerate states belonging to the energy eigenvalue $E_2^{(0)} = -E_1^{(0)}/4$, where $E_1^{(0)}$ is the absolute value of energy of the ground state. I choose to use eigenvectors of the total angular momentum as a basis, so these states are

$$|1\rangle \equiv \left|2, \frac{1}{2}, 0, -\frac{1}{2}\right\rangle, \quad |2\rangle \equiv \left|2, \frac{1}{2}, 0, \frac{1}{2}\right\rangle, \quad |3\rangle \equiv \left|2, \frac{3}{2}, 1, -\frac{3}{2}\right\rangle,$$

$$|4\rangle \equiv \left|2, \frac{3}{2}, 1, \frac{3}{2}\right\rangle, \quad |5\rangle = \left|2, \frac{1}{2}, 1, -\frac{1}{2}\right\rangle, \quad |6\rangle = \left|2, \frac{3}{2}, 1, -\frac{1}{2}\right\rangle,$$

$$|7\rangle = \left|2, \frac{3}{2}, 1, \frac{1}{2}\right\rangle, \quad |8\rangle = \left|2, \frac{1}{2}, 1, \frac{1}{2}\right\rangle. \tag{14.42}$$

To enumerate these states, I introduced for them a simplified notation $|i\rangle$, where $i = 1, 2, \cdots 8$, but it shall be understood that this numeration is quite arbitrary and is needed only to index the elements of the 8×8 perturbation matrix. The total perturbation operator in this case consists of three terms: the relativistic correction, Eq. 14.28, the spin–orbit correction, Eq. 14.5, and the Zeeman term, Eq. 14.35. The first two of these and the J_z portion of the Zeeman energy are diagonal in the basis of vectors given in Eq. 14.42. The respective diagonal elements (expectation values, really) have been calculated using the same basis for the relativistic correction in Eq. 14.32 and for spin–orbit Hamiltonian in Eqs. 14.26 and 14.27, and their sum, which is perfectly suitable for our goals here, was found in Eq. 14.33. Combining Eq. 14.33 with the results for $\hat{J}_z$ presented in Eq. 14.36, I can write for the complete diagonal portion of the perturbation matrix, $H_{pert}^{(diag)}$:

$$\left(H_{pert}^{(diag)}\right)_{i,j} = \left[-\frac{\left[E_n^{(0)}\right]^2}{\mu c^2}\left(\frac{2n}{j + \frac{1}{2}} - \frac{3}{2}\right) + \mu_B B m_J\right]\delta_{i,j}, \tag{14.43}$$

where i and j correspond to the numeration scheme introduced in Eq. 14.42. For instance, the element $i = j = 1$ corresponds to quantum numbers $n = 2, j = 1/2$, $l = 0$, and $m_J = -1/2$, so that the respective matrix element is

$$\left(H_{pert}^{(diag)}\right)_{11} = -\frac{\left[E_2^{(0)}\right]^2}{\mu c^2}\left(4 - \frac{3}{2}\right) - \frac{1}{2}\mu_B B =$$

$$-\frac{5}{2}\frac{\left[E_2^{(0)}\right]^2}{\mu c^2} - \frac{1}{2}\mu_B B. \tag{14.44}$$

Similarly you can find all other diagonal matrix elements of $H_{pert}^{(diag)}$ (I will leave the derivation to you as an exercise):

$$\left(H_{pert}^{(diag)}\right)_{22} = -\frac{5}{2}\frac{\left[E_2^{(0)}\right]^2}{\mu c^2} + \frac{1}{2}\mu_B B; \quad \left(H_{pert}^{(diag)}\right)_{33} = -\frac{1}{2}\frac{\left[E_2^{(0)}\right]^2}{\mu c^2} - \frac{3}{2}\mu_B B,$$

$$\left(H_{pert}^{(diag)}\right)_{44} = -\frac{1}{2}\frac{\left[E_2^{(0)}\right]^2}{\mu c^2} + \frac{3}{2}\mu_B B; \quad \left(H_{pert}^{(diag)}\right)_{55} = -\frac{5}{2}\frac{\left[E_2^{(0)}\right]^2}{\mu c^2} - \frac{1}{2}\mu_B B,$$

$$\left(H_{pert}^{(diag)}\right)_{66} = -\frac{1}{2}\frac{\left[E_2^{(0)}\right]^2}{\mu c^2} - \frac{1}{2}\mu_B B; \quad \left(H_{pert}^{(diag)}\right)_{77} = -\frac{1}{2}\frac{\left[E_2^{(0)}\right]^2}{\mu c^2} + \frac{1}{2}\mu_B B,$$

$$\left(H_{pert}^{(diag)}\right)_{88} = -\frac{5}{2}\frac{\left[E_2^{(0)}\right]^2}{\mu c^2} + \frac{1}{2}\mu_B B. \tag{14.45}$$

Now, let me turn to a more interesting task of finding the matrix of the operator $\hat{S}_z$ in the same basis. To this end I will again invoke the representation of $|n, j, l, m_J\rangle$ vectors in terms of $|n, l, m\rangle |m_s\rangle$ vectors using Clebsch–Gordan expansion, Eqs. 9.93 and 9.94. Specializing these equations first to the specific case of vectors $|1\rangle$, $|2\rangle$, $|3\rangle$, and $|4\rangle$ defined in Eq. 14.42, I find that one of the two terms in Eq. 9.93 or 9.94 vanishes for all these vectors leaving me with

$$|1\rangle = |2, 0, 0\rangle |\downarrow\rangle, \ |2\rangle = |2, 0, 0\rangle |\uparrow\rangle, \tag{14.46}$$

$$|3\rangle = |1, -1\rangle |\downarrow\rangle, \ |4\rangle = |1, 1\rangle |\uparrow\rangle. \tag{14.47}$$

The situation becomes more interesting for vectors $|5\rangle - |8\rangle$. Vectors $|5\rangle$ and $|8\rangle$ are characterized by $j = 1/2, l = 1$, and $m_J = \mp 1/2$ and are obtained, therefore, from Eq. 9.94

$$|5\rangle = \frac{1}{\sqrt{3}} \left(\sqrt{2}\,|1,-1\rangle\,|\uparrow\rangle - |1,0\rangle\,|\downarrow\rangle \right), \tag{14.48}$$

$$|8\rangle = \frac{1}{\sqrt{3}} \left(|1,0\rangle\,|\uparrow\rangle - \sqrt{2}\,|1,1\rangle\,|\downarrow\rangle \right), \tag{14.49}$$

while vectors $|6\rangle$ and $|7\rangle$ correspond to $j = 3/2$, $l = 1$, and $m_J = \mp 1/2$, and for them Eq. 9.93 yields

$$|6\rangle = \frac{1}{\sqrt{3}} \left(|1,-1\rangle\,|\uparrow\rangle + \sqrt{2}\,|1,0\rangle\,|\downarrow\rangle \right), \tag{14.50}$$

$$|7\rangle = \frac{1}{\sqrt{3}} \left(\sqrt{2}\,|1,0\rangle\,|\uparrow\rangle + |1,1\rangle\,|\downarrow\rangle \right). \tag{14.51}$$

The first thing to notice is that vectors $|1\rangle - |4\rangle$ are eigenvectors of the operator $\hat{S}_z$, and, therefore, all nondiagonal matrix elements involving these vectors vanish. In other words it means that the first four columns of matrix $(S_z)_{ij}$ consist of all zeroes with an exception of the elements $(S_z)_{ii}$. Since $\hat{S}_z$ is a Hermitian operator, the same is true for the first four rows as well. To give you an idea about how to compute the diagonal elements in these rows/columns, I compute $(S_z)_{11}$:

$$(S_z)_{11} = \langle\downarrow|\,\langle 2,0,0|\,\hat{S}_z\,|2,0,0\rangle\,|\downarrow\rangle = \langle\downarrow|\,\hat{S}_z\,|\downarrow\rangle = -\frac{\hbar}{2}$$

and leave it up to you to show that other diagonal elements are $(S_z)_{22} = (S_z)_{44} = \hbar/2$, while $(S_z)_{33} = -\hbar/2$. The only nondiagonal elements of $(S_z)_{ij}$ can be found in columns 5 through 8 formed by matrix elements $\langle i|\,\hat{S}_z\,|5\rangle$, $\langle i|\,\hat{S}_z\,|6\rangle$, $\langle i|\,\hat{S}_z\,|7\rangle$, and $\langle i|\,\hat{S}_z\,|8\rangle$ and the respective rows. I will begin with $\hat{S}_z\,|5\rangle$, which is easily computed to be

$$\hat{S}_z\,|5\rangle = \frac{\hbar}{2\sqrt{3}} \left(\sqrt{2}\,|1,-1\rangle\,|\uparrow\rangle + |1,0\rangle\,|\downarrow\rangle \right).$$

This vector is clearly orthogonal to vectors $|1\rangle$ through $|4\rangle$ as well as to $|7\rangle$ and $|8\rangle$ because they all contain basis vectors characterized by quantum numbers, among which at least one (l, m, or m_s) is different from the respective numbers appearing in vector $|5\rangle$. The only non-zero matrix elements in the fifth column are $\langle 6|\,\hat{S}_z\,|5\rangle$ and $\langle 5|\,\hat{S}_z\,|5\rangle$:

$$\langle 6|\,\hat{S}_z\,|5\rangle = \frac{\hbar}{6} \left(\langle\uparrow|\,\langle 1,-1| + \sqrt{2}\,\langle\downarrow|\,\langle 1,0| \right) \left(\sqrt{2}\,|1,-1\rangle\,|\uparrow\rangle + |1,0\rangle\,|\downarrow\rangle \right) = \frac{\sqrt{2}}{3}\hbar$$

$$\langle 5|\,\hat{S}_z\,|5\rangle = \frac{\hbar}{6} \left(\sqrt{2}\,\langle\uparrow|\,\langle 1,-1| - \langle\downarrow|\,\langle 1,0| \right) \left(\sqrt{2}\,|1,-1\rangle\,|\uparrow\rangle + |1,0\rangle\,|\downarrow\rangle \right) = \frac{1}{6}\hbar.$$

Obviously $\langle 5|\hat{S}_z|6\rangle = \langle 6|\hat{S}_z|5\rangle$, and by the same token, the only other non-zero elements are

$$\langle 6|\hat{S}_z|6\rangle = \langle 8|\hat{S}_z|8\rangle = -\frac{1}{6}\hbar; \quad \langle 7|\hat{S}_z|7\rangle = \frac{1}{6}\hbar$$

$$\langle 7|\hat{S}_z|8\rangle = \langle 8|\hat{S}_z|7\rangle = \frac{\sqrt{2}}{3}\hbar.$$

Now let me put all these matrix elements together into the respective matrix representing the contribution of spin $\frac{eB}{2m_e}\hat{S}_z$ to the perturbation:

$$\frac{eB}{2m_e}(S_z)_{ij} = \mu_B B \begin{bmatrix} -\frac{1}{2} & 0 & 0 & 0 & 0 & 0 & 0 & 0 \\ 0 & \frac{1}{2} & 0 & 0 & 0 & 0 & 0 & 0 \\ 0 & 0 & -\frac{1}{2} & 0 & 0 & 0 & 0 & 0 \\ 0 & 0 & 0 & \frac{1}{2} & 0 & 0 & 0 & 0 \\ 0 & 0 & 0 & 0 & \frac{1}{6} & \frac{\sqrt{2}}{3} & 0 & 0 \\ 0 & 0 & 0 & 0 & \frac{\sqrt{2}}{3} & -\frac{1}{6} & 0 & 0 \\ 0 & 0 & 0 & 0 & 0 & 0 & \frac{1}{6} & \frac{\sqrt{2}}{3} \\ 0 & 0 & 0 & 0 & 0 & 0 & \frac{\sqrt{2}}{3} & -\frac{1}{6} \end{bmatrix}. \qquad (14.52)$$

While the idea of dealing with a 8×8 matrix might be terrifying, the matrix appearing in Eq. 14.52 has a very special structure, in which small groups of elements along its main diagonal are surrounded by zeroes from all sides and are, thereby, separated from other elements. Such groups in the first four rows and columns consist just of single elements on the main diagonal, then you see the group of four elements in the sixth and seventh rows and the columns with the same numbers, which appear as 2×2 matrix surrounded by zeroes on all sides, and, finally, the similar structure appears in the lower right corner of the matrix. Matrix having such structure are called block-diagonal, and what makes them special is that each block can be considered independently of the others and treated accordingly. To see what it means, imagine that this matrix is multiplied by a column with eight elements labeled as a_i with i changing from one to eight. In the resulting column, the first four elements will appear just by themselves, not mixed with any other elements, elements a_5 and a_6 will only mix with each other , and the same is true for elements a_7 and a_8. Now if this resulting column is equated to another column to form a system of equations, the entire system will disintegrate into four single independent equations, and a pair of the systems of equations contains only two coupled coefficients. Apparently, instead of having to solve a system of eight equations in eight variables, one is left with four independent equations of a single variable and two pairs of the equations involving only two variables each.[7] As a

---

[7]Gosh, I am repeating myself, aren't I? I said exactly the same words in Chap. 9, but well, it is all for your benefit.

result, if initially an eight-variable problem might appear quite insurmountable for an analytical solution, the block-diagonal structure of the matrix makes it easily solvable even by a high school student. The matrix $H_{ij}^{(pert)}$ representing the entire perturbation with relativistic, spin–orbit, and Zeeman terms included is obtained by adding a diagonal matrix described in Eq. 14.43 through Eq. 14.45 with Eq. 14.52:

$$H_{ij}^{(pert)} =$$

$$
\begin{bmatrix}
-5\epsilon_{so} - \epsilon_Z & 0 & 0 & 0 & 0 & 0 & 0 & 0 \\
0 & -\epsilon_{so} + \epsilon_Z & 0 & 0 & 0 & 0 & 0 & 0 \\
0 & 0 & -5\epsilon_{so} - 2\epsilon_Z & 0 & 0 & 0 & 0 & 0 \\
0 & 0 & 0 & -\epsilon_{so} + 2\epsilon_Z & 0 & 0 & 0 & 0 \\
0 & 0 & 0 & 0 & -5\epsilon_{so} - \frac{1}{3}\epsilon_Z & \frac{\sqrt{2}}{3}\epsilon_Z & 0 & 0 \\
0 & 0 & 0 & 0 & \frac{\sqrt{2}}{3}\epsilon_Z & -\epsilon_{so} - \frac{2}{3}\epsilon_Z & 0 & 0 \\
0 & 0 & 0 & 0 & 0 & 0 & -\epsilon_{so} + \frac{2}{3}\epsilon_Z & \frac{\sqrt{2}}{3}\epsilon_Z \\
0 & 0 & 0 & 0 & 0 & 0 & \frac{\sqrt{2}}{3}\epsilon_Z & -5\epsilon_{so} + \frac{1}{3}\epsilon_Z
\end{bmatrix}
$$

$$(14.53)$$

where I introduced notations

$$\epsilon_{so} = \frac{\left[E_2^{(0)}\right]^2}{2\mu c^2}; \quad \epsilon_Z = \mu_B B$$

to fit this matrix on the page.

The block-diagonal form of the matrix allows me to immediately claim that vectors

$$\left|2, \frac{1}{2}, 0, -\frac{1}{2}\right\rangle = |2, 0, 0\rangle\,|\downarrow\rangle, \quad \left|2, \frac{1}{2}, 0, \frac{1}{2}\right\rangle = |2, 0, 0\rangle\,|\uparrow\rangle,$$

$$\left|2, \frac{3}{2}, 1, -\frac{3}{2}\right\rangle = |1, -1\rangle\,|\downarrow\rangle, \quad \left|2, \frac{3}{2}, 1, \frac{3}{2}\right\rangle = |1, 1\rangle\,|\uparrow\rangle$$

are eigenvectors of the entire perturbation matrix, while the corresponding eigenvalues yield four energy levels of the hydrogen with corrections due to relativistic and magnetic field effects valid to the first order in the perturbation. These levels split off the initial degenerate level characterized by $n = 2$ hydrogen energy level and are distinct for different values of quantum numbers j, l, and m_J. Using the same order of these numbers as in the designation of the states, I can write

$$E_{2,1/2,0,-1/2} = E_2^{(0)} - 5\epsilon_{so} - \epsilon_Z, \tag{14.54}$$

$$E_{2,1/2,0,1/2} = E_2^{(0)} - \epsilon_{so} + \epsilon_Z, \tag{14.55}$$

$$E_{2,3/2,1,-3/2} = E_2^{(0)} - 5\epsilon_{so} - 2\epsilon_Z, \tag{14.56}$$

$$E_{2,3/2,1,3/2} = E_2^{(0)} - \epsilon_{so} + 2\epsilon_Z. \tag{14.57}$$

The next four energy levels splitting off our $n = 2$ unperturbed value cannot be assigned a number j because they originate from the superposition of states with different values of this number. They, however, still can be assigned numbers $l = 1$ and $m_J = -1/2$ for levels arising from the states in the fifth and sixth rows/columns and $l = 1$ and $m_J = 1/2$ for energies originating from rows/columns 7 and 8. To find these energies, I have to solve two pairs of eigenvalue equations:

$$-\left(5\epsilon_{so} + \frac{1}{3}\epsilon_Z\right)a_5 + \frac{\sqrt{2}}{3}\epsilon_Z a_6 = E_{2,sup,1,-1/2}a_5$$

$$\frac{\sqrt{2}}{3}\epsilon_Z a_5 - \left(\epsilon_{so} + \frac{2}{3}\epsilon_Z\right)a_6 = E_{2,sup,1,-1/2}a_6 \qquad (14.58)$$

and

$$\left(-\epsilon_{so} + \frac{2}{3}\epsilon_Z\right)a_7 + \frac{\sqrt{2}}{3}\epsilon_Z a_8 = E_{2,sup,1,1/2}a_7$$

$$\frac{\sqrt{2}}{3}\epsilon_Z a_7 + \left(-5\epsilon_{so} + \frac{1}{3}\epsilon_Z\right)a_8 = E_{2,sup,1,-1/2}a_8, \qquad (14.59)$$

where coefficients a_5 and a_6 determine the structure of eigenvectors $|2, sup, 1, -1/2\rangle$:

$$|2, sup, 1, -1/2\rangle = a_5\,|5\rangle + a_6\,|6\rangle \equiv a_5\left|2, \frac{1}{2}, 1, -\frac{1}{2}\right\rangle + a_6\left|2, \frac{3}{2}, 1, -\frac{1}{2}\right\rangle,$$

corresponding to eigenvalues, $E_{2,sup,1,-1/2}$, while coefficients a_7, a_8 define eigenvectors $|2, sup, 1, 1/2\rangle$:

$$|2, sup, 1, 1/2\rangle = a_7\,|5\rangle + a_8\,|6\rangle \equiv a_7\left|2, \frac{1}{2}, 1, \frac{1}{2}\right\rangle + a_8\left|2, \frac{3}{2}, 1, \frac{1}{2}\right\rangle,$$

where in place of index j in the eigenvectors and the eigenvalues, I placed the abbreviation *sup* reminding that the corresponding vectors represent superposition states which are uncertain values of j. The eigenvalues are found as zeroes of the determinants

$$\begin{Vmatrix} 5\epsilon_{so} + \frac{1}{3}\epsilon_Z + E_{2,sup,1,-1/2} & -\frac{\sqrt{2}}{3}\epsilon_Z \\ -\frac{\sqrt{2}}{3}\epsilon_Z & \epsilon_{so} + \frac{2}{3}\epsilon_Z + E_{2,sup,1,-1/2} \end{Vmatrix}$$

for Eq. 14.58 and the determinate

$$\begin{Vmatrix} \epsilon_{so} - \frac{2}{3}\epsilon_Z + E_{2,sup,1,-1/2} & -\frac{\sqrt{2}}{3}\epsilon_Z \\ -\frac{\sqrt{2}}{3}\epsilon_Z & 5\epsilon_{so} - \frac{1}{3}\epsilon_Z + E_{2,sup,1,-1/2} \end{Vmatrix}$$

for Eq. 14.59. Solving the corresponding quadratic equations, I find

$$E_{2,sup_{1,2},1,-1/2} = E_2^{(0)} - 3\epsilon_{so} - \frac{\epsilon_Z}{2} \pm \sqrt{4\epsilon_{so}^2 - \frac{2}{3}\epsilon_{so}\epsilon_Z + \frac{\epsilon_Z^2}{4}} \qquad (14.60)$$

$$E_{2,sup_{1,2},1,1/2} = E_2^{(0)} - 3\epsilon_{so} + \frac{\epsilon_Z}{2} \pm \sqrt{4\epsilon_{so}^2 + \frac{2}{3}\epsilon_{so}\epsilon_Z + \frac{\epsilon_Z^2}{4}}. \qquad (14.61)$$

In the limit of zero magnetic field, both these expressions together with all other energies in Eqs. 14.54 through 14.57 are reduced to only two values $-5\epsilon_{so}$ and $-\epsilon_{so}$ as predicted in Eq. 14.33 for $j = 1/2$ and $j = 3/2$ restoring the degeneracy with respect to l and m_J, characteristic for the fine structure of the hydrogen. The structure of Eqs. 14.60 and 14.61 provides the quantitative criterion for the weak field and strong field limits discussed in Sects. 14.2.1 and 14.2.2: the former is defined by condition $\epsilon_Z \ll \epsilon_{so}$ and the latter by the opposite inequality. Expanding Eqs. 14.60 and 14.61 in a power series with respect to ϵ_Z/ϵ_{so} or $\epsilon_{so}/\epsilon_Z/$ correspondingly, and keeping only linear terms, one can reproduce results of the corresponding sections. I will let you confirm this fact as an exercise. Coefficients a_5 through a_7 are found by substituting the energy eigenvalues into the respective equations, but the resulting expressions are rather cumbersome and not too informative, so I will refrain from showing them here. In the weak and strong field limits, you will be asked to derive them as an exercise.

14.3 Problems

Problem 166 Derive Eq. 14.20.

Problem 167 Consider a hydrogen atom in vacuum and evaluate all energy levels constituting its fine structure corresponding to the principal quantum number $n = 3$. Which of these energy eigenvalues remain degenerate, and what is the degree of degeneracy? Determine the frequencies of light required to observe transitions from the ground state of hydrogen to each of these states.

For Sect. 14.2

Problem 168 Verify Eq. 14.33.

Problem 169 Consider the optical spectra associated with transition between ground state of the hydrogen atom and the energy levels characterized by $n = 3$. Assume that the observations are conducted in the magnetic field $B = 10^{-2}T$, and evaluate the wavelengths of light corresponding to each transition assuming that you are in the regime of the weak magnetic field. Repeat these calculations for magnetic field $B = 10^2 T$ using the results derived for the strong field limit. Can you relate each of the lines found in the weak field limit to the corresponding lines in the strong field limit?

Problem 170 Verify Eqs. 14.46 and 14.47.

Problem 171 Verify that vectors $|5\rangle$, $|6\rangle$, $|7\rangle$, and $|8\rangle$ defined in Eqs. 14.46–14.49 are orthogonal to the vector obtained by applying operator $\hat{S}_z$ to vector $|3\rangle$.

Problem 172 Demonstrate that Eqs. 14.60 and 14.61 for the energy eigenvalues reproduce results obtained in the limits of weak and strong magnetic fields in the linear in parameters ϵ_Z/ϵ_{so} or ϵ_{so}/ϵ_Z approximation, correspondingly.

Problem 173 Find coefficients a_5 through a_8 in Eqs. 14.58 and 14.59 in the weak and strong field limits using the expressions for the eigenvalues derived in the previous problem.

Problem 174 Find the energy levels arising due to Zeeman splitting of $n = 3$ eigenvalue of the hydrogen in the presence of the spin–orbit and relativistic corrections in the intermediate field regime. (Hint: This is a long problem which involves having to deal with an 18×18 perturbation matrix, which, however, hopefully will have a block-diagonal structure allowing this problem to be solved.)

Chapter 15
Emission and Absorption of Light

15.1 Time-Dependent Perturbation Theory

It is no secret that quantum mechanics grew up to a large extent out of efforts to understand emission and absorption spectra of atoms. Indeed, the famous Planck distribution was introduced to explain the spectrum of the black-body radiation (not to be confused with black hole, of course), and one of Bohr's postulates dealt explicitly with conditions for the emission or absorption of light by atoms. It is not surprising, therefore, that one of the first problems studied by Paul Dirac in his seminal 1926 paper on "new" quantum mechanics[1] (as opposed to the pre-1925 "old" quantum theory based on Bohr–Sommerfeld quantization principle) was the problem of interaction between light and atoms. This problem belongs to a broad class of problems, in which the Hamiltonian of a system can be presented as a sum of the "unperturbed" Hamiltonian $\hat{H}_0$ and a perturbation $\hat{V}$. However, unlike perturbations considered in Chap. 13, operator $\hat{V}$ is now allowed to depend on time so that we end up with the problem involving a time-dependent Hamiltonian:

$$\hat{H} = \hat{H}_0 + \hat{V}(t). \tag{15.1}$$

A time dependence of the Hamiltonian is a big deal because it does not just completely change the way we must approach a problem, it changes the questions that must be asked. To some extent this issue has already been discussed in Sect. 10.2, where I dealt with a two-level system interacting with a periodic electric field and introduced the ideas of quantum transitions and their probabilities that replaced eigenvalues and eigenvectors as main objects of study. If you do not remember what I am talking about, please, go back to Sect. 10.2 for a brief refresher.

---

[1]P.A.M. Dirac, On the theory of quantum mechanics. Proc. R. Soc. A **112**, 661 (1926).

© Springer International Publishing AG, part of Springer Nature 2018
L.I. Deych, *Advanced Undergraduate Quantum Mechanics*,
https://doi.org/10.1007/978-3-319-71550-6_15

The simple model considered in Sect. 10.2 as well as its more generic extension presented by Eq. 15.2 makes an important assumption about the perturbation operator $\hat{V}(t)$. It is supposed that the perturbation potential is due to interaction with some external objects or fields, which are not a part of the system described by Hamiltonian 15.2. Whatever parameters characterizing the perturbation appear in $\hat{V}(t)$, they are assumed to be fixed by some external conditions and do not change due to interaction with the atom. For instance, if the perturbation is an electric field of an electromagnetic wave, this field is taken as being emitted by an external source (laser, lamp, sun, etc.), and any possible changes in it caused by the interaction with the atom are neglected.[2] This assumption is not always valid. For instance, if you would want to describe the lasing phenomenon, you would have to complement Schrödinger or Heisenberg equations for the atom with Maxwell equations for the field that would include the dipole moment of the atoms as a source term. In this approach the atom and the electromagnetic field are treated self-consistently, as an interacting system with interdependent dynamics, with the only difference that light is considered as a classical (not quantum) object. While it is an improvement compared to the initial assumption of the independent electromagnetic field, it is still not sufficient to describe such effects as the finite laser linewidth or spontaneous emission. Dirac understood the shortcomings of this approach very well, and, therefore, just a year after publishing the 1926 paper, he produced another publication, where he described the electromagnetic field as a set of quantized dynamical variables and solved the light–atom interaction problem treating both atom and the field quantum mechanically.[3] Unfortunately, even the semiclassical theory of atom–light interaction, leave alone its full quantum treatment, is beyond the scope of this book, so you will have to wait till you are ready for a more advanced quantum mechanics course to learn about Dirac's derivation of the rate of spontaneous emission as well as about the further development of Dirac's work in Wigner–Weisskopf theory of spontaneous emission. You will also have to refer to more serious books on laser theory to learn about the Schawlow–Townes formula for the fundamental laser linewidth. Here and for now, you are stuck with the simplest version of the theory of interaction between quantum systems and light.

In general, the spectrum of the Hamiltonian $\hat{H}_0$ in Eq. 15.1 might contain both discrete and continuous segments with an infinite number of eigenvalues and eigenvectors. When the number of the unperturbed states is increased beyond the two, even the rotating wave approximation, which I had to use in Sect. 10.2 to solve the two-level model, is not going to help me much. Therefore, several approximation schemes have been invented to deal with this problem. One of the most popular of them is the time-dependent perturbation theory allowing to determine the evolution of a state of a system due to a time-dependent perturbation. Having found the time

---

[2]If you are wondering what kind of changes the atom can impose on the field, here are two examples: (a) absorption by atoms can change its amplitude (or the number of photons if you prefer quantum language), and (b) atoms can emit light at the same frequency as the incident field resulting in the increase of its amplitude.

[3]P.A.M. Dirac, The quantum theory of the emission and absorption of radiation. Proc. R. Soc. A **114**, 243 (1927).

dependence of the state, you would be able to determine the probability distributions and expectation values of any quantity of interest.

I will begin assuming that I know all eigenvalues E_n and eigenvectors $|\alpha_n\rangle$ of the unperturbed Hamiltonian $\hat{H}_0$:

$$\hat{H}_0 |\alpha_n\rangle = E_n |\alpha_n\rangle . \tag{15.2}$$

The lower index n here might actually be a composite index consisting of multiple elements, such as principal, azimuthal, and magnetic quantum numbers for a hydrogen atom, or a set of numbers n_x, n_y, n_z characterizing the states of a particle in a three-dimensional potential well, and may include a spin magnetic number as well. Correspondingly all summations over n appearing below imply summations over all respective indexes, and all vectors $|\alpha_n\rangle$ are assumed to be normalized and orthogonal to each other, even if they belong to the same degenerate energy eigenvalue. Using these states as a basis, I can present an arbitrary time-dependent vector $|\psi(t)\rangle$ as

$$|\psi(t)\rangle = \sum_n c_n(t)e^{-iE_n t/\hbar} |\alpha_n\rangle . \tag{15.3}$$

Equation 15.3 is a simple generalization of Eq. 10.13 from Sect. 10.2, where I have already explained the meaning of coefficients c_n and exponential factors $\exp(-iE_n t/\hbar)$. It should be noted that Eq. 15.3 implies that all eigenvectors of $\hat{H}_0$ belong to the discrete spectrum. If this is not the case, the sum over n has to be complemented by an integral over the continuous eigenvalues. Alternatively, I can always turn a continuous spectrum into a quasi-continuous by imposing periodic boundary conditions as explained in Sect. 11.5. Substituting this expression to the Schrödinger equation

$$i\hbar d |\psi\rangle /dt = \hat{H} |\psi\rangle$$

and taking into account Eqs. 15.1 and 15.2 yield the system of equations for the unknown coefficients c_n:

$$i\hbar \left(-\sum_n \frac{iE_n}{\hbar} c_n(t)e^{-iE_n t/\hbar} |\alpha_n\rangle + \sum_n \frac{dc_n(t)}{dt} e^{-iE_n t/\hbar} |\alpha_n\rangle \right) =$$

$$\sum_n c_n(t)e^{-iE_n t/\hbar} \hat{H}_0 |\alpha_n\rangle + \sum_n c_n(t)e^{-iE_n t/\hbar} \hat{V} |\alpha_n\rangle \rightarrow$$

$$\sum_n E_n c_n(t)e^{-iE_n t/\hbar} |\alpha_n\rangle + i\hbar \sum_n \frac{dc_n(t)}{dt} e^{-iE_n t/\hbar} |\alpha_n\rangle =$$

$$\sum_n E_n c_n(t)e^{-iE_n t/\hbar} |\alpha_n\rangle + \sum_n c_n(t)e^{-iE_n t/\hbar} \hat{V} |\alpha_n\rangle \Rightarrow$$

$$i\hbar \sum_n \frac{dc_n(t)}{dt} e^{-iE_n t/\hbar} |\alpha_n\rangle = \sum_n c_n(t)e^{-iE_n t/\hbar} \hat{V} |\alpha_n\rangle .$$

Premultiplying the last expression by the bra vector $\langle \alpha_m |$ and using the orthogonality of the basis vectors, I generate the following system of equations for coefficients c_n:

$$i\hbar \frac{dc_m(t)}{dt} = \sum_n V_{mn} e^{i\omega_{mn}t} c_n, \tag{15.4}$$

where ω_{mn}, called *transition frequency*, is defined as

$$\omega_{mn} = \frac{E_m - E_n}{\hbar}. \tag{15.5}$$

V_{mn} is again the same perturbation matrix element $V_{mn} = \langle \alpha_m | \hat{V} | \alpha_n \rangle$, which was introduced earlier in Chap. 13. Equation 15.4 is a generalization of the system of equations, Eqs. 10.16 and 10.17, to the case of multiple states and an arbitrary perturbation potential.

In the most generic case, this is a system of infinitely many linear differential equations of the first order, which is equivalent to the initial Schrödinger equation and still cannot be solved exactly. The advantage of this representation over the initial abstract form of the Schrödinger equation is that all the complexity of the perturbation potential, which is not directly concerned with its time dependence (e.g., its dependence on coordinate, momentum, or spin operators), is coded into the corresponding matrix elements, which can be computed using any of the available presentations for the corresponding operators. To make this argument clearer, imagine the original Schrödinger equation in the position representation, when it takes the form of a differential equation in partial derivatives describing dependence of the wave function upon at least four variables (time plus three coordinates). The transition to the equations for the coefficient c_n given by Eq. 15.4 eliminates the spatial coordinates, which are integrated out and are hidden in the definition of the matrix elements V_{mn}.

The price for this is, of course, that now, instead of one equation, you have to deal with a system of infinitely many, albeit much simpler, equations. However, there are several tools, which you can use to make the problem manageable. One of them— restricting the number of states included in the expansion, Eq. 15.3 (and, hence, the number of equations in the system)—has been already demonstrated in Sect. 10.2. In principle, if needed, this approach can be extended to include as many states as necessary so that the resulting finite system of equations can be solved at least numerically with the help of a computer. In this section, however, I will introduce a different method of solving Eq. 15.4 based upon the assumption that the perturbation operator is, in some sense, small. This method, which was first used in the already mentioned famous 1926 paper by P. Dirac, and many times since, is responsible for the most important results in quantum theory of light–atom interaction.

Now, back to Eq. 15.4. To develop the perturbation approach to this equation, I will again pull out of a perturbation operator a formal "strength" parameter ϵ:

$\hat{V} \to \epsilon\hat{V}$, which I will use for the same bookkeeping purposes as in Chap. 13. Now, I will present an arbitrary coefficient c_m as a power series of the form:

$$c_m = c_m^{(0)} + \epsilon c_m^{(1)} + \epsilon^2 c_m^{(2)} \cdots , \tag{15.6}$$

where the first, zero-order, term reproduces the coefficients corresponding to whatever state the system would be in the absence of the perturbation, the second, first-order, term introduces corrections linear in the perturbation operator, the next one adds corrections quadratic in the perturbation, and so on and so forth. Substituting this expression into Eq. 15.4, I generate an equation containing terms with various powers of ϵ:

$$i\hbar \left(\frac{dc_m^{(0)}}{dt} + \epsilon \frac{dc_m^{(1)}}{dt} + \epsilon^2 \frac{dc_m^{(2)}}{dt} + \cdots \right) =$$

$$\epsilon \sum_n V_{mn} e^{i\omega_{mn}t} \left(c_n^{(0)} + \epsilon c_n^{(1)} + \epsilon^2 c_n^{(2)} \cdots \right).$$

This equation can only be satisfied for any arbitrary value of ϵ if and only if terms with equal powers of ϵ on the left-hand and the right-hand sides of this equation are individually equal to each other. Equating these terms yields the following set of equations:

$$i\hbar \frac{dc_m^{(0)}}{dt} = 0,$$

$$i\hbar \frac{dc_m^{(1)}}{dt} = \sum_n V_{mn} e^{i\omega_{mn}t} c_n^{(0)}, \tag{15.7}$$

$$\vdots$$

$$i\hbar \frac{dc_m^{(r)}}{dt} = \sum_n V_{mn} e^{i\omega_{mn}t} c_n^{(r-1)},$$

$$\vdots$$

The first equation simply states that the zero-order coefficients $c_m^{(0)}$ do not depend on time, the second equation defines the first-order coefficients $c_m^{(1)}$ in terms of $c_m^{(0)}$, the next equation defines $c_m^{(2)}$ in terms of $c_m^{(1)}$, and each next coefficient $c_m^{(r)}$ is defined in terms of coefficients of the preceding order $c_m^{(r-1)}$. Thus, starting with $c_m^{(0)}$, which are supposed to be known, one can, in principle, solve each of the equations in this sequence and find the coefficients $c_n^{(r)}$ in any order of the perturbation.

However, in order to actually proceed with this plan, I need additional information, namely, I've got to know the state of the system at some predetermined time instant t_0. Mathematically speaking, what I need are initial conditions for the expansion coefficients c_m, determined by the concrete physical context of the problem. As it has already been explained in Sect. 10.2, in order to provide meaningful physical interpretation to calculations with time-dependent Hamiltonians, one has to assume that the perturbation has a beginning and that it also has to end (doesn't this statement apply to everything in life?). Then the state of the system in the absence of the perturbation, characterized by zero-order coefficients $c_m^{(0)}$, defines natural initial conditions for $c_m(t)$: $c_m(t_0) = c_m^{(0)}$. This, in turn, means that all higher-order coefficients $c_m^{(r)}$ to $c_m(t)$ must vanish at $t = t_0$. Strictly speaking both *on* and *off* times might not always be rigorously defined because switching-on and switching-off of the perturbation is never instantaneous. Still, in some situations, assigning an exact value to t_0, which is usually chosen to be $t_0 = 0$, is justifiable and is frequently used in practical calculations. In other instances it might make more sense to assume that the perturbation grows gradually from zero and assign $t_0 = -\infty$. The particular choice usually depends on the problem at hand, but in what follows, unless I explicitly state otherwise, it will be assumed that the perturbation is being turned on instantaneously and choose $t_0 = 0$.

Now, once the initial conditions for all differential equations in Eq. 15.7 are specified, they can be solved quite easily by simple integration of both sides with respect to time. The simplest way to incorporate the initial conditions into the solution is to write them down as definite integrals with lower limit set to 0 and upper limit to t:

$$c_m^{(0)} = const \tag{15.8}$$

$$c_m^{(1)}(t) = \frac{1}{i\hbar} \sum_n c_n^{(0)} \int_0^t d\tau\, V_{mn}(\tau)\, e^{i\omega_{mn}\tau} \tag{15.9}$$

$$\vdots$$

$$c_m^{(r)}(t) = \frac{1}{i\hbar} \sum_n \int_0^t d\tau\, V_{mn}(\tau)\, e^{i\omega_{mn}\tau} c_n^{(r-1)}(\tau) \tag{15.10}$$

$$\vdots$$

It is quite obvious that this form of the solution ensures that all $c_m^{(r)}$ with $r > 0$ vanish at $t = 0$ as required by the initial conditions. Equation 15.9 presents a ready-to-use solution for the first-order correction to the time-dependent state vector $|\alpha(t)\rangle$. As an illustration I will also derive the final expression for the second-order correction, but I am not going to make much use of it in what follows. Equation 15.10 adapted to $r = 1$ yields

$$c_m^{(2)}(t) = \frac{1}{i\hbar} \sum_n \int_0^t d\tau \, V_{mn}(\tau) \, e^{i\omega_{mn}\tau} c_n^{(1)}(\tau).$$

Substitution of Eq. 15.9 yields

$$c_m^{(2)}(t) = \left(\frac{1}{i\hbar}\right)^2 \sum_n \sum_p c_p^{(0)} \int_0^t d\tau \int_0^\tau d\tau_1 V_{mn}(\tau) \, V_{np}(\tau_1) \, e^{i\omega_{mn}\tau} e^{i\omega_{np}\tau_1} \qquad (15.11)$$

where I made necessary changes in the summation indexes and integration variables to be able to write down the expression as a double sum and a double integral. It is not too difficult to write down an expression for an arbitrary coefficient $c_m^{(r)}(t)$, which will obviously include r summations and integrations; I will let you enjoy figuring it out on your own as an exercise.

If we are only interested in the effects of the first order (*linear*) in the perturbation potential, Eq. 15.9 contains all the general information we need. Equation 15.11 obviously describes second order or quadratic in perturbation effects, the next term will yield third-order effects, and so on. In the context of the light–matter interaction, you can say that Eq. 15.9 describes linear optical effects, while other equations correspond to nonlinear phenomena of the second, third, or even higher orders. Nonlinear optics is a fascinating field, but it is impossible to embrace the unembraceable, so I will have to reluctantly limit the scope of this chapter by the first-order corrections only.

In order to proceed any further, I need to specify the set of initial coefficients $c_m^{(0)}$ and the actual time dependence of the perturbation potential. Obviously, there exist infinite possibilities with respect to both of these, so I will focus on the situations which are most relevant for typical experiments. While lately experimentalists learned how to create complex superposition states of atoms, in most practical situations, you will deal with an initial state represented by a simple eigenvector of the unperturbed Hamiltonian. In this case all coefficients $c_m^{(0)} \equiv c_m(0)$ with $m \neq m_0$, where $|\alpha_{m_0}\rangle$ is a vector representing the initial state of the system, are equal to zero, while $c_{m_0}^{(0)} \equiv c_{m_0}(0) = 1$. Consequently, the summation over n in Eq. 15.9 now vanishes, and one ends up with the following:

$$c_m^{(1)}(t) = \frac{1}{i\hbar} \int_0^t d\tau \, V_{mm_0}(\tau) \, e^{i\omega_{mm_0}\tau}. \qquad (15.12)$$

Going back to Eq. 15.3, I find the first-order approximation for the time-dependent state of the system:

$$|\psi(t)\rangle = e^{-iE_{m_0}t/\hbar} |\alpha_{m_0}\rangle +$$

$$\frac{1}{i\hbar} e^{-iE_m t/\hbar} \sum_m \int_0^t d\tau V_{mm_0}(\tau) e^{i\omega_{mm_0}\tau} |\alpha_m\rangle . \qquad (15.13)$$

If the perturbation is switched off at $t = t_f$, and a measurement of energy is carried out immediately afterward, the probability that the system is found in the stationary state $|\alpha_m\rangle$ is given by

$$p_{mm_0} = \frac{1}{\hbar^2} \left| \int_0^{t_f} d\tau V_{mm_0}(\tau) e^{i\omega_{mm_0}\tau} \right|^2 . \qquad (15.14)$$

The double-index notation of the probability is a reminder of the fact that this probability is conditional (predicated on a system being initially at the state $|\alpha_{m_0}\rangle$) and can be interpreted as a probability of transition between the initial and final states. It shall be noted that the requirement that the perturbation ends at $t = t_f$ shall not be understood too literally. It is sufficient if the act of measurement, taking place at this instant, disrupts the "natural" dynamics of the state prescribed by the perturbation drastically enough so that the entire evolution begins from the very beginning. An example of such a disruption can be the act of spontaneous emission accompanied by the transition of the atom to one of the lower energy states, which will restart the perturbation-induced dynamics. If the final state $|\alpha_m\rangle$ belongs to a non-degenerate eigenvalue of $\hat{H}_0$, it is sufficient to measure the spectral intensity of the spontaneous emission to identify it. If, however, $|\alpha_m\rangle$ belongs to a degenerate eigenvalue, additional characteristics such as polarization, angular distribution, dependence on magnetic field, etc. might be required to achieve the unique identification of the final state. If, however, you are only interested in the probability of a specific energy eigenvalue, you can find it by summing up p_{mm_0} over all states belonging to a given degenerate energy level.

There are several important models of the time dependence of the perturbation matrix elements $V_{mm_0}(t)$ which needs to be considered. However, I will begin with a simple toy example of a "pulse" perturbation described by the following time dependence:

$$V_{mm_0}(t) = v_{mm_0}\theta(t)e^{-t/t_0}, \qquad (15.15)$$

where $\theta(t)$ is a step function, which is equal to unity for positive values of the argument and vanishes for the negative ones. Substituting Eq. 15.15 into Eq. 15.12, I get

$$c_m^{(1)}(t) = \frac{v_{mm_0}}{i\hbar} \int_0^t d\tau e^{(i\omega_{mm_0} - 1/t_0)\tau} = \frac{v_{mm_0}}{i\hbar} \frac{1}{i\omega_{mm_0} - 1/t_0} e^{(i\omega_{mm_0} - 1/t_0)\tau}\Big|_0^t =$$

$$\frac{v_{mm_0}}{i\hbar} \frac{e^{(i\omega_{mm_0} - 1/t_0)t} - 1}{i\omega_{mm_0} - 1/t_0}$$

and

$$p_{mm_0} = \left|c_m^{(1)}(t_f)\right|^2 = \frac{|v_{mm_0}|^2}{\hbar^2} \frac{\left(e^{(i\omega_{mm_0} - 1/t_0)t_f} - 1\right)\left(e^{(-i\omega_{mm_0} - 1/t_0)t_f} - 1\right)}{\omega_{mm_0}^2 + 1/t_0^2} =$$

$$\frac{|v_{mm_0}|^2}{\hbar^2} \frac{1 + e^{-2t_f/t_0} - 2e^{-t_f/t_0}\cos\left(\omega_{mm_0}t_f\right)}{\omega_{mm_0}^2 + 1/t_0^2}.$$

There are several lessons to be learned from this example. First, the transition probability decreases as the spectral distance between the initial and final states exemplified by ω_{m,m_0} increases. This fact can be used as a justification for limiting the number of states included into consideration, in particular, for the two-level model studied in Chap. 10. Second, the number of states δN with appreciable transition probabilities is determined by the product of the spectral distance between two adjacent states $\delta\omega_m = |E_m - E_{m-1}| = |\omega_{m,m_0} - \omega_{m-1,m_0}|$ and the time scale of the exponential decay of the perturbation t_0: $\delta N \approx (\delta\omega_m t_0)^{-1}$. Indeed, as long as $|\omega_{m,m_0}| \ll 1/t_0$, the transmission probabilities remain virtually independent of ω_{m,m_0}, and the probability starts decreasing only as $|\omega_{m,m_0}|$ exceeds $1/t_0$.

This result also has another important interpretation. The temporal Fourier transform $\tilde{V}_{mm_0}(\omega)$ of the perturbation matrix element gives

$$\tilde{V}_{mm_0}(\omega) = \frac{v_{mm_0}}{\sqrt{2\pi}} \int_0^\infty e^{i\omega t} e^{-t/t_0} dt = -\frac{v_{mm_0}}{\sqrt{2\pi}} \frac{1}{i\omega - 1/t_0},$$

and the corresponding power spectrum $\left|\tilde{V}_{mm_0}(\omega)\right|^2$, which determines relative contribution of different frequency component into the perturbation potential, is

$$\left|\tilde{V}_{mm_0}(\omega)\right|^2 = \frac{|v_{mm_0}|^2}{2\pi} \frac{1}{\omega^2 + 1/t_0^2}.$$

This is a so-called Lorentzian function, which has its maximum at $\omega = 0$ and is reduced by a factor of 2 when $\omega = 1/t_0$. For this reason, $1/t_0$, called half-width at half-maximum (HWHM), can be considered as a measure of the breadth of the spectrum of the perturbation. Then condition $|\omega_{m,m_0}| \ll 1/t_0$ means that the largest effect the perturbation has on those energy levels, which fall within the spectral range of the perturbation potential. Finally, one can see that in the limit $t_f \to \infty$, the transition probability becomes proportional to the power spectrum of the perturbation matrix elements at frequency ω_{mm_0}:

$$p_{mm_0} = \frac{2\pi}{\hbar^2} \left| \tilde{V}_{mm_0} (\omega_{mm_0}) \right|^2 .$$

15.2   Fermi's Golden Rule

Fermi's golden rule is one of the most frequently used (and often overused) results of the first-order perturbation theory. It appears in several different reincarnations in a variety of situations and deserves an entire section devoted to it. It all begins with a problem of finding transition probabilities due to a monochromatic perturbation.

15.2.1   First-Order Transmission Probability in the Case of a Monochromatic Perturbation

While a time dependence in the form of a simple trigonometric function

$$V_{mm_0} (t) = 2v_{mm_0} \theta(t) \cos \Omega t \tag{15.16}$$

is the simplest one, it generates transition probabilities that are not quite trivial to interpret, and dealing with them requires certain caution. Matrix elements v_{mm_0} in Eq. 15.16 are assumed to obey condition $v_{mm_0} = v_{mm_0}^*$ to ensure that the entire matrix $V_{mm_0} (t)$ is Hermitian. The choice of the cosine function to represent the time dependence is absolutely arbitrary; the same results would follow if I chose to deal with a sine function instead. The factor 2 is introduced also for convenience and amounts to a redefinition of the remaining matrix elements.

To compute the coefficients $c_m^{(1)} (t)$, it is convenient to rewrite the perturbation in terms of exponential functions, which results in the following integral:

$$c_m^{(1)}(t) = \frac{v_{mm_0}}{i\hbar} \int_0^t d\tau \left(e^{i\Omega\tau} + e^{-i\Omega\tau} \right) e^{i\omega_{mm_0}\tau} =$$

$$\frac{v_{mm_0}}{i\hbar} \left[\frac{\exp\left[i\left(\Omega + \omega_{mm_0}\right)t\right] - 1}{i\left(\Omega + \omega_{mm_0}\right)} + \frac{\exp\left[i\left(-\Omega + \omega_{mm_0}\right)t\right] - 1}{i\left(-\Omega + \omega_{mm_0}\right)} \right] =$$

$$\frac{2v_{mm_0}}{i\hbar}\left[\exp\left[\frac{i\left(\Omega+\omega_{mm_0}\right)t}{2}\right]\frac{\exp\left[\frac{i\left(\Omega+\omega_{mm_0}\right)t}{2}\right]-\exp\left[-\frac{i\left(\Omega+\omega_{mm_0}\right)t}{2}\right]}{2i\left(\Omega+\omega_{mm_0}\right)}+\right.$$

$$\left.\exp\left[\frac{i\left(-\Omega+\omega_{mm_0}\right)t}{2}\right]\frac{\exp\left[\frac{i\left(-\Omega+\omega_{mm_0}\right)t}{2}\right]-\exp\left[-\frac{i\left(-\Omega+\omega_{mm_0}\right)t}{2}\right]}{2i\left(-\Omega+\omega_{mm_0}\right)}\right]=$$

$$\frac{2v_{mm_0}}{i\hbar}\left[\exp\left[\frac{i\left(\Omega+\omega_{mm_0}\right)t}{2}\right]\frac{\sin\frac{\left(\Omega+\omega_{mm_0}\right)t}{2}}{\Omega+\omega_{mm_0}}+\right.$$

$$\left.\exp\left[\frac{i\left(-\Omega+\omega_{mm_0}\right)t}{2}\right]\frac{\sin\frac{\left(-\Omega+\omega_{mm_0}\right)t}{2}}{-\Omega+\omega_{mm_0}}\right].$$

This expression can be rewritten in terms of the so-called *sinc* function defined as

$$sinc(x) = \frac{\sin x}{x}. \tag{15.17}$$

This name is an abbreviation of the full Latin name *sinus cardinalis*; it appears quite often in the theory of signal processing, information theory, and, as you can see, in calculations of transition probabilities with the harmonic perturbation. Its main features are the main maximum at $x = 0$, where it takes value of unity, and equidistant zeroes at $x_n = \pi n$, $n = \pm 1, \pm 2, \cdots$. The zeroes separate secondary maximums (or minimums) with ever-decreasing absolute values. Among other important properties of this function is the following integral:

$$\int_{-\infty}^{\infty} sinc^2(x)dx = \pi. \tag{15.18}$$

It is also useful to take a closer look at a so-called rescaled sinc function defined as $sinc(ax)$ and considered as a function of x. What is interesting about this function is that the distance between its zeroes given by

$$x_{n+1} - x_n = \frac{\pi}{a}$$

goes to zero as parameter a tends to infinity. It means that while the value of the function at $x = 0$ remains unity, the function becomes more and more "compressed" around this point as a is increasing (see Fig. 15.1, where I plotted function $sinc(ax)$

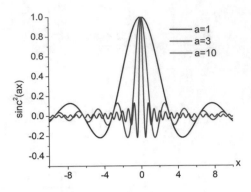

Fig. 15.1 Function $sinc(ax)$ for three values of parameter a

a=1
a=3
a=10

for several values of a). In terms of this function, the expression for the coefficients $c_m^{(1)}$ can be written down as

$$
c_m^{(1)}(t) = \frac{v_{mm_0}}{i\hbar}t\left[\exp\left[\frac{i\left(\Omega + \omega_{mm_0}\right)t}{2}\right] sinc\left(\frac{(\Omega + \omega_{mm_0})t}{2}\right) +\right.
$$
$$
\left.\exp\left[\frac{i\left(-\Omega + \omega_{mm_0}\right)t}{2}\right] sinc\left(\frac{(-\Omega + \omega_{mm_0})t}{2}\right)\right].
$$
(15.19)

Properties of the sinc function offer us a valuable lesson in connection with Eq. 15.19. Since $\Omega + \omega_{mm_0}$ and $-\Omega + \omega_{mm_0}$ never turn zero for the same combinations of Ω and ω_{mm_0}, the corresponding terms in this equation are always disparate in their magnitudes, and the disparity grows stronger with the passage of time. It is clear from the latest remark that I am treating the combination of frequencies as the argument of the *sinc* function, while time t plays the role of the scaling parameter a.

The argument of the *sinc* function in the first term of Eq. 15.19 turns zero when the external frequency Ω is

$$
\Omega = -\omega_{mm_0},
$$
(15.20)

which is only possible if ω_{mm_0} is negative because the frequency Ω is defined as a positive quantity. Negative ω_{mm_0} corresponds to $E_m < E_{m_0}$, i.e., to transitions from the higher energy state to lower energy states, and authors of many textbooks at this point add that it, obviously, corresponds to a transition in which light is being emitted. This conclusion is usually justified by appealing to the energy conservation principle—if the energy of the atom decreases, it must go somewhere, and if the perturbation involved in the transition is an electromagnetic wave, it is quite natural to assume that this is where the energy goes. I would like to emphasize, however, that the approach used to derive Eq. 15.19 is based on the assumption that the perturbation potential is not affected by the state of the atom, so that any processes involving emission or absorption of light, rigorously speaking, cannot be described within this approach. Having cautioned you, now I have to admit being a bit facetious because as long as an interaction with the electromagnetic field is the only interaction the atom can be engaged in, this conclusion is, indeed, true. It can be confirmed directly by a complete theory explicitly taking into account the absorption and emission of light by the atom.

The argument of the *sinc* function in the second term of Eq. 15.19 turns zero when

$$\Omega = \omega_{mm_0}, \tag{15.21}$$

i.e., for transitions from a lower energy state to the higher energy states. Using again energy conservation arguments, you can argue that since the energy required for an atom to make this transition can only come from the perturbation, this process must correspond to absorption of light if the perturbation is again provided by an incident electromagnetic wave. Regardless of which of the conditions, Eq. 15.20 or 15.21, is true, you need to keep only one term in Eq. 15.19. Thus, limiting consideration only to coefficients for which $\omega_{mm_0} > 0$ (and if m_0 is the ground state, it would include all the coefficients, since in this case only transitions to higher states are possible), we can simplify the expression for $c_m^{(1)}(t)$ to

$$c_m^{(1)}(t) = \frac{v_{mm_0}}{i\hbar} t \exp\left[\frac{i(-\Omega + \omega_{mm_0})t}{2}\right] sinc\left(\frac{(-\Omega + \omega_{mm_0})t}{2}\right), \tag{15.22}$$

which yields the transition probability

$$p_{mm_0} = \left|c_m^{(1)}(t)\right|^2 - \frac{|v_{mm_0}|^2}{\hbar^2} t^2 sinc^2\left(\frac{\omega_{mm_0} - \Omega}{2}t\right). \tag{15.23}$$

However, the derivation of this formula is not the entire story. The real story here is the dependence of the transition probability on time and the external frequency detuning $\Omega - \omega_{mm_0}$. When this detuning is equal to zero (people say in this case that the perturbation is *in resonance* with the transition), the probability grows quadratically with time. This growth imposes the upper limit on how long the perturbation can be allowed to act before the approximation used in this approach loses its validity. Using the applicability condition $p_{mm_0} \ll 1$, I obtain for the time limit

$$t_f \ll t^* = \frac{\hbar}{|v_{mm_0}|}. \tag{15.24}$$

It is interesting to compare the expression for t^* with Eq. 10.24 for the Rabi frequency Ω_R of a two-level system. With obvious changes in the notations ($|v_{mm_0}| \to \mathcal{E}$), Eq. 15.24 can be recast as $t_f \ll 1/\Omega_R$ allowing to claim that the first-order perturbation theory remains valid as long as time t_f remains much smaller than the period of the Rabi oscillations defined by the perturbation matrix element

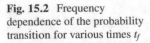

Fig. 15.2 Frequency dependence of the probability transition for various times t_f

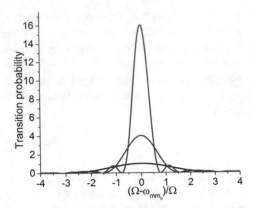

$|v_{mm_0}|$. The quadratic raise of the resonance transition probability with time is not the whole story, again. The second effect occurring simultaneously is the shrinkage with time of the interval of frequencies corresponding to the nonvanishing transition probabilities. This interval, for instance, can be estimated as the spectral distance between the resonance $\Omega_r = \omega_{mm_0}$ and the frequency $\Omega_z - \omega_{mm_0} = 2\pi/t$, where the *sinc* function becomes equal to zero for the first time. Beyond this point even formally non-zero values of the probability are too small to be taken seriously (see Fig. 15.2). With increasing t this spectral interval decreases as $1/t$, so that even though the maximum value of the probability grows as t^2, the area covered by the function $p_{mm_0}(\Omega - \omega_{mm_0})$ increases only as t (*hight* × *width*). This area is expressed mathematically as an integral of the $p_{mm_0}(\Omega - \omega_{mm_0})$ with respect to $\Omega - \omega_{mm_0}$, and while you are scratching your head trying to understand why it even matters to me, let me try to explain. Consider the following integral:

$$\int_{-\infty}^{\infty} sinc^2\left(\frac{ax}{2}\right) dx = \frac{2\pi}{a}, \tag{15.25}$$

where I used Eq. 15.18 and the obvious substitution of variables. This result means that

$$\lim_{a \to \infty} \left(\frac{a}{2\pi} \int_{-\infty}^{\infty} sinc^2\left(\frac{ax}{2}\right) dx \right) = 1,$$

which indicates that function $a\, sinc^2(ax/2)/(2\pi)$ in the limit $a \to \infty$ behaves essentially as Dirac's delta-function—it approaches infinity as its argument goes to zero and vanishes for all other argument values (see Fig. 15.2), and its integral is equal to unity:

$$\lim_{a \to \infty} \frac{a}{2\pi} sinc^2\left(\frac{ax}{2}\right) = \delta(x). \tag{15.26}$$

Identifying now x with $\Omega - \omega_{mm_0}$ and a with t, I can rewrite Eq. 15.23 in the form

$$p_{mm_0} = \frac{2\pi \, |v_{mm_0}|^2}{\hbar^2} t\delta \left(\Omega - \omega_{mm_0} \right),$$

which approximates the transition probability for times t_f large enough to neglect the values of *sinc* function for all frequencies except of the resonance one, and at the same time small enough to still satisfy inequality in Eq. 15.24. Sometimes people write these results in terms of energy levels rather than transition frequency, making argument of the δ-function into $\hbar\Omega - E_m + E_{m_0}$. Taking advantage of the identity $\delta(x/a) = a\delta(x)$, you can rewrite the probability transition as

$$p_{mm_0} = \frac{2\pi \, |v_{mm_0}|^2}{\hbar} t\delta \left(\hbar\Omega - E_m + E_{m_0} \right). \tag{15.27}$$

The linear dependence of this probability upon time allows introducing a transition rate $R_{mm_0} = p_{mm_0}/t$, which expresses the constant (independent of time) probability of transition per unit time:

$$R_{mm_0} = \frac{2\pi \, |v_{mm_0}|^2}{\hbar} \delta \left(\hbar\Omega - E_m + E_{m_0} \right). \tag{15.28}$$

Equation 15.28 is one of the several expressions for what is known in quantum mechanics as Fermi's golden rule. The problem with this result is that despite of its fame and years of successful use, it is not that easy to make sense of it. Indeed, according to Eq. 15.28, the transitions are possible only if the argument of the delta-function is exactly equal to zero—a condition which is experimentally impossible to achieve. Even from the theoretical viewpoint, this expression is dubious because delta-function as a mathematical construct only makes sense inside of an integral with another normal function. You might try to save the situation by assuming that the whole business of replacing the *sinc* function with the delta-function was wrongheaded, and if we just stayed with the *sinc* function, all problems could have been avoided. Unfortunately, this idea does not save the situation for several reasons. First, with increasing time, the width of the main maximum of the *sinc* function objectively grows too small to have experimental relevance, and the use of the delta-function simply emphasizes this real fact (experimentally, starting with some time t, the difference between the zero width of the delta-function and finite but tiny width of the *sinc* function becomes unnoticeable). Second, avoiding transition to the delta-function, you would not be able to reveal the important feature of the time dependence of the transition probability, namely, its linear rather than quadratic dependence on time, understood in some integrated sense. The transition from quadratic time dependence at small t to linear dependence as t grows is not a regular crossover from one asymptotic form of a function to another—it is actually a transition in interpreting the transition probability from a regular function of frequency to what mathematicians call *a distribution*—a mathematical entity which can only have sense when used inside of an integral. The transition from the *sinc* to the delta-function simply illuminates this quite real and experimentally relevant change in the nature of the frequency dependence of the transition probability.

Thus, the only way to make sense of Eq. 15.28 is to find what this transition rate can be integrated with. It is quite clear that within the current picture of a monochromatic perturbation acting on an atom with a discrete spectrum, this goal cannot be achieved, which means that the perturbation theory is not really suitable for dealing with such systems. Indeed, you can remember that in the extreme model of the atom with only two energy levels solved in Sect. 10.2 without reliance on the perturbation theory, we have not encountered any of the problems plaguing us here. But do not worry, you have not just wasted an hour of your life reading the lead-up to Eq. 15.28. This result is still useful, but you need to learn when and how to use it, and this is what I am going to show to you in the next subsections.

15.2.2   A Non-monochromatic Noncoherent Perturbation

To justify the starting discussion of the transition probabilities with the simple monochromatic (described by sin or cos functions) perturbation, I could remind you that any arbitrary time-dependent perturbation can be expended into a linear combination of monochromatic functions. This expansion is well known as a Fourier series, if the initial function is periodic, or a Fourier integral, if it is not. I will not assume periodicity of the perturbation potential and will present it as

$$V_{mm_0}(t) = v_{mm_0} \int_{-\infty}^{\infty} F(\Omega) e^{-i\Omega t} d\Omega, \tag{15.29}$$

where $F(\Omega)$ is the Fourier transform of the function $f(t)$ prescribing the time dependence of the perturbation potential.[4] Since the integration with respect to time required by Eq. 15.12 can be carried out independently of the integration over Ω, I can simply substitute Eq. 15.22 into Eq. 15.29 to get the result for the first-order transition amplitude:

$$c_m^{(1)}(t) = \frac{v_{mm_0}}{i\hbar} t \int_{-\infty}^{\infty} d\Omega \, F(\Omega) \exp\left[\frac{i(-\Omega + \omega_{mm_0})t}{2}\right] \times$$

$$sinc\left(\frac{(-\Omega + \omega_{mm_0})t}{2}\right). \tag{15.30}$$

---

[4]Equation 15.29 obviously implies that time dependence of the perturbation potential appears as a product of a time-dependent function and a time-independent operator, which is the case in many practically important situations.

Note that since Eq. 15.29 does not formally contain $\exp(i\Omega t)$ terms which appeared in Eq. 15.16 in order to maintain the Hermitian nature of the perturbation, I can use Eq. 15.22 omitting the second nominally emission-related term, appearing in the treatment of the monochromatic perturbation. Because the integration over Ω includes both positive and negative values, this term is automatically included in Eq. 15.30. The hermiticity of the perturbation is now ensured by the condition $F(\Omega) = F^*(-\Omega)$, which guarantees that $f(t)$ is a real-valued function. This condition is a known result from the theory of Fourier transforms, but you can easily verify it yourself.

The corresponding expression for the transition probability

$$
p_{mm_0} = \frac{|v_{mm_0}|^2}{\hbar^2} t^2 \int_{-\infty}^{\infty} F(\Omega_1) F^*(\Omega_2) \exp\left[\frac{i(\Omega_2 - \Omega_1)t}{2}\right] \times
$$

$$
sinc\left(\frac{(-\Omega_1 + \omega_{mm_0})t}{2}\right) sinc\left(\frac{(-\Omega_2 + \omega_{mm_0})t}{2}\right) d\Omega_1 d\Omega_2 \tag{15.31}
$$

features the integration of the sinc function over the external frequency, which I sought to find, but I got more than I have bargained for—two integrals instead of one and two external frequencies $\Omega_{1,2}$ (I had to introduce two separate integration variables to turn a product of two integrals into a double integral). To move forward with this expression, I have to make additional assumptions about the perturbation, and I will begin by presenting $F(\Omega)$ in the amplitude–phase form

$$
F(\Omega) = |F(\Omega)| e^{i\varphi(\Omega)}
$$

where I introduced its absolute value $|F(\Omega)|$ and phase $\varphi(\Omega)$. You can think of Eq. 15.29 as a superposition of monochromatic waves each with its own phase, and as your experience with wave superposition might tell you, the result depends a lot on the relation between the phases of the waves being added. In a simplest case, we distinguish between two extremes—the coherent and incoherent superposition. In the former case, the relation between the phases is fixed and the intensity of the resultant wave produces a characteristic interference pattern. Incoherent superposition occurs when the phases of the individual waves are not fixed and change randomly so that all interference terms dependent on the phase difference average out to zero. The resulting intensity becomes merely the sum of intensities of the individual waves with all traces of interference pattern washed out.

With this in mind, I am going to suppose that Eq. 15.29 represents an incoherent superposition in which the phase difference $\varphi(\Omega_1) - \varphi(\Omega_2)$ is a random quantity fluctuating between values of 0 and 2π. This quantity appears in Eq. 15.31 in the form $\exp[\varphi(\Omega_1) - \varphi(\Omega_2)]$, which vanishes upon "averaging" over the phases unless $\Omega_1 = \Omega_2$. If we dealt with a sum over discrete frequencies, I would have assigned the value of unity to this expression at $\Omega_1 = \Omega_2$, but since I am

dealing with an integral over continuously varying quantity, it is more appropriate
to prescribe that

$$\overline{\exp\left[\varphi\left(\Omega_1\right) - \varphi(\Omega_2)\right]} = \delta\left(\Omega_1 - \Omega_2\right), \tag{15.32}$$

where the line above the exponential signifies "averaging" over the phases. Substituting this expression into Eq. 15.31, you end up with the following:

$$p_{mm_0} = \frac{|v_{mm_0}|^2}{\hbar^2} t^2 \int\limits_{-\infty}^{\infty} d\Omega \, |F(\Omega)|^2 \, sinc^2\left(\frac{(-\Omega + \omega_{mm_0})\, t}{2}\right). \tag{15.33}$$

I will not be surprised if you are not completely satisfied with my "derivation" of
the last expression and will think of it as a bit of a hand-waving type. I would agree
as this can hardly be called a derivation at all: the nature of the averaging procedure
was not explained, and Eq. 15.32 was essentially postulated not derived. However,
the result, Eq. 15.33, looks quite reasonable and in sync with our intuition about
incoherent superpositions. Indeed, the story, which this expression tells, makes total
sense: each frequency component of the perturbation potential generates a transition
probability as given by Eq. 15.23 corrected for the spectral intensity $|F(\Omega)|^2$. The
total probability due to all frequency components is just a sum (well, in this case, it is
an integral) of the probabilities caused by each spectral component independently.
The incoherency supposition is in this context equivalent to the assumption that
each spectral component of the perturbation induces transitions independently of
the others. Of course, this result can be derived and formulated in a more rigorous
manner, but I hope you will forgive me for the decision to spare you the technical
details. Those who are not still completely satisfied can google the Wiener–Khinchin
theorem and enjoy the ride. In the meantime, I will replace the *sinc* function by a
delta-function following the rule established above and transforming Eq. 15.33 into

$$p_{mm_0} = \frac{2\pi\,|v_{mm_0}|^2}{\hbar^2} t \int\limits_{-\infty}^{\infty} d\Omega \, |F(\Omega)|^2 \, \delta\left(-\Omega + \omega_{mm_0}\right) =$$

$$\frac{2\pi\,|v_{mm_0}|^2}{\hbar^2} \, |F\left(|\omega_{mm_0}|\right)|^2 \, t. \tag{15.34}$$

Equation 15.34 and the corresponding formula for the transmission rate

$$R_{mm_0} = \frac{2\pi\,|v_{mm_0}|^2}{\hbar^2} \, |F\left(|\omega_{mm_0}|\right)|^2, \tag{15.35}$$

unlike Eq. 15.28, do make total sense and can be put to immediate use. Equation 15.35 can be considered as Fermi's golden rule formulated for an incoherent
perturbation, where the transition rate is determined by the spectral intensity of the

perturbation at the transition frequency. You shall also notice that these equations describe both absorption and emission transitions as exemplified by my use of $|\omega_{mm_0}|$. Indeed, if $\omega_{mm_0} > 0$ (absorption), the zero of the argument of the delta-function in Eq. 15.34 occurs at positive values of Ω, while if $\omega_{mm_0} < 0$, it happens at negative Ω, which are also present in the respective integral. However, property $F(\Omega) = F^*(-\Omega)$ required to keep perturbation Hermitian also yields $F(-\Omega)F^*(-\Omega) = F^*(\Omega)F(\Omega)$ ensuring that the spectral density is an even function of the frequency $|F(-\Omega)|^2 = |F(\Omega)|^2$. Therefore, the use of the absolute value symbol in its argument is not really necessary as transition rates for emission and absorption between the same pairs of states are obviously equal to each other.

15.2.3 Transitions to a Continuum Spectrum

Fermi's golden rule in its delta-functional form can also be used in a meaningful way when either the final or initial (or both) state participating in the transition belongs to a continuous spectrum. Such situation arises, for instance, in the case of ionizing transitions, when an electron is kicked out from its initial discrete energy level to the spectral continuum, where its wave function is no longer localized. We call this process ionization because the electron in such a final state has a non-zero probability to get infinitely far away from the remaining ion, so it does not "belong" to it anymore. Optical (caused by interaction with light) transitions in semiconductors present an example when both initial and final states of the transition belong to a continuous spectrum. In this section I will consider the transitions between discrete and continuous states, leaving the semiconductor case for a special treatment in the section on light absorption.

When dealing with transitions to the states of the continuous spectrum, the first thing you need to realize is that a probability of transition to a state with an exact energy eigenvalue does not make sense anymore, and you need to discuss this situation in terms of probability density. The actual transition probability can only be then defined as an integral of the probability density over some finite energy interval. This is nothing new, of course, as we have already discussed the conversion of probabilities to probability densities in Sects. 2.3 and 3.3.5 when exploring differences between observables with discrete and continuous eigenvalues.

So, my mission now is to turn transition probability $p_{m_0 m}$ into an appropriate probability density, and to accomplish this, I will rely upon the old trick of forced discretization, which I have used in Sect. 11.5 on a system of non-interacting electrons. However, the concrete realization of the discretization procedure depends on how the states of the continuous spectrum are described. I will consider two different ways to deal with them. If, as often is the case, the unperturbed Hamiltonian $\hat{H}_0$ commutes with the operators of the angular momentum $\hat{L}^2$ and $\hat{L}_z$, its eigenvectors can be characterized by azimuthal and magnetic quantum numbers L and M even if the eigenvector belongs to a continuous spectrum. These states, then, are still

characterized by three discrete indexes (including spin) and a single continuously
changing parameter $k = \sqrt{2m_eE}/\hbar$. This parameter appears in the radial equation
for the electron in a central potential in the form of dimensionless coordinate $\rho = kr$
in Eq. 8.12, Chap. 8. You can easily demonstrate it by replacing the dimensionless
parameters ς and $\epsilon_{n,l}$ appearing in that equation by their expressions in terms of
real physical coordinate r and energy $E_{n,l}$. In order to discretize this parameter, I
have to impose boundary conditions on the radial function $R_l(kr)$: $R_l(kD) = 0$,
where $D \to \infty$. Sparing you the details of the derivation, it can be shown that this
boundary condition results in the following allowed values of k_j: $k_j = \pi j/D + \delta_l/D$,
where δ_l is an extra phase dependent on the azimuthal quantum number l, which you
do not have to worry about, and $j = 1, 2, 3 \cdots$. Now the expression for the transition
probability p_{m_0m} can be rewritten as

$$p_{m_0;m',k_j} = \frac{2\pi \left| v_{m_0;m',k_j} \right|^2}{\hbar} t\delta \left(\hbar\Omega - E_{m',k_j} + E_{m_0} \right),$$

where index m_0 includes principal, azimuthal, magnetic, and spin quantum numbers
of the initial states, while the principal number in m' is replaced by a wave number
k_j. To turn this probability into the probability density, I will perform the following
standard series of steps:

$$\Delta p_{m_0;m',k_j} = \frac{2\pi \left| v_{m_0;m',k_j} \right|^2}{\hbar} t\delta \left(\hbar\Omega - E_{m',k_j} + E_{m_0} \right) \Delta j =$$

$$\frac{2D \left| v_{m_0;m'}(k) \right|^2}{\hbar} t\delta \left(\hbar\Omega - E_{m'}(k) + E_{m_0} \right) \Delta k =$$

$$\frac{2D \left| v_{m_0;m'}(k) \right|^2}{\hbar} t\delta \left(\hbar\Omega - E_{m'}(k) + E_{m_0} \right) \frac{dk}{dE} \Delta E =$$

$$\frac{2D \left| v_{m_0;m'}(k) \right|^2}{\hbar^2} \sqrt{\frac{m_e}{2E}} t\delta \left(\hbar\Omega - E_{m'}(k) + E_{m_0} \right) dE.$$

The first line in this expression is obtained by multiplying the initial transition
probability by Δj, which is equal to unity, so this step does not really change
anything. In the second line, I used the quantization rule obtained from the boundary
conditions (note: the phase δ_j, which I asked you not to worry about, disappeared)
to replace Δj with Δk and upgraded k from a discrete lower index to the continuous
argument of the matrix element. Finally I converted Δk into ΔE using the relation
$k = \sqrt{2m_eE}/\hbar$ to compute the derivative $dk/dE = \sqrt{m_e/(2E\hbar^2)}$. The entire
expression in front of dE is the probability density, and the actual probability to
undergo a transition to a state with any energy within a certain interval would be
normally given by an integral over this interval. However, the presence of the delta-
function in the probability density makes transition possible only to the states with

$$E_{tr} = \hbar\Omega + E_{m_0},$$

with the corresponding probability given by the integral over arbitrary energy interval as long as it includes E_{tr}. Thus, the rate of transitions from the initial state $|m_0\rangle$ to the final state $|m', E_{tr}\rangle$ is equal to

$$R_{m_0;m'}(\Omega) = \frac{2D|v_{m_0;m'}(\hbar\Omega + E_{m_0})|^2}{\hbar^2}\sqrt{\frac{2m_e}{\hbar\Omega + E_{m_0}}}. \qquad (15.36)$$

If the perturbation potential does not depend on the spin of the electron, the perturbation matrix element is diagonal with respect to the spin quantum numbers of the initial and final states. If, also, you do not really care about the spin state of your electron, the transition rate needs to be summed over the spin states, which, in this case, produces simply an extra factor of two in the transition rate. Then the spin-indifferent transition rate becomes

$$R_{m_0;m'}(\Omega) = \frac{4D|v_{m_0;m'}(\hbar\Omega + E_{n_0})|^2}{\hbar^2}\sqrt{\frac{2m_e}{\hbar\Omega + E_{m_0,L}}},$$

where the indexes m' and m_0 are now missing the spin quantum numbers.

In the approach leading to Eq. 15.36, the wave functions of the position representation were defined using the spherical coordinates. While this approach is useful in some circumstances, the wave functions even of a free particle in the spherical coordinates have a pretty complicated form. Thus, in many practically relevant situations, it is more convenient to use Cartesian coordinates instead. In this case the states of the continuous spectrum can be characterized by a regular wave vector k with the energy $E(k)$ defined in a standard free particle-like way. If we are dealing with an electron moving in a Coulomb potential, like in a hydrogen atom problem, the respective wave functions (remember, I am talking about the continuous spectrum here, and we did not deal with it in Chap. 8) are called Coulomb wave functions, and they are too mathematically complex to be bothered with here. These functions look like modified plane waves and asymptotically approach the latter as the energy and/or the distance from the nucleus grows. And luckily enough, large distance asymptotic behavior is all I need in order to introduce a quasi-continuous discrete description of these functions. To this end I can go back to the same periodic boundary conditions used in Sect. 11.5 and introduce quantized components of the wave vector as $k_{m_{1,2,3}} = 2\pi m_{1,2,3}/L$ with the set of numbers $m_{1,2,3}$ comprising the composite index m in the expressions for the probability distribution. Then I can convert the transition probability into a transition probability distribution using the same trick as before by introducing

$$1 = \Delta m_1 \Delta m_2 \Delta m_3 = \frac{L^3}{(2\pi)^3}\Delta k_x \Delta k_y \Delta k_z$$

into

$$\triangle p_{m_0 m} = p_{m_0 m} \triangle m_1 \triangle m_2 \triangle m_3 = \frac{L^3}{(2\pi)^3} p_{m_0,s}(\mathbf{k}) \triangle k_x \triangle k_y \triangle k_z, \qquad (15.37)$$

where in the last expression I replaced the discrete index m with the set of quasi-continuous variables k_i while keeping the quantum spin number s (the only remaining discrete characteristics of the final state) as an index. Incorporating Eq. 15.27 I can transform it into the following expression for the transition probability density:

$$dp_{m_0,s}(k_1, k_2, k_3) = \frac{2\pi}{\hbar} t \frac{L^3}{(2\pi)^3} |v_{m_0,s}(\mathbf{k})|^2 \delta(\hbar\Omega - E_{k,s} + E_{m_0}) d^3k. \qquad (15.38)$$

The delta-function in this expression limits the non-zero values of the probability density to what is called the "equienergetic" surface $E_{k,s} = \hbar\Omega + E_{m_0}$. A further development of this expression depends on the shape of this surface. I will only consider here the simplest and, therefore, the most popular case of spherical equienergetic surfaces, which correspond to the free-particle relation between energy and the wave vector:

$$E = \frac{\hbar^2}{2m_e}\left(k_x^2 + k_y^2 + k_z^2\right).$$

In this case, it is convenient to replace the Cartesian components of the wave vector k_x, k_y, and k_z with their spherical counterparts k, θ, and φ, where $k = \sqrt{k_x^2 + k_y^2 + k_z^2}$ and angles θ and φ define the direction of the wave vector with respect to some chosen coordinate axes. Transforming the Cartesian volume element $dk_x dk_y dk_z$ (i.e., the volume in the space of wave vectors) into the spherical one

$$dk_x dk_y dk_z = k^2 \sin\theta d\theta d\varphi dk,$$

I can rewrite Eq. 15.38 as

$$dp_{m_0,s} = \frac{2\pi}{\hbar} t \frac{L^3}{(2\pi)^3} |v_{m_0,s}(k, \theta, \varphi)|^2 \delta(\hbar\Omega - E_{k,s} + E_{m_0}) k^2 \sin\theta d\varphi d\theta dk.$$

Using the relation between k and energy E, it is more convenient to rewrite the latest expression in terms of probability density for the latter. Assuming that energy of the particle in the continuous spectrum does not depend on spin and using $k^2 = 2m_e E/\hbar^2$ and $dk = \sqrt{m_e/(2\hbar^2 E)}dE$, you can obtain

$$dp_{m_0,s} = \frac{2\pi}{\hbar} t \frac{L^3}{(2\pi)^3} \frac{\sqrt{2}m_e^{3/2}}{\hbar^3} \sqrt{E} \, |v_{m_0,s}(k,\theta,\varphi)|^2 \times$$

$$\delta(\hbar\Omega - E + E_{m_0}) \sin\theta d\varphi d\theta dE.$$

Further proceeding follows almost the same line of arguments, as those presented when deriving Eq. 15.36 with one exception. In Eq. 15.36 there was a single continuously distributed variable (E to be sure), so once I integrated it out, the remaining expression became the actual probability for the remaining discrete characteristics of the final state m'. Here E is just one of the three continuous parameters defining the final state, and while the delta-function again takes care of E making its value definite and equal to E_{tr}, after integrating it out, you still remain with a differential probability $dp_{m_0,s}(E_{tr})$ defining the probability of the electron's after-transition wave number to lie within an infinitesimally small solid angle $d\Omega = \sin\theta d\varphi d\theta$ around the direction specified by angles θ and φ. Now defining the corresponding transition rate density $R_{m_0,s}(\theta,\varphi)$ as a probability of transition per unit time per unit solid angle

$$R_{m_0,s}(\theta,\varphi) \equiv \frac{dp_{m_0,s}(E_{tr})}{t d\Omega},$$

I find for it

$$R_{m_0,s}(\theta,\varphi) = \frac{2\pi}{\hbar} \frac{L^3}{(2\pi)^3} \frac{\sqrt{2}m_e^{3/2}}{\hbar^3} \sqrt{E_{tr}} \, |v_{m_0,s}(E_{tr},\theta,\varphi)|^2. \tag{15.39}$$

This result describes spin and angular distribution of electrons kicked out from an atom as the result of ionization for various orientations of the electron spin. Experimentally, this quantity can be measured by placing spin-sensitive electron detectors at different angles and counting the number of electrons ejected under the action of a perturbation per unit time by an ensemble of non-interacting atoms in different directions. If the presence of the quantization volume L^3 (which is a pretty arbitrary quantity) in this expression gives you jitters, fear not. It will be canceled by the $L^{-3/2}$ normalization factor in the wave function of the continuous spectrum when you compute the perturbation matrix element $v_{m_0,s}(E_{tr},\theta,\varphi)$.

Now, if you are not interested in the direction of the propagation of the electron after ionization or its spin state, and only want to know the total transition rate R_{m_0} from the initial state $|\alpha_{m_0}\rangle$ (you can call it the rate of ionization), you will integrate $R_{m_0,s}(\theta,\varphi)$ over the entire solid angle and sum it up over the spin states:

$$R_{m_0} = \frac{2\pi}{\hbar} \frac{L^3}{(2\pi)^3} \frac{\sqrt{2}m_e^{3/2}}{\hbar^3} \sqrt{E_{tr}} \int_0^\pi d\theta \sin\theta \times$$

$$\int_0^{2\pi} d\varphi \left(|v_{m_0,\uparrow}(E_{tr},\theta,\varphi)|^2 + |v_{m_0,\downarrow}(E_{tr},\theta,\varphi)|^2 \right).$$

If the perturbation matrix element does not depend on the angles and the spin, then the integration over angles combined with the summation over spin yields the trivial factor 8π so that the expression for the ionization rate can be written down as

$$R_{m_0} = \frac{2\pi}{\hbar} |v_{m_0}(E_{tr})|^2 g(E_{m_0} + \hbar\Omega), \qquad (15.40)$$

where $g(E)$ is the same density of states for a free electron, which was defined in Eq. 11.56. Equation 15.40 is often passed for one of the reincarnation of Fermi's golden rule, but one would be wise to remember about the limitations involved in its derivation: angle and spin independence of the perturbation matrix element. I will return to this issue in the section on light absorption and emission.

I will complete this section by an example of calculations based on Eq. 15.39.

Example 31 (Ionization of a Hydrogen Atom) Find the angular distribution of the electrons ejected from the ground state of hydrogen atom by a uniform oscillating electric field $\mathcal{E} = \mathcal{E}_0 \cos \Omega t$, assuming that the frequency Ω satisfies the condition $\hbar\Omega \gg 13.6\,\text{eV}$. This condition ensures that the final state of the electron, indeed, belongs to the continuum spectrum and that its energy is high enough to approximate the respective wave functions by a plane wave.

Solution

The perturbation operator corresponding to the uniform electric field can be written down as

$$\hat{V} = e\boldsymbol{r} \cdot \mathcal{E}(t) = e\hat{r}\mathcal{E}_0 \cos \Omega t, \qquad (15.41)$$

where e is the absolute value of the electron's charge. A comparison with Eq. 15.16 allows me to identify the matrix elements $v_{m_0 m}$ as

$$v_{m_0,s}(\boldsymbol{k}) = \frac{e}{2}\sqrt{\frac{1}{\pi a_B^3 L^3}} \iiint d^3 r\, e^{-r/a_B} \boldsymbol{r} \mathcal{E}_0 e^{i\boldsymbol{k}\boldsymbol{r}}, \qquad (15.42)$$

where I substituted the wave function of the ground state of the hydrogen atom from Chap. 8, which represents the initial state, and the box normalized plane wave for the final state from Sect. 11.5.

The structure of the hydrogen ground state wave function, which depends only on the absolute value of the position vector $r = |\boldsymbol{r}|$, indicates that you can save time and effort by using a spherical coordinate system to calculate this integral. However, in order to take advantage of this realization, you, first, have to choose the directions of the axes that you will use to define the spherical coordinates. Actually, there is only one axis which you need to worry about—the polar (Z)-axis. When making such a decision, it is wise to consider first if there are any constant vectors in the integrand (position vector r does not count—it is an integration variable). In Eq. 15.42 there are two such vectors: the wave vector of the final state k and the external electric field E_0. These two vectors define a plane, and it is quite obvious that choosing it as one of the coordinate planes will simplify the task (both vectors will immediately lose one of their coordinates, and two coordinates to worry about are always better than three, right?). Now, the last question is: which of these two vectors choose to define the direction of the polar axis? My bet would be on k since this choice, while making $r \cdot \mathcal{E}_0$ more complex, would simplify expression $k \cdot r$, and it is better to have a more complex expression as a factor than as an argument of the exponential function. Thus, settling on the direction of the Z-axis along k, and choosing $\mathcal{E}$ to belong to $X - Z$ plane, I can immediately express $k \cdot r$ and $r\mathcal{E}_0$ in the corresponding spherical coordinates r, θ, and φ. With this choice of coordinate axes, I have

$$k \cdot r = kz = kr \cos \theta$$

and

$$r\mathcal{E}_0 = x\mathcal{E}_0 \sin \theta_E + z\mathcal{E}_0 \cos \theta_E = r\mathcal{E}_0 (\sin \theta \cos \varphi \sin \theta_E + \cos \theta \cos \theta_E),$$

where θ_E is the angle between the vector of the electric field and the wave vector k (see Fig. 15.3). Finally, introducing the volume element in the spherical coordinates, I have the following for the matrix element:

$$v_{m_0,s}(k) = \frac{e\mathcal{E}_0}{2} \sqrt{\frac{1}{\pi a_B^3 L^3}} \int_0^\infty dr r^2 \int_0^\pi d\theta \sin \theta \times$$

$$\int_0^{2\pi} d\varphi e^{-r/a_B} r (\sin \theta \cos \varphi \sin \theta_E + \cos \theta \cos \theta_E) e^{ikr \cos \theta}. \qquad (15.43)$$

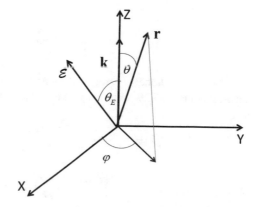

Fig. 15.3 The choice of the coordinate system for calculations of the integral in Eq. 15.42

First, note that the first term in the integral contains the factor $\cos\varphi$, which after integration with respect to the azimuthal angle φ yields zero: $\int_0^{2\pi} d\varphi \cos\varphi = 0$. The remaining integral

$$v_{m_0,s}(\theta_E) = \frac{e\mathcal{E}_0 \cos\theta_E}{2}\sqrt{\frac{1}{\pi a_B^3 L^3}} \int_0^\infty dr r^3 e^{-r/a_B} \int_0^\pi d\theta \sin\theta \cos\theta e^{ikr\cos\theta} \int_0^{2\pi} d\varphi,$$

where I replaced argument $\mathbf{k}$ with θ_E reflecting that this is the only characteristic of $\mathbf{k}$, which this matrix element depends upon, takes a bit more work. The integration with respect to φ is, of course, trivial and produces 2π, while the integral with respect to the polar angle requires the standard substitution of the variable $s = \cos\theta$, which yields

$$\int_0^\pi d\theta \sin\theta \cos\theta e^{ikr\cos\theta} = \int_{-1}^1 ds s e^{ikrs} = \frac{e^{ikr} + e^{-ikr}}{ikr} + \frac{e^{ikr} - e^{-ikr}}{k^2 r^2} =$$

$$\frac{1}{ikr}\left(1 - \frac{1}{ikr}\right) e^{ikr} + \frac{1}{ikr}\left(1 + \frac{1}{ikr}\right) e^{-ikr}.$$

The remaining integral over r is the sum of two integrals:

$$v_{m_0,s}(\theta_E) = \frac{e\mathcal{E}_0\pi\cos\theta_E}{ik}\sqrt{\frac{1}{\pi a_B^3 L^3}} \times \left[\int_0^\infty dr r^2 \left(1 - \frac{1}{ikr}\right) e^{-r(1/a_B - ik)} + \right.$$

$$\left. \int_0^\infty dr r^2 \left(1 + \frac{1}{ikr}\right) e^{-r(1/a_B + ik)}\right] =$$

$$16ie\mathcal{E}_0\pi\cos\theta_E \sqrt{\frac{1}{\pi a_B^3 L^3}} \frac{a_B^5 k}{\left(1 + a_B^2 k^2\right)^3} =$$

$$16i\sqrt{\frac{\pi a_B}{L^3}} e\mathcal{E}_0 a_B^3 \cos\theta_E \frac{k}{\left(1 + a_B^2 k^2\right)^3},$$

which I, to be honest, computed using Mathematica©. Now, this matrix element must go to Eq. 15.39, where it yields (taking into account the summation over spin)

$$R_{m_0}(\theta_E) = 2\frac{L^3}{4\pi^2}\frac{\sqrt{2}m_e^{3/2}}{\hbar^4}\sqrt{E_{tr}}\frac{256 k^2}{\left(1 + a_B^2 k^2\right)^6}\frac{\pi a_B^7}{L^3} e^2 \mathcal{E}_0^2 \cos^2\theta_E =$$

$$\frac{64}{\pi}\frac{(e\mathcal{E}_0 a_B)^2}{\hbar}\frac{2m_e a_B^2}{\hbar^2}\cos^2\theta_E\frac{(k_{tr} a_B)^3}{\left(1 + k_{tr}^2 a_B^2\right)^6}, \tag{15.44}$$

where $k_{tr} = \sqrt{2m_e E_{tr}}/\hbar$. Substituting the expression for the Bohr radius from Eq. 8.8 into $2m_e a_B^2/\hbar^2$, you will immediately see that this factor is just $1/|E_g|$, where E_g is the ground state energy of hydrogen, which in this example plays the role of E_{m_0}—the energy of the initial state. Thus, the expression for $R_{m_0}(\theta_E)$ can now be rewritten as

$$R_{m_0}(\theta_E) = \frac{64}{\pi} \frac{(e\mathcal{E}_0 a_B)^2}{\hbar\,|E_g|} \cos^2 \theta_E \frac{(k_{tr}a_B)^3}{\left(1 + k_{tr}^2 a_B^2\right)^6}.$$

The angular dependence of this transition rate is given by the factor $\cos^2 \theta_E$, which indicates that the majority of the electrons propagate after the ionization in the direction of the field (not a big surprise here, of course), while the ionization probability in the direction perpendicular to the field is zero. To get the complete ionization rate regardless of the direction, one needs to integrate this result with respect to the angles θ_E and φ_E, remembering, of course, that the integration occurs over a solid angle $\sin\theta d\theta d\varphi$. This involves computing the following expression:

$$\int_0^{2\pi} d\varphi_E \int_0^{\pi} d\theta_E \sin\theta_E \cos^2\theta_E = 2\pi \int_{-1}^{1} ds s^2 = \frac{4\pi}{3},$$

so that the total ionization rate becomes

$$R_{m_0} = \frac{256}{3} \frac{(e\mathcal{E}_0 a_B)^2}{\hbar\,|E_g|} \frac{(k_{tr}a_B)^3}{\left(1 + k_{tr}^2 a_B^2\right)^6}. \tag{15.45}$$

The dimensionless parameter $k_{tr}a_B$ can be written down in the form

$$k_{tr}a_B = \frac{\sqrt{2m_e\left(\hbar\Omega + E_g\right)a_B^2}}{\hbar} = \sqrt{\frac{\hbar\Omega + E_g}{|E_g|}} = \sqrt{\frac{\hbar\Omega}{|E_g|} - 1}, \tag{15.46}$$

which reveals that the ionization is only possible if $\hbar\Omega > |E_g|$, where this parameter remains real. The dependence of the ionization rate upon frequency is determined by the factor

$$\beta = \frac{(k_{tr}a_B)^3}{\left(1 + k_{tr}^2 a_B^2\right)^6}.$$

Plotting β as a function of the dimensionless parameter ka_B, you will notice that this factor exhibits two different modes of behavior: growth as $(k_{tr}a_B)^3$ for $k_{tr}a_B \ll 1$ and a much faster decrease as $(k_{tr}a_B)^{-9}$ for $k_{tr}a_B \gg 1$ (see Fig. 15.4). (Actually, you could have noticed this without any plots, by simply analyzing this function in two different limits, but it is always nice to have something to look at.) The transition

Fig. 15.4 Dependence of the ionization rate on the dimensionless parameter ka_B

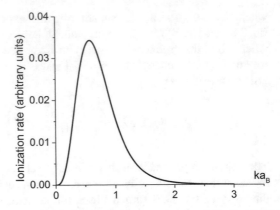

between these two regimes occurs at $k_{tr}a_B = 1/\sqrt{3}$, where the function $\beta(k)$ has a maximum. This corresponds to the external frequency Ω_{max}:

$$\hbar\Omega_{max} = \frac{4\,|E_g|}{3}.$$

This condition means that the maximum efficiency of ionization occurs at such frequency of oscillations, which excite electrons to the energy equal in magnitude to the third of the energy $E_{tr} = |E_g|/3$.

15.3   Semiclassical Theory of Absorption and Emission of Light

15.3.1   Interaction of Charged Particles with Electromagnetic Radiation

In the introductory physics courses, you learned that a particle with charge q moving with instantaneous velocity v in electric and magnetic fields $\mathcal{E}$ and B experiences a force described by the well-familiar Lorentz expression (in SI units):

$$F = q\mathbf{E} + q\,\mathbf{v} \times \mathbf{B}.$$

Unfortunately, this simple equation is ill-suited for application in quantum mechanics, which is based on the Hamiltonian representation of all interactions. Fortunately, it is possible to describe the interaction between a charged particle and electromagnetic field remaining firmly grounded in the Hamiltonian approach. However, to get there I have to remind you a few facts from electromagnetism first.

You know that in the electrostatics and magnetostatics, when an electric field can only be created by charges, and a magnetic field arises only in the presence of an electric current, both fields can be expressed in terms of their respective potential functions. The electric field is related to a scalar potential $V(r)$ as $\mathcal{E} = -\nabla V(r)$, while the magnetic field can be found from a vector potential $A\,(r)$ as $B = \nabla \times A$, where ∇ is a well-known vector Del or nabla operator. Both these potentials are not defined uniquely: you can add any constant term to a scalar potential and any vector function expressible as a gradient to the vector potential without changing the fields (the last statement follows from the operator identity $\nabla \times \nabla f = 0$).

When electric and magnetic fields are allowed to change with time, you are moving into the realm of electrodynamics, where everything gets intertwined and the electric field can be generated by a time-dependent magnetic field and vice versa. One of the immediate consequences of this is a possibility for electric and magnetic fields to sustain each other without the need for any charges or currents. Do not get me wrong here, though: some charges or currents are needed someplace to generate the fields, but the point is that these fields can break out free of their initial sources and travel through space without any charges or currents in the form of electromagnetic waves. Of course, I am sure you know all this, but it is still useful to recall these simple facts to create a proper context for what follows. While the transition to electrodynamics does not change the relation between the magnetic field and the vector potential, the electric field now acquires an additional contribution in the form

$$\mathcal{E} = -\nabla V(r) - \partial A/\partial t.$$

The appearance of the vector potential in the expression for the electric field reflects the possibility for the generation of this field by the time-dependent magnetic field (this explains the time derivative). In the region free of charges, the scalar potential term can be eliminated using the non-uniqueness of the potentials, so that both electric and magnetic fields are expressed in terms of a single vector potential:

$$\mathcal{E} = -\partial A/\partial t; \; B = \nabla \times A, \tag{15.47}$$

which can also be made to satisfy the condition $\nabla \cdot A = 0$. This condition is called *gauge-fixing* condition, and the potential defined this way is said to be in the Coulomb or transverse gauge.

Now I can approach the main issue of concern—the Hamiltonian formulation of the particle–field interaction in classical physics. It turns out that the classical Lorentz force on a particle interacting with the electromagnetic field in the absence of charges and currents can be reproduced from the Hamiltonian equations if the Hamiltonian of the charged particle is written down as

$$H = \frac{[p - qA(r,t)]^2}{2m_e} \equiv \frac{\sum_j \left(p_j - qA_j\right)\left(p_j - qA_j\right)}{2m_e} \tag{15.48}$$

where in the last expression I used the definition of the magnitude of a vector in terms of its Cartesian components. To verify this fact I just need to write down the classical Hamiltonian equations as given by Eqs. 7.8 and 7.7:

$$\frac{dr_i}{dt} = \frac{\partial H}{\partial p_i} = \frac{p_i - qA_i}{m_e}, \tag{15.49}$$

$$\frac{dp_i}{dt} = -\frac{\partial H}{\partial r_i} = \frac{q}{m_e} \sum_j (p_j - qA_j) \frac{\partial A_j}{\partial r_i}. \tag{15.50}$$

Differentiating the first of these equations with respect to time and using the time derivative of the momentum from the second equation, I obtain

$$\frac{d^2 r_i}{dt^2} = -\frac{q}{m_e} \frac{dA_i}{dt} + \frac{1}{m_e} \frac{q}{m_e} \sum_j (p_j - qA_j) \frac{\partial A_j}{\partial r_i}.$$

Multiplying both sides of this equation by m_e and noting that the velocity of the particle $v_i = dr_i/dt = (p_i - qA_i)/m_e$, I can rewrite this as

$$m_e \frac{d^2 r_i}{dt^2} = -q \frac{dA_i}{dt} + q \sum_j v_j \frac{\partial A_j}{\partial r_i}. \tag{15.51}$$

Now you need to pay attention. The time derivative of the vector potential appearing in Eq. 15.51 is a *full* derivative (straight d) as opposed to the partial derivative (round ∂) appearing in Eq. 15.47. The difference is significant because the vector potential is a function of the coordinate of the particle, and it changes with time for two reasons: one is its explicit time dependence, and the other is due to the motion of the particle. In other words, when computing this time derivative, you must treat the position vector r as a function of time. Accordingly, you have to write

$$\frac{dA_i}{dt} = \frac{\partial A_i}{\partial t} + \sum_j \frac{\partial A_i}{\partial r_j} \frac{dr_j}{dt} = \frac{\partial A_i}{\partial t} + \sum_j v_j \frac{\partial A_i}{\partial r_j},$$

where I introduced the particle's velocity $v_j = dr_j/dt$. Taking this into account, Eq. 15.51 takes the following form:

$$m_e \frac{d^2 r_i}{dt^2} = -q \frac{\partial A_i}{\partial t} - q \sum_j v_j \frac{\partial A_i}{\partial r_j} + q \sum_j v_j \frac{\partial A_j}{\partial r_i} =$$

$$-q \frac{\partial A_i}{\partial t} + q \sum_j v_j \left(\frac{\partial A_j}{\partial r_i} - \frac{\partial A_i}{\partial r_j} \right). \tag{15.52}$$

Comparing this with Eq. 15.47, you can immediately identify the first term on the right as $q\mathcal{E}$, which is the electric field component of the Lorentz force. In order to prove that the second term reproduces the contribution of the magnetic field to the force, note that using identity $A \times (D \times C) = D\,(A \cdot C) - (A \cdot D)\,C$, I can transform $v \times B$ as

$$v \times B = v \times (\nabla \times A) = \nabla\,(v \cdot A) - (v\nabla) \cdot A.$$

Rewriting this expression in the component form

$$(v \times B)_i = \frac{\partial}{\partial r_i} \sum_j v_j A_j - \sum_j v_j \frac{\partial A_i}{\partial r_j} =$$

$$\sum_j v_j \left(\frac{\partial A_j}{\partial r_i} - \frac{\partial A_i}{\partial r_j} \right),$$

you see that it exactly coincides with the second term in Eq. 15.52. Now you can take a breath: we have shown that the Hamiltonian given by Eq. 15.48 indeed correctly reproduces the equation of motion for a charged particle in a field of an electromagnetic wave. If, in addition to the electromagnetic field, the particle also has a regular potential energy of interaction, for instance, with another charged particle, the total quantum Hamiltonian capable of handling this situation becomes

$$\hat{H} = \frac{[\hat{p} - qA(\hat{r}, t)]^2}{2m_e} + \hat{V}(r), \tag{15.53}$$

where the classical coordinate and momentum are replaced by the corresponding operators, while the vector potential remains a regular classical quantity. In many problems involving the interaction between quantum systems and radiation, the term proportional to the square of the vector potential can be neglected, in which case the Hamiltonian can be presented as

$$\hat{H} = \frac{\hat{p}^2}{2m_e} + \frac{e}{2m_e}\,(\hat{p}A(\hat{r}, t) + A(\hat{r}, t)\hat{p}) + \hat{V}(r), \tag{15.54}$$

where I replaced the generic charge q with the negative electron charge $-e$, but, most importantly, restrained from equating expressions $\hat{p}A(\hat{r}, t)$ and $A(\hat{r}, t)\hat{p}$, which, while natural in classical description, would be in general wrong in the quantum case due to non-commutativity of momentum and position operators. However, using the gauge-fixing condition for the vector potential, the Hamiltonian 15.54 can still be simplified. Indeed, consider the expression $\hat{p}A(\hat{r}, t)f(r)$, where $f(r)$ is an arbitrary function, using the position representation for the momentum and coordinate operators:

$$\hat{p}A(\hat{r}, t)f(r) = -i\hbar\nabla\left[A(\hat{r}, t)f(r)\right] =$$
$$-i\hbar A(\hat{r}, t)\nabla\left[f(r)\right] - i\hbar f(r)\nabla\left[A(\hat{r}, t)\right].$$

In the Coulomb gauge, the last term in the above expression disappears, allowing me to replace $\hat{p}A(\hat{r}, t)$ with $A(\hat{r}, t)\hat{p}$ and present the Hamiltonian as

$$\hat{H} = \frac{\hat{p}^2}{2m_e} + \frac{e}{m_e}A(\hat{r}, t)\hat{p} + \hat{V}(r). \tag{15.55}$$

Assuming that the vector potential $A(r, t)$ describes a plane linearly polarized electromagnetic wave, I can choose it to have the following form:

$$A(r, t) = -w\frac{\mathcal{E}_0}{\Omega}\cos(kr - \Omega t), \tag{15.56}$$

where w is a unit vector defining the polarization of the wave. To satisfy the condition $\nabla \cdot A = 0$, w must be perpendicular to the propagation direction of the wave k. According to Eq. 15.47, the chosen form of the vector potential describes a wave with the electric field:

$$\mathcal{E} = \mathcal{E}_0 w \sin(kr - \Omega t). \tag{15.57}$$

With the vector potential in the form of Eq. 15.56, the perturbation operator becomes

$$\hat{V} = -\frac{e}{2m_e}\frac{\mathcal{E}_0}{\Omega}\left[e^{i(kr-\Omega t)} + e^{-i(kr-\Omega t)}\right](w \cdot \hat{p}). \tag{15.58}$$

My goal in this section is to apply the perturbation theory for the time-dependent Hamiltonians to optical transitions between discrete states of a hydrogen-like atom due to an incoherent radiation. This job can be made significantly easier if I take into account that the characteristic scale of the wave functions, which I will be using to compute the perturbation matrix elements, is determined by the Bohr radius a_B, which for a hydrogen atom in vacuum is about 5.3×10^{-2} nm. According to Eq. 8.8, it might become two or three orders of magnitudes larger for excitons in semiconductors, but even in this case, it is a far cry from the wavelength λ of visible radiation, which is of the order of 400–700 nm. This means that over the distances, when the wave functions in the perturbation matrix elements have any substantial magnitude, the field of the electromagnetic wave barely changes. Consequently I can safely replace $\exp(\pm i\mathbf{k}\mathbf{r})$ with unity, which is just the first term in the expansion of this exponent with respect to a small parameter $\mathbf{k} \cdot \mathbf{r} \sim 2\pi a_B/\lambda \ll 1$ making the perturbation operator much simpler:

$$\hat{V} = -\frac{e}{2m_e} \frac{\mathcal{E}_0}{\Omega} \left[e^{i\Omega t} + e^{-i\Omega t} \right] (\mathbf{w} \cdot \hat{\mathbf{p}}).$$ (15.59)

This approximation is called a dipole approximation for reasons which will become clear later. Comparing Eq. 15.59 with Eq. 15.16, I can identify the matrix element $v_{m_0 m}$ appearing in all formulas for the transition probability as

$$v_{m_0 m} = -\frac{e}{2m_e} \frac{\mathcal{E}_0}{\Omega} (\mathbf{w} \cdot \hat{\mathbf{p}})_{m_0 m},$$ (15.60)

where

$$(\mathbf{w} \cdot \hat{\mathbf{p}})_{m_0 m} = \langle \alpha_{m_0} | \mathbf{w} \cdot \hat{\mathbf{p}} | \alpha_m \rangle$$ (15.61)

is the matrix element of the component of the momentum operator in the direction of polarization of the incident light.

The computation of this matrix element in the position representation would involve differentiating the wave functions of the final state, which is not a terribly difficult thing to do but is better avoided if possible. And it can be, indeed, avoided with the help of one of the magic tricks from the arsenal of quantum mechanics. Recall a commutator of the unperturbed Hamiltonian with the position operator, which I computed in the section on Ehrenfest theorem, Eq. 4.18. Trivial generalization of that result to all three components of the position vector allows to present it as

$$\left[\hat{\mathbf{r}}, \hat{H}_0 \right] \equiv i\hbar \hat{\mathbf{p}}/m_e.$$

Using this result to replace the momentum operator in Eq. 15.61, I can turn it into

$$(\mathbf{w} \cdot \hat{\mathbf{p}})_{m_0 m} = \frac{m_e}{i\hbar} \langle \alpha_{m_0} | \mathbf{w} \cdot \left(\hat{\mathbf{r}} \hat{H}_0 - \hat{H}_0 \hat{\mathbf{r}} \right) | \alpha_m \rangle =$$

$$\frac{m_e}{i\hbar} \langle \alpha_{m_0} | \mathbf{w} \cdot \hat{\mathbf{r}} \hat{H}_0 | \alpha_m \rangle \quad \frac{m_e}{i\hbar} \langle \alpha_{m_0} | \hat{H}_0 \mathbf{w} \cdot \hat{\mathbf{r}} | \alpha_m \rangle.$$

Acting with the Hamiltonian to the right in the first term, and to the left in the second one, I can present this expression as

$$(\mathbf{w} \cdot \hat{\mathbf{p}})_{m_0 m} = \frac{m_e}{i\hbar} (E_m - E_{m_0}) \langle \alpha_{m_0} | \mathbf{w} \cdot \hat{\mathbf{r}} | \alpha_m \rangle,$$

reducing the matrix element of the momentum operator to that of the position operator. Substituting this result into Eq. 15.60, I find

$$v_{m_0m} = \frac{e}{2} \frac{i\mathcal{E}_0}{\hbar\Omega} (E_m - E_{m_0}) \langle \alpha_{m_0} | \boldsymbol{w} \cdot \hat{\boldsymbol{r}} | \alpha_m \rangle . \tag{15.62}$$

It will be useful to recall now that the transition probability, which I intend to calculate, contains the delta-function $\delta (\hbar\Omega - E_m + E_{m_0})$, which ensures that for any Ω, only those transitions have a non-zero chance to occur for which $E_m - E_{m_0} = \hbar\Omega$. This observation allows further simplification of Eq. 15.62 by canceling the energy difference $E_m - E_{m_0}$ in its numerator and the frequency of the perturbation Ω in the denominator. This yields for the matrix element:

$$v_{m_0m} = \frac{ie\mathcal{E}_0}{2} \langle \alpha_{m_0} | \boldsymbol{w} \cdot \hat{\boldsymbol{r}} | \alpha_m \rangle . \tag{15.63}$$

The reduction of the momentum matrix element to this form is, of course, very useful technically, but it is also quite remarkable from a more philosophical, if you want, point of view. To completely appreciate a significance of this result, take a look at the expression for the electric field corresponding to the chosen form of the vector potential, Eq. 15.57. If you were to neglect the $\boldsymbol{k} \cdot \boldsymbol{r}$ in it, you would end up with a spatially uniform electric field, which can be associated with a potential energy:

$$V = -e\boldsymbol{E} \cdot \boldsymbol{r} = \frac{ie\boldsymbol{E}_0 \cdot \boldsymbol{r}}{2} \left(e^{-i\Omega t} - e^{i\Omega t} \right) \tag{15.64}$$

and which would generate the time-independent perturbation matrix element v_{m_0m} of exactly the form given in Eq. 15.63. (Note that v_{m_0m} is defined as a coefficient in front of $\exp(-i\Omega t)$ term.) Now, think about it: I started out with the perturbation expressed in terms of a vector potential introduced to describe a solenoidal (with zero divergence) electric field, which can only be created by a changing magnetic field, and ended up with the perturbation expressed in terms of a scalar potential reserved for the potential electrostatic field. It might appear that physics works in mysterious ways, but the only mystery here is the beautiful interconnectedness and self-consistency of the various parts of the machinery of electrodynamics and quantum mechanics derived from the unified structure of the world.

A more practically minded person could say, of course, that since this result is only valid in the approximation $\boldsymbol{k} \cdot \boldsymbol{k} \ll 1$, it merely means that locally (over distances much smaller than the wavelength of the electromagnetic wave) there is not much difference between potential electric fields created by electrical charges and solenoidal fields arising due to the Faraday effect.

Now, if you recall that I am talking here about an electron interacting with an equal positive electric charge chosen to serve as an origin of the coordinate system, you can easily see that expression $e\boldsymbol{r}$ represents a dipole moment of this system of charges and that the perturbation potential is just the energy of a dipole in a uniform electric field. This is the reason why we use the term "dipole approximation" to characterize the reduction of the perturbation matrix elements to the form of Eq. 15.63.

Having derived the expression for the matrix element v_{m_0m}, I can consider a few particular examples of interaction between light and matter.

15.3.2   Absorption and Emission of Incoherent Radiation by Atoms

15.3.2.1   Absorption and Stimulated Emission

When discussing the interaction between light and electrons in atoms or semi-conductors, one has to consider three different processes: absorption, stimulated emission, and spontaneous emission. Spontaneous emission refers to the process of emission of light in the absence of any external electromagnetic field, which could "stimulate" the electron to drop from a higher energy level to a lower one. Because of the spontaneous emission, the eigenvectors of the electron's Hamiltonian (with exception of those that correspond to the ground state) do not actually represent true stationary states—an electron does not stay in these states forever, eventually undergoing a transition to the ground state and emitting light.

This circumstance makes it obvious that the theory of atomic energy levels and the stationary states developed so far is not quite complete. What is missing in this story is an interaction of electrons with a quantized electromagnetic field responsible for the phenomenon of spontaneous emission. The true stationary state must be formed by a combination of vectors belonging to a tensor product of electron and electromagnetic field vector spaces, such that the total energy corresponding to such a state would remain constant: a decrease of electron energy due to transition to the lower energy state shall be compensated by an increase of the energy of the electromagnetic field due to emission of light. Unfortunately, the complete theory of interaction between atoms and quantized electromagnetic field is outside the scope of this book as I already have mentioned before.

Thus, while I wouldn't be able to resist the temptation of demonstrating beautiful thermodynamic arguments of Albert Einstein that allowed him to deduce the rate of spontaneous emission without any knowledge of quantum electrodynamics, my main focus in this chapter would be on absorption and stimulated emission in the presence of classical incoherent radiation.

Within the quantum framework, the process of stimulated emission arises naturally as a consequence of the hermiticity requirement of operators representing observables (this is the reason why I had to introduce both $\exp(i\Omega t)$ and $\exp(-i\Omega t)$ terms in Eq. 15.16). However, before the advent of the full quantum theory, the idea of stimulated emission appeared quite bizarre. Indeed, think about it: an electron in an excited state is illuminated by light, and instead of jumping even to the higher energy level absorbing this light, it jumps to the lower level emitting even more light! The concept of stimulated emission was introduced by Einstein when he analyzed thermodynamic equilibrium between light and atoms—it turned out that without this process, the equilibrium cannot exist.

On second thought, however, it might appear surprising that the existence of this process has not been foreseen earlier. After all, remaining completely within the classical picture, one can think of a classical dipole interacting with radiation, in which two oppositely charged particles are connected by an elastic string. This system can both absorb and emit radiation. Indeed, on one hand, the harmonic oscillations excited by the incident radiation take energy away from light resulting in its absorption, while on the other hand, an oscillating dipole emits electromagnetic waves which carry away its energy damping the oscillations. This radiation is stimulated by the incident wave, and while quantitatively it might differ significantly from quantum stimulated emission, conceptually it is a similar process, and it might seem a bit surprising that it took that long to recognize its existence.

Well, it does not do us any good second-guessing our forebears, so let's go back to our immediate task. The transition rates for absorption and stimulated emission, which are equal to each other, are described by Eq. 15.35. If you think of the incident electric field $\mathcal{E}(t)$ as of incoherent "superposition" of monochromatic components with continuously distributed frequencies, you can identify $\mathcal{E}_0$ with $F(\Omega)$ appearing in Eq. 15.29 and $\mathcal{E}_0^2$ with $|F(\Omega)|^2$ in Eq. 15.35. When dealing with the finite or discrete number of incoherent waves, you could find the total energy density of the field by simply adding energy densities of the individual waves. In the case of continuous spectral distribution, the total finite energy density of the field is divided between the infinite numbers of infinitesimally small spectral intervals $d\Omega$. The contribution of each such interval can be described as $u(\Omega)\,d\Omega$, where $u(\Omega)$ is the time-averaged spectral energy density of the corresponding spectral component, which for a wave propagating in vacuum is given by the following expression well known in introductory electrodynamics (I am using SI units here):

$$u(\Omega) = \frac{1}{2}\varepsilon_0 \mathcal{E}_0^2(\Omega).$$

Taking into account Eq. 15.63, I can now present the expression for the transition rate as

$$R_{m_0m} = \frac{2\pi}{\hbar^2}\frac{e^2}{4}\,|\mathbf{w}\cdot\langle\alpha_{m_0}|\,\hat{\mathbf{r}}\,|\alpha_m\rangle|^2\,\mathcal{E}_0^2\,(\omega_{m,m_0}) =$$

$$\frac{\pi u\,(\omega_{m,m_0})}{\varepsilon_0\hbar^2}\,|\mathbf{w}\cdot\langle\alpha_{m_0}|\,e\hat{\mathbf{r}}\,|\alpha_m\rangle|^2. \qquad (15.65)$$

If the incident light has a definite polarization (noncoherent polarized light can be created using a polarizing element), Eq. 15.65 provides a full description for the corresponding transition rate. If, however, the light is not only incoherent but also unpolarized, some additional efforts are still required. In an unpolarized light, the polarization vectors w must be considered as a random quantity, and the completion of the calculation of the transition rate in this situation demands the averaging of Eq. 15.65 over the polarization directions.

To perform the averaging, I, first of all, must recall that an unpolarized light is a mixture of two independent polarizations, each appearing in the mix with equal probability $1/2$. To simplify the arguments, I will think of them as two linear polarizations characterized by mutually perpendicular unit vectors w_1 and w_2 with random but mutually correlated orientations (they must always remain orthogonal to each other). Therefore, I need to take the average over these two vectors and integrate it over their directions.

The direction of both polarization vectors must be perpendicular to the wave vector k. This requirement creates a one-to-one correspondence between w_1, w_2 and the propagation direction of the wave. Consequently, the integration of the transition rate over k automatically averages it over all allowed directions of w_1 and w_2. The resulting polarization-averaged transition rate can be presented as

$$R_{m_0m} = \frac{\pi u\left(\omega_{m,m_0}\right)}{\varepsilon_0 \hbar^2} \frac{1}{8\pi} \oiint d\Omega_k \left(|w_1 \cdot \langle \alpha_{m_0}| e\hat{r} |\alpha_m\rangle|^2 + \right.$$

$$\left. |w_2 \cdot \langle \alpha_{m_0}| e\hat{r} |\alpha_m\rangle|^2 \right), \qquad (15.66)$$

where the integration is over the surface of the constant k, and it is assumed that all directions of k are equally likely. The factor $1/4\pi$ normalizes the uniform distribution of k, so that $(1/4\pi) \oiint d\Omega_k = 1$. To carry out the integration, let me first note that $|w_{1,2} \cdot \varsigma(m, m_0)|^2$, where $\varsigma(m, m_0)$ stays for the vector dipole matrix element[5]

$$\varsigma(m, m_0) \equiv |\langle \alpha_{m_0}| e\hat{r} |\alpha_m\rangle|, \qquad (15.67)$$

does not depend on the choice of coordinate axes used to define the components of the vectors in this expression. Using this freedom to simplify it, I choose the direction of k as the Z-axis of a coordinate system. This choice makes defining polarization vectors $m_{1,2}$ almost trivial: recalling that they must be perpendicular to each other and to k, you can choose them to be directed along the arbitrarily chosen X- and Y-coordinate axes of this system. Now you can immediately have

$$|w_1 \cdot \varsigma(m, m_0)|^2 + |w_2 \cdot \varsigma(m, m_0)|^2 = |\varsigma_x(mm_0)|^2 + |\varsigma_y(mm_0)|^2 =$$

$$\varsigma^2(mm_0) \sin^2 \theta \cos^2 \varphi + \varsigma^2(mm_0) \sin^2 \theta \sin^2 \varphi = \varsigma^2(mm_0) \sin^2 \theta, \qquad (15.68)$$

---

[5] Since I assumed linear polarization, vectors $w_{1,2}$ are real, allowing me to take them outside of the absolute value sign.

where θ is the angle between $\mathbf{k}$ and a vector $\varsigma\,(m, m_0)$, while angle φ defined the direction of this vector in the X–Y plane.[6]

Since the result expressed by Eq. 15.68 is presented in the form independent of the choice of the coordinate system (angle θ is an objectively existing geometrical characteristic independent of the choice of coordinates), I could now move to different coordinate axes, if it suited me better. Indeed, the next step of the averaging procedure involves integration over directions of the wave vectors, so it does not make sense to use $\mathbf{k}$ as a choice for the polar axis. And allowing vector $\varsigma\,(m, m_0)$ to have this honor will make the integration just a bliss. Angle θ in Eq. 15.68 now becomes one of the spherical coordinates of the wave vector and thereby an integration variable. Substituting Eq. 15.68 to the integral in Eq. 15.66, I get

$$
R_{m_0 m} = \frac{\pi u\,(\omega_{m,m_0})\,\varsigma^2\,(mm_0)}{\varepsilon_0 \hbar^2}\,\frac{1}{8\pi} \oiint d\Omega_k \sin^2\theta =
$$

$$
\frac{\pi u\,(\omega_{m,m_0})\,\varsigma^2\,(mm_0)}{\varepsilon_0 \hbar^2}\,\frac{1}{8\pi} \int\limits_0^{\pi} d\theta \sin\theta \int\limits_0^{2\pi} d\varphi \sin^2\theta =
$$

$$
\frac{\pi\,[\varsigma\,(mm_0)]^2}{3\varepsilon_0 \hbar^2}\,u\,(\omega_{m,m_0})\,, \qquad (15.69)
$$

where I used the spherical surface element $d\Omega_k = d\varphi d\theta \sin\theta$.

This result describes the transition rate between a pair of states of a single atom, which can be either related to emission (if $\omega_{m,m_0} < 0$) or absorption (if $\omega_{m,m_0} > 0$) of radiation. It is obvious that since $u\,(\omega_{m,m_0})$ is an even function of frequency (it has to be if we believe in time-reversal symmetry[7]), the transition rates for both emission and absorption are equal to each other. A word of warning: the statement of equal transition probabilities refers to transitions involving the same pair of states with initial and final states reversed. If, however, you are looking at an "atom" in some initial state $|\alpha_{m_0}\rangle$ and are interested to know if the perturbation is more likely to cause upward or downward transitions, you can easily see that these probabilities do not have to be equal. Indeed, in this case, you are dealing with transitions characterized by the same initial and two different final states, so that they might have different transition frequencies ω_{m_1,m_0} and ω_{m_2,m_0} and different matrix elements.

---

[6]One needs to be a bit careful here because the original dipole matrix element vector $\varsigma\,(m, m_0)$ can be complex.

[7]Time-reversal symmetry means that if you change t to $-t$, nothing must change. Wave equations of electromagnetic waves are of the second order with respect to time derivative, and, therefore, they are obviously time-reversal symmetric. The situation is more complex with the Schrödinger equation, which are of the first order in time derivative, but you will have to wait for a more advanced course on quantum mechanics to dig into this issue.

In general when discussing quantum transitions and their probabilities, there is "a must": you must clearly understand which phenomenon you are dealing with. This exhortation is especially important if one or both states connected by the transition are degenerate, in which case Eq. 15.69 has to be used cautiously and the validity of the statement about equal transition rates for reversed transitions depends upon a situation you are dealing with.

The first question that must be asked is how much control do you have about the initial state, meaning do you know in which of the degenerate states with a given energy your system is? If the answer is "yes," then you can proceed to the next question, but if the answer is "no," you have to assume that all of the degenerate states are possible and, in the absence of any reasons to give preference to one state or another, that they can occur with equal probabilities. In this case the transition rate given by Eq. 15.69 must be averaged over the initial states. The next question you must ask is: what do you know (or want to know) about the final state? For instance, if you are only concerned with the total absorption or stimulated emission rate in the presence of unpolarized and incoherent radiation (strictly speaking, this is the only situation for which Eq. 15.69 is good for), then you have to sum up the rate given by this equation over all degenerate final states. In this case the total transition rate between two energy levels (note the careful choice of the language: now I am talking about the transition between the *energy levels* rather than between *states*) can be written down as

$$R_{E_{m_0}, E_m} = \frac{1}{g_{m_0}} \sum_{m_0} \sum_{m} R_{m_0 m} \tag{15.70}$$

where g_{m_0} is the degree of degeneracy of the initial energy level reflecting the procedure of averaging over the initial states. In the case of the reversed transition, the similar expression takes the form of

$$R_{E_m, E_{m_0}} = \frac{1}{g_m} \sum_{m} \sum_{m_0} R_{m, m_0}, \tag{15.71}$$

where g_m is the degeneracy of $|\alpha_m\rangle$, which is now the initial state. Since generally $g_m \neq g_{m_0}$, the two total energy-level-to-energy-level rates $R_{E_{m_0}, E_m}$ and $R_{E_m, E_{m_0}}$ are not equal to each other even if particular state-to-state rates $R_{m_0 m}$ are.

The situation might change if you are interested in the absorption or emission of light of a particular polarization (experimentally you can express your interest by inserting a polarizer between the light source and the atom). In this case you, first, must forgo the averaging over polarizations, go back to Eq. 15.65, and determine which of the perturbation matrix elements are different from zeroes for each of the degenerate state. If the polarization allows you to initiate a transition to just one particular state, you can congratulate yourself: you found a way to optically control a quantum state of your system. If you end up with more than one "allowed" transition (the one for which the perturbation matrix element does not vanish), you still might have to average over the initial states and sum up over the final states, however, including only the states connected by allowed transitions.

To prevent your head from spinning too much, I will illustrate these points with an example. Let $|\alpha_{m_0}\rangle$ be the ground state of the electron in a hydrogen atom, which is twice degenerate with respect to the spin magnetic number m_s, and assume that $|\alpha_m\rangle$ is its first excited state, which is eightfold degenerate. The composite indexes m_0 and m in this case consist of the principal number n, orbital angular momentum number l, and orbital and spin magnetic numbers m_l and m_s. The initial state $|\alpha_{m_0}\rangle$ represents one of two states $|1, 0, 0, 1/2\rangle$ or $|1, 0, 0, -1/2\rangle$, while $|\alpha_m\rangle$ can be one of eight states $|2, l, m_l, \pm 1/2\rangle$ with $l = 0$, $m = 0$ or $l = 1$, $m = -1, 0, 1$. As usual, I am using here the standard nomenclature for the states of a hydrogen atom $|n, l, m_l, m_s\rangle$, where the first is the principal quantum number, followed by the orbital angular momentum number, and then followed by orbital and spin magnetic numbers. Now, assuming that you want to compute the absorption rate for the unpolarized light with any of $|\alpha_{m_0}\rangle$ as the initial state and any of $|\alpha_m\rangle$ as the final, you have to consider the following:

$$R_{E_1,E_2} = \frac{1}{2} \Big[R_{\uparrow;0,0\uparrow} + R_{\uparrow;0,0\downarrow} + R_{\downarrow;0,0\uparrow} + R_{\downarrow;0,0\downarrow} +$$

$$\sum_{m_l=-1}^{1} \left(R_{\uparrow;1,m_l\uparrow} + R_{\uparrow;1,m_l\downarrow} + R_{\downarrow;1,m_l\uparrow} + R_{\downarrow;1,m_l\downarrow} \right) \Big].$$

To make notations less cumbersome, I am showing here only the spin index for the initial state and have dropped the principal quantum number for both initial and final states. Factor $1/2$ here is the degeneracy degree of the initial state, and the expression includes the summation over different values of the spin variable of the initial state. You also see the sum over all possible values of the quantum numbers characterizing the final states—this takes into account the probabilities of transitions to all states with the same final energy. If the perturbation potential does not depend on spin, this expression can be simplified since the matrix elements with different spin magnetic numbers are all equal to each other. In this case, you have for the total absorption transition rate

$$R_{E_1,E_2} = 2 \left(R_{1,0,0;2,0,0} + \sum_{m_l=-1}^{1} R_{1,0,0;2,1,m_l} \right), \tag{15.72}$$

where now I suppressed the index for the spin magnetic number (because nothing depends on it) and restored all other indexes for both initial and final states. For the opposite emission transition from $n = 2$ energy level to the ground state, the similar expression for R_{E_2,E_1} is

$$R_{E_2,E_1} = \frac{1}{8} \left(R_{\uparrow;0,0\uparrow} + R_{\uparrow;0,0\downarrow} + R_{\downarrow;0,0\uparrow} + R_{\downarrow;0,0\downarrow} \right.$$

$$\left. \sum_{m_l=-1}^{1} R_{\uparrow;1,m_l\uparrow} + \sum_{m_l=-1}^{1} R_{\uparrow;1,m_l\downarrow} + \sum_{m_l=-1}^{1} R_{\downarrow;1,m_l\uparrow} + \sum_{m_l=-1}^{1} R_{\downarrow;1,m_l\downarrow} \right) =$$

$$\frac{1}{4} \left(R_{1,0,0;2,0,0} + \sum_{m_l=-1}^{1} R_{1,0,0;2,1,m_l} \right). \qquad (15.73)$$

Further computation of these transition rates requires dealing with the matrix elements between different hydrogen-like states, and I will put it off for now to keep the suspense on. As a compensation, I will, instead, find for you the absorption and stimulated emission rates for one-dimensional harmonic oscillator with mass μ and charge e oscillating at frequency ω_0 along the Z-axis (no degeneracy there!). In this case the dipole matrix element vector ς (mm_0) defined in Eq. 15.67 has just a single component $\varsigma_z (mm_0) = \langle \alpha_{m_0} | e\hat{z} | \alpha_m \rangle$, where $|\alpha_m\rangle \equiv |m\rangle$ now is an eigenvector of a harmonic oscillator corresponding to the eigenvalue $E_m = \hbar\omega_0 (m + 1/2)$. Using Eq. 7.48 for the coordinate matrix elements, I can write

$$\langle m_0 | e\hat{z} | m \rangle = \begin{cases} e\sqrt{\frac{\hbar m_0}{2\mu\omega_0}} \delta_{m,m_0-1} & m = m_0 - 1 \text{ (emission)} \\ e\sqrt{\frac{\hbar (m_0+1)}{2\mu\omega_0}} \delta_{m_0,m-1} & m = m_0 + 1 \text{ (absorption)}. \end{cases} \qquad (15.74)$$

Apparently transitions can occur from any state $|m\rangle$ only to adjacent states $|m \pm 1\rangle$ with the transition frequency in both cases being $|\omega_{m,m_0}| = |\omega_0 (m_0 + 1/2) - \omega_0 (m_0 \pm 1 + 1/2)| = \omega_0$. Then the stimulated emission and absorption rates for downward and upward transitions between, say, states $|m_0\rangle$ and $|m_0 - 1\rangle$ become

$$R_{m_0m_0-1} = R_{m_0-1m_0} = \frac{\pi e^2 u(\omega_0)}{3\varepsilon_0\varepsilon_m\hbar^2} \frac{\hbar m_0}{2\mu\omega_0} =$$

$$\frac{\pi e^2 (E_{m_0} - \hbar\omega_0/2)}{6\varepsilon_0\varepsilon_m\hbar^2\omega_0^2\mu} u(\omega_0), \qquad (15.75)$$

where I replaced the initial state number m_0 with the value of the corresponding energy level E_{m_0}. As a side note: to find transition rates between states $|m_0\rangle$ and $|m_0 + 1\rangle$, one would need to replace in Eq. 15.75 m_0 to m_0+1, so for the given initial state $|m_0\rangle$, the probability of absorption (transition to $|m_0 + 1\rangle$) is larger than the probability of emission (transition to $|m_0 - 1\rangle$) by $(\pi e^2 u(\omega_0))/(2\varepsilon_0\hbar\omega_0\mu)$. This is a useful illustration to a general statement that equality of emission and absorption probabilities refers to transitions with reversed direction between the same pair of states. Probabilities of upward or downward transitions from a given initial state can be and often are different.

The results obtained so far describe a single quantum system (atom, oscillator, etc.) in a given initial state. However, in most situations, in which these results might be of interest, you have to deal with an ensemble of systems, not necessarily all in the same initial state. In the presence of a radiation field, some of these atoms would undergo upward transitions absorbing light; others will emit light while lowering their energy. What we are interested in most of the time is the net result: will there be more upward or downward transitions resulting in an overall absorption or emission of light? The answer to this question depends not only on the transition rates but also upon the number of atoms in different initial states. Below I am going to discuss this situation using a simplified model as per Einstein with radiation described as a collection of Einstein's light quanta—photons.

If the spectral energy density of the incident radiation $u(\omega)$ is peaked at some frequency ω_{max} and falls off within some frequency range comparable to the distance between energy levels of an atom, I can approximate atoms by two-level systems with only two energy levels. Let's assume that per unit volume there is a stationary (independent on time) number N_a of atoms in the lower energy state $|a\rangle$ with respective energy eigenvalue E_a and N_b atoms in the higher energy state $|b\rangle$ with corresponding eigenvalue $E_b > E_a$ such that $E_b - E_a \approx \hbar\omega_{max}$. Since each atom in state $|a\rangle$ absorbs photons at rate R_{ab} (I assume here that each upward transition corresponds to absorption of one photon), the total number of photons absorbed per unit time would be $N_a R_{ab}$, and taking into account that each absorbed photon carries energy $\hbar\omega_{ba} = E_b - E_a$, I conclude that the total absorbed power is $\hbar\omega_{ba} N_a R_{ab}$. Similar arguments yield for the emitted power $\hbar\omega_{ba} N_b R_{ba}$, so that the total change in the energy density of the incident radiation u becomes

$$\frac{du}{dt} = \hbar\omega_{ba} \left(N_b - N_a\right) R_{ab} \tag{15.76}$$

(emission increases the radiation energy and absorption decreases it). So, the overall result depends on the relative number of atoms in the lower and higher energy states: if $N_b < N_a$, the system absorbs light; in the opposite case, the system generates light. Normally, for the system in the thermodynamic equilibrium, more atoms stay in the lower energy state so that absorption of light is the expected normal behavior of any collection of atoms, electrons, etc. However, by using some clever tricks (which I cannot discuss here), it is possible to create a situation in which $N_b > N_a$ and the system becomes a net emitter of light. This situation is quite unusual and cannot exist for systems in thermodynamic equilibrium; systems with this property are said to demonstrate "population inversion." Population inversion is the main condition for operation of lasers and was realized first in ammonia molecules independently by American physicist Charles H. Townes and two Russian physicists, Nikolay Basov and Aleksandr Prokhorov, for which they all shared the 1964 Nobel Prize in Physics.

However, one question here just begs to be answered. A thoughtful reader could say: "OK, let's assume that in the absence of light there is, indeed, some distribution of atoms in the lower and higher energy states. But, once light starts generating those upward transitions, this distribution will start changing, and since the transition rates for upward and downward transitions are the same, eventually we must come to a situation in which $N_b = N_a$ and no net absorption or emission will take place." I will present this argument in a more formal way in the next section, and here I will only mention that it misses an important point. Stimulated emission is not the only process by which an atom can drop from the higher energy level to a lower one. Other possibilities are spontaneous emission, which will be discussed in the next section, and so-called non-radiative processes, when an atom gives away its energy to something else than radiation—elastic wave or kinetic energy of atoms in a gas, or what else. The presented calculation of the absorption coefficient makes sense only if these processes of energy "dissipation" occur faster than the optical transitions from the lower to higher energy level. Only in this case one can really talk about stationary distribution of atoms in various energy levels appearing in Eq. 15.76. An important lesson of this discussion is that even though we compute the absorption rate considering atomic transitions, the actual absorption (the dissipation of energy) occurs only when the atom gives off this energy in a form which can no longer be converted back into radiation. If you are wondering how can I claim that Eq. 15.76 describes the absorption of light without considering all these processes in a more or less explicit form, the answer is that the assumption that the atoms are in thermal equilibrium ensures that all these dissipation processes are accounted for. This is also a beautiful illustration of the universality of thermodynamic equilibrium: it does not matter how the system comes to equilibrium: once it is there, its properties are described by the corresponding distribution functions.

I will conclude this discussion with a derivation of expression for an absorption coefficient of light, which is defined based on the consideration of a beam of light of intensity I propagating in some well-defined direction. If z is the coordinate in the direction of the propagation of the beam, the absorption coefficient κ is defined via the relation

$$I(z) = I_0 e^{-\kappa z}.$$ (15.77)

Differentiation of this expression with respect to the coordinate yields

$$\kappa = -\frac{1}{I}\frac{dI}{dz}.$$

Taking into account that the distance dz light travels in time $dt = n_m dz/c$, where n_m is the refractive index of the medium where light propagates, it can be rewritten as

$$\kappa = -\frac{n_m}{cI}\frac{dI}{dt}.$$

Remembering that the energy density of light is related to its intensity as $I = n_m c u$, I can turn this into

$$\kappa = -\frac{n_m}{cu}\frac{du}{dt} = \hbar\omega_{ba}\left(N_b - N_a\right)B_{ab}$$ (15.78)

where I presented the transition rate as

$$R_{ab} = B_{ab}u(\omega_{ba}) \tag{15.79}$$

with B_{ab} being

$$B_{ab} = \frac{\pi \, [\varsigma \, (mm_0)]^2}{3\varepsilon_0\hbar^2}. \tag{15.80}$$

Using Eq. 15.78 I can formulate for you a verbal definition of the absorption coefficient not related to the specific representations of intensity as in Eq. 15.77:

$$\text{absorption coefficient} = \frac{\text{energy absorbed/unit time/unit volume}}{\text{total incident intensity}}. \tag{15.81}$$

You probably wonder if it was really necessary to introduce new quantities such as B_{ab} or I am doing this just to annoy you. Well, I could have done without this coefficient, but it is really, really important, partially for historical reasons—it was introduced by Einstein and is still called the Einstein B coefficient. You see, we feel so much reverence for the guy that we do not even dare to change the notation he used in his original 1916 paper (more than a hundred years ago!). Actually, there are two such coefficients B_{ab} and B_{ba}, defined as factors in front of the energy density term u in the corresponding expressions for the transition rates. In the absence of degeneracy in the initial states $B_{ab} = B_{ba}$, otherwise comparing Eqs. 15.70 and 15.71, you can show that $g_a B_{ab} = g_b B_{ba}$. Besides being of historical and sentimental value, these coefficients also play a more serious role allowing to separate in the transition rate the effects dependent on the properties of the "atoms" from those dependent on the characteristics of the radiation. The usefulness of this feature will become especially clear in the next section.

15.3.2.2 Spontaneous Emission Without Quantum Electrodynamics: Einstein Meets Planck

Having gotten through the previous section, you might experience a vague and uneasy feeling that you are being duped. You might be thinking: how can one assume that the number of atoms in different states $N_{a,b}$ have stationary (time-independent) values and at the same time argue about emission and absorption processes, which obviously change them? Well, this is a legitimate question that deserves to be answered. The answer as you will see in a minute will not only justify arguments leading to Eqs. 15.76 and 15.78, but it will also demonstrate to you one of the most beautiful results of early quantum mechanics produced by Einstein as early as in 1916.

First, I will drop the offending assumption that $N_{a,b}$ are constant and will try to actually describe their time dependence. Here is what I know: transitions $|a\rangle \rightarrow |b\rangle$

result in a decrease of N_a and simultaneous increase of N_b. The reverse transitions $|b\rangle \rightarrow |a\rangle$ obviously depopulate state $|b\rangle$ and populate state $|a\rangle$. The rate of all these changes is proportional to the rate of the respective transition R_{ab} and the number of atoms in the corresponding states at any given time:

$$\frac{dN_a}{dt} = -N_a B_{ab} u + N_b B_{ba} u \tag{15.82}$$

$$\frac{dN_b}{dt} = -N_b B_{ba} u + N_a B_{ab} u. \tag{15.83}$$

N_a and N_b now are instantaneous numbers of atoms in the respective states per unit volume, and I replaced the transition rates with their expressions in terms of the Einstein coefficients B_{ab} and B_{ba} to emphasize the presence of the radiation spectral energy density u. If you add these two equations, you get

$$\frac{d(N_a + N_b)}{dt} = 0,$$

which simply tells you that the total number of particles in both states remains constant. This is a somewhat trivial result reflecting the fact that an atom does not get vanquished while undergoing a transition from one state to another. The assumption of time-independent numbers $N_{a,b}$ amounts to a statement that these equations have stationary solutions in which the processes of population and depopulation of various states perfectly balance each other. In principle, Eqs. 15.82 and 15.83 do have a stationary solution in which time derivatives on the left-hand side vanish and $N_a B_{ab} = N_b B_{ba}$. In the absence of degeneracy, the Einstein coefficients cancel out, and we find that the stationary solution with constant N_a and N_b is possible only if these numbers are equal to each other. This cannot be right, of course, and for many reasons. For instance, if this were true, then Eq. 15.78 would predict zero absorption in all cases, which is nonsense. Obviously, something is missing in these equations, and a bit of rumination over them can bring you a happy guess: all processes included in the equations are stimulated (proportional to the energy density of the field). But what about the spontaneous emission—the one which is responsible for bringing atoms to their ground state spontaneously without any prompting by the already present field? This process would increase the number of atoms N_a in the lower energy state at the expense of N_b, which would obviously decrease. Since it is the atoms in the higher energy state, which undergo this spontaneous transition, its rate must be proportional to N_b:

$$\left(\frac{dN_a}{dt}\right)_{sp} = A_{ab} N_b$$

where following Einstein, I introduced a new coefficient A_{ab} called the Einstein A coefficient. Note the absence of proportionality to the energy density u: the rate of this process does not depend on the existence of any radiation. Adding this term to Eq. 15.82, I obtain a corrected version of this equation:

$$\frac{dN_a}{dt} = -N_a B_{ab} u + N_b B_{ba} u + A_{ab} N_b. \tag{15.84}$$

Now the stationary state with time-independent values of the population numbers emerges if

$$N_b (A_{ab} + B_{ba} u) = N_a B_{ab} u. \tag{15.85}$$

This result does not look that egregious as the previous attempt—the ratio N_b/N_a now depends on the radiation energy density and three coefficients, two of which we know, but the third one is yet an enigma. This is as far as I can go on without any additional assumptions about the atoms and the radiation.

So, now imagine that the atoms are enclosed in a box maintained at fixed temperature T from which no atoms or radiation can escape. If you wait long enough, all the atoms and the radiation will come to a thermal equilibrium with each other and the walls of the box so that they all can be characterized by a common temperature T. Elementary statistical mechanics tells me that a probability for an atom (or any other system) to have energy E is proportional to $\exp(-E/k_B T)$, where k_B is the Boltzmann constant. Applying this to our system of two-level atoms, I can write $N_a \propto g_a \exp(-E_a/k_B T)$ and $N_b \propto g_b \exp(-E_b/k_B T)$, where $g_{a,b}$ are degrees of degeneracy of the corresponding level (if there are several states with the same energy, then the total probability for the system to have this energy must be multiplied by the number of such states). Then I can write that

$$\frac{N_a}{N_b} = \frac{g_a}{g_b} \exp\left(\frac{E_b - E_a}{k_B T}\right) = \frac{g_a}{g_b} \exp\left(\frac{\hbar \omega_{ba}}{k_B T}\right). \tag{15.86}$$

Using Eq. 15.85 to express the electromagnetic spectral energy density in terms of the population numbers $N_{a,b}$, I have

$$u = \frac{N_b A_{ab}}{N_a B_{ab} - N_b B_{ba}} = \frac{A_{ab}}{\frac{N_a}{N_b} B_{ab} - B_{ba}}.$$

Substitution of Eq. 15.86 yields

$$u = \frac{g_b A_{ab}}{g_a B_{ab} \exp\left(\frac{\hbar \omega_{ba}}{k_B T}\right) - g_b B_{ba}}. \tag{15.87}$$

You know that the energy density for radiation in the thermal equilibrium has been found by Max Planck in 1900. This was the first formula in this book, Eq. 1.1, but for your convenience, I will reproduce it here:

$$u = \frac{\hbar\omega^3}{\pi^2 c^3} \frac{1}{\exp\left(\frac{\hbar\omega}{k_B T}\right) - 1}.$$

Equation 15.87 must agree with Eq. 1.1 because they describe the same quantity. It is encouraging that the exponential terms in the denominator of both expressions are identical, and I can make their entire denominators coincide by requiring that $g_a B_{ab} = g_b B_{ba}$. Isn't it remarkable that this reproduces the relation between the Einstein coefficients that I derived just a few lines above using nothing but honest to G-d quantum mechanical calculations of the transition rates? It turns out that the thermodynamics demands the same. Now, all that is left to make Eq. 15.87 look exactly like Eq. 1.1 is to require that

$$\frac{A_{ab}}{B_{ba}} = \frac{\hbar\omega^3}{\pi^2 c^3}, \tag{15.88}$$

which expresses the unknown coefficient A_{ab} representing the rate of spontaneous emission in terms of the known quantity B_{ba}. Using Eq. 15.80 I finally find for A_{ab}

$$A_{ab} = \frac{[\varsigma\,(mm_0)]^2\,\omega^3}{3\pi\varepsilon_0 \hbar c^3}, \tag{15.89}$$

which agrees exactly with the result obtained by Dirac in his 1927 paper from his quantum theory of electromagnetic field.

This is the point where you are supposed to feel genuinely amazed. Indeed, the spontaneous emission is a result of quantum fluctuations of the electromagnetic field, and I always maintained that it cannot be described without quantization of the field. And here we are: we found this coefficient using nothing but thermodynamics and regular quantum mechanics with classical electromagnetic field. This looks almost like a miracle and is an evidence of deep connections not just between different laws of physics but between different layers of reality described by them.

I shall complete this section by computing the power emitted by a quantum harmonic oscillator via the process of spontaneous emission and comparing this result with the emission of a classical oscillator. Using Eq. 15.89 for the rate of the corresponding transmissions and Eq. 15.67 to compute the required dipole matrix elements, I find for the emitted power

$$P_{sp} = \hbar\omega_0 A_{ab} = \frac{e^2 \hbar m_0 \omega_0^3}{6\pi\varepsilon_0 \mu c^3}. \tag{15.90}$$

To compare this with a power emitted by a classical harmonic oscillator, consider a free (not driven) oscillator with initial energy E_{cl}. One can use a well-known form of electrodynamics Larmor's formula for a power emitted by a charged particle moving with acceleration $\boldsymbol{a}$:

$$P_{cl} = \frac{e^2 a^2}{6\pi\varepsilon_0 c^3}. \tag{15.91}$$

Acceleration of a harmonic oscillator is related to its displacement z as $a = -\omega_0^2 z$, so that the expression for the power becomes

$$P_{cl} = \frac{e^2 z^2 \omega_0^4}{6\pi\varepsilon_0 c^3}.$$

This expression has to be averaged over the period of oscillations, which, as you hopefully know,[8] yields $\overline{z^2} = z_0^2/2$, where z_0 is the oscillator's amplitude related to its energy as $z_0^2 = 2E_{cl}/\mu\omega_0^2$. Thus, the final expression for the power of the oscillator's radiation becomes

$$P_{cl} = \frac{e^2 \omega_0^2 E_{cl}}{3\pi\varepsilon_0 c^3 \mu}. \tag{15.92}$$

Replacing the term $\hbar m_0 \omega_0$ with $E_{m_0} - \hbar\omega_0/2$ in Eq. 15.90, you can see that the classical and quantum results for the power agree with an accuracy to the term $-\hbar\omega_0/2$, which can be neglected in the classical limit $E_{m_0} \gg \hbar\omega_0/2$. At the same time, this term makes sure that there is no spontaneous emission from the ground state.

15.3.3   Optical Transitions in Semiconductors

Optical transitions in semiconductors are another important area of application of Fermi's golden rule. An important peculiarity of the energy spectrum of electrons in semiconductors is that the continuous spectrum of the allowed energy values is broken into a sequence of regions separated by intervals of energy values containing no states whatsoever—so-called forbidden bands. The actual number of allowed bands depends on the properties of atoms (mostly their atomic number) constituting a semiconductor, but as far as interaction with light is concerned, only two are of real importance: the last completely "filled" band and the first "empty" band (the existence of such "full" and "empty" bands is what distinguishes semiconductors and dielectrics from metals). The concept of the full or empty bands is closely

---

[8]For those who forgot: $z(t) = z_0 \cos\omega_0 t$, and $\overline{\cos^2(\omega_0 t)} = (1/T)\int_0^T \cos^2(\omega_0 t)\,dt = 1/2$, where $T = 2\pi/\omega_0$.

related to the discussion of many-particle fermion states we had in Sect. 11.5 (it might make sense for you to return to this section). Recall that the Pauli principle requires the ground state of a many-fermion system to involve as many single-particle orbitals as there are particles. These orbitals, however, belong to energy eigenvalues, which are distributed between the bands, and when orbitals from the lowest bands are all used up, and there are still unassigned electrons, you have to go to the next band, and this process repeats itself until all electrons are accounted for. If the highest energy level assigned to the last electron lies somewhere in the middle of a band so that a spectral distance to the next available level is infinitesimally small (remember energy levels inside each band form a continuous spectrum), the material behaves as a metal (it takes a very small amount of energy, which can be provided by a static electric field to move an electron to a higher energy state where it can conduct current). If, however, the last filled level is also the last level in a band so that the next available state belongs only to the next band, the material behaves as a semiconductor or a dielectric. There is no sharp boundary between the two— the difference is in the magnitude of the band gap—for semiconductors it is usually much smaller and can range between 0.23 eV in *InSb* and 3.44 eV in *ZnO*, while band gaps of materials usually perceived as dielectric are much higher (e.g., the band gap of silica is 8.9 eV). The last filled band in a semiconductor is called valence band, and the first empty band is called conduction band. These names reflect the fact that valence electrons are usually firmly attached to an atom and cannot conduct electricity, while electrons excited to a conduction band become freer of their ions and can respond to an applied electric field. In a way, you can still think of this process as "ionization," which I discussed in Sect. 15.2.3. The difference is that the valence band electrons also have a continuous spectrum and the conduction band electrons are not that free and cannot be simply described by free-particle wave functions.

The curious band structure of electron energy levels in semiconductors is the direct consequence of the periodicity in the arrangement of atoms in these materials. Periodicity means that one can find a small group of atoms contained in such a box that the entire structure can be reproduced by translating this box by one of three non-collinear vectors a_i (I am going to assume for simplicity that they all are mutually perpendicular and can be used, therefore, to define directions of Cartesian coordinate axes). These vectors define the periodicity of the structure in the sense that if you take a coordinate of an arbitrary atom and add one of the vectors a_i, you will get a coordinate of another atom of exactly the same element in the same position with respect to other atoms, see Fig. 15.5.

I am not going to torture you with any derivations related to the theory of wave functions describing the stationary states of electrons in such a periodic environment and their corresponding energy eigenvalues. What you need to know is that the wave functions representing states in both valence and conduction bands have the following form:

$$\psi_{v,c}\left(k_{v,c},r\right) = \frac{1}{\sqrt{V}}u_{v,c}\left(r\right)e^{ik_{v,c}r}. \tag{15.93}$$

Fig. 15.5 Example of a
periodic arrangement of
atoms in what is called a
crystal lattice

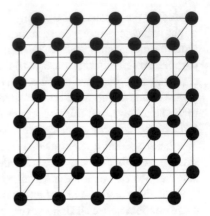

These states are characterized by a discrete index, signifying a particular band
to which they belong, and by quasi-continuous wave vectors $k_{v,c}$, discretized by
the same periodic boundary conditions as in Eq. 11.45. For the purpose of this
discussion, I need only two bands represented in Eq. 15.93 by indexes v and
c corresponding to the valence and conduction bands. Functions $u_{v,c}(r)$ in this
equation are periodic: $u_{v,c}(r + a_i) = u_{v,c}(r)$, and V is the normalization volume
defined by the same boundary conditions. For small enough $k_{c,v}$ ($k_{c,v} \cdot a_i \ll 1$),
the energy of a single electron in the conduction and valence bands can often be
approximated by

$$E_c = E_g + \frac{\hbar^2 k_c^2}{2m_c} \tag{15.94}$$

$$E_v = -\frac{\hbar^2 k_v^2}{2m_v}. \tag{15.95}$$

These formulas look somewhat like expressions for kinetic energy of a free
particle but with two significant differences: the energy of the electron in the
conduction band does not go to zero as $k_c \rightarrow 0$ because of the extra term E_g,
and the energy of the states in the valence band is negative and decreases with
increasing k_v. Both these peculiarities reflect the presence of the gap E_g in the
spectrum of energies, and the supposition (valid for many semiconductors) that
states with $k_{v,c} = 0$ correspond to the maximum and the minimum energies in the
valence and conduction bands correspondingly. If I choose the zero level of energy
to coincide with the maximum energy in the valence band, then the lowest energy
in the conductance band must be E_g, which appears in Eq. 15.94. This choice also
automatically makes all energies in the valence band negative, and they become
more negative the farther away from $k_v = 0$ you go. Parameters m_c and m_v have the
dimensions of mass and appear in the place of a particle's mass in many quantum
mechanical formulas, but they are not equal to the regular electron mass m_e. These

parameters are called effective masses and are usually much smaller than m_e with m_c often less than m_v.

With some preliminary groundwork laid down, I can start talking about optical transitions between states in the valence and conductance bands (these are the only relevant transitions for light with frequencies between infrared and visible). From a general point of view, this situation is an extension of the discussion of Sect. 15.2.3 in two directions: first, now both initial and final states belong to continuous spectra, and second, we are dealing with an ensemble of fermions rather than with a single particle. First of these circumstances means that the conversion of the transition probability into the probability density started out in Eq. 15.37 and continued in Eq. 15.38 must now include two sets of quasi-continuous wave numbers for the valence and conduction bands:

$$dp_{mm_0} = p_{mm_0} \Delta m_{0_1} \Delta m_{0_2} \Delta m_{0_3} \Delta m_1 \Delta m_2 \Delta m_3 = \quad (15.96)$$

$$\left[\frac{L^3}{(2\pi)^3} \right]^2 p_{s_0,s} \left(k^{(c)}, k^{(v)} \right) \Delta k_1^{(c)} \Delta k_2^{(c)} \Delta k_3^{(c)} \Delta k_1^{(v)} \Delta k_2^{(v)} \Delta k_3^{(v)}.$$

The many-body nature of the semiconductor transitions manifest itself first of all via the Pauli principle affecting the occupation of single-particle orbitals by identical fermions even if you can neglect the Coulomb interaction between them (see Chap. 11 if you need to refresh your memory of these ideas). Because of the Pauli principle, any single-particle orbitals can be either occupied (included into a determinant presenting a wave function of a many-particle electron state) or empty (not part of the determinant). Neglecting the interaction between the electrons, you can still describe the transitions in terms of the single-particle orbitals, but you have to take into account that a transition from $\left| \alpha_{m_0}^{(v)} \right\rangle$ in the valence band to $\left| \alpha_m^{(c)} \right\rangle$ in the conduction band is now contingent upon two conditions: first, that the initial state is occupied and, second, that the final state is empty (you cannot put a second fermion in a state which is already occupied since trying to add to a Slater determinant on another row consisting of orbitals already present there will render the whole thing equal to zero). This circumstance turns probability dp_{mm_0} into a *conditional* probability, and according to the general theorem of the probability theory, in order to find the actual probability of the transition, you must multiply dp_{mm_0} by probabilities that both conditions are realized simultaneously. To make this correction, I need two probability distributions: one, $f_v(k_v)$, is the probability of a given orbital in the valence band to be occupied, and the other, $f_c(k_c)$, contains the same information about the conduction band. In the continuous limit, both these distribution functions become *probability densities*. Since a given orbital can be either occupied or empty, the probability of it being empty is obviously $1 - f_{v,c}(k_{v,c})$ for both valence and conduction bands. Since any orbital can contain either one or zero electrons, integration over the orbitals weighted with the respective probability densities is equivalent to the summation over all electrons, so that the total transition probability p_{vc} from any valence band orbital to any conduction band orbital can be found as

$$
p_{vc} = \left[\frac{L^3}{(2\pi)^3} \right]^2 \sum_{s_0,s} \int p_{s_0,s} \left(k_c, k_v \right) f_v \left(k_v \right) \left(1 - f_c \left(k_c \right) \right) d^3 k_v d^3 k_c, \tag{15.97}
$$

where indexes s_0,s indicate the initial and final spin states of the electron. This expression contains two factors with very distinct physical meaning. Factor $p_{s_0,s} \left(k_c, k_v \right)$ is your regular quantum transition probability, which we have been discussing in this chapter all along. Factors $f_v \left(k_v \right) \left(1 - f_c \left(k_c \right) \right)$ express a different kind of uncertainty, which is also inherent to the system under consideration but has nothing to do with quantum mechanics. It simply reflects the fact that in a system comprising many particles, detailed information about any particular particle is inaccessible. This situation is typical for all many-particle systems regardless of their quantum or classical nature. In classical statistical mechanics, the lack of knowledge of coordinates and momentums of individual particles is manifested in the form of Maxwell–Boltzmann distribution, while in the quantum case, the similar role is played by functions $f_{v,c} \left(k_{v,c} \right)$, which have the form of the famous Fermi–Dirac distribution:

$$
f_{v,c}(k) = \frac{1}{\exp\left(\frac{E_{v,c}(k) - \mu}{k_B T} \right) + 1}.
$$

Parameter μ in this expression is called a chemical potential and is essentially a normalization parameter found by requiring that the integral of $f(k)$ over all k yields the total number of particles in the system. I am going to spare you from the task of deriving this function (you can easily find several different derivations if you want), and I will not be using it in what follows. But I do want to note that as temperature T approaches absolute zero, the Fermi–Dirac function turns into a step function, which takes values equal to one for $E < \mu$, and zero for $E > \mu$. Indeed, in the former case, the argument of the exponential function is negative so that it vanishes in the limit $T \to 0$, while in the latter case, this argument is positive, making the exponential function infinite when the temperature goes to zero and vanquishing, thereby, the entire function f. Such a distribution describes the ground state of the many-fermion system with all states in the valence band filled with probability equal to unity and all states in the conduction band remaining empty. It does not mean, however, that parameter μ is necessarily equal to the energy of the last filled orbital. It would have been true for metals, but in semiconductors, the presence of the gap pushes the value of μ toward the middle of the band gap where no single-particle states can exist.

With the help of Eq. 15.38, I can present the total rate of transitions from the valence to conduction band as

$$R_{vc} = \frac{2\pi}{\hbar} \left[\frac{L^3}{(2\pi)^3} \right]^2 \sum_{s_0,s} \int |v_{s_0,s}(k_v, k_c)|^2 f_v(k_v)(1 - f_c(k_c)) \times$$

$$\delta[\hbar\Omega - E_c(k_c) + E_v(k_v)] d^3k_v d^3k_c, \qquad (15.98)$$

where the only remaining discrete indexes refer to the two-spin states of the electrons. If the interaction between light and electrons is spin independent as is often the case, the perturbation matrix element is diagonal with respect to spin indexes and can be presented as $v_{s_0,s}(k_v, k_c) = \delta_{s_0,s} v(k_v, k_c)$. In this case, the summation over the spin variables is reduced to an extra factor of two in the expression for the transition rate:

$$R_{vc} = \frac{4\pi}{\hbar} \left[\frac{L^3}{(2\pi)^3} \right]^2 \int |v(k_v, k_c)|^2 f_v(k_v)(1 - f_c(k_c)) \times$$

$$\delta[\hbar\Omega - E_c(k_c) + E_v(k_v)] d^3k_v d^3k_c. \qquad (15.99)$$

Since the matrix elements $v(k_v, k_c)$ are the same for the transitions in both directions, I can easily turn this expression into the rate for the reverse transitions (from the conduction to the valence band) by simply interchanging the distribution functions $f_v(k_v)$ and $f_c(k_c)$:

$$R_{cv} = \frac{4\pi}{\hbar} \left[\frac{L^3}{(2\pi)^3} \right]^2 \int |v(k_v, k_c)|^2 f_c(k_c)(1 - f_v(k_v)) \times$$

$$\delta[\hbar\Omega - E_c(k_c) + E_v(k_v)] d^3k_v d^3k_c. \qquad (15.100)$$

It follows immediately from these equations that at zero temperature, when all orbitals in the valence band are filled with probability one, and all orbitals in the conduction band are empty, the probability distribution functions become $f_v = 1, f_c = 0$, and one can generate only absorbing transitions from the valence to the conduction band. Indeed, if the system is already in the ground state, it cannot further lower its energy by emitting light, be it a single electron or a whole bunch of them filling the entire valence band. The net transition rate for the non-zero temperature $R_{net} = R_{vc} - R_{cv}$ is given by

$$R_{net} = \frac{4\pi}{\hbar} \left[\frac{L^3}{(2\pi)^3} \right]^2 \int |v(k_v, k_c)|^2 (f_v(k_v) - f_c(k_c)) \times$$

$$\delta[\hbar\Omega - E_c(k_c) + E_v(k_v)] d^3k_v d^3k_c. \qquad (15.101)$$

Normally, the occupation probability in the valence band is larger than in the conduction band $f_v(k_v) > f_c(k_c)$ so that the net flow of electrons is from the valence to conductance band, which results in net absorption of light. If, however, you manage somehow to achieve the reverse relation (a population inversion), the system will become a net emitter of radiation. I think I have already mentioned that a population inversion is one of the preconditions for achieving lasing.

What is left for me now is to compute the transition matrix elements $v(k_v, k_c)$ and evaluate the integrals over the wave numbers. To achieve this goal, I first need, according to Eqs. 15.63 and 15.93, to calculate the vector dipole matrix element:

$$d(k_v, k_c) = \frac{e}{L^3} \int e^{-i(k_v - k_c)r} u_v^*(r)(r) u_c(r) d^3r, \tag{15.102}$$

which determines $v(k_v, k_c)$ via

$$v(k_v, k_c) = \frac{i\mathcal{E}_0}{2} w \cdot d(k_v, k_c). \tag{15.103}$$

Figuring out the integral in Eq. 15.102 takes effort and imagination, so fasten your seat belt—you are in for a ride.

Remember that semiconductors are built by periodic repetition of a single box— elementary cell? Well, each cell can be characterized by a position vector R_n of its center (or one of the corners—does not matter) defined by three integer numbers (relative to the same common origin as the integration variable r, see Fig. 15.5):

$$R_{n_1,n_2,n_3} = n_1 a_1 + n_2 a_2 + n_3 a_3, \tag{15.104}$$

where n_i changes between 0 and $N_i = L/a_i$, a presumed integer, and is used to enumerate all these cells. You can think of the entire integration volume in Eq. 15.102 as being filled by $N_c = L^3/(a_1 a_2 a_3) = N_1 N_2 N_3$ non-overlapping (and leaving no gaps) identical cells of volume $v = a_1 a_2 a_3$. The integral, then, can be presented a sum of integrals over each of these cells as

$$d(k_v, k_c) = \frac{e}{L^3} \sum_{n_1}^{N_1} \sum_{n_2}^{N_2} \sum_{n_3}^{N_3} \int_{v_n} e^{-i(k_v - k_c)r} u_v^*(r) r u_c(r) d^3r, \tag{15.105}$$

where v_n is the region inside nth cell. (Here and elsewhere I will sometimes use $n \equiv \{n_1, n_2, n_3\}$ as a compound index for the sake of brevity.) If you feel uneasy about this trick, consider its one-dimensional version: an integral from 0 to L can be presented as $\int_0^L f(x)dx = \int_0^a f(x)dx + \int_a^{2a} f(x)dx + \cdots \int_{(n-1)a}^L f(x)dx$, where $L = na$. Next, take an arbitrary point inside a cell and define its position relative to its center by vector ϱ, such that you can present the point's position vector r as $r = \rho + R_n$ (see Fig. 15.6).

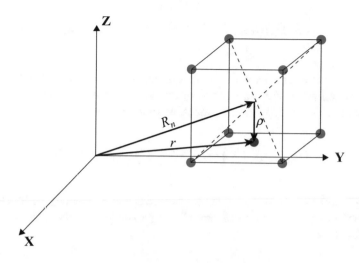

Fig. 15.6 Vectors involved in the transition from integration over an entire volume to the integration over the elementary cell represented by a single square

Substituting this into Eq. 15.105 and taking advantage of the periodicity of functions $u_{v,c}(r)$: $u_{v,c}(\rho + R_n) = u_{v,c}(\rho)$, I obtain

$$d(k_v, k_c) = \frac{e}{L^3} \sum_{n_1=0}^{N_1} \sum_{n_2=0}^{N_2} \sum_{n_3=0}^{N_3} e^{-i(k_v - k_c)R_n}$$

$$\times \int_{v_n} e^{-i(k_v - k_c)\rho} u_v^*(\rho)(\rho + R_n) u_c(\rho) d^3\rho =$$

$$\frac{e}{L^3} \sum_{n_1} \sum_{n_2} \sum_{n_3}^{N_1 \, N_2 \, N_3} \int_{v_n} e^{-i(k_v - k_c)\rho} u_v^*(\rho) \rho u_c(\rho) d^3\rho +$$

$$\frac{e}{L^3} \sum_{n_1} \sum_{n_2} \sum_{n_3}^{N_1 \, N_2 \, N_3} e^{-i(k_v - k_c)R_n} R_n \int_{v_n} e^{-i(k_v - k_c)\rho} u_v^*(\rho) u_c(\rho) d^3\rho.$$

The last line in this expression vanishes because the wave functions $\psi_{c,v}(r)$, representing eigenvectors of a Hermitian operator, are orthogonal.[9] The integral in the second line

---

[9]For especially pedantic and inquisitive readers, I will note: yes you are right, the orthogonality condition must include integration over the entire volume, while the integral I claim is vanishing includes integration only over a single cell. However, you might also notice that such integrals over each elementary cells are all equal to each other, and the integral over the entire volume is their sum. Thus, the orthogonality condition in this case can be naturally formulated for the inner product defined over an integral over a single cell.

$$\tilde{d}\left(k_v, k_c\right) = \int_{v_n} e^{-i(k_v - k_c)\rho} u_v^*\left(\rho\right) \rho u_c\left(\rho\right) d^3\rho,$$

which I will call a reduced dipole matrix element, is taken over a volume of the elementary cell and is the same for all of them. Thus, it can be factored out of the sum over the cells, so that $d\left(k_v, k_c\right)$ takes the following form:

$$d\left(k_v, k_c\right) = \frac{e}{L^3}\tilde{d}\left(k_v, k_c\right)\sum_{n_1}^{N_1}\sum_{n_2}^{N_2}\sum_{n_3}^{N_3} e^{-i(k_v - k_c)R_n}.$$

All what is left now is to figure out the sum over the elementary cells. Using the representation of R_n in terms of periodicity parameters of the crystal structure, Eq. 15.104, I can write

$$\sum_{n_1}^{N_1}\sum_{n_2}^{N_2}\sum_{n_3}^{N_3} e^{-i(k_v - k_c)R_n} =$$

$$\sum_{n_1=0}^{N_1} e^{-i\left(k_{vx} - k_{cx}\right)a_1 n_1} \sum_{n_2=0}^{N_2} e^{-i\left(k_{vy} - k_{cy}\right)a_2 n_2} \sum_{n_3=0}^{N_3} e^{-i\left(k_{vz} - k_{cz}\right)a_3 n_3},$$

where I assumed that vectors a_1, a_2, and a_3 point along X-, Y-, and Z-coordinate axes correspondingly. Each of these sums is a geometric progression with the first term equal to unity and coefficient $q_i = e^{-i\Delta k_i a_i}$, $(\Delta k_i = k_{v_i} - k_{c_i})$, with i taking values x, y, and z. A well-known formula for the sum of the geometric progression yields

$$\sum_{n_i=0}^{N_i} e^{-i\Delta k_i a_i n_i} = \frac{1 - e^{-i\Delta k_i a_i N_i}}{1 - e^{-i\Delta k_i a_i}} = \frac{e^{-i\Delta k_i a_i N_i/2}}{e^{-i\Delta k_i a_i/2}}\frac{\sin\frac{\Delta k_i a_i N_i}{2}}{\sin\frac{\Delta k_i a_i}{2}},$$

and after combining all these sums, the matrix element becomes

$$d\left(k_v, k_c\right) = \frac{e}{L^3}\tilde{d}\left(k_v, k_c\right)\prod_{i=1}^{3}\frac{e^{-i\Delta k_i a_i N_i/2}}{e^{-i\Delta k_i a_i/2}}\frac{\sin\frac{\Delta k_i a_i N_i}{2}}{\sin\frac{\Delta k_i a_i}{2}}. \tag{15.106}$$

For the transition probabilities and rates, I need $\left|d\left(k_v, k_c\right)\cdot w\right|^2$ (in case you forgot w is the polarization vector of the electromagnetic wave):

$$\left|d\left(k_v, k_c\right)\cdot w\right|^2 = \frac{e^2}{L^6}\left|\tilde{d}\left(k_v, k_c\right)\cdot w\right|^2\prod_{i=1}^{3}\frac{\sin^2\frac{\Delta k_i a_i N_i}{2}}{\sin^2\frac{\Delta k_i a_i}{2}}.$$

Function

$$F_N\left(\Delta k_i a_i\right) = \frac{\sin^2 \frac{\Delta k_i a_i N_i}{2}}{\sin^2 \frac{\Delta k_i a_i}{2}}$$

appears quite often in physics, for instance, in the theory of a diffraction lattice. For large values of N_i, it is characterized by a maximum at $\Delta k_i = 0$ where the function takes the value of N_i^2. The larger the parameter N_i, the narrower the peak becomes and the more of the area bounded by the function lies in the small vicinity of $\Delta k_i = 0$. For small Δk_i, I can replace the sin function in the denominator with its arguments reducing $F_N\left(\Delta k_i a_i\right)$ to your old acquaintance, the *sinc* function that you first encountered during the derivation of Fermi's golden rule in Sect. 15.2.1:

$$F_N\left(\Delta k_i a_i\right) = \frac{\sin^2 \frac{\Delta k_i a_i N_i}{2}}{\left(\frac{\Delta k_i a_i}{2}\right)^2} = N_i^2 sinc^2\left(\frac{\Delta k_i a_i N_i}{2}\right).$$

Using Eq. 15.26 with x identified with $\Delta k_i a_i$ and a with N_i, I can write

$$\lim_{N_i \to \infty} F_N\left(\Delta k_i a_i\right) = 2\pi N_i \delta\left(\Delta k_i a_i\right) = \frac{2\pi N_i}{a_i}\delta\left(\Delta k_i\right), \qquad (15.107)$$

where at the last step I used a known scaling property of the delta-functions: $\delta(ax) = \delta(x)/a$ (I apologize for having too many as with different meanings in the last three lines and hope you won't let yourself get confused by them). Taking into account Eq. 15.107, and identifying $\delta\left(\Delta k_x\right)\delta\left(\Delta k_y\right)\delta\left(\Delta k_z\right)$ with $\delta\left(\mathbf{k}_v - \mathbf{k}_c\right)$ (recall Eq. 2.39), I find for $\left|\mathbf{d}\left(\mathbf{k}_v, \mathbf{k}_c\right)\cdot\mathbf{w}\right|^2$:

$$\left|\mathbf{d}\left(\mathbf{k}_v, \mathbf{k}_c\right)\cdot\mathbf{w}\right|^2 = \frac{(2\pi)^3 N_c}{\upsilon L^6}e^2\left|\tilde{\mathbf{d}}\left(\mathbf{k}_v, \mathbf{k}_c\right)\cdot\mathbf{w}\right|^2\delta\left(\mathbf{k}_v - \mathbf{k}_c\right). \qquad (15.108)$$

Take a breath and pat yourself on the shoulder—you wandered into a rugged terrain and came back mentally in one piece. From this point on, it will be all down the slope. First, use Eq. 15.103 to find the perturbation matrix element $\left|\upsilon\left(\mathbf{k}_v, \mathbf{k}_c\right)\right|^2$, substitute it into Eq. 15.101 for the net transition rate, and get rid of the integration over one of the wave numbers using the delta-function:

$$R_{net} = \frac{4\pi}{\hbar}\left[\frac{L^3}{(2\pi)^3}\right]^2\frac{(2\pi)^3 N_c}{\upsilon L^6}\frac{\mathcal{E}_0^2}{4}e^2\times$$

$$\int\left|\tilde{\mathbf{d}}\left(\mathbf{k}_v, \mathbf{k}_c\right)\cdot\mathbf{w}\right|^2\left(f_v\left(\mathbf{k}\right) - f_c\left(\mathbf{k}\right)\right)\delta\left[\hbar\Omega - E_c\left(\mathbf{k}\right) + E_v\left(\mathbf{k}\right)\right]d^3k.$$

Now you can go back to Eqs. 15.94 and 15.95 and write R_{net} as

$$R_{net} = \frac{N_c e^2 \mathcal{E}_0^2}{8\pi^3 \hbar \upsilon} \int \left| \tilde{d}(k,k) \cdot w \right|^2 (f_v(k) - f_c(k)) \times$$

$$\delta \left[\hbar \Omega - E_g - \frac{\hbar^2 k^2}{2m_c} - \frac{\hbar^2 k^2}{2m_v} \right] d^3 k =$$

$$\frac{N_c e^2 \mathcal{E}_0^2}{8\pi^3 \hbar \upsilon} \int \left| \tilde{d}(k,k) \cdot w \right|^2 (f_v(k) - f_c(k)) \times$$

$$\delta \left[\hbar \Omega - E_g - \frac{\hbar^2 k^2}{2\mu} \right] d^3 k, \tag{15.109}$$

where I introduced the reduced effective mass μ, $1/\mu = 1/m_c + 1/m_v$, and remembered that the distribution functions depend only on the magnitude of the wave number k by the way of the corresponding energies E_v and E_c. The last comment is important as it allows me to carry out the integration over directions of k in Eq. 15.109 using the spherical coordinate representation for k. Neglecting a weak k—dependence of the matrix element $\tilde{d}(k,k)$—and replacing it with the constant vector $\tilde{d}$, I can evaluate Eq. 15.109 to be

$$R_{net} = \frac{N_c e^2 \mathcal{E}_0^2}{2\pi^2 \hbar \upsilon} \left| \tilde{d} \cdot w \right|^2 \int (f_v(k) - f_c(k)) \delta \left[\hbar \Omega - E_g - \frac{\hbar^2 k^2}{2\mu} \right] k^2 dk,$$

where I took into account that the integral over angular variables yields 4π. To evaluate the remaining integral over k, I make a substitution of variable $\epsilon = \hbar^2 k^2 / 2\mu$ that yields

$$R_{net} = \frac{N_c e^2 \mathcal{E}_0^2}{2\pi^2 \hbar \upsilon} \left| \tilde{d} \cdot w \right|^2 \frac{1}{2} \left(\frac{2\mu}{\hbar^2} \right)^{3/2} \int (f_v(k) - f_c(k)) \delta \left[\hbar \Omega - E_g - \epsilon \right] \sqrt{\epsilon} d\epsilon =$$

$$\frac{N_c e^2 \mathcal{E}_0^2 \mu^{3/2}}{\sqrt{2} \pi^2 \hbar^4 \upsilon} \left| \tilde{d} \cdot w \right|^2 \left[f_v(\hbar \Omega - E_g) - f_c(\hbar \Omega - E_g) \right] \sqrt{\hbar \Omega - E_g} \, \theta(\hbar \Omega - E_g),$$

where the step function $\theta(x)$ indicates impossibility to fulfill the condition $\hbar \Omega - E_g = \epsilon > 0$ unless $\hbar \Omega > E_g$. This condition reflects a physically quite obvious circumstance: it is impossible to cause a transition from the valence to the conductance band if the energy of a respective photon is less than the energy required to bridge the band gap. Light with frequencies not satisfying this condition does not interact with electrons of the semiconductor merely passing through. Therefore, semiconductors remain almost transparent to frequencies smaller than $E_g/\hbar$. (Almost because there are always other mechanisms for light absorption not accounted for by this model.)

The transition rate in this expression is proportional to the number of elementary cells in the structure, i.e., to its volume. A more physically relevant quantity would be a number of transitions per unit time per unit volume, which is

$$\frac{R_{net}}{L^3} = \frac{e^2 \mathcal{E}_0^2 \mu^{3/2}}{\sqrt{2}\pi^2\hbar^4 \upsilon^2} \left|\tilde{d} \cdot w\right|^2 \times$$

$$\left[f_v\left(\hbar\Omega - E_g\right) - f_c\left(\hbar\Omega - E_g\right)\right]\sqrt{\hbar\Omega - E_g}\,\theta\left(\hbar\Omega - E_g\right), \qquad (15.110)$$

where I replaced $N_c/L^3 = 1/\upsilon$. This result now can be used to find an expression for the absorption coefficient in the semiconductor, but I leave it for you as an exercise. Final remark: if you are dealing with an unpolarized light, the averaging over polarization might be needed, which as we know will result in the replacement $\left|\tilde{d} \cdot w\right|^2 \rightarrow \left|\tilde{d}\right|^2/3$.

15.3.4 Selection Rules: Dipole Matrix Elements Made Easier

Dipole matrix elements $d_{m_0,m} = \langle\alpha_{m_0}|\,er\,|\alpha_m\rangle$ play the key role in determining probabilities of various optical transitions, and their computation constitutes the largest chunk of what researchers studying emission and absorption spectra of various quantum systems do. Quite often knowing the symmetry of the unperturbed Hamiltonian, you can determine if a matrix element in question vanishes or has a non-zero value without any actual calculations. Obviously, this knowledge saves lots of time and effort, and this is why physicists pay a great deal of attention to so-called selection rules. These are the rules which allow you to determine the matrix element of which transitions have a finite (non-zero) value and for which it vanishes. In the former case, the transitions are called dipole-allowed, and in the latter, they are called dipole-forbidden. Usually, if you go beyond the dipole approximation, the new perturbation matrix element might have a non-zero albeit a rather small value, making the probability of such transitions small but still different from zero.

Derivation of the general selection rules is quite involved and requires knowledge of mathematics (theory of groups) which is not expected from the readers of this book. However, selection rules for dipole transitions between states represented by hydrogen-like wave functions can be derived using more elementary methods. In this section I'll show how this can be done, and in the end, I will fulfill my promise and finish the computation started in Eqs. 15.72 and 15.73.

The selection rules for $d_{m_0,m}$ are determined by the symmetry of this expression with respect to inversion and rotations. As I have already explained at least twice in other parts of the book, in order to utilize the symmetry, you have to consider how constituent parts of the matrix element—the state vectors and the respective operator (in this case $\hat{r}$)—change when a symmetry transformation is performed.

If these changes keep the resulting matrix element unchanged, then there is nothing you can say about it—it might or might not be equal to zero, but you will have to actually compute the integrals to find out its actual value. If, however, the changes of the vectors and the operator are such that the resulting matrix element, which must remain invariant, appears to change, you would encounter a paradox, which can only be resolved if the matrix element in question is zero.

The first of the selection rules can be derived by considering the effect of inversion—the symmetry operation described by parity operator $\hat{\Pi}$. I have already analyzed the behavior of matrix elements with respect to inversion in Sect. 7.1.2 and the beginning of Chap. 10, so here I will only remind you of the results. Since the position operator $\hat{r}$ (and the momentum operator $\hat{p}$) are odd operators (they change sign under the action of $\hat{\Pi}$; see Eqs. 5.25 and 5.26), the corresponding matrix element will vanish unless it is calculated between states with different parity. In application to the hydrogen-like atoms, this selection rule means that the dipole-allowed transitions are only possible between states with odd and even values of the orbital quantum numbers l. This follows from the properties of the spherical harmonics, which are even for even l and odd when l is odd; see Sect. 5.1.4.

The rest of the selection rules are determined by the rotational symmetry of the system, but a direct application of the transformation properties of the eigenvectors with respect to rotation is again outside of the mathematics background expected from the readers. Therefore, I will address this problem relying directly on the properties of the spherical harmonics and operators of the angular momentum. So, to make sure that we all are on the same page, let me reiterate: I am interested in the properties of the matrix elements of the form

$$\langle n_0, l_0, m_0 | \, \boldsymbol{r} \, | n, l, m \rangle \,,$$

where n, l, m are standard notations for principal, orbital, and magnetic quantum numbers correspondingly (obviously, the absence of the electric charge in this expression is inconsequential). I am looking for relationships between orbital and magnetic quantum numbers of the initial and final states, which would ensure that the matrix element does not vanish because of the symmetry requirements. I hope you will not get confused by my use of m to represent two different things: sometimes it is a generic index enumerating various states of the system, and sometimes m is a magnetic quantum number defining an eigenvalue of the operator $\hat{L}_z$. I hope that you will be able to differentiate between the two usages of this symbol from the context in which it is used.

The required relations between magnetic quantum numbers m_0 and m are easiest to establish. Indeed, these numbers appear in the matrix elements only in the form $\exp\left[i\left(m - m_0\right)\varphi\right]$, which is a part of the spherical harmonics $Y_{l,m}\left(\theta, \varphi\right)$, multiplied by x-, y-, or z-component of the position vector $\boldsymbol{r}$. Therefore I can write for the Cartesian components of the dipole matrix element:

$$d_x \propto \int\limits_0^{2\pi} \cos\varphi\, e^{i(m-m_0)\varphi}\, d\varphi \propto \delta_{m-m_0+1} + \delta_{m-m_0-1},$$

$$d_y \propto \int\limits_0^{2\pi} \sin\varphi\, e^{i(m-m_0)\varphi}\, d\varphi \propto \delta_{m-m_0+1} - \delta_{m-m_0-1},$$

$$d_z \propto \int\limits_0^{2\pi} e^{i(m-m_0)\varphi}\, d\varphi \propto \delta_{m-m_0},$$

where I omitted the parts of the integrals that do not contain the azimuthal angle φ (those bothersome factors with radial functions and associated Legendre functions and even $\cos\theta$ or $\sin\theta$ parts of the position vector components) and took into account that an integral of $\exp(im\varphi)$ over the interval $[0, 2\pi]$ with integer m is zero unless $m = 0$. These results tell me that the selection rules for different components of the dipole matrix element are different: for x- and y-components, the matrix elements are not zero only if the magnetic numbers of the initial and final states differ by one, $|m - m_0| = 1$, while for the z-component, it is not zero only if $m = m_0$. This difference manifests itself experimentally by the different reactions of the atom to light of different polarizations and propagation directions.

At this point a thoughtful reader should say: "Wait a second, professor! If the system is completely isotropic as a hydrogen atom is, then it should make no difference how we choose *all* axes of the coordinate system, including its Z-axis. In this sense, a light of any linear polarization can be made polarized along any axis—we could choose a Z-axis to go along the propagation direction of the wave, and X-axis (or Y-axis) along the electric field of the wave. This choice would make the wave induce transitions between states with magnetic numbers different by one. If, however, you, on a whim, decide to direct a Z-axis along the electric field, then the induced transitions would be between the states with the same magnetic number. And how does all this make any sense? The nature wouldn't care about the choices of the coordinate system we make, would it?" This, of course, would be a very fair question, but I do have a very good answer to it. First, you need to remember that all states with the same magnetic number have the same energy, they are degenerate. Therefore, the transitions with or without change in m would occur at the same frequency, and you cannot distinguish between them in the absorption or emission spectra. Second, the definition of m and of the state with given m is directly attached to the direction of the polar axis. Changing the direction of the latter, obviously, would not change the actual quantum state (the nature indeed cares very little about our choices of the coordinate systems), but it will change your description of it: the same state, which in one system is an eigenvector of the operator $\hat{L}_z$ characterized by a certain value of m, will now be described as a superposition of eigenvectors of the operator $\hat{L}_{\bar{z}}$, defined with respect to a new polar axis, with different magnetic numbers.

In order to really observe the difference between transitions with or without change of magnetic number, you would need to spoil the symmetry enough to lift the degeneracy of the states, but not too much so as not to destroy the dependence of the wave functions upon φ. This can be achieved, for instance, by placing an atom in a uniform magnetic field. The resulting Hamiltonian as you know still commutes with L_z, if the Z-axis is chosen along the magnetic field, so the eigenvectors of the latter remain eigenvectors of the Hamiltonian and retain their $\exp(im\varphi)$ behavior. The m-selection rules, which were based completely on this dependence on φ, remain valid even in this case, but the direction of the Z-axis is now not arbitrary but set by the magnetic field. By directing the incident light along the field (its polarization will obviously be in X- or Y-directions), you can observe its absorption due to $m \to m\pm1$ transitions. The transition between the states with the same magnetic numbers can be observed using light polarized along the magnetic field.

The m-selection rules can also be derived in a more elegant way without reliance on a direct computation of integrals in the position representation. It is easy to see that operators $\hat{L}_z$ and $\hat{z}$ commute ($\hat{L}_z$ contains only operators $\hat{x}, \hat{y}$ and correspondingly $\hat{p}_x$ and $\hat{p}_y$, all of which commute with $\hat{z}$). So take the identity $\hat{L}_z\hat{z} = \hat{z}\hat{L}_z$ and multiply from the left by the bra version of the eigenvector of $\hat{L}_z$ $\langle l_1, m_1|$ and from the right by its ket version $|l_2, m_2\rangle$:

$$\langle l_1, m_1| \hat{L}_z\hat{z} |l_2, m_2\rangle = \langle l_1, m_1| \hat{z}\hat{L}_z |l_2, m_2\rangle .$$

Using hermiticity of $\hat{L}_z$, you can apply it to the bra vector on the left-hand side of this expression and to the ket vector on its right-hand side:

$$\hbar m_1 \langle l_1, m_1| \hat{z} |l_2, m_2\rangle = \hbar m_2 \langle l_1, m_1| \hat{z} |l_2, m_2\rangle .$$

This expression can only be true if either $m_1 = m_2$ or $\langle l_1, m_1| \hat{z} |l_2, m_2\rangle = 0$, which is exactly our selection rule for the z-component of the dipole matrix element. This analysis shows a way to derive the selection rule for two other components of the dipole matrix elements, but I will leave it for you as an exercise.

I will complete this section by deriving the last set of the selection rules establishing limitations on the orbital quantum numbers l of the initial and final states of the dipole-allowed transitions. This would require analyzing integrals with respect to the polar angle θ, which can be written in the following form:

$$d_{x,y} \propto \int_0^\pi P_{l_0}^{m_0}(\cos\theta) \sin\theta P_l^{m_0\pm1}(\cos\theta) \sin\theta d\theta \qquad (15.111)$$

$$d_z \propto \int_0^\pi P_{l_0}^{m_0}(\cos\theta) \cos\theta P_l^{m_0}(\cos\theta) \sin\theta d\theta. \qquad (15.112)$$

The structure of these expressions is as follows: the associated Legendre functions are, of course, the θ-dependent part of the spherical harmonics for the initial and final states with m-selection rules incorporated, the $\sin\theta$ or $\cos\theta$ factor between them represent the θ-dependent part of the components of the position vector expressed in the spherical coordinates, and, finally, the last $\sin\theta$ factor is a part of the spherical differential volume element dV. A convenient way to compute these integrals is to note that trigonometrical functions can be expressed in terms of the associated Legendre polynomials as $\cos\theta = P_1^0(\cos\theta)$, $\sin\theta = -P_1^1(\cos\theta)$, turning integrals in Eqs. 15.111 and 15.112 into integrals of the product of three associated Legendre polynomials. These integrals can be evaluated using a widely known (among a narrow circle of people) so-called Gaunt's formula. I am not going to reproduce it here because of its rather cumbersome form and easy online availability (if you are interested, you can find it on a Wikipedia page for associated Legendre polynomials). Moreover, for the examples you will actually be dealing with in this book, it is easier to compute these integrals from scratch. However, I will use the conditions on numbers l and m, specified by this formula, which ensure that the respective integral does not vanish. First, it is required that the largest of m be equal to the sum of two other magnetic numbers. In our case this condition is consistent with m-selection rules, which is easily verifiable by looking at the corresponding equations. In Eq. 15.111 the $\sin\theta$ corresponds to $m = 1$, which is a perfect match for the m-values of two other Legendre polynomials, while in Eq. 15.112 the $\cos\theta$ corresponds to $m = 0$, again in agreement with the already established selection rules. Second, it is required that the l numbers of the Legendre functions obey the "triangle rule," in which one of the numbers is less than the sum of two others and larger than their difference (like for the sides of a triangle). In our case it means

$$l + 1 \geq l_0 \geq l - 1,$$

leaving me with only three possibilities $l_0 = l$ or $l_0 = l \pm 1$. The first of them is excluded by the parity argument, so that you are left with the last of the selection rules: dipole transitions are only allowed between the states with azimuthal quantum numbers differing by unity, $|l - l_0| = 1$, which is, of course, consistent with the inversion symmetry-based requirement that initial and final states have different parities. If you are wondering if this result can be obtained in a different way without a referral to an obscure formula too big even to be displayed here, the answer is yes. One can derive it using arguments similar to the one I showed for m-selection rules, but it involves computing commutator $\left[\hat{L}^2, \left[\hat{L}^2, r\right]\right]$ of a rather obscure and non-intuitive origin, so the whole approach does not look that elegant anymore. You can find it in a popular quantum mechanics textbook by D.J. Griffiths.[10] Another way to derive this result is based on the Wigner–Eckart theorem, which deals with

---

[10]D.J. Griffiths, *Introduction to Quantum Mechanics*, 2nd edn. (Cambridge University Press, Cambridge, 2016).

the generic properties of matrix elements involving eigenvectors of the angular momentum, but exposing you to it could be construed as a cruel and unusual punishment.

Now I can finish computing the dipole matrix elements appearing in Eqs. 15.72 and 15.73, which are

$$\varsigma_1 = \langle 1, 0, 0 | e\hat{r} | 2, 0, 0 \rangle ,$$

$$\varsigma_2 = \langle 1, 0, 0 | e\hat{r} | 2, 1, -1 \rangle , \ \varsigma_3 = \langle 1, 0, 0 | e\hat{r} | 2, 1, 0 \rangle , \ \varsigma_4 = \langle 1, 0, 0 | e\hat{r} | 2, 1, 1 \rangle .$$

By the parity argument (or l-selection rule), you can immediately conclude that $\varsigma_1 = 0$, while all other matrix elements might have nonvanishing values. For ς_2 and ς_4, the m-selection rule yields the non-zero values for their x- and y-components, while for ς_3 only the z-component does not vanish. Now, let's go ahead and compute them using the known expressions for the hydrogen wave functions from Chap. 8. I will need the following (assuming $Z = 1$ and vacuum):

$$|1, 0, 0\rangle = \frac{1}{\sqrt{\pi}} \left(\frac{1}{a_B} \right)^{3/2} e^{-r/a_B} ,$$

$$|2, 1, \pm 1\rangle = \frac{1}{8\sqrt{\pi}} \left(\frac{1}{a_B} \right)^{3/2} \frac{r}{a_B} e^{-r/2a_B} \sin\theta e^{\pm i\varphi} ,$$

$$|2, 1, 0\rangle = \frac{1}{4\sqrt{2\pi}} \left(\frac{1}{a_B} \right)^{3/2} \frac{r}{a_B} e^{-r/2a_B} \cos\theta ,$$

which gives me

$$\varsigma_{2x} = \frac{e}{8\pi a_B^4} \int_0^\infty dr r^4 e^{-3r/2a_B} \int_0^\pi d\theta \sin^3\theta \int_0^{2\pi} d\varphi \cos\varphi e^{-i\varphi} =$$

$$\frac{1}{8\pi a_B^4} \frac{256 a_B^5}{81} \frac{4}{3}\pi = \frac{128}{243} e a_B ,$$

$$\varsigma_{2y} = \frac{e}{8\pi a_B^4} \int_0^\infty dr r^4 e^{-3r/2a_B} \int_0^\pi d\theta \sin^3\theta \int_0^{2\pi} d\varphi \sin\varphi e^{-i\varphi} =$$

$$-i\frac{1}{8\pi a_B^4} \frac{256 a_B^5}{81} \frac{4}{3}\pi = -i\frac{128}{243} e a_B ,$$

and

$$|\varsigma_2|^2 = |\varsigma_{2x}|^2 + |\varsigma_{2y}|^2 \approx 0.55 e^2 a_B^2 .$$

Now, you can easily notice (using the expressions for the corresponding wave functions) that $\varsigma_4 = \varsigma_2^*$, which means that x-components for both matrix elements are identical, while their y-components are complex conjugates of each other and $|\varsigma_4|^2 = |\varsigma_2|^2$. And, finally, the last of the non-zero elements

$$\varsigma_{3_z} = \frac{e}{4\sqrt{2}\pi a_B^4} \int_0^\infty dr r^4 e^{-3r/2a_B} \int_0^\pi d\theta \sin\theta \cos^2\theta =$$

$$\frac{256 a_B^5}{81} \frac{2}{3}\pi = \frac{64}{243} e a_B$$

$$|\varsigma_3|^2 \approx 0.14 e^2 a_B^2,$$

and I can finish the computation of the transition rates in this example. For upward (absorbing) transitions

$$R_{E_1,E_2} = 2\frac{\pi u\left(\omega_{E_1,E_2}\right) e^2 a_B^2}{3\varepsilon_0\hbar^2}(0.55 \times 2 + 0.14) = 2.48\frac{\pi u\left(\omega_{E_1,E_2}\right) e^2 a_B^2}{3\varepsilon_0\hbar^2},$$

while the rate for downward transitions due to stimulated emission is eight times smaller due to the difference between degrees of degeneracy of the initial and final states for each direction of the transitions.

15.4 Problems

For Sect. 15.1

Problem 175 Consider a one-dimensional harmonic oscillator (mass m_e, frequency ω_0) acted upon by a uniform force with time dependence of the form

$$F(t) = F_0 \tau \delta\left(t - t_0\right).$$

1. Assuming that the oscillator is initially in a ground state, find the probability that it will be found in an arbitrary state $|n\rangle$ at time $t > t_0$ using first and second orders of the time-dependent perturbation.
2. Solve Eq. 15.4 exactly (not using the perturbation theory) with the same initial condition.

Problem 176 Consider an electron in a one-dimensional infinite potential well $V(z)$ subjected to additional potential of the form

$$V(x,t) = \begin{cases} Fz\frac{t}{\tau} & 0 < t < \tau \\ Fz & t > \tau \end{cases}.$$

1. Find a probability of transition between a state belonging to the second energy level and states belonging to the first and the third energy levels.
2. Write down an expression for the time-dependent state of the electron in the first order of the perturbation theory assuming that initially it is in the ground state, and use it to compute expectation values of coordinate and momentum of the particle (keep only linear in the perturbation terms).

Problem 177 Derive an expression for $c_m^{(r)}$—an arbitrary-order approximation for the time-dependent expansion coefficients $c_n(t)$.

Problem 178 Consider an electron in the ground state of a three-dimensional symmetrical infinite potential well of width a in all dimensions subjected to a perturbation:

$$V(t) = F_0 (x + y + z) \, \theta(t) \, \theta(\tau - t).$$

Find a probability that after the perturbation is over, you will observe an emission from this electron at a frequency $\omega_{21} = (E_2 - E_1)/\hbar$, where E_2 and E_1 are the second and the first lowest energy eigenvalues of the electron. Do not forget that the eigenvalue E_2 is degenerate. For which τ will this probability be largest?

Problem 179 Repeat Problem 178, assuming that the electron is in an infinite spherical potential well, while the perturbation has the same form.

Problem 180 Vectors $|\alpha_m\rangle$ in the derivation of the transition probabilities might represent states of many-particle systems as well. Thus, consider a system of two non-interacting electrons allowed to move only in z-direction in a one-dimensional potential well of width a. They are subjected to a pulse of the uniform electric field:

$$\mathcal{E} = \mathcal{E}_0 \cos \Omega t \theta(t) \, \theta(\tau - t).$$

1. Assuming that the electrons are initially in the ground state, find a probability of transition to the lowest in energy excited state. Do not forget that electrons are fermions and that they have spin.
2. How will the answer change if somehow spins of both electrons are always kept in the spin-up state?

Problem 181 Consider a pure spin $1/2$ (no orbital degrees of freedom to worry about) placed in a uniform constant magnetic field B in the direction of the Z-axis. The spin is also subjected to a weak "rotating" magnetic field $B_\perp(t) = b_0 \left(e_x \cos \Omega t + e_y \sin \Omega t \right)$, where $e_{x,y}$ are unit vectors in the directions of X- and Y-axes. Considering the "rotating" field as a perturbation,

1. Find the time-dependent spinor describing the quantum state of this spin in the first order of the time-dependent perturbation theory, and compute the expectation values of all three spin components.

2. Assuming that the perturbing field was turned off at time $t = T$ and z-component of the field is measured, find the probabilities of various outcomes. Repeat the same if the x-component is measured.

Analyze the obtained results as a function of parameter ΩT. In all cases assume that initially the spin was in its ground state.

Problem 182 Consider a hydrogen atom initially in the ground state subjected to a perturbation:

$$V = \mathcal{E}_0 z e^{-|t|/\tau}.$$

Assuming that "initially" corresponds to $t_0 = -\infty$, modify Eq. 15.12 to account for this new value of t_0 and compute probabilities to find the atom in the ground state at $t \to \infty$ in the first order of the perturbation theory.

For Sect. 15.2.1

Problem 183 Compare the result of the first-order perturbation theory for the transition probability in the case of monochromatic perturbation (Eq. 15.23) with the exact solution for the two-level system from Chap. 10. Derive an approximation for Eq. 10.32 valid for short time intervals, and compare it with the results of the perturbation theory. State the condition to which the duration of the perturbation must obey for the probability expression to yield a reliable result.

Problem 184 Assuming monochromatic perturbation, derive an expression for the transition probability in the second order of the perturbation theory. Analyze its behavior at $t \to \infty$.

Problem 185 Re-derive Eq. 15.23 assuming that the perturbation operator has a time dependence in the form of the sine instead of cosine function.

Problem 186 Consider a spin $3/2$ particle in a uniform magnetic field in z-direction: $\boldsymbol{B} = B_0 \boldsymbol{e}_z$. At time $t = 0$ the particle is subjected to a time-dependent magnetic field in the y-direction: $\boldsymbol{B}_1 = b_0 \sin \Omega t$. Assuming that the particle is initially in the ground state, find the transition probabilities for all three excited states expressed in the delta-functional form of Fermi's golden rule.

Problem 187 Consider a one-dimensional harmonic oscillator with time-dependent frequency:

$$\omega = \omega_0 \left(1 + \Delta \cos \Omega t\right).$$

1. Present the Hamiltonian of the system in the form $\hat{H}_0 + \hat{V}(t)$, and identify the form of the time-dependent perturbation Hamiltonian. Rewrite it using lowering and raising operators.
2. Assuming that the oscillator is initially (at $t = 0$) in the nth state, determine transitions into which states are possible and find the corresponding probabilities in the limit of small t and $t \to \infty$.
3. Write down the time-dependent state of the oscillator, and compute the expectation value of the Hamiltonian in it.

Problem 188 Consider a system with a quasi-monochromatic perturbation operator:

$$\hat{V}(t) = \hat{v}e^{-\epsilon t} \cos \Omega t \theta(t).$$

1. For the perturbation of this form, derive the expressions for the transition probabilities in the first order of the perturbation theory for an arbitrary time t.
2. Find the limiting value of these probabilities in the limit $t \to \infty$.
3. Consider the behavior of these limiting values when $\epsilon \to 0$. Compare the results with the delta-functional form of Fermi's golden rule.

For Sect. 15.2.3

Problem 189 Consider a particle in a one-dimensional δ-functional potential well:

$$V = -\alpha_0 \left[1 + \Delta \cos\left(\Omega t\right) \theta(t)\right] \delta(x).$$

Find the probability that the perturbation will kick out the particle from its ground state. Do not assume that the state in the continuum are just plane waves—find the true scattering states in the delta-functional potential.

Problem 190 Consider a gas consisting of non-interacting potassium atoms with concentration $n_{Na} = 10^{12} \, \mathrm{m}^{-3}$. First, find the electron configuration of a potassium atom (distribute all electrons among available orbitals) and, assuming that a lone s-electron in the last shell behaves as an electron in a quasi-hydrogen atom, determine its energy. The gas is being exposed to a uniform time-periodic electric field of frequency Ω which ionizes the atoms by kicking out this electron to the continuous segment of the spectrum. Find for which frequency Ω the number of free electrons generated per second per unit volume is largest, and find this number.

For Sect. 15.3.1

Problem 191 Sometimes the dipole matrix elements in the expression for the transition rate vanish, in which case other weaker interactions between light and atom start playing a role. One of them is the magnetic dipole interaction described by a perturbation term:

$$\hat{V} = \frac{e}{2m_e} B(0, t) n_B \cdot \left(\hat{L} + 2\hat{S} \right),$$

where n_B and $B(0, t)$ are the direction and magnitude of the magnetic field component of the electromagnetic wave, which I described in Eq. 15.56 by its vector potential. The magnetic dipole approximation consists in taking the value of the field at $r = 0$. It can be shown that the part of this operator containing the orbital angular momentum $\hat{L}$ originates from the first correction to the dipole approximation (linear in the kr term in the expansion of the exponent $\exp(ikr)$), but I will spare you the derivation. This expression is similar to the Zeeman term in atomic Hamiltonian, Eq. 14.34, but, unlike the latter, is time-dependent.

1. Using Eq. 15.56 find the expression for the magnetic field in this wave and relate vector n_B to the polarization vector w.
2. Derive a general expression for the perturbation matrix element $v_{m_0,m}$ for the magnetic dipole perturbation term (do not specify the nature of the initial and final states). Index m here (and in the next problem) is just a generic index enumerating states, and not a magnetic quantum number.

Problem 192 Derive a dipole transition matrix element $v_{m_0,m}$ if the incident electromagnetic radiation is described by the vector potential of the form

$$A(r, t) = -\frac{\mathcal{E}_0}{\Omega} \left[A_x e_x \cos(kz - \Omega t) + A_y e_y \cos(kz - \Omega t + \varphi) \right],$$

where $e_{x,y}$ are unit vectors and $\sqrt{A_x^2 + A_y^2} = 1$. Find a dependence of the matrix elements of the parameters φ and $A_{x,y}$. (Hint: Present the trigonometric functions in the complex exponential form.)

For Sect. 15.3.2.1

Problem 193 Using Eqs. 15.70 and 15.71, demonstrate that Einstein coefficients B_{ab} and B_{ba} are related to each other via $g_a B_{ab} = g_b B_{ba}$, where $g_{a,b}$ are degrees of degeneracy of the initial and final states $|a\rangle$ and $|b\rangle$.

Problem 194 Find Einstein's B_{ab} and B_{ba} coefficients for transitions between the second and third excited levels of a particle in a symmetric three-dimensional infinite cubic potential well. Find the transition rates, taking into account the degeneracy of both the final and the initial states.

Problem 195 Consider an electromagnetic wave propagating through a gas of "pure" spins $1/2$ placed in the constant uniform magnetic field $\boldsymbol{B}$ with concentration N spins per unit volume. This situation can be realized, for instance, in the case of an electron in an atom in a ground state if the frequency of the wave is too low to cause any transitions except of those between the spin states. If this is the case, you can ignore all quantum numbers of the electron and treat it as a "pure" spin. So the magnetic field of the electromagnetic wave induces transitions between two energy levels corresponding to eigenvectors of the component of the spin operator in the direction of the permanent magnetic field $\boldsymbol{B}$. Assume that the spins are in the thermal equilibrium, meaning that the ratio of the number of spins in the higher energy state N_b to those in the lower energy state is

$$
\frac{N_b}{N_a} = \frac{\exp\left(-\dfrac{E_b}{k_B T}\right)}{\exp\left(-\dfrac{E_a}{k_B T}\right)}.
$$

Determine the absorption coefficient in this system. Assuming that the magnitude of the permanent magnetic field is $1\,T$, and the concentration of spins $N = 10^{15}$ per unit volume, find the numerical value of the absorption coefficient as a function of temperature. How does this absorption behave as the temperature goes up?

Problem 196 Find the stimulated emission and absorption transition rates between the second and third excited states of a three-dimensional isotropic harmonic oscillator (mass μ, frequency ω). Do not forget to take into account the degeneracy of both states. For numerical estimates, take μ to be equal to the reduced mass of a molecule of CO_2, and for frequency $\omega = 7.7 \times 10^7$ rad/s. Assume that the molecule is exposed to the black-body radiation at temperature $T = 300$ K.

Problem 197 Find Einstein B coefficients for a magnetic dipole transition (see Problem 191) between states of hydrogen atom $\left|2, \frac{3}{2}, 1\right\rangle$ and $\left|2, \frac{1}{2}, 1\right\rangle$ split by the spin–orbit interaction (see Sect. 14.1), where the nomenclature for the states is in $|n, j, l\rangle$, where n is the principal quantum number and j and l are the total and orbital angular momentum numbers.

For Sects. 15.3.2.2–15.3.4

Problem 198 An electron in an infinite one-dimensional potential well is excited to the third energy level and is left alone to return to the ground state by the way of spontaneous emission. This can be done via two different routes: $|3\rangle \rightarrow |1\rangle$ and

$|3\rangle \to |2\rangle \to |1\rangle$, where I used notation $|n\rangle$ to designate the state belonging to the corresponding energy level.

1. At which frequencies can the spontaneously emitted light be observed?
2. What is the spontaneous emission rate for each frequency?
3. Find the rate at which the number of atoms in the ground state increases.
   Hint: Remember that if a random event is a sequence of two independent events occurring consequently (one after the other), the total probability is the product of the probabilities of the individual events. If, however, an event can occur via several alternative paths, the total probability is the sum of probabilities of each alternative.

Problem 199   Consider a hydrogen atom excited to $|3, 2, -1\rangle$ and $|3, 1, -1\rangle$ states, where a standard notation for hydrogen stationary states is used (spin–orbit interaction is neglected).

1. Describe all possible channels by which the atom can decay to the ground state, and find probabilities for each channel.
2. Find the lifetime of these states. (Do not forget to consider all possible ways for these states to decay, but take into account that once the atom transitioned from its initial state, it is no longer in that state, so for the lifetime calculations, only the first steps in each chain are relevant.)

Problem 200   Using Eq. 15.110 and the definition of absorption coefficient, Eq. 15.81, derive an expression for the latter for a semiconductor. Assume that you are dealing with *GaAs* (band gap $E_g = 1.43\,\mathrm{eV}$, effective mass in conductance band $m_c = 0.067 m_e$, effective mass in the valence band $m_v = 0.45 m_e$, where m_e is a regular mass of a free electron) at temperature $T = 300\,\mathrm{K}$. Find the numerical value of the absorption coefficient for light at frequency $\hbar\Omega = 1.5 E_g$.

Problem 201   Derive the magnetic quantum number selection rules for light polarized in the x- and y-directions using commutation relations between operators $\hat{x}$, $\hat{y}$ and $\hat{L}_z$.

Problem 202   Determine the selection rules for magnetic dipole transitions between hydrogen-like states with its given azimuthal, magnetic, and spin quantum numbers. Show that the magnetic dipole transitions between the states with different principal numbers are forbidden. (Hint: For this you would have to consider the radial part of the matrix element integrals.)

Problem 203   When a dipole matrix element vanishes because of the dipole selection rules and the rate of the corresponding transition goes to zero, it does not mean that such transitions could never occur, but one has to go beyond the dipole approximation in order to find the mechanism responsible for such so-called *forbidden* transitions. By keeping the linear in **kr** term in the expansion of the vector potential in Eq. 15.56, one would find two new transition mechanisms: the magnetic dipole transition (Problems 191 and 202) and a so-called quadrupole transition whose contribution to the perturbation potential can be written down as

$$\hat{V} = \frac{e\mathcal{E}_0}{2m_e\Omega} \left(\sum_{i,j} Q_{ij}w_ik_j \right) \left[e^{i\Omega t} + e^{-i\Omega t} \right],$$

where

$$Q_{ij} = r_ir_j - \frac{1}{3}\delta_{ij}r^2$$

is a so-called quadrupole moment—a 3×3 matrix with zero trace: $\sum_i Q_{ii} = 0$. In the definition of the quadrupole moment, $r_{1,2,3}$ corresponds to the Cartesian components of the position vector x, y, z, while $r = \sqrt{x^2 + y^2 + z^2}$ is its magnitude. Correspondingly, the matrix element, which determines the fate of all transitions, becomes in this case

$$v_{m_0m} = \frac{e\mathcal{E}_0}{2m_e\Omega} \left(\sum_{i,j} \langle \alpha_{m_0}| Q_{ij} |\alpha_m\rangle w_ik_j \right).$$

Determine the selection rules for the quadrupole-allowed transitions assuming hydrogen-like initial and final states.

Problem 204 Consider a transition between the hydrogen states $|3, 2, 0\rangle$ and $|1, 0, 0\rangle$, which is dipole-forbidden, and find the quadrupole moment-mediated transition rate due to spontaneous emission for this transition. Compare it with a dipole-allowed transition between states $|3, 1, 0\rangle$ and $|1, 0, 0\rangle$.

Chapter 16
Free Electrons in Uniform Magnetic Field: Landau Levels and Quantum Hall Effect

The Zeeman effect, which we discussed in some detail in Sect. 14.2, originates from the interaction between the magnetic moment of an electron bound to an atom and a uniform magnetic field. Experimentally this effect is often observed in atomic gases but can also manifest itself with bound electrons in semiconductors and dielectrics. What is important is that the quantum states of the electrons in all these cases belong to discrete energy eigenvalues. In metals and in the conduction band of semiconductors, on the other hand, the energy levels of electrons belong to the continuum spectrum, and in some instances, electrons can even be treated as free particles. The interaction between such unbound, almost free electrons and the uniform magnetic field results in some fascinating effects which had played and are still playing an important role in physics.

Among older phenomena owing their existence to this interaction are the Hall effect (the emergence of an electric current in the direction perpendicular to the electric field), the de Haas–van Alphen effect (oscillations of magnetization of the metal with increasing magnetic field), and the Shubnikov–de Haas effect (oscillations of conductivity with the magnetic field). The Hall effect, which is a purely classical phenomenon, was discovered by American physicist Edwin Hall in 1879. It is interesting to note that Hall first worked as a high school principal before embarking on Ph.D. studies in physics, which he did at Johns Hopkins University. The discovery of the effect, which now bears his name, was part of his doctoral work. He became a professor of physics at Harvard University in 1895, but did not produce anything even remotely as significant as his student discovery. Both de Haas effects owe their existence to the quantum nature of electrons and were discovered in 1930 in the laboratory of Dutch physicist Wander Johannes de Haas. Pieter M. van Alphen was at that time de Haas' student, while Lev Shubnikov was a Soviet physicist working with de Haas as a visiting scholar. After his return to the Soviet Union, Shubnikov was falsely accused of espionage and in 1937 executed by the People's Commissariat for Internal Affairs (NKVD)—Soviet analog of Nazi Gestapo (Secret State Police).

© Springer International Publishing AG, part of Springer Nature 2018
L.I. Deych, *Advanced Undergraduate Quantum Mechanics*,
https://doi.org/10.1007/978-3-319-71550-6_16

One of the most fascinating and fairly recent magnetic field effects discovered in the system of unbound electrons is the quantum Hall effect and even more exotic fractional quantum Hall effects. The former was discovered by German physicist K. von Klitzing in 1980, who showed experimentally that the so-called Hall conductance (I will explain what it means later, have patience) in a two-dimensional electron gas is independent of the geometry and details of the structure and is an integer multiple of a universal parameter e^2/h (e is the fundamental charge, and h is the Planck's constant without the bar). Plotted as a function of the magnetic field, the Hall conductance (or its inverse, called Hall resistance) is seen as a series of plateaus separated by finite jumps with a magnitude of e^2/h (see Fig. 16.1, where the lower curve gives an example of the Shubnikov–de Haas oscillations in a two-dimensional system).

The significance of this discovery was almost immediately recognized by the physics community, and von Klitzing was awarded the 1985 Nobel Prize in Physics.

In 1982 Horst Störmer, Daniel Tsui, and Arthur Gossard, American physicists[1] working at that time at Bell Labs, made an even more unexpected discovery—the Hall conductivity, which is a rational fractional multiple (as opposed to the integer multiple) of e^2/h. This phenomenon was called a fractional Hall effect, and it still

Fig. 16.1 Experimental curves for the Hall resistance and Ohmic resistivity of a heterostructure as a function of the magnetic field at a fixed carrier's density at temperature $T = 0.8$ mK (from K. von Klitzing, The quantum Hall effect. Rev. Modern. Phys. **58**(3) (1986), reprinted with permission)

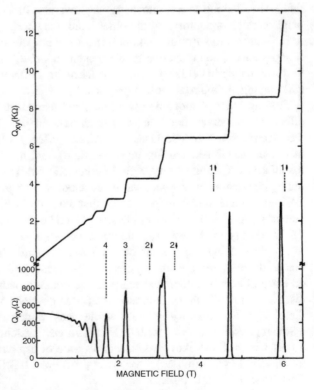

---

[1] Störmer was born in Germany and got his Ph.D. in France, while Tsui was born in China to a family of farmers, did undergraduate work at Lutheran Augustana College in Illinois, and got his Ph.D. at the University of Chicago.

defies full theoretical explanation. The only thing clear from the very beginning was that the model of non-interacting electrons, successful in the explanation of "normal quantum Hall effect," is not applicable here. Just a few months after the experimental discovery, another American theoretical physicist Robert Laughlin pulled out of thin air an exotic many-electron wave function describing a new kind of ground state of interacting fermions which can only exist in two-dimensional systems. This ground state has a bunch of mysterious properties which can be interpreted by introducing particles with fractional charge and fractional (neither boson nor fermion) statistics.[2] While Laughlin did not really derive his new wave function, but rather guessed it, the guess was so successful and generated so many new ideas and concepts that Laughlin together with Tsui and Störmer was awarded the 1998 Nobel Prize in Physics.

In the foundation of all these phenomena, from the old de Haas–Shubnikov oscillations to the newest concepts of the fractional quantum Hall physics, lies Landau quantization—the emergence of quantized energy levels in the spectrum of an otherwise free particle interacting with a uniform magnetic field. In this chapter I will introduce you to the basic physics of Landau levels and will also scratch a little bit at the surface of the regular quantum Hall effect. However, even a cursory discussion of de Haas effects, especially of the fractional Hall conductivity, lies way outside of the scope of this book.

Quantization of energy eigenvalues of a free electron under an influence of a uniform magnetic field was first predicted by Soviet physicist Lev Landau. He is probably one of the best-known Soviet physicists, famous for his seminal contributions to many areas of physics, and winner of the 1962 Nobel Prize in Physics for his theory of superconductivity. It is much less known that he was destined to repeat the fate of Lev Shubnikov and was saved only by a direct personal intervention of Pyotr Kapitsa—the most powerful Soviet physicist of that time, the discoverer of superfluidity, a Nobel Prize winner, and the director of the main physics research institute in the Soviet Union. Landau was arrested in 1938 for comparing Stalin's and Hitler's regimes and spent a year in the infamous NKVD Lubyanka prison. It took Kapitsa two personal requests and the threat of resignation to secure Landau's eventual release in 1939.

---

[2]These are not real "particles" of course and are usually called quasiparticle. They present a convenient theoretical model useful for the description of ground state properties of strongly interacting electrons.

16.1   Classical Mechanics of a Charged Particle in Crossed Uniform Electric and Magnetic Fields

To create an appropriate context for the story of the life of a quantum electron in the uniform magnetic field, I will begin by reminding you of the properties of classical electrons.

16.1.1   Cyclotron Motion in the Uniform Magnetic Field

A particle with electric charge q moving in the static magnetic field $\boldsymbol{B}$ with velocity $\boldsymbol{v}$ experiences the Lorentz force

$$\boldsymbol{F} = q\boldsymbol{v} \times \boldsymbol{B},$$

which is perpendicular to both the field and the velocity. The resulting motion is a combination of a uniform drift with constant speed $v_{\parallel} = v \cos\theta$ in the direction of the field (θ is the angle between $\boldsymbol{B}$ and the velocity) and the circular motion with constant angular velocity $\Omega_L = v_{\perp}/R$ in the plane perpendicular to the field, where $v_{\perp} = v \sin\theta$ and R is the radius of the circular trajectory. The part of this statement referring to the motion in the direction parallel to $\boldsymbol{B}$ is rather obvious—there is no force component in $\boldsymbol{B}$ direction, but to feel at ease with the other part describing rotation in the plane perpendicular to $\boldsymbol{B}$, you would have to dig out of your memory the long-forgotten concept of the centripetal acceleration. (I am trying to avoid having to formally solve equations of motion here because there is no fun in it.) So, centripetal acceleration is always perpendicular to the velocity changing its direction without affecting its magnitude and therefore "forcing" the particle to follow a circular path. Preservation of the magnitude of the velocity v can also be understood as a consequence of the fact that Lorentz force, which is perpendicular to the velocity, does not generate any work, making kinetic energy and, respectively, magnitude of the velocity, conserving quantities. This whole arrangement is shown in Fig. 16.2 for your enjoyment.

The radius of the circle R is determined by Newton's second law

$$\frac{m_e v_{\perp}^2}{R} = qvB \Rightarrow$$

$$R = \frac{m_e v_{\perp}}{qB} = \frac{m_e v \sin\theta}{qB}.$$

Fig. 16.2 A charged particle in the uniform magnetic field: velocity, trajectory, and the system of coordinates

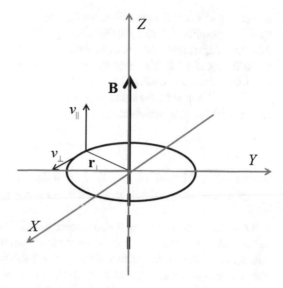

It is remarkable that the angular velocity Ω_L (frequency) of this motion, called Larmor[3] frequency

$$\Omega_L = \frac{v_\perp}{R} = \frac{qB}{m_e},$$

is independent of the magnitude or direction of the particle's velocity (this result must appear rather strange in the limit $v_\perp \to 0$, when radius R goes to zero, and there is no circular motion, but it is what it is). This type of motion is often called cyclotron motion because it is exploited in a particle-accelerating device, called a cyclotron.

Introducing a coordinate system with Z-axis along the field, I can describe the particle's motion by projecting its position vector onto the coordinate axes:

$$x = x_0 + \frac{m_e v \sin\theta}{qB} \cos\Omega_L t \qquad (16.1)$$

$$y = y_0 - \frac{m_e v \sin\theta}{qB} \sin\Omega_L t \qquad (16.2)$$

$$z = z_0 + v \cos\theta t, \qquad (16.3)$$

---

[3] Sir Joseph Larmor (1857–1942) was a Northern Irish physicist famous for the discovery of Lorentz transformations 2 years before Lorentz and 8 years before Einstein. He also discovered the effects of time dilation and length contraction but believed that they are real material changes in length rather than pure kinematic effects. He believed in ether and did not believe in relativity, both special and general. Still, he held a post of Lucasian Professor of Mathematics at Cambridge University, which was established in 1663(!), and was held before him by Newton and after him by Dirac. At the time of writing, this post is held by Stephen Hawking.

where x_0 and y_0 are the coordinates of the center of the circular trajectory determined by initial conditions and z_0 is the initial position of the particle on the Z-axis. (Note that time-dependent terms in equations for x-coordinate and y-coordinate must have opposite signs since if one of the coordinates of the rotating vector increases, the other one must decrease.) Substituting $t = 0$ in these equations, you will find that y_0 is, indeed, equal to the value of y at $t = 0$. At the same time, the relation between x_0 and $x(0)$ is more complex:

$$x_0 = x(0) - p_\perp/qB, \tag{16.4}$$

so that the position of the center of the trajectory in the x direction depends on the component of the particle's momentum in the plane perpendicular to the magnetic field.

One important thing you should notice about these expressions is that coordinates in the plane normal to the field demonstrate the time dependence typical for a harmonic oscillator, while the motion along the field is the one of a free particle. You will see that this connection with the harmonic oscillator on the one hand and the free particle on the other hand persists even in the quantum description of this phenomenon.

16.1.2 Classical Motion in Crossed Electric and Magnetic Fields

To make life a bit more interesting, I will now, in addition to the magnetic field $\boldsymbol{B}$, throw in a uniform electric field $\boldsymbol{\mathcal{E}}$ perpendicular to $\boldsymbol{B}$. Assuming that the electric field is in the positive direction of the Y-axis, I can write Newton's second law equations, describing the particle's motion in this case as

$$m_e \frac{d^2x}{dt^2} = qv_yB, \tag{16.5}$$

$$m_e \frac{d^2y}{dt^2} = -qv_xB + q\mathcal{E}, \tag{16.6}$$

$$m_e \frac{d^2z}{dt^2} = 0. \tag{16.7}$$

It would be natural for you to assume that the electric field, which exerts a force and generates an extra acceleration parallel to the Y-axis, would completely destroy the picture of the motion described in the previous section. Well, watch my hands. I begin by introducing a new coordinate

$$x' = x - v_Et, \tag{16.8}$$

where the constant parameter v_E is yet to be determined. This change of the variable does not affect the equation for the x-component of the acceleration at all (after two differentiations the extra term with v_E vanishes). However, substituting this relation to the equation for the y-component of the acceleration and taking into account that $v_x = dx/dt = dx'/dt + v_E$, I get

$$m_e \frac{d^2y}{dt^2} = -qv'_x B - qv_E B + q\mathcal{E},$$

where $v'_x \equiv dx'/dt$. Now prepare to be amazed: by choosing $v_E = \mathcal{E}/B$, I eliminate the electric field from the equations completely. The system of equations, Eqs. 16.5–16.7, expressed in terms of coordinates $x', y,$ and z, looks exactly the same as if the electric field were not present at all. So much for the "extra" acceleration in the y direction! Consequently, I can use the expressions for the coordinates given in Eqs. 16.1–16.3, replacing x with x' and then using Eq. 16.8 to bring back the original x-coordinate. As a result, equations for y- and z-components do not change at all, while the equation for the x-coordinate becomes

$$x = x_0 + \frac{\mathcal{E}}{B}t + \frac{m_e v \sin\theta}{qB} \cos\Omega_L t.$$

Ponder about this result a bit and enjoy its counterintuitive appeal: an electric field in the y direction generates a uniform displacement of charged particles in the perpendicular direction! You can understand this result qualitatively by noting that transition to the new coordinate x' is essentially the Galileo transformation to a new reference frame moving with speed v_E in the x direction. Since the Lorentz force depends on the velocity of the particle, transition to the new reference frame generates an additional contribution to the force, which, given the right choice of the v_E, cancels the electric force.

Now, let me assume that a particle enters the region of electric and magnetic field in the direction perpendicular to $\mathbf{B}$, allowing me not to worry about the particle's displacement in the z direction. Also assume that instead of just one particle, there is a whole bunch of them but that I can neglect their mutual repulsion. Imagine now that I install a small particle detector of length d in the y direction and width w in the z direction (a bar in Fig. 16.3) perpendicular to the X-axis.

Fig. 16.3 Schematic of electron motion in crossed electric and magnetic fields. Magnetic field (together with Z-axis) points out of the page

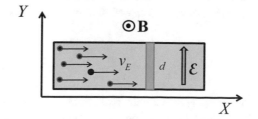

This detector counts the number of particles crossing it during some time τ. Apparently this number is equal to the number of particles inside the volume limited by the detector in the Y–Z plane and by the distance the particles can travel during this time, which can be defined as

$$\Delta x = x(t + \tau) - x(t) = \frac{\mathcal{E}}{B}\tau + \frac{m_e v \sin\theta}{qB}\left(\cos\Omega_L t - \cos\Omega_L\tau\right).$$

Now you need to note that as time τ grows, the first term in this expression grows along with it, while the second term oscillates, remaining limited in value by the radius R of the cyclotron orbit. Thus, for time τ larger than R/v_E, you can drop the oscillating term and count the number of particles crossing the detector as

$$N(\tau) = n_q \frac{\mathcal{E}}{B}\tau \times d \times w,$$

where n_q is the number of particles per unit volume. Consequently, the total charge per unit time per unit area, which not quite accidentally is what people call the current density j, can be found as

$$j = \frac{eN}{\tau \times d \times w} = \frac{en_q}{B}\mathcal{E}. \tag{16.9}$$

Current density j described by Eq. 16.9 is unusual in many respects. First, it describes a current flowing perpendicular to the electric field—the phenomenon described in the beginning of this chapter as the Hall effect. To emphasize this unusual feature, we call the coefficient of proportionality between the current density and the electric field in Eq. 16.9 as the Hall (as opposed to the regular Ohmic) conductivity and use special two-index notation σ_{xy} for it:

$$\sigma_{xy} = \frac{qn_e}{B}. \tag{16.10}$$

One index in σ_{xy} relates to the direction of the field and the other to the direction of the current.

Another important feature of the Hall conductivity is that it has a finite value in the absence of any dissipation mechanisms which are absolutely necessary for establishing a regular stationary flow of charges. The dissipation processes, which are always present in real materials, of course modify the Hall conductivity as

$$\sigma_{xy} = \frac{qn_e}{B}\frac{1}{1 + \left(\frac{m_e}{qB\tau_r}\right)^2},$$

where τ_r is the characteristic time during which a particle loses a significant portion of its energy. Equation 16.10 emerges from the last expression in the limit $\tau_r \to \infty$. You will have a chance to derive this result as an exercise.

Finally, you might notice that the Hall conductivity presented by Eq. 16.10 is linear in the particle's charge, while the standard Ohmic conductivity is proportional to q^2. This is an immensely important difference because particles with charges of opposite signs would generate the Hall current flowing in opposite directions. By observing the direction of the Hall current, one can find out the charge of the current-carrying particles. Thanks to this quirk of the Hall conductivity, experimentalists were able to confirm a theoretical prediction that the electric current in semiconductors can be carried not only by negatively charged electrons but also by positively charged "holes" giving experimental legitimacy to this purely theoretical construct.

16.2 Quantum Theory of Electron's Motion in a Uniform Magnetic Field

16.2.1 Landau Quantization

A quantum theory of electron's motion in a magnetic field begins as any quantum theory—with a classical Hamiltonian for the same problem. The Hamiltonian that have been introduced in Sect. 15.3.1, Eq. 15.53, is of a general enough form so that I can use it here with minimal alterations. Since now I am dealing with static fields, the vector potential appearing in Eq. 15.53 is time-independent and can, therefore, define only the magnetic field. A static electric field has to be described by a traditional scalar potential $\mathcal{E} = -\nabla V$, which can be added to the Hamiltonian as a potential energy. I will need to introduce the electric field only later in the section on quantum Hall effect, but what I need right now is the term describing the interaction between the magnetic field and the spin of the electron.[4] Consequently, I can present the Hamiltonian of a free electron in a static magnetic field as

$$H = \frac{[\hat{p} + e A(r)]^2}{2m_e} + g\frac{\mu_B}{\hbar} B \cdot \hat{S}, \qquad (16.11)$$

where I took into account that electrons have the negative charge $-e$. The last term in this expression is the spin Hamiltonian, defined in Chap. 9, Eq. 9.28, with a slight modification: I replaced factor 2 in this Hamiltonian with a more generic quantity, the so-called spin g-factor. This factor determines the relation between the spin operator and the electron's magnetic moment and is equal to 2 only approximately.

---

[4]When considering the interaction of an atom with electromagnetic wave as in Sect. 15.3.1, I did not have to worry about the interaction between the magnetic field of the wave and the spin because normally such an interaction would be extremely weak. In the case considered in this chapter, the magnetic field can be strong enough to make this interaction relevant.

More accurate relativistic calculations first performed by American physicist Julian Schwinger in 1948 gave for this factor the approximate value 2.002319. New York City residents might be thrilled to find out that Schwinger, who is considered to be one of the most important theoretical physicists of the twentieth century, was born in the City and attended Townsend Harris High School located about 50 meters from my office in the Science Building at Queens College. He started his undergraduate education at City College of New York (Queens College has not yet existed at that time) but transferred later to Columbia University. Experimental verification of the value of the g-factor, which was found to agree with the theoretical prediction with an astonishing accuracy, was one of the triumphs of quantum electrodynamics.

Now I can explore how quantum effects change the story of the electron's life in the magnetic field. To do that, I need, first of all, to choose a form for the vector potential which would reproduce the uniform magnetic field in the direction of the Z-axis of my preferred coordinate system. It is important to realize that the choice is not unique: there exist infinitely many different vector potentials generating the same magnetic field (in science speak it is called gauge invariance). However, not all vector potentials are created equal, even if they do reproduce the same field and describe the same physical reality. Some are more convenient to work with, others are not so much, and different potentials, while providing equivalent descriptions of the world, might emphasize its different aspects. Here I will choose the vector potential in the form

$$A = \tau_y x B \tag{16.12}$$

called *Landau gauge*, where τ_y is the unit vector in the directions of the Y-axis. This expression describes a vector potential with a non-zero y-component (which is its sole non-zero component) linearly dependent on the x-coordinate and independent of the y-coordinate. Of course, there is no inherent preference of one coordinate in the plane perpendicular to the magnetic field over the other, and I could have chosen a potential with a sole non-zero component in the x direction, which would depend on the y coordinates, but it will not really change the description to any significant degree.

Substituting Eq. 16.12 into the Hamiltonian, I get

$$\hat{H} = \frac{\left[\hat{p} + e\tau_y x B\right]^2}{2m_e} = \frac{\hat{p}_x^2}{2m_e} + \frac{\hat{p}_z^2}{2m_e} + \frac{\left[\hat{p}_y + exB\right]^2}{2m_e} + g\frac{\mu_B}{\hbar} B \hat{S}_z. \tag{16.13}$$

You can easily check that this Hamiltonian commutes with operators $\hat{S}_z$, $\hat{p}_y$, and $\hat{p}_z$ (but not with $\hat{p}_x$ because of the x-dependent term in the Hamiltonian). This means that the eigenvectors of the Hamiltonian, which needs to be considered as spinors defined in the tensor product of orbital and spin spaces, must have the following form:

$$|\psi\rangle = |p_y\rangle |p_z\rangle |m_s\rangle |\varphi\rangle, \tag{16.14}$$

where $|p_{y,z}\rangle$ are eigenvectors of the corresponding momentum operators, $|m_s\rangle$ is an eigenvector of $\hat{S}_z$ with m_s being the magnetic spin number, and $|\varphi\rangle$ is the only unknown vector state, which needs to be determined from the corresponding Schrödinger equation. Substituting Eq. 16.14 into

$$\left[\frac{\hat{p}_x^2}{2m_e} + \frac{\hat{p}_z^2}{2m_e} + \frac{[\hat{p}_y + exB]^2}{2m_e} + g\frac{\mu_B}{\hbar}B\hat{S}_z \right] |\psi\rangle , = E|\psi\rangle$$

I obtain the equation for the remaining unknown vector $|\varphi\rangle$

$$\left[\frac{\hat{p}_x^2}{2m_e} + \frac{[p_y + exB]^2}{2m_e} \right] |\varphi\rangle = \left(E - \frac{1}{2}g\mu_B m_s B - \frac{p_z^2}{2m_e} \right) |\varphi\rangle , \tag{16.15}$$

where eigenvalues p_y, p_z, and $\hbar m_s/2$ replace the corresponding operators. Factoring eB out of the respective term in the Hamiltonian, recognizing that $e^2 B^2/m_e$ can be rewritten as $m_e\Omega_L^2$, and introducing the notation

$$x_c = -p_y/eB, \tag{16.16}$$

I can rewrite Eq. 16.15 as

$$\left[\frac{\hat{p}_x^2}{2m_e} + \frac{1}{2}m_e\Omega_L^2 (x - x_c)^2 \right] |\varphi\rangle = \left(E - \frac{1}{2}g\mu_B m_s B - \frac{p_z^2}{2m_e} \right) |\varphi\rangle . \tag{16.17}$$

Do you remember I promised you that connection to the harmonic oscillator problem found in the classical case will persist in quantum treatment as well? So, here it is, consider the promise fulfilled: Eq. 16.17 describes the harmonic oscillator problem with frequency Ω_L, mass m_e, but with the zero point of the potential shifted by x_c (see Chap. 7). In the position representation, this shift can be easily corrected by introducing the new coordinate $x' = x - x_c$ which obviously does not change the eigenvalues and allows to use the well-known wave functions representing the eigenvectors expressed in terms of the new coordinate. Using the results from Chap. 7, I can write now that

$$E_{n,m_s}(p_z) - \frac{1}{2}g\mu_B m_s B - \frac{p_z^2}{2m_e} = \hbar\Omega_L\left(n + \frac{1}{2} \right) \Rightarrow$$

$$E_{n,m_s}(p_z) = \hbar\Omega_L\left(n + \frac{1}{2} + \frac{g\mu_B m_e}{2\hbar e}m_s \right) + \frac{p_z^2}{2m_e}, \tag{16.18}$$

where I replaced eB/m_e with Ω_L. Energy eigenvalues are now determined by one continuous number p_z and two discrete indexes n and $m_s = \pm 1$. They (the eigenvalues) can be thought of as a collection of parabolic bands formed by a

Fig. 16.4 Energy eigenvalues of the electron in a uniform magnetic field. Each parabola represents a Landau "band" corresponding to a particular value of n. The units for both energy and p_z are chosen arbitrarily, so do not pay any attention to the actual numbers on the axes. The point of this figure is to show the general qualitative trend in the behavior of the Landau bands

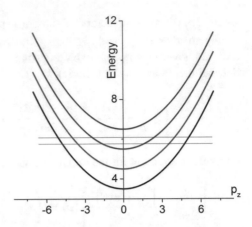

continuously changing p_z, each corresponding to a particular value of n and one of two values of the spin magnetic number m_s (Fig. 16.4). Due to dependence of the energy on p_z, direct manifestations of the Landau quantization in the spectrum of the system are washed out. Indeed, for any arbitrary chosen energy value, you can find multiple states belonging to it (see Fig. 16.4) so that the allowed energy values are continuously distributed despite Landau quantization. The spectrum of the system can become truly discrete only if the electrons are confined to a two-dimensional plane (which experimentalists can routinely do these days), so that the z-component of the momentum p_z vanishes from Eq. 16.18. The complete discretization of the electron's energies in this case is the main reason why all these exciting quantum phenomena that I mentioned earlier occur only in a two-dimensional electron gas. Washing out of the discrete structure of the energies does not mean, however, that Landau quantization does not have any interesting consequences even in the system of regular three-dimensional electrons. One can observe its traces in such phenomena as de Haas–van Alphen and Shubnikov–de Haas effects. I will return to this issue later, after I will have finished with the energy levels $E_{n,m_s}(p_z)$.

If we are dealing with truly free (not interacting with anything but the magnetic field) electrons so that the mass term m_e in the kinetic energy part of the Hamiltonian is just the genuine electron mass, then the factor $\hbar e/m_e$ is exactly twice of the Bohr magneton μ_B, so that Eq. 16.18 simplifies to

$$E_{n,m_s}(p_z) = \hbar\Omega_L\left(n + \frac{1}{2} + \frac{g}{4}m_s\right) + \frac{p_z^2}{2m_e}. \tag{16.19}$$

If in addition you neglect the deviations of the g-factor from 2, you end up with energy levels presented by the even simpler expression

$$E_{n,m_s}(p_z) = \hbar\Omega_L\left(n + \frac{m_s + 1}{2}\right) + \frac{p_z^2}{2m_e}. \tag{16.20}$$

Since m_s only takes two values 1 and -1, $(m_s + 1)/2$ can only be equal to 0 (if $m_s = -1$) or 1 if $m_s = 1$. The immediate consequence of this is that the energy band corresponding to a given n and $m_s = 1$ is exactly the same as the band corresponding to $n+1$ and $m_s = -1$: $E_{n,\uparrow} = E_{n+1,\downarrow}$, making each Landau band double degenerate. Here I decorated the lower index in the notation for energy with up and down arrows instead of numerical values of m_s, 1 and -1, correspondingly, to make the notation visually more striking. The only band, which remained non-degenerate, is the lowest one corresponding to $n = 1$, $m_s = -1$ and described by the energy $E_{1,\downarrow} = p_z^2/2m_e$. This expression looks very much like the energy of a one-dimensional free particle, and if you wish to think of this as of exact cancelation of the magnetic forces on the electron due to its orbital motion and its spin, you can do that, but do not take this analogy too seriously and too far.

The situation changes, however, if you are dealing with electrons in real metals or semiconductors, where they are not quite free as they interact with periodic potential of positively charged ions. Most of the current understanding of semiconductor physics is based on the idea that the interaction with this potential can be accounted for (approximately, of course) by simple replacement of the actual electron's mass m_e by an effective mass m_{eff} (I have already mentioned that in Sect. 15.3.3 on optical transitions in semiconductors). Usually, effective mass is much smaller than the mass of a free electron, for instance, in $GaAs$—one of the most studied and used semiconductors $m_{eff} = 0.067m_e$. This effective mass replaces m_e in the kinetic energy term of Eq. 16.11, but it does not affect the expression for Bohr magneton in the spin-related contribution for the Hamiltonian. You might wonder why we are treating different parts of the Hamiltonian differently and if it can be justified in any way. Actually, there is a good reason for doing this, and while a rigorous explanation of this fact would be a bit over your heads, I can offer a hand-waving argument in favor of this approach. The electron's mass in the kinetic energy describes the reaction of the electron's orbital motion, characterized by the de Broglie wavelength, with ions. This wavelength typically covers several hundred ions, so that electrons feel their potential averaged over large distances, and the effective mass emerges as a result of this averaging process. Spin, and the associated magnetic moment of the electron, on the other hand, is a local characteristic of the electron not subjected to any averaging procedure and remains, therefore, the same as for a free electron. So, if I present the effective mass as $m_{eff} = \kappa m_e$, I can rewrite Eq. 16.18 as

$$E_{n,m_s}(p_z) = \hbar \Omega_L \left(n + \frac{1}{2} + \kappa \frac{g}{4} m_s \right) + \frac{p_z^2}{2m_e}. \tag{16.21}$$

The degeneracy between different Landau bands has all but vanished, and the spin-related separation between bands with the same n is equal to

$$E_{n,\uparrow} - E_{n,\downarrow} \approx \kappa \hbar \Omega_L,$$

where I approximated g with 2. For $GaAs$ this separation is more than ten times smaller than the separation between the bands originating from Landau levels with adjacent values of n, so that Landau bands in this case appear in closely positioned pairs corresponding to the bands with the same n but different orientation of the spin.

Before continuing, let me take a step back and make a detour into an idealized world of a genuinely free two-dimensional electrons with g-factor exactly equal to 2, described by Eq. 16.20 without the term $p_z^2/2m_e$. This model, while not quite realistic, is still a useful toy model which allows you a half glimpse into some of the coolest ideas of modern physics in a relatively safe and unintimidating setting. The main remarkable feature of this model is that its Hamiltonian can be presented in the following form:

$$\hat{H} = \hat{Q}^2, \tag{16.22}$$

where

$$\hat{Q} = \frac{1}{\sqrt{2m_e}} \left[\hat{\sigma}_x \hat{p}_x + \hat{\sigma}_y \left(\hat{p}_y + eB\hat{x} \right) \right].$$

Operators $\hat{\sigma}_{x,y}$ here are Pauli matrices introduced in Eqs. 9.15–9.17. To make it easier, I will remind you those of their properties, which I need in order to continue: (1) the square of any Pauli matrix is a unity matrix, $\hat{\sigma}_{x,y,z}^2 = \hat{I}$, and (2) different Pauli matrices anticommute, $\hat{\sigma}_x \hat{\sigma}_y + \hat{\sigma}_y \hat{\sigma}_x = 0$. To prove Eq. 16.22, you just need to square $\hat{Q}$ and use these properties of the Pauli matrices along the way:

$$\hat{Q}^2 = \frac{1}{2m_e} \left[\hat{\sigma}_x^2 \hat{p}_x^2 + \hat{\sigma}_y^2 \left(\hat{p}_y + eB\hat{x} \right)^2 + \right.$$

$$\hat{\sigma}_x \hat{\sigma}_y \hat{p}_x \left(\hat{p}_y + eB\hat{x} \right) + \hat{\sigma}_y \hat{\sigma}_x \left(\hat{p}_y + eB\hat{x} \right) \hat{p}_x \Big] =$$

$$\frac{1}{2m_e} \left[\hat{p}_x^2 + \left(\hat{p}_y + eB\hat{x} \right)^2 + \right.$$

$$eB \left(\hat{\sigma}_x \hat{\sigma}_y \hat{p}_x \hat{x} + \hat{\sigma}_y \hat{\sigma}_x \hat{x} \hat{p}_x \right) \Big] =$$

$$\frac{1}{2m_e} \left[\hat{p}_x^2 + \left(\hat{p}_y + eB\hat{x} \right)^2 - i\hbar eB\hat{\sigma}_x \hat{\sigma}_y \right].$$

Here terms containing commuting operators $\hat{p}_x \hat{p}_y$ vanish thanks to anticommutativity of the Pauli matrices, which is also responsible for generating the canonical commutator $[\hat{x}, \hat{p}_x] = i\hbar$ in the term mixing momentum and coordinate operators. Finally, you can verify by direct calculations that $\hat{\sigma}_x \hat{\sigma}_y = i\hat{\sigma}_z$, and replacing $(\hbar/2)\hat{\sigma}_z$ with the spin operator $\hat{S}_z$, you will complete the proof.

Operator $\hat{Q}$ deserves a more close inspection because its eigenvectors are also the eigenvectors of the Hamiltonian, while the square of its eigenvalues are supposed to yield the energy levels. Recalling the actual form of the Pauli matrices, I can rewrite $\hat{Q}$ as a two-by-two matrix:

$$\hat{Q} = \frac{1}{\sqrt{2m_e}} \left(\begin{bmatrix} 0 & \hat{p}_x \\ \hat{p}_x & 0 \end{bmatrix} + \begin{bmatrix} 0 & -i\left(\hat{p}_y + eB\hat{x}\right) \\ i\left(\hat{p}_y + eB\hat{x}\right) & 0 \end{bmatrix} \right) =$$

$$\frac{1}{\sqrt{2m_e}} \begin{bmatrix} 0 & \hat{p}_x - i\left(\hat{p}_y + eB\hat{x}\right) \\ \hat{p}_x + i\left(\hat{p}_y + eB\hat{x}\right) & 0 \end{bmatrix}.$$

The combination of operators arising in the off-diagonal elements of this matrix looks somewhat familiar, so it makes sense to take a look at their commutator:

$$\left[\hat{p}_x + i\left(\hat{p}_y + eB\hat{x}\right), \hat{p}_x - i\left(\hat{p}_y + eB\hat{x}\right)\right] =$$
$$2i\left(\hat{p}_y + eB\hat{x}\right)\hat{p}_x - 2i\hat{p}_x\left(\hat{p}_y + eB\hat{x}\right) =$$
$$2ieB\left(\hat{x}\hat{p}_x - \hat{p}_x\hat{x}\right) = -2\hbar eB.$$

The important result here is that the commutator is just a regular constant and not an operator. Taking advantage of this fact, I can now define new operators

$$\hat{a} = \frac{\hat{p}_x - i\left(\hat{p}_y + eB\hat{x}\right)}{\sqrt{2\hbar qB}} \tag{16.23}$$

and

$$\hat{a}^\dagger = \frac{\hat{p}_x + i\left(\hat{p}_y + eB\hat{x}\right)}{\sqrt{2\hbar qB}} \tag{16.24}$$

which have exactly the same commutation relation as raising and lowering operators in the harmonic oscillator problem: $\left[\hat{a}, \hat{a}^\dagger\right] = 1$. Isn't it curious how the ears of the harmonic oscillator problem are sticking out everywhere we look? In fact, it is getting even curiouser and curiouser if you rewrite operator $\hat{Q}$ in terms of $\hat{a}$ and $\hat{a}^\dagger$:

$$\hat{Q} = \sqrt{\frac{\hbar qB}{m_e}} \begin{bmatrix} 0 & \hat{a}^\dagger \\ \hat{a} & 0 \end{bmatrix} = \sqrt{\hbar\Omega_L} \begin{bmatrix} 0 & \hat{a} \\ \hat{a}^\dagger & 0 \end{bmatrix}, \tag{16.25}$$

which yields the Hamiltonian in the form

$$\hat{H} = \hat{Q}^2 = \hbar\Omega_L \begin{bmatrix} 0 & \hat{a} \\ \hat{a}^\dagger & 0 \end{bmatrix} \begin{bmatrix} 0 & \hat{a} \\ \hat{a}^\dagger & 0 \end{bmatrix} =$$

$$\hbar\Omega_L \begin{bmatrix} \hat{a}\hat{a}^\dagger & 0 \\ 0 & \hat{a}^\dagger\hat{a} \end{bmatrix} = \hbar\Omega_L \begin{bmatrix} \hat{a}^\dagger\hat{a} + 1 & 0 \\ 0 & \hat{a}^\dagger\hat{a} \end{bmatrix} =$$

$$\hbar\Omega_L \left(\hat{a}^\dagger\hat{a} \begin{bmatrix} 1 & 0 \\ 0 & 1 \end{bmatrix} + \begin{bmatrix} 1 & 0 \\ 0 & 0 \end{bmatrix} \right) =$$

$$\hbar\Omega_L \left(\hat{a}^\dagger\hat{a} + \frac{1}{2} \right) \begin{bmatrix} 1 & 0 \\ 0 & 1 \end{bmatrix} + \frac{1}{2}\hbar\Omega_L \begin{bmatrix} 1 & 0 \\ 0 & -1 \end{bmatrix}. \tag{16.26}$$

The first term in this expression is a spin-independent Hamiltonian of the harmonic oscillator, while the second term is our familiar spin Hamiltonian in the magnetic field. This result ascertains the role of operators $\hat{a}$ and $\hat{a}^\dagger$ as lowering and raising operators which generate the excited states of the electron in the magnetic field from the ground state in exactly the same manner as they did for a regular harmonic oscillator. More specifically if $|\varphi_n\rangle$ and $|\varphi_{n+1}\rangle$ are normalized eigenvectors representing orbital components of states corresponding to nth and $n+1$st Landau levels, one can write $\hat{a}^\dagger |\varphi_n\rangle = \sqrt{n+1}\,|\varphi_{n+1}\rangle$ and $\hat{a}\,|\varphi_{n+1}\rangle = \sqrt{n+1}\,|\varphi_n\rangle$.

To get a bit more insight into properties of operator $\hat{Q}$, consider a spinor

$$\begin{bmatrix} 0 \\ |\varphi_0\rangle \end{bmatrix}$$

describing the zeroth Landau level with the "spin-down" spin state. Applying $\hat{Q}$ to it, you get

$$\hat{Q}\begin{bmatrix} 0 \\ |\varphi_0\rangle \end{bmatrix} = \sqrt{\hbar\Omega_L}\begin{bmatrix} 0 & \hat{a} \\ \hat{a}^\dagger & 0 \end{bmatrix}\begin{bmatrix} 0 \\ |\varphi_0\rangle \end{bmatrix} = \sqrt{\hbar\Omega_L}\begin{bmatrix} \hat{a}\,|\varphi_0\rangle \\ 0 \end{bmatrix} = 0$$

(the lowering operator acting on the ground state yields zero), which agrees with the previously obtained result that the state of the electron with $n = 0$ and $m_s = -1$ in the magnetic field has zero energy. If, however, you consider the action of this operator on, let's say state $|\varphi_0\rangle\,|\uparrow\rangle$, you will get

$$\hat{Q}\begin{bmatrix} |\varphi_0\rangle \\ 0 \end{bmatrix} = \sqrt{\hbar\Omega_L}\begin{bmatrix} 0 & \hat{a} \\ \hat{a}^\dagger & 0 \end{bmatrix}\begin{bmatrix} |\varphi_0\rangle \\ 0 \end{bmatrix} = \sqrt{\hbar\Omega_L}\begin{bmatrix} 0 \\ \hat{a}^\dagger\,|\varphi_0\rangle \end{bmatrix} = \sqrt{\hbar\Omega_L}\begin{bmatrix} 0 \\ |\varphi_1\rangle \end{bmatrix}.$$

In English it means that operator $\hat{Q}$ flips the spin of the electron whereby lowering its energy by $\hbar\Omega_L$ while simultaneously lifting it to the next Landau level increasing its energy by the same amount, thereby generating a pair of degenerate states discussed in Eq. 16.19.

Now, if you recall the general discussion of the degenerate eigenvalues, you will remember that a degeneracy can often be associated with the set of mutually commuting operators. In the case under consideration here, it is the operator $\hat{Q}$ together with the Hamiltonian that forms such a set. An interesting peculiarity of this example, distinguishing it from all other examples of commuting operators, is that here operator $\hat{Q}$ mixes spatial–temporal and spinor segments of the total tensor product space. By itself it is not big news—you observed this phenomenon when we discussed spin–orbital coupling in Sect. 14.1. What is new here is that the mixing of orbital and spin degrees of freedom occurs in the absence of any spin–orbital interaction. This mixing is a primitive pedestrian version of something called supersymmetry: a symmetry transformation connecting particles with half-integer spins, the fermions, with integer spin particles—bosons. In our example, the harmonic oscillator represents the bosonic degree of freedom (remember, harmonic oscillator lowering–raising operators appear also as creation–annihilation operators of photons which are bosons, Sect. 7.3), while the spin degrees of freedom play the role of the fermionic degrees of freedom. The eigenvalues and eigenvectors of operator $\hat{Q}$ can shed more light on the connection between orbital and spin states in this problem, and I hope you will have fun figuring them out by yourself.

OK, the detour is over, I hope you enjoyed it, but let's get back to the main business of this chapter. Based on Eq. 16.17, I can write down the position representation for the $|\varphi\rangle$ component of the eigenvector of my Hamiltonian, which is just a wave function of a harmonic oscillator centered at $x = x_c$ instead of zero:

$$\varphi_n(x) \equiv \langle x|\,\varphi_n\rangle = \frac{1}{\sqrt{2^n n!\,\xi\,\sqrt{\pi}}}\exp\left(-\frac{(x-x_c)^2}{2\xi^2}\right)\mathcal{H}_n\left(\frac{x-x_c}{\xi}\right), \qquad (16.27)$$

where the characteristic length ξ known as *magnetic length* is

$$\xi = \sqrt{\frac{\hbar}{m_e\Omega_L}} = \sqrt{\frac{\hbar}{eB}}. \qquad (16.28)$$

The total wave function $\psi\,(\boldsymbol{r})$ includes two free particle-like exponents $\exp\left(ip_z z + ip_y y\right)$ reflecting the free particle-like motion in the z and y directions. The corresponding position probability density $|\psi\,(\boldsymbol{r})|^2$ looks like a slab with uniform probability distribution along the Y and Z axes while exponentially decaying for $x > x_c$ and $x < x_c$, Fig. 16.5.

In addition to the trivial free particle-like dependence of the wave function on p_y, it also shows another, much less trivial role played by this parameter, which, via Eq. 16.16, determines the position of the maximum of the distribution along the X-

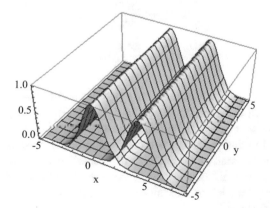

Fig. 16.5 Probability density $|\psi\,(\boldsymbol{r})|^2$ of the electron's position corresponding to the oscillator ground state as a function of two coordinates, x and y. If you can imagine a plot in a four-dimensional space, it would show that the probability distribution behaves in the z direction in the same way as in the y direction. Two different surfaces in the figure correspond to two different values of p_y, illustrating its effect on the center of the probability density

axis. This fact is reflected in Fig. 16.5 where two different plots correspond to two different values of p_y, generating two different values of x_c.

In classical picture, this maximum would correspond exactly to the center of the particle's trajectory, which you can see comparing Eqs. 16.4 and 16.16. It is important to note at this point that the energy eigenvalues do not depend on p_y, indicating their significant degree of degeneracy. This degeneracy is what distinguishes an electron in the magnetic field from the simple harmonic oscillator problem.

16.2.2   Degeneracy of the Landau Levels and the Density of States

The degeneracy of the energy levels determines the number of single-particle orbitals available for constructing many particle states in a many-fermion system and is expressed in terms of the density of state—a notion which has been already discussed in the previous chapters in connection with the concept of Fermi energy (Sect. 11.5) and optical transitions to (or between) states of the continuous spectrum. Our normal practice employed to compute the density of states consists in confining the particles inside a large but final volume and impose boundary conditions on the wave function. However, when dealing with Landau levels, this procedure acquires several unusual twists. First, while I can and should impose the periodic boundary conditions on the wave function in y and z directions, the wave function vanishes in the x directions, so no boundary conditions for the x-coordinate are actually required. As a result, the boundary conditions allow me to discretize the two continuously changing quantities:

$$p_z = \frac{2\pi \hbar b_z}{L_z},$$

$$p_y = \frac{2\pi \hbar b_y}{L_y},$$

where $b_{x,y} = \pm 1, \pm 2 \cdots$, while the third quantum number defining a state of the electron, integer n from the harmonic oscillator problem, is already discrete. In order to find how many states with distinct values of p_y correspond to any given energy value of n, I need to find the restriction on the number of allowed p_y. The periodic boundary conditions are not too helpful in this regard as they just make p_y discrete but put no bounds on their values. Also, since the energy does not depend on p_y, I cannot use our regular trick of looking for the number of states corresponding to a specified energy interval.

To figure all this out, let's take a look at what is happening to the system as the absolute value of p_y increases. Equation 16.16 indicates that as p_y increases or decreases, the center of the probability distribution in the x directions moves further away from zero toward correspondingly negative or positive coordinates. However, in the system of a finite size, it cannot move too far because at some point you will be pushing the system outside of the quantization box, which is not allowed. This restriction sets a limit for the possible value of x_c: $x_c < L_x$.[5] This is where the periodic boundary condition discretizing p_y comes in handy, because it also automatically discretizes parameter x_c, which now only takes values

$$x_c = \frac{2\pi\hbar b_y}{L_y eB}.$$

The condition $x_c < L_x$ limits the number N_{n,p_z} of the allowed states with given n and p_z as

$$\frac{2\pi\hbar b_y}{L_y eB} < L_x \Rightarrow b_y^{max} = \frac{eBL_xL_y}{2\pi\hbar},$$

so that the p_y-related degeneracy of the Landau levels is

$$G = \frac{eBL_xL_y}{2\pi\hbar}. \tag{16.29}$$

This expression is sometimes rewritten as

$$G = \frac{\Phi}{\Phi_0}, \tag{16.30}$$

where $\Phi_0 = h/e$ (note the rare appearance of the Planck's constant without the bar) is called *fundamental quantum of flux*, while $\Phi = BL_xL_y$ is the magnetic flux through the region of area $A_\perp = L_xL_y$.

These expressions tell you how many states with given n and all possible p_y exist in the system occupying the surface area $A_\perp$ normal to the magnetic field. If the particle's motion is limited to a two-dimensional plane, then this is all what you need to know. In the two-dimensional case, which is of main interest in quantum Hall effect studies, the momentum's component p_z does not exist, the respective wave function loses the factor $\exp(ip_zz/\hbar)$, and the spectrum of eigenvalues becomes purely discrete with energies given by the one-dimensional oscillator formula

---

[5]This condition is obviously just an approximation because the wave function in the x direction has finite size, and requiring that its center remains inside the box does not make the probability of the particle described by the harmonic oscillator wave functions to be outside of the box to vanish automatically. The choice of x_c to define the number of states is, in this sense, rather arbitrary (you could choose $x_c + \Delta x$, where Δx is the uncertainty of the coordinate or something else for that matter). However, the error which I make here is of the order of $\Delta x/L_x$ and becomes negligibly small as L_x grows. The count of the number of states derived by this heuristic method is actually confirmed by a more rigorous (and immensely more complicated) mathematical approach to the problem.

$$E_{n,m_s} = \hbar \Omega_L \left(n + \frac{1}{2} + \kappa \frac{g}{4} m_s \right). \tag{16.31}$$

In the case of discrete levels, the concept of the density of states, which was designed to count states in the continuous spectrum, is not really needed because we can simply count all the discrete states belonging to a given degenerate level, and this is what Eq. 16.30 yields.[6] If you wonder where this degeneracy comes from, please recall that you are dealing with a two-dimensional problem which requires two quantum numbers to fully characterize a state. This second dimension manifests itself as a degeneracy of the harmonic oscillator energy levels with respect to the position of the minimum of the harmonic oscillator potential making each Landau level to comprise G number of states, the same for all levels.

In the three-dimensional case, which might currently be out of vogue, but is of interest for more old-fashioned manifestations of Landau quantization, such as de Haas–van Alphen and Shubnikov–de Haas effects, there is an extra step I have to make. Taking into account the quasi-continuous distribution of p_z, I am back to the density of state computation mentality trying to find the number of states within an energy interval ΔE between the values E and $E + \Delta E$. To figure out the answer, I need to go back to Fig. 16.4 and pay attention to the two horizontal lines drawn in there. These lines delineate an energy interval ΔE that appeared in the question posed above. You can see that several Landau bands characterized by their corresponding distinct n-numbers and spin indexes contribute their states to this interval. In order to find a contribution from a single band, consider

$$\Delta p_z^{(n,m_s)} = 2 \frac{d \sqrt{2m_e \left[E - \hbar \Omega_L \left(n + \frac{1}{2} + \kappa \frac{g}{4} m_s \right) \right]}}{dE} \Delta E =$$

$$\frac{\sqrt{2m_e}}{\sqrt{E - \hbar \Omega_L \left(n + \frac{1}{2} + \kappa \frac{g}{4} m_s \right)}} \Delta E, \tag{16.32}$$

where the factor of 2 accounts for positive and negative values of the momentum. The periodic boundary conditions tell me that the number of states Δb_z corresponding to a given interval $\Delta p_z^{(n,m_s)}$ is

$$\Delta b_z = \frac{L_z}{2\pi \hbar} \Delta p_z^{(n,m_s)},$$

---

[6]One can associate a delta-function $\delta(E - E_n)$ with each level and the identity expression $\sum_n N_n \delta(E - E_n)$ as a total density of states. This identification makes sense if you integrate it over any final energy interval to get the total number of states in it.

which together with Eq. 16.32 yields

$$\Delta b_z^{(n,m_s)} = \frac{L_z\sqrt{2m_e}}{2\pi\hbar\sqrt{E - \hbar\Omega_L\left(n + \frac{1}{2} + \kappa\frac{g}{4}m_s\right)}}\Delta E. \tag{16.33}$$

Apparently, this expression makes sense only if $E - \hbar\Omega_L\left(n + \frac{1}{2} + \kappa\frac{g}{4}m_s\right) > 0$, which cuts the number of bands that at a given E can donate their states at

$$n_{max} = \left[\frac{E}{\hbar\Omega_L} - \frac{1}{2} - \kappa\frac{g}{4}\right]$$

where I took into account that for a given n, the higher lying band corresponds to $m_s = 1$. The square brackets in this expression indicate that the number inside needs to be *rounded down* to the nearest integer. To account for the degeneracy in the plane normal to the field, the result in Eq. 16.33 must be multiplied by G. Adding together the contributions from all bands with $n \le n_{max}$, I can finally get the answer to my burning question: the total number of states ΔN with energies between E and $E + \Delta E$ is

$$\Delta N = \frac{e\sqrt{2m_e}B}{4\pi^2\hbar^2}L_xL_yL_z\sum_{n=0}^{n_{max}}\sum_{m_s}\frac{1}{\sqrt{E - \hbar\Omega_L\left(n + \frac{1}{2} + \kappa\frac{g}{4}m_s\right)}}\Delta.E$$

A more convenient version of this expression that does not require you to keep track of n_{max} uses the step function $\theta\left[E - \hbar\Omega_L\left(n + \frac{1}{2} + \kappa\frac{g}{4}m_s\right)\right]$, which automatically limits the summation to only those n for which its argument is positive, making the expression for the density of states more convenient for analysis:

$$g(E) = \Delta N/\Delta E =$$

$$\frac{e\sqrt{2m_e}B}{4\pi^2\hbar^2}L_xL_yL_z\sum_{n=0}^{\infty}\sum_{m_s}\frac{\theta\left[E - \hbar\Omega_L\left(n + \frac{1}{2} + \kappa\frac{g}{4}m_s\right)\right]}{\sqrt{E - \hbar\Omega_L\left(n + \frac{1}{2} + \kappa\frac{g}{4}m_s\right)}}. \tag{16.34}$$

If, as is often the case, the spin-related term in the expression for energy can be neglected, the summation over the spin number m_s simply yields an extra factor of two in the density of states.

16.2.3 Fermi Energy of a Gas of Non-interacting Electrons in Magnetic Field

Now I am going to use the results of the previous section to revisit the concept of the Fermi energy first discussed in Sect. 11.5 for a gas of free electrons. The Fermi energy is the main characteristic of the many-electron ground state and as such affects the most practically important properties of electronic systems such as conductivity, heat capacity, magnetization, etc. It would be interesting, therefore, to see how the magnetic field and Landau quantization affect it.

I will begin with a two-dimensional case, when all what you have to worry about are different harmonic oscillator states and their degeneracy. I will also neglect the spin contribution to the energy of Landau levels, in which case the total number of states at a given Landau level is found by multiplying Eq. 16.29 by two. Then, the total number of states in Landau levels with $n \leq n_{max}$ for a given magnetic field B is found as

$$(n_{max} + 1) \frac{eBL_xL_y}{\pi\hbar}$$

(remember, the first value of n is zero). If the number N_e of electrons in the system obeys inequality

$$n_{max}\frac{eBL_xL_y}{\pi\hbar} < N_e < (n_{max} + 1)\frac{eBL_xL_y}{\pi\hbar},$$

then the $n_{max} - 1$ Landau level is the last one completely filled, and n_{max}-th level is partially filled, so that the Fermi energy is

$$E_F = \hbar\Omega_L \left(n_{max} + \frac{1}{2} \right). \tag{16.35}$$

Introducing the filling fraction v (the fraction of the total number of single-electron orbitals actually used to form a many-electron state) of the n_{max}-th level, I can relate the total number of electrons in the system to n_{max} and v:

$$N_e = (n_{max} + v)\frac{eBL_xL_y}{\pi\hbar}. \tag{16.36}$$

Now, let's see what will happen as I gradually increase the magnetic field. Two things are taking place simultaneously. First, Ω_L and, therefore, the energy of each Landau level grow linearly with B, resulting in the initial increase of the Fermi level. At the same time, the degree of degeneracy of each Landau level grows, so does their capacity, and so the number of filled single-electron orbitals at the last partially filled level decreases until the n_{max}-th level becomes devoid of any electrons. When this happens, the last filled orbital now corresponds to the $n_{max} - 1$-th level, resulting in a drop in the Fermi energy by $\hbar\Omega_L$. Further increase in the field again produces growth of E_F due to the growth of Ω_L, until the filling fraction v of the $n_{max} - 1$-st level drops to zero, resulting in a subsequent drop in the Fermi energy again by $\hbar\Omega_L$. This pattern of linear growth accompanied by sudden drops repeats itself until the magnetic field becomes so large that you would only need states belonging to

$n = 0$ Landau level to accommodate all electrons in the system. This happens when B obeys inequality

$$\frac{eBL_xL_y}{\pi\hbar} > N_e$$

or

$$B > \frac{N_e}{L_xL_y}\frac{\pi\hbar}{e},$$

which can also be recast in the form

$$\Phi > N_e\Phi_0.$$

This pattern can be seen in a more quantitative manner by solving Eq. 16.36 for n_{max}:

$$n_{max} = \left[\frac{N_e}{L_xL_y}\frac{\pi\hbar}{eB}\right]$$

where $[\cdots]$ again means dropping the fractional part of the number inside the brackets. Then the Fermi energy can be written down as

$$E_F = \frac{\hbar eB}{m_e}\left(\left[\frac{N_e}{L_xL_y}\frac{\pi\hbar}{eB}\right] + \frac{1}{2}\right). \tag{16.37}$$

These expressions clearly show the two tendencies in the magnetic field dependence of the Fermi energy, which are also illustrated in Fig. 16.6.

The Fermi energy in the three-dimensional case is obtained by integrating the density of states over all energies from 0 to E_F, which results in equation

Fig. 16.6 Magnetic field dependence of the Fermi energy for a two-dimensional electron gas. Magnetic field is in the units of Φ/Φ_0, and the Fermi energy is in the units of $\hbar\Omega_L$

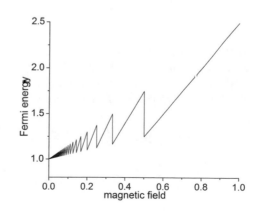

$$\frac{e\sqrt{2m_e B}}{2\pi^2\hbar^2}L_xL_yL_z\sum_{n=0}\int_0^{E_F}dE\frac{\theta\left[E-\hbar\Omega_L\left(n+\frac{1}{2}\right)\right]}{\sqrt{E-\hbar\Omega_L\left(n+\frac{1}{2}\right)}}=N_e.$$

The integral in this expression is easily computed yielding

$$\frac{e\sqrt{2m_e B}}{\pi^2\hbar^2}L_xL_yL_z\sum_{n=0}\sqrt{E_F-\hbar\Omega_L\left(n+\frac{1}{2}\right)}=N_e,$$

where I took into account that the step function in this expression effectively limits the lower limit of integration for each term in the sum over n to value $\hbar\Omega_L\left(n+\frac{1}{2}\right)$. Also it is worth noting that after the integration over energy, you are essentially ending up with the sum of $p_z^{(n)}$. This conforms with a well-known fact that in one-dimensional systems the magnitude of the particle's momentum at a given energy E is equal to the total number of states with energies less than E. The computation of the resulting sum over n is not a trivial matter, and I will not attempt it here (I think I made you suffer quite enough even without it), but I would like to notice that the qualitative pattern of dependence of E_F versus magnetic field is similar: the energy of the last filled level grows with B, while the filling factor of the respective bands decreases, until they become empty, and the Fermi level drops to the lower band corresponding to $n-1$. Thus, again one shall expect the oscillatory dependence of the Fermi energy upon the magnetic field.

I cannot feel completely satisfied unless the total energy of the ground state is computed, and I am going to do it here for the two-dimensional electrons neglecting the spin contribution to the energy of the Landau levels. In the system of non-interacting particles, the total energy is simply equal to the sum of the single-particle energies of all filled orbitals, so all what I need is a bit of bookkeeping. Taking into account Eq. 16.31 together with Eq. 16.29, I can write for the total energy

$$E_g^{N_e}=2G\hbar\Omega_L\left[\sum_{n=0}^{n_{max}-1}\left(n+\frac{1}{2}\right)+\nu\left(n_{max}+\frac{1}{2}\right)\right],$$

where the first term in the square brackets counts all the completely filled Landau levels up to $n_{max}-1$ and the second term accounts for the contribution of the partially filled n_{max} level with the filling fraction ν defined by Eq. 16.36. Recalling the formula for the sum of the arithmetic progression, I have

$$E_g^{N_e}=2G\hbar\Omega_L\left(\frac{n_{max}\left(n_{max}-1\right)}{2}+\frac{n_{max}}{2}+\nu\left(n_{max}+\frac{1}{2}\right)\right)=$$

$$G\hbar\Omega_L\left(n_{max}^2+2\nu\left(n_{max}+\frac{1}{2}\right)\right)=$$

$$G\hbar\Omega_L\left[\left(n_{max}+\nu\right)^2+\nu-\nu^2\right].$$

Using Eq. 16.36 to express

$$n_{max} + \nu = \frac{\pi \hbar N_e}{eBL_xL_y} \equiv \frac{N_e}{2G}, \tag{16.38}$$

I can rewrite the total ground state energy as

$$E_g^{N_e} = \frac{\hbar \Omega_L N_e^2}{4G} + \hbar \nu \left(1 - \nu\right) G\Omega_L =$$

$$\frac{\hbar}{4} \frac{eB}{m_e} \frac{2\pi \hbar^2 N_e^2}{eBL_xL_y} + \hbar \nu \left(1 - \nu\right) \frac{eB}{m_e} \frac{eBl_xL_y}{2\pi \hbar} =$$

$$\frac{\pi \hbar^2 N_e^2}{2m_e L_x L_y} + \nu \left(1 - \nu\right) \frac{e^2}{2\pi m_e} B^2 L_x L_y. \tag{16.39}$$

The first term in this expression does not depend on the magnetic field and can be shown to be the ground state energy of the two-dimensional gas of free electrons (see the problem section in this chapter). If you are confused about the quadratic dependence of this contribution on N_e (you shall rightfully expect a linear dependence in a system of independent particles), note the denominator of the respective expression contains L_xL_y—the area of the sample. Thus, the field-independent part of the total energy is actually proportional to $N_e\sigma_e$, where σ_e is the surface concentration of the electrons, i.e., it is, indeed, linear in N_e as expected and similar to the case of three-dimensional electrons without the field, Eq. 11.57. The magnetic field dependence of $E_g^{N_e}$ is again determined by two factors: oscillations of the filling factor between values 0 and 1 with period $1/B$ (check out Eq. 16.38 and note the $1/B$ factor) and a smooth increase in the energy as B^2 (see Fig. 16.7). The first tendency is a direct consequence of Landau quantization as it reflects changes in the occupation of Landau levels with increasing B, while the second tendency can be explained using the classical argument: the energy of a magnet in the magnetic field is $\propto \mu B$, where the magnetic dipole, which is zero in the absence of B, is itself proportional to the field.

Fig. 16.7 Magnetic field dependence of the ground state energy of the two-dimensional electron gas

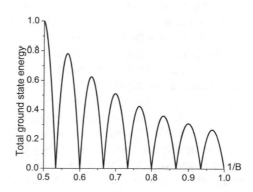

The oscillations of the ground state energy manifest themselves in oscillations of immediately observable quantities such as magnetization (magnetic moment per unit area). Keeping only terms linear in the magnetic field, the magnetization can be found as

$$M = -\frac{1}{L_x L_y}\partial E_g/\partial B = -\nu\,(1-\nu)\,\frac{e^2}{\pi m_e}B. \tag{16.40}$$

This result means that the two-dimensional gas of free electrons can be characterized by magnetic permeability μ, defined as a proportionality coefficient between magnetization and magnetic field

$$\mu = -\nu\,(1-\nu)\,\frac{e^2}{\pi m_e} \tag{16.41}$$

which exhibits periodic oscillations as a function of $1/B$. The negative sign of the permeability indicates that the gas of free electrons is a diamagnetic material, in which the magnetic field induces magnetization in the direction opposite to that of the field. The diamagnetic properties of free electrons were also discovered by Lev Landau in 1930.

16.3   Quantum Hall Effect

I will complete this chapter by a brief excursion into a huge area of the quantum Hall effect, which is a quantum version of motion of a two-dimensional electron in crossed electric and magnetic fields. From the onset I have to warn you that here I am going just to scratch a bit the surface of this field, and by no means you shall assume that this section contains anything even remotely resembling a complete theory of this phenomenon. What I am going to present here is just some very basic, albeit fundamental, underlying reasons for its existence.

The Hall effect (classical or quantum), as you already know, consists in the appearing of an electric current in a crossed electric and magnetic fields in the direction perpendicular to both of them. Computing the electric current in honesty cannot be done within the formalism of the pure quantum mechanics considered in this book. It would require taking into account the temperature-dependent statistical distribution of the electrons among various single-particle orbitals and its dependence on the electric and magnetic fields. You will also have to introduce a variety of dissipative processes, without which a finite conductivity simply cannot exist (go back to Sect. 16.1.2 and Problem 1 in the exercise section of this chapter), and dealing with dissipative processes within the quantum formalism is very far from being a trivial task. All of this will take us into the realm of quantum kinetics, which is way outside of the scope of this text. Fortunately, for us, however, the Hall conductivity does need dissipation to acquire a finite value as you have already seen in the classical treatment of the Hall effect in Sect. 16.1.2. As a result I can get away with neglecting the possible dissipative effects, which is, of course, a gross oversimplification but will give us at least some hints at the underlying physics of the quantum Hall phenomenon. If in addition I assume that the electrons are in their many-particle ground state (zero temperature), I can avoid dealing with any temperature-dependent complications.

In this simplified treatment, the flow of the electrons can be interpreted in terms of the flow of the probability density. Indeed, consider an electron in a single-particle orbital described by a wave function $\psi_\sigma(r)$, where σ is a composite index numerating the orbitals. The measured charge density of these electrons $\rho_\sigma(r)$ is directly related to the position probability distribution $|\psi_\sigma(r)|^2$: $\rho_\sigma(r) = e|\psi_\sigma(r)|^2$, and if there are many electrons occupying multiple states, the total charge density at a given point is given by $\rho_{tot} = e\sum_\sigma |\psi_\sigma(r)|^2$. Taking into account the continuity equation for the probability density, Eq. 5.45, I can rewrite it in the form of the equation for the charge density

$$\frac{\partial \rho_{tot}}{\partial t} = -e\sum_\sigma \nabla \cdot j_\sigma,$$

where j_σ is the probability current density for a given orbital. Consequently, the total electric current density with contributions from all electrons in occupied single-particle orbitals can be found as

$$j_{tot} = e\sum_\sigma \nabla \cdot j_\sigma. \tag{16.42}$$

To lay down some groundwork and as a basis for comparison, I will begin by evaluating the current density j_σ in the absence of the electric field. It is quite tempting to just rush ahead and use Eq. 5.42 with the wave function given by Eq. 16.14 to do the job, and I wouldn't blame if you did that, but unfortunately it would have been wrong. You shall realize it right away if you recall that Eq. 5.42 was derived using the Schrödinger equation with kinetic energy operator $\hat{K} = \hat{p}^2/2m_e$ and velocity $\hat{v} = \hat{p}/m_e$. In the presence of the magnetic field, however, the kinetic energy is given by

$$\hat{K} = \frac{[p + eA(r)]^2}{2m_e},$$

and the velocity operator is now

$$v = \frac{[p + eA(r)]}{m_e},$$

so I will have to re-derive the formula for the probability current density to reflect this new reality.

If, on a hunch, you replace the velocity operator in Eq. 5.43 with the new expression, the probability current density reemerges in the following form:

$$j = \frac{1}{2m_e} \left[\Psi^* (r,t) (p + eA(r)) \Psi (r,t) + \Psi (r,t) \left[(p + eA(r)) \Psi (r,t) \right]^* \right].$$

(16.43)

As it turns out, it is a good guess, and this is indeed the correct expression for j. However, hunch or no hunch, to know that you have the right result for certain, you still have to derive it. This is done by quite literally repeating the same steps that resulted in Eq. 5.42, but with a new form of the kinetic energy. Ignoring any potential energy, which as you know cancels anyway, I start with

$$i\hbar \frac{\partial \Psi(r,t)}{\partial t} = \frac{\left[-i\hbar \nabla + eA(r) \right]^2}{2m_e} \Psi (r, t)$$

$$-i\hbar \frac{\partial \Psi * (r,t)}{\partial t} = \frac{\left[i\hbar \nabla + eA(r) \right]^2}{2m_e} \Psi * (r, t).$$

Expanding the kinetic energy term and ignoring the real-valued $e^2 A^2 / m_e$, which will cancel in the same way as the potential energy, I have

$$i\hbar \Psi^* \frac{\partial \Psi}{\partial t} = -\frac{\hbar^2}{2m_e} \Psi^* \nabla^2 \Psi (r, t) - \frac{ie\hbar}{2m_e} \Psi^* \left[\nabla (A\Psi) + A\nabla \Psi \right]$$

$$-i\hbar \Psi \frac{\partial \Psi^*}{\partial t} = -\frac{\hbar^2}{2m_e} \Psi \nabla^2 \Psi^* (r, t) + \frac{ie\hbar}{m_e} \Psi \left[\nabla (A\Psi^*) + A\nabla \Psi^* \right].$$

You might want to note that unlike the similar calculation in Sect. 15.3.1 I did not use the fact that for the vector potential in the Coulomb gauge ($\nabla \cdot A = 0$) $\nabla (A\Psi) = A\nabla \Psi$, keeping these two terms as they were. Subtracting the second line from the first, I get

$$\frac{\partial |\Psi(r,t)|^2}{\partial t} = -\frac{i\hbar}{2m_e} \left[-\Psi^* \nabla^2 \Psi + \Psi \nabla^2 \Psi^* \right] -$$

$$\frac{e}{2m_e} \left[\Psi^* \nabla (A\Psi) + \Psi^* A \nabla \Psi + \Psi \nabla (A\Psi^*) + \Psi A \nabla \Psi^* \right].$$

The first line here is the same as in Eq. 5.42 and can be written down as

$$\frac{i\hbar}{2m_e} \nabla \cdot \left(\Psi^* \nabla \Psi - \Psi \nabla \Psi^* \right),$$

and in the second line, as you can easily see by reverse engineering, the result is equal to

$$-\frac{e}{2m_e}\nabla \cdot \left(\Psi^* A\Psi + \Psi A\Psi^*\right).$$

Combining these two expressions, I again end up with the continuity equation

$$\frac{\partial |\Psi(r,t)|^2}{\partial t} = -\nabla \cdot j$$

with the probability current now defined as

$$j = -\frac{i\hbar}{2m_e}\Psi^*\nabla\Psi + \frac{i\hbar}{2m_e}\Psi\nabla\Psi^* + \frac{e}{2m_e}\Psi^* A\Psi + \frac{e}{2m_e}\Psi A\Psi^* =$$

$$\frac{1}{2m_e}\left[\Psi^*\left(-i\hbar\nabla + eA\right)\Psi + \Psi\left(i\hbar\nabla + eA\right)\Psi^*\right].$$

Replacing $-i\hbar\nabla$ with p and $i\hbar\nabla\Psi^*$ with $(p\Psi)^*$, you can immediately see that this result reproduces Eq. 16.43.

Now you can take a few minutes to feel proud about your amazing hunch and your power of intuition, but do not get carried away with it. Besides, the main work is still ahead of you: the goal was to compute the probability current in the absence of the electric field and with the magnetic field presented by the vector potential in the Landau gauge, Eq. 16.12. The expressions for the components of current density vector in this case become

$$j_x = \frac{-i\hbar}{2m_e}\left[\Psi^*\frac{\partial\Psi}{\partial x} - \Psi\frac{\partial\Psi^*}{\partial x}\right] \tag{16.44}$$

$$j_y = \frac{1}{2m_e}\left[\Psi^*\left(-i\hbar\frac{\partial\Psi}{\partial y} + eBx\Psi\right) + \Psi\left(i\hbar\frac{\partial\Psi^*}{\partial y} + eBx\Psi^*\right)\right]. \tag{16.45}$$

For the wave function in the form

$$\Psi(r,t) = \frac{1}{\sqrt{L_y}}e^{-iE_n t/\hbar}e^{ip_y y/\hbar}\varphi_n(x - x_c),$$

where E_n is the energy of the corresponding Landau level, while $\varphi_n(x - x_c)$ is the wave function of the displaced oscillator as defined by Eq. 16.27, with x_c given by Eq. 16.16, the x-component of the current density obviously vanishes, thanks to "real-valuedness" of $\varphi_n(x - x_c)$. As for the y-component, you can easily get

$$j_y(n, p_y) = \frac{1}{m_e L_y}\left(p_y + eBx\right)|\varphi_n(x - x_c)|^2 = \frac{eB}{m_e L_y}(x - x_c)|\varphi_n(x - x_c)|^2 \tag{16.46}$$

where in the last step I used Eq. 16.16 for the center of the wave function. So, in the absence of the electric field at each value of coordinate x, there exists a non-zero density of the probability current. The direction of this current changes sign however when the coordinate x changes from $x < x_c$ to $x > x_c$. We are interested in the total current, though, which is normally defined as an integral of a current density over a surface perpendicular to it. In the two-dimensional case, however, the surface should be replaced with a one-dimensional line, which in the case under consideration is just the X-axis. The total electric current, I, is then defined as

$$I = e \int_{-\infty}^{\infty} dx j_y(x) dx = \frac{e^2 B}{m_e L_y} \int_{-\infty}^{\infty} dx \, (x - x_c) \, |\varphi_n(x - x_c)|^2$$

and is proportional to the expectation value of the coordinate $\tilde{x} = x - x_c$ in the $|p_y\rangle |\varphi_n\rangle$ state of the electron. Obviously, this expectation value is equal to zero, so that the total current in the absence of the electric field vanishes, as expected.

Now, let's add the uniform electric field $\mathcal{E} = \tau_x \mathcal{E}_0$ into the picture by going back to Hamiltonian 16.11 with scalar potential $V(x) = -x\mathcal{E}_0$. Since this potential does not depend on the y-coordinate, the separation of the wave function into the product $|p_y\rangle |\varphi_n\rangle$ still works with the equation for $\varphi_n(x)$, now taking the form of

$$\left[\frac{\hat{p}_x^2}{2m_e} + \frac{1}{2} m_e \Omega_L^2 (x - x_c)^2 + e\mathcal{E}_0 x \right] \varphi_n(x) = E_n \varphi_n(x).$$

Obviously, this equation is a simple rehash of Eq. 16.17 with the addition of the scalar potential term $-eV(x)$ and a slight obvious change of notation. To get a handle on this equation, let me play with the potential energy term:

$$\frac{1}{2} m_e \Omega_L^2 (x - x_c)^2 + e\mathcal{E}_0 x = \frac{1}{2} m_e \Omega_L^2 \left(x^2 - 2xx_c + \frac{2e\mathcal{E}_0}{m_e \Omega_L^2} x + x_c^2 \right) =$$

$$\frac{1}{2} m_e \Omega_L^2 \left(x^2 - 2x\tilde{x}_c + \tilde{x}_c^2 - \tilde{x}_c^2 + x_c^2 \right) = \frac{1}{2} m_e \Omega_L^2 (x - \tilde{x}_c)^2 + \frac{1}{2} m_e \Omega_L^2 \left(x_c^2 - \tilde{x}_c^2 \right),$$

where I defined

$$\tilde{x}_c = x_c - \frac{e\mathcal{E}_0}{m_e \Omega_L^2} = x_c - \frac{m_e \mathcal{E}_0}{eB^2}. \tag{16.47}$$

Thus, it is still the same old harmonic oscillator problem

$$\left[\frac{\hat{p}_x^2}{2m_e} + \frac{1}{2} m_e \Omega_L^2 (x - \tilde{x}_c)^2 \right] \varphi_n(x) = \left[E_n - \frac{1}{2} m_e \Omega_L^2 \left(x_c^2 - \tilde{x}_c^2 \right) \right] \varphi_n(x),$$

where the electric field does two things: (1) it shifts (again) the zero of the quadratic potential to new position $\widetilde{x}_c$, and (2) it changes the energy of the Landau levels to a new value defined by

$$E_n = \frac{1}{2}m_e\Omega_L^2\left(x_c^2 - \widetilde{x}_c^2\right) + \hbar\Omega_L\left(n + \frac{1}{2}\right).$$

Using Eqs. 16.16 and 16.47, you can find

$$\frac{1}{2}m_e\Omega_L^2\left(x_c^2 - \widetilde{x}_c^2\right) = \frac{1}{2}\frac{e^2B^2}{m_e}\left(2x_c\frac{m_e\mathcal{E}_0}{eB^2} - \frac{m_e^2\mathcal{E}_0^2}{e^2B^4}\right) =$$

$$e\mathcal{E}_0 x_c - \frac{1}{2}\frac{m_e\mathcal{E}_0^2}{B^2} = -\frac{\mathcal{E}_0 p_y}{B} - \frac{1}{2}\frac{m_e\mathcal{E}_0^2}{B^2}.$$

Then the expression for the energy levels becomes

$$E_n = \hbar\Omega\left(n + \frac{1}{2}\right) + \frac{\mathcal{E}_0 p_y}{B} + \frac{1}{2}\frac{m_e\mathcal{E}_0^2}{B^2}. \tag{16.48}$$

While the constant $m_e\mathcal{E}_0^2/2B^2$ term in this expression can be eliminated by a different choice of zero of the electric field potential, the term proportional to p_y is physically significant. It indicates that the electric field lifts the degeneracy of the Landau levels with respect to p_y and turns them into bands. The dependence on p_y means that the electrons are now propagating in the y direction with the group velocity

$$v_D = \frac{\partial E_n}{\partial p_y} = \frac{\mathcal{E}_0}{B} \tag{16.49}$$

which coincides with classical velocity v_E characterizing the drift of the center of the classical cyclotron orbit. Apparently, Eq. 16.49 represents a quantum reincarnation of that classical result. It is also interesting that the dependence of the energy upon momentum in this case is linear instead of the expected quadratic one. For one, it means that the phase velocity of these electrons coincides with their group velocity, and the packet formed by the corresponding wave functions would not broaden (unlike the case of free electron packets). Essentially, these electrons behave as light with the exception that their energy does not go to zero when $p_y = 0$.

This result, while curious in itself, is of little significance for the calculations of the electric current. Equations 16.44 and 16.45 are not affected by the electric field (remember, any real-valued potential vanishes from the definition of the probability current), so the only change I need to incorporate comes from the wave function. Since the x-dependent part of the wave function is still real-valued, the x-component of the current still vanishes, while Eq. 16.46 for the y-component now acquires the form

$$j_y(n, p_y) = \frac{eB}{m_e L_y}(x - x_c)|\varphi_n(x - \widetilde{x}_c)|^2.$$

Now I want you to pay attention: the center of wave function in the presence of the electric field is shifted from zero by $\widetilde{x}_c$, while the $(x - x_c)$ factor in it still contains the old shift x_c. This simple circumstance makes all the difference, because now the total current does not vanish:

$$I_{n,p_y} = \frac{e^2 B}{m_e L_y}\int\limits_{-\infty}^{\infty} dx\,(x - x_c)|\varphi_n(x - \widetilde{x}_c)|^2 =$$

$$\frac{e^2 B}{m_e L_y}\int\limits_{-\infty}^{\infty} dx\,(x - \widetilde{x}_c)|\varphi_n(x - \widetilde{x}_c)|^2 + \frac{e^2 B}{m_e L_y}(\widetilde{x}_c - x_c)\int\limits_{-\infty}^{\infty} dx\,|\varphi_n(x - \widetilde{x}_c)|^2.$$

The first term in the second line of this expression is still zero, of course, but the integral in the second term is just the normalization integral and is equal to unity. Thus, I have for the total electric current in the y direction, perpendicular to the electric and magnetic fields:

$$I_{n,p_y} = -\frac{e^2 B}{m_e L_y}\frac{m_e \mathcal{E}_0}{eB^2} = -\frac{e\mathcal{E}_0}{BL_y}.$$

What is important here is that I_{n,p_y} does not depend on either p_y or n, and, therefore, in order to find the total current as defined in Eq. 16.42, I simply need to multiply this result by the total number of states occupied by electrons. The contribution from a single Landau level to the current is

$$I_n = I_{n,p_y} G = -\frac{e\mathcal{E}_0}{BL_y}\frac{eBL_x L_y}{2\pi\hbar} = -\frac{e^2}{h}\mathcal{E}_0 L_x,$$

and now if there are exactly n filled Landau levels, the total current becomes

$$I_{tot} = -n\frac{e^2}{h}\mathcal{E}_0 L_x.$$

$\mathcal{E}_0 L_x$ is obviously a potential difference across the sample in the direction of the field, and expression $I_{tot}/\mathcal{E}_0 L_x$ is the quantized Hall conductance σ_{xy} of the system

$$\sigma_{xy} = -n\frac{e^2}{h}. \tag{16.50}$$

You might feel a bit puzzled as why I did not multiply the degeneracy factor G by two as I did when deriving the density of states or the ground state energy. The thing is that even if the spin contribution to the Landau level is relatively small, at low temperatures, when the quantum Hall effect is observed, the Landau levels corresponding to different spin orientations remain well separated from each other, so that each individual level does not have the spin-related degeneracy. But, when counting the total number of filled levels, the spin contribution must be counted so that n in Eq. 16.50 is not the same n as in Eq. 16.31 for Landau levels. (I am sorry but I have a limited number of symbols to be used in the formulas.) This n counts all filled Landau levels, including those with different values of the spin number m_s.

As the magnetic field shifts the position of the Fermi energy such that instead of n filled Landau levels, there are only $n - 1$, the conductivity changes abruptly by a universal amount e^2/h in complete agreement with the experiment. What this simple calculation does not explain is why the conductivity does not change when a Landau level is only partially filled and the number of the occupied states changes with the magnetic field. In other words, this theory explains the quantum jumps of the conductivity, but does not explain the plateaus observed in the experiment. But this is a different story for a different time and a different book. And there are plenty of those if you are interested.

16.4 Problems

Problem 205 Consider a charged particle moving in uniform and time-independent electric and magnetic fields which are perpendicular to each other. The particle is also subjected to a damping force which can be written down as

$$F_d = -\frac{m_e}{\tau_r} v,$$

where m_e is, as always, the particle's mass, τ_r is the relaxation parameter, and v is the particle's velocity.

1. Write down the equations of motion for this particle including the damping force.
2. Do not attempt to solve these equations in general case, but see if there is a solution with time-independent velocity v (set $dv/dt = 0$).
3. Now imagine that you have N_e electrons per unit volume, and using the found solutions, derive the expressions for normal Ohmic conductivity (current in the direction of the electric field) and Hall conductivity (current perpendicular to both electric and magnetic fields). (Hint: Compute the corresponding current densities first.)

Problem 206 Using Eqs. 16.1 and 16.2, show that the center of the classical cyclotron trajectory, x_0 and y_0, obeys the following relations:

$$x_0 = x - \frac{1}{eB}\left(p_y + eA_y\right).$$

Problem 207 Assume that instead of the vector potential given in Eq. 16.12, you use the vector potential of the form $A = -\tau_x yB$:

1. Show that this vector potential describes the same magnetic field as the one used in the text of the chapter.
2. Write down the Schrödinger equation for the electron in the magnetic field using this vector potential and determine the eigenvalues and find the wave functions representing the eigenvectors of the corresponding Hamiltonian.
3. Derive the expression for the degree of degeneracy of a single Landau level using this new vector potential.

Problem 208 Consider operators

$$\hat{\Pi}_x = \hat{p}_x + eA_x$$

$$\hat{\Pi}_x = \hat{p}_y + eA_y,$$

where $A_{x,y}$ are x- and y-components of an arbitrary vector potential defining a magnetic field B. Show that the commutator $\left[\hat{\Pi}_x, \hat{\Pi}_y\right]$ is equal to

$$\left[\hat{\Pi}_x, \hat{\Pi}_y\right] = -i\hbar eB_z,$$

where B_z is the z-component of the magnetic field represented by the vector potential A. (Hint: Use the position representation for the momentum operator.)

Problem 209

1. Using operators $\hat{\Pi}_{x,y}$ introduced in the previous problem, construct the following operators:

$$\hat{a} = \frac{1}{\sqrt{2e\hbar B}}\left(\hat{\Pi}_y + i\hat{\Pi}_x\right)$$

$$\hat{a}^\dagger = \frac{1}{\sqrt{2e\hbar B}}\left(\hat{\Pi}_y - i\hat{\Pi}_x\right)$$

and find their commutator $\left[\hat{a}, \hat{a}^\dagger\right]$.
2. Assuming that $A_z = 0$, present the Hamiltonian given by Eq. 16.11 in terms of operators $\hat{a}$ and $\hat{a}^\dagger$. Comment on the result.

Problem 210 Find the total ground state energy of a two-dimensional gas of free electrons (introduce the periodic boundary conditions and find the density of states; using Pauli principle, find the Fermi energy and the total energy of the many-electron ground state).

Problem 211 Generalize Eqs. 16.39, 16.36, and 16.39 taking into account the interaction of the spin of the electrons with the magnetic field. (Hint: The degeneracy with respect to p_y does not change in the presence of the spin, but the expression for

G in Eq. 16.29 must now be divided by two since spin variables are now counted explicitly. While the energy levels are now labeled by two numbers, you can still organize them in the ascending order and assign a single number to each energy.)

Problem 212 Find the eigenvalues and eigenvectors of operator $\hat{Q}$ defined in Eq. 16.25. Hint: Consider a superposition of a spinor corresponding to Landau level $n - 1$ and spin up with a spinor corresponding to Landau level n and then spin down (both these states correspond to the same eigenvalue of the Hamiltonian equation 16.11).

Problem 213 Find an alternative expression for the potential describing the uniform electric field in the quantum Hall problem which would eliminate the term

$$\frac{1}{2} \frac{m_e \mathcal{E}_0^2}{B^2}$$

in Eq. 16.48 for the Landau energy levels in the presence of the electric field.

Index

Printed in the United States
By Bookmasters